Lecture Notes in Computer Science 13155

More information about this subseries at https://link.springer.com/bookseries/7407

Yongxuan Lai · Tian Wang · Min Jiang ·
Guangquan Xu · Wei Liang ·
Aniello Castiglione (Eds.)

Algorithms and Architectures for Parallel Processing

21st International Conference, ICA3PP 2021
Virtual Event, December 3–5, 2021
Proceedings, Part I

Editors

Yongxuan Lai
Xiamen University
Xiamen, China

Tian Wang
Beijing Normal University
Zhuhai, China

Min Jiang
Xiamen University
Xiamen, China

Guangquan Xu
Tianjin University
Tianjin, China

Wei Liang
Hunan University
Changsha, China

Aniello Castiglione
University of Naples Parthenope
Naples, Italy

ISSN 0302-9743 ISSN 1611-3349 (electronic)
Lecture Notes in Computer Science
ISBN 978-3-030-95383-6 ISBN 978-3-030-95384-3 (eBook)
https://doi.org/10.1007/978-3-030-95384-3

LNCS Sublibrary: SL1 – Theoretical Computer Science and General Issues

This Springer imprint is published by the registered company Springer Nature Switzerland AG
The registered company address is: Gewerbestrasse 11, 6330 Cham, Switzerland

Preface

On behalf of the Conference Committee we welcome you to the proceedings of the 2021 International Conference on Algorithms and Architectures for Parallel Processing (ICA3PP 2021), which was held virtually during December 3–5, 2021. ICA3PP 2021 was the 21st in this series of conferences (started in 1995) that are devoted to algorithms and architectures for parallel processing. ICA3PP is now recognized as the main regular international event that covers the many dimensions of parallel algorithms and architectures, encompassing fundamental theoretical approaches, practical experimental projects, and commercial components and systems. This conference provides a forum for academics and practitioners from countries around the world to exchange ideas for improving the efficiency, performance, reliability, security, and interoperability of computing systems and applications.

A successful conference would not be possible without the high-quality contributions made by the authors. This year, ICA3PP received a total of 403 submissions from authors in 28 countries and regions. Based on rigorous peer reviews by the Program Committee members and reviewers, 145 high-quality papers were accepted to be included in the conference proceedings and submitted for EI indexing. In addition to the contributed papers, eight distinguished scholars, Yi Pan, Daqing Zhang, Yan Zhang, Shuai Ma, Weijia Jia, Keqiu Liu, Yang Yang, and Peng Cheng, were invited to give keynote lectures, providing us with the recent developments in diversified areas in algorithms and architectures for parallel processing and applications.

We would like to take this opportunity to express our sincere gratitude to the Program Committee members and 160 reviewers for their dedicated and professional service. We highly appreciate the six track chairs, Ding Wang, Songwen Pei, Zhiming Luo, Shigeng Zhang, Longbiao Chen, and Feng Wang, for their hard work in promoting this conference and organizing the reviews for the papers submitted to their tracks. We are so grateful to the publication chairs, Yang Wang, Carmen De Maio, Donglong Chen, and Yinglong Zhang, and the publication assistants for their tedious work in editing the conference proceedings. We must also say "thank you" to all the volunteers who helped us in various stages of this conference. Moreover, we are so honored to have many renowned scholars be part of this conference. Finally, we would like to thank all speakers, authors, and participants for their great contribution and support to make ICA3PP 2021 a success!

December 2021

Min Jiang
Aniello Castiglione
Guangquan Xu
Wei Liang
Jean-Luc Gaudiot
Yongxuan Lai
Tian Wang

Organization

General Co-chairs

Jean-Luc Gaudiot	University of California, Irvine, USA
Yongxuan Lai	Xiamen University, China
Tian Wang	Beijing Normal University and UIC, China

Program Co-chairs

Min Jiang	Xiamen University, China
Aniello Castiglione	University of Naples Parthenope, Italy
Guangquan Xu	Tianjin University, China
Wei Liang	Hunan University, China

Track Chairs

Ding Wang	Nankai University, China
Songwen Pei	University of Shanghai for Science and Technology, China
Zhiming Luo	Xiamen University, China
Shigeng Zhang	Central South University, China
Longbiao Chen	Xiamen University, China
Feng Wang	Wuhan University, China

Local Co-chairs

Cheng Wang	Huaqiao University, China
Liang Song	Xiamen University, China

Publication Chairs

Yan Wang	Xiamen University of Technology, China
Carmen De Maio	University of Salerno, Italy
Donglong Chen	Beijing Normal University-Hong Kong Baptist University United International College (UIC), China
Yinglong Zhang	Minnan Normal University, China

Publicity Co-chairs

Fan Lin	Xiamen University, China
Longbiao Chen	Xiamen University, China
Saiqin Long	Xiangtan University, China
Zhetao Li	Xiangtan University, China

Steering Committee

Yang Xiang	Swinburne University of Technology, Australia
Kuan-Ching Li	Providence University, Taiwan, China

Program Committee

A. M. A. Elman Bashar	Plymouth State University, USA
Jiahong Cai	Hunan University of Science and Technology, China
Yuanzheng Cai	Xiamen University, China
Jingjing Cao	Wuhan University of Technology, China
Zhihan Cao	Huaqiao University, China
Arcangelo Castiglione	University of Salerno, Italy
Lei Chai	Beihang University, China
Chao Chen	Chongqiing University, China
Donglong Chen	Beijing Normal University-Hong Kong Baptist University United International College (UIC), China
Haiming Chen	Ningbo University, China
Juan Chen	Hunan University, China
Kai Chen	Institute of Information Engineering, Chinese Academy of Sciences, China
Lifei Chen	Fujian Normal University, China
Rongmao Chen	National University of Defense Technology, China
Shuhong Chen	Guangzhou University, China
Xiaoyan Chen	Xiamen University of Technology, China
Xu Chen	Sun Yat-sen University, China
Yu Chen	Wuhan University of Technology, China
Yuanyi Chen	Zhejiang University, China
Yuxiang Chen	Huaqiao University, China
Lin Cui	Jinan University, China
Haipeng Dai	Nanjing University, China
Hong-Ning Dai	Lingnan University, China
Xia Daoxun	Guizhou Normal University, China

Himansu Das	KIIT Deemed to be University, India
William A. R. De Souza	Royal Holloway, University of London, UK
He Debiao	Wuhan University, China
Xiaoheng Deng	Central South University, China
Chunyan Diao	Hunan University, China
Qiying Dong	Nankai University, China
Fang Du	Ningxia University, China
Guodong Du	Xiamen University, China
Xin Du	Fujian Normal University, China
Shenyu Duan	Shanghai University, China
Xiaoliang Fan	Xiamen University, China
Xiaopeng Fan	Shenzhen Institute of Advanced Technology, Chinese Academy of Sciences, China
Yongkai Fan	China University of Petroleum, Beijing, China
Zipei Fan	University of Tokyo, Japan
Fei Fang	Hunan Normal University, China
Tao Feng	Sichuan University, China
Virginia Franqueira	University of Kent, UK
Anmin Fu	National University of Defense Technology, China
Bin Fu	Hunan University, China
Zhipeng Gao	Beijing University of Posts and Telecommunications, China
Debasis Giri	Maulana Abul Kalam Azad University of Technology, India
Xiaoli Gong	Nankai University, China
Ke Gu	Changsha University of Science and Technology, China
Zonghua Gu	University of Missouri, USA
Gui Guan	Nanjing University of Posts and Telecommunications, China
Yong Guan	Iowa State University, USA
Zhitao Guan	North China Electric Power University, China
Guangquan Xu	Tianjin University, China
Ye Guixin	Northwest University, China
Chun Guo	Institute of Information Engineering, Chinese Academy of Sciences, China
Guijuan Guo	Huaqiao University, China
Jun Guo	Quanzhou University of Information Engineering, China
Shihui Guo	Xiamen University, China
Zehua Guo	Beijing Institute of Technology, China
Yu Haitao	Guilin University of Technology, China

Dezhi Han	Shanghai Maritime University, China
Jinsong Han	Zhejiang University, China
Song Han	Zhejiang Gongshang University, China
Yulin He	Shenzhen University, China
Alan Hong	Xiamen University of Technology, China
Haokai Hong	Xiamen University, China
Zhenzhuo Hou	Peking University, China
Donghui Hu	Hefei University of Technology, China
Yupeng Hu	Hunan University, China
Yuping Hu	Guangdong University of Finance and Economics, China
Qiang-Sheng Hua	Huazhong University of Science and Technology, China
Weh Hua	University of Queensland, China
Yu Hua	Huazhong University of Science and Technology, China
Chenxi Huang	Xiamen University, China
Haiyang Huang	Huaqiao University, China
Jiawei Huang	Central South University, China
Jing Huang	Hunan University, China
Weihong Huang	Hunan University of Science and Technology, China
Xinyi Huang	Fujian Normal University, China
Zhou Jian	Nanjing University of Posts and Telecommunications, China
Fengliang Jiang	Longyan University, China
Wanchun Jiang	Central South University, China
Wenjun Jiang	Hunan University, China
He Jiezhou	Xiamen University, China
Zhengjun Jing	Jiangsu University of Technology, China
Xiaoyan Kui	Central South University, China
Xia Lei	China University of Petroleum, China
Chao Li	Shanghai Jiao Tong University, China
Dingding Li	South China Normal University, China
Fagen Li	University of Electronic Science and Technology of China, China
Fuliang Li	Northeastern University, China
Hui Li	Guizhou University, China
Jiliang Li	University of Goettingen, Germany
Lei Li	Xiamen University, China
Tao Li	Nankai University, China
Tong Li	Nankai University, China

Chengying Mao	Jiangxi University of Finance and Economics, China
Mario Donato Marino	Leeds Beckett University, UK
Yaxin Mei	Huaqiao University, China
Hao Peng	Zhejiang Normal University, China
Hua Peng	Shaoxing University, China
Kai Peng	Huaqiao University, China
Li Peng	Hunan University of Science and Technology, China
Yao Peng	Northwest University, China
Zhaohui Peng	Shandong University, China
Aneta Poniszewska-Maranda	Lodz University of Technology, Poland
Honggang Qi	University of Chinese Academy of Sciences, China
Tie Qiu	Tianjin University, China
Dapeng Qu	Liaoning University, China
Zhihao Qu	Hohai University and Hong Kong Polytechnic University, China
Yang Quan	Huaqiao University, China
Abdul Razaque	International IT University, Kazakhstan
Chunyan Sang	Chongqing University of Posts and Telecommunications, China
Arun Kumar Sangaiah	VIT University, India
Shanchen Pang	China University of Petroleum, China
Yin Shaoyi	Paul Sabatier University, France
Hua Shen	Hubei University of Technology, China
Meng Shen	Beijing Institute of Technology, China
Huibin Shi	Nanjing University of Aeronautics and Astronautics, China
Liang Shi	East China Normal University, China
Peichang Shi	National University of Defense Technology, China
Liang Song	Xiamen University, China
Tao Song	China University of Petroleum, China
Song Han	Zhejiang Gongshang University, China
Riccardo Spolaor	University of Oxford, UK
Chunhua Su	Osaka University, Japan
Bingcai Sui	National University of Defense Technology, China
Nitin Sukhija	Slippery Rock University of Pennsylvania, USA
Bing Sun	Huaqiao University, China
Yu Sun	Guangxi University, China

Sheng Wen	Swinburne University of Technology, Australia
Stephan Wiefling	Bonn-Rhein-Sieg University of Applied Sciences, Germany
Di Wu	Deakin University, Australia
Hejun Wu	Sun Yat-sen University, China
Qianhong Wu	Beihang University, China
Shangrui Wu	Beijing Normal University, China
Xiaohe Wu	Hunan University of Science and Technology, China
Zhongbo Wu	Hubei University of Arts and Science, China
Bin Xia	Nanjing University of Posts and Telecommunications, China
Guobao Xiao	Minjiang University, China
Lijun Xiao	Guangzhou College of Technology and Business, China
Wenhui Xiao	Central South University, China
Yalong Xiao	Central South University, China
Han Xiaodong	Minjiang University, China
Fenfang Xie	Sun Yat-sen University, China
Guoqi Xie	Hunan University, China
Mande Xie	Zhejiang Gongshang University, China
Songyou Xie	Hunan University, China
Xiaofei Xie	Nanyang Technological University, Singapore
Yi Xie	Sun Yat-sen University, China
Zhijun Xie	Ningbo University, China
Peiyin Xiong	Hunan University of Science and Technology, China
Dejun Xu	Xiamen University, China
Jianbo Xu	Hunan University of Science and Technology, China
Ming Xu	Hangzhou Dianzi University, China
Wenzheng Xu	Sichuan University, China
Zhiyu Xu	NSCLab, Australia
Zichen Xu	Nanchang University, China
Zisang Xu	Changsha University of Science and Technology, China
Xiaoming Xue	City University of Hong Kong, China
Xingsi Xue	Fujian University of Technology, China
Changcai Yang	Fujian Agriculture and Forestry University, China
Chao-Tung Yang	Tunghai University, Taiwan, China
Dingqi Yang	University of Macau, China
Fan Yang	Xiamen University, China
Fengxiang Yang	Xiamen University, China

Guisong Yang	University of Shanghai for Science and Technology, China
Hao Yang	Yancheng Teachers University, China
Hui Yang	National University of Defense Technology, China
Lan Yang	Quanzhou University of Information Engineering, China
Lvqing Yang	Xiamen University, China
Mujun Yin	Huaqiao University, China
Haitao Yu	Guilin University of Technology, China
Sheng Yu	Shaoguan University, China
Shuai Yu	Sun Yat-sen University, China
Zhiyong Yu	Fuzhou University, China
Liang Yuzhu	Huaqiao University, China
Tao Zan	Longyan University, China
Yingpei Zeng	Hangzhou Dianzi University, China
Bingxue Zhang	University of Shanghai for Science and Technology, China
Bo Zhang	Shanghai Normal University, China
Chongsheng Zhang	Henan University, China, China
Haibo Zhang	University of Otago, New Zealand
Hong-Bo Zhang	Huaqiao University, China
Jia Zhang	Jinan University, China
Jingwei Zhang	Guilin University of Electronic Technology, China
Jun Zhang	Dalian Maritime University, China
Mingwu Zhang	Hubei University of Technology, China
Qiang Zhang	Central South University, China
Shaobo Zhang	Hunan University of Science and Technology, China
Shengchuan Zhang	Xiamen University, China
Shiwen Zhang	Hunan University, China
Tianzhu Zhang	Nokia Bell Labs, USA
Yi Zhang	Xiamen University, China
Yilin Zhang	Huaqiao University, China
Baokang Zhao	National University of Defense Technology, China
Bowen Zhao	Singapore Management University, Singapore
Jinyuan Zhao	Changsha Normal University, China
Liang Zhao	Shenyang Aerospace University, China
Sha Zhao	Zhejiang University, China
Wan-Lei Zhao	Xiamen University, China
Yongxin Zhao	East China Normal University, China

Qun-Xiong Zheng	Institute of Information Engineering, Chinese Academy of Sciences, China
Ping Zhong	Central South University, China
Binbin Zhou	Zhejiang University City College, China
Qifeng Zhou	Xiamen University, China
Teng Zhou	Shantou University, China
Wei Zhou	Guilin University of Electronic Technology, China
Xinyu Zhou	Jiangxi Normal University, China
Haibin Zhu	Nipissing University, Canada
Shunzhi Zhu	Xiamen University of Technology, China
Weiping Zhu	Wuhan University, China
Xiaoyu Zhu	Central South University, China
Zhiliang Zhu	East China Jiaotong University, China
Haodong Zou	Beijing Normal University-Hong Kong Baptist University United International College (UIC), China
Yunkai Zou	Nankai University, China

Reviewers

Yingzhe He	Peifu Han	Shihang Yu
Zhiyu Wang	Xuefei Wang	Lei Huang
Wang Yin	Wen Dong	Shuqi Liu
Jingyi Cui	Zhengbo Han	Yawu Zhao
Xingda Liu	Jincai Zhu	Feiyu Jin
Guohua Xin	Chen Qi	Yunfeng Huang
Zhangyan Yang	Na Zhao	Fei Zhu
Yuexin Zhang	Bingxuan Li	Shengmin Xu
Chenkai Tan	Zhixiu Guo	Yao Hu
Hongpeng Bai	Xu Wu	Weijing You
Lixiao Gong	Yu Pen	Rui Liu
Xue Li	Jian Yin	Fan Meng
Wenqing Lei	Xinjun Pei	Haiyuan Gui
Xue Hu	Xiaoquan Zhang	Guangjing Huang
Meiqi Feng	Yihao Lin	Qingze Fang
Victor Chiang	Shiqiang Zheng	Hualing Ren
Xin Ji	Huifang Zeng	Ziya Chen
Pengqu Yan	Yixiang Hu	Ruixiang Luo
Xinru Ding	Xincao Xu	Wenxuan Wei
Yumei Li	Jiahao Zhao	Zhiming Lin
Haodong Zhang	Lulu Cao	Xin He
Zhe Chen	Jiahui Yu	Guoyong Dai
Gang Shen	Faquan Chen	Zhiyuan Wang
Hongfei Shao	Xue Zhai	Yajing Xie

Hanbin Hong
Min Wu
Siyao Chen
Kaiyue Zhang
Xiaohai Cai
Zhiwen Zhang
Tieqi Shou
Liqiang Xu
Chenhui Lu
Hang Zhu
Jiannan Gao
Hang Zhou
Haoyang An
Yang Liu
Peiwei Hu
Xiaotong Zhou
Qiang Tang
Chang Yue
Miaoqian Lin
Yudong Li
Zijing Ma
Zhankai Li
Chengyao Hua
Qingxuan Wang
Yi Wang
Mingyue Li
Songsong Zhang
Jingwei Jiang
Meijia Xu
Shuhong Hong
Anqi Yin
Shaoqiang Wu
Shuangjiao Zhai
Fei Ma
Shiya Peng
Kedong Xiu
Shengnan Zhao
Chuan Zhao
Bo Zhang
Baozhu Li
Ning Liu
Wang Yang
Xiaohui Yang
Zihao Dong
Sijie Niu
Kun Ma
Dianjie Lu
Ziqiang Yu
Lizhi Peng
Yilei Wang
Zhang Jing
Tian Jie
Jian Zhao
Hui Li
Yan Jiang
Minghao Zhao
Lei Lyu
Yanbin Han
Yanlin Wu
Lingang Huang
Mingwei Lin
Wenxiang Wang
Xinqiang Ye
Songyi Yang
Cancan Wang
Xingbao Zhang
Yafeng Sun
Li Lin
Jinxian Lei
Wentao Liu

Contents – Part I

Software Systems and Efficient Algorithms

Contents – Part II

Data Science

Edge Computing and Edge Intelligence

Blockchain Systems

Contents – Part III

Service Dependability and Security Algorithms

Software Systems and Efficient Algorithms

Deep Learning Models and Applications

CRFST-GCN: A Deeplearning Spatial-Temporal Frame to Predict Traffic Flow

Chunyan Diao[1], Dafang Zhang[1(✉)], Wei Liang[2,3], Kuan-Ching Li[4], and Man Jiang[5]

[1] College of Computer Science and Electronic Engineering, Hunan University, Changsha 410082, China
{cydiao,dfzhang}@hnu.edu.cn

[2] School of Computer Science and Engineering, Hunan University of Science and Technology, Xiangtan 411201, China
wliang@hnust.edu.cn

[3] Hunan Key Laboratory for Service Computing and Novel Software Technology, Xiangtan 411201, China

[4] Department of Computer Science and Information Engineering (CSIE), Providence University, Taichung 43301, Taiwan
kuancli@pu.edu.tw

[5] School of Architecture and Planning, Hunan University, Changsha 410082, China
jiangman@hnu.edu.cn

Abstract. Long-term traffic prediction has tremendous significance for ITS intelligent traffic management security problem. An accurate forecast can improve the efficiency of traffic management and reduce traffic accidents. Thus, the complex dynamic time-space cycle of traffic flow data makes traffic prediction a considerable challenge. Although the existing graph convolution method can capture the correlation between nodes, but never proposed to capture the similarity between the hidden layers of the graph convolution. This paper combines time, space, occupancy, and other related factors to propose a unique multi-period conditional random field (CRF) graph convolution model to accurately predict long-term traffic flow (CRFST-GCN). First, divide the data into three independent fields: trend, day, and week, and then input the data into CRFGCN frame to effectively extract spatial features. The convolution module captures the time-series relationship. Finally, it is verified on two real data sets that our proposed model effectively extracts similarities, and the results show that the model is 40 % more accurate than traditional methods during peak hours.

Keywords: Conditional random field · Deep learning · Graph convolution network · Spatial-temporal frame · Traffic flow prediction

1 Introduction

As a new diagram of cloud computing, Internet of Things (IoT) [1] computing has attracted many attention on optimize its performance [2], security and

Y. Lai et al. (Eds.): ICA3PP 2021, LNCS 13155, pp. 3–17, 2022.
https://doi.org/10.1007/978-3-030-95384-3_1

intelligence [3]. With the advanced of this field, in recent years, many countries have been committed to vigorously developing intelligent transportation systems (ITS) to help achieve efficient traffic management. As an important part of improving intelligent transportation system, traffic forecasting can not only assist traffic management departments in guiding vehicles more rationally, improving the operating efficiency of the road network, to reduce the probability of accidents and improve public safety, but also effectively provide users with a more effective travel service experience. Therefore, traffic flow forecasting has received more and more attention from all walks of life.

Sensors are the important part of the IoT network, which can realize the collection, transmission, and calculation of data [4]. As the reliability of IoT devices such as sensors and cameras increases [5], these advanced equipment and the security of blockchain [6] provides a convenient data acquisition, a large number of rich data for accurate traffic prediction offers a new direction, also has exposed many problems at the same time; usually, data acquisition device installed on the fixed position, only partial data, from the point of the data content, traffic flow data has certain cyclical trends, Periodicity is more pronounced at specific times and locations. In essence, traffic flow forecasting is a typical time-space forecasting problem, but the flow data itself presents complex non-linearities and uncertainties of external environmental influences. This makes accurate prediction of traffic flow a major challenge for researchers, fortunately, with various sensor cameras and other equipment collecting a large amount of rich data. It provides a new direction for accurate traffic forecasting.

Traditional machine learning methods such as history average (HA) [7], vector autoregression(VAR) [8], autoregression integrates moving average (ARIMA) [9], etc., only consider time-series relationships and fail to capture complex spatial relationships, resulting in the inability to extract spatial features. Machine learning methods, such as k-nearest neighbor classification K-Nearest Neighbor (KNN) [10] and support vector machines (SVM) method [11], can model more complex data, but they require detailed feature engineering, resulting in poor prediction results. The data in the IoT is large, and it is difficult to process data accurately [12]. Deep learning is an algorithm that uses artificial neural network as the architecture to perform characterization learning on data [13]. As deep learning-based methods have shown advantages in capturing Spatial-temporal relationships, researchers have gradually applied deep learning to spatial-temporal data prediction. Recurrent neural networks (RNNs) can capture the features of time series, and Convolution neural networks (CNNs) can capture the similarity features of spatial relationships. Therefore, the method of combining CNN and RNN shows a powerful advantage in Spatial-temporal prediction. However, when faced with a large-scale long-period traffic forecast, the problem of small gradients will lead to loss of prediction accuracy and the inability to deal with emergencies. Long Short-Term Memory (LSTM) effectively solved the gradient problems [14], but the complexity is large, Gate Recurrent Unit (GRU) [15] designed to simplify LSTM. With the advent of Graph convolution network (GCN), the research of traffic prediction has entered a new stage.

Graph Convolutional Neural Networks perform convolution operations on graph data. Unlike general data, graph data has similar information between different nodes. Therefore, it is important to retain this similar information in the hidden layer of the graph convolutional neural network. However, the existing works cannot do this. On the other hand, forcing a hidden layer to store information is challenging.

Based on the existing shortcomings and challenges, this paper proposes a CRF based multi-branch Spatio-temporal graph convolution framework to improve the similarity of the hidden layers of graph convolutional networks and improve prediction efficiency. The main contributions of this paper are as follows:

- This paper proposes a multi-branch space-time framework. Three identical modules capture the long and short cycle dependencies from the local and the whole, respectively.
- A CRF-based graph convolution model is designed to effectively extract the correlation between hidden layers in the graph convolution process and improve the prediction accuracy.
- A large number of experiments have been carried out on two real data sets to prove that the method proposed in this paper is superior to the existing state-of-the-art baselines in many indicators.

The remaining of this paper is organized as follows. First, Sect. 2 reviews the related works of literature, Sect. 3 introduces the notations and problem definition, and Sect. 4 presents the Proposed Method. Then, Sect. 5 shows the experiments, and finally, the concluding remarks and future directions are drawn in Sect. 6.

2 Related Work

2.1 Graph Convolution

In recent years, graph convolution has been widely used in various graph structure tasks. The classic algorithm [16] defines the CNN signal as the Laplacian matrix of the graph, and obtains an efficient depth frame by learning a large number of convolutional layers with independent input parameters. [17] uses a semi-supervised graph data learning method to stimulate the convolution structure through the first-level approximation of the spectrogram convolution, effectively learning the structure of the graph and the hidden representation of node features. The Cluster-GCN algorithm based on graph clustering structure can satisfy the training based on sgd [18]. Representative works based on graph attention. Graph attention networks (GATs) can obtain different weights of nodes by stacking the features of nodes and their neighborhoods without knowing the structure of the graph, and use node feature information to efficiently generate node embeddings [19]. The GraphSAGE induction framework has also made great contributions to the attention mechanism [20].

2.2 Spatial-Temporal Graph Convolutional Network

Basically, loop detectors and cameras are two common sensors for traffic sensing and monitoring. For example, [21] proposed ACSNet to solve traffic density estimation under low-resolution surveillance, and [22] combined density regression and lstm for spatial-temporal prediction. Another option is to forecast traffic flow via road side sensors data. Neural networks based on recursion have received great attention in various spatial-temporal tasks due to their good performance. However, this method is only effective for short-term smooth dynamic data. Specifically, [23] proposed a deep residual learning framework. [24]designed an ST-ResNet model based on residual convolution unit to predict crowd flow. [25] also proposed a method of fusing CNN and LSTM to predict traffic, while simultaneously modeling spatial and temporal dependencies, to further proposed a spatial-temporal dynamic taxi demand prediction network, which can dynamically learn the similarity between locations. [26] later proposed a traffic prediction gated graph convolutional network based on this method, but the model did not consider the dynamic spatial-temporal correlation of traffic data. [27] proposed a spatial-temporal graph convolutional neural network (STGCN) to predict traffic status. It can use spatial features in complex traffic networks to capture temporal features at the same time, but when the expansion rate increases, STGCN and GraphWaveNet [28] method may cause the loss of local information. STSGCN captures the local spatial-temporal correlation through spatial-temporal sub-graphs [29]. In the case of a lot of missing data, it will cause the loss of global information [30]. Spatial-temporal fusion graph neural networks for traffic flow forecasting (SFTGN) fusion of graphs of different time periods [31], while learning hidden spatial-temporal features, solves the long and short cycle dependence, and effectively extracts global and local features. Traffic Flow Forecasting with Spatial-Temporal Graph Diffusion Network (ST-GDN) adopts multi-scale attention and temporal layer aggregation, embeds contextual information in the perception based on granular representation[32], which can effectively retain global information. DCRNN [33] uses diffusion graph convolution to capture the spatial correlation on the spatial map at each time step, [34] proposed two attention mechanisms in the image description task, and used a visual method to visually demonstrate the effects of the attention mechanism. In order to classify graph nodes, [19] used a self-attention layer to process graph structure data through a neural network in 2018, and achieved the most advanced results. In order to predict time series, [35] proposed a multi-level attention network in 2017 to adaptively adjust the correlation between multiple geographic sensor time series. However, due to the need to train a separate model for each time series, this is time-consuming in practice. [36] proposed a spatial-temporal graph convolutional neural network (ASTGCN) based on the attention mechanism. However, Graph Attention Networks (GATs) cannot guarantee to retain the structure of the node distribution before the nonlinear activation function, and the similarity may not be retained after the convolution operation.

3 Preliminaries

In this section, we denote the road network as an undirected graph $G = (V, E, A)$, V is the set of traffic nodes, E is the edge of nodes i which indicate the connection between nodes, A is the adjacent matrix of graph represents all eigenvalues of nodes at time t. We definite as all eigenvalues of all nodes at time t, c is the frequency sample measures. Table 1 shows some notations and descriptions in our task.

Table 1. Notations and Description in CRFST-GCN

Notation	Description
$G =< V; E >$	Undirected graph
V	Finite set of N nodes
E	Connection edges of nodes
A	Adjacent matrix of graph
$x_t^{i,c}$	The cth eigenvalue of node i at time t
x_t^i	All eigenvalues of node i at time t
X_t	All eigenvalues of all nodes at time t
$\mathcal{N}_i$	Neighourhood node set of node i
$\mathcal{X}_t$	Trend segment
$\mathcal{X}_d$	Daily segment
$\mathcal{X}_w$	Weekly segment

3.1 Problem Definition

Given X, to predict the traffic Y of all nodes on the entire transportation network in the next time period. We formulate the prediction problem in Eq. (1),

$$Y = \mathcal{F}_\Theta(\mathcal{X}, A) \tag{1}$$

4 Proposed Method

4.1 CRFST-GCN Framework

From an intuitive point of view, traffic congestion is gradual and the congestion at the last moment will impact the next or even several moments. When congestion occurs in a particular place (peak period, accident, emergency, etc.), it will quickly cause congestion to nearby location. In addition, the trajectory of the flow also shows regular periodicity, for example, the morning peak and the evening peak hours follow daily cyclical patterns. The flow on Saturdays and Sundays will also show a different pattern from working days, thus forming a weekly cycle. Therefore, this paper proposes a multi-branch Spatial-temporal graph convolutional network, which models the adjacent time, daily period, and weekly period to capture the long and short periodic characteristics. In particular, the three modules shared the same structure and finally merged through a fully connected layer. The frame shows in Fig. 1.

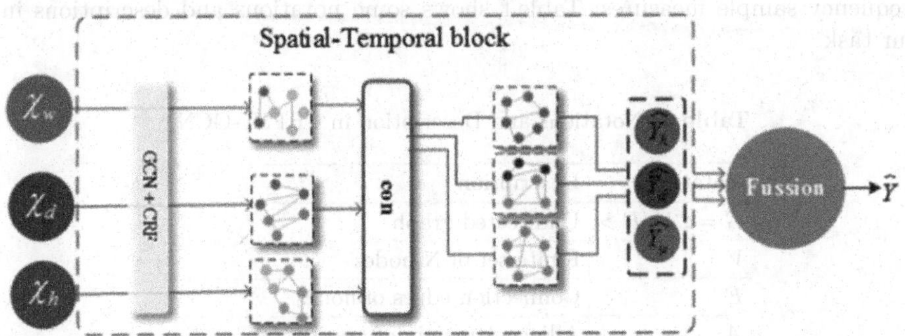

Fig. 1. CRFST-GCN frame,$\mathcal{X}_h$, $\mathcal{X}_d$, $\mathcal{X}_w$ capture the long and short cycle dependence respectively. GCN captures the spatial relationship, the CRF layer captures the graph convolution hidden layer similarity, con captures the time dimension of the long- and short-term periodic relationship, and the fully connected layer outputs the prediction results.

4.2 GCN on Spatial

The forecasting process can be regarded as a causal function, predicting the flow rate in a period of time in the future based on the current observation data. The spectral theory maps two-dimensional data to graph convolutional networks. GCN processes irregular graph data by decomposing the Laplacian matrix. Based on this, we regard transportation networks as graph structures and graph signals as nodes and mapped the traffic network to GCN to represent the spatial relationship of the traffic network. By analyzing the Laplacian matrix and its eigenvalues to retrieve the properties of the graph [37] proposed, the process of realizing frequency domain filtering is as follows:

$$g_\theta \star G(x) = g_\theta(L)x = Ug_\theta(\Lambda)U^T x \tag{2}$$

The eigenvalue decomposition of the Laplace matrix is $L = U\Lambda U^T$. U is the Fourier basis, and U is the diagonal matrix. Obviously that they assume that the graph signal is multi-dimensional, and the oblique parameters are learnable. In this way, each graph needs to be decomposed by L, and the complexity will increase accordingly [38]. Therefore, this paper selects Chebyshev polynomials to approximate and effectively solve the problem,

$$g_\theta *_G x = g_\theta(\mathbf{L})x = \sum_{k=0}^{K-1} \theta_k T_k(\widetilde{\mathbf{L}})x, \tag{3}$$

where $\theta \in R^K$ is a vector parameter of polynomial coefficients, $\tilde{L} = 2(L - I_N)/\lambda_{\max}$, where $\lambda_{\max}$ represents the maximum eigenvalue of L. T_k is Chebyshev polynomial. Adopt Chebyshev polynomial approximate expansion to solve the formula for extracting neighbors from 0 to K order around each node in the graph through the convolution kernel. [39] proposed an effective variant of the convolution neural network that directly operates on the graph, The local first-order approximation of spectrogram convolution can encourage the excellence of convolution structure. In this section, we propose a multi-layer GCN, the stratification breeding rule is:

$$H^{(l)} = \mathrm{ReLU}\left(\hat{A}H^{(l-1)}W^{(l-1)}\right), \tag{4}$$

where ReLU is the active action and $W^{(l-1)}$ is the learned parameters in the $(l-1)$-th layer. $\hat{A}$ is defined as:

$$\hat{A} = \tilde{D}^{-\frac{1}{2}}\tilde{A}\tilde{D}^{-\frac{1}{2}} \tag{5}$$

4.3 Spatial CRF-GCN Layer

In spatial dimension, the influence of traffic conditions between different nodes is highly dynamic. Compared with general data, the graph between different other nodes hides parallel information, so it is challenging to maintain this information in the GCN hidden layer. [40] presents a flexible CRF layer that can directly extract hidden similarity features between nodes in the GCN process, Fig. 2 shows that similar nodes are closer after embedding CRF.

In this paper, in order to maintain the similarity in the propagation of graph neural network, we borrow the idea from CRF which can capture the pair-wise relation. In specific, node representation of hidden layer $H^{(l)}$ can be regarded as random variable $\{H_i^{(l)}\}$, in which $H_i^{(l)}$ corresponding to node i. These random variable are under condition of $\{O_i^{(l)}\}$, where $O_i^l = \mathcal{F}_l(\hat{A}, H^{(l-1)}, W^{(l-1)})$. Based on these, the CRF apply in this paper is formulate as:

$$P(H^{(l)}|O^{(l)}) = \frac{1}{Z(O^{(l)})}\exp(-E(H^{(l)}|O^{(l)})), \tag{6}$$

where $Z(\cdot)$ is normalize factor, $E(\cdot)$ is energy function. Otherwise, the distance between node i and j will be enlarged. Subsequently, similar nodes are mapped to parallel position in hidden space. Finally, the energy function can be define as:

$$E(H_i^{(l)}|O_i^{(l)}) = \alpha\left\|H_i^{(l)} - O_i^{(l)}\right\|_2^2 + \beta\sum_{j\in\mathcal{N}_i} g_{ij}\left\|H_i^{(l)} - H_j^{(l)}\right\|_2^2, \tag{7}$$

in order to flexibly control the two components, we introduce two non-negative parameters α and β. The similarity fraction g_{ij} is computed by:

$$g_{ij} = \exp\left(\frac{\alpha O_j^{(l)} O_i^{(l)T}}{\left\|O_j^{(l)}\right\|_2\left\|O_i^{(l)}\right\|_2}/\sigma^2\right), \tag{8}$$

Obviously, the Eq. 7 can also forces the approximation of $O_i^{(l)}$ and $H_i^{(l)}$. Compared to [5], the pairwise components play a role in regularization.

After defining Eq. 7, we acquire a renew rule for the target representation $H^{(l)}$ by

$$\left(H_i^{(l)}\right)^k = \frac{\alpha O_i^{(l-1)} + \beta \sum_{j \in N_i} g_{ij} \left(H_i^{(l-1)}\right)^{k-1}}{\alpha + \beta \sum_{j \in N_i} g_{ij}}, \tag{9}$$

where N_i is the neighborhood of node i, and $\alpha > 0$ and $\beta > 0$ are used to process the balance of [6], respectively.

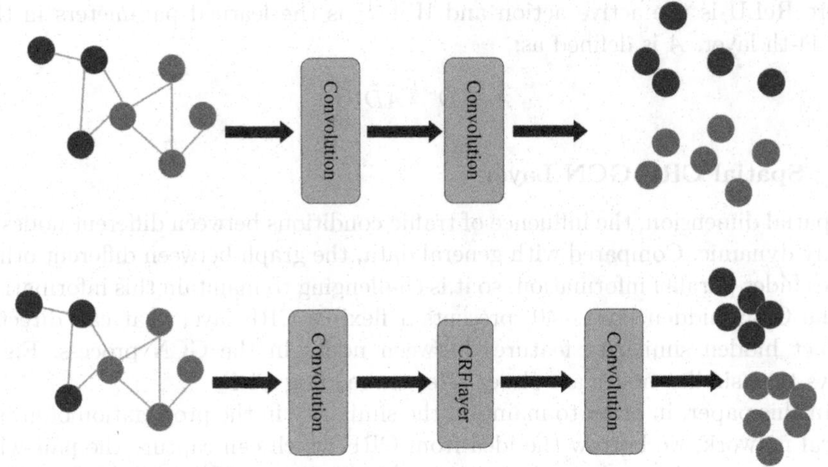

Fig. 2. Comparing the graph convolution neural network with the CRF layer, the designed CRF layer can better maintain the similarity of the output characteristics.

4.4 Temporal Convolution

After capturing the neighboring information of the node in the spatial dimension through graph convolution operation, Then, through a standard convolution layer in the time dimension, the information on the adjacent time slices of the signal of a node is updated by merging. Then, the operation of the first level in the time dimension is expressed as:

$$\mathcal{X}_h^{(l)} = \mathrm{Re}\,LU\left(\Phi * \left(\mathrm{ReL}\,U\left(g_{\theta * G}\widehat{\mathcal{X}}_h^{(l-1)}\right)\right)\right) \in \mathbb{R}^{C_r \times N \times T_r}, \tag{10}$$

where $*$ denotes a standard convolution operation, Φ is the temporal dimension convolution kernel, and ReLU is the activation function.

4.5 Multi-scale Fusion

In different time segment, daily and weekly components have different influences on traffic, thus the output paraments is different during fusion the three components and need to be learned from historical data. The final fusion result is:

$$\widehat{\mathbf{Y}} = \mathbf{W}_h \odot \widehat{\mathbf{Y}}_h + \mathbf{W}_d \odot \widehat{\mathbf{Y}}_d + \mathbf{W}_w \odot \widehat{\mathbf{Y}}_w, \tag{11}$$

where $\odot$ denotes inner product and $\mathbf{W}_h$, $\mathbf{W}_d$, and $\mathbf{W}_w$ denote trainable linear transformation weights. The prediction process of the entire framework is shown in Algorithm 1.

Algorithm 1: Algorithm of CRFST-GCN block

```
Input: num_of_timesteps, num_of_vertices, graph_signal_matrix,
       points_per_hour.
Output: Result.
for time_step in range(num_of_timesteps): do
    graph_signal = x[:, :, :, time_step]
    output self.name_scope()
    for k in range(crf.K): do
        graph_signal = x[:, :, :, time_step]
        T_k= crf.cheb_polynomials[k]
        T_k_with_crf= T_k * spatial_crf
        theta_k= crf.Theta.data()[k]
        rhs= nd.batch_dot(T_k_with_crf.transpose(),graph_signal)
        Result = nd.relu(nd.concat(*outputs, dim=-1));
    end
end
return Result;
```

5 Experiments

We have carried out the experimental process on two real highway traffic data sets, and a large number of comparative experiments show that the prediction accuracy of the proposed model is the highest.

5.1 Evaluation Metrics

Three commonly used indicators are considered, the Root Mean Squared Errors (RMSE), Mean Absolute Errors (MAE), and Mean Absolute Percentage Error (MAPE), to evaluate the performance of all compared models as follows:

– **MAE**

$$\mathrm{MAE} = \frac{1}{n}\sum_{i=1}^{n}|x_i - \hat{x}_i| \tag{12}$$

– **RMSE**

$$\mathrm{RMSE} = \sqrt{\frac{1}{n}\sum_{i=1}^{n}(x_i - \hat{x}_i)^2} \tag{13}$$

– **MAPE**

$$\mathrm{MAPE} = \frac{100\%}{n}\sum_{i=1}^{n}\left|\frac{\hat{x}_i - x_i}{x_i}\right| \tag{14}$$

where n is the number of test samples.

5.2 Experimental Datasets

We perform experiments on PeMSD4 and PeMSD8 datasets. The Caltrans Performance Measurement System collects the datasets (PeMS) 111 in real-time every 30 s, composed of 39000 detectors on the California highways, USA. The total flow, average occupancy, and average speed are considered as prediction indexes in this experiment. The training and test sets are the first 50 days and the remaining data, respectively. Details of the statistical properties of the two datasets introduced above are available in Table 2.

Table 2. Statistics of PEMSD4 and PEMSD8 datasets

Data	#Nodes	#Edges	#Timesteps	Time ranges
PEMSD4	307	340	16992	1/1/2018-2/28/2018
PEMSD8	170	277	17856	1/7/2016-31/8/2016

This algorithm uses linear interpolation to complete the data, minimizes the error through backpropagation, and implements the CRFST-GCN model on the MXNET-based framework, the hardware environment is a single desktop device running Windows 10 Professional 64-bit version having an Intel Core i5-4460 processor with a CPU 3.2 GHz. The convolution kernel size is 64, K = 3, and the learning rate is 0.0001.

5.3 Performance Evaluation

To compare our model with eight baselines on PeMSD4 and PeMSD8. Table 3 shows the average performance of predict the traffic flow in the next hour. the method CRFST-GCN get the best performance in all evaluation. The low accuracy of traditional time series is due to the limitations of time series-based forecasts in obtaining spatial correlation. In addition, the LSTM model does not take into account the dynamic spatial similarity and the transition of periodic changes, resulting in poor performance. The excellent performance of this model proves the effectiveness of the periodic system shunt mechanism in capturing the similarity of dynamic time and space. STGCN predicts the traffic flow from spatial and temporal and gets an excellent result. But there is also a limitation in GCN on capturing spatial structure. ASTGCN catches the fine-grained spatial-temporal relation through the attention mechanism, but the active line function is before convolution, resulting in the loss. Diffusion graph convolution is used to model the spatial-temporal correlation on each time step and then put them into temporal series components GRU, finally getting the predicted value by encoding and decoding structure. The proposed method CRFST-GCN can capture the similarity of the hidden layer during the convolution processes, which can result in the accurate spatial correction.

Table 3. Comparison of the average performance of different methods on PeMSD4 and PeMSD8.

Model	PEMSD4			PEMSD8		
	MAE	RMSE	MAPE	MAE	RMSE	MAPE
HA [7]	38.03	59.24	27.88	34.86	52.04	24.07
VAR [8]	24.44	37.76	17.27	19.21	29.74	13.09
LSTM [14]	25.72	39.67	17.71	20.40	31.55	12.77
GRU [15]	25.80	39.67	17.40	20.48	31.54	12.83
DCRNN [33]	21.08	32.74	14.62	16.10	25.00	10.28
STGCN [27]	22.89	35.39	14.94	19.00	28.83	11.78
ASTGCN [36]	22.69	34.65	16.23	18.97	28.39	12.70
STSGCN [29]	21.19	33.65	13.90	17.13	26.86	10.96
CRFST-GCN (Ours)	20.82	31.13	14.94	15.24	24.23	10.19

Figure 3 depicts the results of different baselines. As the length of the prediction increases, the prediction error also increases. In addition, in short interval prediction, methods such as LSTM and GRU only use time correlation as an indicator, and the prediction accuracy is high. However, with the increase of the prediction length, the prediction accuracy decreases sharply. In contrast, in Fig. 5, the decline rate of method VAR is slower than these methods, mainly because VAR can also take into account more critical temporal and spatial correlations in long-term forecasts. However, when the scale of traffic network becomes increasingly prominent, the more time series considered in the model, the greater the prediction error of VAR. In Fig. 4 that its performance on PeMSD4 is worse than that on PeMSD8. The error of the deep learning method increases slowly with the increase of the prediction length. In short, the overall degree of completion is better.

In addition, the performance of ASTGCN is greater than that of STGCN, indicating that the energy change of the data stream can be controlled by multi-level attention. However, without any attention mechanism, the CRFST-GCN model proposed in this paper has achieved superior results over the most

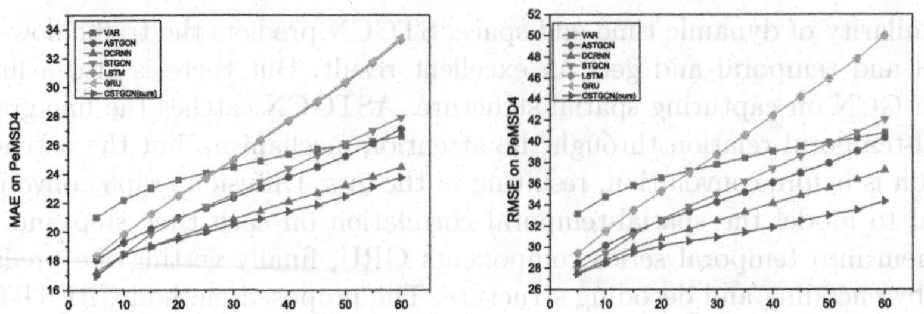

Fig. 3. Performance Evaluation with increasing prediction interval on MAE.

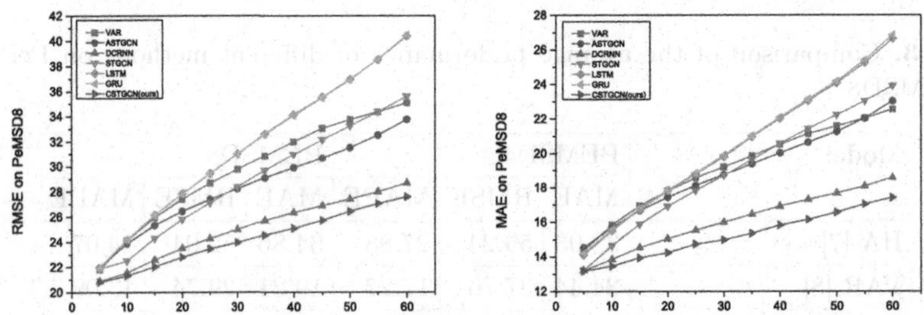

Fig. 4. Performance Evaluation with increasing prediction interval on RMSE.

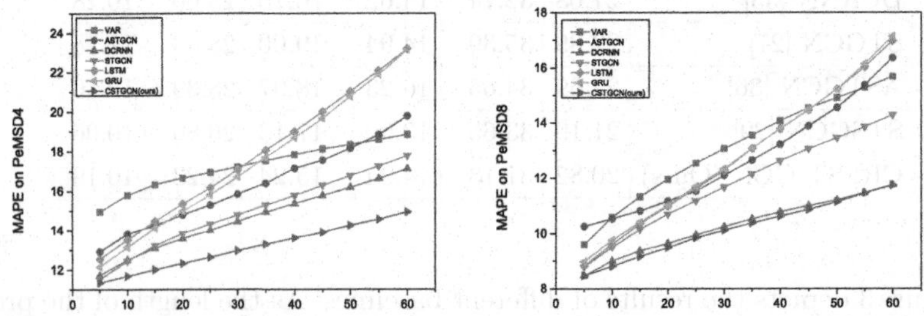

Fig. 5. Performance Evaluation with increasing prediction interval on MAPE.

advanced baselines, which proves the advantages of the proposed model in describing the characteristics of spatial-temporal data.

6 Conclusion

This paper proposes a flexible CRFST-GCN model and has achieved the best results in all evaluation indicators. It proves the advantages of the multi-segment model processing method in capturing spatial relationships. The prediction accuracy is also affected by objective external factors, so in future work, we will consider embedding more external factors such as weather and temperature to improve the prediction accuracy.

Acknowledgement. This work is supported by the National Natural Science Foundation of China under Grants 61976087, 62072170, 61872130, and 61872138, the Fundamental Research Funds for the Central Universities under Grant 531118010527, the Science and Technology Key Projects of Hunan Province (No.2022GK2015) and the Hunan Provincial Natural Science Foundation of China (No.2021JJ30141).

References

1. Liang, W., Long, J., Li, K.-C., Xu, J., Ma, N., Lei, X.: A fast defogging image recognition algorithm based on bilateral hybrid filtering. ACM Trans. Multimedia Comput. Commun. Appl. **17**(2), 1–16 (2021). https://doi.org/10.1145/3391297
2. Xu, J., et al.: NFMF: neural fusion matrix factorisation for QoS prediction in service selection. Connect. Sci. **33**, 1–16 (2021)
3. Liang, W., Li, Y., Xu, J., Qin, Z., Li, K.C.: QoS prediction and adversarial attack protection for distributed services under DLaaS. IEEE Trans. Comput., 1–14 (2021)
4. Liang, W., Xie, S., Long, J., Li, K.-C., Zhang, D., Li, K.: A double puf-based rfid identity authentication protocol in service-centric internet of things environments. Inf. Sci. **503**, 129–147 (2019). https://www.sciencedirect.com/science/article/pii/S0020025519305857
5. Liang, W., Ning, Z., Xie, S., Hu, Y., Lu, S., Zhang, D.: Secure fusion approach for the internet of things in smart autonomous multi-robot systems. Inf. Sci. **579**, 468–482 (2021)
6. Liang, W., Zhang, D., Lei, X., Tang, M., Li, K.C., Zomaya, A.Y.: Circuit copyright blockchain: blockchain-based homomorphic encryption for IP circuit protection. IEEE Trans. Emerg. Topics Comput. **9**(3), 1410–1420 (2020)
7. Garrow, D.: Odd deposits and average practice. a critical history of the concept of structured deposition. Arch. Dial. **19**(2), 85–115 (2012)
8. Zivot, E., Wang, J.: Vector autoregressive models for multivariate time series. In: Modeling Financial Time Series with S-Plus®, pp. 385–429 (2006)
9. Williams, B.M., Hoel, L.A.: Modeling and forecasting vehicular traffic flow as a seasonal arima process: theoretical basis and empirical results. J. Transp. Eng. **129**(6), 664–672 (2003)
10. Denoeux, T.: A k-nearest neighbor classification rule based on dempster-shafer theory. In: Classic Works of the Dempster-Shafer Theory of Belief Functions, pp. 737760. Springer, Heidelberg (2008). https://doi.org/10.1007/978-3-540-44792-4_29
11. Joachims, T.: Making large-scale SVM learning practical. Technical report (1998)
12. Liang, W., Xie, S., Zhang, D., Li, X., Li, K.: A mutual security authentication method for RFID-PUF circuit based on deep learning. ACM Trans. Internet Technol. **22**, 1–20 (2020)
13. Liang, W., Xiao, L., Zhang, K., Tang, M., He, D., Li, K.-C.: Data fusion approach for collaborative anomaly intrusion detection in blockchain-based systems. IEEE Internet Things J., 1 (2021)
14. Hochreiter, S., Schmidhuber, J.: Long short-term memory. Neural Comput. **9**(8), 1735–1780 (1997)
15. Cho, K., et al.: Learning phrase representations using rnn encoder-decoder for statistical machine translation. arXiv preprint arXiv:1406.1078 (2014)
16. Bruna, J., Zaremba, W., Szlam, A., LeCun, Y.: Spectral networks and locally connected networks on graphs. arXiv preprint arXiv:1312.6203 (2013)

17. Kipf, T.N., Welling, M.: Semi-supervised classification with graph convolutional networks. arXiv preprint arXiv:1609.02907 (2016)
18. Chiang, W.-L., Liu, X., Si, S., Li, Y., Bengio, S., Hsieh, C.-J.: Cluster-GCN: an efficient algorithm for training deep and large graph convolutional networks. In: Proceedings of the 25th ACM SIGKDD International Conference on Knowledge Discovery & Data Mining, pp. 257–266 (2019)
19. Veličković, P., Cucurull, G., Casanova, A., Romero, A., Lio, P., Bengio, Y.: Graph attention networks. arXiv preprint arXiv:1710.10903 (2017)
20. Hamilton, W.L., Ying, R., Leskovec, J.: Inductive representation learning on large graphs. In: Proceedings of the 31st International Conference on Neural Information Processing Systems, pp. 1025–1035 (2017)
21. Ge, Z., Li, Y., Liang, C., Song, Y., Zhou, T., Qin, J.: Acsnet: adaptive cross-scale network with feature maps refusion for vehicle density detection. In: IEEE International Conference on Multimedia and Expo (ICME) 2021, pp. 1–6 (2021)
22. Zhang, S., Wu, G., Costeira, J.P., Moura, J.M.: FCN-RLSTM: deep spatio-temporal neural networks for vehicle counting in city cameras. In: Proceedings of the IEEE International Conference on Computer Vision, pp. 3667–3676 (2017)
23. He, K., Zhang, X., Ren, S., Sun, J.: Deep residual learning for image recognition. In: Proceedings of the IEEE Conference on Computer Vision and Pattern Recognition, pp. 770–778 (2016)
24. Zhang, J., Zheng, Y., Qi, D., Li, R., Yi, X., Li, T.: Predicting citywide crowd flows using deep spatio-temporal residual networks. Artif. Intell. **259**, 147–166 (2018)
25. Yao, H., Tang, X., Wei, H., Zheng, G., Li, Z.: Revisiting spatial-temporal similarity: a deep learning framework for traffic prediction. Proc. AAAI Conf. Artif. Intell. **33**(01), 5668–5675 (2019)
26. Kong, X., Xing, W., Wei, X., Bao, P., Zhang, J., Lu, W.: STGAT: spatial-temporal graph attention networks for traffic flow forecasting. IEEE Access **8**, 134363–134372 (2020)
27. Yu, B., Yin, H., Zhu, Z.: Spatio-temporal graph convolutional networks: a deep learning framework for traffic forecasting. arXiv preprint arXiv:1709.04875 (2017)
28. Wu, Z., Pan, S., Long, G., Jiang, J., Zhang, C.: Graph wavenet for deep spatial-temporal graph modeling. arXiv preprint arXiv:1906.00121 (2019)
29. Song, C., Lin, Y., Guo, S., Wan, H.: Spatial-temporal synchronous graph convolutional networks: a new framework for spatial-temporal network data forecasting. Proc. AAAI Conf. Artif. Intell. **34**(01), 914–921 (2020)
30. Chen, X., Liang, W., Xu, J., Wang, C., Li, K.-C., Qiu, M.: An efficient service recommendation algorithm for cyber-physical-social systems. IEEE Trans. Netw. Sci. Eng., 1 (2021)
31. Li, M., Zhu, Z.: Spatial-temporal fusion graph neural networks for traffic flow forecasting. arXiv preprint arXiv:2012.09641 (2020)
32. Zhang, X., et al.: Traffic flow forecasting with spatial-temporal graph diffusion network (2020)
33. Lin, Z., Feng, J., Lu, Z., Li, Y., Jin, D.: Deepstn+: context-aware spatial-temporal neural network for crowd flow prediction in metropolis. Proc. AAAI Conf. Artif. Intell **33**(01), 1020–1027 (2019)
34. Xu, K., et al.: Show, attend and tell: neural image caption generation with visual attention. In: International Conference on Machine Learning, PMLR, pp. 2048–2057 (2015)
35. Liang, Y., Ke, S., Zhang, J., Yi, X., Zheng, Y.: Geoman: multi-level attention networks for geo-sensory time series prediction. IJCAI **2018**, 3428–3434 (2018)

36. Guo, S., Lin, Y., Feng, N., Song, C., Wan, H.: Attention based spatial-temporal graph convolutional networks for traffic flow forecasting. Proc. AAAI Conf. Artif. Intell **33**(01), 922–929 (2019)
37. Shuman, D.I., Narang, S.K., Frossard, P., Ortega, A., Vandergheynst, P.: The emerging field of signal processing on graphs: Extending high-dimensional data analysis to networks and other irregular domains. IEEE Signal Processing Mag. **30**(3), 83–98 (2013)
38. Henaff, M., Bruna, J., LeCun, Y.: Deep convolutional networks on graph-structured data. arXiv preprint arXiv:1506.05163 (2015)
39. Simonovsky, M., Komodakis, N.: Dynamic edge-conditioned filters in convolutional neural networks on graphs. In: Proceedings of the IEEE Conference on Computer Vision and Pattern Recognition, pp. 3693–3702 (2017)
40. Gao, H., Pei, J., Huang, H.: Conditional random field enhanced graph convolutional neural networks. In: Proceedings of the 25th ACM SIGKDD International Conference on Knowledge Discovery & Data Mining, pp. 276–284 (2019)

BFR-RetinaNet: An Improved RetinaNet Model for Vehicle Detection in Aerial Images

J. I. N. Zhang[1,3], Meng Luo[2](✉), Cheng Sun[2], and Peiqi Qu[1]

[1] College of Information Science and Engineering, Hunan Normal University, Changsha, China
qpeggy@hunnu.edu.cn

[2] College of Mathematics and Statistics, Hunan Normal University, Changsha, China
luomeng@hunnu.edu.cn

[3] School of Computer and Communication Engineering, Changsha University of Science and Technology, Changsha, China

Abstract. Vehicle detection in aerial images has been applied in many fields and attracted more and more scholars' attention. In the task, the objects are multi-directional and arranged densely, the background information is complex, and the scale of the object is different. To achieve better detection performance, an improved detection model BFR-RetinaNet is proposed, which is based on the single-stage object detection model RetinaNet. BFR-RetinaNet optimizes vehicle positioning by adding a directional anchor box regression branch. Simultaneously, the model introduces a balanced feature pyramid structure to enhance the extraction of features and reduce the interference of complex backgrounds. The experimental results on the aerial dataset show that the precision, recall, and average precision of the proposed model have been improved to varying degrees. It achieves 86.2% Precision, 98.4% Recall, and 90.8% Average Precision (AP), which is 7.96, 0.45, and 0.58 points higher than R-RetinaNet.

Keywords: Vehicle detection · Aerial imagery · Oriented bounding box · RetinaNet · Feature pyramid network

1 Introduction

In the field of computer vision, object detection is a basic and important visual recognition task. It mainly includes two tasks. One is the positioning task, which is to accurately locate the instance object on the tested image. The general positioning is determined in the form of rectangular boxes. The other is the classification task for determining the

This work is partially supported by the Open Research Project of the State Key Laboratory of Industrial Control Technology (No. ICT2021B10), the Natural Science Foundation of Hu-nan Province (2021JJ30456), the Open Fund of Science and Technology on Parallel and Distributed Processing Laboratory (WDZC20205500119), the Hunan Provincial Science and Technology Department High-tech Industry Science and Technology Innovation Lead-ing Project (2020GK2009) and the Scientific and Technological Progress and Innovation Program of the Transportation Department of Hunan Province (201927).

Y. Lai et al. (Eds.): ICA3PP 2021, LNCS 13155, pp. 18–32, 2022.
https://doi.org/10.1007/978-3-030-95384-3_2

specific category of the object, such as common people, animals, vehicles, etc. Object detection provides the most basic information for computer vision, that is, what kind of object is in and where [1].

In the early stage, the process of object detection consists of three consecutive steps: generate candidate regions; extract feature vectors; and return the region category. In the first stage, search for the position of the detected object through the sliding window. Simultaneously, to obtain object information of multiple scales and different aspect ratios, the tested image is usually scaled; in the second stage, the information acquired by the sliding window is extracted to a fixed-length feature vector; finally, the feature vector is sent to a specific classifier, and the object category is returned.

Subsequently, breakthrough progress has been made in object detection with the rapid development of deep learning. At this stage, object detection methods are roughly divided into two major branches: two-stage detection methods and single-stage detection methods [2]. The two-stage detection method will firstly frame the candidate regions with coarse precision, and then perform detailed classification and regression of the candidate regions, such as R-CNN [3], Fast R-CNN [4], Faster R-CNN [5], etc.; the single-stage detection method, abandoning the region recommendation step, treating the object detection as a regression task, and directly classifying and regressing the object, such as SSD [6], DSSD [7], YOLO (You Only Look Once) [8], YOLO9000 [9], YOLOv3 [10], etc.

Generally, a two-stage detector can achieve higher accuracy requirements, and a single-stage detector can achieve faster detection. Some scholars have studied the balance between the accuracy of the two-stage detection model and the speed of the single-stage detection model. Lin et al. proposed the RetinaNet [11], which achieves high accuracy and excellent speed at the same time. They proposed the focus loss function and constructed a new single-stage detection framework RetinaNet based on the novel loss function. Based on this, the paper will be based on the RetinaNet model which has certain advantages in accuracy and speed.

Recently, vehicle detection in aerial images becomes a hot issue under the development of drone technology and the application of intelligent transportation systems. Zhao Shuang [12] et al. proposed an improved YOLOv2 algorithm, which constructs a three-step method of cropping, detection, and stitching to overcome the difficulty of large aerial images, and fuses different features to address the problem of small object detection in vehicle detection; Lu Bo [13] et al. proposed an improved YOLOv3-tiny aerial vehicle detection method. This method incorporates the BiFPN network pyramid structure and proposes a new up-sampling structure, which makes the model achieve great improvement in the terms of the speed and accuracy for the task of vehicle detection in aerial images.

Compared with object detection in general images, vehicle detection in aerial images must overcome the following difficulties:

- The direction of the detected object is arbitrary. UAV aerial images acquire a wide range of scene information through the top view, ensuring the integrity of real-world information. At the same time, due to the particularity of the bird's-eye view, the detected objects are arranged densely and have a specific rotation angle, which brings greater challenges to object detection.

- The scale of the object varies greatly. Aerial images are taken by drones at high altitudes. There are differences in the size of the image obtained at different flying heights. The same object may have different sizes, which leads to higher requirements for object location and recognition.
- The aerial image has a complicated background. Compared with general images, UAV aerial images have a larger size, contain richer scene information, and cover more complex background information. This undoubtedly adds negative noise and increases the difficulty of object detection.

Therefore, we propose an improved RetinaNet, which combines a rotating branch and a balanced feature pyramid structure. For the multi-directional problem of the detected object, the model introduces a rotation prediction branch, which increases the angle information of the positioning object, greatly reduces the invalid background information; Facing the scale changes of the objects and the background interference of large scenes, the model uses a balanced feature pyramid module. The low-level feature map of the feature pyramid contains rich details, which is conducive to the recognition and positioning of small objects; and the high-level feature extracts abstract semantic information, which is convenient for the model to detect large objects. The introduction of the balanced feature pyramid structure combines the low-level and high-level feature maps and solves the problem of feature information imbalance. It eliminates the difficulty of object detection at different scales and reduces the interference of invalid background. Experimental results show that the performance of our model has achieved great improvements for vehicle detection in aerial images.

The main work of this paper is as follows:

- The single-stage object detection model RetinaNet is introduced into the field of vehicle detection in aerial images. From the visualized experimental results, it is necessary to enhance the position ability in RetinaNet.
- For the positioning problem of the original RetinaNet, this paper adds a regression branch to get the rotation angle of the object, which expands the object positioning space, from the original horizontal anchor box positioning space to an arbitrary rotation angle anchor box positioning space.
- To solve the multi-scale and complex background problems, this paper introduces a balanced feature pyramid structure and finds a better solution to the use of feature information at different levels. Based on fully fusing the feature information of adjacent layers, the cross-layer feature information can be used.

2 RetinaNet

The RetinaNet detector is a single-stage detector, which is composed of three main parts: a feature extraction network, a feature pyramid network, a classification sub-network, and a regression sub-network. The structure of the network is shown in Fig. 1 (the feature extraction network takes the stage of s1, s2, and s3 as an example). The feature extraction structure sends the original image to a series of convolutional neural networks, and the mapped feature map is obtained through calculation; then, the feature pyramid network merges the features of different layers in the previous stage, and outputs the adjusted feature maps; Finally, the classification and regression sub-networks respectively return the detected classification category and the bounding box position.

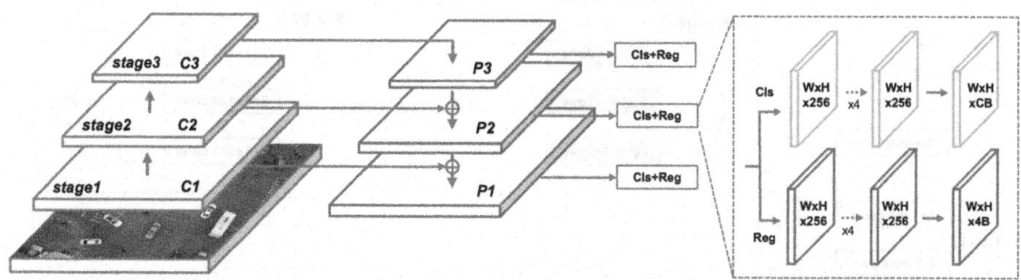

Fig. 1. The network architecture of RetinaNet

2.1 Feature Extraction Network

A series of breakthroughs made by deep convolutional neural networks, such as LeNet [14], AleNet [15], VGGNet [16], etc., have shown that the depth of the network is an important factor affecting the recognition results. However, in [17], as the network deepens, the related experiments show that the recognition model cannot keep the recognition accuracy gradually increasing but will experience saturation of recognition performance or even a decline in recognition accuracy. The residual neural network solves the gradient degradation problem by introducing a residual learning module, so the network can obtain a stable and improved recognition performance while increasing the depth.

Therefore, the current mainstream feature extraction networks all select the deep residual neural network [17]. The deep residual neural network is formed by stacking residual blocks. Different kinds of residual blocks and different numbers of residual blocks can be stacked to form residual networks of different depths. Common residual neural networks from shallow to deep are ResNet18, ResNet34, ResNet50, ResNet101, and ResNet152. Considering the complexity of the network and the feature extraction performance of the network, we use ResNet50 to extract the fundamental features, which not only guarantees the feature extraction ability of the network model, but also does not make the network model too complicated.

The ResNet50 network consists of a 7×7 convolution at the head, four sets of convolutions in the middle, and a fully connected layer in the last stage. The four sets of convolutions in the middle part are stacked by 3, 4, 6, and 3 convolution blocks. Each convolution block is composed of three convolutions of 1×1, 3×3, and 1×1 connected in sequence. Since the ResNet50 network was applied to the field of recognition, the fully connected layer maps the features to the corresponding category activation values, and this paper focuses on the object detection, so the last fully connected layer of ResNet50 is discarded directly and the modified structure is shown in Fig. 2.

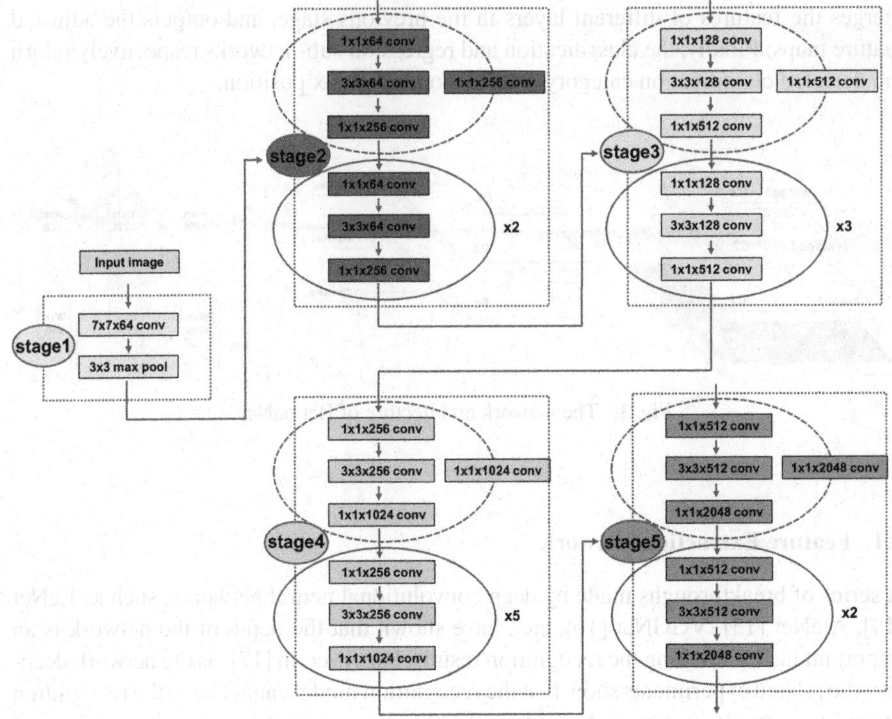

Fig. 2. The network architecture of ResNet50

2.2 Feature Pyramid Network

The object detection and recognition of different sizes in the real world have always been a basic problem. The feature pyramid network realizes the fusion of different feature maps through a bottom-up, top-down, and horizontal connection structure. It can establish a feature pyramid with rich semantics on a single-scale image, and then solve the problem of identifying objects with different sizes.

The RetinaNet builds Feature Pyramid Network (FPN) [18] based on the feature extraction network ResNet. In the bottom-up stage, the model selects the feature maps from the outputs of the last three groups of convolutional networks in the middle part of ResNet; in the lateral connection stage, the network sends feature maps to 1 × 1 convolution respectively and obtain feature maps with 256 output channels; in the top-down stage, the network sequentially merges the expanded feature maps to the next layer through superposition operation, and connects the output results to 3 × 3 convolutions; finally, it obtains the feature maps from P3 to P5. The P6 is obtained by layer C5 after 3 × 3 convolution operation, and the P7 is obtained by 3 × 3 convolution and ReLU activation operation from P6. The complete network structure is shown in Fig. 3.

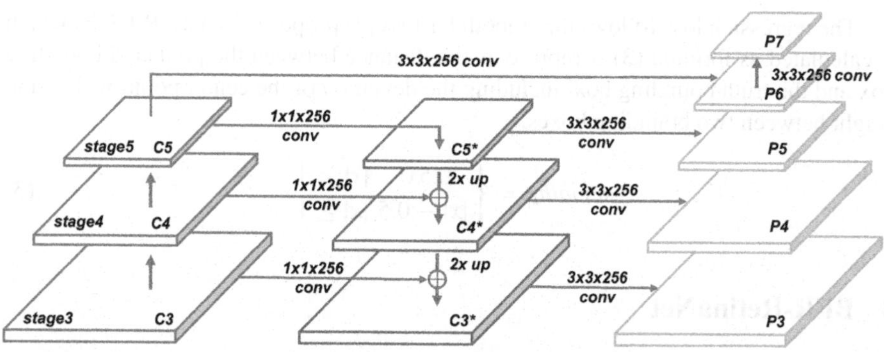

Fig. 3. The network architecture of FPN

2.3 Classification and Regression Subnets

The classification and regression tasks of the RetinaNet fall into the classification and regression sub-networks. The classification sub-network is responsible for predicting the category probability of each anchor box. The structure is shown in Fig. 1. At the end of each feature map from the feature fusion network, four 3x3 convolutional networks are connected in sequence, and then follows a 3×3 convolutional network with CxB (C represents the number of classification categories, B represents the number of preset bounding boxes at each location) channels, and finally, the activation function is passed in to obtain the category probability value of each bounding box.

The regression sub-network returns the offset value of the predict bounding box and the ground-truth bounding box. The structure is shown in Fig. 1. The regression sub-network is parallel to the classification sub-network, the same four convolutional layers are connected in sequence. Since the learning offset value of the regression sub-network is four dimensions, a convolutional network with a channel number of 4xB and a convolution kernel of 3×3 is connected later, and finally, the position offset value of each anchor box is calculated. These two sub-networks share the same structure but use separate model parameters.

2.4 Loss Function

The training loss value of the RetinaNet network is composed of classification loss and regression loss. The calculation method of classification loss selects the focal loss, and its calculation is as formula (1). Taking the two-classification model as an example, multiple classifications can be extended in parallel. p_t is an indicator function, and its value can be expressed by formula (2). When the tested category is true, the probability value p_t is taken p; when the tested category is false, the probability value p_t is taken $1 - p$, a_t, γ represents the weighting factor. According to [11], we set $a_t = 0.25$ and $\gamma = 2$.

$$FL(p_t) = -a_t(1 - p_t)^{\gamma} \log(p_t) \tag{1}$$

$$p_t = \begin{cases} p & y = 1 \\ 1 - p & y \neq 1 \end{cases} \tag{2}$$

The regression loss follows the smooth L1 loss [4] proposed by Fast R-CNN, which is calculated as formula (3), x represents the distance between the predicted bounding box and the truth bounding box, including the deviation of the center point, width, and height between two bounding boxes.

$$smooth_{l_1} = \begin{cases} 0.5x^2 & |x| < 1 \\ |x| - 0.5 & |x| \geq 1 \end{cases} \tag{3}$$

3 BFR-RetinaNet

BFR-RetinaNet keeps the overall framework of the original RetinaNet and focuses on the feature fusion and anchor box regression. As shown in Fig. 4, in the feature fusion stage, the model introduces a balanced feature pyramid structure; in the regression branch stage, the model adds the angle component for predicting the anchor box.

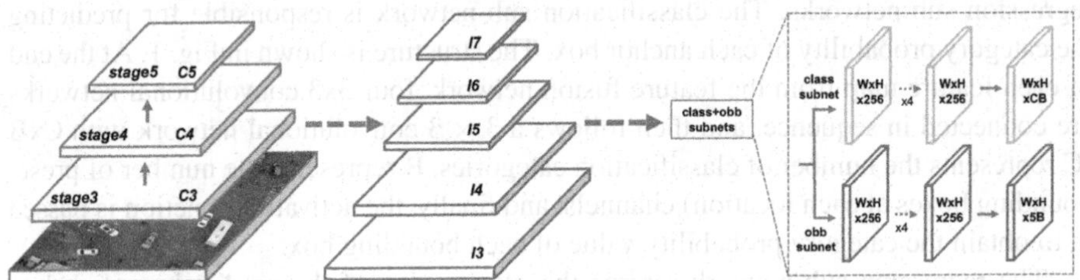

Fig. 4. The network architecture of BFR-RetinaNet

3.1 Balanced Feature Pyramid

In the feature extraction network, shallow features usually contain more specific and detailed content, while deep features usually contain more abstract semantic information. By fusing different levels of information, the model's detection results will make great progress, such as FPN [18], PANet [19]. For the detection effect, whether it is a large feature map or a small feature map, they can all play an active role. However, how to fully use the information of all layers to achieve a better fusion effect is a question. Pang et al. and proposed a novel Balanced Feature Pyramid Network (BFPN) [20] based on FPN.

In Fig. 5, it is obvious that the BFPN consists of four parts: extraction, integration, refinement, and superposition. Firstly, to fuse the feature maps of each layer extracted from the FPN, they are adjusted to the middle layer size through adaptive pooling or nearest-neighbor interpolation; then, the adjusted feature maps of each layer are linearly averaged to obtain the integration layer. Specifically, the calculation method is shown in formula (4).

$$I = \frac{1}{L}\sum_{l_{min}}^{l_{max}} P_l \tag{4}$$

P_l denotes the feature map of the level l, the value of L is the number of all fused feature layers; then, the integration layer is sent to the non-local module [21] to refine the feature layer; finally, adjust the refined feature layer to the original different size through adaptive pooling or nearest-neighbor interpolation calculation, then add the original input feature maps to the refined different feature maps for getting the integration layer with different size.

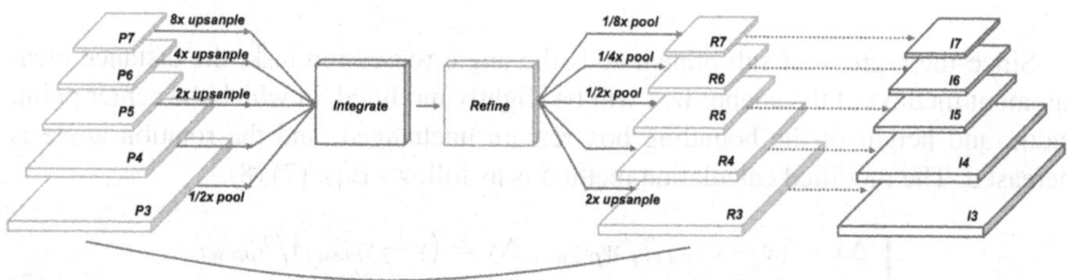

Fig. 5. The network architecture of BFPN

3.2 Oriented Anchor Box Regression

In the regression subtask of the RetinaNet, to accurately predict the location of the objects, the regression branch learns the distance between the predict bounding box and the preset bounding box, and gradually approaches the distance between the truth bounding box and the preset bounding box. Then the center point, width, and height of the predicted bounding box are calculated. The calculation of the distance between the predicted bounding box and the preset bounding box is shown in the following formula (5). Similarly, the distance between the truth bounding box and the preset bounding box is calculated as formula (6). The vector (x, y, w, h) denotes the bounding box, which includes the center point coordinates, width, and height.

$$\begin{cases} \Delta x = (x - x_{preset})/w_{preset}, \ \Delta y = (y - y_{preset})/h_{preset} \\ \Delta w = \log(w/w_{preset}), \qquad \Delta h = \log(h/h_{preset}) \end{cases} \tag{5}$$

$$\begin{cases} \Delta x^* = (x_{truth} - x_{preset})/w_a, \ \Delta y^* = (y_{truth} - y_{preset})/h_{preset} \\ \Delta w^* = \log(w_{truth}/w_{preset}), \ \Delta h^* = \log(h_{truth}/h_{preset}) \end{cases} \tag{6}$$

For detection tasks in general scenarios, the rectangular box predicted by the regression sub-branch of the RetinaNet can achieve good positioning performance. However, as for the detection task in aerial imagery, the detected object has the characteristics of multi-directional and is arranged tightly. So, the rectangular box locates the object not well.

As shown in Fig. 6, a horizontal rectangular box often contains a lot of invalid background information. Therefore, for obtaining more accurate positioning information, we introduce the rotation angle regression task in the regression subtask and replace the horizontal anchor box with the directional anchor box, so that the regression sub-branch predicts the object with the rotation angle.

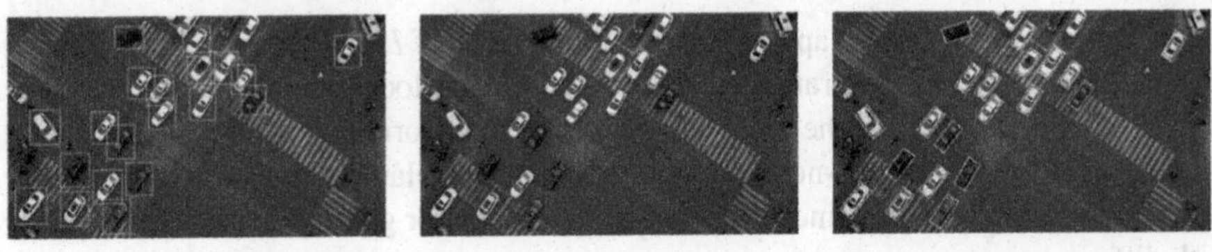

Fig. 6. Comparison of horizontal bounding boxes and oriented bounding boxes

Since the regression sub-branch adds the angle regression task, the distance measurement method of the anchor box will be slightly modified, in which the center point, width, and height of the bounding box remain unchanged, and the rotation angle is increased. The modified calculation method is as follows Eqs. (7) (8).

$$\begin{cases} \Delta x = (x - x_{preset})/w_{preset}, & \Delta y = (y - y_{preset})/h_{preset} \\ \Delta w = \log(w/w_{preset}), & \Delta h = \log(h/h_{preset}) \\ \Delta\theta = \theta - \theta_{preset} \end{cases} \tag{7}$$

$$\begin{cases} \Delta x^* = (x_{truth} - x_{preset})/w_{preset}, & \Delta y^* = (y_{truth} - y_{preset})/h_{preset} \\ \Delta w^* = \log(w_{truth}/w_{preset}), & \Delta h^* = \log(h_{truth}/h_{preset}) \\ \Delta\theta^* = \theta_{truth} - \theta_{preset} \end{cases} \tag{8}$$

The direction bounding box can be represented by (x, y, w, h, θ), which is the center coordinate, the width, the height, and the rotation angle respectively. Then the distance between the predicted direction bounding box and the preset bounding box can be expressed as $(\Delta x, \Delta y, \Delta w, \Delta h, \Delta\theta)$, and the distance between the true value direction anchor box and the preset anchor box is $(\Delta x^*, \Delta y^*, \Delta w^*, \Delta h^*, \Delta\theta^*)$. It is worth mentioning that the preset anchor box generated by the network is still horizontal, but it does not affect the directional anchor box regression training. The preset horizontal anchor box can be regarded as a directional anchor box with a rotation angle of $-\pi/2$.

4 Experiments

4.1 Dataset

We select an aerial vehicle dataset UAV ROD for experiments. The UAV ROD dataset [22] contains 1577 images, a total of 30090 vehicle instances, and each image contains 19 vehicle instances on average. The division of the UAV ROD dataset is described exhaustively in Table 1. The trainset contains 1150 images and a total of 18626 vehicle instances; the testset contains 427 images and a total of 11464 vehicle instances. Different from the instance labeling of the general dataset, the UAV-ROD dataset is labeled with the open-source tool roLabelImg. The label of each vehicle instance includes the center point coordinates, width, height, and rotation angle, which provides the true value of the anchor box during the model training.

Table 1. UAV ROD dataset

	Trainset	Testset	Dataset
Image quantity	1150	427	1577
Instances quantity	18626	11464	30090
Avg. instances quantity	16	27	19
Image size	1920 × 1080 2720 × 1530	1920 × 1080 2720 × 1530	1920 × 1080 2720 × 1530

4.2 Training and Testing

In our experiments, the used deep learning framework is Pytorch 1.1.0, the computer hardware configuration is NVIDIA GTX 2080Ti-11G GPU, and the operating system is Ubuntu 18.04.

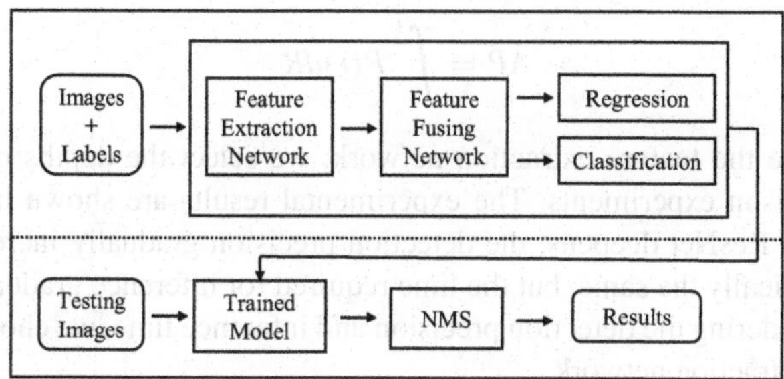

Fig. 7. Diagram of the model training and testing process

As shown in Fig. 7, during the process of training, the images and labels files of the trainset are sent to the improved RetinaNet network. After multiple epochs of iterative training, the model becomes stable and the training process ends; in the test phase, the images of testset are sent to the trained model, the predicted results are output, and then filter through non-maximum suppression, finally, the vehicle position coordinates and category probabilities predicted by the model are obtained.

The realization of the object detection model is supported by an open-source object detection code library AerialDetection [23]. The basic parameter settings are shown in Table 2. The training epoch of the model is set to 12. Since there is only one GPU, the learning rate is linearly adjusted to 0.0025, and other parameter settings remain.

Table 2. Training parameter settings

Parameter	Epoch	Batch size	Learning rate	Momentum
value	12	2	0.0025	0.9

4.3 Evaluation Metrics and Experimental Results

The evaluation indicators of this article are Precision (Pre), Recall (Rec), and Average Precision (AP). The specific calculation is shown in the following formula (9–11). TP, FP, and FN are the confusion matrix parameters of the detection result. The values of Precision, Recall, and Average Precision are all proportional to the detection effect, that is, the higher value, the better performance. The Precision can describe the recognition accuracy of the model, and the Recall can evaluate the recall rate of the model. Normally, the Precision is inversely proportional to the Recall, that is, as the value of precision increases, the recall rate gradually decreases. The Average Precision is an evaluation index that balances Precision and Recall.

$$pre = \frac{TP}{TP + FP} \tag{9}$$

$$\mathrm{Re}c = \frac{TP}{TP + FN} \tag{10}$$

$$AP = \int_0^1 P(r)dR \tag{11}$$

To determine the feature extraction network, we select the depths of 34, 50, and 101 for comparison experiments. The experimental results are shown in Table 3. As the depth of the ResNet deepens, the detection precision gradually increases, and the recall rate is basically the same, but the time required for inference gradually increases. Therefore, considering the detection precision and inference time, we choose ResNet50 as the feature extraction network.

Table 3. Comparison of the backbone

Model	Backbone	Pre	Rec	Time
RetinaNet	ResNet34	32.15	**66.47**	32.78
	ResNet50	34.32	66.11	39.81
	Resnet101	**37.7**	66.38	**51.52**

The experimental results of our model are shown in Table 4. After adding the direction anchor box regression branch, the value of Precision, Recall, and Average Precision make a great increase. In the original RetinaNet model without rotation regression branch, the predicted results are all horizontal anchor boxes, resulting in a low level of the intersection ratio between the predicted and the truth anchor box. Under the same intersection ratio threshold evaluation condition, the values of indicators will naturally be greatly reduced. After introducing the balanced feature pyramid, the precision of the BFR-RetinaNet increases by 7.96 points, and the recall rate and average accuracy achieve a slight improvement. By comparing the experimental results, the improved model in this paper has certain advantages.

Table 4. Comparison of the improved model

Model	Pre	Rec	AP
RetinaNet	34.32	66.11	45.26
R-RetinaNet	78.19	97.94	90.24
BFR-RetinaNet(ours)	**86.15**	**98.39**	**90.82**

4.4 Visual Analysis of Experimental Results

To express the improvement of vehicle detection performance of BFR-RetinaNet more intuitively, we select aerial images in different scenarios in the testset, visualize the predicted bounding boxes of vehicle detection, and constitute a comparison of detection results. As shown in Fig. 8, each row of the contrast image matrix represents a different actual scene, from top to bottom: traffic main road, traffic branch road, public parking apron, and road parking apron. Each column represents the visualization results of different detection models, which are the original image, the RetinaNet, the R-RetinaNet, and the BFR-RetinaNet in order from left to right.

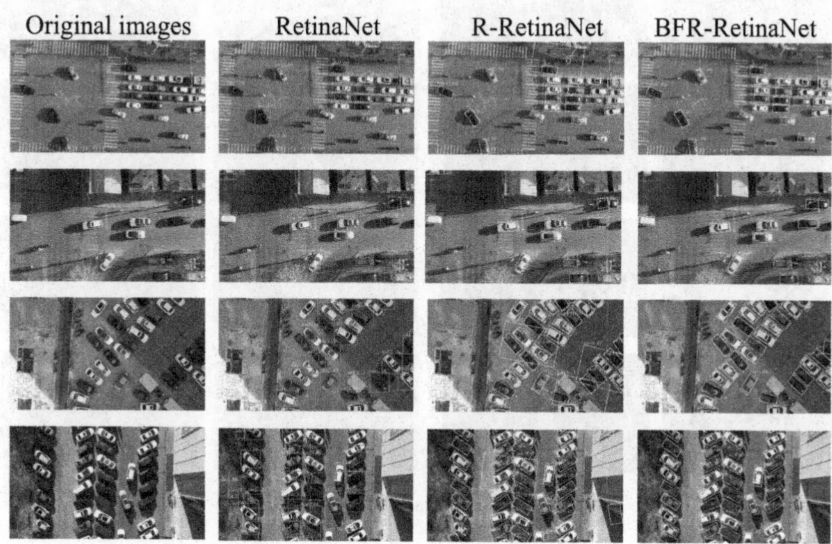

Fig. 8. Visualization of the results of different models

It can be seen from the comparison of the image matrix that the BFR-RetinaNet has achieved good detection results in four different detection scenarios. Regarding the improvement of the rotation regression branch, the comparison between the second column and the third column shows that in the general sparse vehicle scene, the RetinaNet model performs well, but as mentioned above, the positioning result returned by the model contains a lot of background information, which is not the optimal positioning result, and there is still room for improvement.

However, in scenes with more vehicles, that is, common parking lot scenes, such as the third and fourth rows. The performance of the RetinaNet model is very bad, and there are quite a lot of missed and false detections. On the one hand, facing the complex background of many ground objects, the model is prone to misdetection. The ground objects that are similar to the contour and color of the vehicle are mistakenly detected. On the other hand, when facing dense scenes, the objects generally have a certain angle. Using a horizontal anchor box for positioning will produce a large amount of overlap and in the later non-maximum suppression operation, seriously missed detection will inevitably occur.

As for the introduction of the balanced feature pyramid, it can be seen from the comparison of the third and fourth columns that the balanced features can filter out some background interference and reduce the false detection, such as the small non-motor vehicle. At the same time, the further improved model also avoids the generation of invalid large-area anchor boxes and shows stronger positioning capabilities in vehicle positioning regression.

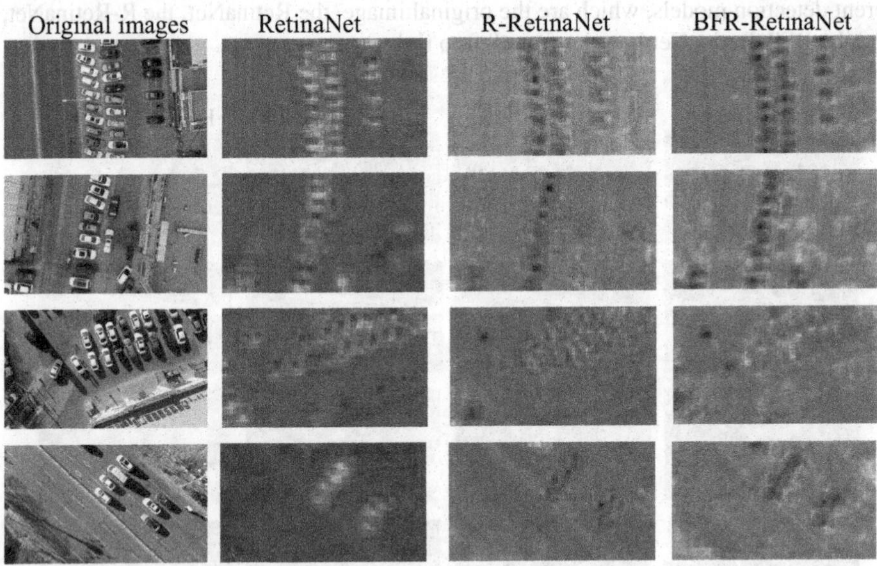

Fig. 9. The comparison of feature maps

As shown in Fig. 9, for exploring the improvement of BFR-RetinaNet, we build a feature map comparison matrix between RetinaNet, R-RetinaNet and BFR-RetinaNet. we take the original layer P_3 and the improved layer I_3 as an example for comparison. Each feature layer contains multiple channels. To visualize the feature layer, we averaged it on the channel dimension to make the dimension one.

From left to right, they represent the output feature maps of the models RetinaNet, R-RetinaNet and BFR-RetinaNet, respectively. It can be seen from the second column that the feature contour extracted by RetinaNet is horizontally rectangular, and adjacent objects are connected without clear boundaries. However, the images in the last two columns show that the contours of the features extracted by the model with the rotating regression branch are clearer and the boundaries are more obvious.

At the same time, the comparison between the three and four columns shows that the BFR-RetinaNet model strengthens the feature extraction of the vehicle significantly and enhances the contour information of the detected object. In addition, the returned feature information is richer and more complete, which is also more beneficial to the classification and positioning in the later stage of the model.

5 Conclusion

We propose an improved detection model BFR-RetinaNet based on the single-stage object detector RetinaNet. The model adds a directional anchor box regression branch, which improves the positioning of the model in the task of vehicle detection for aerial images and reduces the high missed detection rate caused by the differences in positioning methods. At the same time, the improved model also introduced the balanced feature pyramid structure, which achieves the fusion and reuse of feature information of different levels and different depths, enhances the feature extraction of vehicle information. Comparing the experimental results, BFR-RetinaNet can achieve better detection results in the vehicle detection task for aerial imagery. In the next step, we will focus on more complex situations and plan to use a multi-step approach to simplify the complexity of vehicle detection.

References

1. Zou, Z.X., Shi, Z.W., Guo, Y.H., et al.: Object detection in 20 years. https://arxiv.org/pdf/1905.05055.pdf. Accessed 2 Sept 2021
2. Wu, X.W., Sahoo, D., Hoi, Sch.: Recent advances in deep learning for object detection. Neurocomputing **396**, 39–64 (2020)
3. Girshick, R., Donahue, J., Darrell, T., et al.: Rich feature hierarchies for accurate object detection and semantic segmentation. In: Proceedings of the 2014 IEEE Conference on Computer Vision and Pattern Recognition, pp. 580–587. IEEE, Piscataway (2014)
4. Girshick, R.: Fast R-CNN. In: Proceedings of the 2015 IEEE International Conference on Computer Vision, pp. 1440–1448. IEEE, Piscataway (2015)
5. Ren, S., He, K., Girshick, R., et al.: Faster R-CNN: towards real-time object detection with region proposal networks. In: Proceedings of the 28th International Conference on Neural Information Processing Systems, pp. 91–99. MIT Press, Cambridge (2015)
6. Liu, W., Anguelov, D., Erhan, D., et al.: SSD: single shot multibox detector. In: Proceedings of the 2016 European Conference on Computer Vision, pp. 21–37. Springer, Cham (2016). Doi: https://doi.org/10.1007/978-3-319-46448-0_2
7. Fu, C.Y., Liu, W., Ranga, A., et al.: DSSD: deconvolutional single shot detector. https://arxiv.org/pdf/1701.06659.pdf. Accessed 2 Sept 2021
8. Redmon, J., Divvala, S., Girshick, R., et al.: You only look once: unified, real-time object detection. In: Proceedings of the 2016 IEEE Conference on Computer Vision and Pattern Recognition, pp. 779–788. IEEE, Piscataway (2016)
9. Redmon, J., Farhadi, A.: YOLO9000: better, faster, stronger. In: Proceedings of the 2017 IEEE Conference on Computer Vision and Pattern Recognition, pp. 7263–7271. IEEE, Piscataway (2017)

10. Redmon, J., Farhadi, A.: YOLOv3: an incremental improvement. https://arxiv.org/pdf/1804.02767.pdf. Accessed 2 Sept 2021
11. Lin, T.Y., Goyal, P., Girshick, R., et al.: Focal loss for dense object detection. In: Proceedings of the 2017 IEEE International Conference on Computer Vision, pp. 2999–3007. IEEE, Piscataway (2017)
12. Zhang, S., Huang, H.Y., Hu, Y.M., et al.: Vehicle detection in satellite imagery based on deep learning. J. Comput. Appl. **39**(S2), 91–96 (2019)
13. Lu, B., Qu, S.J.: Vehicle detection method in aerial images based on BiFPN and improved Yolov3-tiny network. J. Chinese Comput. Syst. **42**(08), 1694–1698 (2021)
14. Lecun, Y., Bengio, Y.: Convolutional networks for images, speech, and time series. In: The Handbook of Brain Theory and Neural Networks, vol. 3361(10), 1995 (1995)
15. Krizhevsky, A., Sutskever, I., Hinton, G.E.: ImageNet classification with deep convolutional neural networks. In: International Conference on Neural Information Processing Systems, pp. 1097–1105. Curran Associates Inc (2012)
16. Simonyan, K., Zisserman, A.: Very deep convolutional networks for large-scale image recognition. https://arxiv.org/pdf/1409.1556.pdf. Accessed 2 Sept 2021
17. He, K., Zhang, X., Ren, S., et al.: Deep residual learning for image recognition. In: Proceedings of the 2016 IEEE Conference on Computer Vision and Pattern Recognition, pp. 770–778. IEEE, Piscataway (2016)
18. Lin, T.Y., Dollár, P., Girshick, R., et al.: Feature pyramid networks for object detection. In: Proceedings of the 2017 IEEE Conference on Computer Vision and Pattern Recognition, pp. 2117–2125. IEEE, Piscataway (2017)
19. Liu, S., Qi, L., Qin, H., et al.: Path aggregation network for instance segmentation. In: Proceedings of the IEEE Conference on Computer Vision and Pattern Recognition, pp. 8759–8768. IEEE, Piscataway (2018)
20. Pang, J., Chen, K., Shi, J., et al.: Libra R-CNN: towards balanced learning for object detection. In: Proceedings of the 2019 IEEE Conference on Computer Vision and Pattern Recognition, pp. 821–830. IEEE, Piscataway (2019)
21. Wang, X., Girshick, R., Gupta, A., et al.: Non-local neural networks. In: Proceedings of the IEEE Conference on Computer Vision and Pattern Recognition, pp. 7794–7803. IEEE, Piscataway (2018)
22. Feng, K.B.: UAV ROD: a car detection dataset. https://github.com/fengkaibit/UAV-ROD. Accessed 2 Sept 2021
23. Ding, J., Xue, N., Xia, G.S., et al.: Object detection in aerial images: a large-scale benchmark and challenges. https://arxiv.org/pdf/2102.12219.pdf. Accessed 2 Sept 2021

Learning Knowledge Graph Embeddings by Multi-Attention Mechanism for Link Prediction

Meihong Wang(✉), Han Li, and Linling Qiu

School of Informatics, Xiamen University, Xiamen, China
wangmh@xmu.edu.cn, {lihan,qiulinling}@stu.xmu.edu.cn

Abstract. Knowledge graphs (KGs) are being widely utilised in many fields of artificial intelligence. However, the lack of relations has become a huge obstacle to their real-life application. Therefore, recent researches in this field tend to focus on link/relation prediction techniques on KGs, but most of them merely learn KG embeddings from the central nodes' neighbourhood in an absolute or unconditional way, fusing a great deal of weak or even useless information. To tackle this issue, we propose a novel end-to-end KG embedding model named **R**elational **G**ated **G**raph **A**ttention ne**T**work (R-GGAT), which learns embeddings by supervising and controlling the node aggregation process by selecting and filtering the aggregated information through its multi-attention mechanism (MAM), in an attempt to fully and effectively mine the underlying semantic information of KGs. Evaluation experiments on several datasets have been performed, where our proposed R-GGAT achieved the elevated performance in link prediction tasks compared to several state-of-the-art baselines. We also carried out ablation study and analysed the changing of node embeddings filtered by the MAM to demonstrate the effectiveness of our model.

Keywords: Knowledge graph embedding · Link prediction · Gating mechanism · Graph attention

1 Introduction

A Knowledge Graph (KG), also known as an Information Graph (IG), is a semantic network that reveals the interrelationship between entities. Its unique graph structure, which is used to store semantic information, enables a human-like analytical and reasoning potential. KGs have been integrated in many fields, including information retrieval [23], recommendation systems [3,22] and natural language processing [8].

A KG usually contains great quantities of facts constructed by billions of real-world entities and relations, which are represented as nodes and edges connecting these nodes, respectively. The interrelationship between each two nodes can be represented as a set of triples. For instance, in Fig. 1(a), the connection between

Y. Lai et al. (Eds.): ICA3PP 2021, LNCS 13155, pp. 33–49, 2022.
https://doi.org/10.1007/978-3-030-95384-3_3

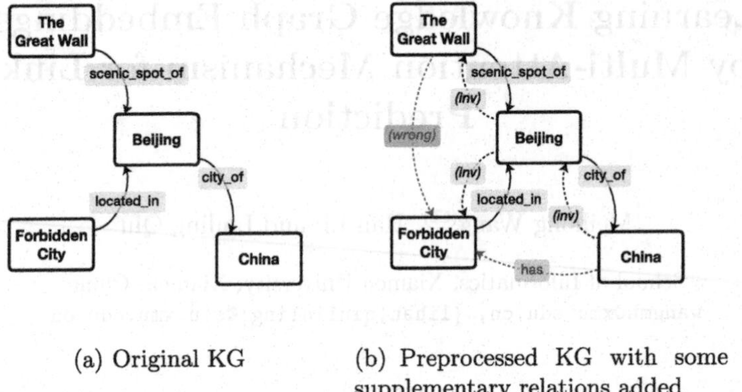

(a) Original KG

(b) Preprocessed KG with some supplementary relations added

Fig. 1. A sample KG which depicts the relations between some places in China. Solid and dashed lines are original and supplementary relations, respectively. Lines labelled '*(inv)*' are inverted relation with respect to its original version in the KG. Dashed lines that coloured green are of reasonable semantic meanings, whilst red dashed lines labelled '*(wrong)*' indicate relations that are invalid or confusing.

Beijing and China can be denoted by a triple (`Beijing`, *`city_of`*, `China`), where *`city_of`* represents the concrete relation in between.

There are many open-access and widely-recognised KGs, such as Freebase [1] and DBPedia [7]. However, because of the openness of the existing KGs, a majority of them are based on limited knowledge sets, which means that they are dynamic and incomplete. Therefore, it is of necessity to supplement a given KG to improve its completeness and interpretability before it can be further applied to other fields of artificial intelligence.

The supplementation of KG is known as Knowledge Graph Completion (KGC). In order to carry out the dominant KGC task, link prediction, knowledge graph embedding (KGE) is put forward to transform entities and relations into low-dimension continuous vector space, while preserving the semantic information that is sealed inside the KG. For example, Fig. 3(a) is the preprocessed version of the KG shown in Fig. 1(b) with corresponding embeddings being associated with each entity and relation.

Recently, with the development of deep learning techniques, neural network models have been applied to KGC tasks. Regrettably, most of the traditional neural network models, such as R-GCN [13], ConvE [5] and ConvKB [12] , merely rely on short-range nodes, viz. the direct neighbours (1-hop neighbourhood) of the central nodes, when learning embeddings. Notwithstanding the attempts made to integrate multi-hop neighbourhood, e.g. the method proposed in KB-GAT [11], a considerable amount of weak or even useless information from a wider range of nodes is still unconditionally aggregated.

We take the multi-hop KG shown in Fig. 1(b) as an example, which was preprocessed on the basis of its original version (Fig. 1(a)) by adding supplementary links (represented by dashed, black and colourful arrows). In fact, some newly-added relations are actually depicting weak or futile connections. For instance, the 2-hop relation from `The Great Wall` to `Forbidden City` is marked 'wrong', as it has confusing semantic meaning which may lead to unexpected aggregation between these two entities, resulting in mistaken reasoning. On the contrary, the 3-hop supplementary relation represented by `(China,` ***has*** `, Forbidden City)` is semantically inferable.

The above example reveals that although multi-hop relations are of significance in a KGE progress, a larger neighbourhood area can be much noisier. Therefore, a KG embedding model should have the ability to evaluate the strength of the multi-hop reasoning relations and entities, and to drop unneeded or useless information after node aggregation. We therefore propose a novel knowledge graph embedding model named *Relational Gated Graph Attention neTwork* (R-GGAT), which is capable of learning embeddings more intelligently by supervising and controlling the information aggregation process by attaching different weights to triples through its neighbouring attention mechanism (NAM), and filtering the aggregated information through its gated attention mechanism (GAM). The NAM and GAM, jointly called the multi-attention mechanism (MAM), make it possible to effectively mine the underlying semantic information of KGs.

Our contributions are three-fold:

- We proposed a novel KG embedding model. To the best of our knowledge, we are the first to introduce MAM into the link prediction task to alleviate the limitation of short-range reasoning and filter information within embedding dimensions.
- To the best of our knowledge, we are the first to develop and integrate the GAM into the field of KGC. Experiments have shown the powerfulness of GAM in embeddings dimension filtering, which enables a more reliable excavation of KG semantic information. We perform experiments for link prediction on several benchmark datasets, where our model achieves better performance over the state-of-the-arts.
- We conducted extensive studies on the GAM's behaviour in representation learning by monitoring the changes of embeddings before and after being processed by the gates. The analyses reveal the way in which GAM learns to refine the embeddings during training.

2 Related Works

A series of KGE models have been used to carry out link prediction tasks, which can roughly be divide into three categories - interpretative distance-based models, semantic information-based models and neural network models.

Interpretative distance-based models, e.g. TransE [2] , use distance measurements as the scoring system for link prediction. Relations and entities are mapped into embeddings by a certain transformation, and a distance function is then used to measure the credibility between each two entities. Notwithstanding the simple and effective translation operation, the overall features they can capture are limited, that is, they learn insufficient representation of the whole KG during their learning process.

Semantic information-based models are those that focus on structural information of KGs, including traditional semantic matching models [17,21], and supplementary information-based models [10,20]. The latter obtained better performance, since they can integrate information from an outside knowledge domain. However, not all structural information is valid, and the external knowledge bases are typically noisy because they are not yet made suitable for many fields [15].

With the rapid development of convolutional neural networks (CNNs), many CNN-based models are widely proposed for link prediction tasks, e.g. ConvE [5] and ConvKB [12]. ConvE uses a 2D-convolution layer to predict missing links, which outperforms 1D-convolution in terms of extracting feature interactions between two embeddings. ConvKB uses 1D-convolution in exchange of the translation technique in TransE, which is sufficient for capturing both the overall structure and the transitional characteristics between entities. These methods, however, treats the knowledge triples individually, without taking into account their neighbouring nodes which are of great significant for knowledge reasoning tasks like KGC.

Recently, the convolution techniques are well integrated with graph-structured data, which enables many graph-based KGE methods to realise their goals. R-GCN [13] extends GCN by modeling multi-relational graph data. SACN [14] proposed an end-to-end structure-aware convolutional network that takes into account node connectivity, attributes and relation types simultaneously, which is also known as a weighted GCN. CompGCN [18] leverages various entity-relation composition operations from KGE techniques and also generalises several multi-relational GCN methods. KB-GAT [11] borrows the idea from GAT [19] by applying the triple-wise attention mechanism in link prediction for KGs. It is the first attempt to learn graph attention-based embeddings that specifically target relation prediction on KGs. This solves the 'one-size-fits-all' flaw of many other graph-based models, by assigning unequal weights to each neighbouring node when they are being aggregated.

Nevertheless, we find that once a weight is calculated for an entity, every component in the embedding associated with the entity is then multiplied by the weight value, before they are aggregated to the central nodes. Intuitively, not all components of the embeddings are important; some of them may be useless for further analysis, but are somehow mistakenly assigned a higher weight during the learning process, and the potential mistakes are accumulated and amplified as training steps. Therefore, in our proposed method, we develop a way to conduct judgement on embedding dimensions either, by assigning different weights to them just as is done to neighbouring entities, and reset those with lower weights with their original states before the embeddings are ready to output.

3 Method

In this section, we first introduce the main symbols and notations involved throughout this study; then we give an overview of our model; finally, we describe and demonstrate the proposed method.

3.1 Notations

In any real-world entity-relation models, any connection can be represented by a triple $t_{htr} = (e_h, r, e_t)$ where e_h and e_t represent the head and tail entities, respectively, and r denotes the type of relation between both entities. Therefore, any KG, which is an abstract of real-world connections, can be represented by a series of triples. A KG is formally symbolised as $\mathcal{G} = (\mathcal{E}, R, \mathcal{T})$ where $\mathcal{T}$, $\mathcal{E}$ and $R \in \{r | (e_h, r, e_t) \in \mathcal{T} \wedge e_h, e_t \in \mathcal{E} \wedge r \in R\}$ represents the triples, entities (nodes) and relations (edges) sets of the knowledge graph, respectively.

3.2 Overview of Our Approach

The end-to-end view of our encoder-decoder model can be seen in Fig. 2.

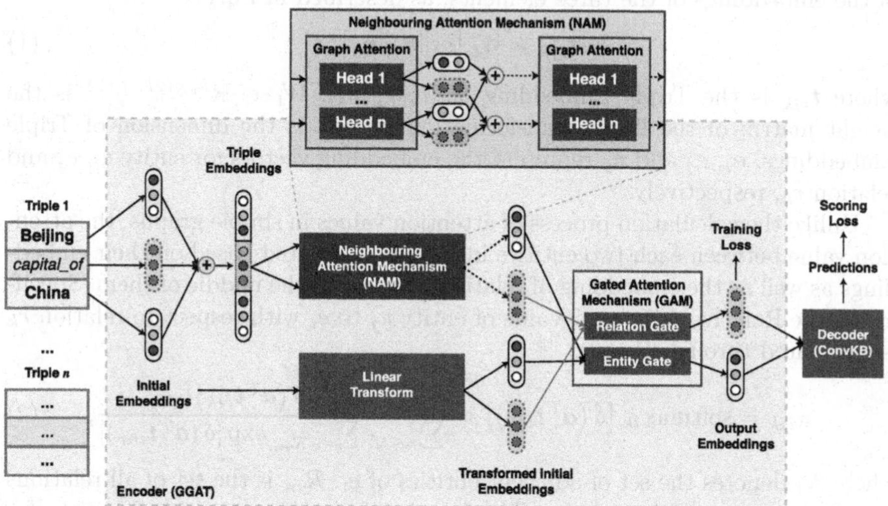

Fig. 2. The overall structure of R-GGAT. To display the whole process more clearly, the structure of NAM has been zoomed in. Solid and dashed lines indicate ordinary data flows and data flows with multiple inside transforms (which are omitted in the figure for simplify), respectively. ⊕ indicates the concatenation of vectors.

3.3 The Encoder

The encoder of our model is used to learn the best embedding vectors of the entities and relations of a graph by the MAM, which we called embedding learning. Apart from a list of training triples, the encoder takes 2 matrices of embeddings as input:

- **Entity Embeddings Matrix** $E \in \mathbb{R}^{N_e \times F_e}$ where N_e is the number of entities and F_e is the dimension of an entity's embedding.
- **Relation Embeddings Matrix** $R \in \mathbb{R}^{N_r \times F_r}$ where N_e is the number of relations and F_e is the embedding dimension of every relation.

Each row in E and R is the Initial Embedding of an entity and a relation, respectively, i.e. $E = \{\boldsymbol{e}_1, \boldsymbol{e}_2, \cdots, \boldsymbol{e}_{N_e}\}^T$, $R = \{\boldsymbol{r}_1, \boldsymbol{r}_2, \cdots, \boldsymbol{r}_{N_r}\}^T$ where $\boldsymbol{e}_i$ is the Initial Embedding of entity e_i, and $\boldsymbol{r}_j$ represents the Initial Embedding of relation r_j.

The NAM. To learn new embedding for an KG entity e_i, all its neighbours, as well as the relations in between, shall be taken into consideration. To this end, representations of all triples starting with the entity e_i shall be obtained in advance. Similar to [11], the vector representation of a triple, or Triple Embedding, is acquired by performing a linear transformation by doing concatenation of the embeddings of the three elements, as described in Eq. (1)

$$\boldsymbol{t}_{ijk} = W_t \left[\boldsymbol{e}_i \| \boldsymbol{e}_j \| \boldsymbol{r}_k\right], \tag{1}$$

where $\boldsymbol{t}_{ijk}$ is the Triple Embedding of (e_i, r_k, e_j). $W_t \in \mathbb{R}^{F_t \times (2F_e + F_r)}$ is the weight matrix of the linear transformation and F_t is the dimension of Triple Embeddings. $\boldsymbol{e}_i$, $\boldsymbol{e}_j$ and $\boldsymbol{r}_k$ represent the embedding vectors for entity e_i, e_j and relation r_k, respectively.

Unlike the calculation process of attention values in simple graphs, the attention value between each two entities in KGs is calculated based on their embeddings as well as the embedding of relation that lies in the middle of them. Specifically, the Relative Attention Value of entity e_j to e_i with respect to relation r_k is calculated through Eq. (2)

$$\alpha_{ijk} = \text{softmax}_{jk}\left[\delta\left(\boldsymbol{a}^T \boldsymbol{t}_{ijk}\right)\right] = \frac{\exp\left[\delta\left(\boldsymbol{a}^T \boldsymbol{t}_{ijk}\right)\right]}{\sum_{n \in \mathcal{N}_i} \sum_{r \in \mathcal{R}_{in}} \exp\left[\delta\left(\boldsymbol{a}^T \boldsymbol{t}_{inr}\right)\right]}, \tag{2}$$

where $\mathcal{N}_i$ denotes the set of adjacent entities of e_i. $\mathcal{R}_{in}$ is the set of all relations that lie between entities i and n. The function $\delta(\cdot)$ is an activation function, for which we adapt LeakyReLU$(\cdot)$ in our implementation.

In Eq. (2), the Relative Attention Value of triple (e_i, r_k, e_j), i.e., $\alpha_{ijk} \in (0, 1)$, is obtained, which reflects the relative significance of e_j to e_i when they are connected via relation r_k, compared to other neighbouring entities and relations of e_i. The resulting New Embeddings is then derived from Eq. (3)

$$\boldsymbol{e}'_i = \sigma\left(\sum_{j \in \mathcal{N}_i} \sum_{k \in \mathcal{R}_{ij}} \alpha_{ijk} \boldsymbol{t}_{ijk}\right); \quad \boldsymbol{r}'_j = \sigma\left(W_R^{(A)} \boldsymbol{r}_j\right), \tag{3}$$

where $W_R^{(A)}$ is a weight matrix which transforms the Initial Embedding of relation r_i ($\boldsymbol{r}_i \in \mathbb{R}^{F_r}$) into another vector space ($\boldsymbol{r}_i'$). $\sigma(\cdot)$ is a non-linearity activation function. In this work, we employ the ELU activate function as $\sigma(\cdot)$. As suggested by [19].

Equations (2) and (3) are calculated by an attention layer. As suggested by [19], an attention layer may either act as a single Graph Attention Layer (GAL) on its own, or become one of the attention heads in a multi-head GAL. For the latter case, H attention layers calculate the embeddings independently, and the resulting embeddings are concatenated eventually, which comes to be the new multi-head embeddings of entities and relations. Specifically, for e_i and r_j, their Multi-head New Embeddings are calculated through Eqs. (4) and (5)[1]

$$\boldsymbol{e}_i' = \Big\Vert_{h=1}^{H} \boldsymbol{e}_i^{(h)\prime} = \Big\Vert_{h=1}^{H} \sigma\left(\sum_{j\in\mathcal{N}_i}\sum_{k\in\mathcal{R}_{i,j}} \alpha_{ijk}^{(h)} \boldsymbol{t}_{ijk}^{(h)}\right) \tag{4}$$

$$\boldsymbol{r}_j' = \Big\Vert_{h=1}^{H} \boldsymbol{r}_j^{(h)\prime} = \Big\Vert_{h=1}^{H} \sigma\left(W_R^{(A)} \boldsymbol{r}_j^{(h)}\right), \tag{5}$$

where $\alpha_{ijk}^{(h)}$ denotes the Relative Attention Value of e_j to e_i via r_k that is calculated in the h-th attention head. $\boldsymbol{t}_{ijk}^{(h)}$, $\boldsymbol{e}_i^{(h)}$ and $\boldsymbol{r}_j^{(h)}$, likewise, are the embedding of (e_i, r_k, e_j), e_i and r_j that are learnt by or input to the same attention head, respectively.

Equations (1) to (5) are collectively called the NAM because of its unique behaviour: it tries to learn new embeddings of entities by accumulating their neighbours', which are weighted by their relative attention values. To this end, it attempts to find the best transformation to produce the right attention values on their neighbours.

The GAM. The transformations in Sect. 3.3 are utilised to learn new embeddings of KG entities and relations. For simplicity, the aforementioned operations are concluded by the two transformations in Equation (6)

$$\begin{cases} \boldsymbol{e}_i' = f_e(\boldsymbol{e}_i) \\ \boldsymbol{r}_j' = f_r(\boldsymbol{r}_j), \end{cases} \tag{6}$$

where $\boldsymbol{e}_i' \in \mathbb{R}^o$ and $\boldsymbol{r}_j' \in \mathbb{R}^o$ are outputs of Eq. (4) and Eq. (5), respectively.

As mentioned in former sections, we introduce a GAM into our encoder to do final filtering of the embeddings. The GAM was developed based on an extension of the idea of the Gated Recurrent Unit (GRU) [4] that was first used in the field of sequence modelling.

There are three gates, namely Update Gate, Reset Gate and Forget Gate, that lies in the terminal side of the encoder. Initially, the input embeddings

[1] As suggested by [19], vector averaging is applied in exchange of concatenation in Eqs. (4)and(5)if it is the end most GAL of the encoder.

are linear transformed into another vector space, to ensure that the two input vectors of the gates are same in dimension number:

$$\begin{cases} \boldsymbol{e}_i^t = W_E \boldsymbol{e}_i \\ \boldsymbol{r}_j^t = W_R \boldsymbol{r}_j, \end{cases} \tag{7}$$

where $W_E \in \mathbb{R}^{o \times F_e}$ and $W_R \in \mathbb{R}^{o \times F_r}$ are weight matrices. $\boldsymbol{e}_i^t$ and $\boldsymbol{r}_j^t$ are called Transformed Initial Embeddings of e_i and r_j. Then, both the (Multi-head) New Embeddings as well as Transformed Initial Embeddings are fed into two foremost gates, i.e. the Update Gate and Reset Gate, at the same time. Specifically, for each entities and e_i and r_i, each gate calculates its Gated Output Vector according to the two embeddings. An instance for the embeddings $\boldsymbol{e}_i'$ and $\boldsymbol{e}_i^t$ to go through the Update and Reset Gates are described in Eqs. (8) and (9)

$$\boldsymbol{z}_i = \sigma \left(W_z \boldsymbol{e}_i' + U_z \boldsymbol{e}_i^t \right) \tag{8}$$

$$\boldsymbol{s}_i = \sigma \left(W_r \boldsymbol{e}_i' + U_r \boldsymbol{e}_i^t \right), \tag{9}$$

where $\boldsymbol{z}_i$ and $\boldsymbol{s}_i$ are the Gated Output Vectors of the Update Reset Gates respectively. W_z, U_z, W_r and U_r are weight matrices to be trained.

The Transformed Initial Embedding, New Embedding, together with the Reset Gate's Gated Output Vector of entity e_i are then merged to produce a New Information Vector through Eq. (10)

$$\boldsymbol{n}_i = \tanh \left[W_n \boldsymbol{e}_i' + U_n \left(\boldsymbol{s}_i \odot \boldsymbol{e}_i^t \right) \right], \tag{10}$$

where $\odot$ is the Hadamard Product of two vectors of the same dimension, and W_n and U_n are weight matrices to be trained. Finally, along with the two Gated Output Vectors, the New Information Vector is then fed into the Forget Gate, to acquire Gated New Embedding ($\widetilde{\boldsymbol{e}_i'}$) of entity e_i.

$$\widetilde{\boldsymbol{e}_i'} = (1 - \boldsymbol{z}_i) \odot \boldsymbol{e}_i^t + \boldsymbol{z}_i \odot \boldsymbol{n}_i \tag{11}$$

Equations (7) to (11) are jointly called the GAM based on its final effect on the embeddings. In Eq. (10), some pieces of initial information contained in $\boldsymbol{e}_i^t$ are ignored and reset with their counterparts in the New Embedding ($\boldsymbol{e}_i'$), when the corresponding dimensions of the reset gate's Gated Output Vector ($\boldsymbol{s}_i$) are close to 0. Besides, as is shown in Eq. (11), the Update Gate determines the amount of information that is adopted from the Transformed Initial Embeddings.

Example. We divide the whole process of the encoder into 2 steps as shown by the example diagrams (Figs. 3(c) and 3(d)). The example is based on the sample KG that was introduced in Sect. 1. Each node will be considered the central node with respect to its neighbours in the whole encoder process; in this example, we select `Forbidden City` as the central node. The initial state of the KG in this example is depicted in Figs. 3(a) and 3(b).

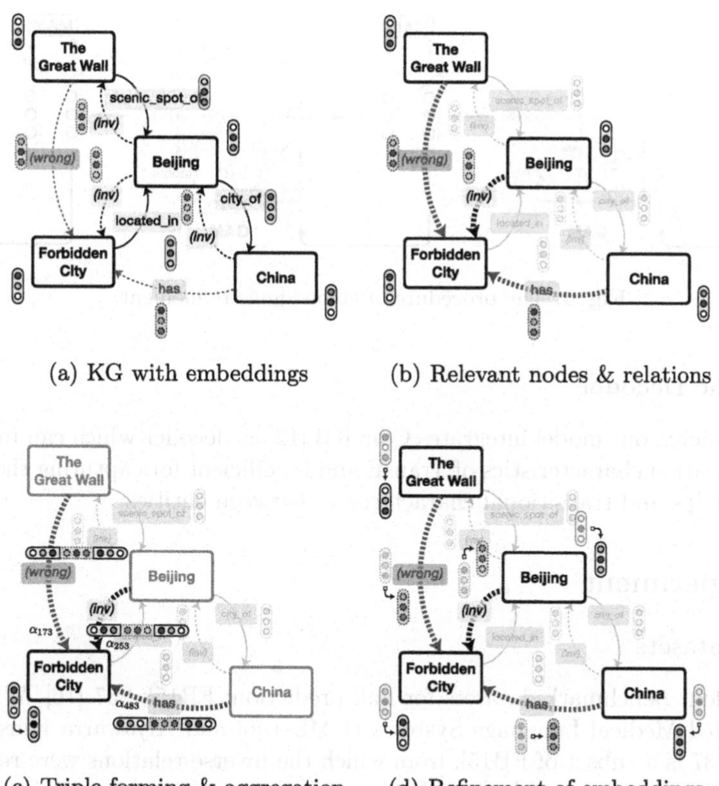

Fig. 3. Initial state and the 2 steps of the encoder. The aggregation process is highlighted in the diagram by thick lines. Irrelevant relations and entities have been masked.

We give details of the two steps as follows.

1. **Triple forming & aggregation (NAM).** This step is performed by the NAM. Embeddings of entities and relations are concatenated according to the definition of triples. The resulting Triple Embedding ($\boldsymbol{t}$) are used by the NAM to calculate attention values, and the embeddings of relevant neighbours are then aggregated to the central node.
2. **Refinement of embeddings (GAM)**. In this step, embeddings of entities and relations are selectively filtered by the GAM. In Fig. 3(d), the GAM is simplified as an arrow with a block attach to its tail.

The detailed procedure of an embedding being altered in step 2 and 3 is depicted in Fig. 4, were use the schematic in [4] to depict the data transformation process in GAM. We use $\boldsymbol{e}_i^{(l)}$ and $\boldsymbol{e}_i^{\prime(l)}$ to represent $\boldsymbol{e}_i$ (Initial Embedding) and $\boldsymbol{e}_i'$ (New Embedding) of the l-th model layer. The procedure is recurrent, i.e. the same process occurs in successive layers, and the output of one layer serves as input of the subsequent layer.

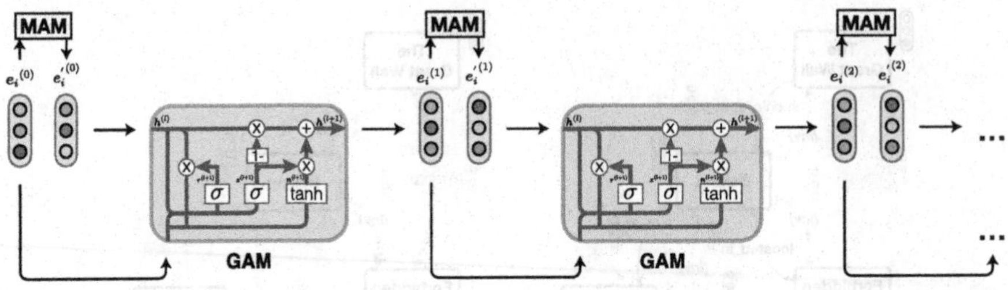

Fig. 4. The procedure of embeddings refinement.

3.4 The Decoder

For simplicity, our model integrates ConvKB [12] as decoder which can maintain the translation characteristics of TransE and is sufficient for capturing the global relationships and transitional characteristics between entities.

4 Experiment

4.1 Datasets

We use four benchmark datasets for link prediction: FB15k-237 [16], WN18RR [5], Unified Medical Language Systems (UMLs) [6] and Alyawarra Kinship [9]. FB15k-237 is a subset of FB15k from which the inverse relations were removed. WN18RR, likewise, is a subset of WN18, where inversion relations were removed from its original version. The UMLs is a digital dataset consists of many health and biomedical vocabularies and standards. Kinship dataset brings together 24 unique names in two families that have equivalent structures. The statistics of these datasets are shown in Table 1.

Table 1. Statistics of datasets.

Name	Entities	Relations	#Train	#Valid	#Test
FB15k-237	14,541	237	272,115	17,535	20,466
WN18RR	40,943	11	86,835	3,034	3,134
UMLS	135	46	5,216	652	661
Kinship	104	25	8,544	1,068	1,074

4.2 Training and Evaluation Protocols

The encoder and decoder are trained separately. We first train the encoder to obtain the best effectiveness on embedding learning; then the embeddings produced by the optimal encoder are used as inputs of the decoder, which is then trained to calculate accurate scores of any triples.

For encoder, the input embeddings for certain datasets are initialised using the embeddings produced by a pre-trained TransE [2] model. Besides, to enable multi-hop reasoning, auxiliary multi-hop relations are added in advance by performing Depth-first Search (DFS) on the original KGs. We preserve the Gated New Embeddings of both entities and relations from the encoder at the end of the final training epoch, which are then fed into decoder as inputs.

For decoder, the set of invalid triples are created by replacing the head and tail entities of each valid triple. Specifically, the invalid entities produced by replacing head and tail are equal in number when sampling invalid triples for loss calculation and model evaluation.

Several hyper-parameters are defined for both encoder and decoder. N/P Ratio is defined as $r = \frac{N_v}{N_i}$, where N_v and N_i are the numbers of invalid triples (negative samples) and valid triples (positive samples) that are to be sampled during the training process. Loss Margin is used to force encoder loss to decrease. Specifically, we set N/P Ratio = 2, learning rate = 1×10^{-3}, dropouts = 0.3 and the number of attention heads = 2 for encoder, and N/P Ratio = 40, learning rate = 1×10^{-3} and weight decay = 1×10^{-5} for decoder. The other hyper-parameters used in the encoder and decoder are listed in Table 2.

Table 2. Training settings of encoder and decoder for the evaluated datasets. "Dim.", "Conv.", "Enc." and "Dec." are abbreviations of Dimension, Convolution, Encoder and Decoder, respectively.

Dataset name	Weight decay	Enc. epochs	Emb. dim.	Margin	Dec. epochs	Dec. dropout	Conv. filters
FB15k-237	1×10^{-5}	3,000	200	1	200	0.3	50
WN18RR	5×10^{-5}	3,600	200	5	200	0	500
UMLS	1×10^{-5}	3,000	200	1	400	0.3	500
Kinship	1×10^{-5}	3,000	400	3	400	0.3	500

4.3 Experimental Results and Analysis

Results. We conducted a series of experiments to verify the effectiveness of our model, where we selected several recently proposed or highly cited, representative models as baselines. The performances are reported using 5 metrics, i.e. Mean Rank (MR), Mean Reciprocal Rank (MRR) and Hits@N where $N \in (1, 3, 10)$. Many performance scores are produced by running the models' authentic implementations in our unified environment, whilst the values of certain models are taken from their original papers or other works. The evaluation results of those being evaluated are shown in Tables 3 and 4.

Table 3. Experimental results on FB15k-237 and WN18RR datasets. The values in **bold** and being underlined are the top-2 scores amongst those being compared. The reported metrics are MR, MRR and Hits@N. For Hits@N we report the percentage of correct ones in the sampled entities whose scores are ranked top N. The results of those models are evaluated in our unified environment except otherwise stated. [†] The results are taken from [5]. [‡] The results are taken from [11].

	FB15k-237					WN18RR				
	MR	MRR	Hits@N (%)			MR	MRR	Hits@N (%)		
			@1	@3	@10			@1	@3	@10
TransE (2013) [2]†	323	0.279	19.8	37.6	44.1	2300	0.243	4.27	44.1	53.2
ConvE (2018) [5]‡	245	0.312	22.5	34.1	49.7	4464	0.456	41.9	47.0	53.1
ConvKB (2018) [12]	216	0.289	19.8	32.4	47.1	2554	0.248	5.82	44.5	55.8
R-GCN (2018) [13]‡	600	0.164	10.0	18.1	30.0	6700	0.123	8.0	13.7	20.7
KB-GAT (2019) [11]	210	0.518	46.0	54.0	62.6	1940	0.440	36.1	48.3	58.1
CompGCN (2020) [18]	207	0.354	26.3	39.0	53.6	3463	**0.465**	**42.9**	47.9	52.9
R-GGAT (ours)	**154**	**0.543**	**46.9**	**58.2**	**68.0**	**1441**	0.448	36.0	**50.4**	**59.8**

Table 4. Experimental results on UMLs and Kinship datasets. The values in **bold** and being underlined are the top-2 scores amongst those being compared. The reported metrics are the identical to Table 3. The results of those models are evaluated in our unified environment except otherwise stated. [†] The results are taken from [11].

	UMLs					Kinship				
	MR	MRR	Hits@N (%)			MR	MRR	Hits@N (%)		
			@1	@3	@10			@1	@3	@10
TransE [2] †	1.77	0.797	64.1	92.1	99.24	6.8	0.309	0.9	64.3	84.1
ConvE [5] †	1.38	0.935	89.8	96.7	99.0	2.03	0.833	73.8	91.7	**98.14**
ConvKB [12] †	1.66	0.785	60.8	96.14	99.25	3.3	0.614	43.62	75.5	95.3
R-GCN [13] †	24.9	0.204	12.6	20.8	32.5	25.92	0.109	3.0	8.8	23.9
KB-GAT [11]	1.13	**0.987**	**98.1**	**99.1**	99.7	1.94	0.904	85.9	94.1	98.0
CompGCN (2020) [18]	1.83	0.891	82.1	95.6	98.1	2.80	0.766	65.1	85.7	95.7
R-GGAT (ours)	**1.12**	0.981	97.1	**99.1**	**99.8**	**1.72**	**0.956**	**93.9**	**96.6**	**98.14**

Here, we provide several analysis concerning the experiment results.

1. Our proposed R-GGAT surpass the state-of-the-arts on several relation prediction datasets at a noticeable margin. In particular, the MR in FB15k-237 dataset is reduced by almost 40%. All metrics on FB15k-237 and Kinship, 2 metrics on WN18RR and UMLs have seen clear improvements of R-GGAT over the competitors, with small, tolerable differences in several metrics on WN18RR and UMLs compared to the highest scores.
2. We noticed that the GCN- or GAT-based models generally achieved better results than other baselines. For instance, KB-GAT and CompGCN show good results on FB15k-237 and WN18RR datasets. Besides, the GAT-based models (KB-GAT and our R-GGAT) are generally more effective than

GCN-based models on nearly all datasets. For example, despite the release dates KB-GAT and CompGCN are similar, the KB-GAT's performances are generally better than CompGCN's. We believe that this can be credit to their ability of unequally neighbour aggregating which filters information in automatic and dynamic ways – this is what a GCN-based model is not good at. Despite seemly poorer performance, we chose CompGCN for comparison as it can act as a newest representative of GCN-based models, thus show clearly the significance of graph attention mechanisms.

3. The UMLs and Kinship experiments' results show minor improvement on our model's performance compared to the experiments performed on the other two datasets. For instance, the R-GGAT's Hits@5 improved by only 0.1% and MR decreased by 1 compared to the second-best model, KB-GAT, on UMLs. However, performances of the existing models on these two datasets have been very high (e.g., the Hits@N on UMLs are as high as 99%), resulting in smaller lifting space and more barriers to make much progress.

Ablation Study. We conduct an ablation study on our model to compare the experimental performance of the encoder with and without a GAM. Specifically, we analysed four evaluation metrics (MR, MRR, Hits@1 and Hits@3), on which we recorded the performance that both R-GGAT and KB-GAT [11] have shown. The analytical charts are shown in Fig. 5.

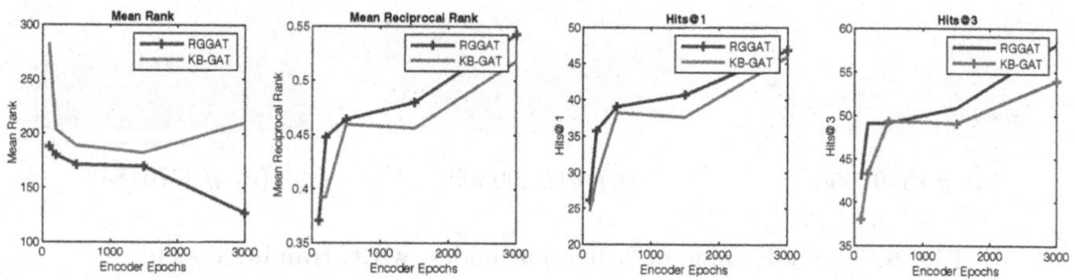

Fig. 5. Value changes of 4 experimental metrics of R-GGAT and KB-GAT.

On all analysed metrics, our R-GGAT outperforms the KB-GAT at noticeable margins at *any* training epochs. Based on the fact that KB-GAT learns embeddings without dropping any data, whilst R-GGAT succeeded in filtering its learning results with the help of GAM, it is true that our proposed GAM does make a huge difference on embedding learning and thus can improve the model performance in a huge step.

Analysis on GAM. As a key component of R-GGAT, GAM plays an important role in embedding filtering. In order to better understand the learning process of our model, we study the changes of embeddings to reveal how GAM affects the representation learning process.

The changes of an embedding can be modelled by comparing its components before and after being filtered by the GAM. Here, the before- and after-GAM versions of $\boldsymbol{e}$ are described as $\boldsymbol{e}_{\text{before}}$ and $\boldsymbol{e}_{\text{after}}$, respectively. We first define two metrics to model the changes of embeddings:

- **The Similarity of Embedding (SoE)**: the similarity between $\boldsymbol{e}_{\text{before}}$ and $\boldsymbol{e}_{\text{after}}$. Here, we use cosine similarity to value the degree of resemblance between $\boldsymbol{e}_{\text{before}}$ and $\boldsymbol{e}_{\text{after}}$.
- **The Difference of the i-th component (DoC$_i$)**: the normalised difference between the i-th component of both $\boldsymbol{e}_{\text{before}}$ and $\boldsymbol{e}_{\text{after}}$, i.e. $\text{DoC}_i = \text{norm}\left(|\boldsymbol{e}_{\text{after}} - \boldsymbol{e}_{\text{before}}|\right)$. We use Min Max Scaling for function $\text{norm}(\cdot)$, i.e. $\text{norm}(\boldsymbol{e}) = \frac{\boldsymbol{e}-\min(\boldsymbol{e})}{\max(\boldsymbol{e})-\min(\boldsymbol{e})}$.

To monitor the whole training process, we trained the encoder on WN18RR dataset for 3,600 epochs, and found that the two aforementioned metrics have dissimilar changing patterns on different nodes. Based on careful observation, we identified three major types of changing pattern and select three representative nodes accordingly. Then, we visualise the SoE and DoC$_i$ values of the three nodes with respect to training epochs (Figs. 6 and 7) to analyse the learning process of GAM.

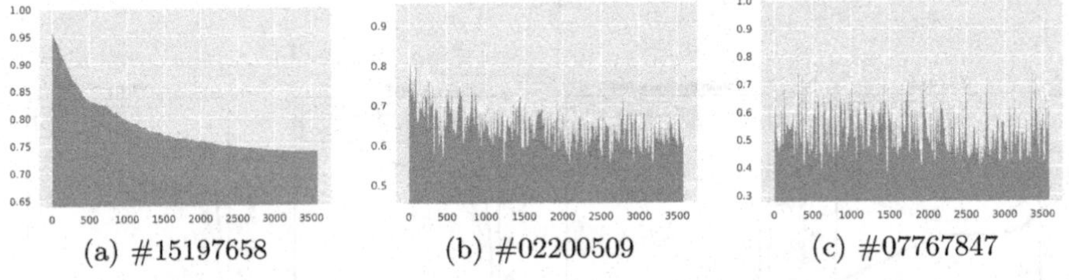

(a) #15197658 (b) #02200509 (c) #07767847

Fig. 6. The SoE value of 3 different nodes. w.r.t. training epoch.

The three charts in Fig. 6 correspond to three types of changing pattern, with each having a uniquely-shaped shaded area:

- **Smooth-decline** (e.g. Node #15197658). The SoE value decreases smoothly as training epoch cumulates.
- **Generally-decline** (e.g. Node #10477077). Despite the volatility, the SoE value has an overall decreasing trend as training epoch cumulates.
- **Generally-stable** (e.g. Node #07767847). Despite the volatility, the SoE value fluctuates up and down within a certain range as training epoch cumulates.

We can observe one common feature of these three changing patterns: despite the difference of how they change, the overall SoE values will eventually become steady as the GAM converges. To learn the possible cause of SoE fluctuation, we randomly select 20 embedding components and visualise their DoC_i values with respect to training epochs using heat maps (Fig. 7).

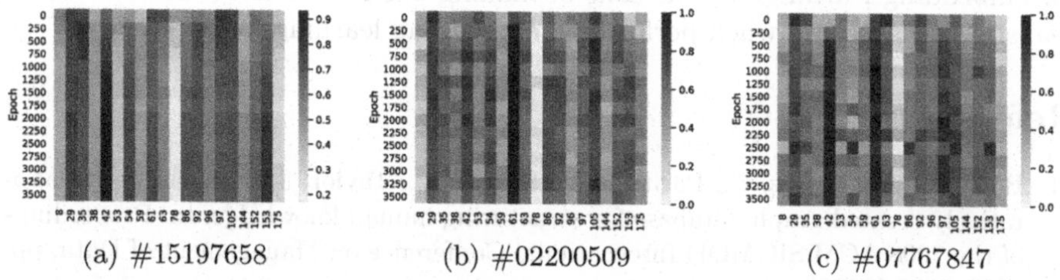

(a) #15197658 (b) #02200509 (c) #07767847

Fig. 7. The DoC_i values of 3 different nodes. w.r.t. training epoch, where $i \in$(8, 29, 35, 38, 42, 53, 54, 59, 61, 63, 78, 86, 92, 96, 97, 105, 144, 152, 153, 175).

By comparing Figs. 6 and 7, we made the following findings and analysis:

1. For nodes that have Smooth-decline SoE value, smooth changes can also be observed in their DoC_i values. For example, in the case of node #15197658, a gradual increase in DoC_{42} and decrease in DoC_8 can be observed in Fig. 7(a).
2. It is obvious in Figs. 7(b) and 7(c) that for nodes with fluctuating SoE, the changes of their DoC_i values are also volatile.
3. DoC_i values of one node can change very differently. For example, in the case of node #15197658, DoC_{42} and DoC_8 change in opposite directions. This indicates that GAM can gradually learns to accept modifications to some components (e.g. e_{78}) whilst forget changes to others (e.g. e_{42}).
4. For a given i, the changes in DoC_i may vary from node to another. For example, DoC_{42} has an overall increasing trend in the case of node #15197658 and #07767847, whilst for node #02200509 it fluctuates all the time. This reveals that a specified component may have varied contributions to different node embeddings.

5 Conclusion and Future Work

In this paper, we propose R-GGAT, a novel KG embedding model based on MAM. By introducing the NAM and GAM, our model can either learn different importance of the neighbouring entities or select and drop the weak or irrelevant information from embeddings, so as to distinguish the characteristics of entities and each dimension of embeddings. Experiments show that our model can perform the KG semantic information mining more effectively, by achieving state-of-the-art performance on several datasets. In addition, our proposed idea is based on the universal graph attention theory which can be used in many other tasks targeting graph-structured data processing.

Our future work will be focusing on a deeper study of the whole process of our model. We plan to integrate an unsupervised triples scoring system into our model, to develop an end-to-end link prediction structure which is more portable and efficient. In addition, we will consider blending external information or prior knowledge with the training process to enable more sufficient information for embedding learning, for the sake of making our model more discriminative, realistic and accurate when performing embedding learning tasks.

References

1. Bollacker, K., Evans, C., Paritosh, P., Sturge, T., Taylor, J.: Freebase: a collaboratively created graph database for structuring human knowledge. In: Proceedings of the 2008 ACM SIGMOD International Conference on Management of Data, pp. 1247–1250 (2008)
2. Bordes, A., Usunier, N., Garcia-Duran, A., Weston, J., Yakhnenko, O.: Translating embeddings for modeling multi-relational data. In: Neural Information Processing Systems (NIPS), pp. 1–9 (2013)
3. Chen, W., Chang, L., Bin, C., Gu, T., Jia, Z.: Jointing knowledge graph and neural network for Top-N recommendation. In: PRICAI 2019: Trends in Artificial Intelligence (2019)
4. Cho, K., et al.: Learning phrase representations using rnn encoder-decoder for statistical machine translation. arXiv preprint arXiv:1406.1078 (2014)
5. Dettmers, T., Minervini, P., Stenetorp, P., Riedel, S.: Convolutional 2D knowledge graph embeddings. In: Proceedings of the AAAI Conference on Artificial Intelligence, vol. 32 (2018)
6. Kok, S., Domingos, P.: Statistical predicate invention. In: Proceedings of the 24th International Conference on Machine Learning, pp. 433–440 (2007)
7. Lehmann, J., et al.: Dbpedia-a large-scale, multilingual knowledge base extracted from wikipedia. Semant. Web **6**(2), 167–195 (2015)
8. Li, S., Huang, Z., Cheng, G., Kharlamov, E., Gunaratna, K.: Enriching documents with compact, representative, relevant knowledge graphs. In: Twenty-Ninth International Joint Conference on Artificial Intelligence and Seventeenth Pacific Rim International Conference on Artificial Intelligence IJCAI-PRICAI-20 (2020)
9. Lin, X.V., Socher, R., Xiong, C.: Multi-hop knowledge graph reasoning with reward shaping. arXiv preprint arXiv:1808.10568 (2018)
10. Ma, L., Sun, P., Lin, Z., Wang, H.: Composing knowledge graph embeddings via word embeddings. arXiv preprint arXiv:1909.03794 (2019)
11. Nathani, D., Chauhan, J., Sharma, C., Kaul, M.: Learning attention-based embeddings for relation prediction in knowledge graphs. In: Proceedings of the 57th Annual Meeting of the Association for Computational Linguistics. Association for Computational Linguistics (2019)
12. Nguyen, D.Q., Nguyen, T.D., Nguyen, D.Q., Phung, D.: A novel embedding model for knowledge base completion based on convolutional neural network. In: Proceedings of the 2018 Conference of the North American Chapter of the Association for Computational Linguistics: Human Language Technologies, vol. 2 (Short Papers), pp. 327–333. Association for Computational Linguistics, New Orleans (2018). https://doi.org/10.18653/v1/N18-2053, https://www.aclweb.org/anthology/N18-2053
13. Schlichtkrull, M., Kipf, T.N., Bloem, P., van den Berg, R., Titov, I., Welling, M.: Modeling relational data with graph convolutional networks. In: Gangemi, A., et al. (eds.) ESWC 2018. LNCS, vol. 10843, pp. 593–607. Springer, Cham (2018). https://doi.org/10.1007/978-3-319-93417-4_38

14. Shang, C., Tang, Y., Huang, J., Bi, J., He, X., Zhou, B.: End-to-end structure-aware convolutional networks for knowledge base completion. In: Proceedings of the AAAI Conference on Artificial Intelligence, vol. 33, pp. 3060–3067 (2019)
15. Stewart, M., Liu, W.: Seq2kg: an end-to-end neural model for domain agnostic knowledge graph (not text graph) construction from text. In: Proceedings of the International Conference on Principles of Knowledge Representation and Reasoning, vol. 17, pp. 748–757 (2020)
16. Toutanova, K., Chen, D., Pantel, P., Poon, H., Choudhury, P., Gamon, M.: Representing text for joint embedding of text and knowledge bases. In: Proceedings of the 2015 Conference on Empirical Methods in Natural Language Processing, pp. 1499–1509 (2015)
17. Trouillon, T., Welbl, J., Riedel, S., Gaussier, É., Bouchard, G.: Complex embeddings for simple link prediction. In: International Conference on Machine Learning, pp. 2071–2080. PMLR (2016)
18. Vashishth, S., Sanyal, S., Nitin, V., Talukdar, P.: Composition-based multi-relational graph convolutional networks. arXiv preprint arXiv:1911.03082 (2019)
19. Veličković, P., Cucurull, G., Casanova, A., Romero, A., Lio, P., Bengio, Y.: Graph attention networks. arXiv preprint arXiv:1710.10903 (2017)
20. Wang, Y., Liu, Y., Zhang, H., Xie, H.: Leveraging lexical semantic information for learning concept-based multiple embedding representations for knowledge graph completion. In: Shao, J., Yiu, M.L., Toyoda, M., Zhang, D., Wang, W., Cui, B. (eds.) APWeb-WAIM 2019. LNCS, vol. 11641, pp. 382–397. Springer, Cham (2019). https://doi.org/10.1007/978-3-030-26072-9_28
21. Yang, B., Yih, W.T., He, X., Gao, J., Deng, L.: Embedding entities and relations for learning and inference in knowledge bases. arXiv preprint arXiv:1412.6575 (2014)
22. Zhu, G., Bin, C., Gu, T., Chang, L., Jia, Z.: A neural user preference modeling framework for recommendation based on knowledge graph. In: PRICAI 2019: Trends in Artificial Intelligence (2019)
23. Zhu, Z., Yu, J., Wang, Y., Sun, Y., Wu, Q.: MUCKO: multi-layer cross-modal knowledge reasoning for fact-based visual question answering. In: Twenty-Ninth International Joint Conference on Artificial Intelligence and Seventeenth Pacific Rim International Conference on Artificial Intelligence IJCAI-PRICAI-20 (2020)

GlowImp: Combining GLOW and GAN for Multivariate Time Series Imputation

Caizheng Liu[1,2], Houquan Zhou[1,2], Zhi Sun[1,2](✉), and Guangfan Cui[2]

[1] Institute of Computing Technology, Chinese Academy of Sciences, Beijing, China
{liucaizheng17b,sunzhi17b}@ict.ac.cn

[2] University of Chinese Academy of Sciences, Beijing, China

Abstract. Multivariate time series data generally contains missing values, which can confound subsequent analysis and compromise downstream applications. Commonly, an imputation method is used to estimate the missing values. The statistical and case deletion type methods can destroy the original data distribution, while RNNs contribute to a higher performance but these model may suffer from bias exploding problem. The method proposed in this paper, GlowImp is a deep sequential latent variable model that combines the advantages of both Glow-VAE with WGAN to learn the actual distribution of the hidden variable space but with high training stability and high-quality samples. Additionally, the framework integrates a variational autoencoder to reduce the dimensionality of the inputs as a way of combatting the high computational costs associated with Glow models. The result is an imputation process with state-of-the-art accuracy (ROC), as demonstrated with two public datasets.

Keywords: Data mining · Multivariate time series imputation · Deep neural network

1 Introduction

Many real-world data can be represented as multivariate time series. Heartbeats, temperature, stock prices, traffic flows - any data that changes value over time, any data captured by a sensor or measured at intervals - all of it can take the form of multivariate time series data. Unfortunately, sensor failures, power failures, human error and a multitude of other reasons frequently mean that time series data is missing some of its values. These gaps in the data obscure the original data distribution, undermining the accuracy of any subsequent analysis and the efficacy of the final application [10]. The most common solution to this problem is to impute the missing data via imputation processes. The recently works are to use the complete data of existing observations to build a model or learn the data distribution, estimate and impute in the missing values of the current time series. Once built, the model then imputes the values that are missing. The commonly-regarded "best" frameworks at present involve RNNs, and

Y. Lai et al. (Eds.): ICA3PP 2021, LNCS 13155, pp. 50–64, 2022.
https://doi.org/10.1007/978-3-030-95384-3_4

GANs [18] with autoencoders [21,34] as a more recent entrant that have been demonstrating state-of-the-art performance. Among these autoencoders methdods, [15] proposed a model based on a deep autoencoder that maps the missing values of multivariate time series data into a continuous low-dimensional hidden space, treating the low-dimensional representation as a Gaussian process. The downside is that the VAE structure can not generate real samples; the closest it can do is generate data that are near to real samples, i.e., the samples are fuzzy. Our innovation, inspired by [11,13,23] and the successful application of the Glow model to image generation, is to use a Glow-based autoencoder as an end-to-end multivariate time series imputation model. Called GlowImp, the first step in the framework is to learn the raw data distribution. Then, the raw data distribution is transformed by a series of reversible functions to derive their corresponding hidden variables (from a one-to-one correspondence). Calculating the logarithmic similarity of the training data is done accurately by selecting the appropriate reversible function transformation. Next, the generator in the filling module of the GAN structure is used to generate an accurate prediction value. Finally, the filling value discriminator evaluates the prediction value and updates the network according to the discriminant result.
In summary, our main contributions are as follows:

- We propose an imputation model for multivariate time series data Glow-based autoencoder. The model combines Glow-VAEs and GANs into a generative model that simultaneously learns to encode, generate and compare dataset samples.
- We design Glow-VAEs that infer the latent variables and evaluate and optimize the exact log-likelihood via a series of reversible function transformations instead of calculating its lower bound.
- Combining Glow-VAEs with GANs to generate the missing values encourages highly diverse data samples and prevents mode collapse of the GANs. It also exploits the VAE's capabilities at dimension reduction to lower the high overall time cost typically associated with Glow models.
- Experimental results on multiple real-world datasets show that the method achieves state-of-the-art performance.

2 Related Work

Our literature review covers work related to dealing with missing values. We begin by covering the traditional methods before moving on to discuss the more recent deep learning methods.

2.1 Traditional Methods

Statistical Methods: Simple statistical imputation methods impute missing values via statistical features, such as mean value [20], median value [1], mode value [14], the last observed valid value [2], etc.

Machine Learning Based Methods: Machine learning approaches include expectation-maximization [29], k-nearest neighbor [24], autoregressive [31], autoregressive vector imputation [5], matrix factorization [7], tensor singular value decomposition [19] and multivariate imputation by chained equations (MICE) [4]. Each of these methods relies on carefully hand-designed features or shared features that are based on the particular type of missing data.

2.2 Deep Learning-Based Methods

RNN-based Methods: With RNN methods, recurrent components are trained on an RNN network equipped with a classification or regression model, which significantly improves accuracy. GRU-D [9]assumes that a missing variable can be computed by combining its corresponding last observed value with the global mean and some decay factors. However, this approach has many drawbacks on general datasets [9]. M-RNN, developed by [33], imputes the missing values via a bi-directional RNN. BRITS is fully based on an RNN structure and is able to perform unidirectional uncorrelated, bidirectional uncorrelated and correlated Recurrent Imputation [8]. Although the RNN yields better high accuracy, the model can suffer from problems with exploding bias [6].

VAE-based Methods: Kingma and Welling [22] proposed stochastic gradient VB, a novel estimator of the variational lower bound, for efficient approximate inference with continuous latent variables. Garnelo et al. [16,17] assume that the data is complete at the time of training and therefore missing values are only imputed during the test period. HI-VAE deals with missing data by defining an ELBO whose reconstruction error term only sums over the observed part of the data on heterogeneous and incomplete data [28]. However, HI-VAE does not exploit temporal information. GP-VAE is a deep probabilistic model that combines a VAE with Gaussian processes to capture the temporal dynamics in time series data [15]. The VAE maps the missing data from the input space into a latent space where the temporal dynamics are modeled by the Gaussian processes.

GAN-based Models: GANs were introduced by [18] as a framework for estimating generative models via an adversarial process. GAN architectures provide the discriminator with information in the form of "hints", which reveal partial information about the missingness in the original sample. In this way, the generator ensures that the samples generated stem from a true data distribution [32]. Luo et al. [25] designed a two-second stage GAN called GRUI. The generator tries to generate a realistic time series from a random noise vector z, while the discriminator tries to distinguish whether the input data is real or fake. Although this adversarial-style structure can lead to a significant increase in accuracy, a two-stage training process adds considerable time and training the "best" matched data is not always a stable undertaking, especially when the input contains random noise. E2E-GAN [26] imputes incomplete time series via an end-to-end strategy. The solution is based on an encoder-decoder GRUI structure as the generator, which can improve the accuracy and stability when training the model. The discriminator consists of a GRUI layer and a fully connected layer working as the encoder. In a recent notable effort, [27] put forward SSGAN, a novel semi-supervised GAN with a generator, a discriminator, and a classifier to predict missing values for partially labeled time series data.

3 Method

3.1 Notations

A multivariate time series $X = \{x_1, x_2, ..., x_n\}$ is a sequence with data observed on n timestamps $T = (t_0, t_1, ..., t_{n-1})$. The i-th observation x_i contains d attributes $(x_i^1, x_i^2, ..., x_i^d)$. Time series X may contain missing values, and a binary mask vector $R^{n \times d}$ is introduced to indicate the missing positions, which is defined as:

$$M_i^j = \begin{cases} 0, & \text{if } x_i^j \text{ is null} \\ 1, & \text{otherwise} \end{cases} \tag{1}$$

If the j-th attribute of x_i is observed, M_i^j is set to 1. Otherwise, M_i^j is set to 0. Figure 1 gives an example of multivariate time series and its corresponding masking vectors.

time series X

5	/	/	/	18
12	32	9	/	76
0	/	24	/	47
x_1	x_2	......		x_5

masking matrix M

1	0	0	0	1
1	1	1	0	1
1	0	1	0	1
m_1	m_2	......		m_5

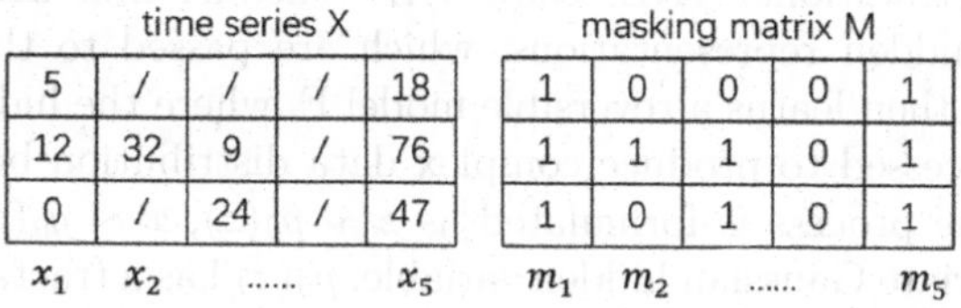

Fig. 1. Multivariate time series and corresponding masking vectors.

3.2 The GlowImp Architecture

To impute reasonable values in place of the missing values, a Glow-VAE/GAN model is trained to learn the original distribution of the dataset. As shown in Fig. 2, the method used is a traditional VAE method designed to make the latent vector of the data coding obey a Gaussian distribution for subsequent data generation. The VAE also optimizes the lower bound of the log-likelihood, i.e., the ELBO (evidence lower bound). It also approximates the distribution of the hidden variable space. This is done by minimizing the KL divergence. However, the model can only approximate the distribution of the hidden variable space, so the data generated will only be fuzzy. As shown in Fig. 3, using a Glow-VAE model instead of a traditional VAE solves this problem. Glow-VAE comprises an encoder, a decoder, and Glow components. Its purpose is to map the data to low-dimensional space and then use the hidden variables to accurately model the real distribution of the data. In other words, its job is to accurately estimate the log-likelihood. The decoder then generates complete multivariate time series data. In the GAN model, a discriminator calculates the difference between the fake data and the real data. The Glow-VAE is trained jointly with the GAN with updates for the generator and discriminator derived through gradient descent.

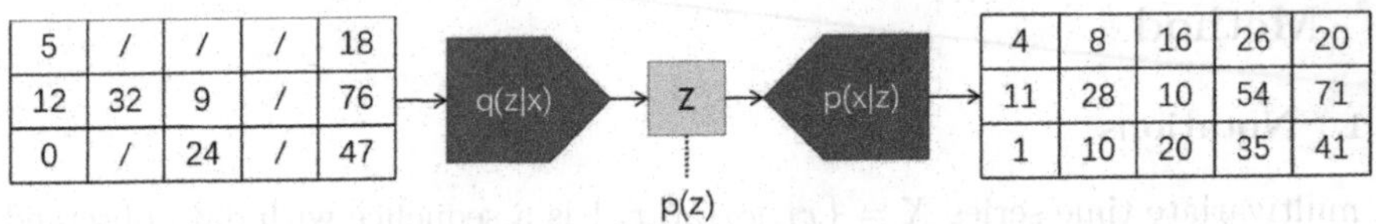

Fig. 2. The structure of the traditional VAE method. VAE method designed to make the latent vector of the data coding obey a Gaussian distribution for subsequent data generation. The VAE also optimizes the lower bound of the log-likelihood, so the data generated will only be fuzzy.

3.3 Glow-VAE

As shown in Fig. 3, Glow-VAE is designed to reconstruct target samples. In addition to mapping the original data to the hidden representations, the encoder also greatly reduces the dimensionality of the data to decrease modeling time. Unlike an encoder in a traditional VAE, Glow-VAEs encoder first learns a Gaussian distribution as hidden representations, which are passed to the Glow model. The Glow model then learns a reversible model F, where the hidden representations can be processed to produce complex data distribution beyond Gaussian distributions. The process is formulated as $z \sim p_\theta(z)$, $x \sim g_\theta(z)$, where z is a complex multivariate Gaussian hidden variable, $p_\theta(z)$ has a tractable density and g_θ is a reversible bijective function that also becomes a bijective function,i.e., $z = f_\theta(x) = g_\theta^{-1}(z)$. Thus, the result obtained is not only accurate, but also can be interpolated. To impute the missing values, it is simply a question of minimizing the following:

$$L_R = E_{z \sim p_\theta(z|x)}[q_\phi(\hat{x} \mid z)] + KL(p_\theta(z \mid x) \| p(z)), \tag{2}$$

$$L_G(D) \cong \frac{1}{N} \sum_{i=1}^{N} - \log p_\theta(x^{(i)}) + c, \tag{3}$$

$$L_{Glow-VAE} = L_R + L_G(D). \tag{4}$$

Where L_R is the loss of VAE, $L_G(D)$ is the loss of Glow model. Given dataset D, $x^{(i)} = x^{(i)} + u$, $u \sim U(0, a)$, $c = -M \cdot \log_a$, $p_\theta(z)$ is a prior distribution. a is determined by the discrimination level of the data and M is the dimensionality of x.

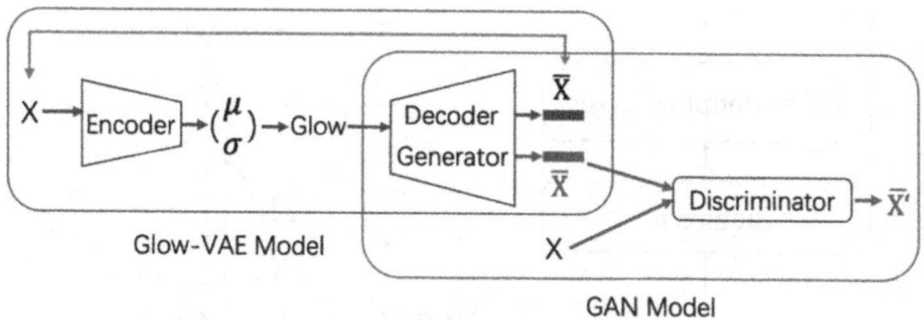

Fig. 3. The GlowImp framework of the multivariate time series missing value imputing model.

3.4 Glow Model Details

Unlike traditional VAEs, the Glow-VAE network structure is reversible. The encoder and decoder actually have similar network structures, except that they are input from different ends when the encoding and decoding functions are performed. As such, a Glow network can be seen as being divided into several large layers, each containing many depth units. As shown in Fig. 4(a), a depth unit consists of two parts, [23].

The first component is the actnorm layer that performs an affine transformation using a scale and bias parameter, similar to batch normalization. These parameters are initialized such that the post-actnorm activation have zero mean and unit variance given an initial mini-batch of data.

The second component is affine coupling layers. With careful function design, a Glow model can be learned that is both tractable and extremely flexible. This property is guaranteed by a series of Jacobi transformations (see [12] for the details of the proof). As shown Fig. 4(b), the affine coupling layers divide the feature vector z passed from actnorm in the channel dimension into two vectors to result in z1 and z2. The length of z1 is half of the feature vector z. The equations in the affine coupling layers are as follows:

$$
\begin{aligned}
y_1 &= z_1, \\
y_2 &= z_2 \odot exp(s(z_1)) + t(z_1), \\
z &= concat(y_1, y_2).
\end{aligned} \tag{5}
$$

Where the s and t denote the nonlinear transformations. Since computing the Jacobian determinant of the coupling layer operation does not involve computing the Jacobian of s or t [12], we made them deep convolutional neural networks in our experiments.

3.5 The GAN Architecture

The GAN consists of a generator (G) and a discriminator (D). G learns a mapping that tries to map the hidden vector from Glow-VAE to a predicted multivariate time series, while D tries to find a mapping that can distinguish between

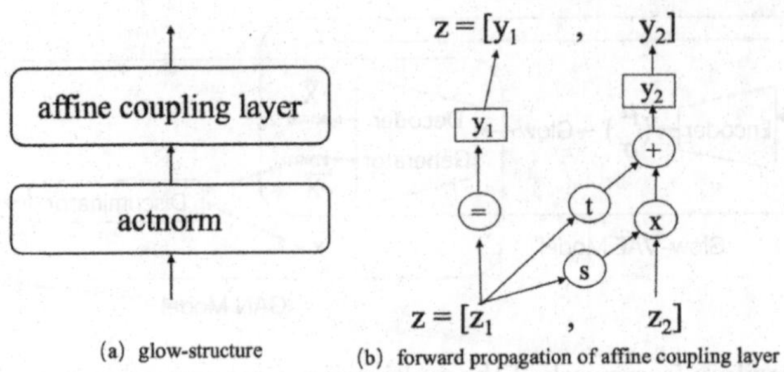

Fig. 4. The structure of the Glow module in the GlowImp framework.

the true and fake generated samples. It is noteworthy that the input of the D contains real, but incomplete, samples and fake, but complete, samples generated by G. Because of the mode collapse problem, a traditional GAN can be hard to train. WGAN is an extension of GAN that uses the Wasserstein distance and is usually easier to train than the original GAN. WGAN can improve stability in the learning stage. It can also escape from mode collapse, making the model easier to optimize. A generative model can be learned to optimize divergence notions beyond the earth mover's divergence. We chose the WGAN by [3] over a traditional GAN. The loss function of WGAN in our model follows:

$$L_{GAN} = \sum_{x \in D} \|x \otimes m - \hat{x} \otimes m\|_2 + \beta(E_{\hat{x} \sim q_\phi(\hat{x}|z)}[f_D(\hat{x})] - E_{x \sim P_r}[f_D(x)]). \quad (6)$$

x represents the input multivariate time series data, and m means the masking matrix. The first expression in Eq. 6 is the masked reconstruction loss that calculates the squared errors between the original sample x and the generated sample. It is noteworthy that we only calculate the non-missing part of the data. The second expression in Eq. 6 computes discriminative loss. β is a hyper-parameter that controls the proportion between the masked reconstruction loss and the discriminative loss.

The overall objective function of our GlowImp model is formulated as:

$$L_{total} = L_{Glow-VAE} + L_{GAN}. \quad (7)$$

4 Experiment

To verify and measure the performance of the GlowImp framework, we imputed missing values in two real-world data sets and compared the imputations to other contemporary methods.

4.1 Dataset Description

KDD CUP 2018 Dataset: The KDD CUP Challenge 2018 dataset[1] contains air quality and weather data collected hourly collected from 20 Jan 2017 to 30 Jan 2018 in Beijing. Each record contains 12 features, including M2.5, PM10, CO, weather, temperature, etc., with roughly 15% of the values missing. We selected 11 of the 12 features and imputed the missing values in batches of one hour and 48 h.

Imputation performance was assessed in terms of root mean squared error (RMSE) and mean absolute error (MAE). RMSE is the mean value of the square root of the error between the predicted value and the true value and is a popular choice for assessing the imputation in multivariate time series. The formulation is as follows:

$$RMSE = \sqrt{\frac{1}{n}\sum_{i=1}^{n}(x-\hat{x})^2}.$$

MAE(Mean Absolute Error) is the absolute residual, which is summed and averaged to get the MAE; it reflects the size of the error between the actual value and the imputed value of the missing value in the multivariate time series. The formulation is as follows:

$$MAE = \frac{1}{n}\sum_{i=1}^{n}|x-\hat{x}|,$$

where x is the actual value , $\hat{x}$ is the imputed value and n is the number of the time internal.

Challenge 2012 Health-care Data: The PhysioNet Challenge 2012 [2] dataset is a public collection of multivariate clinical time series data. The set consists of around 8,000 records gathered from intensive care units. Each record is a multivariate time series of about 48 h containing 42 features, including age, gender, height, heart rate, glucose levels, etc. Roughly 80.67% of the values are missing, which is too many to directly impute with any real level of accuracy. However, 554 patients are labeled as either alive (negative mortality) or dead (positive mortality), which is enough data to impute values for the remaining records and train a model in a (binary) mortality prediction task. For a fair comparison, we imputed the values using each of the different models but used the same classifier to perform the predictions. Performance was evaluated in terms of AUC (area under ROC curve) and Accuracy (ACC), calculated as:

$$ACC = \frac{TP+TN}{TP+TN+FP+FN},$$

[1] KDDCUP. Available on: http://www.kdd.org/kdd2018/, 2018.

[2] Predicting in-hospital mortality of icu patients. Available on: The physionet/computing in cardiology challenge 2012. In CinC (2012).

where TP is true positive, TN is true negative, FP is false positive and FN is false negative. ACC(Accuracy) refers to the proportion of correct predictions in all prediction structures. AUC is an indicator to measure the performance of a multivariate time series missing value imputing model.

4.2 Comparison Methods

We compared GlowImp to eight current imputation methods. A brief description of each follows.

- Statistical imputation methods [30], where the missing values are simply replaced with zero, mean and median.
- KNN [24], which uses a k-nearest neighbor algorithm to find neighboring data. Unobserved data is imputed as the weighted average of k neighbors.
- MF [7], which factorizes an incomplete matrix into low-rank matrices to fill the missing values.
- SVD [19], which uses matrix completion through iterative singular value decomposition (SVD).
- GP-VAE [15], a method that combines ideas from VAEs and Gaussian processes to capture temporal dynamics for time series imputation.
- BRITS [8], one of methods that uses bidirectional RNN to impute time series.
- GRUI [25], which uses a based two-stage GAN to impute missing values.
- E2E-GAN [26], one of the state-of-the-art methods. It relies on an end-to-end GAN network to impute the missing values in time series data.

4.3 Implementations Details

All experiments were conducted under the same conditions with the same settings. The hardware platform was an Intel i7 9700k PC with 48GB memory and an NVIDIA GTX 1080 8 GB video card. The deep learning framework was PyTorch 1.7 and TensorFlow 1.15.0. The Glow model contained five actnorm and affine coupling layers.

To maintain the same number of training and test samples as the contemporary method, the dataset was split with 80% of the records used for the training set and the remaining 20% used for the test set. All values are normalized in the range of 0 to 1. For training, 50% of the KDD data in the training set was deleted. When testing, we deleted between 10% and 80% of the KDD data, tested each method at a range of levels of missing data. For PhysioNet dataset, we imputed the values using each of the different models but used the RNN to perform the predictions.

4.4 Experiment Results

KDD Dataset Results: The results with the KDD dataset and a missing value ratio of 10% appear in Table 1. Here, GlowImp returned significantly fewer errors than the other methods in RMSE and MAE.

Table 1. The RMSE and MAE results of the GlowImp and other methods on the KDD datasets(lower is better)

Method	Zero	Mean	Median	KNN	MF	SVD	GP-VAE	BRITS	GRUI	E2E-GAN	GlowImp
RMSE	1.009	1.009	1.079	0.853	0.683	1.553	0.994	0.196	0.179	0.153	**0.136**
MAE	1.004	1.005	1.038	0.924	0.826	1.246	0.989	0.168	0.142	0.114	**0.089**

Generally, the proportion of missing data in multivariate time series may be uncertain. Data with a higher proportion of missing values means that it is more difficult to correctly predict the missing values. Therefore, the ability to correctly predict data with different missing ratios is an important basis for evaluating model performance. To assess the frameworks with different levels of missing data, we then conducted the same experiment with the BRITS, GRUI, E2E-GAN and GlowImp, varying the ratios of missing values from 10% to 80% in steps of 10%. The results are shown in Fig. 5. Again, GlowImp returned the fewest errors.

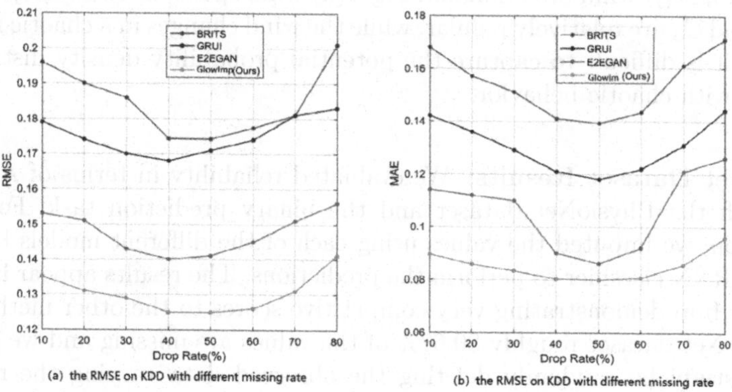

Fig. 5. The RMSE and MAE results of the GlowImp and other methods on the KDD dataset(the lower, the better).

Our third set of experiments was an ablation study, designed to assess the contribution of the Glow module. It comprised four tests: the first with the Glow module, the second where we simply removed the Glow module; a third where we replaced the Glow module with a 5-layer fully connected layer; and the last where we replaced the Glow module with a five-layer nonlinear fully connected layer. All tests were conducted with a range of missing value ratios. Figure 6 shows the results. The tests with the Glow module returned substantially fewer errors, verifying Glow's contribution to the framework.

Figure 7 and Fig. 8 show the imputation results of Tongzhou and Mentougou districts, respectively. The blue point is the ground true time series and the read curve is the imputed values. We could see that GlowImp could capture the evolution trend and impute the missing values pretty well. The GlowImp could

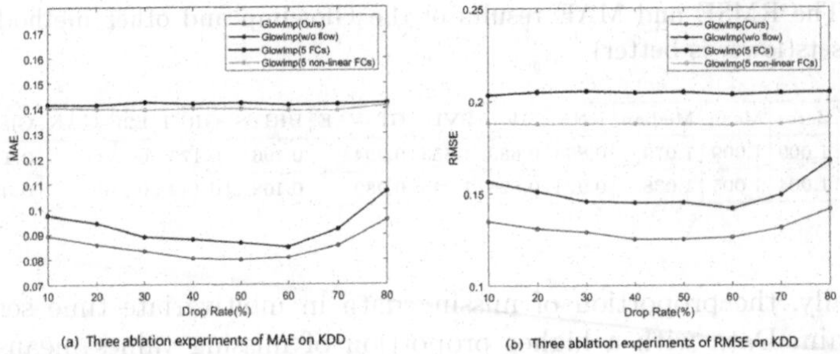

Fig. 6. Three ablation experiments of the GlowImp framework on RMSE and MAE on KDD dataset(the lower, the better).

capture the potential probability density distribution of multivariate time series and make full use of the interactive information of multivariate time series. We can observe that in Fig. 7(b) NO_2 and Fig. 7(c) O_3 maintain better performance than the Fig. 7(j) wind direction and Fig. 7(k) wind speed, as the change pattern of NO_2 and O_3 are relatively regular, while the wind changes in a chaotic manner, verifying it is difficult to capture the potential probability density distribution perfectly with chaotic behavior.

PhysioNet Dataset Results: We evaluated reliability in terms of ACC and AUC with the PhysioNet dataset and the binary prediction task. For a fair comparison, we imputed the values using each of the different models but used the same RNN classifier to perform the predictions. The results appear in Fig. 9, with GlowImp demonstrating very competitive scores to the other methods.

PhysioNet dataset roughly 80.67% of the values are missing and we conduct ablation study by randomly deleting the observed data varying the ratios of missing values about 80% and 90%. It comprised three tests: the first with the Glow module, the second where we replaced the Glow module with a 5-layer fully connected layer; and the last where we replaced the Glow module with a five-layer nonlinear fully connected layer. Table 2 and Table 3 shows the results. The tests with the Glow module returned higher ACC and AUC, verifying Glow's contribution to the framework.

Table 2. The AUC results of the GlowImp and other methods on the PhysioNet datasets

Method	5 FCs	5 non-FCs	Ours(GlowImp)
80%	0.751	0.872	**0.881**
90%	0.737	0.859	**0.864**

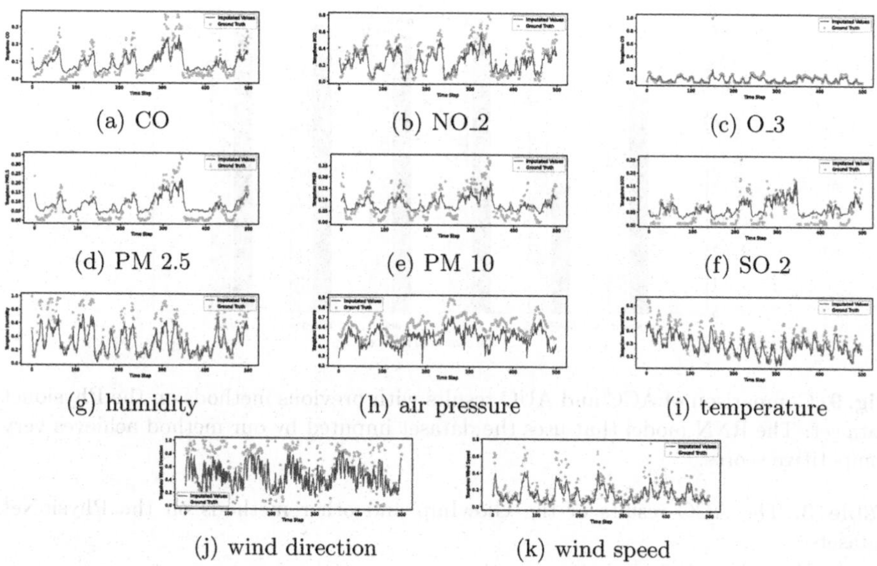

Fig. 7. The KDD dataset imputed results. The ground true(blue) and the imputed values(red) in Tongzhou, Beijing of KDD dataset. The X axis indicates time step. The Y axis indicates imputed values. (Color figure online)

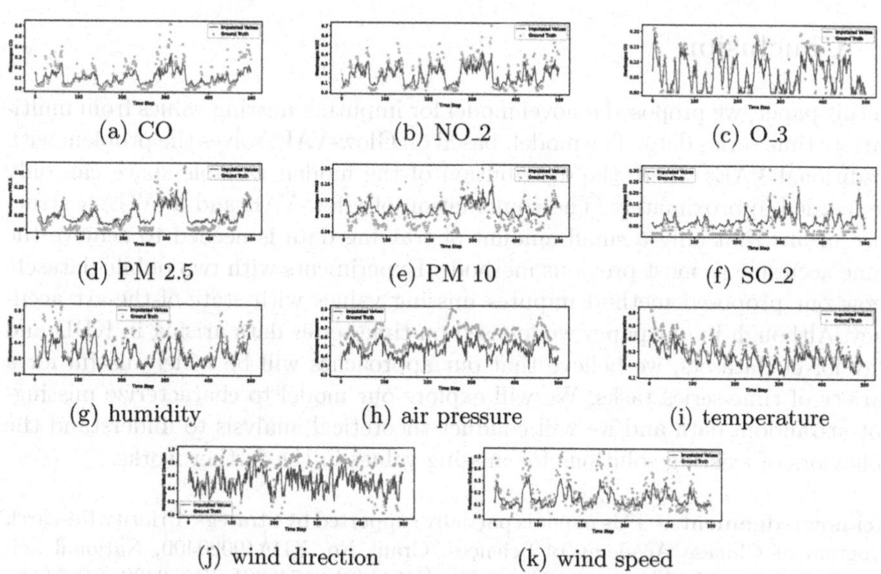

Fig. 8. The KDD dataset imputed results. The ground true(blue) and the imputed values(red) in Mentougou, Beijing of KDD dataset. The X axis indicates time step. The Y axis indicates imputed values. (Color figure online)

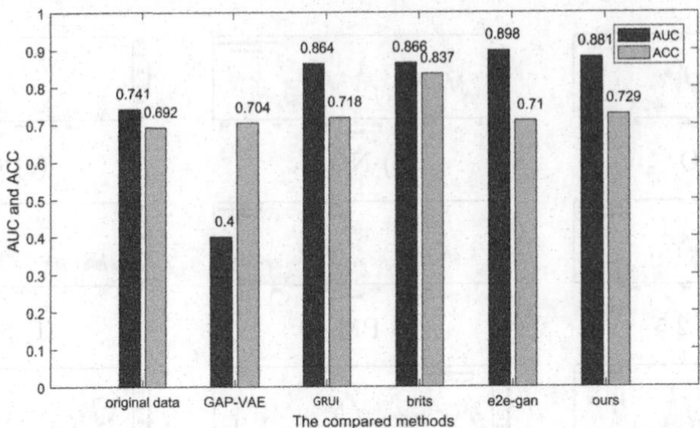

Fig. 9. Comparison of ACC and AUC results with previous methods on the Physionet data set. The RNN model that uses the dataset imputed by our method achieves very competitive scores.

Table 3. The ACC results of the GlowImp and other methods on the PhysioNet datasets

Method	5 FCs	5 non-FCs	Ours(GlowImp)
80%	0.718	0.722	**0.729**
90%	0.671	0.699	**0.711**

5 Conclusion

In this paper, we proposed a novel model for imputing missing values from multivariate time series data. The model, based on Glow-VAE, solves the problem with traditional VAEs where the distribution of the hidden variable space can only be learned approximately. The combination of Glow-VAE and a WGAN structure means that only a small amount of training data is needed to achieve the same accuracy as most previous methods. Experiments with two public datasets show our proposed method imputes missing values with state-of-the-art accuracy. Although in our paper we focused on time-series data arising in KDD and PhysioNet datasets, we believe that our approaches will be widely useful for a variety of time-series tasks. We will explore our model to characterize missing-not-at-random data and we will conduct theoretical analysis to understand the behaviors of existing solutions for missing values in the further works.

Acknowledgements. This paper is partially supported by Strategic Priority Research Program of Chinese Academy of Sciences, Grant No. XDA19020400, National Science Foundation of China under Grant No. U1911401,91746301, 61772498, 61872206, 61802370.

References

1. Acuña, E., Rodriguez, C.: The treatment of missing values and its effect on classifier accuracy. In: Classification, Clustering, and Data Mining Applications, pp. 639–647 (2004)
2. Amiri, M., Jensen, R.: Missing data imputation using fuzzy-rough methods. Neurocomputing **205**, 152–164 (2016)
3. Arjovsky, M., Chintala, S., Bottou, L.: Wasserstein generative adversarial networks. In: Proceedings of the 34th International Conference on Machine Learning, vol. 70, pp. 214–223 (2017)
4. Azur, M.J., Stuart, E.A., Frangakis, C., Leaf, P.J.: Multiple imputation by chained equations: what is it and how does it work? Int. J. Methods Psychiatr. Res. **20**(1), 40–49 (2011)
5. Bashir, F., Wei, H.L.: Handling missing data in multivariate time series using a vector autoregressive model-imputation (VAR-IM) algorithm. Neurocomputing **276**, 23–30 (2018)
6. Bengio, S., Vinyals, O., Jaitly, N., Shazeer, N.: Scheduled sampling for sequence prediction with recurrent neural networks. In: Proceedings of the 28th International Conference on Neural Information Processing Systems, vol. 1, pp. 1171–1179 (2015)
7. Li, C., et al.: Recommending missing sensor values. In: 2015 IEEE International Conference on Big Data (Big Data), pp. 381–390 (2015)
8. Cao, W., Wang, D., Li, J., Zhou, H., Li, L., Li, Y.: BRITS: bidirectional recurrent imputation for time series. In: Advances in Neural Information Processing Systems, vol. 31 (2018)
9. Che, Z., Purushotham, S., Cho, K., Sontag, D., Liu, Y.: Recurrent neural networks for multivariate time series with missing values. Sci. Rep. **8**(1), 6085 (2018)
10. Cheema, J.R.: A review of missing data handling methods in education research. Rev. Educ. Res. **84**(4), 487–508 (2014)
11. Dinh, L., Krueger, D., Bengio, Y.: Nice: non-linear independent components estimation (2015)
12. Dinh, L., Sohl-Dickstein, J., Bengio, S.: Density estimation using real NVP. CoRR abs/1605.08803 (2016)
13. Dinh, L., Sohl-Dickstein, J., Bengio, S.: Density estimation using real NVP (2017)
14. Donders, A.R.T., van der Heijden, G.J., Stijnen, T., Moons, K.G.: Review: a gentle introduction to imputation of missing values. J. Clin. Epidemiol. **59**(10), 1087–1091 (2006)
15. Fortuin, V., Baranchuk, D., Raetsch, G., Mandt, S.: GP-VAE: deep probabilistic time series imputation. In: Proceedings of the Twenty Third International Conference on Artificial Intelligence and Statistics, vol. 108, pp. 1651–1661 (2020)
16. Garnelo, M., et al.: Conditional neural processes. In: Proceedings of the 35th International Conference on Machine Learning. Proceedings of Machine Learning Research, vol. 80, pp. 1704–1713 (2018)
17. Garnelo, M., et al.: Neural processes. arXiv:1807.01622 [cs, stat] (2018)
18. Goodfellow, I.J., et al.: Generative adversarial nets. In: Proceedings of the 27th International Conference on Neural Information Processing Systems, vol. 2, pp. 2672–2680 (2014)
19. He, J., Sun, G., Zhang, Y., Geng, T.: Data recovery in heterogeneous wireless sensor networks based on low-rank tensors. In: 2016 IEEE Symposium on Computers and Communication (ISCC), pp. 616–620 (2016)
20. Kantardzic, M.: Data Mining: Concepts. Methods, and Algorithms, Models (2011)

21. Kingma, D.P., Welling, M.: Auto-encoding variational bayes (2014)
22. Kingma, D.P., Welling, M.: Auto-encoding variational bayes. In: 2nd International Conference on Learning Representations, ICLR 2014, Banff, AB, Canada, 14–16 April 2014, Conference Track Proceedings (2014)
23. Kingma, D.P., Dhariwal, P.: Glow: generative flow with invertible 1×1 convolutions. In: Advances in Neural Information Processing Systems, vol. 31. Curran Associates, Inc. (2018)
24. Liew, A.W.C., Law, N.F., Yan, H.: Missing value imputation for gene expression data: computational techniques to recover missing data from available information. Brief. Bioinf. **12**(5), 498–513 (2011)
25. Luo, Y., Cai, X., Zhang, Y., Xu, J., Yuan, X.: Multivariate time series imputation with generative adversarial networks. In: Proceedings of the 32nd International Conference on Neural Information Processing Systems, pp. 1603–1614 (2018)
26. Luo, Y., Zhang, Y., Cai, X., Yuan, X.: E^2GAN: end-to-end generative adversarial network for multivariate time series imputation. In: Proceedings of the Twenty-Eighth International Joint Conference on Artificial Intelligence, IJCAI-19, pp. 3094–3100 (2019)
27. Miao, X., Wu, Y., Wang, J., Gao, Y., Mao, X., Yin, J.: Generative semi-supervised learning for multivariate time series imputation. In: Proceedings of the AAAI Conference on Artificial Intelligence, vol. 35, no. 10, pp. 8983–8991 (2021)
28. Nazabal, A., Olmos, P.M., Ghahramani, Z., Valera, I.: Handling Incomplete Heterogeneous Data using VAEs. arXiv:1807.03653 [cs, stat] (2020)
29. Nelwamondo, F.V., Mohamed, S., Marwala, T.: Missing data: a comparison of neural network and expectation maximization techniques. Curr. Sci. **93**(11), 1514–1521 (2007)
30. Rubinsteyn, A., Feldman, S.: fancyimpute: an imputation library for python (2016). https://github.com/iskandr/fancyimpute
31. Sridevi, S., Rajaram, S., Parthiban, C., SibiArasan, S., Swadhikar, C.: Imputation for the analysis of missing values and prediction of time series data. In: 2011 International Conference on Recent Trends in Information Technology (ICRTIT), pp. 1158–1163 (2011)
32. Yoon, J., Jordon, J., van der Schaar, M.: GAIN: missing data imputation using generative adversarial nets. In: Proceedings of the 35th International Conference on Machine Learning, vol. 80, pp. 5689–5698 (2018)
33. Yoon, J., Zame, W., Schaar, M.: Multi-directional recurrent neural networks: a novel method for estimating missing data (2017)
34. Zhao, J., Kim, Y., Zhang, K., Rush, A.M., LeCun, Y.: Adversarially regularized autoencoders (2018)

Accurate Indoor Localization Using Magnetic Sequence Fingerprints with Deep Learning

Xuedong Ding, Minghua Zhu(✉), and Bo Xiao(✉)

MOE Engineering Research Center for Software/Hardware Co-Design Technology and Application, East China Normal University, Shanghai 200062, China
{mhzhu,bxiao}@sei.ecnu.edu.cn

Abstract. Magnetic field fingerprinting has been an interesting topic in indoor localization researches because of its advantages of being ubiquitous, energy-efficient and infrastructure-free. Most existing indoor magnetic field-based positioning methods use the raw three-dimensional magnetic field strength obtained by the magnetic sensor built in smartphones. However, they have to overcome the problem of ambiguity that originates from the nature of geomagnetic data, especially in the large-scale environment. In this paper, we first expand the dimension of magnetic data elements, and a sliding window mechanism is designed to construct magnetic sequence fingerprints to increase the distinguishability of magnetic field fingerprints. Moreover, an accurate indoor positioning model combining the advantages of one-dimensional convolutional neural network and long short-term memory network is designed to automatically learn the mapping between ground-truth positions and magnetic sequence fingerprints. To demonstrate the effectiveness of our proposed method, we perform a comprehensive experimental evaluation on three real-world datasets, and the results show that the proposed approach can remarkably improve positioning performance compared with other methods.

Keywords: Indoor localization · Magnetic field · Magnetic sequence fingerprints · Smartphone · Deep learning

1 Introduction

With the rapid development of mobile technology, accurate and pervasive location-based service significantly facilitates and enriches daily life. The proliferation of modern smartphones and their wide usage in daily life motivate the rapid development of LBS (Location-Based Services) that could provide precise location information for the user, both outdoor and indoor. The outdoor location can be obtained by GPS (Global Positioning System) with high accuracy. However, its signals cannot cover many indoor scenarios, such as shopping malls, libraries, and museums. Therefore, indoor localization has become a hot research area.

In order to achieve robust meter-level accuracy of indoor location-based services, many types of indoor positioning methods have been proposed, including

Y. Lai et al. (Eds.): ICA3PP 2021, LNCS 13155, pp. 65–84, 2022.
https://doi.org/10.1007/978-3-030-95384-3_5

Bluetooth [12], Wi-Fi [15] and CSI (Channel State Information) [11]. Unfortunately, these positioning signals have inherent limitations: Bluetooth-based positioning requires extra infrastructure; Wi-Fi-based positioning can only provide rough location estimation due to its signal fluctuation; CSI can provide high precision positioning only in a small-range environment because it needs to sample a large number of dense training points.

Magnetic field-based indoor localization has been actively studied in recent years since its signal is ubiquitous, and it incurs almost no additional energy consumption and requires no extra infrastructure. In addition, researchers find that the magnetic field strength (MFS), determined by the geomagnetic field and building's iron structures such as steel frames and electrical appliances, is sufficiently stable in indoor environments [4]. So far, various indoor localization approaches (see Sect. 2 for a review) have been presented that take advantage of magnetic field data to locate a person in the indoor environment. However, existing magnetic field-based positioning methods using smartphone are limited by two factors. First, the magnetic field dataset collected from smartphones is a collection of three-dimensional vectors with great ambiguity, especially in a large-scale indoor environment with a similar structure. Second, the existing magnetic field-based positioning models cannot effectively extract enough features from raw magnetic sequence data.

In this paper, we propose an accurate magnetic field-based indoor localization system with a hybrid neural network that combines a one-dimensional convolutional neural network (1D-CNN) and a long short-term memory (LSTM) network. In order to reduce the ambiguity and improve the uniqueness of magnetic field fingerprints, we expand the raw three-dimensional magnetic data elements to five-dimensional. In addition, we observed in the experiment that although the magnetic field strength may fluctuate at a point, the magnetic sequence of the same path has the same trend. Therefore, a sliding window mechanism is designed to construct the magnetic sequence fingerprints (MSFs) based on the transformed magnetic data elements. In order to obtain the optimal positioning model, we divide the model training into two steps: the Adam optimizer first to converge the model quickly and then the stochastic gradient descent (SGD) optimizer to make the model better.

The contributions of this research can be summarized as follows.

1. An indoor localization model called CNN-LSTM is proposed that combines the advantages of 1D-CNN and LSTM and uses magnetic field fingerprints alone to realize localization in the indoor environment.
2. A novel fingerprint construction method is devised, which first expands the dimension of magnetic data elements and then uses a sliding window mechanism to construct magnetic sequence fingerprints.
3. The proposed approach is tested on three real-world datasets, and experimental results show that the proposed positioning method significantly surpasses the existing approaches in terms of mean positioning error and cumulative error distribution.

The remainder of the paper is arranged as follows. Section 2 surveys related work on indoor positioning with the magnetic field or deep learning. Section 3 provides our proposed system architecture, data collection and preprocessing, MSFs construction. The detailed process of the proposed positioning model is described in Sect. 4. And Sect. 5 describes how the experimental campaign was set and conducted, and results analysis. Section 6 concludes the paper. Some related acronyms are listed in Table 1.

Table 1. List of important abbreviations in alphabetical order

Abbr.	Definition	Abbr.	Definition
1D	One-dimensional	CNN	Convolutional Neural Network
LSTM	Long Short-term Memory	MFS	Magnetic Field Strength
MSF	Magnetic Sequence Fingerprint	RNN	Recurrent Neural Network

2 Related Work

The request for pervasive indoor LBS (Location-Based Services) has spurred the development of efficient indoor positioning techniques. In this section, we mainly review two types of indoor localization systems, i.e., magnetic field-based and deep learning-based systems.

Earth magnetic field has been proven to be useful in indoor positioning because it is natural and ubiquitous. After Suksakulchai [14] realized that the magnetic field disturbances could be used for indoor localization, Chung et al. [5] and Grand et al. [8] used the magnetic field signal as fingerprint for indoor localization. In order to make better use of the information of the magnetic field, different magnetic field features (e.g., kurtosis, mean and slope) were tested to achieve room-level accuracy [7]. The work [6] studied the magnetic field intensity and direction distribution features for constructing magnetic maps. The improved work [13] proposed a feature distinguishability measurement technique to evaluate the performance of different feature extraction methods for magnetic fingerprints. Unfortunately, the study [3] found that the magnetometers in smartphones are vulnerable to a few factors such as user's postures and walking speed, which causes the magnetic field strength corresponding to a location often shift in time or exhibit local distortions, thus greatly limits the positioning performance of existing methods rely on raw magnetic field strength.

Recently, many researchers use deep learning technology for indoor positioning to improve positioning accuracy. There are three types of deep networks used for indoor localization, including deep autoencoder networks, deep convolution neuron networks (CNN), and LSTM networks [18]. DeepFi was the first work based on autoencoder to use CSI amplitudes for indoor positioning [16]. Moreover, deep autoencoder networks-based indoor localization systems with Bluetooth Low Energy (BLE) [20] and Wi-Fi [1] have also been proposed. CiFi was the first system that incorporated a deep CNN for indoor localization [17]. Immediately afterwards, AMID [10] utilized a convolution neural network (CNN) for analyzing magnetic field features and a multi-layer perceptron (MLP) for magnetic landmark classification. In addition, MINLOC [2] designed multiple

CNN models combining with a voting mechanism to improve positioning performance. The proposed DeepML [19] system was the first to employ deep LSTM with magnetic and light bimodal data for indoor localization. In order to save the workforce, in reference [4], Chiang et al. built a robotic platform for magnetic field data collection, and they studied some data augmentation mechanisms to facilitate the RNN-LSTM model for improving the positioning accuracy. However, many of the above-mentioned works require a large amount of data for training to predict user position. Additionally, heterogeneous smartphones and the different walking speeds of users reduce the scalability of these methods.

We, therefore, seek to minimize those drawbacks by using magnetic data combined with deep learning to better perform indoor localization. First, a novel fingerprint construction method is proposed to construct MSFs with high distinguishability, and then these MSFs are used to train the CNN-LSTM model to predict the user location accurately. Furthermore, our system has good scalability, because it has achieved considerable positioning performance in three different typical experimental environments, using different smartphones, and with different walking speeds of users.

3 System Architecture

3.1 General Overview

The architecture of the proposed indoor positioning system is presented in Fig. 1. The main processes of the system contains magnetic data collection and preprocessing phase, model training phase and online localization phase. The details of these phases are described in the Sect. 3.2, Sect. 3.3 and Sect. 4.

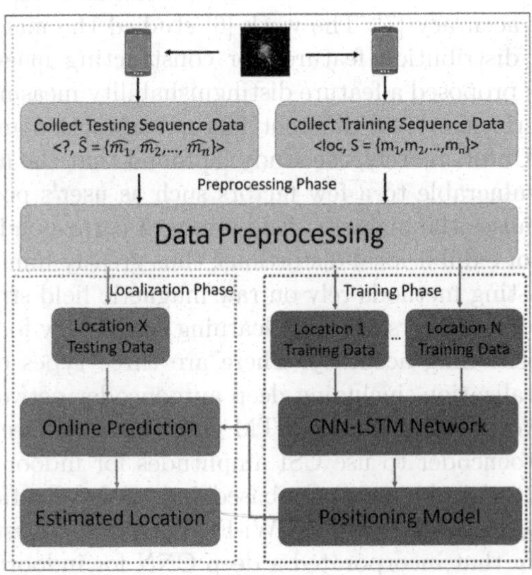

Fig. 1. The indoor positioning system architecture

3.2 Magnetic Data Collection and Preprocessing

The magnetic data collection and preprocessing process consists of data collection, magnetic data elements expansion, magnetic sequence data scaling, noise reduction, and data normalization.

Magnetic Data Collection. There is a large error in the point-based positioning due to the following two factors: 1) the MFS at a given indoor location is a 3-D vector in space that varying similarly with near location; 2) different orientation or postures of mobile phone lead to different MFS readings at the same location. Moreover, we collected three magnetic sequences for the same space trajectory at different times, which are shown in Fig. 2. We could observe that there may be some shifts in the magnetic sequence at different times, but the changing trend is basically the same. Therefore, this study adopts a fast continuous collection method to collect magnetic sequence data on multiple paths in the indoor environment, which can well reflect the distribution of the indoor magnetic field and has good uniqueness. Figure 3 shows the procedure of the collection of magnetic data on an indoor path. The distance between every two reference points on the experimental path is 0.5 m. Since the sampling frequency 50 Hz and the walking speed is 1 m/s, the magnetic sequence between the two reference points includes 25 magnetic data.

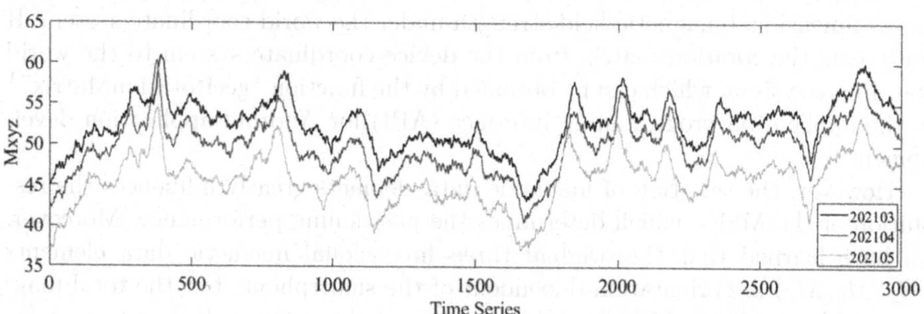

Fig. 2. Three magnetic sequences of the same space trajectory at different times

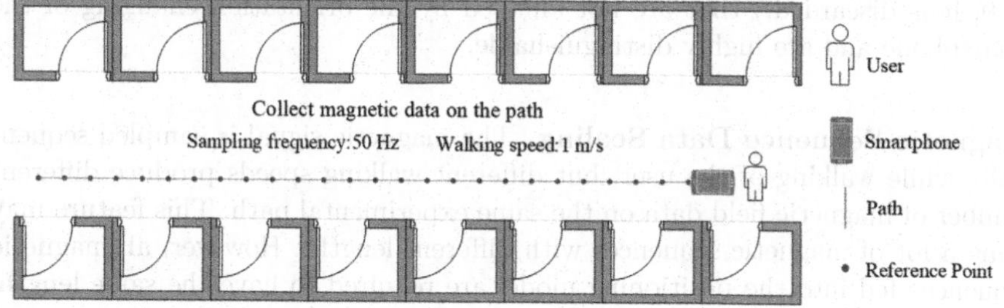

Fig. 3. Collection of magnetic data

Magnetic Data Elements Expansion. As shown in Fig. 4(a), the magnetic field information collected by the smartphone is represented by three-axis data (M_x, M_y, M_z). When we synthesize them, we can represent the rich magnetic field information, *e.g.*, total magnetic field strength M_{xyz} and horizontal component M_{xy}. The magnetic data obtained by the built-in magnetic sensor of the smartphone is based on the device coordinate system of the mobile phone, and the direction of the mobile phone held on the user is arbitrary in the walking, so the device coordinate system is changing. This study converts the measured magnetic data from the device coordinate system to the world coordinate system which is fixed (Their definitions are shown in Fig. 4(b) and Fig. 4(c), respectively.). The triaxial component ($M_x^{'}$, $M_y^{'}$, $M_z^{'}$) of the magnetic field in the world coordinate system is obtained. These elements can be calculated by the Eq. (1):

$$\begin{gathered} M_{xy} = \sqrt{M_x^2 + M_y^2}, \\ \begin{pmatrix} M_x^{'} \\ M_y^{'} \\ M_z^{'} \end{pmatrix} = R \begin{pmatrix} M_x \\ M_y \\ M_z \end{pmatrix}, \\ M_{xyz} = \sqrt{M_x^2 + M_y^2 + M_z^2}. \end{gathered} \tag{1}$$

where (M_x, M_y, M_z) represents the three-dimensional magnetic field strength obtained by the smartphone's magnetic sensor, ($M_x^{'}$, $M_y^{'}$, $M_z^{'}$) represents the three-dimensional magnetic field strength under the world coordinate system. R represents the rotation matrix from the device coordinate system to the world coordinate system, which can be obtained by the function "getRotationMatrix"[1] in the application programming interface (API) for Android application development.

However, the selection of magnetic data elements greatly influences the resolution of the MSFs, which determines the positioning performance. Moreover, we have learned that the original three-dimensional magnetic data elements (M_x, M_y, M_z) are orientation-dependent of the smartphone, but the total magnetic field magnitude M_{xyz} is stable and orientation-independent of the smartphone. Therefore, this study uses five-dimensional magnetic data elements (M_{xy}, M_z, $M_y^{'}$, $M_z^{'}$, M_{xyz}) (Because the value of the $M_x^{'}$ component is always close to 0, it is discarded.) that are not affected by the orientation changing of the smartphone and are highly distinguishable.

Magnetic Sequence Data Scaling. The magnetic signal is sampled sequentially while walking of the user, but different walking speeds produce different number of magnetic field data on the same experimental path. This feature may bring a lot of magnetic sequences with different lengths. However, all magnetic sequences fed into the positioning model are required to have the same length.

[1] See the help document of Android developer.http://developer.android.com/reference/packages.html..

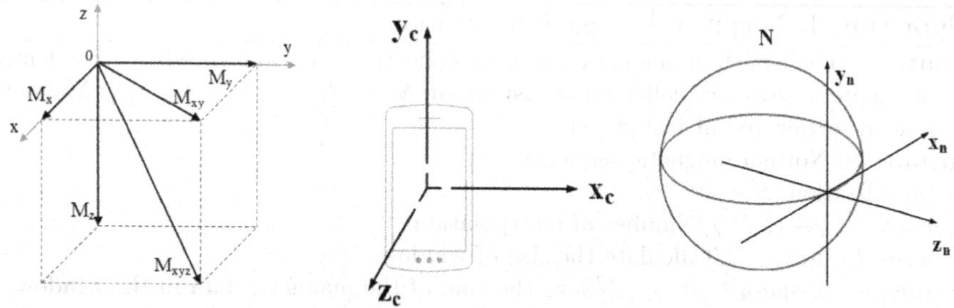

(a) Information of the magnetic field (b) Device coordinate system (c) World coordinate system

Fig. 4. Information of the magnetic field and the definition of coordinate systems

Therefore, user speed is one of the main issues when using a sequence fingerprint for indoor localization.

In this study, we use interpolation and deletion for getting the same number of magnetic field data at different walking speeds. For magnetic sequences obtained in fast walking, we propose an interpolation algorithm named Neighbor Average Interpolation, as shown in Algorithm 1. For magnetic sequences obtained in slow walking, we delete redundant data in the magnetic sequences at an equal interval d. This interval d is calculated by Eq. (2).

$$d = \frac{l_S}{l_S - l_N} \tag{2}$$

where l_S and l_N are the length of magnetic sequence collected at slow speed and normal speed, respectively.

Noise Reduction. Low-quality magnetometers built in smartphones produce a lot of noise. To reduce the noise, we use a smoothing filter (Median Average Filtering) with a particular window size, and the window size is 25 (Because the magnetic sequence between the two reference points includes 25 magnetic data.). Median average filtering is a combination of arithmetic average filtering and median filtering, whose principle is to remove a maximum value and a minimum value from the data in the window and then average the remaining data.

Data Normalization. In this paper, we adopt Min-Max normalization, which is formulated by Eq. (3), to convert the input data to the range of 0 to 1.

$$z_{nor} = \frac{z - z_{min}}{z_{max} - z_{min}} \tag{3}$$

where z_{nor}, z, z_{max} and z_{min} represent normalized data, data, maximum and minimum, respectively.

Algorithm 1. Neighbor Average Interpolation

Input: l_N: The length of magnetic sequence collected at normal speed; l_F: The length of magnetic sequence collected at fast speed; $S = \{S_1, \cdots, S_i, \cdots, S_{l_F}\}$: Magnetic sequence collected at fast speed.
Output: $\hat{S}$: Normal magnetic sequence.

Initialization: $\hat{S} \leftarrow \emptyset$
$num \leftarrow l_N - l_F$ //Number of interpolation.
$w_s \leftarrow l_F / num$ //Calculate the size of window.
$sum_L \leftarrow 0$, $sum_R \leftarrow 0$ //Save the sum of the magnetic data in the window.
for $i = 1$ to num **do**
  //Calculate the sum of the magnetic data in the left window.
  $sum_L \leftarrow \sum_{j=(i-1)*w_s+1}^{i*w_s} S_j$
  //Add the magnetic data in the left window to $\hat{S}$.
  $Add\ \{S_{(i-1)*w_s+1}, \dots, S_{i*w_s}\}$ to $\hat{S}$
  $size \leftarrow w_s$
  //Judge whether the right window exceeds the range of the sequence.
  if $(i+1) * w_s \geq l_F$ **then**
    //Calculate the sum of the magnetic data in the right window.
    $sum_R \leftarrow \sum_{k=i*w_s+1}^{l_F} S_k$
    $size \leftarrow l_F - (i * w_s)$
  else
    $sum_R \leftarrow \sum_{k=i*w_s+1}^{(i+1)*w_s} S_k$
  end if
  $avg \leftarrow (sum_L + sum_R)/(w_s + size)$
  $Add\ avg$ to $\hat{S}$ //Take the average as a new interpolation and add it to $\hat{S}$.
end for
return $\hat{S}$

3.3 Magnetic Sequence Fingerprints Creation

The essential quality of a fingerprint is its distinguishability or uniqueness. Increasing the length of the magnetic sequence to represent a location will increase the uniqueness, but it will boost the localization time which affects the real-time performance of the positioning system. On the other hand, smaller length reduces the time but affects the fingerprint uniqueness and ultimately degrades the localization accuracy. Hence, a proper fingerprint size is very critical. We analyzed the magnetic sequences with different lengths to ensure that MSFs can reach sufficiently distinguishable and found that MSFs are distinguishable with the length of 75 (Each MSF includes 3 reference points, and each reference point contains 25 magnetic data.).

Figure 5 shows that comparison of MSFs with different lengths of 25, 50, 75, and 100 in four adjacent positions (P_1, P_2, P_3, P_4). It can be observed that MSFs with the length of 25 are almost similar and make the distinction very difficult. It is clear that the MSFs of P_1 and P_2 become dissimilar, but the MSFs of P_3 and P_4 look almost identical when the length is 50. They become distinguishable with the length of 75. Therefore, we use MSFs with the length of 75 to train the CNN-LSTM model used in our study.

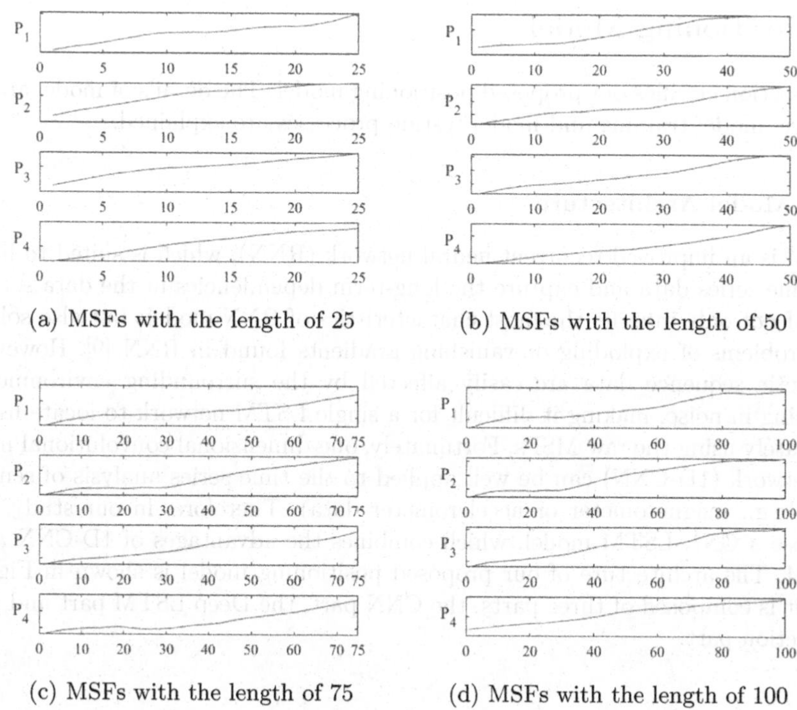

(a) MSFs with the length of 25 (b) MSFs with the length of 50

(c) MSFs with the length of 75 (d) MSFs with the length of 100

Fig. 5. Comparison of MSFs with different lengths in four adjacent positions

In this paper, we design a sliding window to cut the magnetic sequences in the same length for every position label. Figure 6 shows an example of how the MSFs creation process. In detail, the magnetic data elements value of the reference points in each sliding window are concatenated sequentially to construct a magnetic sequence, which is labelled by the location coordinates of the endpoint (see the Example in the Fig. 6). Moreover, the size of the sliding window is 75, and the sliding step is 25.

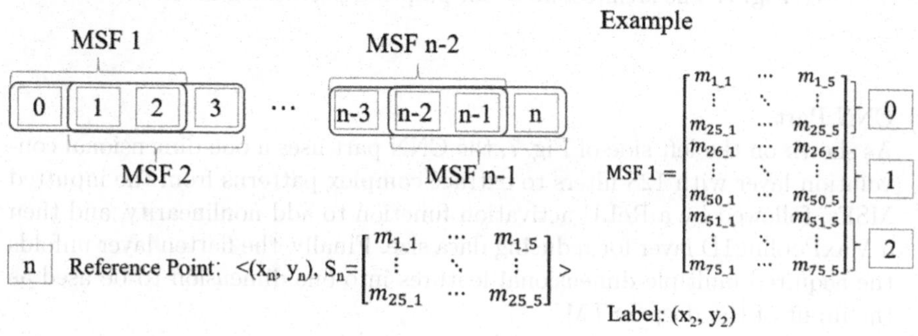

Fig. 6. The magnetic sequence fingerprints creation process

4 Positioning Model

This section clarifies our proposed positioning model. The details of model architecture, model training and model testing processes are explained.

4.1 Model Architecture

LSTM is an improved recurrent neural network (RNN), which is suited to handle time-series data and capture the long-term dependencies in the data series. LSTM not only inherits the most characteristics of RNN models but also solves the problems of exploding or vanishing gradients found in RNN [9]. However, magnetic sequence data are easily affected by the surrounding environment, resulting in noise, making it difficult for a single LSTM network to locate users accurately using the raw MSFs. Fortunately, one-dimensional convolutional neural network (1D-CNN) can be well applied to the time series analysis of sensor data (e.g., magnetometer or accelerometer data). Therefore, in our study, we propose a CNN-LSTM model, which combines the advantages of 1D-CNN and LSTM. The architecture of our proposed positioning model is shown in Fig. 7. And it is composed of three parts, the CNN part, the Deep LSTM part and the Prediction part.

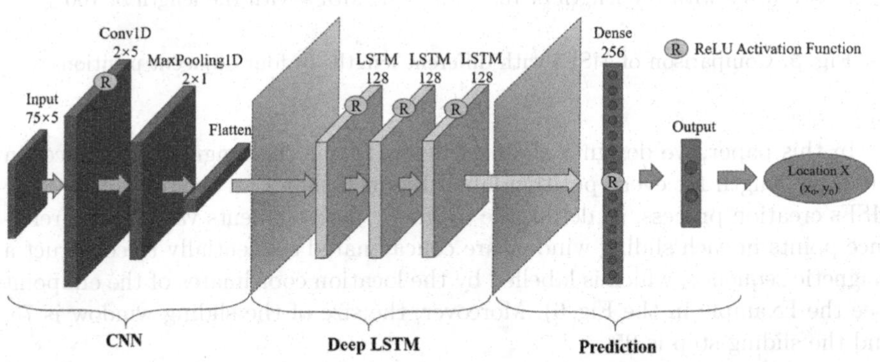

Fig. 7. The architecture of our proposed positioning model

1) CNN Part
As shown on the left side of Fig. 7, the CNN part uses a one-dimensional convolution layer with 128 filters to extract complex patterns from the inputted MSFs, followed by a ReLU activation function to add nonlinearity, and then a MaxPooling1D layer for reducing data size. Finally, the flatten layer unfolds the acquired multiple dimensional features into one dimension to be used as the input of the deep LSTM.

The theory of the CNN part is formulated by Eq. (4) and the process is shown in Fig. 8.

$$\mathbf{z}_t = Maxpooling1D(ReLU(\mathbf{W}\mathbf{s}_t + \mathbf{b})) \tag{4}$$

where $\mathbf{s}_t$ and $\mathbf{z}_t$ are the input and output of the CNN part, respectively, $\mathbf{W}$ and $\mathbf{b}$ are the weight and bias vectors of the convolution layer, ReLU(•) is the rectified linear unit, which is the activation function defined as

$$ReLU(x) = \begin{cases} x & x > 0 \\ 0 & x \leq 0 \end{cases} \tag{5}$$

The ReLU(•) function has the advantages of sparse representation, high gradient propagation efficiency and low computational complexity. And Maxpooling1D is the pooling operation, which leverages a max filter to filter the sub-regions of the initial feature map and takes the most obvious feature of that region, creating a new output feature map.

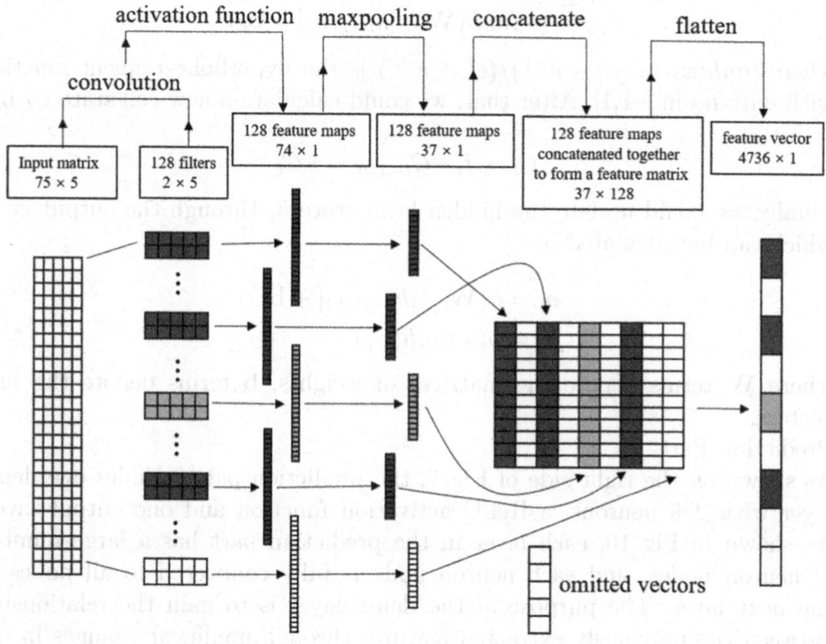

Fig. 8. The process of the CNN part

2) Deep LSTM Part
As shown in the middle of Fig. 7, the deep LSTM part including three LSTM layers and each LSTM layer is followed by a ReLU activation function to add nonlinearity. And what needs to be explained is that each LSTM layer has 128 neurons.

The key idea of LSTM is to use the forget gate to decide what information to be removed and use the input gate to determine what information to be added to the cell state. As shown in Fig. 9, the LSTM model comprises the time-series input x_t, cell state C_t, temporary cell state $\widetilde{C}_t$, hidden layer state h_t, forget gate f_t, input gate i_t, and output gate o_t. The first step in LSTM is to decide what information to be thrown away from the previous cell state C_{t-1}. This action is achieved by the forget gate which is formulated by

$$f_t = \sigma(\mathbf{W}_f \cdot [h_{t-1}, x_t] + \mathbf{b}_f) \tag{6}$$

where $\sigma(x) = 1/(1 + e^{-x})$ is the sigmoid function with outputs in [0,1]. Therefore, the value of f_t close to 0 means to throw away all information, while close to 1 means to retain all information. Then, we want to determine what information we would store in the cell state C_t. This procedure can be done by the input gate, which is defined as

$$\begin{aligned} i_t &= \sigma(\mathbf{W}_i \cdot [h_{t-1}, x_t] + \mathbf{b}_i) \\ \widetilde{C}_t &= tanh(\mathbf{W}_c \cdot [h_{t-1}, x_t] + \mathbf{b}_c) \end{aligned} \tag{7}$$

where $tanh(x) = (e^x - e^{-x})/(e^x + e^{-x})$ is the hyperbolic tangent function with outputs in [–1,1]. After that, we could calculate a new cell state C_t by

$$C_t = f_t * C_{t-1} + i_t * \widetilde{C}_t \tag{8}$$

Finally, we could update the hidden layer state h_t through the output gate, which can be calculated by

$$\begin{aligned} o_t &= \sigma(\mathbf{W}_o \cdot [h_{t-1}, x_t] + \mathbf{b}_o) \\ h_t &= o_t * tanh(C_t) \end{aligned} \tag{9}$$

where $\mathbf{W}$ terms denote the matrices of weights, $\mathbf{b}$ terms denote the bias vectors.

3) Prediction Part
As shown on the right side of Fig. 7, the prediction part includes one dense layer with 256 neurons, a ReLU activation function and one output layer. As shown in Fig. 10, each layer in the prediction part has a large number of neuron nodes, and each neuron node is fully connected to all nodes in the next layer. The purpose of the dense layer is to gain the relationship between the previously extracted features through nonlinear changes in the dense layer and finally map them to the output space. Therefore, the function of the prediction part is to convert the output of deep LSTM into the location coordinates of the user.

4.2 Model Training

In this paper, the training on the proposed positioning model is carried out using the MSFs. In order to improve the generalization ability of the model,

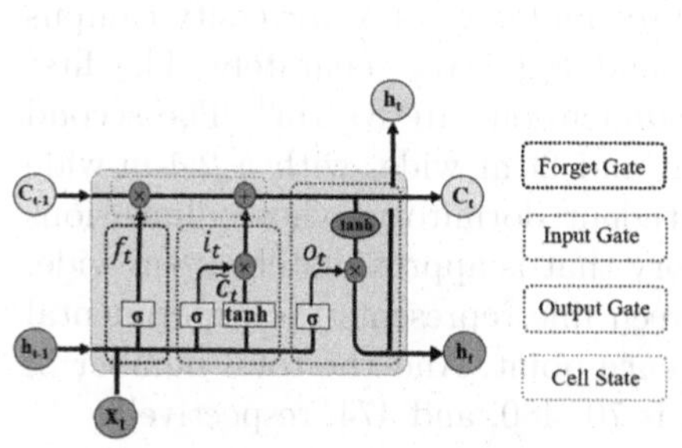

Fig. 9. The structure of LSTM

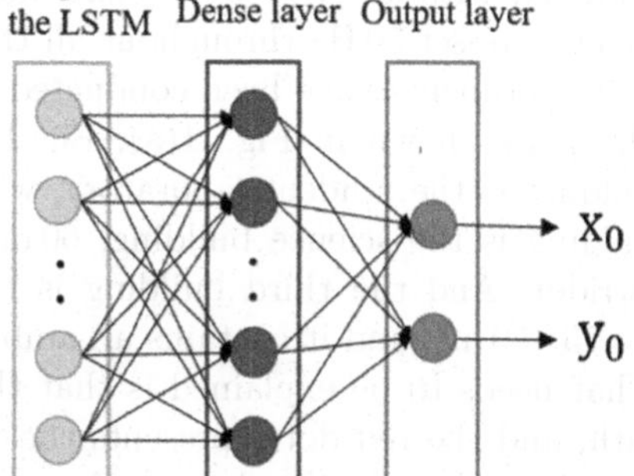

Fig. 10. The structure of the prediction part

we divide the collected data (We collect 230 samples (Each sample includes 25 magnetic data.) for every reference point on each experimental path.) into a training dataset and a validation dataset, train the model on the training dataset and validate the model on the validation dataset, where the training dataset accounts for 90% and the validation dataset accounts for 10%. Furthermore, the model training process is divided into two steps to find the best hyperparameters of the model. First of all, the Adam optimizer is used to train the model to reduce the training loss fast so that the model converges quickly. Next, the fine-tuned SGD optimizer is used to train the model for making the model better.

4.3 Model Testing

In the online testing phase, newly collected magnetic field data (We collect 26 samples (Each sample includes 25 magnetic data.) for every reference point on each experimental path.) from a smartphone are first processed in the data preprocessing module. Then the newly MSFs are constructed and fed into the trained positioning model for testing the performance.

5 Experiments and Results

In this section, the evaluation of the proposed system is demonstrated. The experiment setup is first described. Then, we explore and evaluate the performance of the proposed indoor localization technique. Finally, we compare our presented solution with other positioning methods.

5.1 Experiment Setup

We developed an android application with Android Studio 4.0.1 on the smartphone (Mi 10 Youth, Beijing, China) for data collection. The Mi 10 Youth is equipped with an AK0991X 3-axis magnetic field sensor, and its sampling frequency are set 50 Hz throughout all the experiments.

Experiments have been conducted in three buildings of a university campus which are shown in Fig. 11(a), Fig. 11(b) and Fig. 11(c) separately. The first building is the student laboratory, whose dimensions are $8 \times 5\,\text{m}^2$. The second building is the science building, 60 m long and 20 m wide, with a 2.4-m-wide corridor. And the third building is the student dormitory, whose dimensions are $75 \times 50\,\text{m}^2$, and it contains an ambulatory that is approximately 1.8 m wide. What needs to be explained is that the green line represents the experimental path, and the red dot represents the reference point. And the total number of reference points in the three environments is 70, 480, and 474, respectively.

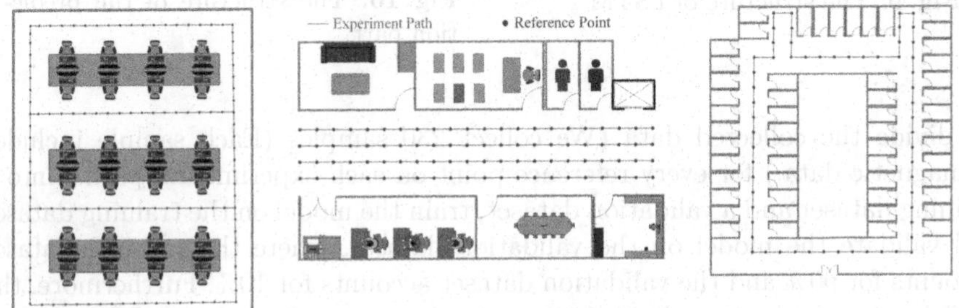

(a) The layout of the student laboratory (b) The layout of the science building (c) The layout of the student dormitory

Fig. 11. The layout of three environments

5.2 Localization Performance Exploration

The key to fingerprint-based indoor location is the identifiability of fingerprints. In this paper, we explore the influence of magnetic data elements and sequence length on the identifiability of MSFs.

Optimal Component of Magnetic Sequence Fingerprint. In this paper, we divided the magnetic data elements into three types of combinations and carried out experiments.

- Class1: (M_x, M_y, M_z)
- Class2: $(M_y^{'}, M_z^{'}, M_{xyz})$
- Class3: $(M_{xy}, M_z, M_y^{'}, M_z^{'}, M_{xyz})$

Figure 12 shows the cumulative distribution function (CDF) and mean distance error of localization errors under different types of MSF in the three experimental environments. Observing the results, we can find that the positioning performance can reach the best when the MSF is composed of five-dimensional magnetic data elements $(M_{xy}, M_z, M_y^{'}, M_z^{'}, M_{xyz})$.

There is no surprising that the MSFs composed of original three-dimensional magnetic data elements (M_x, M_y, M_z) have poor positioning accuracy, because raw magnetic readings are highly orientation-dependent. Although, the MSFs composed of the transformed magnetic data elements $(M_y^{'}, M_z^{'}, M_{xyz})$ have a good feature is that independent on the orientation of the mobile phone. Unfortunately, this kind of MSF contains few features leading to low distinguishability, reducing positioning accuracy. To sum up, the MSFs composed of five-dimensional magnetic data elements $(M_{xy}, M_z, M_y^{'}, M_z^{'}, M_{xyz})$ are optimal, which are orientation-independent of the mobile phone and have high distinguishability.

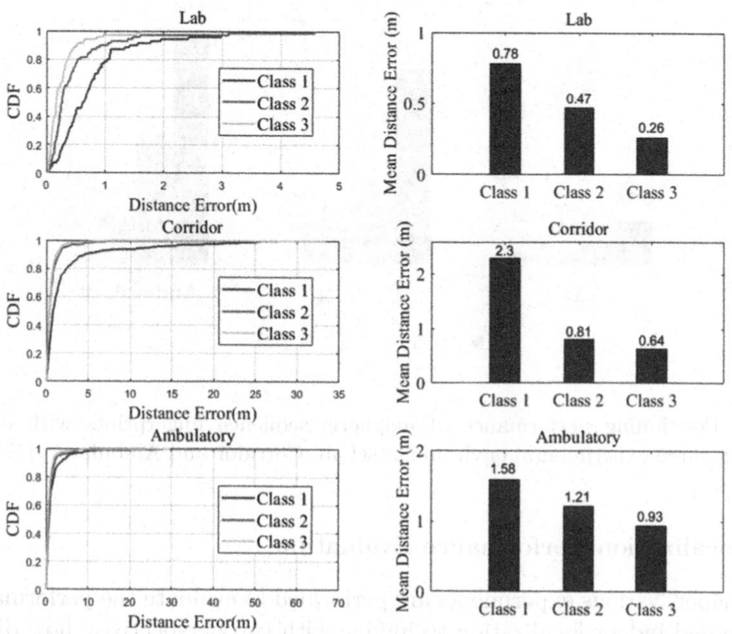

Fig. 12. Positioning performance of different types of magnetic sequence fingerprints in three experimental environments(Lab, Corridor and Ambulatory)

Optimal Length of Magnetic Sequence Fingerprint. In this section, we evaluated the influence of MSF with different lengths on positioning accuracies. We carried out experiments on MSFs with lengths of 25, 50, 75, and 100. And the mean distance errors in the three experimental environments are shown in Fig. 13. From this figure, we observe that the positioning performance increases gradually when the MSF length increases from 25 to 75 and then drops when the MSF length is greater than 75. Therefore, the best results are 0.26 m for lab,

0.64 m for corridor and 0.94 m for ambulatory when the MSF length equals to 75. Another observation is the mean positioning error degrades significantly when the MSF length is shorter. This is no surprising that since there are few magnetic field data about each MSF, leading the advantage of sequence fingerprint-based indoor positioning can be ignored. However, the mean positioning error also degrades significantly when the MSF length is relatively large. The reason may be that too much magnetic field data in each MSF, which leads to overfitting.

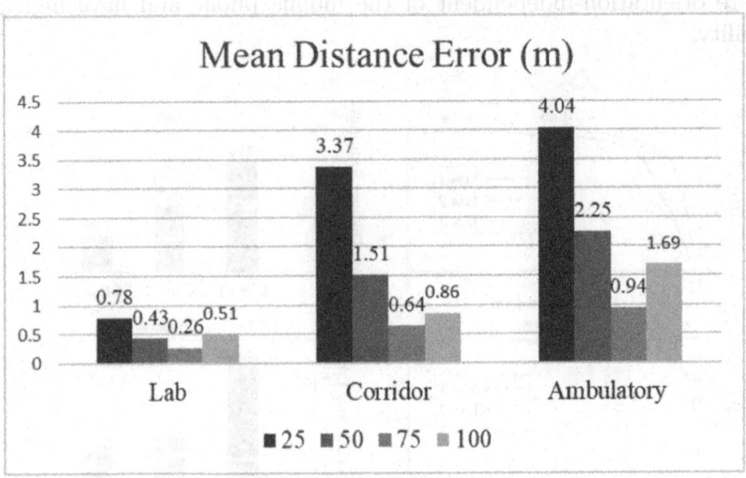

Fig. 13. Positioning performance of magnetic sequence fingerprints with different lengths in three experimental environments(Lab, Corridor and Ambulatory)

5.3 Localization Performance Evaluation

In this paper, various experiments are performed to evaluate the performance of the proposed indoor localization technique with two perspectives: how different walking speeds of the user can influence the positioning performance; and the effect of device diversity on positioning performance.

Impact of User Speed. In this section, we conduct experiments on the magnetic sequences obtained at three different speeds (Normal, Slow, Fast). Figure 14 shows the results obtained by the positioning model under different walking speeds of the user in the corridor. In general, the proposed method has obtained good results in terms of the mean localization error, i.e., 0.68 m for normal speed, 0.77 m for slow speed, and 0.84 m for fast speed. And the results also reveal that the positioning performance under slow speed is slightly better than that under fast speed. The reason may be that more original magnetic field information can be read at a slow speed.

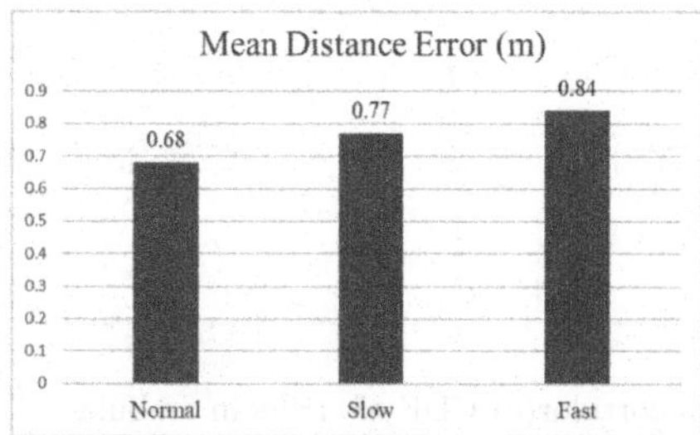

Fig. 14. Positioning performance under different walking speeds

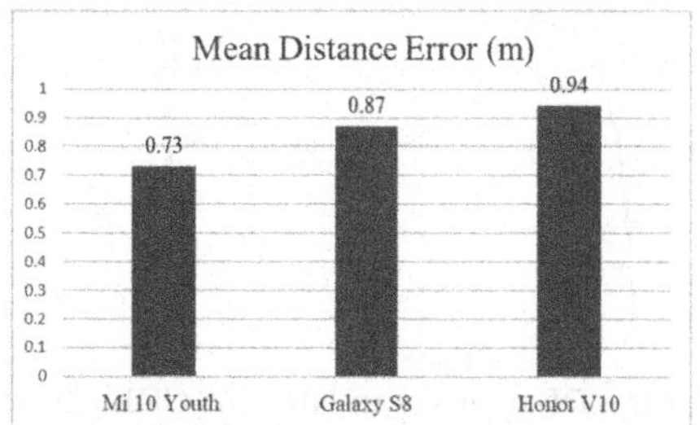

Fig. 15. Positioning performance of using different smartphones

Impact of Device Diversity. In order to verify the influence of device diversity on the proposed positioning model, we exerted three different smartphones (Mi 10 Youth, Galaxy S8 and Honor V10) to obtain magnetic information on the same path in the corridor. The built-in magnetic sensors of these smartphones are shown in Table 2. And the result of Fig. 15 reveals that the effect of device diversity on positioning performance is acceptable.

Table 2. Built-in magnetometer of different smartphones

Smartphones	Mi 10 Youth	Galaxy S8	Honor V10
Magnetometer	AK0991X	AK09916C	AK09918

5.4 Localization Performance Comparison

Based on the prediction accuracy results, we choose the best prediction model for each test environment. To obtain the positioning error, we calculate the Euclidean distance between reference points and predicted points. In order to demonstrate the utility of our proposed model, we conduct the performance comparison between our method and the existed works, such as RNN-LSTM [4] and DeepML [19].

(1) The cumulative distribution function (CDF) of localization error
Fig. 16 plots the CDFs of the localization errors in the lab, corridor and ambulatory, respectively. In the lab, as shown in Fig. 16(a), in 90% case, the positioning errors of CNN-LSTM (our method) are less than 0.53 m, while RNN-LSTM and DeepML are less than 1.13 m and 2.01 m, respectively. In the corridor, as shown in Fig. 16(b), for over 80% of the test spots, the errors of our method are less than 0.8 m. However, RNN-LSTM and DeepML have errors of 1.77 m and 1.14 m under the same experimental conditions. In the ambulatory, as shown in Fig. 16(c), our scheme has a distance error of 0.88 m for 70% of the test spots in this more complex indoor environment. Meanwhile, the distance error of RNN-LSTM and DeepML is 1.02 m and 1.2 m, respectively.

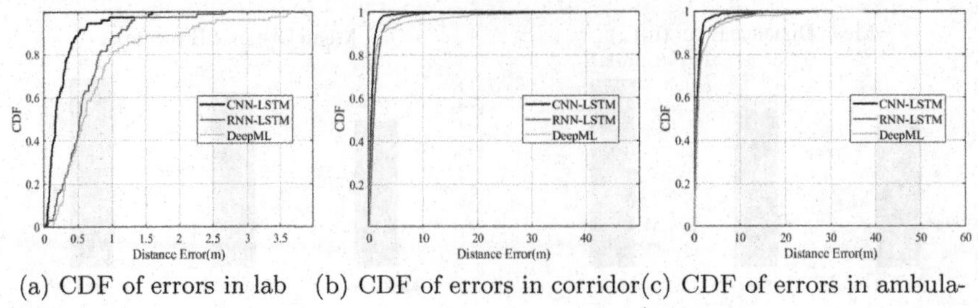

(a) CDF of errors in lab (b) CDF of errors in corridor (c) CDF of errors in ambulatory

Fig. 16. The CDFs in the different indoor environments

(2) Mean distance error
Fig. 17 gives the mean distance error obtained by the proposed algorithm CNN-LSTM (our method), RNN-LSTM [4] and DeepML [19]. As shown in the figure, in the lab, CNN-LSTM achieves the mean error of 0.26 m, which outperforms RNN-LSTM and DeepML by more than 0.35 m and 0.57 m, and the gain is about 57%, 68%, respectively. Moreover, in the corridor, the mean error of our approach is 0.64 m, which is about 47% and 59% gain than RNN-LSTM and DeepML, respectively. What's more, in the ambulatory scenario, which is a large-scale indoor environment with a similar structure, the mean accuracy of our approach is 0.93 m, which is about 34% and 50% gain than RNN-LSTM and DeepML, respectively.

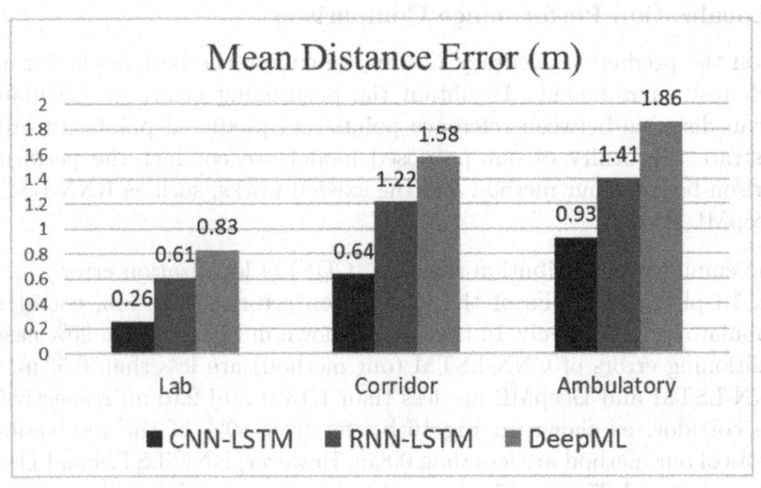

Fig. 17. The comparison of mean distance error

6 Conclusion

This paper proposes an accurate indoor positioning system utilizing deep learning. It introduces a novel fingerprint representation scheme that firstly converts raw three-dimensional magnetic field signals into five-dimensional and designs a sliding window mechanism to construct MSFs. Compared with the existing positioning methods, our system designs a CNN-LSTM model, which combines the advantages of 1D-CNN and LSTM. It can also automatically learn the mapping between ground-truth positions and MSFs, which assures the achieving of high positioning accuracies. The objective of the proposed method is to improve the performance and robustness of magnetic field-based localization system. Extensive experiments are performed to evaluate the performance of the proposed approach. Results demonstrate that CNN-LSTM can localize a user within 0.9 m with an accuracy of 70% under three experimental environments, different smartphones and different user speeds. For future work, we plan to further enhance indoor positioning performance by the proposed magnetic sequence fingerprints, and enable real-time indoor magnetic-based positioning by implementing our method using cloud-edge-end cooperation architecture.

Acknowledgement. This work is supported by the National Key Research and Development Program of China (2020YFB2104202).

References

1. Abbas, M., Elhamshary, M., Rizk, H., Torki, M., Youssef, M.: Wideep: Wifi-based accurate and robust indoor localization system using deep learning. In: 2019 IEEE International Conference on Pervasive Computing and Communications, PerCom, Kyoto, Japan, 11–15 March 2019, pp. 1–10. IEEE (2019). https://doi.org/10.1109/PERCOM.2019.8767421
2. Ashraf, I., Kang, M., Hur, S., Park, Y.: MINLOC: magnetic field patterns-based indoor localization using convolutional neural networks. IEEE Access **8**, 66213–66227 (2020). https://doi.org/10.1109/ACCESS.2020.2985384
3. Chen, Y., Zhou, M., Zheng, Z.: Learning sequence-based fingerprint for magnetic indoor positioning system. IEEE Access **7**, 163231–163244 (2019). https://doi.org/10.1109/ACCESS.2019.2952564
4. Chiang, T.H., Sun, Z.H., Shiu, H.R., Lin, K.C.J., Tseng, Y.C.: Magnetic field-based localization in factories using neural network with robotic sampling. IEEE Sensors J. **20**(21), 13110–13118 (2020). https://doi.org/10.1109/JSEN.2020.3003404
5. Chung, J., Donahoe, M., Schmandt, C., Kim, I., Razavai, P., Wiseman, M.: Indoor location sensing using geo-magnetism. In: Agrawala, A.K., Corner, M.D., Wetherall, D. (eds.) Proceedings of the 9th International Conference on Mobile Systems, Applications, and Services (MobiSys 2011), Bethesda, MD, USA, 28 June–01 July 2011, pp. 141–154. ACM (2011). https://doi.org/10.1145/1999995.2000010
6. Frassl, M., Angermann, M., Lichtenstern, M., Robertson, P., Julian, B.J., Doniec, M.: Magnetic maps of indoor environments for precise localization of legged and non-legged locomotion. In: 2013 IEEE/RSJ International Conference on Intelligent Robots and Systems, Tokyo, Japan, 3–7 November 2013, pp. 913–920. IEEE (2013). https://doi.org/10.1109/IROS.2013.6696459

7. Galván-Tejada, C.E., García-Vázquez, J., Brena, R.F.: Magnetic field feature extraction and selection for indoor location estimation. Sensors **14**(6), 11001–11015 (2014). https://doi.org/10.3390/s140611001
8. Grand, E.L., Thrun, S.: 3-axis magnetic field mapping and fusion for indoor localization. In: IEEE International Conference on Multisensor Fusion and Integration for Intelligent Systems, MFI 2012, Hamburg, Germany, 13–15 September 2012, pp. 358–364. IEEE (2012). https://doi.org/10.1109/MFI.2012.6343024
9. Hochreiter, S., Schmidhuber, J.: Long short-term memory. Neural Comput. **9**(8), 1735–1780 (1997). https://doi.org/10.1162/neco.1997.9.8.1735
10. Lee, N., Ahn, S., Han, D.: AMID: accurate magnetic indoor localization using deep learning. Sensors **18**(5), 1598 (2018). https://doi.org/10.3390/s18051598
11. Li, T., Wang, H., Shao, Y., Niu, Q.: Channel state information-based multi-level fingerprinting for indoor localization with deep learning. Int. J. Distrib. Sens. Netw. **14**(10) (2018). https://doi.org/10.1177/1550147718806719
12. Pusnik, M., Galun, M., Sumak, B.: Improved bluetooth low energy sensor detection for indoor localization services. Sensors **20**(8), 2336 (2020). https://doi.org/10.3390/s20082336
13. Shao, W., et al.: Location fingerprint extraction for magnetic field magnitude based indoor positioning. J. Sensors **2016**, 1945695:1–1945695:16 (2016). https://doi.org/10.1155/2016/1945695
14. Suksakulchai, S., Thongchai, S., Wilkes, D.M., Kawamura, K.: Mobile robot localization using an electronic compass for corridor environment. In: Proceedings of the IEEE International Conference on Systems, Man & Cybernetics: "Cybernetics Evolving to Systems, Humans, Organizations, and their Complex Interactions", Sheraton Music City Hotel, Nashville, Tennessee, USA, 8–11 October 2000, pp. 3354–3359. IEEE (2000). https://doi.org/10.1109/ICSMC.2000.886523
15. Wang, R., Li, Z., Luo, H., Zhao, F., Shao, W., Wang, Q.: A robust wi-fi fingerprint positioning algorithm using stacked denoising autoencoder and multi-layer perceptron. Remote Sens. **11**(11), 1293 (2019). https://doi.org/10.3390/rs11111293
16. Wang, X., Gao, L., Mao, S., Pandey, S.: Deepfi: deep learning for indoor fingerprinting using channel state information. In: 2015 IEEE Wireless Communications and Networking Conference, WCNC 2015, New Orleans, LA, USA, 9–12 March 2015, pp. 1666–1671. IEEE (2015). https://doi.org/10.1109/WCNC.2015.7127718
17. Wang, X., Wang, X., Mao, S.: Cifi: deep convolutional neural networks for indoor localization with 5 ghz wi-fi. In: 2017 IEEE International Conference on Communications (ICC), pp. 1–6 (2017). https://doi.org/10.1109/ICC.2017.7997235
18. Wang, X., Wang, X., Mao, S.: RF sensing in the internet of things: a general deep learning framework. IEEE Commun. Mag. **56**(9), 62–67 (2018). https://doi.org/10.1109/MCOM.2018.1701277
19. Wang, X., Yu, Z., Mao, S.: Indoor localization using smartphone magnetic and light sensors: a deep LSTM approach. Mob. Netw. Appl. **25**(2), 819–832 (2020). https://doi.org/10.1007/s11036-019-01302-x
20. Xiao, C., Yang, D., Chen, Z., Tan, G.: 3-D BLE indoor localization based on denoising autoencoder. IEEE Access **5**, 12751–12760 (2017). https://doi.org/10.1109/ACCESS.2017.2720164

Wasserstein Graph Auto-Encoder

Yan Chu, Haozhuang Li, Hui Ning(✉), and Qingchao Zhao(✉)

Harbin Engineering University, 150001 Harbin, China
{ninghui,zhaoqc418}@hrbeu.edu.cn

Abstract. Conventional Graph Auto-Encoder-based(GAE) minimizes the variational lower bound by Kullback-Leibler divergence in link prediction task and forces a variational posterior distribution for each node to match the prior Gaussian distribution. During the matching process, different variational posterior distributions may intersect. The nodes located at the intersection are subjected to multiple variations which is not suitable for graph reconstruction. We propose Wasserstein Graph Auto-Encoder (WGAE) to minimize the penalty in the Wasserstein distance between the model distribution and the target distribution, which encourages continuous mixed distributions to match the prior distribution. The latent codes of different nodes are far away from each other in the mixed distribution, which is conducive to reconstruction. In addition, gradients are employed to update the mean and variance of the Gaussian distribution in Variational Graph Auto-Encoder(VGAE), which may fall into a kind of local optimization. For each node, the variational posterior distribution is always equal to the prior distribution, and the latent vector is treated as a kind of noise and be ignored by the model. WGAE uses a d-1 dimensional manifold distribution on a d-dimensional hypersphere von Mises-Fisher distribution, controlled by mean direction and concentration parameter. For the prior distribution and posterior distribution, different concentration parameters can be selected and set as hyper-parameters to avoid the local optimum. The von Mises-Fisher distribution also learns a richer topology in non-linear dimensionality reduction. Extensive experiments show that the accuracy of our proposed WGAE in the Cora data set is 2% higher than those of the GAE and VGAE, and 3% higher in the Citeseer data set.

Keywords: Link prediction · Graph auto-encoder · Wasserstein distance · von Mises-Fisher distribution

1 Introduction

A graph or network is the data structure that appears in various real-world applications, such as the social network and co-authorship network, etc. These networks contain not only complex structural information between nodes but

Supported by National Natural Science Foundation of China under Grant No. 61771155. Haozhuang Li and Qingchao Zhao are the co-first authors.

Y. Lai et al. (Eds.): ICA3PP 2021, LNCS 13155, pp. 85–99, 2022.
https://doi.org/10.1007/978-3-030-95384-3_6

also rich content information for each node. Link prediction is an important topic in the field of graph research. In real life, the link prediction task has a wealth of applications, such as interactions between proteins [1], movie recommendations [2], citations of academic papers [3], recommendation of web hyperlinks [4], etc.

In recent years, due to the rapid development of Graph Convolutional Networks (GCN) [5,6] framework has been widely used in link prediction [7,8]. However, in some methods based on GCN, the task in link prediction cannot achieve high accuracy because of following shortcomings. First, most of these models are improved based on the VGAE. They use the same regularization–KL divergence for each node to generate a variational posterior distribution to force the matching of the prior distribution. Different variational posterior distributions may intersect, and latent codes of different nodes located at the intersection are affected by multiple generation distributions. This is not conducive to reconstruction. In addition, the prior distribution and variational posterior distribution in VGAE follow the Gaussian distribution. When the model is optimized, it may fall into the local optimum. The variational posterior is always equal to the prior for each node. The latent vector is treated as a kind of noise and will be ignored by the model.

In this paper, we propose a novel graph auto-encoder model Wasserstein Graph Auto-Encoder for link prediction. It is a framework for unsupervised learning of graph structure data, which is composed of an encoder and a decoder. This model will encourage continuous mixed distributions (the marginal distribution of conditional distributions) to match the prior distributions, latent codes of different nodes are far away from each other in the mixed distribution, which is conducive to graph reconstruction and can improve the accuracy of link prediction. We use a d-1 dimensional manifold distribution on a d-dimensional hypersphere, vMF distribution. It is controlled by mean direction and concentration parameter. We can regard the concentration parameter as a hyper-parameter and set different values for the prior distribution and the conditional distribution (the marginal distribution is required to be equal to the prior distribution) to avoid falling into the local optimum structurally. Compared with Gaussian distribution, vMF can make the model learn a richer topology in non-linear dimensionality reduction. Our contributions are summarized as follows:

1. We effectively applied the Wasserstein distance to the GAE model and constructed the WGAE model. The WGAE model uses a continuous mixed distribution to match the prior distribution. The latent codes of different nodes are far away from each other in the mixed distribution, which is conducive to reconstruction. For any cost function c, our model minimizes the optimal transmission $W_c(P_G, P_A)$ between the generated distribution P_G and the original data distribution P_A.
2. We use a new potential distribution in the model–vMF distribution. Different concentration parameter values κ for the prior distribution and the conditional distribution are set to prevent the local optimum in the structure. Compared with the Gaussian distribution, the model of the vMF distribution can also learn a richer topology in the process of non-linear dimensionality reduction.

3. Experiments have shown that the proposed WGAE model is superior to the current GAE, VGAE and their variants on the benchmark data sets of link prediction.

The rest of the article is organized as follows. Firstly, we discuss the related work of link prediction, and then we give the notation and model framework. In Sect. 4, we describe the encoder and decoder and optimization of the model. Then the experimental results are given in Sect. 5. Finally, we come to the conclusions.

2 Related Work

As the classic topic in data mining, the link prediction methods are usually divided into three categories: classic link prediction methods, graph embedding link prediction methods and graph network link prediction methods.

2.1 Classical Link Prediction Methods

Classical link prediction algorithms are mainly based on structural similarity methods. This kind of method is relatively simple and has been studied in depth. The idea is that the more similar the structure of the two nodes, the greater the possibility of connecting edges. Methods based on structural similarity are usually divided into two categories: local similarity indicators and global similarity indicators. The local similarity index only considers some of the most basic structural information between nodes. Such methods usually have relatively low complexity and low prediction efficiency. Such methods include Common Neighbour (CN) [9], Admic-Adar (AA) [10], Resource Allocation (RN) [11], etc. The global similarity index usually takes all the paths between these two nodes as structural information. The accuracy of this method is improved compared with that of the local part, but it also increases the computational complexity. Such approaches include Katz [12] and so on.

2.2 Graph Embedding Link Prediction Methods

In recent years, with the development of deep learning, many learning methods based on network representation have emerged. These methods are also called graph embedding methods. The basic idea of this type of method is to learn a mapping relationship and then map the graph data to a vector in a low-dimensional space. Then the mapping relationship is continuously optimized so that the low-dimensional vector representation can reflect the topological structure, node content and other edge information in the graph. We employ these low-dimensional vectors to complete some downstream tasks. Such algorithms include DeepWalk [13], node2vec [14], LINE [15], etc. Compared with classical link prediction algorithms, this type of algorithm has improved link prediction accuracy. However, it also has limitations: this type of method lacks supervision information in training, which makes the obtained nodes and tasks insufficiently related. Moreover, the global topological structure and random walk algorithm are used in the calculation, which significantly increases the complexity of the calculation.

2.3 Graph Convolutional Network Link Prediction Methods

GCN generates a node representation by aggregating its characteristics and the characteristics of neighboring nodes, which is used in node classification tasks. Afterward, Kif et al. used GCN as the encoder of the Variational Auto-encoder (VAE) and proposed GAE and VGAE [7]. VGAE learns the mean μ and variance σ represented by the low-dimensional vector of the node through GCN and then uses the decoding Reconstruction graph. The loss function of the GAE model measures the error between the generated graph and the original input graph. Since the error between the generated posterior distribution and the prior distribution is not considered, the accuracy of the link prediction task can not be high. VGAE uses KL to measure the error between distributions based on GAE, which slightly improves link prediction accuracy. Based on GAE and VGAE, Shirui Pan et al. proposed Adversarially Regularized Graph Auto-Encoder (ARGA) and Adversarially Regularized Variational Graph Auto-Encoder (ARVGA) models [16]. The model introduced an adversarial training module to distinguish potential codes from real prior distributions or the graph auto-encoder. These two models improve link prediction accuracy, and the model still uses KL divergence to minimize the lower bound of variation during optimization.

We propose the link prediction model WGAE, which combines the Wasserstein distance and graph neural network model to minimize the penalty in the form of Wasserstein distance between the model distribution and the target distribution. WGAE encourages the employment of continuous mixed distributions to match the prior distribution, which is conducive to reconstruction. In addition, vMF distribution can also learn a richer topology in non-linear dimensionality reduction. We verified the effectiveness of the model in citation networks.

3 Notations and Framework

3.1 Notations

Throughout the paper, the symbols we used are shown in Table 1.

3.2 Preliminaries

Wasserstein Distance. The Optimal Transportation (OT) [17,18] is often used to resolve differences in probability distributions. Suppose P_X and P_G are two probability distributions on R^n, the Kantorovich formula can be used to express the OT of P_X and P_G as:

$$W_c(P_X, P_G) = \inf_{\pi \in \Pi(M \sim P_X, N \sim P_G)} \mathbb{E}_{(M,N)\sim\pi}[c(M, N)] \tag{1}$$

Table 1. Commonly used symbols.

Notations	Descriptions
G	An undirected graph
V,v	V is a set of nodes in the graph, a node v ∈ V
E,e_{ij}	E is a set of edges in graph, an edge e_{ij} ∈ E
A	The adjacency matrix of graph
D	The degree matrix of **A**
n	The number of nodes, n=|V|
d	The dimension of a node feature vector
$\mathbf{X} \in R^{n \times d}$	The feature vector of a graph
$\mathbf{x_i} \in \mathbf{X}$	The feature vector of each node v
Z	The set of latent variables
$\mathbf{z_i}$	Low-dimensional vector obtained by node mapping

Here $c(X, Y)$ is any measurable cost function, the joint π of P_X and P_G is the probability distribution on $R^n \times R^n$, with marginal distributions P_X and P_G, $\Pi(P_X, P_G)$ means The set of all joint distributions $c(X, Y) = d^p(X, Y)(p \geq 1)$, the p-th root of W_c is called p-Wasserstein distance [19]. We use 1-Wasserstein distance, referred to as Wasserstein distance for convenience later.

Wasserstein distance can be interpreted as the minimum cost of transporting one probability distribution to another. Wasserstein also has some advantages that other divergence cannot achieve. Wasserstein distance can naturally measure the distance between discrete and continuous distributions. When Wasserstein distance measures the distance between distributions, it can also provide a solution for transforming one distribution into another.

Kantorovich-Rubinstein (KR) duality [20] provides a new method for evaluating the Wasserstein distance between distributions. The dual form is

$$W(P_X, P_G) = \sup_f \{\mathbb{E}_{x \sim P_X}[f(x)] \mathbb{E}_{x \sim P_G}[f(x)]\} \tag{2}$$

$$s.t. f(x_a) f(x_b) \leq \|x_a x_b\|, \forall x_a, \forall x_b \tag{3}$$

von Mises-Fisher Distribution. The von Mises-Fisher (vMF) distribution[21] is the maximum entropy distribution on the surface of the hyper-sphere, and its parameters are the mean direction μ and the isotropic concentration κ. The probability density function of the n-dimensional unit vector x is:

$$p(x \mid \mu, \kappa) = C_n(\kappa) \exp(\kappa \mu^T x) \cdot, C_n(\kappa) = \frac{\kappa^{n/21}}{(2\pi)^{n/2} I_{n/21}(\kappa)} \tag{4}$$

$x, \mu \in S^{n1}$. I_v represents the first kind of modified Bessel function of order v. The greater the concentration parameter κ, the more concentrated the probability distribution is around the direction μ, where $\kappa \geq 0$. x is unimodal for the distribution of $\kappa > 0$, and $\kappa = 0$ is uniformly distributed on the sphere.

vMF distribution is suitable for directional data modeling, because directional data can be objectively mapped to unit vectors. The probability density function of vMF is the part of the unit hyper-sphere centered on the origin of the probability density function of the multivariate normal distribution.

Prior distribution and posterior distribution use vMF distribution. The prior distribution is a uniform distribution on the unit hyper-sphere (κ=0). For the posterior distribution, it is a distribution where κ is a fixed hyper-parameter.

3.3 Framework

The WGAE model is composed of an encoder and a decoder, which is shown in Fig. 1.

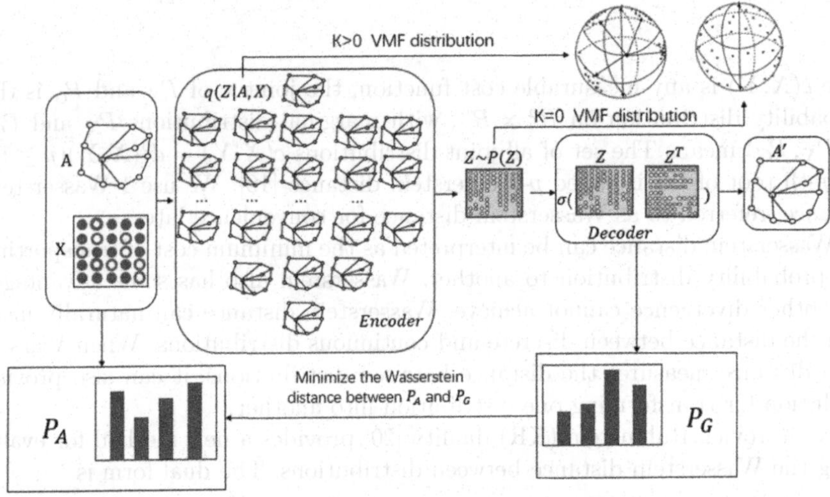

Fig. 1. The architecture of the Wasserstein Graph Auto-Encoder. This is a graph convolutional encoder that uses the graph structure **A** and the node content matrix **X** to generate an embedding **Z**, and then uses **Z** to reconstruct **A**. WGAE minimizes a penalized form of the Wasserstein distance between the generated distribution P_G and the original data distribution P_A. Prior distribution $P(\mathbf{Z})$ adopts vMF distribution with $\kappa = 0$, conditional distribution $Q(\mathbf{Z} \mid \mathbf{A}, \mathbf{X})$ adopts vMF distribution with $\kappa(\kappa > 0)$ as the hyper-parameter.

The model employs the topological structure of the adjacency matrix **A** and the node features of the feature matrix **X** as a combined input to obtain a latent representation **Z**, and then reconstructs an **A** from **Z** through the decoder. There may be noise and some missing information for the recovered **A**. Since the adjacency matrix of the graph represents the similarity or association between the node pairs, the decoder of link prediction usually uses the inner product of the learned latent representation, $\langle \mathbf{Z}, \mathbf{Z}^{\mathrm{T}} \rangle$. For the optimization of the model, we use

Wasserstein distance. Wasserstein distance is often used to measure the distance between two distributions. The regularization method requires the continuous mixing distribution (the edge distribution of the conditional distribution) to match the prior distribution, so that latent codes of different nodes are far away from each other in the mixing distribution, which is more conducive to reconstruction.

4 Methodology

The structure of the Wasserstein Graph Auto-Encoder is specified. Firstly, we give the encoder and decoder to show how our auto-encoder uses graph structure to reconstruct node features. Secondly, we describe the optimization and inference of the proposed model and give the optimization formula. Finally, the algorithm of the WGAE model is presented.

4.1 Wasserstein Graph Auto-Encoder WGAE

Encoder Part. Here we will define a inference model of WGAE formula as follows:

$$q(\mathbf{Z} \mid \mathbf{X}, \mathbf{A}) = \prod_{i=1}^{n} q\left(\mathbf{z}_i \mid \mathbf{X}, \mathbf{A}\right) \tag{5}$$

$$q\left(\mathbf{z}_i \mid \mathbf{X}, \mathbf{A}\right) = q\left(\mathbf{z}_i \mid \mu_i, \kappa\right) = C_n(\kappa) \exp\left(\kappa \mu_i^T \mathbf{z}_i\right), C_n(\kappa) = \frac{\kappa^{\frac{n}{2}1}}{(2\pi)^{\frac{n}{2}} J_{\frac{n}{2}1}(\kappa)} \tag{6}$$

In Eq. 5, $\mu = \mathrm{GCN}_\mu(\mathbf{X}, \mathbf{A})$ is the matrix of the average vector μ_i. Here κ is the concentration around μ.Two-level GCN is defined as $\mathrm{GCN}(\mathbf{X}, \mathbf{A}) = \widetilde{\mathbf{A}}\,\mathrm{ReLU}\left(\widetilde{\mathbf{A}}\mathbf{X}\mathbf{W}^{(0)}\right)\mathbf{W}^{(1)}$, Weight matrix $\mathbf{W}^{(i)}$. ReLU is the activation function, $\widetilde{\mathbf{A}} = \mathbf{D}^{\frac{1}{2}}\mathbf{A}\mathbf{D}^{\frac{1}{2}}$is a symmetric normalized adjacency matrix.

Decoder Part. Our decoder is used to reconstruct the graph structure $\widetilde{\boldsymbol{A}}$. This is more flexible than the reconstructed graph structure $\widetilde{\boldsymbol{A}}$ and the feature matrix $\mathbf{X}$ of the graph, because even if there is no available information, our model can still work. Our decoder is used to predict whether there is a link between two nodes. In our experiment, the generative model is given by graph embedding training:

$$p(\widetilde{\boldsymbol{A}} \mid \boldsymbol{Z}) = \prod_{i=1}^{n}\prod_{j=1}^{n} p\left(\widetilde{\boldsymbol{A}}_{ij} \mid \mathbf{z}_i, \mathbf{z}_j\right) \tag{7}$$

$$p\left(\widetilde{\boldsymbol{A}}_{ij} = 1 \mid \mathbf{z}_i, \mathbf{z}_j\right) = \mathrm{sigmoid}\left(\mathbf{z}_i^\top, \mathbf{z}_j\right) \tag{8}$$

Among them, $\widetilde{\boldsymbol{A}}_{ij}$ is an element of $\mathbf{A}$.

Graph Auto-Encoder Model. For graph embedding Z and reconstructed adjacency matrix $\widetilde{A}$ can be expressed as:

$$\widetilde{A} = \text{sigmoid}\left(ZZ^{\top}\right), Z = q(Z \mid \mathbf{X}, \mathbf{A}) \tag{9}$$

4.2 Model Optimization and Inference

For the Wasserstein graph auto-encoder, our error comes from the deviation between the real data distribution P_A and the latent variable distribution P_G obtained from the prior distribution $P(\mathbf{Z})$ and the generated distribution $P(\mathbf{A}|\mathbf{Z})$. Most models are difficult or even impossible to calculate when calculating the deviation. Especially when the real data distribution is unknown and P_G is parameterized by a deep neural network.

Wasserstein distance is a method often used to measure the difference between two probability distributions. We first give the latent variable distribution P_G, which is obtained in two steps. The first step is to sample from the prior distribution P($\mathbf{Z}$), which is fixed in the latent space, to obtain a $\mathbf{z}_i$ ($\mathbf{z}_i \in \mathbf{Z}$) and the second step is to map $\mathbf{z}_i$ to the graph Structure $\mathbf{A}_{ij}$ ($\mathbf{A}_{ij} \in \mathbf{A}$). The density distribution of P_G can be obtained:

$$p_G(\mathbf{A}_{ij}) = \int_Z p_G(\mathbf{A}_{ij} \mid \mathbf{z}_i)\, p_z(\mathbf{z}_i)\, dz, \forall \mathbf{A}_{ij} \in \mathbf{A} \tag{10}$$

The generation distribution $P(\mathbf{A} \mid \mathbf{Z})$ here can deterministically exist a mapping $\mathbf{A_{ij}} = G(z)$ under the condition of a given mapping G: $Z \rightarrow A$ so we can get a straightforward form by mapping G to find the Wasserstein distance. Originally, we needed to find two marginal distributions about π (Eq. 1). One distribution is based on the real data distribution P_A, and the other distribution is based on the latent variable distribution P_G. Here we only need to find a conditional distribution $Q(\mathbf{Z} \mid \mathbf{A}, \mathbf{X})$, make the marginal distribution of this conditional distribution $Q(\mathbf{Z}) = \mathbb{E}_{A \sim P_A}[Q(\mathbf{Z} \mid \mathbf{A}, \mathbf{X})]$ the same as our prior distribution $P(\mathbf{Z})$.

The Wasserstein distance between P_G and P_A above can be calculated with the following formula:

$$\inf_{\pi \in \Pi(M \sim P_A, N \sim P_G)} \mathbb{E}_{(M,N)\sim\pi}[c(M,N)] = \inf_{Q(\mathbf{Z})=P(\mathbf{Z})} \mathbb{E}_{P_A}\mathbb{E}_{Q(\mathbf{Z}|\mathbf{A},\mathbf{X})}[c(\mathbf{A}, G(\mathbf{Z}))] \tag{11}$$

Q($\mathbf{Z}$) in Eq. 11 is the marginal distribution of $\mathbf{Z}$ when M obeys P_A and Z obeys $Q(\mathbf{Z} \mid \mathbf{A}, \mathbf{X})$.

In this way, we only need to optimize $Q(\mathbf{Z} \mid \mathbf{A}, \mathbf{X})$ when optimizing, instead of optimizing the coupling between M and N mentioned above. Of course, this problem will still be limited, and it will not be possible to get an accurate numerical solution. We can add a penalty factor to the target expression to relax the restriction on $Q(\mathbf{Z})$, and add a penalty to the target expression. The final formula is:

$$L_{WGAE}(P_A, P_G) = \inf_{Q(z|A,X)\in\Omega} \mathbb{E}_{P_A}\mathbb{E}_{Q(\mathbf{Z}|\mathbf{A},\mathbf{X})}[c(\mathbf{A}, G(\mathbf{Z}))] + \sigma \cdot \mathcal{D}_Z(Q(\mathbf{Z}), P(\mathbf{Z})) \tag{12}$$

where Ω is any non-parameter set of the probabilistic encoder, The penalty factor D_Z is the arbitrary divergence between $Q(\mathbf{Z})$ and $P(\mathbf{Z})$, and σ is a hyper-parameter that can be adjusted.

To prevent falling into the local optimal problem, we use vMF distributions with different density parameters to solve the problem in our experiments. The density parameter κ of the prior distribution $P(\mathbf{Z})$ is set to 0, and then the density parameter κ of the conditional distribution $Q(\mathbf{Z} \mid \mathbf{A}, \mathbf{X})$ is set as a hyper-parameter, so that the two distributions will not always be equal to avoid the in-applicability of the model Latent variables. Therefore, $P(\mathbf{Z}) = p(\mathbf{Z} \mid \mu, 0)$, $Q(\mathbf{Z} \mid \mathbf{A}, \mathbf{X}) = p(\mathbf{Z} \mid \mu, \kappa)$ in the above Eq. 4 obeys the vMF distribution.

For the penalty factor $D_Z(Q(\mathbf{Z}), P(\mathbf{Z}))$, we adopt a method based on Maximum Mean Discrepancy (MMD). For a positive definite reproducing core $\lambda(\cdot)$: $\mathbf{Z} \times \mathbf{Z} \rightarrow R$, there is the following formula:

$$\mathrm{MMD}_\lambda(P(\mathbf{Z}), Q(\mathbf{Z})) = \left\| \int_Z \lambda\left(\mathbf{z}_{i},\right) dP(\mathbf{Z})\left(\mathbf{z}_i\right) \int_Z \lambda\left(\mathbf{z}_{i},\right) dQ(\mathbf{Z})\left(\mathbf{z}_i\right) \right\|_{\mathcal{H}_\lambda} \quad (13)$$

H_s in the formula is the Reproducing kernel Hilbert Space (RKHS) of real-valued function mapping $\mathbf{Z}$ to R. If s in the formula represents a feature, then MMD_λ can be defined as a metric, which can be used as a measure of divergence. MMD is an unbiased U statistical estimator, which can be used in conjunction with the stochastic gradient descent (SGD) method. The maximum average difference has better performance in matching high-dimensional distributions, so this kind of penalty factor can be used.

Algorithm 1. Wasserstein Graph Auto-Encoder

Initialization:

G = {V, E}:a Graph with links and features

α:Regularization coefficient

λ: characteristic positive-definite kernel

κ: The density parameter of the conditional distribution

Initialize the parameters of the encoder Q_Φ, decoder P_Γ

While(Φ,Γ) not converged do

Sample{$\mathbf{A}_{1j}, \ldots, \mathbf{A}_{nj}, \mathbf{x}_1, \ldots, \mathbf{x}_n$}from the training set

Sample{$\mathbf{z}_1, \mathbf{z}_2, \mathbf{z}_3, \ldots, \mathbf{z}_n$} from the prior P($\mathbf{Z}$)

Sample $\tilde{\mathbf{z}}_i$ from $Q_\phi\left(\mathbf{Z} \mid \mathbf{A}_{ij}, \mathbf{x}_i\right)$ for i=1,,n

Update Q_Φ and P_Γ by descending:

$\frac{1}{n} \sum_{i=1}^{n} c\left(\mathbf{A}_{ij}, \mathbf{x}_i, P_\Gamma\left(\tilde{\mathbf{z}}_i\right)\right) + \frac{\alpha}{n(n-1)} \sum_{\ell \neq m} \lambda\left(\mathbf{z}_\ell, \mathbf{z}_m\right)$

$+\frac{\alpha}{n(n-1)} \sum_{\ell \neq m} \lambda\left(\tilde{\mathbf{z}}_\ell, \tilde{\mathbf{z}}_m\right) \frac{2\alpha}{n^2} \sum_{\ell, m} \lambda\left(\mathbf{z}_\ell, \tilde{\mathbf{z}}_m\right)$

end while

Algorithm 1 is the proposed WGAE algorithm. Given a graph G, the regularization coefficient α, the characteristic positive definite kernel λ, the density parameter κ of the conditional distribution, sample the values of the adjacency matrix and the characteristic matrix from the training set. We also need to sample from the prior distribution, use the encoder to calculate $\widetilde{z}_i$ according to the values sampled in the training set, and update our model itself through the formula gradient until the encoder and decoder converge.

5 Experiments

We have conducted experiments on benchmark data sets to extensively validate our WGAE model by comparing it with GAE, VGAE and their variants.

5.1 Data Sets

We have conducted experiments on three citation network data sets: Cora, Citeseer and Pubmed [3]. In link prediction experiments, these data sets are widely used. In these data sets, the network nodes are scientific publications, and the edges of the network are the citation relationships among them. We summarize the statistics of the data sets in Table 2.

Table 2. Statistics of the data set used in the experiment.

DATASET	#NODES	#LINKS	#FEATURES	#CONTENT WORDS
Cora	2,708	5,429	1,433	3,880,564
Citeseer	2,327	4,732	3,703	12,274,336
Pubmed	19,717	44,338	500	9,858,500

5.2 Evaluation Metrics

We evaluate performance based on two metrics, Area Under Curve (AUC) and Average Precision (AP), which are also used for link prediction in [7,16]. AUC is defined as the area enclosed by the coordinate axis under the Receiver Operating Characteristic (ROC) curve. The value of this area will not be greater than 1, and the ROC curve is located above the $y=x$ line, so the value of AUC is between 0.5 and 1. The closer the AUC value is to 1, the higher the accuracy of this algorithm will be. AP refers to the area under the Precision-Recall (PR) curve, and the same value will not be greater than 1. In the link prediction algorithm, the higher the AP value is, the higher the accuracy will be.

5.3 Baselines

We have selected several link prediction models and WGAE models as comparison experiments.

Spectral Clustering(SC) [22]: is a clustering method based on the graph theory. The basic idea is to use the similarity matrix of the sample data to perform the clustering of the eigenvectors obtained after matrix decomposition.

DeepWalk [13]: is the data mining algorithm, which combines the classical random walk method based on graphs and the skip-gram model in natural language processing, and uses the cooccurrence relationship among nodes in the graph to learn the vector representation of nodes.

GAE [7]: It is an unsupervised graphic data frame based on auto-encoder. Use the graph topology and content information to learn the low-dimensional vector representation of the section, and then reconstruct the graph.

VGAE [7]: is an unsupervised graphics data structure framework based on variational auto-encoders. Use the graph topology and content information to learn the low-dimensional vector representation of nodes, and then reconstruct the graph.

ARGA [16]: Improved based on GAE, and introduced an adversarial training module in training.

ARVGA [16]: Improve based on VGAE, and introduce adversarial training modules in training.

VGAE+vMF: Improve based on VGAE and replace the Gaussian distribution with the vMF distribution.

WGAE+Gaussian(GS): Improve based on VGAE to minimize the optimal transmission $W_c(P_G, P_A)$ between the generated distribution P_G and the real data distribution P_A.

WGAE: Improve based on VGAE by using a Wasserstein distance and the vMF prior distribution instead of the Gaussian distribution.

5.4 Experimental Parameter Settings

In all experiments, we train for 200 iterations using Adam [23] with a learning rate of 0.01 for our WGAE model. And we use 32-dimensional hidden layers and 16-dimensional latent variables. We follow the training scheme in [7,16]: For the links in all data sets, we randomly divide them into the training set, validation set and test set according to the ratio of 85:5:10; We deleted 15% of the links while retaining the node features to train the model. The validation and test sets are derived from deleted edge and node pairs. We run our model 10 times on each data set, and use its result and error as the final result. For the conditional distribution $Q(Z \mid A, X)$, the hyper-parameter $\kappa = 176$ can reach the optimum in all data sets. The hyper-parameter σ in the penalty factor was tried, and it was found that when $\sigma = 5$, it can reach the optimal value for all data sets. For baseline SC, DW (DeepWalk), GAE* (input feature not used), VGAE* (input feature not used), GAE, VGAE, ARGA and ARVGA, we set them according to their descriptions in the paper.

5.5 Results and Analysis

The details of the experimental results on the link prediction are shown in Table 3. The best performance is marked in bold. It can be seen from the table that our VGAE+vMF is better than the GAE model in the Cora and Citeseer

Table 3. AUC and AP results of different algorithms on the data sets.

Approaches	Cora		Citeseer		Pubmed	
	AUC	AP	AUC	AP	AUC	AP
SC [22]	84.6 ± 0.01	88.5 ± 0.00	80.5 ± 0.01	85.1 ± 0.01	84.3 ± 0.03	87.8 ± 0.01
DW [19]	83.1 ± 0.01	85.0 ± 0.00	80.5 ± 0.02	83.6 ± 0.01	84.4 ± 0.00	84.1 ± 0.00
GAE* [22]	84.3 ± 0.02	88.1 ± 0.01	78.7 ± 0.02	84.1 ± 0.02	82.2 ± 0.01	87.4 ± 0.00
VGAE* [22]	84.0 ± 0.02	87.7 ± 0.01	78.9 ± 0.03	84.1 ± 0.02	82.7 ± 0.01	87.5 ± 0.01
GAE [22]	91.0 ± 0.02	92.0 ± 0.03	89.5 ± 0.04	89.9 ± 0.05	96.4 ± 0.00	96.5 ± 0.00
VGAE [22]	91.4 ± 0.01	92.6 ± 0.01	90.8 ± 0.02	92.0 ± 0.02	94.4 ± 0.02	94.7 ± 0.02
ARGE [23]	92.4 ± 0.003	93.2 ± 0.003	91.9 ± 0.003	93.0 ± 0.003	96.8 ± 0.001	97.1 ± 0.001
ARVGE [23]	92.4 ± 0.004	92.6 ± 0.004	92.4 ± 0.003	93.0 ± 0.003	96.5 ± 0.001	96.8 ± 0001
VGAE+vMF	92.1 ± 0.01	92.5 ± 0.03	90.5 ± 0.02	91.3 ± 0.03	95.9 ± 0.03	96.7 ± 0.02
WGAE+GS	92.5 ± 0.02	93.1 ± 0.03	92.6 ± 0.02	92.9 ± 0.02	95.6 ± 0.01	96.2 ± 0.03
WGAE	**93.1 ± 0.04**	**93.8 ± 0.03**	**93.0 ± 0.02**	**93.3 ± 0.01**	**96.9 ± 0.02**	**97.2 ± 0.03**
Increase	0.7	1.2	0.6	0.3	0.1	0.1

data sets, and worse than GAE in the Pubmed data set. This may be because the PubMed data set is relatively large, and the Gaussian distribution is easier to adjust parameters than the vMF distribution. WGAE+Gaussian is better than the VGAE model on the three data sets. It can be seen that Wasserstein distance is more conducive to graph reconstruction than KL divergence. Compared with VGAE that includes node characteristics, WGAE+vMF has increased AUC and AP by about 1.5% on Cora and Citeseer, and raised by 2.5% in Pubmed data set. Compared with the ARGA and ARVGA models, our model has increased by about 1% in the Cora data set. Although ARVGA has introduced an adversarial training module, it still uses KL divergence to measure the variational posterior distribution and prior distribution during optimization. This further shows that Wasserstein distance is better than KL divergence in graph reconstruction.

5.6 Hyper-parameter κ in $Q(\mathbf{Z} \mid \mathbf{A}, \mathbf{X})$

We set the density parameter κ as a hyper-parameter to prevent falling into local optimum. Our prior distribution adopts a uniform distribution with $\kappa = 0$, and the selection of the density parameter κ in the conditional distribution is shown in Table 4.

We used different κ to run 10 times on each data set, and got the results in Table 4. The best performance is marked in bold.

5.7 Training Performance Analysis

To better verify our experimental results, we train our model, GAE and VGAE models on the Cora data set for 200 epochs at the same time to obtain the AUC and AP. The results are shown in Fig. 2(a) and Fig. 2(b).

The performances of GAE and VGAE in the Fig. 2 are carried out entirely according to the description in [7]. It can be seen that our model does not surpass

Table 4. AUC and AP results of different κ on the data sets.

κ	Cora		Citeseer		Pubmed	
	AUC	AP	AUC	AP	AUC	AP
$\kappa = 1$	91.81 ± 0.04	91.93 ± 0.06	91.23 ± 0.02	92.28 ± 0.05	94.63 ± 0.03	94.98 ± 0.03
$\kappa = 50$	92.52 ± 0.03	92.97 ± 0.03	91.85 ± 0.02	92.56 ± 0.03	95.21 ± 0.02	95.33 ± 0.02
$\kappa = 100$	92.75 ± 0.02	93.16 ± 0.02	92.52 ± 0.02	92.77 ± 0.01	95.84 ± 0.03	95.72 ± 0.02
$\kappa = 150$	92.86 ± 0.03	93.32 ± 0.03	92.93 ± 0.03	92.98 ± 0.03	96.12 ± 0.02	96.35 ± 0.02
$\kappa = 175$	93.01 ± 0.02	93.73 ± 0.03	93.08 ± 0.03	93.27 ± 0.03	96.83 ± 0.02	97.13 ± 0.03
$\kappa = 176$	$\mathbf{93.14 \pm 0.02}$	$\mathbf{93.78 \pm 0.02}$	$\mathbf{93.15 \pm 0.02}$	$\mathbf{93.34 \pm 0.02}$	$\mathbf{97.04 \pm 0.02}$	$\mathbf{97.18 \pm 0.02}$
$\kappa = 177$	93.05 ± 0.02	93.70 ± 0.02	93.04 ± 0.02	93.25 ± 0.02	96.71 ± 0.02	97.00 ± 0.02
$\kappa = 200$	92.82 ± 0.03	93.28 ± 0.02	92.86 ± 0.03	93.07 ± 0.03	96.35 ± 0.02	96.84 ± 0.02

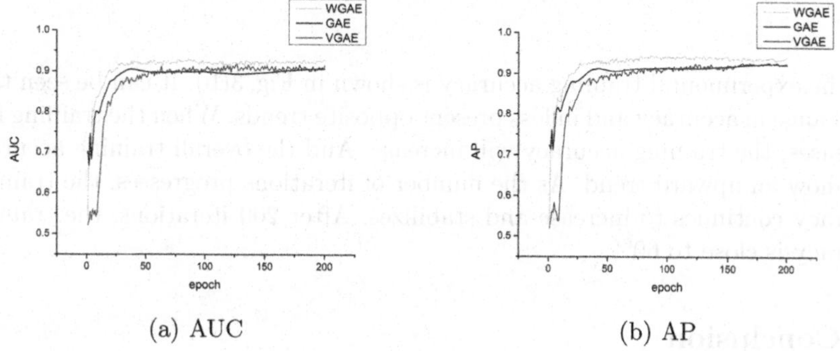

(a) AUC (b) AP

Fig. 2. Experimental results of 200 iterations of WGAE, GAE and VGAE on the Cora data set. (a)AUC score; (b) Average Precision score.

the GAE and VGAE models at a certain moment. Our model started to be higher than the performance of GAE and VGAE in the first few epochs, and kept it until the end of the experiment. This experiment shows that our WGAE model is more suitable for link prediction tasks than GAE and VGAE. From Fig. 2, we can see that our model converges faster than VGAE. It may be because the continuous mixed posterior distribution is used to match the prior distribution in our model. The posterior distribution of each sample is required to match the prior distribution in VGAE, which slows down the convergence speed of VGAE. And at the same time, our model also converges faster than the GAE model. Our model uses the vMF distribution, we only train the direction parameter μ, and in the Gaussian distribution, we need to train μ and σ.

In Fig. 3(a), we show the loss changes of the WGAE model during training. It can be seen that the loss of our model was almost 0.80 at the beginning, and it showed a linear decline in the first few iterations. After the 50-th iteration, the downward trend of loss became moderate. As the number of iterations progresses, the loss gradually becomes more stable. After 200 iterations, it was reduced from 0.80 to 0.40.

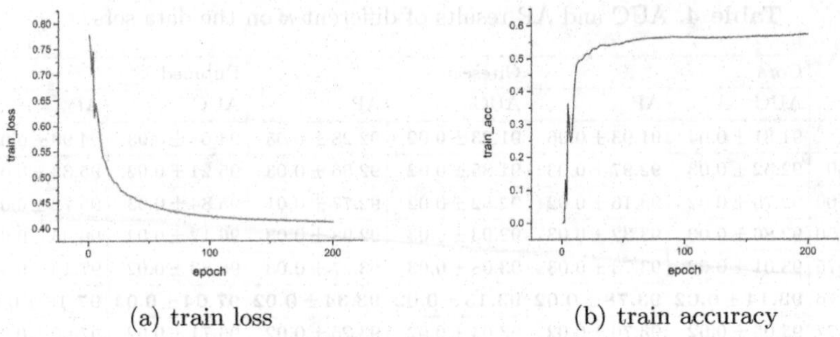

(a) train loss (b) train accuracy

Fig. 3. The training loss and training accuracy change of the WGAE model on the Cora data set for 200 epochs. (a)train loss; (b) train accuracy.

The experimental training accuracy is shown in Fig. 3(b). It can be seen that the change in accuracy and in loss present opposite trends. When the training loss decreases, the training accuracy will increase. And the overall training accuracy will show an upward trend. As the number of iterations progresses, the training accuracy continues to increase and stabilizes. After 200 iterations, the training accuracy is close to 60%.

6 Conclusion

This paper proposes a novel generative model–WGAE. The WGAE model minimizes the penalty in the form of Wasserstein distance between the model distribution and the target distribution. WGAE encourages the use of continuous mixed distribution to match the prior distribution. The latent codes of different nodes are far apart in the mixed distribution, which is conducive to reconstruction. In WGAE, a d–1 dimensional distribution on the d-dimensional hypersphere is used–vMF distribution, which can set different density parameters for the prior distribution and the conditional distribution and avoid the local optimum. Experimental results have shown that our WGAE is superior to GAE, VGAE and their variants in link prediction tasks.

References

1. Nickel, M., et al.: A review of relational machine learning for knowledge graphs. In: Proceedings of the IEEE, vol. 104, no. 1, pp. 11–33 (2015)
2. Bennett, J., Lanning, S., et al.: The netflix prize. In: Proceedings of KDD Cup and Workshop, New York, USA, pp. 35 (2007)
3. Prithviraj, S., et al.: Collective classification of network data. In: AI Magazine, pp. 93–106 (2008)
4. Pei, H., Wei, B., Chang, K.C.C., Lei, Y., Yang, B.: Geom-gcn: geometric graph convolutional networks. In: International Conference on Learning Representations, Addis Ababa, Ethiopia (2020)

5. Kipf, T.N., Welling, W.: Semi-supervised classification with graph convolutional networks. In: International Conference on Learning Representations, Toulon, France. (2017)
6. Wu, Z., et al.: A comprehensive survey on graph neural networks. IEEE Trans. Neural Netw. Learn. Syst. **32**, 4–24 (2020)
7. Kipf, T.N., Welling, M.: Variational graph auto-encoders. arXiv: 1611.07308 (2016). https://arxiv.org/abs/1611.07308
8. Zhang, M., Li, P., Xia, Y., Wang, K., Jin, L.: Revisiting graph neural networks for link prediction (2020). arXiv preprint arXiv:2010.16103
9. Newman, M.: Clustering and preferential attachment in growing networks. Phys. Rev. E **64**, 025102 (2001)
10. Adamic, L.A., Adar, E.: Friends and neighbors on the web. In: Social Networks, pp. 211–230 (2003)
11. Zhou, T., Lü, L., Zhang, Y.C.: Predicting missing links via local information. Eur. Phys. J. B **71**, 623–630 (2009)
12. Katz, L.: A new status index derived from sociometric analysis. Psychometrika **18**(1), 39–43 (1953)
13. Perozzi, B., Al-Rfou, R., Skiena, S.: DeepWalk: online learning of social representations. In: ACM (2014)
14. Grover, A., Leskovec, J.: node2vec: Scalable feature learning for networks. In: ACM (2016)
15. Tang, J., et al.: LINE: large-scale information network embedding. In: International Conference on World Wide Web WWW (2015)
16. Pan, S., Hu, R., Long, G., Jiang, J., Yao, L., Zhang, C.: Adversarially regularized graph auto-encoder for graph embedding. In: International Joint Conference on Artificial Intelligence, Stockholm, Sweden, pp. 2609–2615 (2018)
17. Villani, C.: Topics in optimal transportation. In: AMS Graduate Studies in Mathematics, p. 370 (2003)
18. Tolstikhin, I., et al.: Wasserstein auto-encoders. arXiv:1711.01558 (2017)
19. Xu, M., et al.: Towards generalized implementation of wasserstein distance in GANs. In: Proceedings of the AAAI Conference on Artificial Intelligence, pp. 10514–10522 (2021)
20. Villani, C.: Optimal transport: old and new. In: Grundlehren der mathematischen Wissenschaften. Springer, Heidelberg (2008). https://doi.org/10.1007/978-3-540-71050-9
21. Nurminen, H., et al.: 3D angle-of-arrival positioning using von Mises-Fisher distribution. In: 21st International Conference on Information Fusion (2018)
22. Liu, T.H.: Leveraging social media networks for classification. In: Data Mining & Knowledge Discovery (2011)
23. Kingma, D., Ba, J.: Adam: a method for stochastic optimization. In: Computer Science (2014)

Fine-Grained Activity Recognition Based on Features of Action Subsegments and Incremental Broad Learning

Shi Chen[1,2], Sheng Wu[1,2], Licai Zhu[1,2(✉)], and Hao Yang[1,2,3,4]

[1] Nanjing Tech University, Nanjing 211816, Jiangsu, China
[2] Yancheng Teachers University, Yancheng 224001, Jiangsu, China
[3] Jiangsu Provincial Key Constructive Laboratory for Big Data of Psychology and Cognitive Science, Yancheng Teachers University, Yancheng 224002, Jiangsu, China
[4] Suzhou Research Institute, University of Science and Technology of China, Suzhou 215123, Jiangsu, China

Abstract. Human activity recognition using MEMS on mobile devices has become one of the most compelling solutions owing to the miniaturization of sensors. A crucial challenge is to recognize precisely activities when they are changing. Sliding window is a type of common methods. However, the interference of historical data in the sliding window is harmful to insight into changing of actions or uncommon behaviors. This paper proposes a fine-grained activity recognition method and designs a corresponding system *farer*. It employs features of action subsegments and incremental broad learning to precisely distinguish the alterations of actions and abnormal movements. Firstly, *farer* achieves the accurate segmentation of activities as data preprocessing. A neighborhood extreme value method (NEV) is adopted to avoid the intervention of peaks and valleys of data. Secondly, the current action is partitioned to fine-grained subsegments to elaborately abstract subtle features. We propose a feature extraction technique based on adjacent difference (FETAD), and furthermore reduce its resulting dimension through the complete two-dimensional principal component analysis (C2DPCA). Finally, broad learning theory is employed to construct the activity recognition model, especially incremental learning for unusual behaviors. Extensive experiments demonstrate that *farer* could accurately recognize activities when they abruptly change, and its performance is considerable stability. Meanwhile, it can quickly establish a valid incremental model that only needs a short sampling time for special activities. The overall accuracy of *farer* is 97.91% with 90.14% for changed activities, which is far superior to the current mainstream methods.

Keywords: Fine-grained recognition · Broad learning · Action subsegments · Incremental model

1 Introduction

Human Activity Recognition (HAR) has been widely applied in some scenarios, e.g. industry and medicine [1]. For instance, it could be used for early warning when factory

Y. Lai et al. (Eds.): ICA3PP 2021, LNCS 13155, pp. 100–114, 2022.
https://doi.org/10.1007/978-3-030-95384-3_7

operators carry out dangerous behaviors, and exhibiting the accuracy degree of mobility impairments patients' performed actions for physical therapy. The basis pattern of HAR is to collect actual data of specific activities and then achieve accurate recognition through comparison and classification. In generally, feature extraction by sliding window is an effective activity recognition paradigm. The basic essence of this type of method is to design a sliding window which contains the perception data of the activity to be recognized, then extract features of the window data to represent the activity, and design an effective classifier for activity recognition. However, in the above special scenarios, most of them require the deployment of dedicated sensing systems, such as IoT devices and sensor networks [2–4], resulting in high cost and poor universality.

The development of wireless communication technology [5] and the miniaturization of sensors [6] have enabled smart devices [7] embedded with many MEMS sensors (such as smart phones, tablets, etc.) to have good data collection capabilities. In the meanwhile, in-depth research in the field of deep learning has further improved the accuracy of activity recognition [8]. However, the popularization of mobile devices puts forward higher requirements for activity recognition in complex scenarios. Particularly, mobile devices must quickly and accurately perceive activity changes, and recognition methods must be oriented to the special behavior habits of different users. The main challenges that activity recognition still faces are as follows: 1) The raw data collected by the accelerometer contains much noise. 2) The influence of historical data on current activity data. 3) Special behavior habits are not easy to accurately recognize.

To address the above challenges, in this paper, we propose a fine-grained activity recognition method and designs a corresponding system *farer*. The system implements effective preprocessing of the source data by smoothing and filtering the sampled data and segmenting activities into individual actions. Then deeply mine the features of actions through fine-grained subsegment division, feature extraction and dimensionality reduction. Finally, an incremental activity recognition model based on broad learning (BL) is constructed to realize accurate activity recognition and satisfy incremental model update. The main contributions of this paper are summarized as follows:

1) Realize accurate segmentation of activities. By smoothing and filtering the source data, a neighborhood extreme value method is proposed to avoid the interference of peaks and valleys.
2) Deeply mine the features of action subsegments. The action data is partitioned to fine-grained subsegments according to changes of acceleration, and a feature extraction technique oriented to adjacent difference is designed.
3) The *farer* system based on BL is designed with stable performance and good practicability. After a series of experimental verifications, the system has a high recognition rate for different activities under the condition of stable activities, with an overall recognition rate of 97.91%. Under the condition of changed activities, the recognition accuracy is 90.14%, far exceeding other methods.

2 Related Work

Recognizing human activities based on sensor data is essentially a pattern classification problem. When using a sliding window for activity recognition, it is required to process the noise generated during the sampling process, select an appropriate sliding window, and design an effective activity classifier. These are all key factors that determine the recognition performance. This section mainly introduces the current research status of three aspects in the case of activity recognition: the processing of data noise, the setting of sliding window and the construction of classifier model.

Since the data collected by the sensor often contains high-frequency noise, it needs to be processed, otherwise it will affect the accuracy of activity recognition. Yang et al. [9] used Gaussian filtering algorithm to eliminate the influence of noise. Garcia-Ceja et al. [10] used the average smoothing method, replacing each original data with the average of two adjacent data points. Khan et al. [11] used a third-order moving average filtering algorithm to remove noise. These methods cannot eliminate the interference extreme points to a great extent.

The collection of human activity data often takes a long time. In the activity recognition stage, the sensor data stream needs to be segmented by windows. Because the fixed-size sliding window segmentation technology is simple to operate, it is adopted by most researches. In the terms of sliding window setting, Fida et al. [12] tested the window of 0.5 s to 3 s, and achieved the best result in 1.5 s. Elsts [13] adopted a 2.56 s sliding window to design the energy-saving activity recognition framework. Shuvo [14] and Xia [15] adopted a sliding window of 2.56 s with a step length of 1.28 s, achieving the recognition accuracy of more than 95%. Cha et al. [16] adopted a window length of 1 to 4 s and found that using 4 s achieved the best accuracy of 96.1%. Pienaar et al. [17] adopted a large window of 10 s with a step length of 1 s to segment data and achieved the recognition accuracy of 94%. The above methods verify that the selection of the action window will greatly affect the recognition performance of the activity. The traditional method of fixed-size sliding window is oriented to the recognition of a single stable activity, while ignoring the change of the activity.

In the terms of classifier construction, researchers manually extract features from activity sensor data and employ them in various traditional machine learning algorithms. Since these data fragments cannot adequately represent complex human activities, they have become the performance bottleneck of the classifier [18]. To further improve the accuracy of activity recognition, methods based on deep learning are employed by more and more people, such as Convolutional Neural Network (CNN) [19], Long Short-Term Memory (LSTM) [20] and their joint improved models CNN-LSTM [21] and ConvLSTM [22]. Although these models show good performance, their designs are more complex. In addition, the amount of calculation is large and hardware requirements are high. More importantly, these models are constructed based on training data, so they have poor perception of activity changes and lack robustness [23].

To solve the problems of accurate recognition and flexibility of activities, we propose effective data preprocessing and employ fine-grained features of action subsegments, which improve the accuracy of activity recognition, especially for changed activities. Incremental model construction based on broad learning is also proposed to realize the incremental update of the special activity behavior, without using the source data to retrain the entire model.

3 Proposed Method

For complex activity situations, we propose a fine-grained activity recognition method based on features of action subsegments and incremental broad learning. The corresponding recognition system named *farer* is also designed. The system contains three sub-modules, namely data preprocessing, feature extraction based on fine-grained subsegment, and incremental recognition model based on broad learning. First, smooth and filter the sampled data, and design a peak and valley recognition method to accurately segment activities into individual actions. Then deeply mine the features of action subsegments through the fine-grained segmentation of the action data, targeted feature extraction and dimensionality reduction. Finally, build an incremental activity recognition model based on broad learning to realize the accurate recognition of activities and satisfy the incremental model update. The framework of *farer* is shown in Fig. 1.

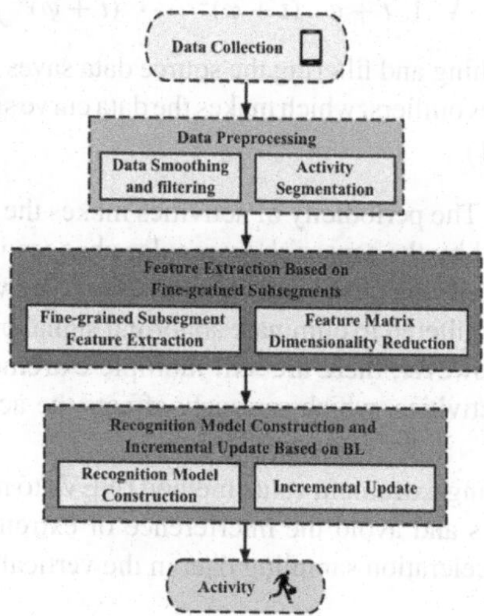

Fig. 1. *farer* framework.

3.1 Data Preprocessing

Data Smoothing and Filtering. According to the travel characteristics of pedestrians, the sensor data changes smoothly in a relatively short period. Since people cannot maintain a fixed posture when traveling, the sensor perceives irregular changes during the collection process, resulting in abnormal points. Without changing the trend of data changes, we employ the neighborhood smoothing method to filter the source data.

Assuming that the sampled data at time t is x_t, its neighborhood interval is $[t-\varphi, t+\varphi]$ and the interval range is $\mu = 2\varphi + 1$. Construct a k-order polynomial to fit the points in the interval. Denote $S_0, S_1, \cdots S_k$ as the polynomial coefficients, then the data x_t can be expressed as

$$x_t = s_0 + s_1 t + s_2 t^2 + \cdots + s_k t^k \tag{1}$$

The least square fitting of the neighborhood interval $X = \left(x_{t-\varphi}, \cdots, x_t, \cdots, x_{t+\varphi}\right)^T$ is calculated as

$$\widehat{X} = P\left(P^T P\right)^{-1} P^T X \tag{2}$$

where,

$$P = \begin{pmatrix} 1 & t-\varphi & (t-\varphi)^2 & \cdots & (t-\varphi)^k \\ \vdots & \vdots & \vdots & \vdots & \vdots \\ 1 & t-1 & (t-1)^2 & \cdots & (t-1)^k \\ 1 & t & t^2 & \cdots & t^k \\ 1 & t+1 & (t+1)^2 & \cdots & (t+1)^k \\ \vdots & \vdots & \vdots & \vdots & \vdots \\ 1 & t+\varphi & (t+\varphi)^2 & \cdots & (t+\varphi)^k \end{pmatrix}$$

The method of smoothing and filtering the source data saves the change information of the signal and eliminates outliers, which makes the data curve smoother. After filtering, the value of x_t is $\widehat{X}(\varphi + 1)$.

Activity Segmentation. The periodicity of activities makes the acceleration data in the vertical direction sampled by the sensor show regular changes in peaks and valleys, so dividing activities by identifying peaks and valleys is the basic way of activity segmentation. The activity data is filtered to eliminate abnormal signal points, making the entire data stream smoother. However, there are still multiple extreme interference points at the peak and valley of activities, which seriously affects the accurate segmentation of activities.

We design a neighboring extremum value method (NEV) to more accurately identify the real peaks and valleys and avoid the interference of extreme points. Denote $X = (x_1, x_2, \ldots, x_n)$ as the acceleration sampling data in the vertical direction. Then,

1. Calculate extreme points.
   Obtain all maximum points $X_{max} = \left(x_{max}^1, x_{max}^2, \ldots\right)$ and minimum points $X_{min} = \left(x_{min}^1, x_{min}^2, \ldots\right)$.
2. Filter extreme interference points.
   We employ the altitude, prominence and isolation of the peaks and valleys to filter the noise at the extreme points, where the altitude refers to the height of the peak or the depth of the valley, the prominence refers to the convexity of the peak or the concavity of the valley, and the isolation is the horizontal distance between two peaks or two valleys. We set the altitude threshold Γ_A, the prominence threshold Γ_P and the isolation threshold Γ_I to eliminate extreme points that do not meet the threshold.
3. Segment activities into individual actions.
   When the vertical acceleration direction is upward and gradually increases from 0, it is defined as the starting point of the action. Then the peak and valley are reached. After reaching valley, when the vertical acceleration direction is downward and decreases to 0, it is defined as the ending point of the action. Thus, the activity data is segmented into individual action data.

Different from the traditional activity recognition, we set the size of the sliding window to the size of the complete action segment, so each window has a different size. We call the sliding window here the action window. The action window only contains the data of the current action, without historical data, and it slides a complete action window every time. Our design avoids the influence of historical data on current data.

3.2 Feature Extraction Based on Fine-Grained Subsegments

Fine-grained Subsegment Feature Extraction. To extract fine-grained features of segmented actions, we design an activity recognition method based on fine-grained subsegments. We perform fine-grained subsegment division of actions to realize fine-grained cognition of actions, that is, the action window is evenly divided into several subsegments. The fine-grained division of the action window shows the change of the behavior state.

In order to fully mine the action characteristics to realize the fine-grained cognition of the action data, we design a feature extraction technique based on adjacent difference (FETAD) for 3 axes acceleration. The change of measured data is relatively stable in a short period due to the continuity of the action. Therefore, when the action is fine-grained divided, the difference in adjacent subsegments changes most smoothly. According to this principle, the steps of FETAD are as follows:

1) The action window is evenly divided into k_f subsegments and the length is l_m. The three-axis data vectors are $G^x_{l_i} = \left[g^x_{(i,1)}, g^x_{(i,2)}, \cdots, g^x_{(i,l_i)}\right]$, $G^y_{l_i} = \left[g^y_{(i,1)}, g^y_{(i,2)}, \cdots, g^y_{(i,l_i)}\right]$, $G^z_{l_i} = \left[g^z_{(i,1)}, g^z_{(i,2)}, \cdots, g^z_{(i,l_i)}\right]$, where $1 \leq i \leq k_f$.
2) Adopt the difference between the data in adjacent subsegments as the data of the previous subsegment, i.e. $G^x_{l_i} = G^x_{l_{i+1}} - G^x_{l_i}$, $G^y_{l_i} = G^y_{l_{i+1}} - G^y_{l_i}$, $G^z_{l_i} = G^z_{l_{i+1}} - G^z_{l_i}$, where $1 \leq i < k_f$. Each axis gets $k_f - 1$ difference vectors.
3) Extract features for each difference vector of each coordinate axis. Denote n_f as the number of features to be extracted. The feature vector of the difference vector is $D_{l_i} = \left[d_i(1), d_i(2), \cdots, d_i(n_f)\right]$.
4) Combine the features of $k_f - 1$ difference vectors on each coordinate axis into a new feature vector. The feature vectors of the three coordinate axes are expressed as $D^x = \left[D^x_{l_1}, D^x_{l_2}, \cdots, D^x_{l_{k_f-1}}\right]^T$, $D^y = \left[D^y_{l_1}, D^y_{l_2}, \cdots, D^y_{l_{k_f-1}}\right]^T$ and $D^z = \left[D^z_{l_1}, D^z_{l_2}, \cdots, D^z_{l_{k_f-1}}\right]^T$.
5) Finally, the feature vectors of the three coordinate axes are combined into a two-dimensional matrix as $D^{xyz} = [D^x, D^y, D^z]^T$. The size is $3 \times [(k_f - 1) \times n_f]$.

Feature Matrix Dimensionality Reduction. To improve the speed of activity recognition, we further extract the effective information of the three-axis features. First, perform feature extraction on the combined matrix D^{xyz} of the x, y and z three axes. Aiming at the obtained two-dimensional feature matrix, we adopt complete two-dimensional principal component analysis (C2DPCA). C2DPCA reduces the dimensionality of the matrix from two aspects: row projection and column projection. Then, flatten the principal component matrix of the three axes to obtain a one-dimensional vector to meet the input requirements of BLS.

Denote D as the feature matrix set of N actions. The feature matrix of the i-th action is $D_i \in R^{p\times q}$, $i \in [1, N]$. To realize the complete two-dimensional principal component analysis of D_i, it needs to be projected from two angels of row and column. The column divergence matrix and row divergence matrix of D are formulated as $G_P = \sum_{i=1}^{N}(D_i - \overline{D})(D_i - \overline{D})^T$ and $G_Q = \sum_{i=1}^{N}(D_i - \overline{D})^T(D_i - \overline{D})$, where $\overline{D} = \frac{1}{N}\sum_{i=1}^{N} D_i$ is the average value of the feature matrix set D.

By choosing proper eigenvectors of the matrices G_P and G_Q, the projection of D_i is as dispersed as possible. The numbers of eigenvalues of G_P and G_Q are calculated as n_P and n_Q. The eigenvalues of G_P and G_Q are sorted in descending order. The eigenvalue set of G_P are $\alpha = [\alpha_1, \alpha_2, \cdots, \alpha_{nP}]$, and the corresponding eigenvector set is $[\mu_1, \mu_2, \cdots, \mu_{nP}]$. The eigenvalue set of G_Q are $\beta = [\beta_1, \beta_2, \cdots, \beta_{nQ}]$, and the corresponding eigenvector set is $[\nu_1, \nu_2, \cdots, \nu_{nQ}]$. Since G_P and G_Q are both non-negative definite matrices, their eigenvalues are not less than zero.

If the first n_1 eigenvalues of the eigenvalue sequence α of G_P satisfy $\sum_{i=1}^{n_1}\alpha_i \ge \partial_P \cdot \sum_{i=1}^{nP}\alpha_i$, where ∂_P is the column hash threshold. Select the eigenvectors corresponding to n_1 eigenvalues to form the column projection of D_i, i.e. $P = [\mu_1, \mu_2, \cdots, \mu_{n_1}]^T$. The size of P is $n_1 \times p$. Similarly, if the first n_2 eigenvalues of the eigenvalue sequence β of G_Q satisfy $\sum_{j=1}^{n_2}\beta_j \ge \partial_Q \cdot \sum_{j=1}^{nQ}\beta_j$, where ∂_Q is the row hash threshold. Select the eigenvectors corresponding to n_2 eigenvalues to form the row projection of D_i, i.e. $Q = [\nu_1, \nu_2, \cdots, \nu_{n_2}]$. The size of Q is $q \times n_2$. P and Q respectively project D_i to obtain the principle component analysis matrix is $H = P \cdot D_i \cdot Q$. The size of H is $n_1 \times n_2$.

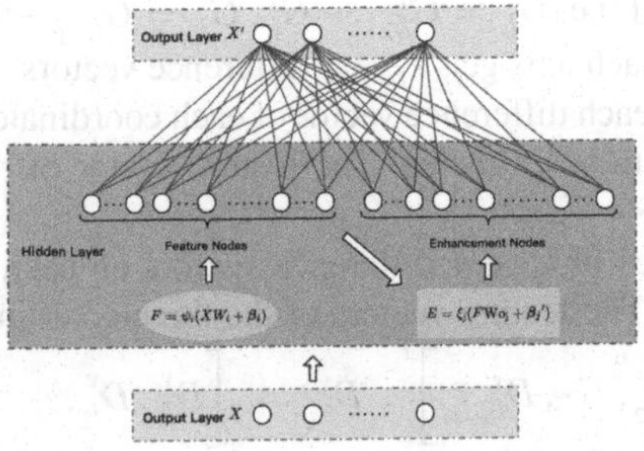

Fig. 2. Broad learning system model.

3.3 Recognition Model Construction and Incremental Update Based on BL

Broad learning system was proposed by Chen [24] in 2018. It is a structure of a single hidden layer including an input layer, a hidden layer and an output layer. The hidden layer includes a feature layer and an enhancement layer. First, the input layer receives the activity features which are extracted by windows. Then, the feature layer linearly maps activity features to construct feature nodes and the enhancement layer employs non-linear activation function for the feature nodes to obtain enhancement nodes. The feature nodes and the enhancement nodes are combined to form the hidden layer matrix. Finally, the output layer obtains the output coefficients by the method of pseudo inverse, and gives the learning results. The model architecture is shown in Fig. 2.

Recognition Model Construction Based on BL. We adopt broad learning to build an activity recognition model. Suppose the training data sample set is $X = [x_1, x_2, \cdots, x_n]^T$ and each sample has m dimensions. The sample category label set is $C = [c_1, c_2, \cdots, c_n]^T$. The construction process of the recognition model is as follows:

1. Input layer.
   Reduce the dimensionality of the action feature matrix and flatten it to obtain targeted one-dimensional data, which is employed as the training sample of the input layer.
2. Feature layer.
   Suppose there are m_g groups of feature mapping and each group has n_g feature nodes. The i-th group of feature mapping is calculated as follows:
   $$F_i = \psi_i(XW_i + \beta_i) \tag{3}$$
   where $1 \leq i \leq m_g$, ψ_i is linear transformation function, W_i is a randomly generated matrix and β_i is a randomly generated vector. The integrated feature node matrix is $F = [F_1|F_2|\cdots|F_{m_g}]$.
3. Enhancement layer.
   The main purpose of the enhancement layer is to increase the non-linear factor of the entire network. Since the feature nodes are all obtained in a linear manner, the BL recognition model introduces enhancement nodes to supplement it. Suppose there are m_e groups of enhancement nodes and each group has n_e enhancement nodes. The j-th group of enhancement nodes is calculated as follows:
   $$E_j = \xi_j(FW_{oj} + \beta_j) \tag{4}$$
   where $1 \leq j \leq m_e$, ξ_i is nonlinear transformation function, W_{oj} is a randomly generated orthogonal matrix and β_j is a randomly generated vector. The integrated feature node matrix is $E = [E_1|E_2|\cdots|E_{m_e}]$.
   Finally, the feature node matrix F and the enhancement node matrix E are integrated to generate the hidden layer input matrix $\Lambda = [F|E]$.
4. Output layer.
   The output layer mainly realizes the mapping from the input matrix of hidden layer to the label matrix. Since the category label matrix is C and the input matrix is Λ, if the mapping matrix Ω satisfies
   $$\Lambda \cdot \Omega = C \tag{5}$$
   Then Ω can be obtained by matrix inversion. However, it should be noted that since Λ is generally not a square matrix, its pseudo-inverse can be solved as
   $$\Lambda^{-1} = \left(\Lambda^T \Lambda + \delta I\right)^{-1} \Lambda^T \tag{6}$$
   where I is the identity matrix, δ is the regularization coefficient and the value of δ is close to zero.

Incremental Update. When the recognition objects are some special users, such as lameness, large swing during the activity, etc., the false alarm rate of the model will increase, resulting in a poor user experience. Simply matching the activity characteristics of a special user to the recognition model makes it difficult to guarantee the recognition rate of the special activity. In addition, it also affects the accurate recognition of the trained actions. On the other hand, if the special user's activity data is added to the source data and the model is retrained, a lot of model construction time is spent. To achieve targeted activity recognition, the recognition model needs to be updated incrementally.

By incrementally fusing the characteristics of user's behavior, we realize effective recognition of personalized activities. Our system *farer* has incremental learning capabilities and can be updated on the trained model without retraining historical data. This method satisfies activity recognition scenarios in more complex situations.

Denote $\dot{X}$ as the special activity data set and $\dot{C}$ as special activity label. The feature node matrix $F_{\dot{X}}$ and the enhancement node matrix $E_{\dot{X}}$ of $\dot{X}$ are given by the random matrix of *farer*. The hidden layer matrix is $\Lambda_{\dot{X}} = \left[F_{\dot{X}}|E_{\dot{X}}\right]$, and its pseudo-inverse is

$$\Lambda_{\dot{X}}{}^{-1} = [\Lambda^{-1} - \phi \cdot \sigma)|\phi] \tag{7}$$

where $\phi = \begin{cases} \omega^{-1} & \omega \neq 0 \\ (\Lambda^{-1}) \cdot \sigma \cdot (I + \sigma^T \cdot \sigma)^{-1} & \omega = 0 \end{cases}, \sigma = \Lambda_{\dot{X}} \cdot \Lambda^{-1}, \omega = \Lambda_{\dot{X}} - \sigma^T \cdot \Lambda$, I is the identity matrix.

After getting the incremental pseudo-inverse, the output of *farer* is calculated as

$$\Omega = \Lambda_{\dot{X}}{}^{-1}\begin{pmatrix} C \\ \dot{C} \end{pmatrix} = \Omega + \phi \cdot (\dot{C} - \Lambda_{\dot{X}} \cdot \Omega) \tag{8}$$

Therefore, the system in this paper can achieve rapid incremental update of the original model through matrix operations.

4 Experiment and Analysis

4.1 Experimental Settings

To verify the recognition advantages of *farer*, we not only compare the classification effect of the traditional machine learning method SVM [25], but also compare the convolutional neural network CNN, LSTM, and their joint model CNN-LSTM and ConvLSTM.

We collect a large amount of three-axis accelerometer data of activities. There are 1556436 pieces of sampling data, and the sampling frequency is set to 180 Hz. There are 7 activity states collected in the experiment, namely trickling, walking, brisk walking, jogging, upstairs, downstairs and jumping.

The action window is divided into 6 subsegments, employing typical features in the activity recognition research, which are the maximum, minimum, average, median, standard deviation, variance, interquartile range, skewness, kurtosis, root mean square, sum, range and entropy. The feature vector size of each coordinate axis is 1×65 (Table 1).

Table 1. Experimental parameter settings.

Parameter meaning	Value	Parameter
Smoothing filter window	51	μ
Polynomial order	3	k
Altitude threshold	0.5	Γ_A
Prominence threshold	1.2	Γ_P
Isolation threshold	55	Γ_I
Number of action windows	6	k_f
Column hash threshold	99.95%	∂_P
Row hash threshold	99.95%	∂_Q
Number of features	13	n_f
regularization coefficient	2^{-30}	δ
Number of feature windows	10	m_g
Number of feature nodes	12	n_g
Number of enhancement windows	1	m_e
Number of enhancement nodes	2000	n_e
Zoom scale	0.8	γ

4.2 Performance Evaluation

System Recognition Accuracy. Table 2 shows the comparison of the activity recognition effect between *farer* and other methods. In the table, SA represents single activities and CA represents changed activities. According to the experimental results, the performance of *farer* is the best, with an overall recognition accuracy of 94.03%, which surpasses other recognition methods. Compared with single activity recognition, the accuracy of *farer* is 97.91%. More importantly, the recognition performance of *farer* is stable. It has high recognition accuracy for different activities, and there is no tendency deviation. In contrast, the recognition rate of other methods is either lower than that of *farer*, or has a higher false alarm rate for certain activities. For example, for downstairs, LSTM achieves 100% recognition accuracy, which exceeds 98.99% of *farer*. However, in the recognition of two more common activities, walking and brisk walking, the recognition rates of LSTM are only 73.98% and 78.20%, while the rates of *farer* are 96.64% and 97.78%.

Comparing the situation of activity changes, the overall recognition rate of *farer* is 90.14%, which is much higher than other recognition methods. This is because other methods are affected by historical data. The historical data leads to a great reduction in recognition accuracy. The sliding window of *farer* only represents the current action, so it effectively avoids the interference of historical data. For example, the recognition accuracy of ConvLSTM which recognizes a single activity more accurately is reduced to 77.78%. LSTM's recognition of downstairs reaches 100%, but the recognition of changed activities is only 67.78%.

Table 2. Comparison of the activity recognition effect.

		farer	SVM	CNN	LSTM	CNN-LSTM	ConvLSTM
SA	Trickling	96.71%	96.58%	99.65%	97.90%	100.00%	100.00%
	Walking	96.64%	93.33%	88.21%	73.98%	89.84%	96.34%
	Brisk Walking	97.78%	93.66%	93.23%	78.2%	92.48%	97.74%
	Jogging	98.01%	98.94%	99.04%	99.05%	99.05%	98.1%
	Upstairs	98.88%	98.23%	94.33%	98.38%	97.17%	94.33%
	Downstairs	98.88%	98.09%	98.72%	100.00%	99.15%	99.15%
	Jumping	100.00%	96.64%	98.29%	99.15%	100.00%	99.15%
	Overall	**97.91%**	96.4%	95.69%	92.33%	96.71%	97.74%
CA	Overall	**90.14%**	72.22%	76.67%	67.78%	81.11%	77.78%
SA/CA average		**94.03%**	84.31%	86.18%	80.06%	88.91%	87.76%

Table 3. Recognition accuracy of different sliding windows for activity scenes.

		Action window	1.28 s	2.56 s	4 s	6 s	8 s	10 s
Stable		97.91%	94.29%	95.41%	97.41%	97.12%	96.52%	97.91%
Changed	S	84.97%	65.71%	63.22%	65.46%	52.78%	37.04%	33.33%
	Q	95.75%	72.97%	69.57%	68.97%	60.53%	39.29%	31.82%

Performance Comparison of Different Windows. The experiment in this section compares the three types of stable activities, activity changes slowly (S) and activity changes quickly (Q), as shown in Table 3. To verify the improvement of the activity recognition rate, we employ window data of different durations, namely 1.28 s, 2.56 s, 4 s, 6 s, 8 s, and 10 s. The sliding step of them is 1/2.

According to the experiment results, when the activity is stable, the recognition ability of action window is slightly higher than that of the fixed-size sliding window. However, when the activity changes, the advantage of the former increases significantly.

As the window duration increases, the historical data has an increased influence on the feature extraction of the activity to be recognized. This results in a sharp drop in the recognition rate for fixed-size sliding windows.

When the activity changes slowly, such as trickling to walking, there is more interference data in the window because of the long switching time, resulting in a lower recognition rate. More importantly, the activity changes slowly, except for reasons of their own behavior habits, mostly because the front and back activities are similar. These changes further increase the difficulty of recognition. When the activity changes quickly, such as walking to upstairs, the difference between the front and back activities is generally large. As the activity changes quickly, the interference data in the window becomes less. The old activity is quite different from the new activity, so the recognition rate is increased.

Performance Comparison of Feature Matrix Dimensions. We adopt FETAD to extract features of the action data, and adopt C2DPCA to reduce the dimensionality of the three-axis feature matrix. According to the above parameter settings, the number of fine-grained subsegments is 6 and the number of features is 13. Thus, the number of features of each axis is $(6 - 1) \times 13 = 65$. The size of action feature matrix is 3×65. We compare different dimensionality reduction results, as shown in Fig. 3.

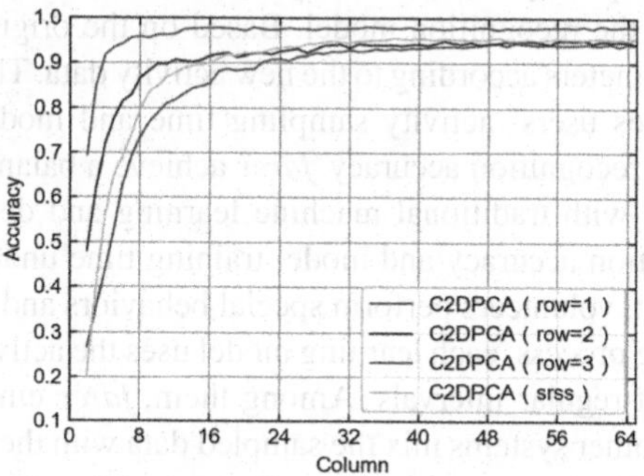

Fig. 3. Dimensionality reduction effect of *farer*.

Table 4. The maximum and minimum accuracy in the dimensionality reduction process.

C2DPCA	Row	Column	Accuracy (Max)	Column	Accuracy (Min)
flatten	1	51	94.39%	2	26.72%
flatten	2	59	95.08%	2	48.10%
flatten	3	61	98.28%	2	69.48%
srss	3	49	95.59%	2	20.22%

Figure 3 shows the recognition accuracy of the three dimensionality reduction curves. The dimensionality of the feature matrix is reduced from 3×65 to $3 \times n$, $2 \times n$, and

$1 \times n$, where $2 \leq n \leq 65$. Experimental results show that the recognition rate of $3 \times n$ is higher than that of the two categories. The experiment in this section further compares the difference between the two processing methods of flattening and square root of sum of squares (srss) when the row dimension is 3 after dimensionality reduction.

Table 4 compares the highest recognition rate and the lowest recognition rate when the feature dimension drops to different sizes, as well as the corresponding row and column values. It can be seen from the table that the row dimension of 3 has the best effect. Especially the lowest recognition rate is much higher than other cases. Considering the recognition accuracy and speed of *farer*, the column hash threshold and row hash threshold are both set to 99.95%. The size of the feature matrix after dimensionality reduction is 3×21. The recognition accuracy of the system is 97.91%.

Performance Comparison of Incremental Update. Some users' behavior habits are different from most people's common activities. For traditional neural network, a lot of special behavior data need to be sampled as new training samples. These new training samples are appended to the original training samples. Then the model is retrained, which takes a long time. Even worse, it requires users to perform special activities for a long time to obtain adequate behavior samples. Obviously, this update method brings a lot of trouble to users and is impractical.

To effectively and quickly recognition special activities of different users, *farer* incrementally updates the recognition model. Based on the original model, it directly updates the model parameters according to the new activity data. The incremental update of *farer* greatly reduces users' activity sampling time and model training time, and ensures the balance of recognition accuracy. *farer* achieve a balance of the three.

We compare *farer* with traditional machine learning and deep learning from the perspective of recognition accuracy and model training time under the same sampling time. In the experiment, volunteers perform special behaviors and acts continuously for 30 *min*. In the sampling process, each learning model uses the activity data of this period to update the model at regular intervals. Among them, *farer* employs an incremental update method, while other systems mix the sampled data with the original training data and retrain models.

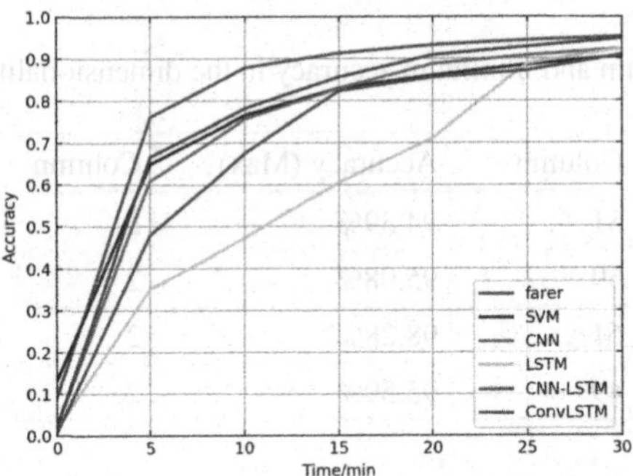

Fig. 4. Recognition accuracy of special activities with different durations

Figure 4 shows the recognition accuracy of special activities of different durations, with an interval of 5 min. It can be seen from the figure that the recognition accuracy of this system is the highest when the sampling time is the same. *farer* is growing faster than other methods.

5 Conclusion

By studying the problem of poor accuracy of activity recognition for sliding windows, we propose a fine-grained activity recognition method, which employs fine-grained subsegments and incremental broad learning. We also design the corresponding activity recognition system *farer*. The system can effectively process the activity data, realize the accurate segmentation of activities and ensure the effectiveness of the action feature extraction. Furthermore, the fine-grained subsegment division is used to dig deeply into the action features and reduce the dimensionality to ensure the recognition rate of the activity. In the process of activity recognition, the incremental recognition model based on broad learning is adopted to learn the activity behavior of special users, which improves the user experience. After a large number of experiments, the performance of *farer* is better than that of other recognition methods. The recognition rates for stable activities and changed activities are 97.91% and 90.14%.

Acknowledgments. This work is supported by National Natural Science Foundation of China under Grant No. 61772448, Natural Science Foundation of Jiangsu Province under Grant No. BK20191481, Natural Science Major Project of the Higher Education Institutions of Jiangsu Province of China under Grant No. 17KJA520006, China Postdoctoral Science Foundation funded project under Grant No. 2019M660132, Jiangsu Province Postdoctoral Research Subsidy Program under Grant No. 2019K123, Project on the Industry-University-Research Cooperation of Jiangsu Province under Grant No. BY2021326, Scientific Research Subsidy of Jiangsu Province under Grant No. BRA2019288, Future Network Scientific Research Fund Project Grant No. FNSRFP-2021-YB-45, Jiangsu Provincial Key Constructive Laboratory for Big Data of Psychology under Grant No. 72591962002G and 72592162003G, Open Foundation of Jiangsu Provincial Key Laboratory of Coastal Wetland Bioresources and Environmental Protection under Grant No. JLCBE14008.

References

1. Jiang, W., Miao, C., Ma, F., et al.: Towards environment independent device free human activity recognition. In: Proceedings of the 24th Annual International Conference on Mobile Computing and Networking, pp. 289–304 (2018)
2. Youke, W., Huang, H., Ningyun, W., Yue Wang, M., Bhuiyan, Z.A., Wang, T.: An incentive-based protection and recovery strategy for secure big data in social networks. Inf. Sci. **508**, 79–91 (2020)
3. Zhao, L.: Novel online sequential learning-based adaptive routing for edge software-defined vehicular networks. IEEE Trans. Wirel. Commun. (2020)
4. Zhao, L., Han, G., Li, Z., Shu, L.: Intelligent digital twin-based software-defined vehicular networks. IEEE Network (2020)

5. Zhao, L., Li, H., Lin, N., Lin, M., Fan, C., Shi, J.: Intelligent content caching strategy in autonomous driving towards 6G. IEEE Trans. Intell. Transp. Syst. (T-ITS) (2021)
6. Wang, T., et al.: Propagation modeling and defending of a mobile sensor worm in wireless sensor and actuator networks. Sensors **17**(1), 139 (2017)
7. Wang, T., Luo, H., Zeng, X., Yu, Z., Liu, A., Sangaiah, A.K.: Mobility based trust evaluation for heterogeneous electric vehicles network in smart cities. IEEE Trans. Intell. Transp. Syst. **22**(3), 1797–1806 (2020)
8. Wang, J., Chen, Y., Hao, S., Peng, X., Hu, L.: 'Deep learning for sensorbased activity recognition: a survey.' Pattern Recognit. Lett. **119**, 3–11 (2019)
9. Yang, J., Bang, W., Choi, E., et al.: A 3D Hand-drawn gesture input device using fuzzy ARTMAP-based recognizer. J. Syst. Cybern. Inf. **4**(3), 1–7 (2006)
10. Garcia-Ceja, E., Brena, R.: Long-term activity recognition from accelerometer data. Procedia Technol. **7**, 248–256 (2013)
11. Khan, A.M., Lee, Y.K., Lee, S., et al.: Accelerometer's position independent physical activity recognition system for long-term activity monitoring in the elderly. Med. Biol. Eng. Comput. **48**(12), 1271–1279 (2010)
12. Fida, B., Bernabucci, I., Bibbo, D., Conforto, S., Schmid, M.: Varying behavior of different window sizes on the classification of static and dynamic physical activities from a single accelerometer. Med. Eng. Phys. **37**(7), 705–711 (2015)
13. Elsts, A., Twomey, N., Mcconville, R., et al.: Energy-efficient activity recognition framework using wearable accelerometers. J. Network Comput. Appl. **168,** 102770 (2020)
14. Hossain Shuvo, M.M., Ahmed, N., Nouduri, K., Palaniappan, K.: A hybrid approach for human activity recognition with support vector machine and 1D convolutional neural network. In: 2020 IEEE Applied Imagery Pattern Recognition Workshop (AIPR), pp. 1–5 (2020)
15. Xia, K., Huang, J., Wang, H.: LSTM-CNN architecture for human activity recognition. IEEE Access **8**, 56855–56866 (2020)
16. Cha, S.H., Seo, J., Baek, S.H., Koo, C.: Towards a well-planned, activity-based work environment: automated recognition of office activities using accelerometers. Build. Environ. **144**, 86–93 (2018)
17. Pienaar, S.W., Malekian, R.: Human activity recognition using LSTM-RNN deep neural network architecture. In: 2019 IEEE 2nd Wireless Africa Conference (WAC), pp. 1–5 (2019)
18. Gao, W., Zhang, L., Teng, Q., et al.: DanHAR: dual attention network for multimodal human activity recognition using wearable sensors. Appl. Soft Comput. **111,** 107728 (2021)
19. Panwar, M., et al.: CNN based approach for activity recognition using a wrist-worn accelerometer. In: Proceedings of EMBC, Seogwipo, South Korea, pp. 2438–2441, July 2017
20. Lee, S.-M., Yoon, S.M., Cho, H.: Human activity recognition from accelerometer data using Convolutional Neural Network. In: 2017 IEEE International Conference on Big Data and Smart Computing (BigComp) (2017)
21. Mutegeki, R., Han, D.S.: A CNN-LSTM approach to human activity recognition. In: 2020 International Conference on Artificial Intelligence in Information and Communication (ICAIIC), pp. 362–366 (2020)
22. Ordóñez, F.J., Roggen, D.: 'Deep convolutional and LSTM recurrent neural networks for multimodal wearable activity recognition.' Sensors **16**(1), 115 (2016)
23. Chen, H., et al.: Assessing impacts of data volume and data set balance in using deep learning approach to human activity recognition. In: 2017 IEEE International Conference on Bioinformatics and Biomedicine (BIBM), pp. 1160–1165 (2017)
24. Fu, Z., He, X., Wang, E., et al.: Personalized human activity recognition based on integrated wearable sensor and transfer learning. Sensors **21**(3), 885 (2021)
25. Hong, J.H., Ramos, J., Dey, A.K.: Toward personalized activity recognition systems with a semipopulation approach. IEEE Trans. HumMach. Syst. **46**(1), 101–112 (2015)

ADFA-LSTM: An Abnormal Trajectory Prediction Method Based on Bionic Neural Network

Yan Wang[1,2](✉), Gang Cui[1], Junqiang Zhou[1], and Zhiheng Han[1]

[1] College of Computer Science, Inner Mongolia University, Hohhot 010021, China
cswy@imu.edu.cn

[2] Inner Mongolia Engineering Laboratory for Cloud Computing and Service Software, Hohhot 010021, China

Abstract. In order to detect the state of vehicle trajectory, an abnormal prediction method of vehicle trajectory based on bionic neural network, called ADFA-LSTM, is proposed according to the temporal characteristics and position relationship of vehicle trajectory. The long-short memory neural network (LSTM) is used to predict the trajectory in the short term based on the characteristics of the vehicle's historical trajectory, and the trajectory prediction results are analyzed to identify the abnormal trajectory. In the process of trajectory prediction, the improved biomimetic firefly algorithm (FA) is used to optimize the prediction the parameters of LSTM model, which improves its processing efficiency and enhances its fitting ability. Through a lot of experimental analysis, the detection effect of the algorithm is verified.

Keywords: Trajectory · LSTM · Firefly algorithm · Anomaly detection

1 Introduction

With the development of Global Positioning System (GPS), wireless technology and location-aware services, a large amount of trajectory data has now been collected. In the field of data mining of moving objects, the problem of trajectory anomaly detection is a hot topic [1]. How to dig out the behaviors and factors related to road traffic safety, and give warnings to the safety behavior of vehicles, so as to effectively reduce traffic violations and reduce the incidence of accidents, has become an urgent problem in the process of modern transportation [2]. According to the driving data of road vehicles, the abnormal identification of the vehicle trajectory is carried out, and the establishment of a driving safety monitoring model can more comprehensively and accurately detect and predict the dangerous state that may cause a traffic accident, and provide a valuable reference for vehicle safety warning. Therefore, in-depth research on driving safety monitoring technology has important significance and use value.

In this paper, we mainly analyze the trajectory data through the bionic deep learning model, identify the abnormal trajectory of the vehicle, and evaluate the driving safety level of the vehicle. The contributions of the paper are summarized as follows. (1)

Y. Lai et al. (Eds.): ICA3PP 2021, LNCS 13155, pp. 115–128, 2022.
https://doi.org/10.1007/978-3-030-95384-3_8

The neural network structure of LSTM model is optimized by the brightest individual disturbance adaptive firefly algorithm (ADFA), and the vehicle trajectory prediction model is established based on it. (2) The trajectory prediction results are analyzed in combination with the trajectory characteristics, and the abnormal trajectory evaluation index is given to judge the situation of vehicle trajectory to identify abnormal trajectories.

2 Related Work

Anomaly detection of traffic trajectory is a hot research topic. Traditional mathematical model and machine learning model are two important methods for anomaly detection of traffic trajectory.

Among them, the traditional mathematical model mostly adopts statistical probability method, and its core idea is multi-detection fusion strategy, which comprehensively judges different motion trajectories, usually including Hidden Markov Statistical Model (HMM), Bayesian decision probability model and Linear regression and function approximation model [3–5]. Zhang et al. measures the similarity between trajectories by calculating the relative similarity, and then uses the Gaussian model to match the new trajectory data to detect whether the trajectory is abnormal or not [6]. However, the correlation between trajectories is not considered, so the efficiency of abnormal detection is not high. Pang et al. adopts a Bayesian deep learning method to effectively predict the trajectory [7]. Due to the difference of features of trajectories, Venkatesan considers the global features of trajectories, including direction, speed and distance. The abnormal trajectories are found by measuring the distance difference, speed change and direction correlation between trajectory points [8]. But the method ignores the local difference between trajectories. Wonjik et al. uses a path prediction method of self-organizing incremental neural network for possible errors, and evaluates the errors effectively [9–11].

Machine learning model has the advantages of fast processing speed and strong learning ability, and it's applicable in processing nonlinear mapping data such as traffic trajectory [12, 13]. Kong etc. comprehensively considers factors such as longitude, latitude and vehicle route, uses the particle swarm optimization algorithm to optimize parameters of support vector machine, and predicts vehicle speed based on this model, and gives regional congestion level [14]. Sovan et al. extracts some features of trajectory points under different tags, and then detects abnormal trajectories in real-time trajectories data [15]. Debat uses a neural network with peak-time-related plasticity rules to train in a supervised manner to predict trajectories [16]. However, these methods simply consider the characteristics of trajectory points, without considering the continuity of trajectory sequence, so the effect is limited. De et al. decompose GPS data into multiple indefinite trajectory sequences, and uses the time series learning ability of B-RNN neural network to identify the overall characteristics of historical trajectory, and the detection effect is better [17].

According to the above-mentioned trajectory anomaly detection work, the current research on trajectory prediction and anomaly detection mainly focuses on the research on trajectory prediction methods, often using empirical value to select feature parameters and forcibly truncate long trajectories, less considering the impact of trajectory factors on the model, such as the problem of factor selection and the problem of variable length trajectory sequence, which have a great impact on the accuracy of model prediction. Therefore, this paper proposes a bionic neural network model based on trajectory timing features. Firstly, the influence degree of each attribute of trajectory on the model is analyzed. Secondly, the embedding layer is used to complete and align the timing trajectory sequence, and then the vehicle trajectory information is predicted by the improved LSTM model. Finally, the anomaly analysis of trajectory data is completed.

3 Trajectory Prediction Analysis Based on ADFA-LSTM

3.1 Definitions

The collection of a series of points formed by moving objects in space motion is called spatial-temporal trajectory, which generally includes longitude and latitude, time, direction, speed, etc. Based on the features of spatio-temporal trajectory, this paper gives the following definitions.

Definition 1 Trajectory (T): The spatio-temporal trajectory sequence T is a set of points with some information such as time stamps and positions, that is, T = $<p_1, p_2, ..., p_i, ...p_n>$. T represents a sequence of trajectory points formed by a vehicle and contains n trajectory points.

Definition 2 Trajectory point (P_i): The point P_i is a position in the trajectory, typically $P_i = <x_i, y_i, s_i, h_i, t_i> (i = 1, 2, 3, \cdots, m)$, x_i, y_i is the spatial coordinate (longitude and latitude), v_i is the speed of P_i, h_i is the height of P_i, t_i is the timestamp.

Definition 3 Trajectory length (T.length): The length of a trajectory refers to the number of trajectory points contained in a trajectory sequence from the start point to the end point.

3.2 Trajectory Prediction Model Based on LSTM

As a variant of RNN neural network, LSTM solves the defect of its long-term dependence by adding four interaction layers, which can perform better in longer time series and has great advantages in processing traffic trajectory. Therefore, this paper selects LSTM model for vehicle trajectory prediction.

Figure 1 shows the trajectory prediction model based on LSTM established in this paper, mainly including four layers of structure: input layer, embedded layer, hidden layer and output layer.

(1) Input layer. The input layer is the input sequence of LSTM network, and the transmitted data is the original vehicle trajectory data collected through GPS. Each trajectory data represents a travel path of a vehicle, which is composed of n $(0 < n \leq T.length_{\max})$ trajectory points, in which each trajectory point contains timestamp, longitude, latitude, height, speed and so on. Each trajectory has different length, so it needs to be aligned.

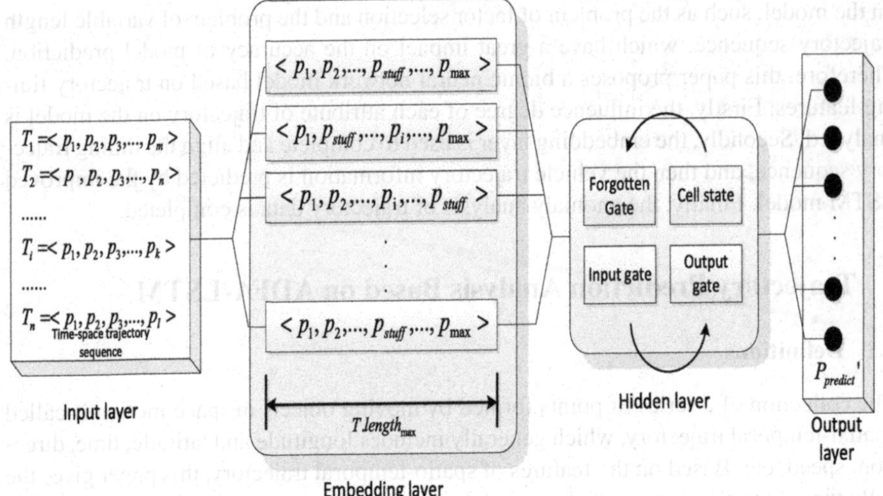

Fig. 1. LSTM trajectory prediction model

(2) Embedded layer. This layer transforms high-dimensional sparse data into low-dimensional dense vectors. In this paper, the embedded layer is used to convert the discrete data of the input layer into a vector with fixed size to solve the input problem of variable length sequence of LSTM model. For LSTM model, the variable-length trajectory cannot be directly trained, so it is necessary to combine the embedding layer to vectorize the trajectory data and convert it into a fixed-length trajectory sequence.

(3) Hidden layer. The training and learning of LSTM for trajectory sequence is mainly carried out in the hidden layer, and the structure includes input gate, output gate, forget gate and cell state.

(4) Output layer. The output layer obtains the position point prediction sequence by processing the data generated by the hidden layer.

In general, the calculation process of LSTM prediction model in this paper can be summarized as follows: combined with the historical trajectory input and current trajectory, the useful information for subsequent trajectory can be transmitted through forgetting and memorizing new information in cell state, while the useless information is discarded.

3.3 Improved Firefly Algorithm Optimizing LSTM Neural Network

Aiming at the defects of slow convergence speed and complex parameter adjustment process in the training of LSTM model, this paper optimizes the network structure by firefly algorithm (FA) to improve the prediction performance. Because the firefly algorithm itself also has the problems of easy falling into local optimization and easy oscillation, this paper establishes an adaptive FA algorithm based on the brightest individual disturbance (ADFA). The improvement of ADFA for FA algorithm mainly includes two aspects:

a) Introduce the update strategy of the brightest individual perturbation position based on the diversity of the population, and enlarge the global search range of the firefly algorithm to prevent falling into the local optimum. Since all fireflies will look for the brightest individual to move in each iteration, the brightest individual will not be updated before the brighter individual appears. Therefore, a disturbance mechanism is added to the brightest individual to actively seek a position with a higher fitness value.

b) Using an adaptive function, the step factor is adaptively changed with the number of iterations, so that the algorithm pays more attention to the global search in the early stage, and pays more attention to the local search in the later stage to solve the problem of continuous oscillation near the optimal solution in the later stage of the search.

Its core processes for LSTM parameter optimization include:

Step 1: Initialize algorithm parameters, including the firefly population size P, problem dimension N, maximum attraction β, individual disturbance factor S, light intensity absorption coefficient γ, initial step factor α, maximum iteration number *max* and search accuracy W;

Step 2: Initialize that individual position of the population, and calculate the individual brightness of the firefly through the objective function;

Step 3: Determine the individual moves in a brighter direction through the brightness difference between the fireflies and the attractiveness value β;

Step 4: Update the position according to the individual disturbance factor S;

Step 5: Recalculate the brightness of the fireflies according to the differences in the positions of the updated individuals;

Step 6: If the maximum number of iterations *max* is reached or the search accuracy W is satisfied, go to step 7; otherwise, continue iteration, use the adaptive function to determine the step size, go to step 3, and start the next search;

Step 7: Output the optimal firefly individual set, that is, the LSTM optimal parameter set.

3.4 ADFA-LSTM Trajectory Prediction Algorithm

Because GPS trajectory data has the characteristics of time and position information before and after, the LSTM neural network model has more advantages than other neural network algorithms, but it still has problems such as slower parameter search and easy fall into local optimality. Therefore, in our paper, ADFA algorithm is used to optimize its parameters, and then the vehicle historical trajectory data is used as input for training to establish a vehicle trajectory prediction model. The model framework of the algorithm is shown in Fig. 2.

The model mainly includes three stages: the first stage mainly selects feature attributes and sequence optimization on the historical trajectory of the vehicle; the second stage optimizes the LSTM model parameters through the ADFA algorithm, and determines the LSTM optimal parameter set based on the brightest individual firefly iterative optimization; in the final stage, train the LSTM model according to the historical trajectory to predict the position at the next moment, and add the predicted data to the data set for continuous position prediction to obtain the final prediction result. According to the above process, the algorithm is described as follows (Table 1):

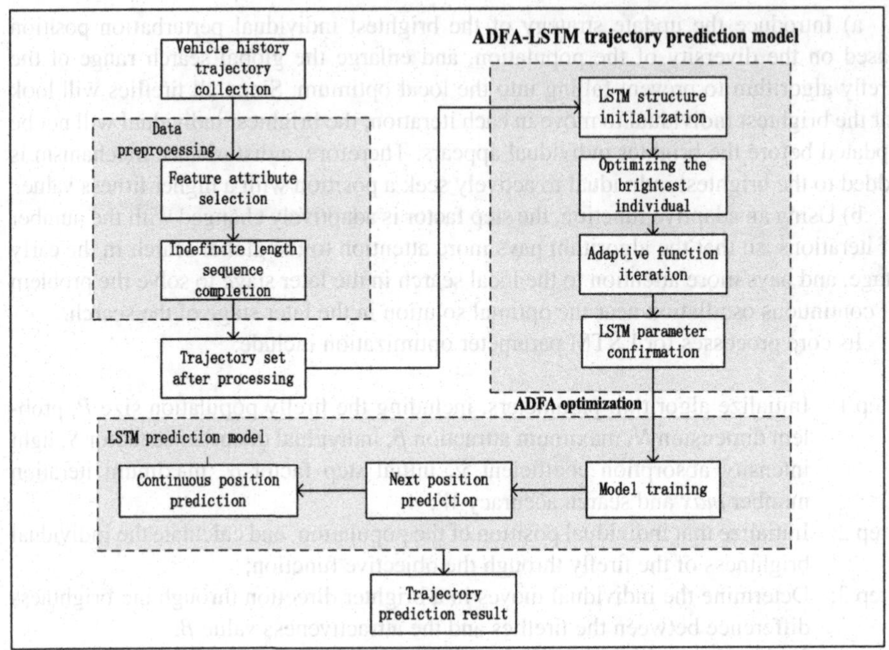

Fig. 2. ADFA-LSTM model frame diagram

Firstly, the trajectory sequence Traj_i is aligned through the embedding interface, and then the preprocessed trajectory sequence is divided into a training set and target set, which are input into LSTM neural network for training. The ADFA algorithm is used to optimize the training parameters. Firstly, find NewBest () is used to implement the brightest individual disturbance location update strategy, and then the FitnessSelf-Function() is used to recalculate the fitness value of the firefly to speed up the location update process until the iteration conditions are met, and the LSTM optimal parameter set is obtained. Finally, the historical trajectory sequence to be predicted is input into the optimized LSTM model to obtain the final trajectory prediction result.

Table 1. Trajectory prediction algorithm based on ADFA-LSTM

Algorithm 3.1 Trajectory Prediction Algorithm Based on ADFA-LSTM

Input: vehicle trajectory sequence set $Traj_{SA}$

Output: Future vehicle location $P_{prediction}$

```
for each Traj_i ∈ Traj_SA do
  Traj_i = embedding(Traj_i) // Time series data completion
  for (k in st: N)
    DataSet trainSet = Traj_i.subTraj(0, k-1);
                                        // Sliding window dynamic training
    DataSet targetSet = Traj_i.subTraj(k-1, k);
    LSTM.train (trainSet,targetSet);
    FA.train ()← LSTM.train ();//LSTM parameter optimization
      While (t < MaxGeneration)
        FA.best ()← FindNewBest ();
                              // The brightest individual perturbation optimization
        FA.update ()← FitnessSelfFunction ();
                                        // fast iteration of adaptive function
        LSTM.train ()←FA.best ();
    end for
end for
DataSet paramSet = embedding ( T_current);
P_prediction = LSTM.learn (paramSet);
```

3.5 Abnormal Trajectory Detection

After the trajectory prediction model is used to obtain the prediction information of the trajectory, the trajectory features are then extracted to analyze the trajectory prediction results, and the abnormal conditions in the trajectory can be judged by detecting the difference between the normal and frequent trajectory. The specific detection ideas are as follows:

(1) Based on the ADFA-LSTM trajectory prediction model, extract the key features of the trajectory to be measured, and establish anomaly detection classification labels based on the trajectory data and road features.
(2) Based on the trajectory characteristic, the anomaly evaluation index is established, and the detection threshold of the evaluation index is set by analyzing the characteristic laws and differences between normal frequent trajectories and different abnormal trajectories.
(3) By comparing the difference between the characteristic index predicted by the trajectory and the threshold value of the trajectory abnormality evaluation index, the abnormality of the trajectory to be measured is analyzed, and the final trajectory abnormality detection result is obtained.

Next, an example is given to explain our estimation method of abnormal trajectory. Through the historical trajectory data, the difference analysis of the evaluation indicators is carried out, and the threshold range of all trajectory abnormal detection evaluation indicators is obtained. In this paper, 50 groups of trajectory segments are selected and brought into the ADFA-LSTM model for training, of which 30 groups are normal trajectories from the historical vehicle trajectory data set collected by GPS; 20 groups are abnormal trajectory segments, including 10 groups of historical vehicle data generated by GPS alarm abnormal trajectory segment (yaw alarm, speeding alarm, etc.), 10 groups are simulated road yaw abnormal trajectory segments. Analyze the characteristic of the trajectories through the value range and the overall trend of the characteristic indicators of different trajectory segments. The error rate and mean square error of the normal trajectory and abnormal trajectory are shown in Fig. 3.

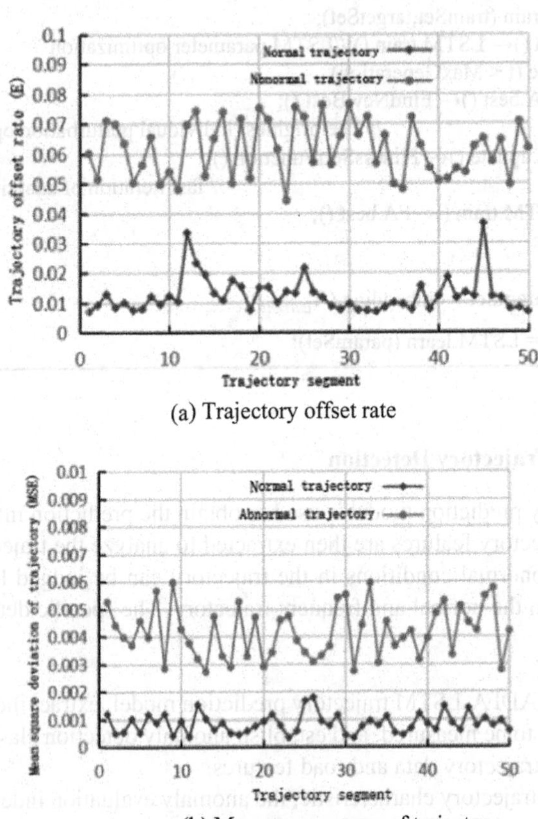

(a) Trajectory offset rate

(b) Mean square error of trajectory

Fig. 3. Threshold range of indicators for different track segments

According to Fig. 3, it can be seen that there are obvious differences in the value ranges of the two types of indicators for the normal trajectory segment and the abnormal trajectory segment. Among them, the value range of the trajectory offset rate (E) of the normal trajectory segment is basically maintained in the interval [0, 0.03], and the value range of the mean square error (MSE) is basically maintained in the range of [0, 0.0015]. The value range of the trajectory deviation rate (E) of the abnormal trajectory segment fluctuates in the interval of [0.04, 0.08], and the value range of the mean square error (MSE) fluctuates in the interval of [0.003, 0.006]. Therefore, it can be assumed that when the trajectory deviation rate E < 0.04 and the mean square error MSE < 0.002, the deviation between the measured trajectory segment and the normal frequent trajectory is small, and the trajectory is judged to be a normal state. When the trajectory deviation rate E > 0.08 or the mean square error MSE > 0.007, it is judged that there may be abnormal data in the trajectory segment. When the value of the trajectory deviation rate E is in the interval [0.04, 0.08] or the value of the mean square error MSE is in the interval [0.002, 0.007], the deviation between the measured trajectory segment and the normal frequent trajectory is large, and the vehicle may have unsafe behavior, which is judged that the trajectory is abnormal.

4 Experiments and Analysis

4.1 Experimental Data Set and Evaluation Criteria

According to the vehicle trajectory data and road characteristics, this paper divides the abnormal trajectory detection results into two categories: data abnormality and trajectory abnormality. Data abnormalities include various GPS data errors caused by the process of collection, transmission and storage; and abnormal trajectories are based on the true and reliable GPS data, and the trajectory characteristics are quite different from the normal frequent trajectory. This difference is certain. To a certain extent, it implies the behavior of the vehicle's abnormal driving.

In this paper, the mean square error (MSE) is used to verify the prediction ability of the model. The smaller the value of MSE is, the better the accuracy of the prediction model is. At the same time, the trajectory deviation rate (E) is introduced and combined with MSE to judge the deviation degree between the predicted position and the normal trajectory, and used as the evaluation index of trajectory anomaly detection.

The formula of trajectory offset rate is as follow:

$$E = \frac{P_{\text{real}} - P_{predict}}{P_{real}/100} \tag{1}$$

The formula of mean square error is as follow:

$$MSE = \frac{1}{N}\sum_{t=1}^{N}\left(P_{real_t} - P_{predict_t}\right)^2 \tag{2}$$

Where N is the number of samples, P_{real} is the normal value, and $P_{predict}$ is the predicted value.

4.2 Test and Analysis of Traffic Trajectory Anomaly Detection Model

This section mainly includes two experiments. The first experiment is the improvement experiment of FA algorithm. The FA algorithm is optimized by different methods and applied to the LSTM prediction model to test the effectiveness of ADFA algorithm. The second experiment is the comparison of ADFA-LSTM and other common LSTM improved algorithms to test the prediction accuracy of the model and verify the recognition ability of the model.

The data used in the experiment in this section is from the GPS trajectory data of operating vehicles collected in the actual project, including about 40,000 trajectories. These data can be used as experimental data in this paper because they represent the route and topological structure of most operating vehicles.

Table 2. Vehicle GPS track data segment

UnitID	GPS Date	GPSTime	Longitude	Latitude	Height (m)	Speed (km/h)
6	20180403	16:59:14	111.546637	41.285371	916	60
6	20180403	16:59:44	111.547987	41.289561	917	61
6	20180403	17:00:14	111.549214	41.294636	912	56
6	20180403	17:00:44	111.548915	41.299101	911	55
6	20180403	17:01:14	111.549035	41.301745	891	65
6	20180403	17:01:44	111.549681	41.303357	889	58

Table 2 shows a data segment of a trajectory generated on April 3, 2018 for a vehicle with GPS equipment No. 6. A complete vehicle trajector contains amount of position points, each of which contains information such as vehicle ID, timestamp, longitude and latitude, altitude and speed.

The first experiment uses different methods to optimize the standard FA algorithm, and applies it to the LSTM prediction model to test the effectiveness of ADFA algorithm to improve the LSTM network structure.

The standard firefly algorithm (FA) mainly includes three steps: population initialization, population flight to brighter individuals and determination of optimal firefly position. However, the fixed step factor and the brightest individual remain unchanged during initialization, which will lead to the oscillation of the result, and it is difficult to achieve the optimal effect. Therefore, this paper adds the adaptive step factor and the brightest individual to optimize the FA algorithm, in which the brightest individual is mainly perturbed once and perturbed N times respectively, and the adaptive step factor is calculated by two adaptive functions, respectively:

$$\alpha(t+1) = \left(1 - \frac{t}{T_{\max}}\right) \cdot \alpha(t) \tag{3}$$

$$\alpha(t+1) = \alpha(t) \cdot \exp\left(-\frac{k \cdot t}{T_{\max}}\right) \tag{4}$$

Formula (3) and Formula (4) represent adaptive function 1 and adaptive function 2 respectively. In this paper, the ADFA algorithm selects the brightest individual disturbance once in combination with adaptive function 2 to optimize the standard firefly. In the comparison experiment, the original FA algorithm is used to establish the FA-LSTM model, and the FFA1-LSTM and FFA2-LSTM models are established based on adaptive functions 1 and 2 respectively. The brightest individual behavior is improved by perturbation once and disturbance N times, and the DFA1-LSTM and DFA2-LSTM models are established respectively. Comparison results of final model prediction mean square error are shown in Fig. 4.

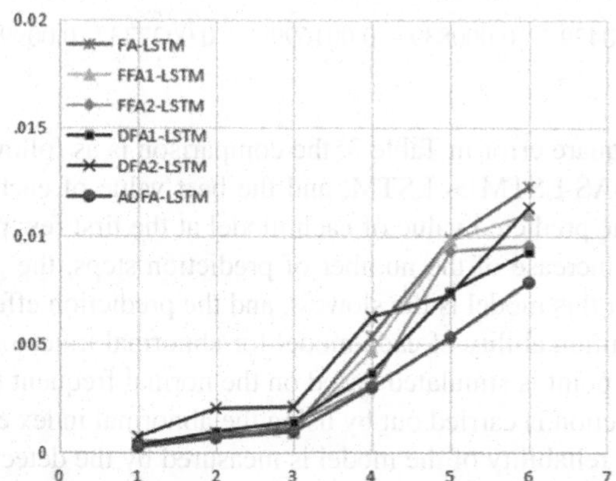

Fig. 4. Comparison of average error of FA algorithm optimization model

According to the growth rate of overall mean square error in Fig. 4, the order is: FA-LSTM > FFA1-LSTM > DFA2-LSTM > FFA2-LSTM > DFA1-LSTM > ADFA-LSTM. It can be seen that the growth rate of the unmodified FA-LSTM model is the fastest, followed by the FFA1-LSTM model and the DFA2-LSTM model with the adaptive function 1 and the brightest individual disturbance N times, and the growth rate of the FFA2-LSTM model and the ADFA1-LSTM model with the adaptive function 2 and the brightest individual disturbance once is relatively slow. It can be seen that the optimization effect of the adaptive function 2 is better than that of the adaptive function 1 after playing the role of the local exploration of the optimal individual. The prediction effect after adding the disturbance factor is obviously smaller than that of the unmodified FA algorithm, and the prediction accuracy of the first disturbance is higher. The ADFA algorithm proposed in this paper has better optimization effect on the LSTM model.

The second experiment is the comparison between ADFA algorithm and other commonly used LSTM improved algorithms, which verifies that the proposed ADFA-LSTM algorithm has better prediction effect, and then tests the recognition ability of each model for abnormal trajectory. In this paper, particle swarm optimization (PSO) and longicorn whisker algorithm (BAS) are selected as the comparison algorithm. In order to verify the effectiveness of the improved algorithm, four models, LSTM, PSO-LSTM, BAS-LSTM and ADFA-LSTM, are established in this paper for experimental test to verify the optimization ability of each algorithm to the model. The mean square error of the experimental results is shown in Table 3.

According to the trend of prediction results, the overall trend of this model is closer to the real value, and the prediction difference is the smallest. According to the accurate

Table 3. Prediction mean square error of LSTM improved model

Mean square error	First point	Second point	Third point	Fourth point	Fifth point	Sixth point
LSTM	0.000587	0.000987	0.001637	0.002867	0.009878	0.025442
ADFA-LSTM	**0.000288**	0.000588	**0.001293**	**0.001723**	**0.003694**	**0.008584**
PSO-LSTM	0.000397	**0.000579**	0.001336	0.003824	0.007331	0.016439
BAS-LSTM	0.000439	0.000839	0.001699	0.002773	0.009967	0.023831

value of the mean square error in Table 3, the comparison is as follows: ADFA-LSTM > PSO-LSTM > BAS-LSTM > LSTM, and the best value of each step represented by bolding is that the predicted value of each model at the first few points is relatively close. But with the increase of the number of prediction steps, the growth rate of the mean square error of this model is the slowest, and the prediction effect is more stable.

Next, the recognition ability of each model for abnormal trajectory is verified. The abnormal trajectory point is simulated based on the normal frequent trajectory, and the abnormal point detection is carried out by using the abnormal index estimation method in Sect. 3.5, and the reliability of the model is measured by the detection accuracy rate of abnormal points and the false detection rate of normal points. The specific formula is as follows:

$$\text{Outlier accuracy} = \frac{\text{Correctly identify the number of abnormal points}}{\text{Total number of abnormal points}} \times 100\% \quad (5)$$

$$\text{False detection rate of normal points} = \frac{\text{Number of false detection points}}{\text{Total number of normal points}} \times 100\% \quad (6)$$

The test results of each model are shown in Table 4.

Table 4. Abnormal trajectory detection results of LSTM improved model

Model name	Track offset rate		Mean square error		Outlier accuracy	False detection rate at normal points
	Normal point	Outliers	Normal point	Outliers		
LSTM	0.036244	0.046825	0.001623	0.003246	77.3%	3.89%
ADFA-LSTM	0.019386	0.053613	0.000541	0.004677	89.1%	1.44%
PSO-LSTM	0.026832	0.050388	0.000977	0.004108	85.8%	1.87%
BAS-LSTM	0.036244	0.049931	0.001038	0.003986	81.4%	2.68%

According to the test results in Table 4, compared with other models, the ADFA-LSTM model in this paper has the lowest evaluation index value (including trajectory deviation rate and mean square error) at the normal point, while the evaluation index value at the abnormal point is higher. Since the predicted value of the model is more biased towards the normal value in the training set, the higher the evaluation index value of the abnormal point, the easier it is to reach the abnormal detection threshold, and then the abnormal trajectory point can be identified earlier and more accurately. The trajectory points have strong recognition ability.

5 Conclusion

With the rapid development of the economy, the number of road traffic mileage and highway density are continuously increasing, and the problems of road traffic safety are becoming more and more prominent. In recent years, the number of casualties caused by traffic problems has consistently ranked first in the world. How to dig out the behaviors and factors related to road traffic safety to realize the early warning of the safety behaviors of vehicles so as to reduce the incidence of traffic violations and accidents, is a very hot topic. In this paper, aiming at the problems caused by the complexity of vehicle GPS trajectory for abnormal trajectory analysis, a LSTM trajectory prediction model based on bionic neural network is proposed. At the same time, in order to solve the shortcomings of LSTM easy to fall into local optimality, an adaptive function is proposed. The Firefly Algorithm (ADFA) perturbed by the brightest individual optimizes the LSTM network structure, and recognizes abnormal trajectories based on the prediction results. Finally, the effectiveness of the proposed method is verified by experiments.

Acknowledgments. This work was supported in part by Natural Science Foundation of China under Grants 62162047, Natural Science Foundation of Inner Mongolia under Grants 2019ZD15, 2019MS06029, Inner Mongolia Science and Technology Plan Project under Grant 2021GG0155, 2019GG372, the Self-topic of Engineering Research Center of Ecological Big Data, Ministry of Education, the Open-topic of Inner Mongolia Big Data Laboratory for Discipline Inspection and Supervision (IMDBD2020012), Inner Mongolia Engineering Laboratory for Cloud Computing and Service Software, Inner Mongolia Key Laboratory of Social Computing and Data Processing, Inner Mongolia Engineering Lab of Big Data Analysis Technology.

References

1. Zhang, T., Song, W., Fu, M., Yang, Y., Wang, M.: Vehicle motion prediction at intersections based on the turning intention and prior trajectories model. IEEE/CAA J. Automatica Sinica **8**(10), 1657–1666 (2021)
2. Sharma, B., Katiyar, V.K., Kumar, K.: Traffic accident prediction model using support vector machines with Gaussian kernel. Adv. Intell. Syst. Comput. **437**, 1–10 (2016)
3. Jirong, X., Yuan, L.: Azimuth-only trajectory detection method based on hidden Markov model. Syst. Eng. Electron. Technol. **38**(07), 1496–1501 (2016)
4. McCarthy, D.M., et al.: Applying Bayesian cognitive models to decisions to drive after drinking. Addiction **116**(6), 1424–1430 (2021)
5. Kim, S., Lee, K.: En-route trajectory prediction via weighted linear regression. J. Korean Soc. Aviat. Aeronaut. **24**(4), 44–52 (2016)

6. Zhang, X., Zhu, Z., Lin, H., et al.: Trajectory prediction algorithm based on Gaussian mixture-variational self-encoder. Comput. Eng. **46**(07), 50–57 (2020)
7. Pang, Y., Zhao, X., Yan, H., Liu, Y.: Data-driven trajectory prediction with weather uncertainties: a Bayesian deep learning approach. Transp. Res. Part C Emerg. Technol. **130**, 103326 (2021)
8. Kanagaraj, V., et al.: Trajectory data and flow characteristics of mixed traffic. Transp. Res. Rec. **2491**(1), 1–11 (2015)
9. Kim, W., Hasegawa, O.: Time Series prediction of tropical storm trajectory using self-organizing incremental neural networks and error evaluation. J. Adv. Comput. Intell. Intell. Inf. **22**(4), 465–474 (2018)
10. Lago-Rodriguez, A., Miall, R.C.: Online visual feedback during error-free channel trials leads to active unlearning of movement dynamics: evidence for adaptation to trajectory prediction errors. Front. Hum. Neurosci. **10**, 472 (2016)
11. Zhang, X., et al.: Evaluation of multi - source forcing datasets for drift trajectory prediction using Lagrangian models in the South China Sea. Appl. Ocean Res. **104**, 1–20 (2020)
12. Sun, S., Chen, J., Sun, J.: Traffic congestion prediction based on GPS trajectory data. Int. J. Distrib. Sens. Netw. **15**(5), 1–18 (2019)
13. Karlaftis, M.G., Vlahogianni, E.I.: Statistical methods versus neural networks in transportation research: differences, similarities and some insights. Transp. Res. Part C Emerg. Technol. **19**(3), 387–399 (2011)
14. Kong, X., Xu, Z., et al.: Urban traffic congestion estimation and prediction based on floating car trajectory data. Fut. Gener. Comput. Syst. **61**, 97–107 (2016)
15. Biswas, R.S., Babu, V.: Anomaly detection via short local trajectories. Neurocomputing **242**, 63–72 (2017)
16. Debat, G., Chauhan, T., Cottereau, B.R., Masquelier, T., Paindavoine, M., Baures, R.: Event-based trajectory prediction using spiking neural networks. Front. Comput. Neurosci. **15**, 658764 (2021)
17. De Brebisson, A., Simon, E., Auvolat, A., et al.: Artificial neural networks applied to taxi destination prediction. Computer Science, pp. 1–18 (2015)

Online Multiple Object Tracking Algorithm Based on Heat Map Propagation

Wei Ding[1,2], Haokai Hong[1,2], Dejun Xu[1,2], and Min Jiang[1,2(✉)]

[1] School of Informatics, Xiamen University, Fujian 361005, China
minjiang@xmu.edu.cn

[2] Key Laboratory of Digital Protection and Intelligent Processing of Intangible Cultural Heritage of Fujian and Taiwan, Ministry of Culture and Tourism, Fujian, China

Abstract. Online multiple object tracking is an important computer vision task with a wide range of application scenarios which integrates target detection, re-ID (Re-identification) and feature association matching. Most of the existing methods have two problems: ignoring the spatio-temporal information, resulting in poor tracking effect, and using two-stage detector makes the tracking speed slow. In this paper, we design an online multiple object tracking algorithm by generating a heat map based on the spatio-temporal information of the tracker, and propagate it to a one-stage detector to improve tracking speed and stabilize detection quality. Without introducing additional regressors, the generation of the propagation heat map are both simply and efficiently. Consequently, the proposed algorithm can achieve a balance between the speed and accuracy, and provides a paradigm for the utilization of the spatio-temporal characteristic information of the one-stage detector. Experiments are carried out on MOT-17 and 2DMOT-15 which verifies that 43.27% and 63.7% improvement in tracking speed is obtained with a small accuracy compromise.

Keywords: Online multiple object tracking · Video object detection · Spatio-temporal information · Heat map propagation

1 Introduction

Online multiple object tracking (OMOT), which refers to a task of continuously tracking multiple objects at the same time, is very common in [5, 18, 29, 30] video surveillance, traffic monitoring, automatic driving and many other real-world applications. OMOT is an attractive and challenging task in computer vision community which integrates three tasks: target detection, re-ID [5] and feature association matching. Specifically, the task requires us to mark the position of the target in the image with a frame similar to the size of the target, and record the motion trajectory of the same target with a specific index in a group of consecutive frames.

In order to obtain the position and trajectory of the tracked target in the online situation, that is, only the video stream data available for the current frame and the past frame, the existing algorithms can be divided into two groups: multiple object

Y. Lai et al. (Eds.): ICA3PP 2021, LNCS 13155, pp. 129–143, 2022.
https://doi.org/10.1007/978-3-030-95384-3_9

tracking algorithm with detector and multiple object tracking algorithm without detector. SORT [18] (Simple Online and Realtime Tracking) is a representative algorithm without detector which uses the existing multiple object detection algorithms (such as [4, 10, 19]) to obtain the position of the target. Then, some association matching strategies are used to obtain the trajectory of the target being tracked. With the development of multi task learning [3, 5, 8, 11, 12, 20], multiple object tracking algorithms with detectors such as Fair MOT [11] have emerged. The integration of appearance feature extraction and target detector enables Fair MOT to significantly improve the speed of tracking multiple object.

However, room for improvement on the OMOT algorithms still remains. The aforementioned algorithms only use image detectors and fail to utilize the spatio-temporal information contained in the tracker. But the video stream data in the multiple object tracking task processes has strong spatio-temporal correlation. Transmitting spatio-temporal information to target detector is the key to improve the performance of the algorithm. However, since the one-stage detector does not display the module of filtering detection box like ROI (Regions of Interest), it is difficult to propagate spatio-temporal information. Therefore, the existing algorithms for propagating spatio-temporal information (such as [13–16]) are almost two-stage. For the two-stage algorithm, the high accuracy is achieved at the cost of significant computing speed, which makes it difficult to transplant to the field of online multiple object tracking with high real-time requirements.

When solving the above problems, we found that the heat map can well spread the spatio-temporal information to the one-stage detector Center-Net, which can also be proved from our experimental results and video target detection algorithm CHP [17] (Center-Net Heatmap Propagation). Therefore, we design a heat map propagation algorithm HPMOT (Heat Map Propagation based Online Tracking) for Fair MOT [11], which is an algorithm with detector. HPMOT not only maintains the advantage that the speed of Fair MOT is less affected by the number of targets, but also reduces redundancy and improves the speed. The main contributions are summarized as follows:

1. The algorithm integrates detection and tracking, and the tracking speed is accelerated with a small accuracy cost by using spatio-temporal information.
2. We use the Kalman filter in the tracking stage to avoid some hyper-parameter adjustment in CHP and there is no additional cost of time and accuracy.
3. The algorithm is simple and scalable to combine with many kinds of detection algorithms. In fact, the proposed heat map generation process can be generated only according to the tracker, thus it is independent of the detection network.

The rest of this paper is organized as follows: the Sect. 2 introduces the relevant work. The details of the proposed algorithm HPMOT is described in the Sect. 3. The Sect. 4 presents the experiments and analyses the results. Finally, the conclusions and future topics are drawn in the Sect. 5.

2 Related Work

2.1 Multiple Object Tracking

Multiple object tracking algorithm can be divided into online algorithm [1, 9, 18, 21, 29–31] and batch algorithm [32–37]. For the batch algorithms, or the offline algorithms, all frames in the video can be received in advance. While for the online algorithms, only the past frames and the current frame are available. This paper focuses on the online multiple object tracking algorithm.

Before deep learning was used in tracking algorithms, most online multiple object detection algorithms assume that the detector is completely reliable. The most representative algorithms are SORT [18] and IOU-Tracker [21] (Intersection Over Union). SORT uses Kalman filter to predict the position of the detection frame of the current frame and associates the prediction frame with the detection frame, scores and weights, then uses weighted Hungary algorithm to match them according to the association graph. IOU-Tracker uses IOU between the detection frames of the past frame and the current frame to weight the association graph, and then uses the Hungarian algorithm for similarly matching.

The main advantages of these methods without using deep networks are simple, easy to use and fast. However, since these algorithms only use the location information of the tracking target and ignore the appearance information (re-ID information), they are difficult to be applied in scenes with dense targets and large motion changes, and easy to lose track targets or track wrong targets. To solve this problem, Xiang et al. [29] consider the online multiple object tracking problem as a Markov decision processes (MDPs). They also use single target tracking and reinforcement learning to determine the emergence and disappearance of each tracking object. Bae et al. [30] applied linear discriminant analysis (LDA) to extract re-ID information of the target to obtain more stable results.

In recent years, more and more methods have used deep learning to strengthen their tracking capabilities. Nicolai Wojke et al. [5] use a small re-ID network to extract the appearance features of each tracking target, and then measure the appearance feature distance and position distance to construct the distance matrix of the bipartite graph, and finally match it with the Hungarian algorithm. There are also a few algorithms [38–40] using more complex association and matching strategies, such as group models and recurrent neural networks to improve the tracking accuracy. However, with the increase of the number of tracking targets, the speed of these algorithms also decrease rapidly.

Benefiting from the development of multi-tasking learning and end-to-end capability promotion of deep learning, the combination of target detector and re-ID feature extractor further improves the tracking speed in a large number of target scenes. For example, Voigtlaender et al. [20], Zhongdao Wang et al. [8] and Yifu Zhang et al. [11] modified the Mask Region-CNN, YOLOv3 [2] (You only look once vision 3) and CenterNet [7], respectively, and increase the speed while ensuring the tracking accuracy by adding the re-ID output. However, these methods don't make full use of spatio-temporal information, which means that there is still room for further improvement.

In addition, in order to make full use of spatio-temporal information, some algorithms such as D&T [6] (Detect to Track and Track to Detect) combining detection and tracking has emerged. Such a combination is inspired from the easy propagation position and optical flow characteristics of the two-stage detector. However, these algorithms can not reason online because most of the two-stage detectors are incapable of using the future frame information in real time.

2.2 Video Target Detection

Generally speaking, there are two kinds of methods for video target detection. The first kind of algorithms includes post-processing, multi frame feature aggregation, tracking feature assisted guidance methods [6, 16] etc. This kind of algorithms improve the detection quality of the current frame through the spatio-temporal information of the front and rear frames globally or within a certain range. The other is to improve the detection speed by reducing the detection redundancy of the current frame by using the information of the past frame. Such as some algorithms [13–15] based on regression prediction frame, optical flow and feature propagation.

Obviously, to track multiple object in an online mode, only the second categories of video target detection algorithm can be adopted. The general optical flow information and feature propagation algorithm is difficult to be used in the one-stage detection algorithm without explicit screening detection frame such as ROI (Regions of Interest) network. Consequently, tracking speed is the bottleneck of this kind of algorithms and speed promotion is needed by avoiding non key frame detection methods as far as possible. The anchor-free detector Center-Net in one stage based on the center point detection method makes it relatively easy to spread the characteristic information of past frames, such as CHP proposed by Zhujun Xu et al. [17]. Based on CHP, using the properties of heat map generated by Center-Net detector and the target trajectory information obtained in the tracking stage, we propose a simple and effective method to accelerate the acquisition of detection results.

3 Proposed Method

This section consists of three parts. The Sect. 3.1 gives the definition of heat map and the generation method of heat map we designed, and then we will present the detail of the proposed method of fusing heat map with the output of detector network. The Sect. 3.3 will describe our algorithm flow in detail.

3.1 Heat Map Generation

A heat map is a matrix that displays data in color changes. As illustrated in Fig. 1, a darker red block in the heat map corresponds to a higher confidence, indicating that the point is more likely to be located in the center of the object.

The heat map in this paper is actually superimposed by the output of the last layer of the detection network (contains the current frame detection information) and the algorithm introduced below (contains spatio-temporal information). The size is defined by expression (1), where W and H are the width and height of the original image, R is the down sampling proportion (the value is 4 in actual use), F is the dimension of

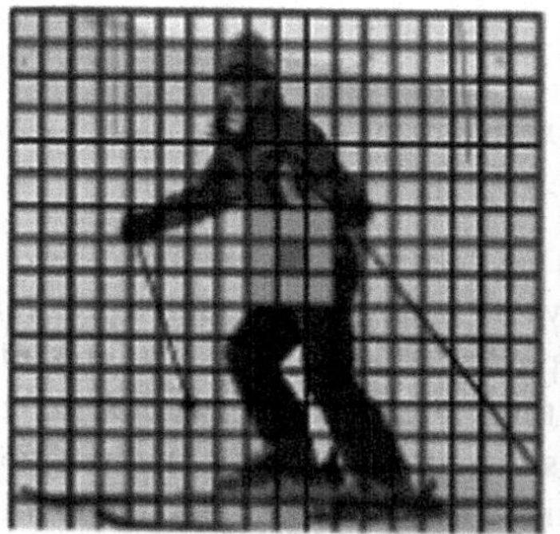

Fig. 1. Example of heat map, On the left, the closer to the center of the target, the redder the color. The right figure shows the detection box according to the center (from [7])

re-ID feature vector, P is the regression dimension of the detection target frame and the confidence prediction of the target at this point. The dimension of P can be represented by 5 dimensions, 1 dimension represents confidence, 2 dimension represents coordinate offset, and 2 dimension represents the width and height of the box.

$$\frac{W}{R} \times \frac{H}{R} \times (F + P) \tag{1}$$

The following part analyzes the value characteristics of the output heat map of the detection network, and explains the way to generate the propagation heat map based on this characteristic and Kalman filter.

In order to spread spatio-temporal information, we generate a heat map in accordance with the size of expression (1) according to the tracking information of the previous frame. The key is to generate the values of each dimension of P.

We first analyze the confidence. In fact, the confidence obeys a Gaussian distribution from the center point of the target to the surrounding area on the heat map. The closer the point is to the target center, the higher its value is. Therefore, the propagation heat map can also be generated in the form of Gaussian distribution to increase the stability of the detection network. This also requires more attention to the center point of the target. To highlight the center point, we suppress the points other than the center point instead of directly assigning the same value.

The suppressed confidence h is defined as follows:

$$h_t^{x',y'} = p_t^{x,y} \alpha^{|x'-x|+|y'-y|} \tag{2}$$

where *x, y* and *x', y'* are the coordinates in the heat map of the previous frame and the current frame respectively, t represents the current frame t, P represents the propagation confidence value of the predicted target, α is defined as the inhibition rate (In this paper, it is taken as 0.95).

Formula (2) shows the value propagated by different distances between each coordinate of the generated heat map and the predicted target, decreasing from the center to the surrounding. For a frame with multiple targets, actual value of h in the generated heat map is:

$$h_t^{x',y'} = \sum_i^n p_{t,i}^{x_i,y_i} \alpha^{|x'-x|+|y'-y|} \tag{3}$$

where n is the number of targets and i is the current target number.

For the coordinates of the confidence value of prediction target propagation, we assign a tracker based on a Kalman filter to predict and update the location information for each stable tracking target. At the same time, a Kalman filter is modeled according to Gaussian distribution. The filter provides the current position, the position change rate and variance measurement of the tracking target. Accordingly, the coordinates of predicted targets can be directly updated by Kalman filter as follows:

$$x_{t,i}, y_{t,i} = KF_i(x_{t-1,i}, y_{t-1,i}, \hat{x}_{t-1,i}, \hat{y}_{t-1,i}) \tag{4}$$

where KF represents the Kalman filter of the target, and $x, y, \hat{x}, \hat{y}$ refers to the state coordinate value of the Kalman filter and the coordinate value detected by the actual detector, respectively.

Besides, since the position change rate and variance of the target are estimated in the filter, we can define the range of confidence propagation according to these two values without adjustment by setting hyperparameter or filling the whole graph according to the suppression rate. It not only avoids the waste of computing resources, but also avoids the attention distracting of the detection network to the background. In this way, we can obtain the confidence propagation formula and the range correction of Eq. (3):

$$p_{t,i}^{x_i,y_i} = h_{t-1}^{\hat{x}_{t-1,i},\hat{y}_{t-1,i}} \tag{5}$$

$$h_t^{x',y'} = \sum_i^n p_{t,i}^{x_i,y_i} \alpha^{|x'-x|+|y'-y|} \left[x', y' \in U\left(x, y, v + \sqrt{\Sigma}\right)\right] \tag{6}$$

where U represents the neighborhood of the predicted target center with respect to the position change rate and standard deviation, (5) indicates that the propagation confidence value comes from the confidence value of the tracking target successfully matched with the detector in the heat map of the previous frame, and (6) indicates that the confidence propagation range is within the domain control range of the tracking target position change rate and standard deviation, rather than unlimited dissemination.

Similarly, Kalman filter can also provide the width and height of the prediction target frame as follows:

$$w_{t,i}, h_{t,i} = KF_i\left(w_{t-1,i}, h_{t-1,i}, \hat{w}_{t-1,i}, \hat{h}_{t-1,i}\right) \tag{7}$$

In addition, we can recover the offset value of the target center coordinate according to the down sampling proportion, that is:

$$\begin{cases} rx_{t,i} = x_{t,i}/R - \lfloor x_{t,i}/R \rfloor \\ ry_{t,i} = y_{t,i}/R - \lfloor y_{t,i}/R \rfloor \end{cases} \tag{8}$$

Finally, we can keep the re-ID eigenvector as it is within the propagation range of confidence, and set the confidence, eigenvector and width and height of detection frame within the non propagation range as 0.

3.2 Heat Map Propagation

In order to propagate the spatio-temporal information extracted in the method described in Sect. 3.1 to the next frame, we need an appropriate strategy to propagate the heat map.

By combing Eqs. (2)–(8) in Sect. 3.1, we can generate a heat map with the same size as expression (1). To facilitate the tracking speed, we directly replace the heat map output by the network with the generated heat map since the size and measurement of the heat map is the same as the network output. The Kalman filter also needs to be updated in the correlation matching stage of Fair MOT, so the generation of heat map by Kalman filter does not bring additional time cost. Using this method, the algorithm speed of multiple object tracking can be greatly improved by alternately generating heat map and network detection at a certain frequency.

The propagation generated heat map can also improve the detection quality and stabilize the re-ID feature of the tracking target. Because even if the tracking target has occlusion and overlap, the center is not completely consistent. Therefore, the heat map can alleviate inaccurate tracking ID switch caused by the instability of the detected re-ID features caused by the occlusion and overlap areas in crowded scenes.

The heat map at the beginning of the second frame can be superimposed output heat map Hdetection from the network and generated heat map Htrack from the Eqs. (2)–(8) by the following formula:

$$hmap_{real} = \begin{cases} \beta hmap_{track} + (1 - \beta)hmap_{detection}(\text{detection frame}) \\ hmap_{track}(\text{accelerated frame}) \end{cases} \tag{9}$$

Where β is the propagation rate. $\beta = 1$ indicates that only the generated heat map is used; $\beta = 0$ indicates that only the detection heat map is used.

3.3 Online Tracking Algorithm Flow

Combining the heat map generation and propagation algorithm proposed above with the detector, we can process each frame of online video stream data in the following way (illustrated by Fig. 2 and Algorithm 1) to achieve online multiple object tracking.

We give the implementation of HPMOT in Algorithm 1. where p_i is the *i*-th frame of the video; $o^i{}_j$ represents the sequence number, width, height, horizontal and vertical coordinates of the *j*-th tracking object in frame *i*; O_i represents the collection of $o^i{}_j$ in frame *i*. *net(x)* refers to the output of the detection network when *x* is input.

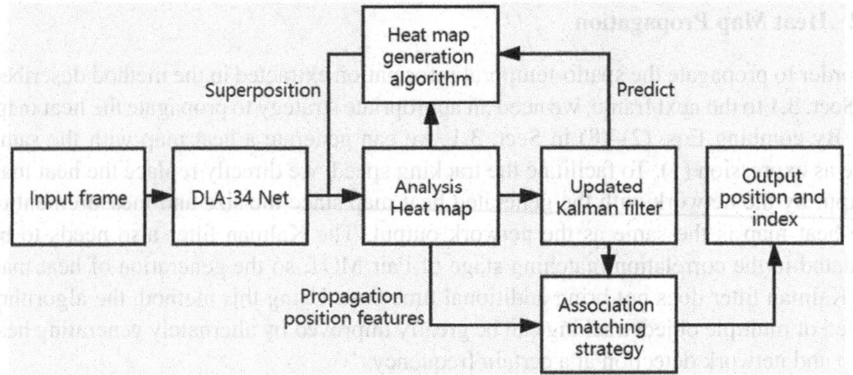

Fig. 2. Flowchart of HPMOT

ALGORITHM 1: HPMOT

Input: The inhibition rate α , The propagation rate. β , Video stream $P=\{p_0, p_1, \cdots\}$

Output: Tracking object collection stream $O=\{O_0, O_1, \cdots\}$

$hmap^0_{real} = net(p_0)$; $hmap_{track} = 0$; O_0 initialize d by $hmap^0_{real}$;

for $p_i \in P$ do

  Calculate the detection heat map, $hmap_{detection} = net(p_i)$;

  Update $hmap^i_{real}$ by (9);

  Update the status of all tracking targets, $O_i = KM(O_{i-1}, hmap^i_{real})$;

  Update $hmap_{track}$ by (2) - (8);

$i = i + 1$;

end for

Return O;

4 Experiments and Analysis

4.1 Datasets and Experiment Settings

The data set used in the experiment is consistent with that used by Fair MOT. Caltech Pedestrian [24], CUHK-SYSU [25], PRW [26] and MOT-17 [27] provide both box and identity annotations, which can be used to train the detection branch and the track branch. ETHZ [22] and City Persons [23] only provide box annotations, so they are only used to train the detection branch. Finally, the self-monitoring strategy is used to train the CrowdHuman [28] dataset containing only the object bounding box. The test set is

2DMOT-15 and half of the MOT-17 training set (this part is removed from the training set in this experiment). The missing data of MOT-17 is not publicly available.

The commonly used metrics of MOT challenge are MOTA, MOTP, MT, ML, IDF1, IDS and FPS. The indication of these metrics are listed in Table 1:

Table 1. Meaning of evaluation indicators (the up arrow indicates that the higher the better, and the down arrow indicates that the lower the better)

MOTA↑	Multiple Object Tracking Accuracy: This measure combines three error sources: false positives, missed targets and identity switches
MOTP↑	Multiple Object Tracking Precision: The misalignment between the annotated and the predicted bounding boxes
MT↑	Mostly tracked targets. The ratio of ground-truth trajectories that are covered by a track hypothesis for at least 80% of their respective life span
ML↓	Mostly lost targets. The ratio of ground-truth trajectories that are covered by a track hypothesis for at most 20% of their respective life span
IDF1↑	ID F1 Score:. The ratio of correctly identified detections over the average number of ground-truth and computed detections
IDSW↓	Number of Identity Switches
FPS↑	Processing speed (in frames per second)

The calculation formula of MOTA, MOTP and IDF1 is as follows:

$$MOTA = 1 - \frac{\sum (FN + FP + IDSW)}{\sum GT} \in (-\infty, 1] \tag{10}$$

$$MOTP = \frac{\sum_{t,i} d_{t,i}}{\sum_{t} c_t} \tag{11}$$

$$IDF1 = \frac{2IDTP}{2IDTP + IDFP + IDFN} \tag{12}$$

$\alpha = 0.95, \beta = 0.5$ is used in the experiment, and the size of the figure is uniformly formatted as 1088×608. Other settings are consistent with Fair MOT.

4.2 Experimental Results and Analysis

In order to verify the acceleration ability of HPMOT algorithm, and test the accuracy maintain ability of HPMOT given a high accuracy original model, experiment is carried out on half of the following MOT-17 data sets in the way of one frame detection + propagation and one frame only propagation (the other half is used for training).

It can be seen from Table 2 that although the accuracy of HPMOT is slightly lower compared with Fair MOT, the FPS is increased by 43.27%, and the number of ID switches

Table 2. Metrics obtained by Fair MOT (upper half) and proposed HPMOT (lower half) in MOT-17-half Dataset

DATA SET	IDF1	MOTP	MOTA	MT	ML	IDSW	FPS
MOT17-02	0.486	0.776	0.445	11	13	93	12.70
MOT17-04	0.817	0.799	0.826	50	2	88	10.56
MOT17-05	0.651	0.804	0.636	19	13	21	16.09
MOT17-09	0.748	0.831	0.716	13	1	14	14.08
MOT17-10	0.710	0.758	0.606	14	6	35	13.12
MOT17-11	0.736	0.845	0.689	17	12	22	13.98
MOT17-13	0.760	0.755	0.621	19	6	26	13.41
AVERAGE	**0.728**	**0.795**	**0.691**	**143**	**53**	299	13.08
MOT17-02	0.511	0.754	0.439	11	13	79	17.24
MOT17-04	0.809	0.792	0.822	49	3	86	13.03
MOT17-05	0.510	0.794	0.341	20	14	20	21.74
MOT17-09	0.644	0.826	0.689	13	1	18	19.84
MOT17-10	0.707	0.754	0.605	13	6	34	19.05
MOT17-11	0.675	0.841	0.639	16	12	25	20.11
MOT17-13	0.717	0.751	0.609	18	5	28	20.96
AVERAGE HPMOT(OURS)	0.704	0.787	0.663	140	54	**290**	**18.74**

has decreased. This is in accord with the original intention of the algorithm, that is, the space-time information is transmitted to the next frame by generating heat map to reduce the amount of detection and accelerate the algorithm. IDF1, MOTP and MOTA decreased only slightly. Besides, By observing ML and MT, we can see that their values are only 1 and 3 worse than Fair MOT, indicating that for the targets accurately tracked by most frames and the targets lost by most frames, reducing detection doesn't lead to significant precision reduction.

In order to verify the ability of the heat map propagation and generation algorithm to speed up and maintain the accuracy given a low accuracy of the original model, further investigation is carried out on 2DMOT-15 data set in the same way:

It can be seen from Table 3 that the quality of the transmitted heat map slightly decreases when the original model performs poorly on the data set, resulting in the decline of the final accuracy. However, even in this case, the number of MT with good target tracking in most frames and ML lost in target tracking in most frames is still similar, and even ID switching is less. It shows that when the detection performance of the

Table 3. Metrics obtained by Fair MOT (upper half) and proposed HPMOT (lower half) in 2DMOT-15 Dataset

DATA SET	IDF1	MOTP	MOTA	MT	ML	IDSW	FPS
VENICE-2	0.498	0.796	-0.314	24	0	67	8.11
KITTI-13	0.453	0.756	-0.825	26	2	23	8.68
KITTI-17	0.596	0.785	0.047	6	0	10	8.19
ETH-BAHNH	0.697	0.818	0.826	176	23	23	8.69
ETH-SUNNY	0.965	0.849	0.957	27	0	0	9.21
PETS09-S2	0.898	0.749	0.894	19	0	10	8.74
CAMPUS	0.742	0.803	0.774	6	0	2	8.84
TUD-STADT	0.816	0.749	0.827	7	0	6	8.86
RUNDLE-6	0.619	0.803	0.689	19	0	44	8.55
RUNDLE-8	0.588	0.773	0.133	25	0	38	8.12
PEDCROSS2	0.759	0.768	0.686	89	24	49	8.42
TUD-STADT	0.816	0.749	0.827	7	0	6	8.85
AVERAGE	**0.658**	**0.787**	**0.463**	**431**	49	278	8.50
VENICE-2	0.499	0.796	0.016	23	0	30	11.15
KITTI-13	0.334	0.757	-0.312	16	1	33	14.48
KITTI-17	0.314	0.775	0.051	2	0	10	15.60
ETH-BAHNH	0.546	0.796	0.511	174	20	17	14.31
ETH-SUNNY	0.778	0.831	0.676	27	0	1	14.22
PETS09-S2	0.598	0.736	0.655	19	0	8	14.69
CAMPUS	0.496	0.792	0.579	7	0	3	14.72
TUD-STADT	0.557	0.736	0.484	8	0	4	14.36
RUNDLE-6	0.586	0.793	0.702	17	0	32	13.58
RUNDLE-8	0.529	0.771	0.611	24	0	22	13.04
PEDCROSS2	0.575	0.755	0.405	88	25	65	14.19
TUD-STADT	0.557	0.736	0.484	8	0	4	14.25
AVERAGE HPMOT(OURS)	0.546	0.777	0.405	413	**46**	**229**	**13.92**

original model is poor, the propagation algorithm can still spread the target information that can be stably detected and tracked. The decline of IDF1 and MOTA shows that for those targets whose detection is not accurate, the propagation algorithm may spread continuation errors, resulting in a gap between the total number and the actual number.

It is worth noting that, MOTP measures the distance between the tracking target box and the actual box (such as the intersection and union ratio of the box), so the value of MOTP is more relevant to the accuracy of the size and position of the tracking target detection frame. Since the wrong detection and wrong propagation targets clearly deviate from the real target, the MOTP is only slightly affected in HPMOT. Moreover, it can be observed from FPS that the speed of the HPMOT is 63.7% higher than Fair MOT In more dense scenes, which also confirms the effectiveness of the algorithm.

5 Conclusion and Future Works

Online multiple object tracking is a task with a wide range of application scenarios, and the use of spatio-temporal information is an important factor to improve its performance. Based on the one-stage detector center net, this paper proposes a heat map propagation algorithm (HPMOT), which effectively leverages the Kalman filter algorithm and avoids the use of additional regression modules. The spatio-temporal information is well exploited in HPMOT and the strength of one-stage detector with strong online capability and fast speed are also achieved. The excellent performance of the algorithm in online multiple object tracking task is verified by a series of experiments. In addition, the autonomy of the proposed heat map generation process to the detection network also enables the algorithm to combine with various promising detection algorithms. We will also try to use Evolutionary Transfer Learning [41–48] to further improve the performance of the algorithm.

Acknowledgement. This work was supported in part by the National Natural Science Foundation of China under Grant 61673328, and in part by the Collaborative Project Foundation of Fuzhou-Xiamen-Quanzhou Innovation Demonstration Zone under Grant 3502ZCQXT202001.

References

1. Hou, X., Wang. Y., Chau, L.: Vehicle tracking using deep sort with low confidence track filtering. In: 2019 16th IEEE International Conference on Advanced Video and Signal Based Surveillance (AVSS), Taipei, Taiwan, pp. 1–6 (2019). https://doi.org/10.1109/AVSS.2019.8909903
2. Redmon, J., Farhadi, A.: YOLOv3: an incremental improvement. arXiv preprint arXiv:804.02767 (2018)
3. Redmon, J., Divvala, S., Girshick, R., Farhadi, A.: You only look once: unified, real-time object detection. In: 2016 IEEE Conference on Computer Vision and Pattern Recognition (CVPR), pp. 779–788 (2016).https://doi.org/10.1109/CVPR.2016.91
4. Liu, W., et al.: SSD: single shot multibox detector. In: Leibe, B., Matas, J., Sebe, N., Welling, M. (eds.) Computer Vision – ECCV 2016: 14th European Conference, Amsterdam, The Netherlands, October 11–14, 2016, Proceedings, Part I, pp. 21–37. Springer, Cham (2016). https://doi.org/10.1007/978-3-319-46448-0_2

5. Wojke, N., Bewley, A., Paulus, D.: Simple online and realtime tracking with a deep association metric. In: 2017 IEEE International Conference on Image Processing (ICIP), pp. 3645–3649. IEEE (2017)
6. Feichtenhofer, C., Pinz, A., Zisserman, A.: Detect to track and track to detect. In: Proceedings of the IEEE International Conference on Computer Vision, pp. 3038–3046 (2017)
7. Zhou, X., Wang, D., Krahenb, P.: Objects as points. arXiv preprint arXiv:1904.07850 (2019)
8. Wang, Z., Zheng, L., Liu, Y., Wang, S.: Towards real-time multiobject tracking. arXiv preprint arXiv:1909.12605 (2019)
9. Zhou, X., Koltun, V., Krahenb, P.: Tracking objects as points. arXiv arXiv:2004.01177 (2020)
10. Bochkovskiy, A., Wang, C.-Y., Liao, H.-Y.M.: YOLOv4: optimal speed and accuracy of object detection. arXiv arXiv:2004.10934 (2020)
11. Zhang, Y., Wang, C., Wang, X., Zeng, W., Liu, W.: FairMOT: on the fairness of detection and re-identification in multiple object tracking. arXiv: Computer Vision and Pattern Recognition (2020)
12. Yu, F., Wang, D., Shelhamer, E., Darrell, T.: Deep layer aggregation. In: 2018 IEEE/CVF Conference on Computer Vision and Pattern Recognition, pp. 2403–2412 (2018). https://doi.org/10.1109/CVPR.2018.00255
13. Zhu, X., et al.: Deep feature flow for video recognition. In: 2017 IEEE Conference on Computer Vision and Pattern Recognition (CVPR), pp. 4141–4150 (2017)
14. Zhang, Z., Cheng, D., Zhu, X., Lin, S., Dai, J.: Integrated object detection and tracking with tracklet-conditioned detection. arXiv arXiv:1811.11167 (2018)
15. Zhu, X., Dai, J., Zhu, X., Wei, Y., Yuan, L.: Towards high performance video object detection for mobiles. arXiv arXiv:1804.05830 (2018)
16. Chen, Y., Cao, Y., Hu, H., Wang, L.: Memory enhanced global-local aggregation for video object detection. In: 2020 IEEE/CVF Conference on Computer Vision and Pattern Recognition (CVPR), pp. 10334–10343 (2020)
17. Xu, Z., Hrustic, E., Vivet, D.: CenterNet heatmap propagation for real-time video object detection. In: Vedaldi, A., Bischof, H., Brox, T., Frahm, J.-M. (eds.) ECCV 2020. LNCS, vol. 12370, pp. 220–234. Springer, Cham (2020). https://doi.org/10.1007/978-3-030-58595-2_14
18. Bewley, A., Ge, Z., Ott, L., Ramos, F., Upcroft, B.: Simple online and realtime tracking. In: ICIP, pp. 3464–3468. IEEE (2016)
19. Ren, S., He, K., Girshick, R., Sun, J.: Faster R-CNN: towards real-time object detection with region proposal networks. IEEE Trans. Pattern Anal. Mach. Intell. **39**(6), 1137–1149 (2017). https://doi.org/10.1109/TPAMI.2016.2577031
20. Voigtlaender, P., et al.: MOTS: multi-object tracking and segmentation. In: CVPR, pp. 7942–7951 (2019)
21. Bochinski, E., Eiselein, V., Sikora, T.: High-speed tracking-by-detection without using image information. In: 2017 14th IEEE International Conference on Advanced Video and Signal Based Surveillance (AVSS), pp. 1–6. IEEE (2017)
22. Ess, A., Leibe, B., Schindler, K., Van Gool, L.: A mobile vision system for robust multi-person tracking. In: CVPR. IEEE, pp. 1–8 (2008)
23. Zhang, S., Benenson, R., Schiele, B.: CityPersons: a diverse dataset for pedestrian detection. In: CVPR, pp. 3213–3221 (2017)
24. Dollar, P., Wojek, C., Schiele, B., Perona, P.: Pedestrian detection: a benchmark. In: CVPR, pp. 304–311. IEEE (2009)
25. Xiao, T., Li, S., Wang, B., Lin, L., Wang, X.: Joint detection and identification feature learning for person search. In: CVPR, pp. 3415–3424 (2017)
26. Zheng, L., Zhang, H., Sun, S., Chandraker, M., Yang, Y., Tian, Q.: Person re-identification in the wild. In: CVPR, pp. 1367–1376 (2017)
27. Milan, A., Leal-Taixe, L., Reid, I., Roth, S., Schindler, K.: Mot16: a benchmark for multi-object tracking. arXiv preprint arXiv:1603.00831 (2016)

28. Shao, S., et al.: CrowdHuman: a benchmark for detecting human in a crowd. arXiv preprint arXiv:1805.00123 (2018)
29. Xiang, Y., Alahi, A., Savarese, S.: Learning to track: online multiobject tracking by decision making. In: ICCV, pp. 4705–4713 (2015)
30. Bae, S.-H., Yoon, K.-J.: Robust online multi-object tracking based on tracklet confidence and online discriminative appearance learning. In: Proceedings of the IEEE Conference on Computer Vision and Pattern Recognition, pp. 1218–1225 (2014)
31. Dicle, C., Camps, O.I., Sznaier, M.: The way they move: tracking multiple targets with similar appearance. In: Proceedings of the IEEE International Conference on Computer Vision, pp. 2304–2311 (2013)
32. Zhang, L., Li, Y., Nevatia, R.: Global data association for multi-object tracking using network flows. In: 2008 IEEE Conference on Computer Vision and Pattern Recognition, pp. 1–8. IEEE (2008)
33. Berclaz, J., Fleuret, F., Turetken, E., Fua, P.: Multiple object tracking using k-shortest paths optimization. IEEE Trans. Pattern Anal. Mach. Intell. **33**(9), 1806–1819 (2011)
34. Zamir, A.R., Dehghan, A., Shah, M.: GMCP-tracker: global multi-object tracking using generalized minimum clique graphs. In: Fitzgibbon, A., Lazebnik, S., Perona, P., Sato, Y., Schmid, C. (eds.) ECCV 2012. LNCS, pp. 343–356. Springer, Heidelberg (2012). https://doi.org/10.1007/978-3-642-33709-3_25
35. Milan, A., Roth, S., Schindler, K.: Continuous energy minimization for multitarget tracking. IEEE Trans. Pattern Anal. Mach. Intell. **36**(1), 58–72 (2013)
36. Wen, L., Li, W., Yan, J., Lei, Z., Yi, D., Li, S.Z.: Multiple target tracking based on undirected hierarchical relation hypergraph. In: Proceedings of the IEEE Conference on Computer Vision and Pattern Recognition, pp. 1282–1289 (2014)
37. Choi, W.: Near-online multi-target tracking with aggregated local flow descriptor. In: Proceedings of the IEEE International Conference on Computer Vision, pp. 3029–3037 (2015)
38. Mahmoudi, N., Ahadi, S.M., Rahmati, M.: Multi-target tracking using CNN-based features: CNNMTT. Multimedia Tools Appl. **78**(6), 7077–7096 (2019)
39. Zhou, Z., Xing, J., Zhang, M., Hu, W.: Online multi-target tracking with tensor-based high-order graph matching. In: 2018 24th International Conference on Pattern Recognition (ICPR), pp. 1809–1814. IEEE (2018)
40. Fang, K., Xiang, Y., Li, X., Savarese, S.: Recurrent autoregressive networks for online multi-object tracking. In: WACV, pp. 466–475. IEEE (2018)
41. Jiang, M., Huang, W., Huang, Z., Yen, G.G.: Integration of global and local metrics for domain adaptation learning via dimensionality reduction. IEEE Trans. Cybern. **47**(1), 38–51 (2017)
42. Jiang, M., Huang, Z., Qiu, L., Huang, W., Yen, G.: Transfer learning based dynamic multi-objective optimization algorithms. IEEE Trans. Evol. Comput. **99**, 1–14 (2017)
43. Jiang, M., Qiu, L., Huang, Z., Yen, G.G.: Dynamic multi-objective estimation of distribution algorithm based on domain adaptation and nonparametric estimation. Inf. Sci. **435**, 203–223 (2018)
44. Jiang, M., Wang, Z., Qiu, L., Guo, S., Gao, X., Tan, K.C.: A fast dynamic evolutionary multiobjective algorithm via manifold transfer learning. IEEE Trans. Cybern. **51**(7), 3417–3428 (2021)
45. Jiang, M., Wang, Z., Hong, H., Yen, G.G.: Knee point based imbalanced transfer learning for dynamic multi-objective optimization. IEEE Trans. Evol. Comput. **25**, 117–129 (2020)
46. Jiang, M., Wang, Z., Guo, S., Gao, X., Tan, K.C.: Individual-based transfer learning for dynamic multi-objective optimization. IEEE Trans. Cyben. **51**, 4968–4981 (2020)

47. Xu, D., Jiang, M., Hu, W., Li, S., Pan, R., Yen, G.G.: An online prediction approach based on incremental support vector machine for dynamic multi-objective optimization. IEEE Trans. Evol. Comput., 1 (2021). https://doi.org/10.1109/TEVC.2021.3115036
48. Wang, Z., Hong, H., Ye, K., Zhang, G.-E., Jiang, M., Tan, K.C.: Manifold interpolation for large-scale multi-objective optimization via generative adversarial networks. IEEE Trans. Neural Netw. Learn. Syst., 1–15 (2021). https://doi.org/10.1109/TNNLS.2021.3113158

Software Systems and Efficient Algorithms

Spatio-Temporal Topology Routing Algorithm for Opportunistic Network Based on Self-attention Mechanism

Xiaorui Wu[1,2], Gang Xu[1,2(✉)], Xinyu Hao[1,2], Baoqi Huang[1,2], and Xiangyu Bai[1,2]

[1] College of Computer Science, Inner Mongolia University, Hohhot 010021, China

[2] Inner Mongolia A.R. Key Laboratory of Wireless Networking and Mobile Computing, Inner Mongolia University, Hohhot, China

csxugang@imu.edu.cn

Abstract. In opportunistic networks, it is difficult to find the best relay node, which not only makes the message transmission inefficient but also wastes resources in the network. Most of the existing routing algorithms based on node similarity often choose relay nodes from a single point of view such as time or space. In this paper, a new neural network architecture is designed, namely, OSAN. It combined with the self- attention mechanism, the spatial and temporal variation characteristics of nodes in two dimensions are taken into account comprehensively to the maximum. Firstly, we divide the opportunistic networks into opportunity network snapshots according to the time window and input each opportunity network snapshot into the spatial structure self-attention layer. At the same time, to capture the interaction between different snapshots, we use the temporal self-attention mechanism. Thus, the Spatio-temporal characteristics of nodes in different snapshots are extracted. Finally, the similarity between nodes is calculated according to the Spatio-temporal characteristics extracted by nodes, so we propose a Spatio-temporal topology routing algorithm in opportunistic networks based on the self-attention mechanism (STSA). The simulation results show that STSA has advantages in several common performance indexes.

Keywords: Opportunistic networks · Dynamic graph embedding · Self-attention mechanism · Node spatio-temporal feature extraction · Routing algorithm

1 Introduction

Opportunistic networks are proposed based on the research of delay-tolerant network (DTN) and mobile ad hoc network MANET [1,2]. Opportunistic networks are defined as: opportunistic network is a kind of self-organizing network that can realize communication without fixed transmission path [3]. In the case

Y. Lai et al. (Eds.): ICA3PP 2021, LNCS 13155, pp. 147–161, 2022.
https://doi.org/10.1007/978-3-030-95384-3_10

of intermittent connectivity, the stored-carry-forward mode is used to complete data forward, which is different from traditional networks. The special characteristics of opportunistic networks can satisfy the application requirements of some areas which cannot meet the condition of a fully connected network in real scenarios. Therefore, in recent years, opportunity networks are increasingly used in real scenarios. Such as data collection from large areas of the wild [4], vehicle network [5], and communication in backward areas [6], in sparsely populated areas such as mountains and deserts, residents can use opportunistic networks to communicate. Due to the characteristics of the opportunistic network, the working mode of side-by-side transmission, the traditional routing algorithms are not suitable for opportunistic networks. Therefore, routing is one of the hot issues of opportunistic networks [7]. To complete the message transmission, how to select the best forwarding time and relay node when the topology changes dynamically with time, and finally determine the appropriate transmission path are the key that researchers need to consider.

This paper proposes STSA routing algorithm, which aggregates the characteristics of neighbor nodes in space and the dynamic changes of nodes in time, so it can effectively capture the intermittent changes of opportunistic networks. Finally, we find the appropriate relay node by calculating the spatio-temporal similarity of nodes, which effectively improves the shortcomings of message transmission success rate, transmission delay, routing overhead, and average hops of nodes.

2 Related Work

At present, some achievements have been made in research on opportunistic network routing based on the similarity between nodes. By calculating the similarity between nodes, we can measure the strength of the relationship between nodes and select the appropriate relay node. Lin et al. and Liu et al. deem that the data packets carried by nodes are the key to calculate the similarity between nodes [8,9]. While Cui et al. defined the node similarity rate by recording the encounter set between nodes and other nodes in the network [10]. Others used node similarity to divide communities, and then designed a routing algorithm based on community division [11,12]. Xiaokaiti et al. used mobile edge computing, combined with network topology attributes and social attributes to measure node similarity to divide communities [13]. Kang et al. comprehensively considered the mobile similarity and social similarity of nodes, and proposed a fuzzy routing and forwarding algorithm based on node similarity, while Mei et al. designed a more effective message transmission strategy on this basis [14,15].

In addition, some algorithms are based on the improvement of traditional routing algorithms. Zhou et al. believed that the traditional routing algorithms were not suitable for opportunistic networks with dynamic network topology, so they optimized and improved the Epidemic algorithm combining with the sleep mechanism [16]. The uneven energy consumption of nodes and the limited residual cache of nodes are also the main factors affecting the opportunistic network routing algorithm, in order to improve these shortcomings, Chen et al. proposed an energy balance and cache optimization routing algorithm based on communication willingness on the basis of Prophet algorithm [17]. Although the performance of the routing algorithms mentioned above is improved, the Spatio-temporal topology between nodes is not well considered. So they don't cope well with the topology change of opportunistic networks.

Combined with the historical connection of nodes in the opportunity network, this paper obtains the topology changes of nodes in two dimensions on different snapshots, with the help of the structural self-attention layer and the temporal self-attention layer. In addition, to adapt to various complex topological changes, we use the multi-head attentional mechanism to capture various implicit features of nodes. Finally, the Spatio-temporal features extracted from the nodes are applied to calculate cosine similarity. We believe that pairs of nodes with high spatial and temporal similarity will meet more frequently in the future.

3 Spatio-Temporal Topology Routing Algorithm Based on Self-attention Mechanism

It is very important to learn the potential representation of nodes in the graph. This representation can preserve the dynamic structure information of the graph and is widely used in many fields, such as node cluster, link prediction, social media recommendation, bioinformatics, and so on. As a new type of network, the opportunistic network changes its network topology with the movement of nodes over time, which directly affects the effective transmission of messages between nodes. To solve this problem, we regard the whole opportunistic network as a dynamic graph, using the method of learning the node representation in the dynamic graph to learn the node representation in the opportunistic networks, and effectively capture the characteristics of nodes under Spatio-temporal evolution. Inspired by DySAT [18], a neural network framework is designed, namely, Opportunistic Self-Attention Networks (OSAN). Specifically, it obtains the representation of nodes through space and time dimensions combined with self-attention mechanism and adds a multi-head attention mechanism on the two layers of structure and time, to capture the multifaceted evolution feature of nodes. Finally, we design the STSA routing algorithm according to the similarity of Spatio-temporal characteristics between nodes.

3.1 The OSAN Framework

The OSAN framework is mainly composed of two modules: the spatial self-attention layer and the temporal self-attention layer. The architecture learns structure embedding on each observable static snapshot, and learns time embedding through multiple static snapshots.

In opportunistic networks, the connection and disconnection between nodes can be regarded as a series of static snapshots.

Definition 1. *(The set of snapshot G)*

$$G = \{g_1, g_2, ..., g_t\} = OSAN(ER|\lambda, \delta) \quad (1)$$

Where ER denotes an encounter record between nodes of the opportunistic network; λ denotes the time window; t is the number of time steps; δ denotes the time slice. The OSAN denotes our framework and extracts snapshots from the ER.

Each snapshot is a record of node encounter information at a certain time interval, and the snapshot set G is a description of the overall dynamic opportunistic networks. For any static snapshot, $g_t = \{V, E_t\}$, V represents all nodes in opportunistic networks, and E_t represents the connection of edges between nodes at time t. Due to the change of the edge set of each static graph, they have an adjacency matrix. The weighted adjacency matrix of g_t at time t is expressed as M_t, where $M_{i,j}$ represents vertex v_i,v_j cumulative connection duration in time t.

The goal of the OSAN framework is to learn the potential representation e_n of each node $v \in V$ in the time step $t = \{t_1, t_2, ..., t_n\}$ by Eq. 2,

Definition 2. *(The spatial-temporal feature of node v)*

$$e_1, e_2, ..., e_n = f(G|\Theta) \quad (2)$$

The OSAN framework ensures that the potential representation can not only obtain the structure of node v in the local snapshot but also retain its time evolution behavior, such as the connection and removal between nodes until time t. The Θ is a parameter which needs to be learned in architecture. The architecture takes λ static snapshots as input, and the output is the node representation of the latest time step. Figure 1 displays the overall framework of OSAN.

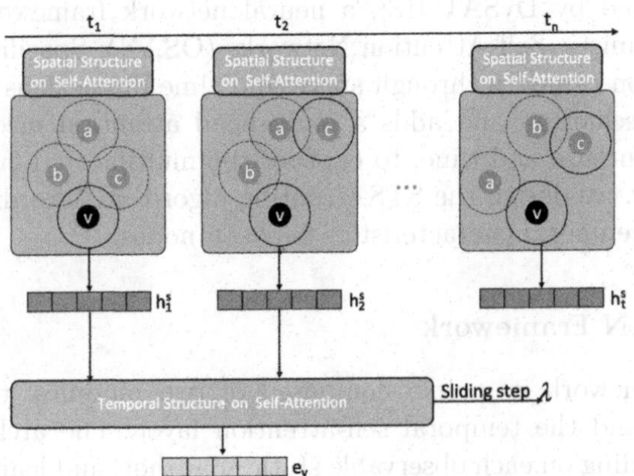

Fig. 1. OSAN architecture

The Self-attention Layer of Spatial Structure. The self-attention layer of spatial structure is mainly used to capture the evolution pattern of opportunistic network in spatial structure. The input for this layer is an adjacency matrix corresponding to static snapshot g_t and an initialized d-dimensional vector for all nodes. Here, we use one-hot encoding to initialize the node v which can be expressed as $X_v \in R^d$, $\forall v \in V$ vector, where d represents the embedding dimension of the input and equals to the total number of nodes. Through the weight of the attention mechanism, we aggregate the nodes' direct neighbors in snapshot g_t. Finally, the potential representation of any node in the graph is the output: $h_v \in R^F, \forall v \in V$. The self-attention of spatial structure is expressed as h_v^s, and the calculation is based on Eqs. 3, 4 and 5.

$$h_v^s = \sigma(\sum u \in N_v, w_{uv} \theta^s X_u) \tag{3}$$

and

$$w_{uv} = \frac{exp(score_{uv})}{\sum_{i \in N_V} exp(score_{iv})} \tag{4}$$

and

$$score_{uv} = \sigma(M_{uv} \cdot a^T \cdot [\theta^s X_u || \theta^s X_v]) \tag{5}$$

Where N_v represents the historical connection node set of node v in a certain time interval; w_{uv} is the attention weight of node u and node v, which is calculated by softmax normalization; $\theta_s \in R^{F \times D}$ is the shared transition matrix of nodes; $\sigma()$ represents an activation function, we use LeakyRELU nonlinear function as the activation function. The $score_{uv}$ is an attention score, which is calculated by Eq. 5. In this formula, the node and the connected node in its snapshot are mapped into the same embedded space. The a is a parameter vector that adjusts the dimension of the aggregation result of the nodes so that it can be linearly transformed with the connection duration M_{uv}. Therefore, the essence of the attention value between a node and its direct neighbor node lies in the connection strength of the edges between them. It can be seen from the above formula 3 that h_v^s weighted aggregates the embedding vector of node v's direct neighbors, and the weight of the edge is the connection time between node v and its neighbors. Therefore, it can be considered that this aggregation preserves the local spatial structure information of node v.

The Self-attention Layer of Temporal Structure. Through the operation of nodes in the structural self-attention layer of OSAN frame, the potential representation of any node in static snapshot g_t contains its nearest neighbor structure, $\{h_v^{s,1}, h_v^{s,2}, ..., h_v^{s,\lambda}\}$. On this basis, to capture the time evolution pattern of opportunistic networks in multiple time steps, we give attention weight to each static snapshot through the time self-attention layer, to selectively save important historical information in features of node, e_v. For any time t in time window λ, we calculate the attention values at all times before t. We obtain the representation of e_v according to Eqs. 6, 7, 8.

$$e_v = \beta_v(X_v\theta_V) \tag{6}$$

and

$$\beta_v^{ij} = \frac{exp(score_v^{ij})}{\sum_{k=1}^{T} exp(e_k^{ij})} \tag{7}$$

and

$$score_v^{ij} = (\frac{(X_v\theta_Q)(X_v\theta_K)^T}{\sqrt{F'}} + M_{ij}^{\lambda}) \tag{8}$$

Where β_v is the attention weight matrix of node v at different times, and its value is determined by the attention $score_v^{ij}$ which normalizes by softmax. The $\theta_Q \in R^{D' \times F'}, \theta_V \in R^{D' \times F'}, \theta_K \in R^{D' \times F'}$ are linear projection matrixs and map query, key, and value to a same space respectively. To learn the representation of node v at time t, the scaled dot product notice form [19] is used in our frame. The $\sqrt{F'}$ is a scaling factor controlling the size of the self-attention score; The $M_{ij}^{\lambda} \in R^{\lambda \times \lambda}$ is a time mask matrix, when $i <= j, M_{ij}^{\lambda} = 0$ otherwise $M_{ij}^{\lambda} = -\infty$. It suggests time only moves forward; The λ is the size of time windows. To save on network overhead, we also slide the time windows λ in every learning time, which can dispose of outdated historical information.

Multiple Head Attention Mechanism. After extracting the node features from the two self-attention layers of spatial structure and temporal structure, a single attention mechanism fully captures the development trend of the node in a single aspect. However, in opportunistic networks, nodes are moving randomly all the time. The static snapshots have many potential development possibilities. To capture the potential properties of nodes in opportunistic networks, we add the multi-head attention mechanism in both spatial structure and temporal structure. We concatenate the node representations in the output layer by each head, as shown by,

$$e_v = Concat(e_v^1, e_v^2, ..., e_v^H) \tag{9}$$

where the H is the number of the head; The $Concat$ is a concatenating operation.

Objective Function. The objective Eq. 10 is designed by [18]. It aggregates the spatial and temporal characteristics of nodes and uses a cross-entropy loss function to ensure that nodes in the connected state have similar representations at each moment.

$$L = \sum_{t=1}^{\lambda} \sum_{v \in V} (\sum_{u \in N_{walk(v)}^t} -log(\sigma(< e_u^t, e_v^t >)) - w_n \cdot \sum_{u' \in P_n^t(v)} log(1 - \sigma(< e_{u'}^t, e_v^t >))) \tag{10}$$

3.2 Spatio-Temporal Topology Routing Algorithm Based on Self-attention Mechanism

To combine the OSAN architecture and the routing algorithm effectively, we designed STSA routing algorithm. The implementation of the algorithm is divided into three stages: initialization stage, recording stage and update stage. The initialization phase is only executed the first time to generate the first snapshot. In the recording stage, the algorithm needs to record the snapshot at each time, and selects the relay node according to the Spatio-temporal characteristics of the previous time window. In the update stage, the node representations are learned by a counted snapshot in a time window, and the node representation is updated. Algorithm 1 shows STSA.

Algorithm 1: Spatio-Temporal Topology Routing Algorithm Based on Self-Attention Mechanism

Input: $G = \{g_0, g_1, g_2 ..., g_n\}, \delta = 1.5h \times 3600, \lambda = 10, M, T$
Output: $r_{v_{next}}$
for *all* $t \in T$ **do**
  if $t < \delta$ **then**
    Recode(g_0)
    continue
  if $t == \delta$ **then**
    Update(g_0)
  $w = (t - \delta)\%(\delta \times \lambda)$
  if $t > \delta$ *and* $w == 0$ **then**
    Update$(g_{\frac{t}{\delta}-\lambda}, g_{\frac{t}{\delta}-\lambda+1}, ..., g_{\frac{t}{\delta}-\lambda+9})$
  Recode$(g_{\lfloor \frac{t}{\delta} \rfloor})$
  for *all* $m \in M$ **do**
    for *all* $n \in N^{m_v}_{g_{\lfloor \frac{t}{\delta} \rfloor}}$ **do**
      if $sim(e_{m_v}, e_{m_d}) < sim(e_n, e_{m_d})$ **then**
        $r_{v_{next}} = e_n$
return $r_{v_{next}}$

Algorithm 1 shows the node learning process and message forwarding strategy. Where M and T represent the message set and simulation end time respectively. Record denoted a function for recording the node encounter and UpdateE_v used to update the representation of node only in the last time slice of a time window. Because the resource in opportunistic networks is very limited, we calculate the similarity between nodes by equation.

$$sim(\boldsymbol{a}, \boldsymbol{b}) = \frac{\boldsymbol{a} \cdot \boldsymbol{b}}{|\boldsymbol{a}||\boldsymbol{b}|}. \tag{11}$$

The e is the embedded representation of nodes, so e_{m_s}, e_{m_d} and e_n respectively represent the characteristics of nodes carrying messages, destination nodes and neighbor nodes. When the node v_a carrying the message meets the node v_b, the cosine similarity between the node v_a, its encounter node v_b and the destination node v_d are calculated. If the similarity between node v_a and destination node v_d is less than that between its encounter node v_b and destination node v_d, the message is forwarded to v_b. The strategy considers that nodes with similar spatial and temporal characteristics have a high probability of meeting each other at the next moment.

4 Simulations and Analysis

4.1 The Parameter Settings of the Simulations

For STSA routing algorithm proposed in the above process, we use open-source simulation tools ONE [20] v1.6.0 to simulate node encounters. The platform provides a variety of mobile models and classical routing algorithms, on this basis, researchers make use of them and gradually improve them. The parameter settings of simulations are shown in Table 1. To better show the performance advantages of STSA, we compare three classical routing algorithms: Epidemic, Prophet and FirstContact. However, they all have some problems, such as poor congestion control and unbalanced energy consumption.

To achieve the purpose of the experiment, the node cache size and simulation time are taken as conditional variables, and finally taking the message transmission success rate, average transmission delay, routing overhead, and average hops as indicators to compare the advantages and disadvantages of different algorithms. These indicators are defined as:

Definition 3. *(The delivery rate) The delivery rate refers to the ratio of the number of successfully delivered messages to the number of all messages in the network, as shown in Eq. 12:*

$$R_d = \frac{N_s}{N_a} \tag{12}$$

Definition 4. *(The average latency) The average delay refers to the time required for all messages in the network to arrive at the destination node from the source node, as shown in the Formula 13:*

$$L_l = \sum_{i=1}^{n} \frac{(t_d - t_s)}{N} \tag{13}$$

Definition 5. *(The overhead rate) The overhead rate refers to the difference between the number of forwarded copies of all messages in the network and the total number of messages successfully delivered to the destination node and then divided by the total number of messages successfully delivered to the destination node as shown in Formula 14:*

$$R_o = \frac{N_r - N_s}{N_s} \tag{14}$$

Definition 6. *(The average hops) The average hops refer to the forwarding times of all messages arriving at the destination node, as shown in Formula 15:*

$$H_a = \frac{H_t}{N_a} \tag{15}$$

Table 1. The parameter settings of simulations.

Parameter	Value
Simulation time(s)	200000
Simulation map	HelsinkMedium
Pedestrian nodes amount (pcs)	40×2
Pedestrian node buffersize (MB)	$5 \sim 25$
Car nodes amount (pcs)	40×1
Car node buffersize (MB)	5
Tram nodes amount (pcs)	2×3
Tram node buffersize (MB)	50

4.2 Snapshot Divide

In opportunistic networks, the transmission of messages depends on the random movement of the node. This opportunistic communication causes the continuous dynamic change of network topology. To facilitate the selection of appropriate relay nodes during message transmission, the initial continuous data set is divided into several discrete opportunistic network snapshots according to time interval δ. Therefore, how to divide snapshots will affect the effect of routing. As shown in Fig. 2, the figure shows the effect of node embedding after snapshot partition at different time intervals. The results show that both accuracy and time complexity increase with the increase of time intervals. In this paper, 1.5h is selected as the division time interval, because the accuracy is high and the time complexity is relatively low.

4.3 Experimental Details

In our experiment, we set $\lambda = 10, \delta = 1.5h$ and epochs, learning rate and finally embedded dimension are set 300, 0.01, and 128 respectively. In addition, we choose the negative sample size of 30 and the random walk length of 20.

4.4 The Comparative Analysis of Results

To show the influence of node cache size on routing algorithm, the node cache is gradually increased from 5 M to 25 M in this paper, Figs. 3, 4, 5 and 6 show the

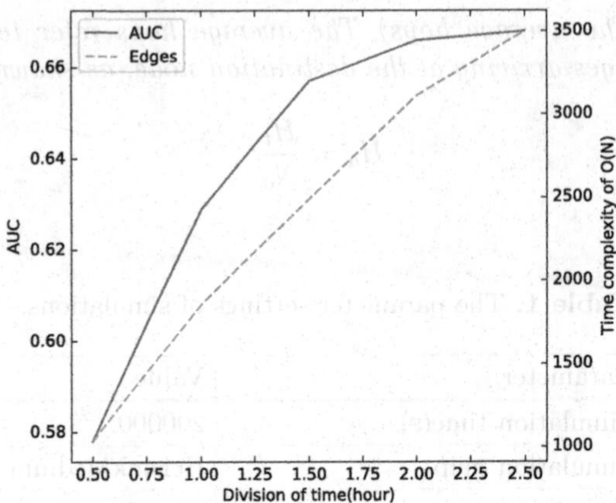

Fig. 2. The effect of node embedding with different time slice

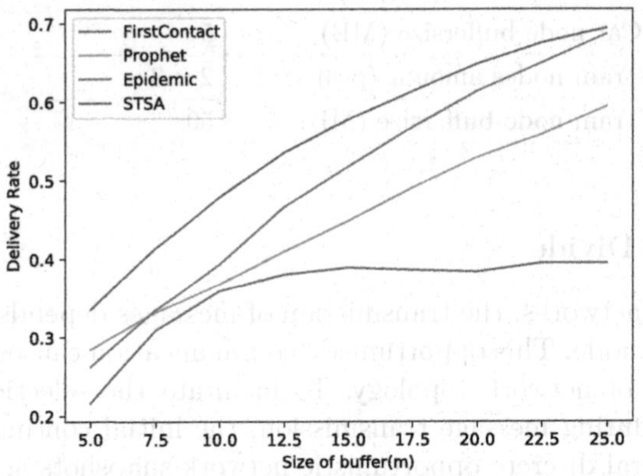

Fig. 3. The effect of buffer size on the delivery rate

performance of four routing algorithms respectively and Fig. 7 shows the impact of simulation time on the delivery rate of four routing algorithms.

Figure 3 shows the changes of message delivery rate of four routing algorithms. With the increase of node cache, the delivery rate of STSA, Epidemic, and Prophet routing algorithms are significantly increased, but STSA algorithm maintains the highest delivery rate as a whole. FirstContact algorithm immediately removes the message from the source node when it is forwarded to the first encountered node, so when the node's cache is increased to 12.5 M, the delivery rate of the algorithm will not change significantly. As the node cache increases, the Epidemic algorithm uses the flooding mechanism to forward more information, so the delivery rate will rise rapidly; Prophet forwards messages only when the source node encounters the destination node, so when the cache increases, Prophet has a higher delivery rate than FirstContact, since Prophet has only one data copy in the whole transmission process, the delivery rate of Prophet will be lower than that of Epidemic. When the cache increases, STSA algorithm forwards more messages and considers the historical topology between the meeting nodes, so its delivery rate is the highest as a whole.

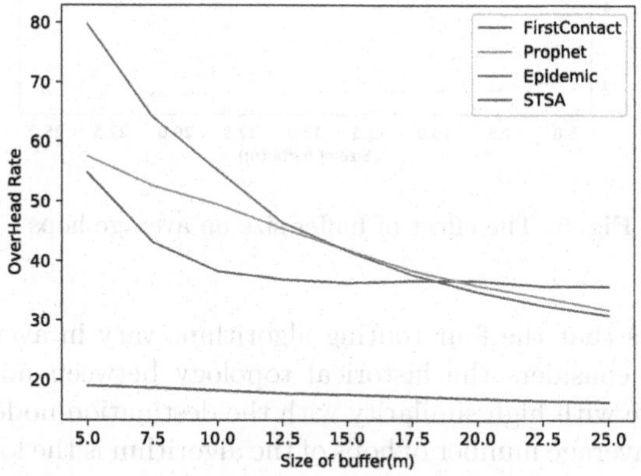

Fig. 4. The effect of buffer size on routing overhead

Figure 4 demonstrates that the four routing algorithms vary in network overhead. STSA algorithm only forwards messages to nodes with high similarity of the destination node, so it maintains a very low routing overhead as a whole. With the increase of node cache size, node cache is no longer a limit factor for network performance, so the network load rates of Epidemic, Prophet, and FirstContact algorithms all show a decreasing trend. Concerning the Epidemic algorithm, more information is forwarded by the flooding mechanism, so the routing overhead is the highest when 5 M–15 M. Prophet algorithm forwards messages to the node with a high meeting probability of the destination nodes, so the routing overhead is lower than the Epidemic algorithm at 5 M–15 M. After the FirstContact algorithm forwards the message to the first encountered node, the message is deleted from the source node, so the routing overhead does not change after 10M.

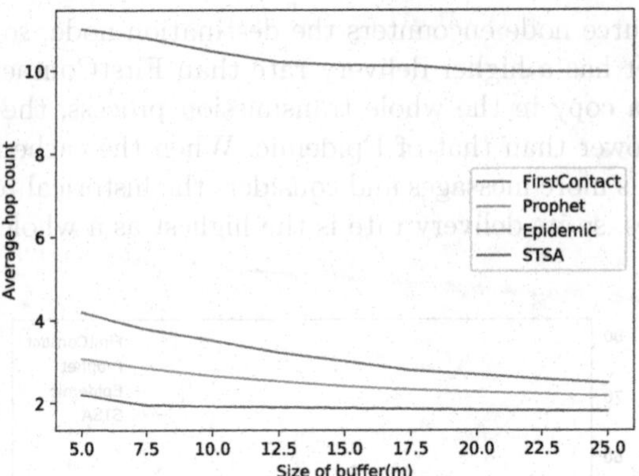

Fig. 5. The effect of buffer size on average hops

Figure 5 shows that the four routing algorithms vary in average hop count. STSA algorithm considers the historical topology between nodes and selects the next-hop node with high similarity with the destination node to forward the messages, so the average number of hops of the algorithm is the lowest as a whole. When the node cache increases, Epidemic algorithm uses the flooding mechanism to forward more messages, while the Prophet algorithm only forwards messages to the node with a high meeting probability of destination nodes, so the average hops of the two algorithms maintain a low and downward trend. FirstContact algorithm removes the message from the source node after forwarding it to the first encountered node, so it has the highest average hop count.

Figure 6 shows the changes in the average transmission delay of the four routing algorithms. With the increase of node cache, the number of messages carried by the node increases, increasing the average delay of all the four routing algorithms, but STSA algorithm is the lowest on the whole. When the node cache increases, Epidemic algorithm uses the flooding mechanism to forward more message copies, so its average delay is lower than FirstContact as a whole. But when the node cache increases to 25 M, their delay are the same. Prophet algorithm forwards messages to the node with a high meeting probability of destination nodes, so it is close to STSA algorithm in terms of transmission delay. STSA algorithm considers the characteristics of nodes in time and can find relay nodes more accurately, so the algorithm has the lowest average delay on the whole. FirstContact algorithm randomly forwards the message to the first encountered node, so it has the highest average latency.

Figure 7 shows the influence of simulation time on the delivery rate of the four routing algorithms. When the simulation time is between 0–5000 s, the delivery rate of the four algorithms increases rapidly. While after 5000 s, the delivery rate of the four routing algorithms will not change significantly. Epidemic algorithm uses a flooding mechanism to forward messages to all encounter nodes. Some nodes receive too many messages and eventually run out of resources and fail to

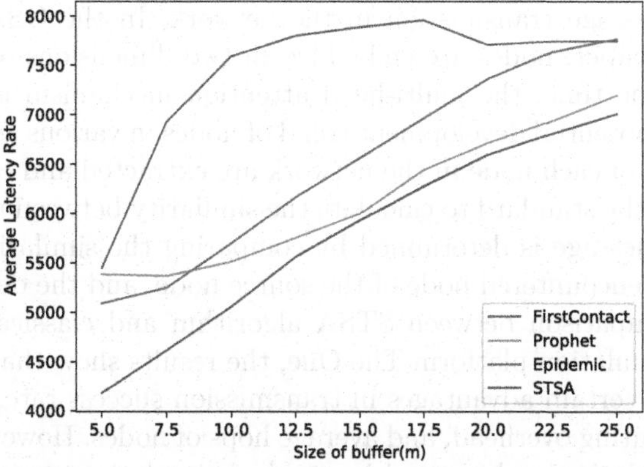

Fig. 6. The effect of buffer size on average delay

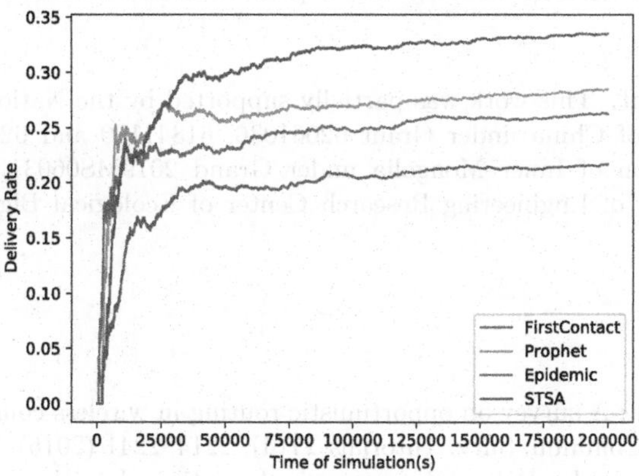

Fig. 7. The effect of simulation time on the delivery rate

deliver the message to the destination node, so its delivery rate is low. Prophet algorithm will forward the message to the node with a high probability of meeting the destination node and selectively transmit the message, so its delivery rate will be slightly higher. When nodes meet, STSA algorithm considers the historical connections between nodes and forwards messages to nodes with high similarity of the destination node, so STSA algorithm has the highest delivery rate. After FirstContact algorithm forwards the message to the first encountered node, it deletes the message from the source node, resulting in the lowest delivery rate of the algorithm.

5 Conclusion

STSA routing algorithm proposed in this paper makes full use of the current popular deep learning and considers the impact of the historical topology between nodes on the message transmission in the network. In the OSAN model constructed in this paper, nodes are embedded in two dimensions of structure and time. At the same time, the multi-head attention mechanism is added to pay attention to the potential development trend of nodes in various aspects, then the intrinsic features of each node in the network are extracted and cosine similarity is finally used as the standard to calculate the similarity between nodes. Whether to forward the message is determined by comparing the similarity between the source node, the encountered node of the source node, and the destination node. Through the comparison between STSA algorithm and classical routing algorithms on the simulation platform The One, the results show that the algorithm we proposed has certain advantages in transmission success rate, average transmission delay, routing overhead, and average hops of nodes. However, the routing algorithm can only be implemented in an ideal simulation environment, so the next step of our work is to improve the learning process of node representation so that the node forwarding strategy and representation learning can be executed in parallel.

Acknowledgment. This work was partially supported by the National Natural Science Foundation of China under Grant 62061036, 61841109 and 62077032, Natural Science Foundation of Inner Mongolia under Grand 2019MS06031, in part by the Self-Open Project of Engineering Research Center of Ecological Big Data, Ministry of Education.

References

1. Chakchouk, N.: A survey on opportunistic routing in wireless communication networks. IEEE Commun. Surv. Tutorials **17**(4), 2214–2241 (2015)
2. Xiao, M., Huang, L.: Delay-tolerant network routing algorithm. J. Comput. Res. Dev. **46**(7), 1065 (2009)
3. Xiong, Y.P., Sun, L.M., Niu, J.W., Liu, Y.: Opportunistic networks. J. Softw. **20**(1), 124–137 (2009)
4. Ehsan, S., et al.: Design and analysis of delay-tolerant sensor networks for monitoring and tracking free-roaming animals. IEEE Trans. Wireless Commun. **11**(3), 1220–1227 (2012)
5. Bitam, S., Mellouk, A., Zeadally, S.: Vanet-cloud: a generic cloud computing model for vehicular ad hoc networks. IEEE Wirel. Commun. **22**(1), 96–102 (2015)
6. Hao, L.I., Chen, Z., Jia, W.U., Software, S.O., University C.S.: Opportunistic network routing algorithm based on community and sociality. Computer Engineering (2015)
7. Ma, H.D., Yuan, P.Y., Zhao, D.: Research progress on routing problem in mobile opportunistic networks. J. Softw. **26**(3), 600–616 (2015)
8. Lin, Y.C., Chen, Z.G., Jia, W.U., S.O. Software, and C. S. University. Routing algorithm for computing similarity between nodes in opportunistic network. Journal of Chinese Computer Systems (2018)
9. Liu, M.Y., Chen, Z.G., Jia, W.U.: Efficient forwarding strategy for computing the cosine similarity of data packets between nodes in opportunistic network. J. Chinese Comput. Syst. **11**(8), 119 (2019)

10. Cui, J., Wu, S., Chang, Y., Huang, D.: Probabilistic routing algorithm based on node similarity rate in opportunity network. J. Chin. Comput. Syst. **42**(3), 6
11. Liang, Z., Yang, F., Jianping, L.I.: Community structure detection based on node similarity in complex networks. J. Comput. Appl. **35**(5), 1213 (2015)
12. Lidong, F.U., Hao, W., Dan, L.I., Fan, L.I.: Community dividing algorithm based on similarity of common neighbor nodes. J. Comput. Appl. **39**(7), 2024 (2019)
13. Xiaokaiti, A., Qian, Y., Wu, J.: Efficient data transmission for community detection algorithm based on node similarity in opportunistic social networks. Complex. 2021, 9928771:1–9928771:18 (2021)
14. Mssn: An attribute-aware transmission algorithm exploiting node similarity for opportunistic social networks. Information (Switzerland) **10**(10), 299 (2019)
15. Fcns: A fuzzy routing-forwarding algorithm exploiting comprehensive node similarity in opportunistic social networks. Symmetry, 10(8) (2018)
16. Zhou, C., Dong, Y., Tian, H.: An opportunistic networks energy-saving routing algorithm based on epidemic and sleeping mechanism. J. Beijing Jiaotong Univ. **43**(2), 18 (2019)
17. Chen, J., Xu, G., Wu, X., Wei, F., He, L.: Energy balance and cache optimization routing algorithm based on communication willingness. In: 2021 IEEE Wireless Communications and Networking Conference (WCNC), pp. 1–6. IEEE (2021)
18. Sankar, A., Wu, Y., Gou, L., Zhang, W., Yang, H.: Dysat: deep neural representation learning on dynamic graphs via self-attention networks. In: Proceedings of the 13th International Conference on Web Search and Data Mining, pp. 519–527 (2020)
19. Vaswani, A., et al.: Attention is all you need. In: Advances in Neural Information Processing Systems, pp. 5998–6008 (2017)
20. Wang, Z., Wang, X.H., Sui, J.Q.: Extending research for one simulator of opportunistic network. Jisuanji Yingyong Yanjiu **29**(1), 272–276 (2012)

TSAEns: Ensemble Learning for KPI Anomaly Detection

Chengyu Wang[1], Tao Yang[1], Jinhua Cui[1(✉)], Yu Li[2], Tongqing Zhou[1], and Zhiping Cai[1]

[1] College of Computer, National University of Defense Technology, Changsha, China
jhcui.gid@gmail.com

[2] Department of Computer Engineering and Microelectronics, Technical University of Berlin, Berlin, Germany

Abstract. Time series anomaly detection is a critical task in the domain of Artificial Intelligence for IT Operations (AIOps). Large companies that provide Internet-based services need to closely monitor massive time-series data from applications and hardware in real-time and provide timely troubleshooting to keep reliable services and smooth business. However, selecting and ensembling diverse detectors for better detection results is challenging due to the complexity of time series data. In this paper, an selection framework for time series anomaly detection, which can select proper detectors for time series data of diverse characteristics according to the detector's performance on historical data. Also, we combine active learning methods to propose unseen samples for labeling, which can significantly alleviate the labeling overhead of operators. Experimental results show the effectiveness of the proposed framework.

Keywords: Time series anomaly detection · Ensemble learning · AIOps · Time series · Anomaly detection

1 Introduction

Time series anomaly detection is critical in the domain of artificial intelligence for IT Operations (AIOps). Large Internet companies need to closely monitor massive time-series data (e.g., cpu usage, response time) in real-time and provide timely troubleshooting to keep reliable services and smooth business, since a short service interruption can cause tremendous financial losses. With the rise of AIOps, dozens of time series anomaly detection algorithms (called detectors for short) have been proposed in recent years [1–10]. Nevertheless, different detectors are designed for time series of different statistical features, and one detector working well for one type of time series may perform poorly for another type of time series [11]. A natural question in AIOps is: how to select proper detectors for massive operational time-series data that may exhibit various statistical patterns?

C. Wang and T. Yang—These authors contribute equally to this work.

Y. Lai et al. (Eds.): ICA3PP 2021, LNCS 13155, pp. 162–177, 2022.
https://doi.org/10.1007/978-3-030-95384-3_11

In practice, selecting appropriate detectors is a hard problem. As reported in [3], various kinds of time series data from different business scenarios have quite different statistical properties (e.g., seasonality), and it is generally agreed that no single detector is general enough for anomaly detection of all kinds of time series data. For example, SPOT [5] does not perform well for seasonal time-series data; and Donut [1] does not work for non-seasonal time-series data. It is extremely difficult, if not impossible, for operators to manually select detectors for massive time-series data. As reported in [12], it is not uncommon that operators give up after a few attempts of testing detectors and settle with static threshold-based detection. In short, how to automatically select and ensemble proper anomaly detectors for anomaly detection on massive time-series data has become a big headache that plagues the whole AIOps industry.

Considering that it is unlikely to detect all types of anomalies via a single detector and existing detectors [1,2] have been validated to be effective for specific types of time series (e.g., seasonal, stable, and unstable) [3], we in this work mainly focus on ensembling different detectors rather than designing a new detector. For this, we need to address the following challenges:

- Different time series data have different types of anomalies, which root from the specific application context. We need to find a suitable method to encode context information in the ensembling learning framework.
- The overhead of labelling time series data is prohibitive. For instance, labelling a year-long time series may take hours [13]. This poses a nontrivial barrier for training supervised ensemble models.

To tackle the above problems, we propose a time series anomaly detection framework, called TSAEns, which can ensembles various detectors for anomaly detection on massive and diverse time-series data. To tackle the first problem, we associate time series with its application context and encode this information in the detector ensembling process. To tackle the second problem, we use active learning to selectively propose unseen samples for labelling. In this way, operators can utilize historical labelled data and only label a small number of proposed samples. Equipped with the above two mechanisms, TSAEns first obtains the performance of all base detectors on the historical (labelled) data. Then it relates the performance of the detectors to the context of time series. In the detection stage, all base detectors run in parallel, and TSAEns automatically infers which detector is most suitable for anomaly detection for given time series under current consideration. Furthermore, TSAEns is extendable since a new detector can be easily integrated into the framework.

To sum up, our main contributions include:

- We propose a time series anomaly detection framework, called TSAEns, which ensembles various detectors for anomaly detection on massive, diverse time-series data. Two critical features of TSAEns, context encoding and active learning-based sample labelling, make it a powerful tool for AIOps.
- We conduct extensive experiments to demonstrate the effectiveness of TSAEns. In addition, we present two case studies to explain how and why TSAEns works.

- We find that the context of time point (i.e., previous time points) can encode critical information that helps ensemble diverse detectors, which has been neglected by most of researchers.

2 Related Work

There are only a limited number of studies on selecting or ensembling anomaly detection methods. Related works can be divided into two categories: selection methods and ensemble methods. The difference between the two is that selection methods select a single detector for the given time series, while ensemble methods use multiple detectors in parallel.

Selection Methods. Selection methods mainly focus on selecting a proper detector and tuning proper parameters for a given time series. In general, these methods consider a time series as a whole, and for a single time series, only one detector will be selected according to the statistical features of the time series. Ying et al. proposed Auto-selector [14], a selection framework that can select a detector based on the features of the given time series. To be detailed, it first extracts features (e.g., variance, auto-correlation, and period) from the given time series via Fast Fourier Transform (FFT), Deseasonality, and Spectral Residual [3]. Then a proper detector is selected based on the extracted features, and the parameters of the selected detector are also estimated at the same time. Although selection methods have achieved promising results, they are sensitive to different anomaly types. It is often that a time series may have different types of anomalies. In this case, a selected detector may accurately detect one type of anomaly but completely miss another type of anomaly, causing a serious problem in service interruption. Thus we argue that only selection is not enough for reliable results. It is better to ensemble complementary detectors to achieve better and reliable detection results.

Ensemble Methods. Ensemble methods mainly focus on fusing the results of different detectors to achieve better results [12,15]. These methods often use base detectors as feature extractor, and apply a classification method to the results from base detectors to detect if there is an anomaly. Laptev et al. proposed EGADS [15], a framework that utilizes a collection of detectors in parallel to detect several types of anomaly. Liu et al. proposed Opprentice [12], an ensemble method that can automatically combine and tune diverse detectors to satisfy operators' accuracy requirements. Opprentice uses multiple detectors in parallel to extract anomaly features, then uses the features and the labels to train a random forest classifier to automatically select the appropriate detector-parameter combinations and the detection threshold. These methods usually achieve better performance than selection methods. Nevertheless, using multiple detectors in parallel has a risk that the result from a poor detector will pollute the final result [14]. Except for the above methods, there are some general ensemble methods like Isolation Forest [16,17], but these methods can not take the context into account. In addition, operators usually do not know labelling what kind of data will benefit the final result. TSAEns includes new mechanisms to alleviate the above problems.

3 Problem Statement

In this section, we first introduce essential definitions and notations. Then we formulate and state the problem we are addressing in this paper based on our definitions.

3.1 Definitions and Notations

Definition 1. (Time Series). Univariate time series is essentially a kind of real valued sequence denoted as:

$$\mathbf{x} = \{x_i\}_{i=1}^N, x_i \in \mathbb{R} \tag{1}$$

where x_i is the value at time index i, N is the length of the time series. Note that we always use bold font to denote a time series (or subsequence) and normal font to denote a time point. We will also use $\mathbf{x}_{m:n} = \{x_i\}_{i=m}^n$ to represent a continuous subsequence of length $n - m + 1$ of time series $\mathbf{x}$. Note that we focus on univariate time series only. Multivariate time series anomaly detection is left as our future work.

Definition 2. (Context). The context of a time point x_i is defined as the previous w time points of x_i, denoted as:

$$\mathbf{c}_i = \mathbf{x}_{i-w:i-1} = \{x_j\}_{j=i-w}^{i-1} \tag{2}$$

where w is the length of the context. The context is used to model dependence between a time point and the previous time points. It is essentially a consecutive time series segment ahead of time point x_i.

Definition 3. (Performance Vector). Performance vector is a vector (of binary values) associated to time point x_i or its context $\mathbf{c}_i$, denoted as $\mathbf{v}_i = \{v_i\}, v_i \in \{0, 1\}$. The length of the vector is the number of base detectors, and each bit of the performance vector indicates whether the corresponding base detector has correctly detected the time point (0 means incorrect and 1 means correct). In the rest of this paper, we will denote the performance vector of x_i by $\mathbf{v}_i$, and the d-th bit of $\mathbf{v}_i$ by $\mathbf{v}_i^d$. For example, a performance vector $\mathbf{v}_i = \{0, 1, 0\}$ means that the 1-th and 3-th base detectors' result on x_i is wrong and the 2-th base detector's result on x_i is correct.

3.2 Problem Statement

As introduced in Sect. 1, we need to address two problems: the **ensemble problem** and the **active learning problem**. The former relates to ensembling base detectors to obtain better detection results; the latter relates to proposing unseen samples for labelling to reduce the overhead of manual labelling.

Ensemble Problem. Now consider a set of unsupervised time series anomaly detectors, each taking as input a point in a time series and outputting an anomaly score of 0 or 1, indicating if the point is anomalous (0 for normal and 1 for anomaly). Then all base detectors run on the time series in parallel. Given a point of time series and its context, our goal is to automatically select which base detector is most likely to correctly detect this point based on the context. Thus the ensemble problem can be formulated as follows:

Given a set of base detectors $\mathcal{D} = \{d_i\}_{i=1}^{M}$ and a time series $\mathbf{x}$, for each time point x_i and its corresponding context $\mathbf{c}_i$, can we approximate conditional probability $p_d(1|\mathbf{c})$, i.e., the probability that base detector d's result is correct when given context $\mathbf{c}$? Once we know $p_d(1|\mathbf{c})$, we can select the proper detector for time point x_i with:

$$d = \underset{d \in \mathcal{D}}{\operatorname{argmax}}\, p_d(1|\mathbf{c}_i). \tag{3}$$

Active Learning Problem. The high cost of manually labelling time series data poses a barrier to training supervised ensemble models. In practice, time series generated from different Internet services may have various statistical features. Thus how to effectively select samples for labelling is a complex problem. Our idea is to utilize the historically labelled dataset, based on which we propose unseen samples for labelling. Leveraging active learning methods, we can significantly reduce the data labelling overhead for massive time-series data.

Active learning focuses on minimizing the labelling overhead of users and improving the accuracy of the prediction model. Most active learning methods use uncertainty as the main criterion to select samples for labelling [7]. Yet, we believe that labelling unseen samples would be more helpful in TSAEns because TSAEns uses a distinctive ensemble mechanism. Thus the active learning problem for our context can be formulated as follows:

Given a set of historically labelled time series $\mathcal{X} = \{\mathbf{x}_i\}$ and a context $\mathbf{c}$, can we approximate a score function $s(\mathbf{c})$ that indicates if there is any context that is similar to the given context? Once we have the score function, we can propose unseen samples with the score function.

4 Design and Implementation of TSAEns

In this section, we first introduce our core idea. Then we give a high-level overview of TSAEns, containing three main stages in the workflow of TSAEns. Finally, we explain the implementation details of TSAEns.

4.1 Core Idea

In practice, operators usually select a detector for a given time series based on the detector's performance on historical data. For example, if the given time series has a strong seasonality, operators may select a detector that has good performance on seasonal data. If some kind of anomaly (e.g., abrupt changes) frequently occurs in the time series, operators may select a detector that has good performance in detecting this kind of anomalies. Nevertheless, if a time series looks unfamiliar to the operators, they will have to label the data and test which detector performs well on this kind of time series. The above industrial practice implies that

Assumption 1 The same detector's performance is similar on similar data (w.r.t. both shape and statistical characteristics).
Assumption 2 For the same time series, the future data and historical data come from the same distribution.

Assumption 2 is a basic assumption in time series anomaly detection. From the perspective of unsupervised anomaly detection, anomaly is defined in contrast to normal data [2], that is, any data that do not follows the distribution of training data can be deemed as anomaly. In the context of IT operation, this assumption is also easy to guarantee, operators can always collect and label training data from the same applications. To take a step back, our experimental results and the practice of operators have proved that these two assumptions are general enough to work most of the time.

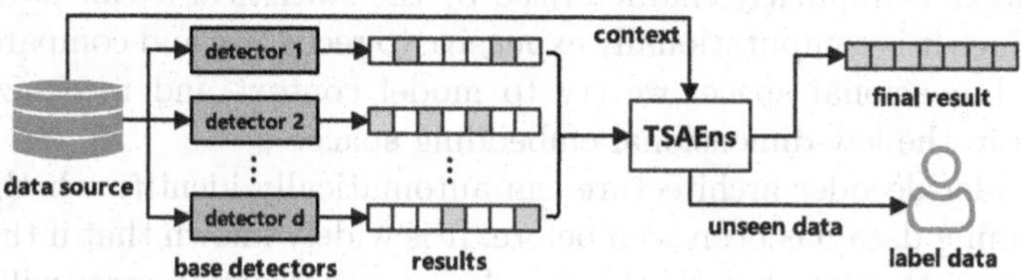

Fig. 1. The core idea of TSAEns. All base detectors run in parallel, only one base detector will be selected to give its result at a time step. When the incoming context is not similar to any context in the historical dataset, TSAEns will notify the operator to label the data.

As is shown in Fig. 1, based on the above analysis, our core idea is to simulate the behaviour of operators. In more detail, we first identify if the context of a time point has been seen before (i.e., similar to any context in a labelled time series dataset). If yes, TSAEns will select a proper detector according to the detectors' historical performance; otherwise, TSAEns will notify operators to label the data. Since the context indicates what type of anomaly is likely to happen, we can infer which detector is most likely to detect anomalies correctly, if any. To model the context (dependence between a time point and previous time points), we use a sliding time window of length w over the time series. Thus a time series can be transformed into a context sequence. Based on the context, we infer which detector is the most suitable for the given time series.

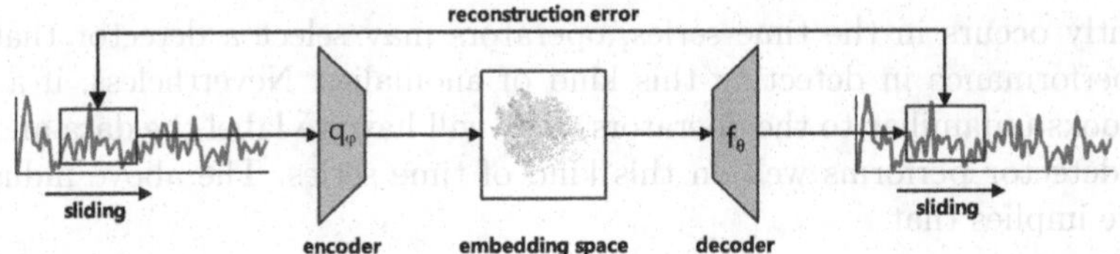

Fig. 2. The architecture of TSAEns. TSAEns uses an encoder-decoder architecture. The encoder is responsible for encoding the context into a low-dimensional embedding space. The decoder is responsibile for reconstructing the context from the embedding.

4.2 Overview of the Design

As shown in Fig. 2, we use an encoder-decoder architecture in TSAEns. We use a sliding window over the time series to retrieve the context for every time point. The encoder is responsible for encoding a context sequence into a low-dimensional embedding space, and the decoder is responsible for reconstructing the context sequence from the embeddings. The rationale for the choice of the encoder-decoder architecture is as follows:

- The context is implicitly characterized by the statistical features of the time series. Since it is computationally expensive to recognize and compare features in high-dimensional space, we try to model context and recognize similar contexts in the low-dimensional embedding space.
- The encoder-decoder architecture can automatically identify whether or not the incoming data has been seen before. It is widely known that if the encoder has not seen the data before, the decoder's reconstruction error will be large; otherwise, the decoder should reconstruct it with a high accuracy [8]. Thus we can utilize this feature to propose unseen data for manual labelling.

In our implementation, we implement the encoder-decoder as a Variational Auto Encoder (VAE) [18], the implementation details are covered in Sect. 4.3. Note that the selection of the encoder-decoder combination is a design choice; other encoder-decoder combinations may be also used in TSAEns.

The workflow of TSAEns consists of three main stages: pre-processing, training, and detection stage.

Pre-processing. In the pre-processing stage, we run all base detectors on the labelled data and get their performance on each labelled time series. Therefore, for each time point, we can obtain its performance vector, indicating whether the detection result of each base detector is correct (0 for wrong and 1 for right). Then the vector is associated with the corresponding context. By querying the context of labelled data, we can infer which detector is most likely to correctly detect if the time point is an anomaly. In this stage, all the base detectors run on the labelled data in an offline fashion. Note that in the detection stage, all base detectors are running in parallel, but only one detector will be selected to give its result at a time.

Training. In the training stage, we first normalize the windowed data by $x_i = \frac{x_i - \mu}{\sigma}$, where μ, σ are the mean and standard variation of the windowed data, respectively, then the encoder and decoder are jointly trained on labelled data, but the labels will not be used in this stage (note that the labels are used for encoding context information, not for training VAE). We use a sliding window of size w that slides over the time series with step length 1 to retrieve context for each time point. In this way, the original time series is transformed into a context sequence. Then we feed the encoder q_ϕ with the time window and reconstruct the time window by the decoder f_θ. The objective function and other details will be introduced in Sect. 4.3. After the training stage, we use the encoder q_ϕ to encode all the labelled data. Thus the embeddings of all labelled data are mapped into the same embedding space. We call the embeddings of all labelled data z map for short. As introduced above, each embedding in the z map is associated with a binary performance vector, indicating whether the detection result of each base detector is correct. The z map will be used as a performance database in the detection stage.

Detection. In the detection stage, all the base detectors are running in parallel, and only one base detector will be selected for a given time point. As is shown in Fig. 3, for each incoming time point x_i, we use the encoder q_ϕ to encode its context $\mathbf{c}_i$, then we query its k nearest neighbors in the z map. After that, we estimate which detector is most likely to correctly detect whether x_i is an anomaly according to the performance vectors associated with the neighbours.

Since the VAE will reconstruct data that it has seen before very well, while failing to do so for data unseen before, we use the reconstruction error as the criterion for proposing unseen samples for manual labelling. Time series with a high reconstruction error is considered as unfamiliar data, and in this case, TSAEns will use a user-specified default detector to detect the given time series and in the meantime notify operators to label the data in the time series.

4.3 Implementation

Inplementation of the Encoder-Decoder. We implement the encoder-decoder as a standard VAE [18]. VAE is a deep Bayesian network that models the dependence between latent variable z and visible variable x. VAE is composed of an encoder-decoder pair. The prior $p_\theta(z)$ is usually chosen to be normal distribution (i.e., $\mathcal{N}(0,1)$). The likelihood $p_\theta(x|z)$ is approximated by the decoder. The true posterior $p_\theta(z|x)$ is intractable, thus we use the encoder to approximate another posterior $q_\phi(z|x)$. In TSAEns, the encoder is essentially the approximation of the posterior $p_\theta(z|x)$ and the decoder is essentially the approximation of the likelihood $p_\theta(x|z)$. The architecture and details of VAE can be found in the related paper [18].

The encoder and the docoder are jointly trained by maximizing the evidence lower bound (ELBO):

$$\log p_\theta(x) \geq \log p_\theta(x) - KL[q_\phi(z|x)||p_\theta(z|x)] \tag{4}$$

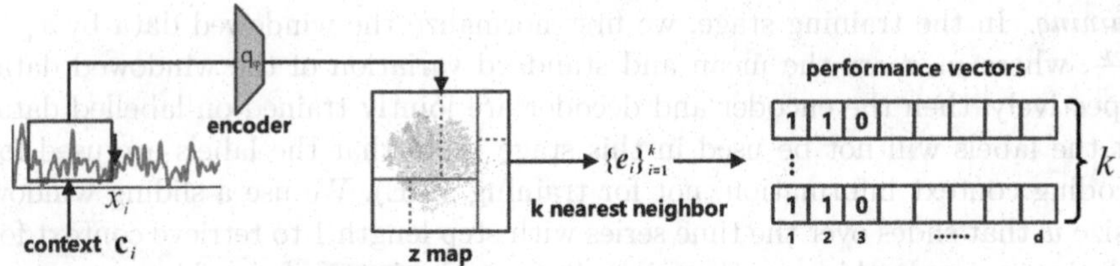

Fig. 3. In the detection stage, TSAEns first encodes a context into an embedding; then it finds k nearest neighbors in the z map and retrieves the associated performance vectors. The z map is implementated as a KD-Tree.

where KL means KL divergence. Thus our objective function is of the following form:

$$\begin{aligned}\mathcal{L}(x) &= \mathbb{E}_{q_\phi(z|x)}[\log p_\theta(x) + \log p_\theta(z|x) - \log q_\phi(z|x)] \\ &= \mathbb{E}_{q_\phi(z|x)}[\log p_\theta(x,z) - \log q_\phi(z|x)] \\ &= \mathbb{E}_{q_\phi(z|x)}[\log p_\theta(x|z) + \log p_\theta(z) - \log q_\phi(z|x)] \qquad (5)\end{aligned}$$

We adopt SGVB [18] as the variational inference algorithm. In addition, we use linear layer to implement the encoder-decoder.

Implementation of z *Map.* Let the embedding space be a K-dimensional space. Searching the nearest k neighbours of an embedding is computationally expensive when the number of embeddings is large. Brute-force search entails calculating the distance between the given point and all other points, which is unacceptable due to the real-time requirements of online anomaly detection. To accelerate the search process, we implement z map as a K-dimensional tree (KD tree) [19]. Note that we use Euclidian distance to measure the distance between embeddings. The KD tree is a structure that stores the instance in K-dimensional space for fast retrieval. It is a particular case of a binary space partition tree. It is mainly applied to the search of multidimensional spatial data (such as range search and nearest neighbor search). More details about the KD tree can be found in the related paper [19].

Implementation of the Conditional Pdf. Here we explain how the conditional $p_d(1|\mathbf{c})$ is approximated. Assume that we have a z map. Let $\mathbf{c}_i$ denote the context of incoming time point x_i. Let e_i denote the embedding of $\mathbf{c}_i$. Let $\mathbf{v}_i$ denote the performance vector associated to x_i. The first step is to find the k nearest neighbor for e_i in the z map. Once we have got the k nearest neighbours, we can approximate the conditional $p_d(1|\mathbf{c}_i)$ for each base dector by:

$$p_d(1|\mathbf{c}) = \frac{\sum_{j=1}^{k} \mathbf{v}_j^d}{k} \qquad (6)$$

where j ranges from 1 to k, $\mathbf{v}_j (j = 1, \ldots, k)$ denote the performance vectors of the k nearest neighbours in the z map.

Implementation of the Score Function s. As we have mentioned above, we use the reconstruction error as the criterion for proposing unseen samples for manual labelling. We use Root Mean Squared Error (RMSE) as the reconstruction error, which is of the following form:

$$score(\mathbf{x}) = ||\mathbf{x} - q_\phi(f_\theta(\mathbf{x}))||_2 = ||\mathbf{x} - \hat{\mathbf{x}}||_2 \tag{7}$$

where $\mathbf{x}$ is the input data, $\hat{\mathbf{x}}$ is the reconstruction of $\mathbf{x}$ (i.e., $q_\phi(f_\theta(\mathbf{x}))$). $||\cdot||_2$ is the l_2-norm, which is equal to the RMSE. In addition, operators can always specify a threshold t as well as a default base detector, for each time point x_i. TSAEns will use the default detector if $score(\mathbf{c}_i) \geq t$. When the accumulated reconstruction error of a time series reaches a certain threshold, TSAEns will inform the operators to manually label the data.

5 Experimental Evaluation

In this section, we first introduce the real-world datasets used in our experiments. Then we introduce the evaluation metrics as well as base detectors in our experiments. Finally, we present our experimental results.

5.1 Datasets and Metrics

For easy reproducibility, we use public datasets in our experiments. There are three widely-used datasets in the time series anomaly detection domain: Yahoo [15], AIOps challenges [20] (we will call it AIOps for short), and NAB [21]. Among these datasets, AIOps and NAB only contain a limited amount of data, and they are not suitable for our experiments (as there is not enough training data). Hence, we only use the Yahoo dataset in our experiments. Yahoo contains four sub-datasets, named A, B, C, and D, respectively. More details of these datasets can be found in the related papers [15,20,21] .

Among these sub-datasets, only dataset A is real, and the other three datasets are synthetic and highly homogeneous with each other. Thus we mainly use datasets A and B in our experiments. The statistics of the sub-datasets A and B are listed in Table 1. For the split of training and test set, we divide each dataset into two parts, which are for training (50%) and evaluation (50%), respectively. Note that the base detectors need to be trained. For each time series, we divide them into two parts, which are used for training (20%) and evaluating (80%), respectively. The split of training and test set for base detectors is illustrated in Fig. 4. We also exchange and shuffle the training and test sets to obtain the average performance of TSAEns.

Table 1. Statistics of the Yahoo dataset.

Dataset	Number of KPI	Average length	Anomaly ratio (%)
A	67	1.414	1.761
B	100	1.420	0.328

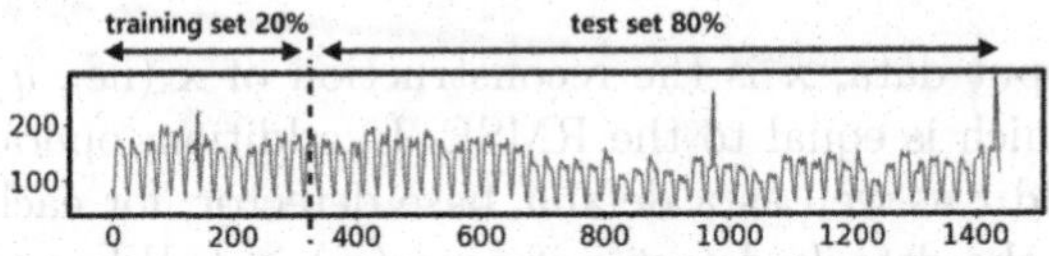

Fig. 4. The split of training set and test set.

We follow the same evaluation metrics adopted in [1–3,9,10,22]. We use F1-score, precision, and recall to evaluate the performance of the proposed method. For the judgment of true positive (TP) number, false positive (FP) number, and false negative (FN) number, we adopt the method used in AIOps challenges [20]. To be detailed, for an anomaly segment, if an algorithm detects any anomaly point, all anomaly points in the segment are counted as being correctly detected. Each anomaly point in the segment is counted as true positive (TP) once. Otherwise, each anomaly point is counted as a false negative (FN). In addition to F1-score assessment, we also evaluate the alarm delay. In practice, the anomaly detector should alarm an anomaly as early as possible. For this purpose, the alarm delay is defined as the distance between the first anomaly point and the first detected point in the anomaly segment.

5.2 Base Detectors and Baseline Methods

We employ four open-source detectors as base detectors in the experiments, which are DDCOL [2], Moving Average (MA) [4], Donut [1], and DSPOT [5], respectively. More details about the base detectors can be found in the related papers [1,2,4,5].

Although there are several ensemble or selection methods in time series anomaly detection, none of them is open-sourced. Thus we introduce two general methods as baseline methods, which are Isolation Forest (IF) and Support Vector Machine (SVM), respectively. Isolation Forest is a strong ensemble method in industrial practice. Following the strategy described in [12], we use base detectors as feature extractors and classify the features by IF and SVM, respectively.

5.3 Overall Performance

Table 2. Experimental results on datasets A and B.

Method	A				B			
	F_1-Score	Precision	Recall	Delay	F_1-Score	Precision	Recall	Delay
MA	0.5272	0.5158	0.6871	0.1423	0.5562	0.5499	0.6800	**0.0**
DDCOL	0.2209	0.2205	0.2633	**0.009**	0.4014	0.3085	0.6000	**0.0**
Donut	0.2747	0.3173	0.3074	**0.009**	0.4723	0.4887	0.9200	49.68
DSpot	0.4975	0.5021	0.6915	7.7410	**0.9009**	**0.8852**	0.9600	0.0733
TSAEns	**0.5628**	**0.5752**	0.6834	0.0593	0.8336	0.8085	0.9417	**0.0**
IF	0.4402	0.3791	**0.7425**	18.83	0.3688	0.3023	**1.0**	49.75
SVM	0.4342	0.4456	0.5810	6.0378	0.8304	0.8396	0.8000	**0.0**

For parameters setting, we set the window size w to 128 and optimize k in $\{10, 30, 50, 100\}$. Since the time series in the Yahoo dataset is relatively short, their seasonal length is short. In this situation, the base detector of the best average performance will always be selected if k is set to a large value. The parameters of IF and SVM are also optimized to get their best performance. We set MA and DSPOT as the default detectors for datasets A and B, respectively, because they have the best average performance on A and B, respectively. The parameters of base detectors are set to the recommended value in their original paper. We did not tune the parameters since parameter tuning is beyond the scope of this paper.

The experimental results are shown in Table 2. TSAEns has the best performance on dataset A compared to any base detectors. It has the best F_1-score and precision on dataset A. TSAEns improves the F-score by 6.75% and precision by 11.52%. For dataset A, TSAEns improves the performance of base detectors significantly but only has a trivial impact on the recall and delay. On the other hand, IF and SVM achive limited results, this phenomenon indicates that it is hard to effectively ensemble base detectors in the absence of context. This result demonstrates that TSAEns can effectively ensemble the base detectors and get better detection results by taking the context into account.

Nevertheless, TSAEns performs not that well on dataset B. Through a close inspection on dataset B, we find this is mainly because dataset B is a synthetic dataset and violates the two assumptions we made in Sect. 3. More specifically, dataset B is generated in a heuristic way. As a consequence, dataset B contains simple patterns and monotonous anomaly types. There are only *sin* and *cos* waves and point anomalies in dataset B, and the anomalies are context-independent (i.e., they occur suddenly without any omen and thus the anomalies has no relationship with its context), thus the context fails to encode what type of anomaly is likely to happen in this situation.

6 Analysis and Interpretation

In this section, we provide an empirical explanation of why TSAEns works well. First, we analyze how well TSAEns can learn the distribution of the data. Then we explain how TSAEns can improve the performance of base detectors.

6.1 Distribution of the Training Data

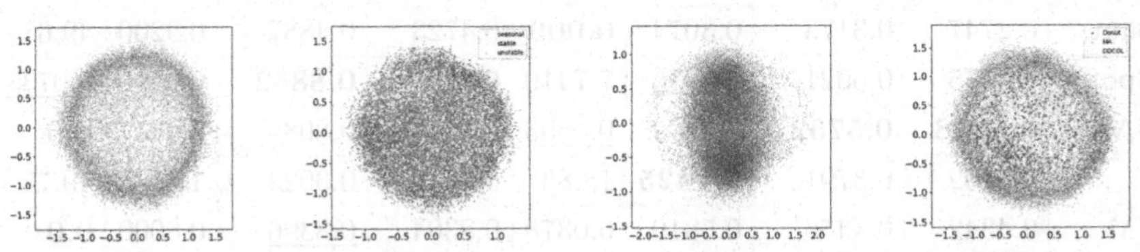

(a) distribution of dataset $\mathcal{A}$ (b) distribution of different data type (c) distribution of dataset $\mathcal{B}$ (d) performance of base detectors

Fig. 5. z map visualization of dataset A. (Color figure online)

Figure 5(a) shows the z map of dataset A. From the figure, we can see that the embeddings are distributed in a circular region in the embedding space with a high density at the boundary area and a low density at the center area. Interestingly, a similar phenomenon has been reported in [1]. This is caused by the distribution of the data itself and by the objective function (ELBO) that pushes dissimilar data away and makes embeddings of seasonal time series circular distributed. In addition, we classify the data in dataset A, mark their embeddings in different colours, and replot the z map. As shown in Fig. 5(b), seasonal data are distributed in the boundary area, and non-seasonal data (e.g., stable and unstable) are pushed to the center area. This experiment indicates that the VAE can effectively capture the distribution of the training data. We also calculate the proportions of all kinds of data in dataset A. As shown in Table 3, seasonal data accounts for 62.2% of the data, while non-seasonal data accounts for only 37.8% of the data, which explains why the density of the marginal area is high and the density of the center area is low.

Table 3. Proportions of all kinds of data in dataset A

Type	Num	Avg length (k)	Ratio (%)
Seasonal	42	1.404	62.2
Stable	15	1.434	22.7
Unstable	10	1.430	15.1

Figure 5(c) shows the z map of dataset B. As illustrated, the distribution of the embeddings is quite different from dataset A. Through an inspection on dataset B, we find that the data in dataset B are all seasonal, but they do not form a ring structure in the embedding space. In contrast, the density of the center area is higher. As mentioned above, the embeddings of seasonal data tend to be circularly distributed, but this experiment shows that if there is no other type of data (e.g., stable, unstable), the formed area does not fit well in a circle. This experiment also proves that the balance of data matters in the training of TSAEns.

6.2 Performance Visualization in z Map

Figure 5(d) shows the z map of dataset A, but it is coloured according to the detectors' detection results on the data. We run all base detectors on data set A and get if their detection results are correct. Then we plot the z map and gray the embeddings on which all detectors' detection results are correct. The embeddings, on which any base detector's detection result is wrong, are coloured in green for MA, blue for DDCOL, and orange for Donut. This figure clearly illustrates the base detectors' performance on dataset A. The density of coloured points indicates the error rate of base detectors. The higher the density, the greater the error rate. For example, those areas with a high density of green points indicate that MA performs poorly on these data. This experiment validates the two assumptions we made in Sect. 3 and shows that TSAEns can effectively learn the performance of base detectors on the training data.

6.3 Effectiveness of Base Detector Inference Mechanism

Figure 6 intuitively shows a time series picked from dataset A (upper) and the corresponding probablity density $p_d(1|\mathbf{c})$ estimated by TSAEns (lower). There are three anomalies (coloured in red) in the time series, the first one is a contextual anomaly [23], the last two are point anomalies.

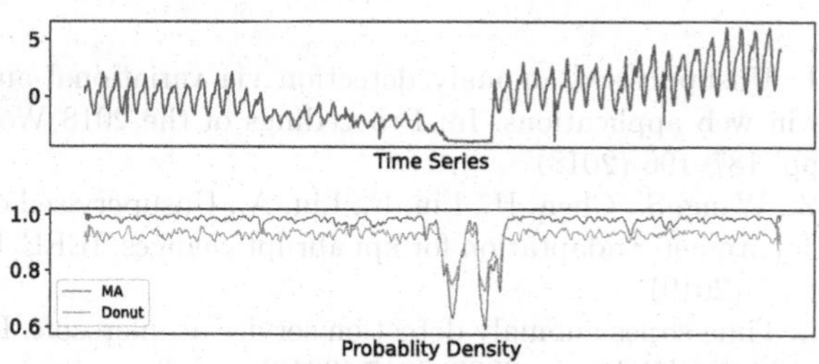

Fig. 6. Probablity density $p_d(1|\mathbf{c})$ estimated by TSAEns. TSAEns successfully identifies the context of contextual and point anomalies.

We employ two base detectors in this experiment, which are MA and Donut, respectively. MA is suitable for detecting point anomalies but has limited power in detecting contextual anomalies; Compared to MA, Donut has a better ability to detect contextual anomalies, but is not as good as MA in detecting point anomalies. As is illustrated, TSAEns successfully identifies the context of different anomalies. $p_{MA}(1|\mathbf{c})$ is higher than $p_{Donut}(1|\mathbf{c})$ most of the time, this is because that the average performance of MA is better than Donut and MA has a lower false positive rate. Furthermore, TSAEns correctly select MA during the point anomalies and Donut during the contextual anomaly. This experiment intuitively shows the effectiveness of the base detector inference mechanism.

7 Conclusion

Time series anomaly detection is critical in the context of AIOps. Monitoring time series of various characteristics and detecting anomalies in real-time is a crucial step to ensure the stability and reliability of Internet services. Although dozens of time series anomaly detectors have been proposed over the years, the selection of proper anomaly detectors has remained unsolved for a long time due to the complex patterns in operational data. To tackle this problem, we propose TSAEns, a framework for time series anomaly detection, which integrates ensemble learning and active learning methods and utilizes historical data to infer which detector is most suitable for a given time series. In addition, by recognizing unseen data and setting a default detector, TSAEns can (1) guarantee the detection performance and (2) propose unseen data for manual labelling to reduce the labelling overhead. In addition, we also find that the context is critical information that helps ensemble the detectors.

Acknowledgment. This work is supported by the National Key Research and Development Program of China (2020YFC2003400), the National Natural Science Foundation of China (62172155, 62102425, 62072465), and the Science and Technology Innovation Program of Hunan Province (2021RC2071).

References

1. Xu, H., et al.: Unsupervised anomaly detection via variational auto-encoder for seasonal kpis in web applications. In: Proceedings of the 2018 World Wide Web Conference, pp. 187–196 (2018)
2. Yu, G., Cai, Z., Wang, S., Chen, H., Liu, F., Liu, A.: Unsupervised online anomaly detection with parameter adaptation for kpi abrupt changes. IEEE Trans. Network Serv. Manage., 1 (2019)
3. Ren, H., et al.: Time-series anomaly detection service at microsoft. In: Proceedings of the 25th ACM SIGKDD, pp. 3009–3017 (2019)
4. Choffnes, D.R., Bustamante, F.E., Ge, Z.: Crowdsourcing service-level network event monitoring. In: Proceedings of the 2010 ACM SIGCOMM, pp. 387–398 (2010)
5. Siffer, A., Fouque, P.-A., Termier, A., Largouet, C.: Anomaly detection in streams with extreme value theory. In: Proceedings of the 23rd ACM SIGKDD International Conference on Knowledge Discovery and Data Mining, pp. 1067–1075 (2017)

6. Yaacob, A.H., Tan, I.K.T., Chien, S.F., Tan, H.K.: Arima based network anomaly detection. In: Second International Conference on Communication Software and Networks 2010, pp. 205–209 (2010)
7. Zhang, X., et al.: Cross-dataset time series anomaly detection for cloud systems. In: 2019 USENIX Annual Technical Conference (2019)
8. Audibert, J., Michiardi, P., Guyard, F., Marti, S., Zuluaga, M.A.: Usad: unsupervised anomaly detection on multivariate time series. In: Proceedings of the 26th ACM SIGKDD, pp. 3395–3404 (2020)
9. Hundman, K., Constantinou, V., Laporte, C., Colwell, I., Soderstrom, T.: Detecting spacecraft anomalies using lstms and nonparametric dynamic thresholding. In: Proceedings of the 24th ACM SIGKDD (2018)
10. Lin, S., Clark, R., Birke, R., Schönborn, S., Trigoni, N., Roberts, S.: Anomaly detection for time series using vae-lstm hybrid model. In: IEEE International Conference on Acoustics, Speech and Signal Processing, pp. 4322–4326. IEEE (2020)
11. Wang, C., Wu, K., Zhou, T., Yu, G., Cai, Z.: Tsagen: synthetic time series generation for kpi anomaly detection. IEEE Trans. Network Serv. Manage. (2021)
12. Liu, D.: Opprentice: towards practical and automatic anomaly detection through machine learning. In: Proceedings of the 2015 Internet Measurement Conference, pp. 211–224 (2015)
13. Zhao, N., Zhu, J., Liu, R., Liu, D., Zhang, M., Pei, D.: Label-less: a semi-automatic labelling tool for kpi anomalies. In: IEEE INFOCOM 2019 - IEEE Conference on Computer Communications, pp. 1882–1890 (2019)
14. Ying, Y., Duan, J., Wang, C., Wang, Y., Xu, B.: Automated model selection for time-series anomaly detection (2020)
15. Laptev, N., Amizadeh, S., Flint, I.: Generic and scalable framework for automated time-series anomaly detection. In: Proceedings of the 21th ACM SIGKDD International Conference on Knowledge Discovery and Data Mining, pp. 1939–1947 (2015)
16. Liu, F.T., Ting, K.M., Zhou, Z.-H.: Isolation forest. In: Eighth IEEE International Conference on Data Mining 2008, pp. 413–422 (2008)
17. Liu, F.T., Ting, K.M., Zhou, Z.H.: Isolation-based anomaly detection. ACM Trans. Knowl. Discovery Data **6**(1), 1–39 (2012)
18. Kingma, D.P., Welling, M.: Auto-encoding variational bayes (2014)
19. Chen, P., Wang, Y.: Optimized kd tree application in instance-based learning. In: 2008 Fifth International Conference on Fuzzy Systems and Knowledge Discovery, vol. 2, pp. 187–191 (2008)
20. Aiops challenge (2018). http://iops.ai/dataset_list/
21. Lavin, A., Ahmad, S.: Evaluating real-time anomaly detection algorithms - the numenta anomaly benchmark. In: 2015 IEEE 14th International Conference on Machine Learning and Applications (ICMLA), pp. 38–44 (2015)
22. Wen, T., Keyes, R.: Time series anomaly detection using convolutional neural networks and transfer learning, arXiv preprint arXiv:1905.13628 (2019)
23. Chandola, V., Banerjee, A., Kumar, V.: Anomaly detection: a survey. ACM Comput. Surv. 41(3), July 2009. doi: 10.1145/1541880.1541882

Towards Transferable Adversarial Examples Using Meta Learning

Mingyuan Fan, Jia-Li Yin, Ximeng Liu(✉), and Wenzhong Guo(✉)

College of Computer and Data Science, Fuzhou University, Fuzhou, China
{x,guowenzhong}@fzu.edu.cn

Abstract. With the discernment of the vulnerability of deep neural networks recently, adversarial attack methods have become one of the hot spots for the security of artificial intelligence technologies. While previous researches can effectively generate adversarial examples in white-box attacks, it remains challenging to transfer these adversarial examples to black-box models, where the attacker has no knowledge about the model structure and parameters. This paper focuses on the transferability of adversarial examples and proposes a novel approach named Model-Agnostic Attack (MAA), in which meta-learning is explored to facilitate the transferability of adversarial examples crafted on vanilla adversarial attacks across diverse black-box models. Specifically, model-agnostic meta-learning, a meta-learning approach, can train a well-generalized model to various unknown tasks and is utilized to alleviate the overfitting problem of adversarial examples for the specified models, so that the adversarial examples can be easily transferred to black-box models. Besides, we highlight that the MAA is a plug-and-play approach and can be effortlessly integrated with any existing technologies to further boost transferability. Extensive experiment results on CIFAR-10 and CIFAR-100 exhibit the superiority of MAA that achieves higher transferability than state-of-the-art methods on average against black-box models.

Keywords: Adversarial attack · Black-box attack · Black-box scenario · Meta learning · Transferability · Transferable adversarial examples · Model-agnostic meta-learning

1 Introduction

Deep neural networks (DNNs), benefited from their superior ability of fitting complex functions, have obtained competitive achievement in a broad spectrum of complicated tasks, particularly in computer vision [33], natural language processing [25], and recommendation systems [3]. However, recent researches [5,6,8,16,17,21,29] have shown that DNNs are exceedingly sensitive to adversarial attacks in various tasks. By maliciously and purposely adding tiny crafted noises towards the sensitive direction of DNNs, such as gradient direction, into the natural examples, the resultant examples,

This paper is supported by organization the National Natural Science Foundation of China (No. 62072109, No. U1804263); Natural Science Foundation of Fujian Province (No. 2021J06013).

Y. Lai et al. (Eds.): ICA3PP 2021, LNCS 13155, pp. 178–192, 2022.
https://doi.org/10.1007/978-3-030-95384-3_12

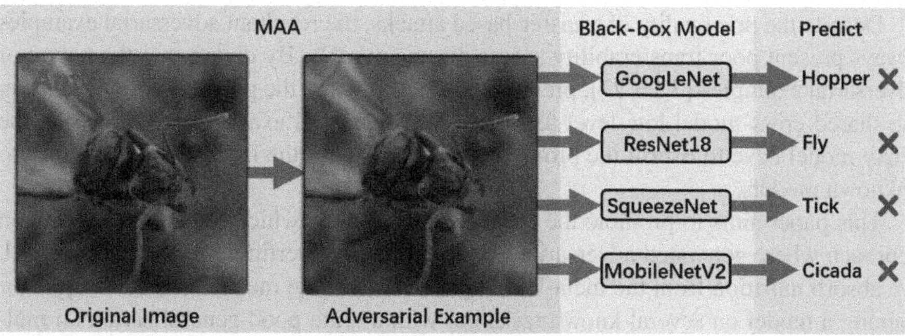

Fig. 1. An example of MAA modification. The original image, the ground-truth label is ant, on the left was randomly extracted from the publicly-used dataset ImageNet, and we crafted the counterpart based on MAA to illustrate. GoogLeNet, ResNet18, SqueezeNet, MobileNetV2 are the black-box models used in the experiment. People would think that there is no any difference between the two pictures, but it was misidentified on the four black-box models.

called adversarial examples, can cause dramatic shifts to the predictions from state-of-the-art models. Such vulnerability of DNNs immediately raises a severe crisis for the applications of DNNs in the physical world and attracts increasing attention from the AI community to the robustness of DNNs.

It is broadly shared that attack and defense play complementary roles in the security of DNNs. In other words, the emergence of threatening attack methods can significantly promote the advancement of defense. Adversarial attack methods [4,10,12,15] focus on the white-box scenario and achieve remarkable attack performance (almost 100% against various common models). Behind the extraordinary effectiveness is that the attackers fully hold all parameters and detailed architecture of the victimized model. However, despite their success, the condition is quite harsh in the real world. For instance, in most cases [1,29,31,32], the attackers only access the labels from the victimized model to inputs (provided by attackers). Hence, white-box attacks suffer from the unpractical assumption and are heavily limited.

In contrast to the white-box scenario, the black-box scenario requires attackers who can only obtain output (score or label) of the victimized model and thus is fairly practical than the white-box scenario. Although the black-box scenario aligns the physical scenario, the available information in the setting is fairly poor and making it challenging. Overall, there are two avenues to conduct black-box attacks: transfer-based attack and query-based attack. Query-based attacks generally require numerous queries [1,28,31] (hundreds to thousands in CIFAR-10 per image) to the victimized model for estimating the gradients to craft the threatening adversarial examples. Such abnormal query phenomenon probably leads to doubt from the defenders and the hefty prices of the attackers. In contrast to query-based attacks, transfer-based attacks desire to simulate the white-box scenario in case the risk of being detected and expensive cost. In detail, it is conducted in the manner where training a proxy model and then harnessing the model to produce transferable adversarial examples across diverse models (with white-box attack methods) to attack the victimized model.

Despite the practicality of transfer-based attacks, the resultant adversarial examples always present poor transferability across diverse models. By diving into the nature of adversarial examples [7,26,27], previous works show that the property originates from the shared cross-model low-level features, and adversarial examples generated on the proxy model have to overfit the proxy model that induces the ill attack performance to unknown models.

This paper aims to promote the transfer-based attack, which requires designing an approach where generated adversarial examples are not overfitting to the proxy model. We absorb nutrition from the meta-learning area, where the methods are developed for training a model on several known tasks but with a quite good generalization on multiple agnostic tasks analogous to known ones. Benefited from the meta-learning area, a novel black-box attack approach in this paper is proposed, named Model-Agnostic Attack (MAA). Specifically, crafting transferable adversarial examples in MAA can be regarded as training a well-generalized model to unknown tasks, which is the consistent target with meta-learning, and in this way, we can adopt advanced meta-learning technologies to avoid overfitting stalemate. In more detail, as shown in Fig. 1, MAA is conducted by training multiple proxy models as known tasks to simulate the meta-learning scene and crafting adversarial examples with meta-learning methods on these known models. Subsequently, the generated adversarial examples have great generalization ability to transfer, or attack successfully, on unknown but similar models so that the preset attack goal is attained. To validate the performance of MAA, we conduct extensive experiments where MAA strikingly surpasses the state-of-the-arts on popular datasets, i.e., CIFAR-10 and CIFAR-100.

- We propose a novel yet effective black-box attack method MAA from a high-level perspective (meta-learning), in which the transferability of adversarial examples can be greatly enhanced via combining vanilla adversarial attack methods and model-agnostic meta-learning.
- Delving the bottom of the black-box attack, we deeply analyze the difference between MAA and the existing method (ensemble attack) and highlight that MAA is generic technology to enhance the performance of existing methods. Moreover, we give insights on how to improve the robustness of DNNs essentially.
- We conduct extensive experiments to validate the performance of MAA, and the results illustrate that MAA has a great improvement on transferability compared to state-of-the-art in two benchmark datasets, namely CIFAR-10 and CIFAR-100.

The remainder of this paper is organized as follows. Section 2 briefly introduces related work, including adversarial examples and meta-learning. Section 3 elaborates the proposed method MAA and its connections to existing approaches. Section 4 conducts extensive experiments to examine the performance of MAA. Section 5 concludes the paper and discusses some future works.

2 Background and Related Work

2.1 Adversarial Attacks

Szegedy et al. [24] found the existence of adversarial examples that can be used to fool DNNs and proposed an optimization method to produce adversarial examples.

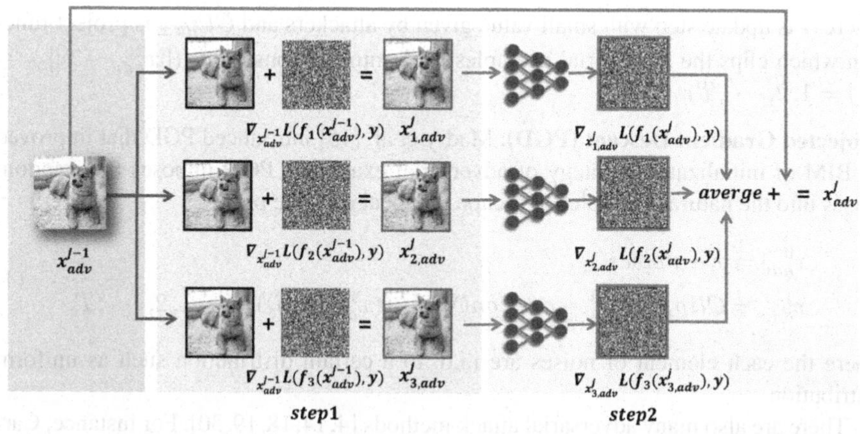

Fig. 2. Running process of MAA.

Although the noises added in the natural examples are imperceptible to humans, it effortlessly misleads the remote target model. Let x and $||x||_p$ denote the natural example with ground-truth label y and the p-norm of x. Given a victimized model f, the attackers desire to craft adversarial example x_{adv} to x and require x_{adv} around ϵ-ball centered at x, i.e., $||x_{adv} - x||_p \leq \epsilon$ (p is commonly set to infinite) that used to measure control the similarity between the natural examples and corresponding adversarial examples. Formally, the process can be formulated as follows:

$$\max_{x_{adv}} f(x_{adv}) \neq y, \quad s.t. \; ||x_{adv} - x||_\infty \leq \epsilon, \tag{1}$$

where $x_{adv} - x$ is adversarial noises for x. But, because the optimization task is difficult to navigate, the majority of existing attack methods adopt the gradient ascent method to solve it approximately. Next, we give a briefly review of adversarial attacks and defenses.

Fast Gradient Sign Method (FGSM). Goodfellow et al. [10] proposed FGSM that crafts the adversarial examples by adding sign of gradient into the natural examples for one step:

$$x_{adv} = x + \epsilon \cdot sign(\nabla_x L(f(x), y)), \tag{2}$$

where $L(\cdot, \cdot)$ is common loss function, e.g., cross-entropy loss function.

Basic Iterative Method (BIM). Kurakin et al. [12] proposed BIM that is a variant of FGSM and perturbs the natural examples for multiple times with small step. In detail, given total perturbation times T, BIM is expressed as follows:

$$\begin{aligned} x_{adv}^0 &= x, \\ x_{adv}^j &= Clip_{x,\epsilon}(x_{adv}^{j-1} + \alpha \cdot sign(\nabla_x L(f(x_{adv}^{j-1}), y))), \; j = 1, 2, \cdots, T \end{aligned} \tag{3}$$

where α is update step with small value given by attackers and $Clip_{x,\epsilon}$ is project function which clips the adversarial examples back into the constraints ($||x^j_{adv} - x||_\infty \leq \epsilon,\ j = 1, 2, \cdots, T$).

Projected Gradient Descent (PGD). Madry et al. [15] introduced PGD that improved on BIM of initialization strategy of adversarial examples. PGD imposes tiny random noises into the natural examples to escape the local extreme point:

$$\begin{aligned} x^0_{adv} &= x + noises, \\ x^j_{adv} &= Clip_{x,\epsilon}(x^{j-1}_{adv} + \alpha \cdot sign(\nabla_x L(f(x^{j-1}_{adv}), y))),\ j = 1, 2, \cdots, T, \end{aligned} \tag{4}$$

where the each element of noises are i.i.d. to a certain distribution such as uniform distribution.

There are also many adversarial attack methods [4,14,18,19,30]. For instance, Carlini and Wagner [4] replaced a new surrogate loss function with the cross-entropy function to enhance the attack performance. Moosavi-Dezfooli et al. [18] assumed that the DNNs are completely linear and then utilize the linear property to generate efficiently adversarial examples.

2.2 Meta Learning

Meta-learning aims to accumulate knowledge acquired from previous tasks for guiding the model to fit in a new task similar to previous tasks quickly. There are two mainstream kinds of meta-learning: optimization-based meta-learning and metric-based meta-learning. Optimization-based meta-learning [2,9,13,20] desires to train a universal model that performs well on new tasks after finetuning with a few examples. Finn et al. [9] suggested adopting the parameters of the model trained in many tasks as initial parameters in new tasks. Metric-based meta-learning [11,22,23] hopes to obtain a great embedding representation for all examples and then predicting the label of an unknown example based on the distance between the example and training example. For instance, Hoffer and Ailon [11] proposed Siamese network in which whether the two examples belong to the same class is done by computing their distance. In Snell et al. [22], the authors constructed distinct prototypes for each class and compared the distance between the input and each prototype to predict.

3 Model-Agnostic Attack

In this section, we will elaborate proposed method. For conciseness, the symbols in Sect. 2 is used.

The adversarial examples with vanilla white-box attacks are prone to overfit the proxy model (causes the poor attack performance against black-box model) because vanilla white-box attacks blend the model-specified noises into the generated adversarial examples. Besides, the model-specified noises are attributed to the intrinsic difference over diverse models. Therefore, the current hurdle is how to reduce the model-specified noises contained in the crafted adversarial noises. It is also equivalent to how

Algorithm 1: MAA

Input: x: the arbitrary natural example given by attackers,
y: ground-truth label of x,
$M = \{f_1, f_2, \cdots, f_n\}$: f_i is i-th proxy model with loss function,
$L(\cdot,\cdot)$: Loss function,
T: total iterations,
ϵ: the size of perturbation,
α, β: hyperparameter.
Output: x_{adv}^T: adversarial example.
Initialize $x_{adv}^0 = x$;
for $j = 1$ *to* T **do**
  Initialize g to empty list;
  for *all* f_i **do**
    compute $noise_i^j = \nabla_{x_{adv}^{j-1}} L(f_i(x_{adv}^{j-1}), y)$;
    let $x_{i,adv}^j = x_{adv}^{j-1} + noise_i^j$;
    compute $\nabla_{x_{i,adv}^j} L(x_{i,adv}^j, y)$ and join it to g;
  update $x_{adv}^j = x_{adv}^{j-1} + \beta \cdot noise$;
  clip x_{adv}^j, $x_{adv}^j = Clip_{x,\epsilon}(x_{adv}^j)$;
return x_{adv}^T.

to align the crafted adversarial noises with model-shared vulnerable direction. Moreover, meta-learning is an effective way to produce an well-generalized model on black-box tasks based on serveral white-box tasks. Therefore, meta-learning approaches can be adopted to relieve the overfitting issue and hence the transferable adversarial examples cross models are crafted in this way. As a famous method in the meta-learning field, model-agnostic meta-learning (MAML) adopts a novel strategy to train a well-generalized model on new tasks. Next, we introduce how to leverage MAML on the adversarial attack to craft transferable adversarial examples.

Assume that there are several models of $f_1, f_2, \cdots, f_n$ that is a mild condition in the black-box scenario [14]. $L(\cdot,\cdot)$ is common loss function and we adopt cross-entropy loss function throughout this paper. As shown in Fig. 2, in each iteration, each model is regarded as a distinct task, and our goal is to produce adversarial examples that can be transferred in the new task. Overall, crafting adversarial examples in MAA is an iteration process, and each iteration consists of two steps. In the first step, the example is fed into all proxy models to craft adversarial noises with identical adversarial attack methods, such as FGSM. Formally, given total iterations T, this step can be expressed as follows:

$$
\begin{aligned}
x_{adv}^0 &= x, \\
noise_i^j &= \nabla_{x_{adv}^{j-1}} L(f_i(x_{adv}^{j-1}), y),\ i = 1, 2, \cdots, n, \\
j &= 1, 2, \cdots, T,
\end{aligned} \tag{5}
$$

where x_{adv}^j and $noise_i^j$ denote the adversarial example crafted in $j-th$ iteration and adversarial noise for $i-th$ proxy model in $j-th$ iteration, respectively. Then, MAA craft adversarial examples for each model (task) separately:

$$x^j_{i,adv} = x^{j-1}_{adv} + \alpha \cdot noise^j_i, \tag{6}$$

where $x^j_{i,adv}$, x^{j-1}_{adv}, and α indicate the adversarial example for $i-th$ proxy model in $j-th$ iteration, crafted adversarial example in $(j-1)-th$ iteration, and hyperparameter given by the attackers.

Next, similar to the first step, these adversarial examples $x^j_{i,adv}$ $(i = 1, 2, \cdots, n)$ are fed into corresponding proxy model. Furthermore, the new adversarial noises (same method in first step, $\nabla_{x^j_{i,adv}} L(x^j_{i,adv}, y)$ $(i = 1, 2, \cdots, n)$) are obtained and then fused to form $noise^j$ that is used to update the x^{j-1}_{adv}. In detail, the procedure can be written as follows:

$$noise^j = \frac{1}{n}\sum_{i=1}^{n} \nabla_{x^j_{i,adv}} L(x^j_{i,adv}, y), \tag{7}$$

Finally, the adversarial example in $i-th$ iteration is crafted:

$$\begin{aligned} x^j_{adv} &= x^{j-1}_{adv} + \beta \cdot noise^j, \\ x^j_{adv} &= Clip_{x,\epsilon}(x^j_{adv}), \end{aligned} \tag{8}$$

where β is hyperparameter given by the attackers and $Clip_{x,\epsilon}(\cdot)$ is used to constrain the difference between the crafted adversarial examples and the natural examples into a certain range. The full algorithm is outlined in Algorithm 1.

Intuitively, MAA first crafts threatening adversarial examples for each base model f_i and then further evaluate the vulnerable directions of these threatening adversarial examples against each base model f_i. Then, the vulnerable directions are collected and blended by average to form the final update direction (adversarial noise) at the iteration. Furthermore, the shared vulnerable directions across diverse models can be maintained, which are probably susceptible to any agnostic black-box model. Meanwhile, the model-specified directions have to weaken due to fuse operation. Consequently, MAA can craft transferable adversarial examples over multiple black-box models based on a handful of proxy models.

Explain Why MAA Works. We introduced how to integrate the MAML into existing methods. But MAML assumes that the known and unknown task, i.e., proxy models and black-box models in context of MAA, should be similar. Next, we explain why the existing models is similar so as to show that MAA can produce transferable adversarial examples across various models. There are three reasons for it. Firstly, it is natural that two similar models should have almost consistent predictions over different inputs. Notice that well-trained models (at least most CNNs such as ResNet, EfficientNet, etc.) achieve great performance on datasets (e.g., higher than 90% accuracy on CIFAR-10), indicating the same predictions from these models to the data over the dataset. Therefore, there is a good similarity between these models. Secondly, we observe that current mainstream models contain many identical modules, such as residual modules widely used in convolutional neural networks. These modules coupled with similar hyperparameters, e.g., comparable learning rates or the size of filters, significantly facilitate the similarity over the models. Thirdly, data is the lifeblood of DNNs, and many DNNs share same or akin training datasets. For example, most of existing models build on pre-trained backbone models (such as ResNet50). Therefore, it is believed arguably that current models are similar.

Comparison to Ensemble Attack. Interestingly, we find that ensemble attack (EA) [14] is sort of similar, but not identical, to MAA. In EA, the loss functions of all proxy models are coarsely integrated as one for producing adversarial examples during each iteration. Intuitively, we plot Fig. 3 to compare the difference of EA and MAA. Formally, the EA can be expressed with our symbols as follows:

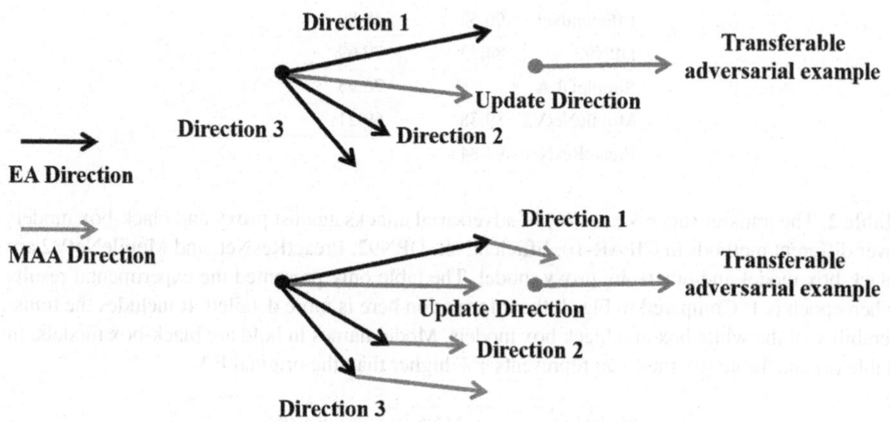

Fig. 3. Difference between EA and MAA.

$$
\begin{aligned}
x_{adv}^{0} &= x, \\
x_{adv}^{j} &= x_{adv}^{j-1} + \frac{1}{n}\sum_{i=1}^{n} \nabla_{x_{adv}^{j-1}} L(f_i(x_{adv}^{j-1}), y), \\
j &= 1, 2, \cdots, T.
\end{aligned} \tag{9}
$$

Compared to EA, in each iteration, MAA first craft threatening adversarial examples and harness them to obtain more deceptive directions than crafted directions by EA. Although the difference introduced by one iteration may be slight, the difference accumulates during iterations and is ultimately non-trivial. Hence, the adversarial examples generated by MAA are more transferable than EA, which is empirically confirmed in the experiments.

4 Experiment Results

4.1 Experiment Setup

Model. We prepared a total of nine models for the experiment and randomly split them into two groups, i.e., proxy models and black-box models. Proxy models included

Table 1. The accuracy of proxy and black-box models. On the left is the name of the model. These models are all trained on the corresponding training set. The two columns on the right are the accuracy of the corresponding model on the test set of CIFAR-10 and CIFAR-100. The accuracy of the most advanced models on CIFAR-10 and CIFAR-100 is about 95% and 80%.

Model	Cifar-10 (%)	Cifar-100 (%)
VGG16	93.69	71.40
ResNet18	95.40	77.15
GoogleNet	95.60	80.01
ShuffleNetV2	89.85	69.70
MobileNet	87.34	64.43
EfficientNet	90.62	68.80
DPN92	84.53	79.69
SimpleDLA	–	76.95
MobileNetV2	89.38	69.21
PreactResNet	94.84	–

Table 2. The transfer success rates(%) of adversarial attacks against proxy and black-box models over different methods in CIFAR-10. EfficientNet, DPN92, PreactResNet, and MobileNetV2 are black-box model and others are proxy model. The table only presented the experimental results when epoch is 1. Compared to Fig. 4, the information here is more detailed. It includes the transferability of the white box and black box models. Model names in bold are black-box models. In Table (a) and Table (b), the $(+x)$ represents $x\%$ higher than the original EA.

Model	MAA (%)	EA (%)
EfficientNet	79.35(+46.96)	32.39
DPN92	77.76(+36.70)	41.06
PreactResNet	75.87(+53.67)	22.20
MobileNetV2	82.04(+47.26)	34.78
VGG16	72.26(+42.04)	30.22
ResNet18	76.68(+49.58)	27.10
GoogleNet	85.67(+54.13)	31.54
ShuffleNetV2	86.75(+43.75)	43.00
MobileNet	87.00(+39.50)	47.50
Mean of all models	80.38(+45.96)	34.42

VGG16, ResNet18, GoogleNet, ShuffleNetV2, and MobileNetV2 to produce transferable adversarial examples, while black-box models included EfficientNet, DPN92, PreactResNet, MobileNetV2 to validate the performance of MAA. These model are trained with common procedure using public code and detailed informations can be refered to the site[1]. Because attacking models with poor performance is useless, the model is required to achieve great performance and we follow the rule. Table 1 presents the accuracy of the models (on the test set) used in this paper. Besides, PreactResNet was

[1] https://github.com/kuangliu/pytorch-cifar/.

Table 3. The transfer success rates(%) of adversarial attacks against proxy and black-box models over different methods in CIFAR-100. EfficientNet, DPN92, SimpleDLA, and MobileNetV2 are black-box model and others are proxy model.

Model	MAA(%)	EA(%)
EfficientNet	93.24(+30.95)	62.29
SimpleDLA	90.37(+38.46)	51.91
DPN92	91.24(+39.38)	51.86
MobileNetV2	92.98(+30.74)	62.24
Average	91.96(+34.88)	57.08
VGG16	89.77(+28.19)	61.58
ResNet18	94.28(+32.76)	61.52
GoogleNet	93.79(+31.71)	62.08
ShuffleNetV2	94.29(+29.41)	64.88
MobileNet	93.60(+21.71)	71.89
Mean of all models	92.62(+31.48)	61.14

replaced with SimpleDLA on CIFAR-100 due to PreactResnet performed poorly on CIFAR-100.

Baseline. We first highlight MAA is additive and can be applied in combination with arbitrary technologies which can boost transferability of adversarial examples, except EA due to the same role played in the algorithm. Therefore, only EA was selected to compare because of the restricted capacity of the paper.

Datasets. We examine the performance of MAA on two widely-used benchmark datasets, namely CIFAR-10 and CIFAR-100, wherein test sets (10000 images) were served as attacked images. For fairness, all experiments with pytorch1.6 and Nvidia GTX 2070 on the Ubuntu 16.02 system.

Attack Setting. The total iterations T for EA and MAA was set to the same value for fair comparison ($T = 1, 3, 5$). The α and β for MAA and EA in this paper were set to $\frac{\epsilon}{T}$.

Metric. To effectively measure the transferability generated adversarial examples, the misclassified rate of models to these generated adversarial examples is selected as the metric of performance.

4.2 Comparison to State-of-the-Art

We evaluated the effectiveness of EA and MAA against diverse black-box models and proxy models in two baseline dataset CIFAR-10 and CIFAR-100, and results were reported in Table 2 and Table 3.

Overall, as shown in Table 2 and Table 3, MAA consistently obtains better transfer results compared EA over various black-box models. As an example of the conclusion, in CIFAR-10, the transferability of adversarial examples generated by MAA

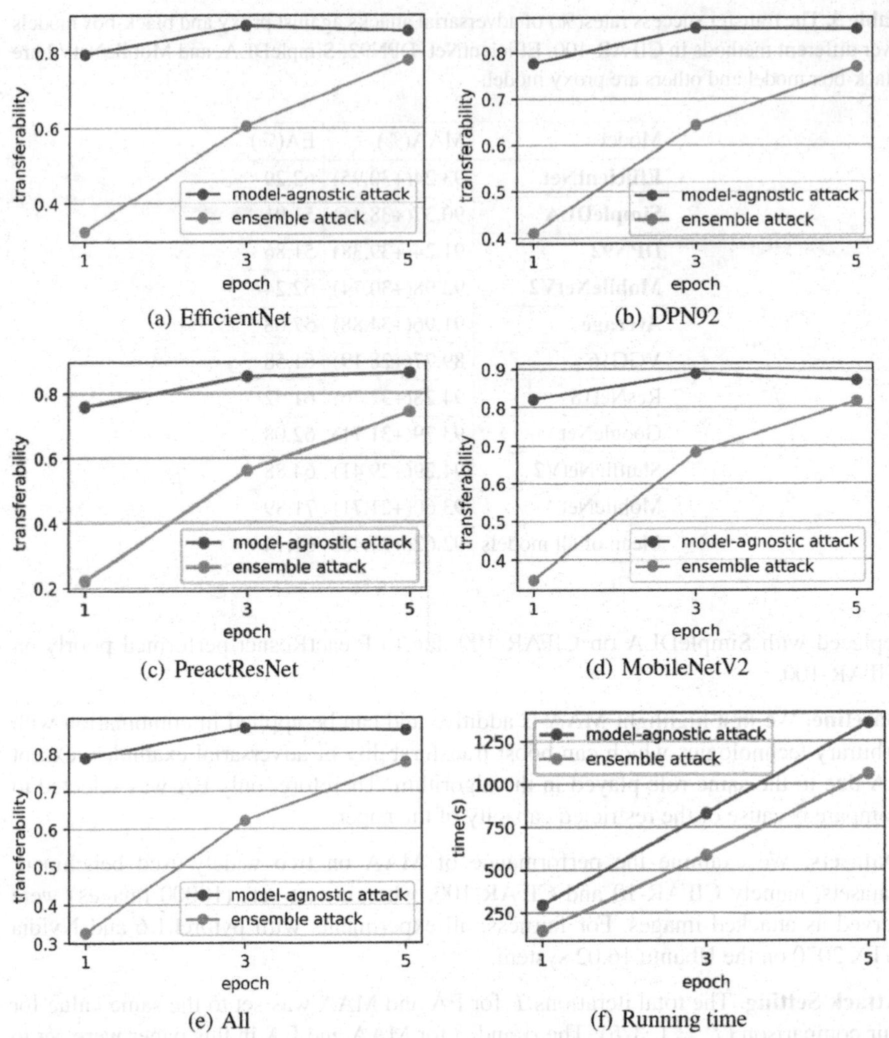

Fig. 4. Transfer result against different black-box models in CIFAR-10. The data used is a test set of CIFAR-10, which has a total of 10,000 images. (e) shows the average transferability against all black-box models at different epochs. The successful transfer of a sample indicates that its adversarial example is misclassified on the specified model. (a-d) shows the transferability against different black-box models. (f) shows the running time of model-agnostic attack and ensemble attack. Finally, here MAA is pure MAA that is no other enhancement measures are added.

significantly surpasses EA by 36.70%–53.67%, which is a considerable improvement compared to EA. Besides, the phenomenon in CIFAR-10 has to happen in CIFAR-100 as shown in Table 3. We also find that the transfer results in CIFAR-100 are strikingly higher than CIFAR-10, and it is natural because the accuracy of models in CIFAR-100 is lower than CIFAR-10.

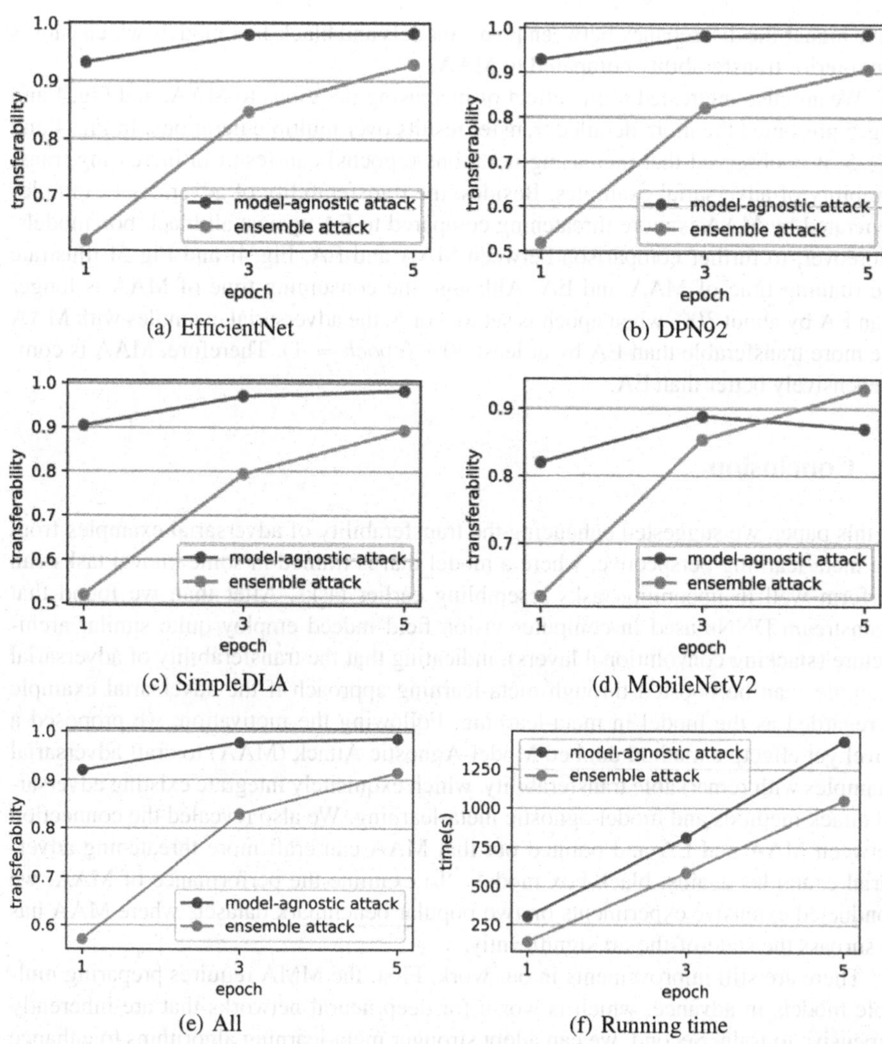

(a) EfficientNet (b) DPN92 (c) SimpleDLA (d) MobileNetV2 (e) All (f) Running time

Fig. 5. Transfer result against different black-box models in CIFAR-100.

Moreover, the transfer results of black-box models only slightly differ from proxy models, which empirically determines that adversarial examples generated by MAA can not overfit proxy models. There is another evidence for it. The transfer results of white-box models not achieve 100% (white-box attacks can easily obtain 100% transfer results) indicate that crafted adversarial examples do not overfit proxy models to a certain extent. Thus, the adversarial examples produced by MAA contain fewer model-specified noises than common adversarial examples so as to have higher transferability across diverse black-box models. In contrast, the adversarial examples crafted by EA have higher the discrepancy between proxy models and black-box models which causes the inferior transferability compared to MAA.

We are also interested in the effect of increasing iterations to MAA, and Fig. 4 and Fig. 5 presented the more detailed transfer results over multiple iterations. In Fig. 4 and Fig. 5, it is observed that increasing iterations (epochs) can result in increasing transferability of adversarial examples. Besides, the transferability of adversarial examples generated by MAA is more threatening compared to EA against all black-box models. Moreover, to further comparison between MAA and EA, Fig. 4f and Fig. 5f illustrate the running time of MAA and EA. Although the consuming time of MAA is longer than EA by about 30% when epoch is set to 3 or 5, the adversarial examples with MAA are more transferable than EA by at least 30% ($epoch = 1$). Therefore, MAA is comprehensively better than EA.

5 Conclusion

In this paper, we suggested enhancing the transferability of adversarial examples from the meta-learning perspective, where a model that is trained in some known tasks can perform well in upcoming tasks resembling earlier tasks. After that, we found that mainstream DNNs used in computer vision field indeed employ quite similar architecture (stacking convolutional layers), indicating that the transferability of adversarial examples can be boosted through meta-learning approach if the adversarial example is regarded as the model in meta-learning. Following the motivation, we proposed a novel yet effective method dubbed Model-Agnostic Attack (MAA) to craft adversarial examples with remarkable transferability, which exquisitely integrate existing adversarial attack methods and model-agnostic meta-learning. We also revealed the connection between MAA and EA and pointed out that MAA can craft more threatening adversarial examples against black-box models. To examine the performance of MAA, we conducted extensive experiments on two popular benchmark datasets where MAA has to surpass the state-of-the-art significantly.

There are still improvements in our work. First, the MMA requires preparing multiple models in advance, which is worse for deep neural networks that are inherently expensive to train. Second, we can adopt stronger meta-learning algorithms to enhance the transferability of adversarial examples further, which is a promising direction. We leave these issues in future works.

References

1. Al-Dujaili, A., O'Reilly, U.M.: Sign bits are all you need for black-box attacks. In: International Conference on Learning Representations (2019)
2. Al-Shedivat, M., Bansal, T., Burda, Y., Sutskever, I., Mordatch, I., Abbeel, P.: Continuous adaptation via meta-learning in nonstationary and competitive environments. arXiv preprint arXiv:1710.03641 (2017)
3. Batmaz, Z., Yurekli, A., Bilge, A., Kaleli, C.: A review on deep learning for recommender systems: challenges and remedies. Artif. Intell. Rev. **52**(1), 1–37 (2018). https://doi.org/10.1007/s10462-018-9654-y

4. Carlini, N., Wagner, D.: Towards evaluating the robustness of neural networks. In: 2017 IEEE Symposium on Security and Privacy (SP), pp. 39–57. IEEE (2017)
5. Chang, H., et al.: A restricted black-box adversarial framework towards attacking graph embedding models. In: Proceedings of the AAAI Conference on Artificial Intelligence, vol. 34, pp. 3389–3396 (2020)
6. Croce, F., Hein, M.: Reliable evaluation of adversarial robustness with an ensemble of diverse parameter-free attacks. In: International Conference on Machine Learning, pp. 2206–2216. PMLR (2020)
7. Dong, Y., Pang, T., Su, H., Zhu, J.: Evading defenses to transferable adversarial examples by translation-invariant attacks. In: Proceedings of the IEEE/CVF Conference on Computer Vision and Pattern Recognition, pp. 4312–4321 (2019)
8. Ebrahimi, J., Rao, A., Lowd, D., Dou, D.: Hotflip: white-box adversarial examples for text classification. In: Proceedings of the 56th Annual Meeting of the Association for Computational Linguistics (Volume 2: Short Papers), pp. 31–36 (2018)
9. Finn, C., Abbeel, P., Levine, S.: Model-agnostic meta-learning for fast adaptation of deep networks. In: International Conference on Machine Learning, pp. 1126–1135. PMLR (2017)
10. Goodfellow, I.J., Shlens, J., Szegedy, C.: Explaining and harnessing adversarial examples. Stat **1050**, 20 (2015)
11. Hoffer, E., Ailon, N.: Deep metric learning using triplet network. In: Feragen, A., Pelillo, M., Loog, M. (eds.) SIMBAD 2015. LNCS, vol. 9370, pp. 84–92. Springer, Cham (2015). https://doi.org/10.1007/978-3-319-24261-3_7
12. Kurakin, A., Goodfellow, I.J., Bengio, S.: Adversarial examples in the physical world. arXiv preprint arXiv:1607.02533 (2016)
13. Lee, Y., Choi, S.: Gradient-based meta-learning with learned layerwise metric and subspace. In: International Conference on Machine Learning, pp. 2927–2936. PMLR (2018)
14. Liu, Y., Chen, X., Liu, C., Song, D.: Delving into transferable adversarial examples and black-box attacks. arXiv preprint arXiv:1611.02770 (2016)
15. Madry, A., Makelov, A., Schmidt, L., Tsipras, D., Vladu, A.: Towards deep learning models resistant to adversarial attacks. arXiv preprint arXiv:1706.06083 (2017)
16. Maho, T., Furon, T., Le Merrer, E.: Surfree: a fast surrogate-free black-box attack. In: Proceedings of the IEEE/CVF Conference on Computer Vision and Pattern Recognition, pp. 10430–10439 (2021)
17. Mao, X., Chen, Y., Wang, S., Su, H., He, Y., Xue, H.: Composite adversarial attacks. In: Proceedings of the AAAI Conference on Artificial Intelligence, vol. 35, pp. 8884–8892 (2021)
18. Moosavi-Dezfooli, S.M., Fawzi, A., Frossard, P.: Deepfool: a simple and accurate method to fool deep neural networks. In: Proceedings of the IEEE Conference on Computer Vision and Pattern Recognition, pp. 2574–2582 (2016)
19. Papernot, N., McDaniel, P., Jha, S., Fredrikson, M., Celik, Z.B., Swami, A.: The limitations of deep learning in adversarial settings. In: 2016 IEEE European Symposium on Security and Privacy (EuroS&P), pp. 372–387. IEEE (2016)
20. Rusu, A.A., et al.: Meta-learning with latent embedding optimization. In: International Conference on Learning Representations (2018)
21. Shukla, S.N., Sahu, A.K., Willmott, D., Kolter, Z.: Simple and efficient hard label black-box adversarial attacks in low query budget regimes. In: Proceedings of the 27th ACM SIGKDD Conference on Knowledge Discovery & Data Mining, pp. 1461–1469 (2021)
22. Snell, J., Swersky, K., Zemel, R.S.: Prototypical networks for few-shot learning. arXiv preprint arXiv:1703.05175 (2017)
23. Sung, F., Yang, Y., Zhang, L., Xiang, T., Torr, P.H., Hospedales, T.M.: Learning to compare: relation network for few-shot learning. In: Proceedings of the IEEE Conference on Computer Vision and Pattern Recognition, pp. 1199–1208 (2018)

24. Szegedy, C., et al.: Intriguing properties of neural networks. arXiv preprint arXiv:1312.6199 (2013)
25. Torfi, A., Shirvani, R.A., Keneshloo, Y., Tavvaf, N., Fox, E.A.: Natural language processing advancements by deep learning: a survey. arXiv preprint arXiv:2003.01200 (2020)
26. Wu, D., Wang, Y., Xia, S.T., Bailey, J., Ma, X.: Skip connections matter: on the transferability of adversarial examples generated with resnets. arXiv preprint arXiv:2002.05990 (2020)
27. Wu, L., Zhu, Z., Tai, C., et al.: Understanding and enhancing the transferability of adversarial examples. arXiv preprint arXiv:1802.09707 (2018)
28. Yan, Z., Guo, Y., Liang, J., Zhang, C.: Policy-driven attack: learning to query for hard-label black-box adversarial examples. In: International Conference on Learning Representations (2020)
29. Yokota, T., et al.: A conformable imager for biometric authentication and vital sign measurement. Nature Electron. 3(2), 113–121 (2020)
30. Yu, Y., Gao, X., Xu, C.Z.: Lafeat: piercing through adversarial defenses with latent features. In: Proceedings of the IEEE/CVF Conference on Computer Vision and Pattern Recognition, pp. 5735–5745 (2021)
31. Zhao, P., et al.: On the design of black-box adversarial examples by leveraging gradient-free optimization and operator splitting method. In: Proceedings of the IEEE/CVF International Conference on Computer Vision, pp. 121–130 (2019)
32. Zhou, M., Wu, J., Liu, Y., Liu, S., Zhu, C.: Dast: data-free substitute training for adversarial attacks. In: Proceedings of the IEEE/CVF Conference on Computer Vision and Pattern Recognition, pp. 234–243 (2020)
33. Zou, Z., Shi, Z., Guo, Y., Ye, J.: Object detection in 20 years: a survey. arXiv preprint arXiv:1905.05055 (2019)

Temporal Convolution Network Based on Attention for Intelligent Anomaly Detection of Wind Turbine Blades

Jianwen Ding, Fan Lin(✉), and Shengbo Lv

School of informatics, Xiamen University, Xiamen 361000, China
iamafan@xmu.edu.cn

Abstract. With the concept of "carbon neutral", wind power industry ushers in a golden period of development. Wind turbine blades are the key components of wind turbine, their normal operation directly affects the operation of the whole wind power generation. The traditional diagnosis method of wind turbine blades needs to disassemble the wind turbine blades. However, due to the potential of long-short-term memory network model and feedforward network based on attention mechanism in anomaly detection and diagnosis, sequence model has become the mainstream of wind turbine blades anomaly detection. Unfortunately, RNN needs to wait for the end of the forward transmission of the previous time step before the forward transmission of the next time step. It can only input all the historical information by default, and can not achieve the fine control of the input information. Therefore, this paper applies TCN-ATT, a temporal convolution network based on attention mechanism, to intelligent anomaly detection of wind turbine blades by combining dilation convolution, causal convolution, the skip connection of residual blocks and attention module. In this method, causal convolution and dilation convolution are used to adapt to the frame of multi-dimensional time series data of wind turbine blades, which provides more flexibility for changing the size of receptive field. At the same time, attention mechanism is added to capture the relevant features in the sequence, so as to improve the classification accuracy of anomaly detection of wind turbine blades. To verify the effectiveness of the system, this paper compares the performance of TCN-ATT with TCN, LSTM, GRU and other methods. The comparative test shows that TCN-ATT has better performance of feature extraction and anomaly detection. In addition, TCN-ATT is applied to multi fault anomaly detection, which also achieves high classification accuracy.

Keywords: Temporal convolution network · Attention · Time series data · Anomaly detection · Wind turbine blades

1 Introduction

Wind turbine blade is an important part of wind turbine, which undertakes the important work of wind power generation. Because the working environment of

Y. Lai et al. (Eds.): ICA3PP 2021, LNCS 13155, pp. 193–209, 2022.
https://doi.org/10.1007/978-3-030-95384-3_13

the wind turbine blades is very complex, it needs to face high-speed wind, shear wind and other bad weather, and the operation of the wind turbine blades is controlled by the spindle and bearing, so the wind turbine blades are prone to abnormal, which may have serious potential consequences. In order to avoid the loss of profits caused by abnormal, power enterprises usually carry out regular preventive maintenance on the wind turbine blades. However, the traditional preventive maintenance is uneconomic and unreasonable, unable to reconcile the contradiction between "insufficient maintenance" and "excessive maintenance" [1]. Based on this, power enterprises now mainly use industrial stethoscope to monitor the running state of the wind turbine blades in real time, and carry out preventive maintenance on the wind turbine blades through various time sequence signals.

At present, deep learning is recognized as a powerful machine learning method. Its strong fault tolerance, robustness, learning prediction ability and parallel processing ability make it play a huge role in the modern industrial environment with many exceptions and complex equipment structure. Intelligent condition monitoring models using deep learning method have been widely used in induction motor [2], planetary gearbox [3], axial piston pump [4] and other systems. In recent years, many excellent deep learning methods have also been applied to wind turbine blades anomaly detection. R rahimilarki et al. implemented a method for diagnosing even small anomalies in the output of wind turbine system based on convolutional neural network [5]. [6] used integration technology, made full use of all features extracted from different hidden layers of deep self encoder network, and constructed an integrated anomaly detection model. D Xian et al. used random forest to screen the variables related to the target variables in SCADA, and used the screened variables and target variables to construct a deep neural network [7]. Inspired by these successful cases, this paper applied intelligent anomaly diagnosis technology based on deep learning to condition monitoring of wind turbine blades. The deep learning method adopted here is temporal convolution network based on attention mechanism.

Temporal convolutional network (TCN) is a network structure based on convolutional neural network, which can process time series data, proposed by Bai and Kolter [8] in 2018. TCN is actually a CNN based method. The basic TCN architecture mainly includes causal convolution, dilation convolution and residual block. Causal convolution considers the causality between variables and the time series characteristics of data, which ensures that the elements in the output sequence can only depend on the elements before it in the input sequence. Dilation convolution provides more flexibility to change the size of receptive field without changing the parameters, which is conducive to the extraction of advanced features. The residual block can solve the problem of gradient vanishing caused by the deepening of the network layer. Yan et al. [9] used TCN for the task of weather forecast, and compared TCN with LSTM, successfully verifying that TCN has great advantages in the prediction task of time series data. Dai et al. [10] proposed a hybrid spatiotemporal graph convolution network H-STGCN, which uses the segment-line-flow-density relationship to convert the upcoming traffic volume into the equivalent travel time. Guirguis et al. [11] recently proposed a novel SELD-TCN architecture for sound events, and claimed that the framework was ahead of the most advanced technology at that time, and the training time was shorter.

Due to the good performance of TCN in processing temporal data, we propose TCN-ATT based on attention mechanism. TCN-ATT is based on the standard TCN, adding the attention module to effectively obtain the global and local contact. Compared with the recurrent neural network (RNN), TCN-ATT has less parameters, low model complexity, strong parallel computing ability and faster training speed. Through TCN network, TCN-ATT overcomes the shortcoming that the attention mechanism cannot learn the order relation in the sequence, and it can also change the input size and filter size, increase the dilation factor and stack more layers, showing greater advantages in time series modeling. TCN-ATT is mainly composed of two modules: temporal convolution module and attention module. Among them, the temporal convolution module adopts one-dimensional full convolution, and adds 0-padding in each layer to achieve the same input length and output length; at the same time, in order to reflect the time sequence characteristics, causal convolution is introduced [12]. When the sequence length is increasing, in order to ensure that historical information is not missed and control the network depth and parameters, the temporal convolution module introduces dilation convolution [13] to expand the receptive field. In addition, in order to prevent the gradient from disappearing due to the increase of network depth, we also add a residual module into the temporal convolution module [14]. The attention module gets the weight coefficient of each key by calculating the similarity between the target vector query and each group of keys, and then gets the final attention value by weighted summation of value, which helps the model to capture important information. Finally, the effectiveness of TCN-ATT model in intelligent detection of single abnormality and multiple faults of fan blades is verified through laboratory experiments, and the performance of TCN-ATT model is compared with that of autoencoder [15], GRU [16], long-short-term memory neural network (LSTM) [17], LSTM with attention mechanism [18] and standard TCN. The results show that TCN-ATT method is superior to other methods in classification accuracy.

Overall, this paper has made the following contributions:

- Abnormal detection of fan blade based on TCN. TCN model was introduced into fan blade anomaly detection task, and three core modules, causal convolution, dilated convolution and residual block, were introduced to help improve the accuracy of fan blade anomaly detection.
- Attention mechanisms. The TCN-ATT model adds an attention module that focuses on more useful features and is flexible enough to capture global and local connections in one step.
- Comprehensive evaluation of TCN-ATT. This paper evaluates that TCN-ATT can achieve high accuracy in both single anomaly detection and multi-fault detection experiments. In addition, the importance of residual block and attention module in TCN-ATT to improve the accuracy of anomaly detection is verified by ablation experiments.

2 Model Architecture

In this section, we introduce the basic architecture of TCN-ATT in detail. In Sect. 2.1, this paper introduces the definition of anomaly detection task of time series data of wind turbine blades, and then describes the processing process of time series data from input to output. Section 2.2 and Sect. 2.3 introduces two modules of TCN-ATT in detail: temporal convolution module and attention module. Section 2.2 introduces three sub modules of temporal convolution module, including causal convolution, dilation convolution and residual module. Section 2.3 deduces the theoretical basis of attention module mathematically.

2.1 Anomaly Detection Task of Wind Turbine Blade

Given a time series $X = \{x_1, x_2, ..., x_t\}$, in which $x_k \in R^m$ $(1 \leq k \leq t)$, representing all kinds of data signals at the kth timestep monitoring by industrial stethoscopes. The task of anomaly detection of time series data of wind turbine blade is to train a sequence model with supervision so that with the input data X, we can accurately judge whether there is an outlier for the corresponding time step y_k in the output time series $Y = \{y_1, y_2, ..., y_t\}$, where $y_k \in \{0, 1\}(1 \leq k \leq t)$. The sequence model we train is shown as follows:

$$Y = SequenceModel(X) \tag{1}$$

The sequence model needs to satisfy two constraints: (1) the model only processes the information forward according to time and the y_t in the time series of the outputs $\{y_1, y_2, ..., y_t\}$ only depends on $\{x_1, x_2, ..., x_t\}$, having nothing to do with the future information $\{x_{t+1}, x_{t+2}, ...,\}$; (2) input sequences of any length need to be mapped to output sequences of the same length. At the same time, we hope that the sequence model can achieve the accuracy of anomaly detection as high as possible.

2.2 Temporal Convolution Module

Based on the two constraints of the sequence model proposed in Sect. 2.1, The temporal convolution module of TCN-ATT adopts one-dimensional full convolution network architecture in order to achieve the principle of timeliness. At the same time, in order to ensure that the output sequence and the input sequence have the same length, zero padding will be applied. Then we take the dilation convolution and causality convolution with different kernel sizes as the hidden layer. After the last hidden layer, the classification layer generates a sequence of categories $\hat{Y} = \{\hat{y_1}, \hat{y_2}, ..., \hat{y_t}\}$. All in all, calculating the intermediate variables of the lth layer at the tth timestep needs to go through three steps, as shown in Fig. 1.

(1) First of all, for the intermediate variables of $(l-1)$th layer $I_t^{l-1} = \{i_1^{l-1}, i_2^{l-1}, \ldots, i_t^{l-1}\}$, go through the causal convolution layer: $C_t^{l-1} = Conv1d(I_t^{l-1})$, where $C_t^{l-1} = \{c_1^{l-1}, c_2^{l-1}, \ldots, c_t^{l-1}\}$ is the output of causal convolution. To keep the input and output of each layer the same length, as shown in the beige square in Fig. 1, we use 0-padding on the left.

(2) The dilation convolution skips some input values according to a certain dilation factor, and applies the filter to the region larger than its size for convolution. By using dilation convolution, the receptive field of the network can be expanded to cover every input in the history. In the TCN-ATT model used in this paper, the size of dilation factor increases exponentially with the increase of network depth, as shown in the mark on the right of Fig. 1. The expansion factor d of the lth layer is 2^l.

(3) Before the eigenvector obtains the intermediate variables of the lth layer $I_t^l = \{i_1^l, i_2^l, \ldots, i_t^l\}$ through the activation function, add the residual block, as shown by the dotted line in Fig. 1. The residual block will train the deep network more effectively. In the residual block, in order to speed up the training speed and solve the problem of gradient disappearance in deep model, skip connections are used.

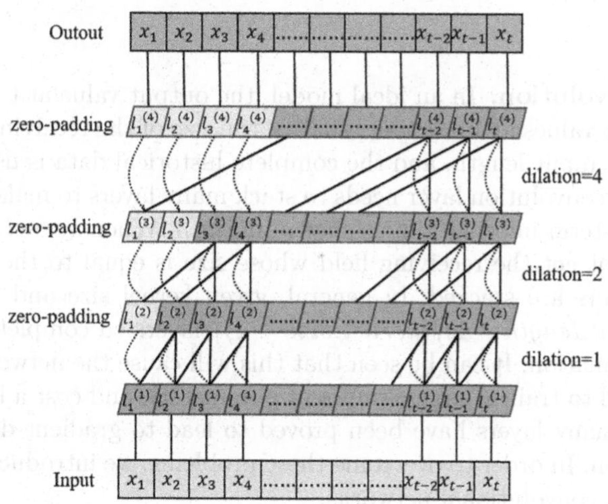

Fig. 1. Temporal convolution module

Causal Convolution. For causality, $I_t^{l-1} = \{i_1^{l-1}, i_2^{l-1}, \ldots, i_t^{l-1}\}$ obtains the output sequence $C_t^{l-1} = \{c_1^{l-1}, c_2^{l-1}, \ldots, c_t^{l-1}\}$ by causal convolution. The kth element of C_t^{l-1} can only depend on the first k elements in I_t^{l-1}. In other words, the elements in the temporal data output by causal convolution can only depend on the elements before it in the input temporal data. According to Sect. 2.1, in order to ensure that the output tensor and the input tensor have the same length and realize causal convolution, we need to carry out zero padding on the left. An example with an input length of 4 and a kernel_size of 2 is shown in Fig. 2. The red square in Fig. 2 is the rightmost output element, which depends on i_t^{l-1} and i_{t-1}^{l-1} in the input sequence. The penultimate output element, the green square in Fig. 2, depends on the i_{t-1}^{l-1} and i_{t-2}^{l-1} in the input sequence. And so on, in order to get the same output length as the input length and obey the causality rule, we need to carry out zero padding on the left side of the input sequence. When the dilation factor is 1, the number of zero padding on the left side of the input sequence is always equal to the kernel size minus 1.

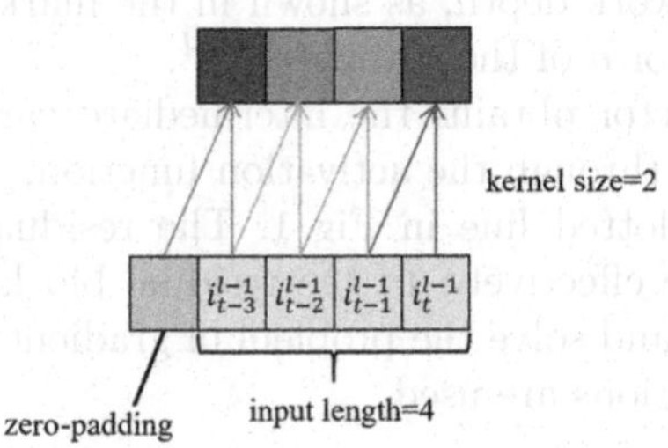

Fig. 2. Causal convolution

Dilation Convolution. In an ideal model, the output value at t time depends on all the input values at $1 \sim t$ time, that is, the size of the receiving field should be equal to the input length, and the complete historical data is used. However, the traditional convolution layer needs to stack many layers to make the network memorize long-term information, as shown in Fig. 3. When $kernel\_size = 2$, the network can not get the receiving field whose size is equal to the input length until three layers are stacked. In general, given kernel size and input length, we need $\lceil (input\_length - 1)/(kernel\_size - 1) \rceil$ layers to completely cover the historical information. It can be seen that this will cause the network to become very deep, need to train a large number of parameters, and cost a lot of time. In addition, too many layers have been proved to lead to gradient degradation of the loss function. In order to overcome these problems, we introduce the concept of dilation into convolutional networks.

The dilation factor d in dilation convolution refers to the distance between the elements of the input sequence. Therefore, the ordinary convolution network can be regarded as a dilation convolution with dilation factor is 1. When using dilation convolution, we will increase the dilation factor exponentially with the increase of network depth, which ensures that the size of the receiving field can be equal to the input length, covering all the historical information. Figure 4 shows a dilated casual convolution with dilation factor $d = 1, 2, 4$ and $kernel\_size = 3$.

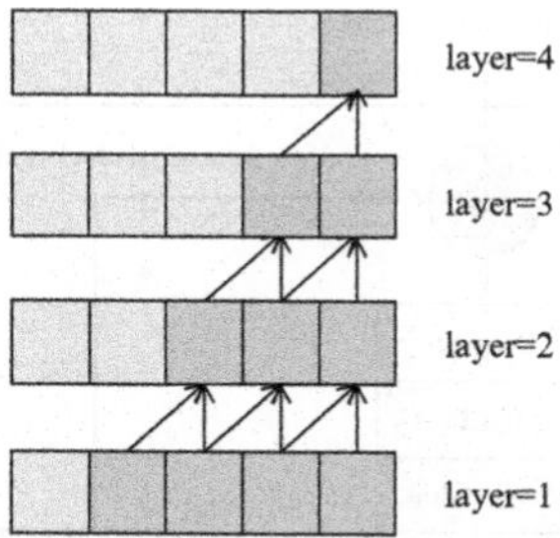

Fig. 3. Ordinary convolution network

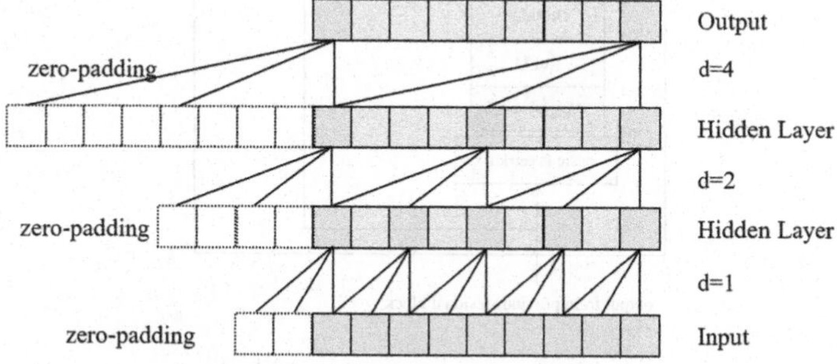

Fig. 4. Dilated casual convolution

Residual Block. S. Bai et al. [19] suggested that some additions should be made to the basic TCN architecture to improve the performance. The most important thing is to change the basic building block of the model from a simple one-dimensional causal convolution layer to a residual block composed of two layers with the same dilation factor and residual connections. The outputs of the two build-up layers are added to the input of the residual block, resulting in the input of the next residual block. At the same time, considering that the first convolution layer of the first residual block and the second convolution layer of the last residual block may have different input and output channel widths, we use 1×1 convolution to adjust the width of the residual tensor.

In addition, in order to normalize the input of hidden layer and counteract the problem of gradient burst, weight normalization is added to each convolution layer. ReLU activation is added to the residual block after the two convolution layers to introduce nonlinearity. Finally, in order to prevent overfitting, dropout is added after each convolution layer of each residual block, and regularization is also introduced. Figure 5 shows the structure of the final residual block.

2.3 Attention Module

In deep learning, generally speaking, the more layers and parameters the model has, the stronger the model is expressive and the more feature information it

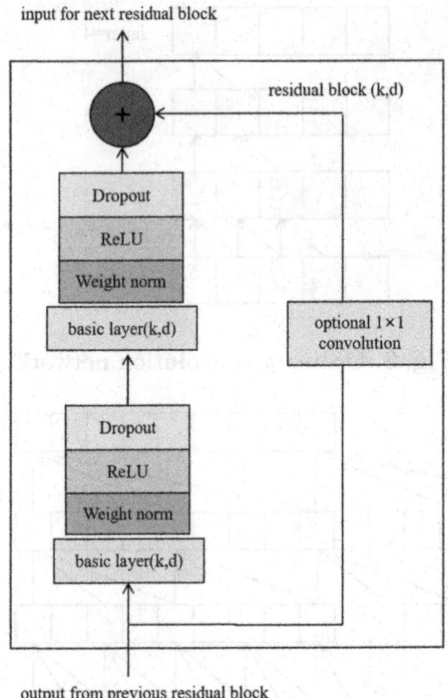

Fig. 5. Structure of residual block

reflects. However, this will cause the model to allocate the same amount of attention to important features and general features, leading to the problem of information overload. Compared with human visual attention mechanism, By scanning the global content, the attention mechanism can get the target area that needs to be focused on, and then invest more attention resources in this area to obtain more detailed information related to the target, while ignoring other irrelevant information, or even filtering out irrelevant information, which can solve the problem of information overload and improve the efficiency and accuracy of task processing.

The basic principle of attention mechanism is to imagine the elements in source as a series of $\langle key, value \rangle$ pairs. At this time, given a certain element query in target, the weight coefficient of each key corresponding value is obtained by calculating the similarity or correlation between query and each key, and then weighted sum the value, and then the final attention value is obtained. So in essence, attention mechanism is to sum up the value of elements in source, and query and key are used to calculate the weight coefficient of corresponding value. The basic formula is as follows, where Len is the length of source:

$$Att(query, source) = \sum_{i=1}^{Len} similarity(query, key_i) * value_i \tag{2}$$

In attention module, attention calculation can be divided into the following three stages:

(1) In the first stage, according to the query and key_i, calculate their similarity and correlation. In this paper, we calculate the similarity by finding the dot product of the two vectors.

$$similarity(query, key_i) = query \cdot key_i \tag{3}$$

(2) In the second stage, the numerical conversion of the score in the first stage can be carried out by adopting the calculation method similar to softmax. On the one hand, the original calculated score can be normalized and sorted into a probability distribution with the sum of the weights of all elements equal to 1. On the other hand, the inherent mechanics of softmax can be used to highlight the weight of important elements.

$$a_i = softmax(sim_i) = \frac{e^{sim_i}}{\sum_{i=1}^{Len} e^{sim_i}} \tag{4}$$

(3) The result a_i calculated in the second stage is the weight coefficient corresponding to $value_i$. In the third stage, the value of attention can be obtained only by carrying out the weighted sum.

$$Att(query, source) = \sum_{i=1}^{Len} a_i * value_i \tag{5}$$

The attention value of the attention module in the TCN-ATT model completely conforms to the above three-stage abstract calculation process, and the specific structure is shown in Fig. 6.

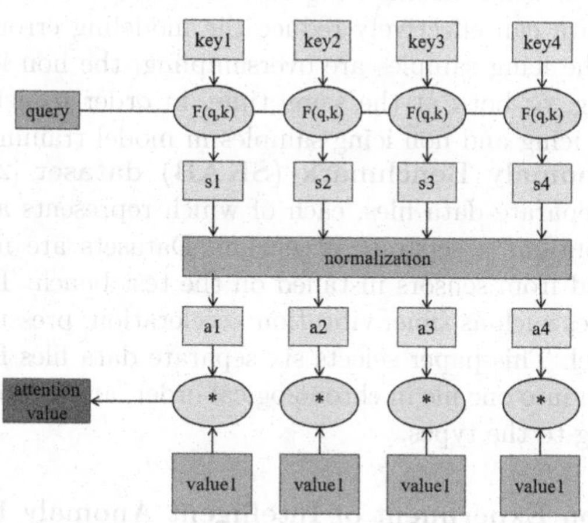

Fig. 6. Computational architecture of attention module

3 Experiments

In order to evaluate the effectiveness of our TCN-ATT model in dealing with the task of intelligent anomaly detection of wind turbine blades, we conducted an experiment on the actual data set of intelligent anomaly detection of wind turbine blades. Section 3.1 introduces the datasets used in the experiments. In Sect. 3.2, we designed the experiment. According to the time series data signal of the wind turbine blades obtained from the industrial stethoscope, we applied the proposed TCN-ATT model to judge whether the wind turbine blades were in an abnormal state at a certain moment, that is, to solve the simple declassification problem. We mainly compared the TCN-ATT model proposed in this paper with the AutoEncoder, GRU, LSTM, LSTM with attention mechanism and standard TCN on the same data set. In Sect. 3.3, we also designed a variety of anomaly detection experiments on wind turbine blades to verify the high efficiency of the TCN-ATT model in the task of multiple anomaly detection. In addition, we use the experimental data in Sect. 3.2 to verify the validity of the residual network in TCN and the TCN-ATT model proposed in this paper in Sect. 3.4.

3.1 Datasets

(1) **Blade icing prediction dataset of 2017 industrial big data:** The experimental data set consists of three parts: complete data set of wind turbine blades (without labels), time data of fault wind turbine blades and time data of normal working wind turbine blades. The complete wind turbine blades data set contains time and other 27 features. Each data in the complete wind turbine blades data set can be judged to be fault, normal or invalid by using the time data of fault wind turbine blades and normal working wind turbine blades. When observing the data set, the class imbalance problem of wind turbine icing data is considered. Generally speaking, resampling data can effectively reduce the modeling error caused by class imbalance. The icing samples are oversampling, the non icing samples are undersampling, or both at the same time, in order to achieve the similar proportion of icing and non icing samples in model training.

(2) **Skoltech Anomaly Benchmark (SKAB) dataset** [20]**:** This dataset contains 35 separate data files, each of which represents a separate experiment and contains a separate exception. Datasets are multivariate time series collected from sensors installed on the test bench. The data set contains 9 features such as time, vibration acceleration, pressure and temperature, and label. This paper selects six separate data files from the dataset, arranges them into one file in chronological order, and relabels the exception tags according to the types.

3.2 Comparison Experiment of Intelligent Anomaly Detection for Wind Turbine Blades

In order to verify the effectiveness of the TCN-ATT model in intelligent anomaly detection of wind turbine blades, this paper uses the 2017 industrial big data blade icing prediction dataset to carry out a comparative experiment, and compares the performance of the TCN-ATT model with that of AutoEncoder [21,22], LSTM [23,24], LSTM with Attention, GRU [25] and standard TCN. The comparison results are shown in Table 1. It can be seen from Table 1 that both the TCN model and the TCN-ATT model proposed in this paper based on TCN have achieved good detection results in the intelligent anomaly detection task of wind turbine blades. Due to the high quality of the experimental dataset used, the TCN-ATT model proposed in this paper can achieve 99.79% accuracy in the intelligent anomaly detection task of the wind turbine blade, and the accuracy rate, recall rate and F1 score can also reach 99%. Compared with the common deep learning method AutoEncoder, LSTM, LSTM with Attention and GRU, which achieved 72.33% accuracy rate, 86.18% accuracy rate, 91.84% accuracy rate and 94.17% accuracy rate respectively, the accuracy of TCN-ATT model increased by 5.62%–27.46%. In addition, it can be seen from Table 1 that, compared with the standard TCN model, the TCN-ATT model proposed in this paper improves the accuracy by 2.39% and the precision, recall rate and F1 score by 2%. Therefore, it can be seen that attention mechanism can focus on more useful features and then improve the accuracy of anomaly detection. Accordingly, TCN-ATT model not only can real-time detect the abnormal state of wind turbine blade effectively, but also due to the flexibility of dilation convolution in TCN-ATT, more layers can be stacked and the memory length of the model can be better controlled, it is to deal with the problem of temporal data signals anomaly detection performance is better than that of common anomaly detection models of deep learning. Through further study, it is found that the TCN-ATT model has similar or even smaller number of parameters compared with the commonly used anomaly detection models [26]. Therefore, with the continuous development of the Internet of Things and edge computing technology, TCN-ATT, a lightweight model, is more suitable for real-time and efficient intelligent anomaly detection of wind turbine blades at the edge end closer to the data source.

Table 1. Comparison experiment results of intelligent anomaly detection for wind turbine blades

Models	Accuracy	Precision	Recall	F1 score
AutoEncoder	0.7233	0.74	0.86	0.80
LSTM	0.8618	0.86	0.86	0.86
LSTM+Attention	0.9184	0.92	0.92	0.92
GRU	0.9417	0.94	0.94	0.94
TCN	0.9740	0.97	0.97	0.97
TCN-ATT	**0.9979**	**0.99**	**0.99**	**0.99**

3.3 Multi Fault Anomaly Detection Experiment of Wind Turbine Blades

The purpose of the multi fault anomaly detection experiment is to test the effectiveness of the proposed TCN-ATT model in the multi fault anomaly detection task. This experiment uses the data from the SKAB dataset collected in the laboratory. In this experiment, six separate data files are selected from the SKAB dataset. There are seven kinds of wind turbine blades states, including normal state and six different faults. The data are mixed in time order to form a new complete data set for experiment. The details of the dataset are shown in Table 2. There are 3636 normal data items and 194, 384, 398, 398, 410 and 401 abnormal data items respectively.

Table 2. Anomaly detection dataset of wind turbine blades with multi faults

Data file	Normal	Fault-1	Fault-2	Fault-3	Fault-4	Fault-5	Fault-6
1	557	194	–	–	–	–	–
2	281	–	384	–	–	–	–
3	739	–	–	398	–	–	–
4	570	–	–	–	398	-	–
5	743	–	–	–	–	410	–
6	746	–	–	–	–	–	401
Total	3636	194	384	398	398	410	401

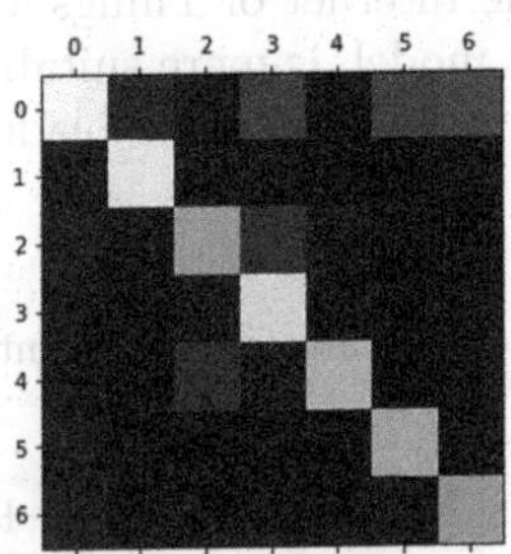

Fig. 7. Confusion matrix of classification results of multi faults anomaly detection for wind turbine blades

The confusion matrix of the classification results of multi-fault anomaly detection for wind turbine blades is shown in Fig. 7, where each row represents the actual anomaly categories, and each column represents the state of wind turbine blades predicted by TCN-ATT. 0 in the figure corresponds to the normal state, and 1–6 corresponds to 6 kinds of anomaly categories respectively. Table 3 lists in detail different abnormal categories of wind turbine blades, classification precision, recall rate and F1 score of TCN-ATT model, as well as the overall classification accuracy of the model in the multi-fault anomaly detection task. As can be seen from Table 3, the overall classification accuracy of the TCN-ATT model proposed in this paper can reach 84.09% in the multi-fault anomaly detection experiment of the wind turbine blades. For the health state of the wind turbine blades and 6 different fault types, the precision rates are respectively 87%, 94%, 81%, 71%, 94%, 74% and 70%, and the precision rates are all up to over 70%, some of the fault categories with obvious anomaly characteristics, such as Fault-4, can even achieve a classification precision of 94%. It can be seen that the TCN-ATT model proposed in this paper can not only simply judge whether the wind turbine blades are in abnormal state, but also effectively detect which abnormal type the wind turbine blades belong to, thus reducing the time and money cost for enterprise maintenance personnel to carry out fault maintenance. In addition, from Fig. 7 and Table 3, we can also see that the characteristics of some abnormal states of wind turbine blades are very similar to the characteristics of normal state or other abnormal states, so it is difficult to accurately distinguish them. For example, the wind turbine blades in the normal state may be predicted to be abnormal in Fault-3, Fault-5 or Fault-6. At this time, in the actual industrial production, we can combine the expert system based anomaly detection technology and knowledge-based anomaly detection technology with TCN-ATT model to form an integrated model of anomaly detection, which makes all kinds of anomaly detection technologies learn from each other, play their own advantages, and make up for the shortcomings, so as to improve the comprehensive performance of the wind turbine blades multi fault anomaly detection system.

Table 3. Experiment results of multi fault anomaly detection for wind turbine blades

Label	Precision	Recall	F1 score
Normal	0.87	0.93	0.90
Fault-1	0.94	0.90	0.92
Fault-2	0.81	0.58	0.68
Fault-3	0.71	0.83	0.76
Fault-4	0.94	0.68	0.79
Fault-5	0.74	0.64	0.69
Fault-6	0.70	0.57	0.63
Total accuracy	0.84087		

3.4 Ablation Experiments

In order to better reflect the efficiency of the proposed TCN-ATT model architecture in the anomaly detection task of wind turbine blades, we designed two ablation experiments on the blade icing prediction dataset of 2017 industrial

Table 4. Comparison experiment results of intelligent anomaly detection for wind turbine blades

Models	Residual block	Attention	Accuracy	Precision	Recall	F1 score
TCN	×	×	0.9591	0.96	0.96	0.96
TCN	√	×	0.9740	0.97	0.97	0.97
TCN-ATT	×	√	0.9639	0.96	0.96	0.96
TCN-ATT	√	√	0.9979	0.99	0.99	0.99

big data. The ablation experiments not only further proved that the Attention module we added would improve the effect of the model to some extent, but also verified the contribution of the residual module of the TCN model and the TCN-ATT model proposed in this paper based on standard TCN to the improvement of model performance by combining features of different levels together to increase the diversity of features. Table 4 shows the experimental results of four varieties of standard TCN models on the prediction dataset of leaf icing in the industrial big data in 2017. The four models used in the ablation experiment are the most basic TCN model without the Attention module and the residual module, and the modified TCN without the Attention module but with the residual module, and the TCN-ATT model proposed in this paper, which includes the Attention module and the residual module, and the basic TCN-ATT model, which includes the Attention module but does not include the residual module, used for comparative experiments.

When verifying the effect of the Attention module on experimental performance, the models are controlled to include or not include the residual module at the same time. We can see from the comparison results of the first and third lines or the second and fourth lines in Table 4 that the Attention module makes the model more focused on finding useful information in the input data that is significantly related to the current output, thus improving the quality of the output. Compared with the models without the Attention module, the models with the Attention module are more focused on finding useful information related to the current output.

The residual module uses the form of skip layer connection to splice the features of different levels together to achieve the purpose of increasing feature diversity and speeding up training, which can well solve the problem of gradient disappearance caused by the increase of network depth. The comparison results of the first and second lines or the third and fourth lines in Table 4 show that, when other structural parameters of the models are of the same value, adding the residual module can improve the detection accuracy of the model in the anomaly detection task of wind turbine blades by about 1.49%–3.4%.

The ablation experiments show that the TCN-ATT model proposed in this paper that adds the residual module and integrates the Attention mechanism, increases the diversity of features, makes the model pay more attention to the features that are more effective for anomaly detection results, which can play a good role in the intelligent anomaly detection task of wind turbine blades in the wind power industry.

4 Conclusion

The health condition of wind turbine blades is related to the normal operation of wind power generation. Based on the time series data signals provided by industrial stethoscope, an intelligent anomaly detection method, TCN-ATT for wind turbine blades based on temporal convolutional network and attention mechanism was proposed. Since the common RNN method needs to consider the future information and takes up a lot of memory when the input sequence is long, in order to solve these problems, this paper applies the TCN-ATT based on attention to the intelligent anomaly detection of wind turbine blades. TCN-ATT model includes temporal convolution module and attention module. There are three core parts in temporal convolution module: causal convolution, dilation convolution and residual block. Because TCN-ATT model uses convolution, it can capture local information, so it is better at capturing temporal dependence, and the dilation convolution module allows TCN-ATT to flexibly adjust the size of receptive field. The attention module is added to focus on the more important features of the detection results to improve the accuracy of anomaly detection. In this paper, the comparative experiment of intelligent anomaly detection and multi-fault anomaly detection of wind turbine blades are designed to verify the detection performance of TCN-ATT model. In the comparative experiment of intelligent anomaly detection of wind turbine blades, the TCN-ATT model was compared with AutoEncoder, LSTM, attention-based LSTM, GRU and standard TCN. The TCN-ATT model has higher accuracy, precision, recall rate and F1 score than other commonly used anomaly detection methods. The average accuracy of anomaly detection is 99.79%, which is 2.39% higher than that of standard TCN with better relative performance. In the multi-fault anomaly detection experiment, the classification accuracy of TCN-ATT model reaches 84.09%, which verifies the good performance of the TCN-ATT model in the presence of multiple anomalies. In addition, we also designed ablation experiments to highlight the role of the Attention mechanism and residual block in TCN-ATT in improving the performance of intelligent anomaly detection of wind turbine blades. In conclusion, the TCN-ATT model proposed in this paper can extract more valuable features from time series data signals and achieve good detection results in the intelligent anomaly detection task of wind turbine blades, which has great potential in the intelligent anomaly detection of wind turbine blades in power system. However, since the experimental data in this paper are all from the laboratory, we hope to evaluate the real-time anomaly detection accuracy of the TCN-ATT model proposed in this paper in the actual fan blade detection work in the future, so as to help wind power enterprises save human and material costs.

References

1. Lu, X., Fan, Y., Qian, K.: Summary and development trend of equipment fault diagnosis technology. Mining Mach. **035**(012), 15–18 (2007). (in chinese)
2. Ince, T., Kiranyaz, S., Eren, L., et al.: Real-time motor fault detection by 1D convolutional neural networks. IEEE Trans. Ind. Electron. 63(11) (2016)
3. Zhao, M., Kang, M., Tang, B., et al.: Deep residual networks with dynamically weighted wavelet coefficients for fault diagnosis of planetary gearboxes. IEEE Trans. Industr. Electron. **65**(5), 4290–4300 (2018)
4. Wang, S., Xiang, J., Zhong, Y., et al.: A data indicator-based deep belief networks to detect multiple faults in axial piston pumps - ScienceDirect. Mech. Syst. Signal Process. **112**, 154–170 (2018)
5. Rahimilarki, R., Gao, Z., Jin, N., et al.: Time-series deep learning fault detection with the application of wind Turbine benchmark. In: 2019 IEEE 17th International Conference on Industrial Informatics (INDIN). IEEE (2019)
6. Liu, Y., Cheng, H., Kong, X., et al.: Intelligent wind turbine blade icing detection using supervisory control and data acquisition data and ensemble deep learning. Energy Sci. Eng. 7(6) (2019)
7. Xian, D., Han, N., Wei, T.: Fault identification of direct drive wind turbine based on deep learning. Renewable Energy Resources (2018)
8. Bai S, Kolter J Z, Koltun V.: An empirical evaluation of generic convolutional and recurrent networks for sequence modeling. CoRR, abs/1803.01271 (2018)
9. Yan, J., Mu, L., Wang, L., et al.: Temporal convolutional networks for the advance prediction of ENSO. Sci. Rep. **10**(1), 8055 (2020)
10. Dai, R., Xu, S., Gu, Q., et al.: Hybrid spatio-temporal graph convolutional network: improving traffic prediction with navigation data. ACM (2020)
11. Guirguis, K., Schorn, C., Guntoro, A., et al.: SELD-TCN: sound event localization & detection via temporal convolutional networks (2020)
12. Oord, A., Dieleman, S., Zen, H., et al.: WaveNet: a generative model for raw audio (2016)
13. Yu, F., Koltun, V.: Multi-scale context aggregation by dilated convolutions. In: ICLR (2016)
14. He, K., Zhang, X., Ren, S., et al.: Deep Residual Learning for Image Recognition. IEEE (2016)
15. Zhang, H., Zhou, Q., Ren, X.: Analog detection for wind turbines based on autoencoder model. Software Guide **018**(009), 158–162 (2019)
16. Chung, J., Gulcehre, C., Cho, K.H., et al.: Empirical evaluation of gated recurrent neural networks on sequence modeling. Eprint Arxiv (2014)
17. Hochreiter, S., Schmidhuber, J.: Long short-term memory. Neural Comput. **9**(8), 1735–1780 (1997)
18. Vaswani, A., Shazeer, N., Parmar, N., et al.: Attention Is All You Need. arXiv (2017)
19. Bai, S., Kolter, J.Z., Koltun, V.: Trellis Networks for Sequence Modeling (2018)
20. Katser, I.D., Kozitsin, V.O.: Skoltech Anomaly Benchmark (SKAB). Kaggle (2020). https://doi.org/10.34740/KAGGLE/DSV/1693952
21. Malhotra, P., Ramakrishnan, A., Anand, G., et al.: LSTM-based Encoder-Decoder for Multi-sensor Anomaly Detection (2016)
22. Fan, H., Zhang, F., Li, Z.: AnomalyDAE: dual autoencoder for anomaly detection on attributed networks. In: ICASSP 2020–2020 IEEE International Conference on Acoustics, Speech and Signal Processing (ICASSP) (2020)

23. Luo, W., Wen, L., Gao, S.: Remembering history with convolutional LSTM for anomaly detection. In: 2017 IEEE International Conference on Multimedia and Expo (ICME). IEEE (2017)
24. Cao V L, Nicolau M, Mcdermott J. A Hybrid Autoencoder and Density Estimation Model for Anomaly Detection[C]// International Conference on Parallel Problem Solving from Nature. Springer, Cham, 2016
25. Tao, X., Peng Y, Zhao F, et al. An Improved Parallel Network Traffic Anomaly Detection Method Based on Bagging and GRU[C]// International Conference on Wireless Algorithms, Systems, and Applications. Springer, Cham, 2020
26. Hao, H., Wang, Y., Xia, Y., et al.: Temporal convolutional attention-based network for sequence modeling (2020)

Error Serial Episodes Discovery from Mobile Payment Log in Distributed ETC

Xiang Li[1], Yawei Zhao[1(✉)], Donghui Li[2], and Wei Guan[3]

[1] School of Engineering Sciences, University of Chinese Academy of Sciences, Beijing, China
lixiang194@mails.ucas.edu.cn, zhaoyw@ucas.edu.cn
[2] Beijing Powershu Technology Co., Ltd., Beijing, China
[3] Beijing Sutong Technology Co., Ltd., Beijing, China

Abstract. The **E**lectronic **T**oll **C**ollection (ETC), as a kind of distributed system, has become more and more mature in application of highways. The mobile payment system, distributed in main network links of ETC, possibly causes the functional failure of entire system if an error occurs. But the error message can be directly reflected in log. Therefore, effective analysis of mobile payment log is crucial to the stability of distributed system. In this paper, we propose a novel analysis method, which is *error serial episodes discovery*. This method converts mobile payment log in ETC into a long event sequence, and improves window truncation by *WINEPI* algorithm to explore the relevance of events in the log. We use real data from mobile payment log in ETC of a Chinese highway company. The experiments indicate that this method obtains a significant rule that *multiple identical events* will cause the error event to occur. This rule, which is found in actual scenarios, will provide a basis for data analysis of the log, and it has been recognized by business experts from relevant institutions.

Keywords: Electronic toll collection · Distributed system · Mobile payment log · Long event sequence · Error serial episodes discovery

1 Introduction

1.1 The Electronic Toll Collection Failure Issue

The coverage of **E**lectronic **T**oll **C**ollection (ETC) in the whole network is an important basis for promoting the large-scale development of intelligent transportation systems. However, it is not easy to monitor and maintain the operating status of a large-scale network system, not to mention that system indicators should be kept to a normal value. ETC as a distributed system contains various aspects, like the *vehicle identification* and the *transaction processing* [12]. An error in one aspect will lead to the failure of entire system. As a crucial aspect, the *mobile payment* including various systems controls the billing of vehicles [3]. Countless traffic jams are caused by errors in that aspect. To this end, maintaining the normal operation of the ETC can not only hold stability of the distributed system, but also reduce traffic congestion and accidents.

Y. Lai et al. (Eds.): ICA3PP 2021, LNCS 13155, pp. 210–221, 2022.
https://doi.org/10.1007/978-3-030-95384-3_14

1.2 Data Analysis of Mobile Payment Log in Electronic Toll Collection

In most cases, system abnormalities are caused by the mobile payment. And log can directly reflect the real-time status of the system. In case of any system errors, data containing errors will be presented in the log in real time.

Accordingly, by monitoring and processing the error data in the log, we are able to get useful conclusions. Traditional methods monitor the frequency of related log error messages by setting thresholds [13]. Similarly, related scholars [9] detect abnormal values of network traffic through clustering. Another new way is to find the useful patterns by *pattern mining*, while finding the events associated with the target events. Mining frequent patterns of telecommunication network alarm sequence is the initial application [10], thus finding the root cause of alarm. This method enables the pattern mining technology to provide a theoretical basis for explaining the occurrence of common related events. The mobile payment log of ETC still allows adopt this method, and this data structure is similar to the structure of event sequence.

Event sequence refers to a long sequence containing a series of events and associated time. Finding frequent patterns in this long event sequence is called *episodes mining* [10]. The main purpose of episodes mining is to discover significant and valuable patterns from existing data, which belongs to the field of pattern mining. The *Apriori* algorithm [1], which pioneered this field, aims to mine frequent item sets in the sales transaction database, but ignores the time sequence between item sets. Based on this, the *sequential pattern mining* task [2] is proposed. Generally, event sequence is regarded as a special sequence containing single information, and the episodes mining can also be used as a special case of sequence pattern mining tasks.

For the event sequence, due to its singularity, there are many design forms for the definition of episodes frequency. Mannila et al. [10] define frequency as the window ratio of episodes in fixed width windows, and propose the *WINEPI* algorithm. After that, related scholars overcome the limitation based on fixed window by holding new concepts and frequency generation methods [6,7]. In applications, pattern mining is utilized to predict user behavior [8], as well as related to the financial field [11]. In the subsequent research, there are more episodes mining algorithms for optimization and compression, yet relatively few for the specific application [4,5].

1.3 Our Contribution

The mobile payment log in ETC can be regarded as a manifestation of a long event sequence. In addition, the data in log are always complicated. It is of great research value if we can understand the data in advance and find the events associated with the error data in the log. According to our existing investigations, the analysis aspect of episodes in event sequence is not enough [4], and valuable rules are rarely found. Furthermore, there is no relevant research on the application of episodes mining technology in the mobile payment log in ETC. Therefore, the main contributions of this paper are as follows:

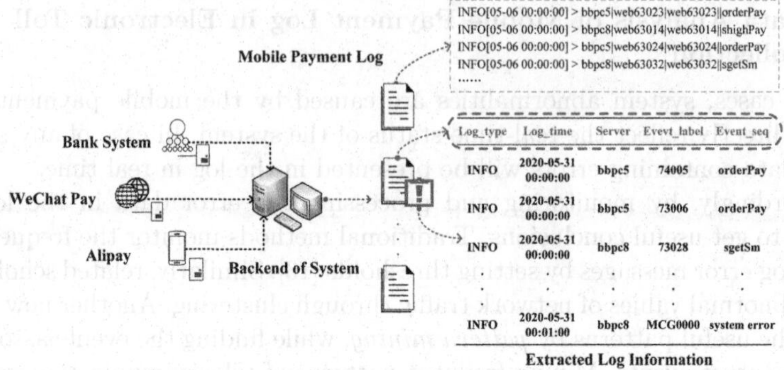

Fig. 1. The generation and extraction process of the mobile payment log in ETC

- A novel method of *error serial episodes discovery* is proposed and applied to analyze errors in the specific mobile payment log of ETC;
- The mobile payment log in ETC is converted into the general form of a long event sequence, then *event window*, *error serial episodes* and *error rules* as new concepts are proposed;
- The mobile payment log in ETC is preprocessed, and the method of window truncation is improved by WINEPI algorithm;
- The significant rules obtained can lay a foundation for subsequent research, and provide early warning information for dstributed ETC and other intelligent transportation systems.

2 Proposed Approach

2.1 Data Characteristics

The generation and extraction process of the mobile payment log information in ETC is shown in Fig. 1. Generally, mobile payment includes several network nodes to complete payment operations, such as *bank*, *WeChat* and *Alipay*, etc. At the same time, the backend of system automatically generates log information, and enables to extract information at any time reflecting the operating status of mobile payment system and the payment status of vehicle. For example, normal events are displayed like *transaction successful* and *the vehicle has not applied for a card* and so on. Normal events are usually abbreviated in the log for brevity, such as *orderPay.* In contrast, if an error occurs in the system, a clear keyword of the error event will be displayed in the log, such as *system error*.

Excluding specific event keywords, the log also includes some information related to the event. The log format generated by the system is unified (the information is divided by "|", as shown in Fig. 1), and the rows contain specific events and related information. However, the information involved in each row may be repeated, causing information redundancy. In order to facilitate subsequent processing, we extract information by setting rules and set column fields.

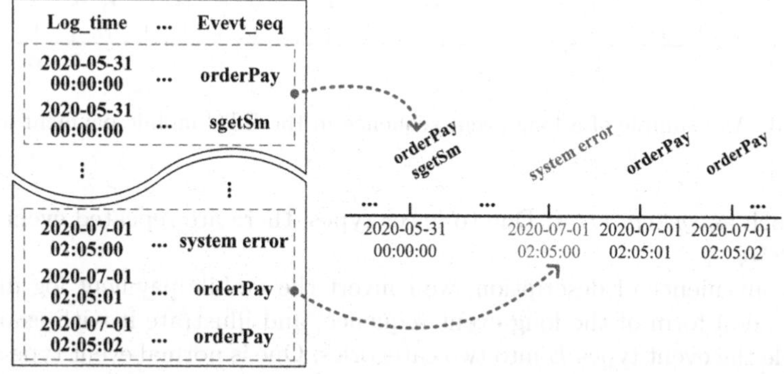

Fig. 2. The mobile payment log in ETC and related long event sequence

The finally extracted log information includes *type of log*, *time of log*, *server called*, *event label* and *event sequence*.

In the following problem statement, we convert the mobile payment log in ETC into the general form of a long event sequence, and give relevant definitions.

2.2 Problem Statement

Long Event Sequence of Mobile Payment Log in ETC. According to the description of Mannila et al. [10], an event is composed of event type and time of occurrence, denoted as a tuple (E_i, t_i), where $E_i \in E$, t_i is the time of corresponding event occurrence (represented by an integer). The event sequence is composed of multiple event tuples, which can be expressed as a triple (s, T_s, T_e). T_s is the sequence start time, T_e is the sequence end time, and s is denoted as:

$$s =< (E_1, t_1), (E_2, t_2), (E_3, t_3)...(E_n, t_n) > \tag{1}$$

where $T_s \leq t_1 \leq ... \leq t_{i-1} \leq t_i \leq t_{i+1} \leq ... \leq t_n \leq T_e$.

Nevertheless, this type of event sequence cannot be directly applied to the mobile payment log in ETC. Through the description of the data characteristics, we summarize several features of the mobile payment log in ETC:

- The mobile payment log generates data all the time, so it will be very long to convert to an event sequence;
- In the mobile payment log, multiple events occur at the same time;
- There are few types of events related to mobile payment;
- Not only multiple types of normal events appear in the mobile payment log, but also different error events.

Figure 2 demonstrates the form of the ETC mobile payment log transformed into a long event sequence. As detailed in Fig. 2, the two events *orderPay* and *sgetSm*, that occurred at the same time are reflected in the same position in the long event sequence. Moreover, this long event sequence also contains error

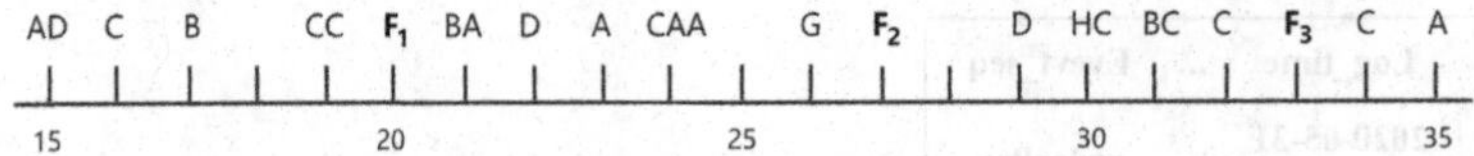

Fig. 3. An example of a long event sequence in the ETC mobile payment log

events such as *system error*. Due to fewer types, there are repeated events such as *orderPay*.

For convenience of description, we convert the mobile payment log in ETC into a general form of the long event sequence, and illustrate it with examples. We divide the event types E into two categories. One is normal event type set N, and the other is error event type set F. For the long event sequence $(s, 15, 35)$, where

$$s = < (A, 15), (D, 15)...(C, 19), (F_1, 20)...(A, 35) > \quad (2)$$

This long event sequence $(s, 15, 35)$ is shown in Fig. 3. Among them, events A, D and C belong to the normal event type set N, while F_1, F_2 and F_3 all belong to the error event type set F. $(A, 15)$ and $(D, 15)$ represent two different events occurring at the same time. $(C, 19)$ and $(C, 19)$ indicate two identical events occurring at the same time. Moreover, there are other repeated events in the sequence. Obviously, the time period contained in the example is relatively short, which will be very long for the event sequence in the log. As we can see, this general form of the long event sequence conforms to the features of the mobile payment log in ETC.

Long Event Sequence Truncation. We consider truncating the long event sequence through some methods. Traditional method is to cut the sequence by setting a time window with a fixed width [10]. Nevertheless, it can be seen from the example in Fig. 3 that the events on the time axis are not evenly arranged. There will be a lot of redundancy if the time window is selected. So we improve the method based on the event window, and the definition is given below.

Definition 1 (Event Window). *For a long event sequence s, the time when the event does not appear is deleted, and all events appearing in the sequence are arranged in order. Then the length of the sequence is total number of all events, which is* $|s|$. *The event window is denoted as* w_E, *which is an intercepted sequence of* l_E *consecutive events (i.e., window width) on the sequence s, where* $l_E < |s|$. *We denote by* $W_E = (w_E, l_E)$ *the set of event windows.*

After truncating the sequence through event windows, we attempt to define the step size of sliding windows. Figure 4 explains two event windows w_{E1} and w_{E2}, where the step size is 1 and the event window width is 7. To present both the start event and the end event of a sequence in windows of the same number, like any other events, it is stipulated that the first window only contains the first event, and the last window only contains the last event. In the following research, the setting of the step size will continue this way. Given a long event

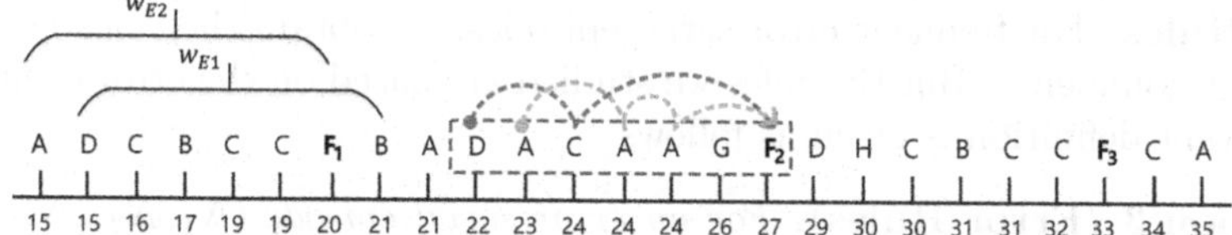

Fig. 4. Event windows and error serial episodes.

sequence s and the width of an event window l_E, the number of event windows in W_E is as follows when the step size is sp:

$$N\_W_E(s, l_E, sp) = \lceil \frac{|s| - 1 + l_E}{sp} \rceil \tag{3}$$

Note that when the number is a floating-point number, it should be rounded up to avoid the incomplete sequence.

Error Serial Episodes Discovery. Episodes include *serial episodes*, *parallel episodes* and *non-serial and non-parallel episodes* [10]. Due to the importance of order, we only discuss serial episodes in the event sequence of log. To explore the rules of error events, error episodes discovery is essential. In this way, we give the following definition.

Definition 2 (Error Serial Episodes). *The normal event type set E and the error event type set F are given for a long event sequence s. An error serial episode is a serial episode in which only the last item contains an error event and the other items are normal events. It is denoted as* $(N_1, N_2 ... N_i, F_i)$, *where* $N_i \in N$ *and* $F_i \in F$.

For an event window in Fig. 4, it is as follows:

$$< (D, 22), (A, 23), (C, 24), (A, 24) ... (F_2, 27) > \tag{4}$$

This event window contains two error serial episodes. One is *several different normal events*: (D, C, F_2), and the other is *multiple identical normal events*: (A, A, A, F_2).

For an error serial episode, we calculate its frequency by the proportion of episodes in the event windows [5]. That is, given a long event sequence s, step size *sp* and the set of event windows W_E, the frequency of an episode α in s is as follows:

$$Freq(s, \alpha, W_E) = \frac{|\{\alpha \; occurs \; in \; W_E\}|}{N\_W_E(s, l_E, sp)} \tag{5}$$

If the frequency of this error serial episode is not less than a frequency threshold, then it is an episode we are interested in.

Error Rules. For frequent error serial episodes, we obtain the rules by calculating the confidence. But the rules we studied are based on the error event, and the relevant definition is given as follows.

Definition 3 (Error Rules). *For an error serial episode* $(N_1, N_2...N_i, F_i)$ *in a long event sequence, an error rule is expressed as* $(N_1, N_2...N_i) \rightarrow (F_i)$*. That is, the antecedent of rule is composed of normal events other than error event, while the consequent only contains the error event.*

Besides, rules are generated by setting a confidence threshold. Given a long event sequence s and the set of event windows W_E, the antecedent of an error rule generated from an error serial episode α is denoted as γ, then the confidence of this rule is as follows:

$$Conf(s, \alpha, W_E) = \frac{Freq(s, \gamma, W_E)}{Freq(s, \alpha, W_E)} \tag{6}$$

2.3 Algorithm for Error Serial Episodes Discovery

For the description of data characteristics and problem statement, we propose a novel method of *error serial episodes discovery* to analyze the errors of mobile payment log in ETC. Error serial episodes discovery is to find those episodes in *Definition 2*.

In the *WINEPI* algorithm, given an event sequence and a set of episodes, the episodes discovery algorithm can be divided into two steps by setting the window width and the frequency threshold: a) Generating a set of candidate episodes; b) Frequent episodes mining. The algorithm relies on a priori that all non-empty subepisodes of frequent episodes must also be frequent [10]. Through this pruning operation, the generation of a large number of candidate episodes can be avoided. Despite this, the step size is constant in this algorithm, which is 1, ignoring the impact of different step sizes on the results. Meanwhile, the generation of rules ignores the order of events. Accordingly, we improve this algorithm by automatically generating the candidate set and truncating the dataset by the event window width and step size. In this way, we can automatically get the set of serial episodes.

Our approach is to firstly convert the mobile payment log of the ETC into a long event sequence. By assigning numbers to the events and taking numbers as the time when the events occurred, we get the long event sequence of log (s, T_s, T_e). Secondly, by the event window in *Definition 1*, we set its width l_E and step size *sp*. These are the inputs of Algorithm 1, that is, the detailed steps of truncating the long event sequence through event windows. In Algorithm 1, we can calculate the number of event windows N_W_E according to (3). For the output of algorithm, the set of event windows W_E is obtained by traversing the event sequence, which is the basis of subsequent research.

The generated candidate episodes are obtained by the set of event windows and their frequencies are calculated according to (5). Next, the frequency threshold is set to get error serial episodes in the form of *Definition 2*. Finally, we can

Algorithm 1. Algorithm for Long Event Sequence Truncation

Input: A Long Event Sequence (s, T_s, T_e), Event Window Width l_E, Step Size sp.
Output: Set of Event Windows W_E.

Initialize the end time of event window $t_e \leftarrow T_s + 1$;
Initialize the start time of event window $t_s \leftarrow t_e - l_E$;
Initialize the set of event windows $W_E \leftarrow \varnothing$;
Calculate the number of event windows N_W_E according to (3);
for $i = 1$ to $|N\_W_E|$ **do**
  $w_E \leftarrow \varnothing$;
  for all events (E_i, t_i) in s **do**
    if $t_s <= t_i$ and $t_e > t_i$ **then**
      $w_E \leftarrow w_E \cup \{E_i\}$;
    end if
  end for
  $t_s \leftarrow t_s + sp$; $t_e \leftarrow t_e + sp$;
  $W_E \leftarrow W_E \cup \{w_E\}$;
end for
return W_E

get error rules by *Definition 3* and their confidence can be calculated by (6). It should be noted that we keep other serial episodes here. In practice, when the dataset is large, error serial episodes discovery can get results quickly. Besides, the generation of rules should pay attention to the occurrence time of events in the sequence.

3 Experiments

3.1 Dataset

We use real data from the mobile payment log in ETC of a Chinese highway company. Due to the huge amount of data, we choose mobile payment log data from May to July 2020, and the data are divided by date.

The probability of error events in the mobile payment log in ETC is extremely low, and generally error events are expressed in Chinese. So it is convenient for us to filter out dates containing Chinese. We finally select a total of eight days. But not all Chinese displays are error events, such as *transaction successful*, *order form is empty*, *vehicle not paid*, etc., thus this part of the information can be deleted. In the end, the only error event we select is *system error*, which is what we really care about. In all eight days of logs, the *system error* only appeared on the day of May 6, 2020 and was concentrated at the beginning. Since the moment when the *system error* event occurred is not affected by the subsequent events, we intercept the time period when the *system error* event occurred.

We finally get 538 pieces of data in the log. Although the dataset is small, it is enough to illustrate the problem. After preprocessing, the dataset is transformed into a long event sequence, which can be used to mine error serial episodes.

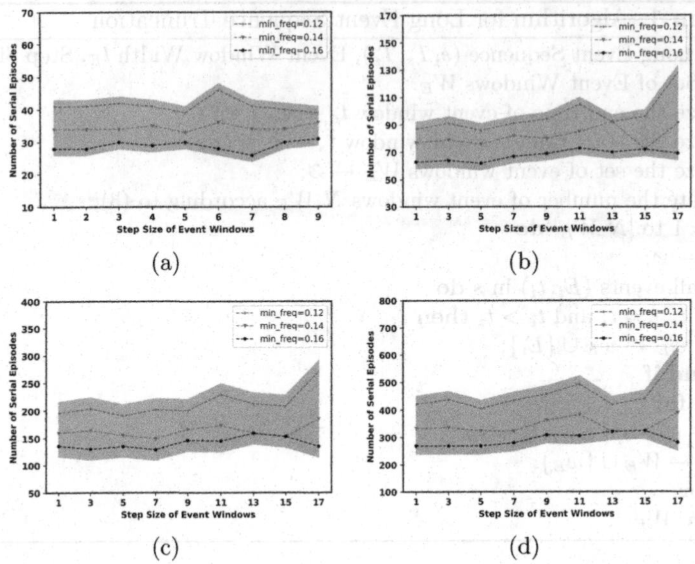

Fig. 5. Distribution law of data when the event window width is (a) 6, (b) 8, (c) 10 and (d) 12.

3.2 Results and Discussion

Data Exploration. When the sequence length is constant, the event window width, step size, and frequency threshold are important parameters that affect the number of serial episodes. In order to better analyze the log data, we first find the serial episodes. We design several comparative experiments to explore the data as shown in Fig. 5.

Figure 5 demonstrates the effect of step size and frequency threshold on the number of serial episodes when the event window width is 6, 8, 10, and 12. In order to mine as many episodes as possible, the frequency threshold (i.e., min_freq) we set is low. In addition, the step size is also selected appropriately according to the event window width, so that the event window width is roughly in the middle of the step size range. From Fig. 5 we can draw the following conclusions:

- When event window width and step size are constant, the number of serial episodes will increase after the increase of frequency threshold;
- When step size and frequency threshold are constant, the number of serial episodes will increase after the increase of event window width;
- When event window width and frequency threshold are constant, the number of serial episodes will fluctuate with the increase of step size, but the overall trend is upward, which is shown in the yellow area of Fig. 5.

The first and second conclusions are in accordance with formulas (2) and (3). For the last conclusion, we consider two cases. One is the effect of the number of windows. When the event window width is constant, the number of windows

Table 1. Form and frequency of the longest error serial episodes (event window width).

Event window width	The longest error serial episodes	Frequency
6	(*sgetSm, sgetSm, system error*)	0.274
8	(*sgetSm, sgetSm, sgetSm, system error*)	0.183
10	(*shighPay, sgetSm, sgetSm, system error*)	0.143
	(*sgetSm, sgetSm, shighPay, system error*)	0.141
	(*sgetSm, sgetSm, sgetSm, system error*)	0.291
	(*orderPay, orderPay, orderPay, system error*)	0.135
12	(*sgetSm, sgetSm, sgetSm, sgetSm, system error*)	0.209
	(*orderPay, orderPay, orderPay, orderPay, system error*)	0.140

will decrease with the increase of step size. So for a certain serial episode, its frequency will increase. If the frequency threshold remains stable, episodes that did not appear before will meet the frequency threshold and then the amount will increase. The second is the effect of episodes types. As the step size is further increased, some episodes will be cut by the windows to reduce episodes types, thus the number of episodes will gradually decrease.

However, it can be found from Fig. 5 that the trend fluctuates locally, while the overall trend is an upward one. The reason is that the episodes in data does not decrease greatly with the increase of step size, so the most critical factor is the change in the number of windows. This is related to the distribution law of the data. Because there are few errors and few kinds of events in the log, there are many repeated events in the data. And the presentation of this distribution law can be derived from the relation between the number of episodes and the step size. As a result, it is of great concern to explore data.

Results of Error Serial Episodes Discovery. According to results of preliminary data exploration, due to fewer error events, too many enlarged windows will make the mined episodes unconvincing. Meanwhile, there are situations where all combinations of episodes that contain fewer events. To this end, we only show the error serial episodes discovery with the most events, that is, the longest one.

In order to consider all the episodes appeared, the experiment set the step size to 1, and the frequency threshold to 0.12. Table 1 illustrates the specific form of the longest error serial episodes as the event window width changes. It can be seen from Table 1 that the longest error serial episodes are mostly caused by the multiple identical normal events, and only there are multiple different normal events in the experiment with an event window width of 10. With the event window width changes, the number of identical events is increasing. Besides, the fact that the frequency of episodes are low accords with the situation that there are few error events.

Results of Error Rules. Through error serial episodes, we can derive error rules. According to the order of confidence, Table 2 demonstrates the rule and corresponds to the episodes in Table 1.

Table 2. Form and confidence of error rules (event window width).

Event window width	Error rules	Confidence
6	$(sgetSm, sgetSm) \rightarrow (system\ error)$	0.606
8	$(sgetSm, sgetSm, sgetSm) \rightarrow (system\ error)$	0.680
10	$(sgetSm, sgetSm, sgetSm) \rightarrow (system\ error)$	0.713
	$(orderPay, orderPay, orderPay) \rightarrow (system\ error)$	0.704
	$(shighPay, sgetSm, sgetSm) \rightarrow (system\ error)$	0.595
	$(sgetSm, sgetSm, shighPay) \rightarrow (system\ error)$	0.575
12	$(orderPay, orderPay, orderPay, orderPay) \rightarrow (system\ error)$	0.778
	$(sgetSm, sgetSm, sgetSm, sgetSm) \rightarrow (system\ error)$	0.706

From the confidence in Table 2, we can see that rules are all generated by multiple identical events. In the experiment with an event window width of 10, the rules containing multiple different normal events have much less confidence than other rules, such as $(shighPay, sgetSm, sgetSm) \rightarrow (system\ error)$. Therefore, we conclude that multiple identical normal events can lead to the occurrence of error event in the log.

Rule Analysis. It can be seen from the results of data exploration that there are many repeated events in the sequence, and this repetitiveness is the reason for this rule. Using this method of episodes mining can not only discover the potential information of the data, but also lay a foundation for the interpretation of subsequent rules. This method can still be applied to other fields.

Furthermore, the rule that the multiple identical events will eventually lead to errors does not only exist in the mobile payment log of ETC. In actual scenarios, we often encounter *overflow* situations. If an event is not responded in time, the system will continue to call the event. If the number of repetitions is too many, it will eventually lead to an error. *Buffer overflow* is also an example of this rule.

4 Conclusion

In this paper, we use the mobile payment log in ETC to mine error serial episodes. Through this method of analysis, we can not only understand the data in advance, but also provide a novel idea for analysis method of log data.

In addition, we obtain significant rules. That is, the multiple identical events will lead to the occurrence of the error event. The actual measurements also confirm our conclusion, and such rules often appear in computer network platforms. Such significant rules can provide support in other areas. For example, after an event associated with an error event is found, the technology behind the event presumably also has problems, which lead to the occurrence of errors. This is helpful for root cause analysis and fault detection of distributed systems. Specially, when building related models, these related events can be regarded as important features in the feature selection stage.

We believe that this method of error serial episodes discovery will open a promising path for future research on log analysis, and this idea will still achieve meaningful results in other fields.

Acknowledgment. This work was supported in part by the National Key Research and Development Program of China under Grants 2020YFC1807104, in part by IBM Visiting Scholar Program and Chinese Academy of Sciences Informatization Project (XXH13504-05).

References

1. Agarwal, R., Srikant, R., et al.: Fast algorithms for mining association rules. In: Proceedings of the 20th VLDB Conference, pp. 487–499 (1994)
2. Agrawal, R., Srikant, R.: Mining sequential patterns. In: Proceedings of the Eleventh International Conference on Data Engineering, pp. 3–14. IEEE (1995)
3. Hari Charan, E.V.V., Pal, I., Sinha, A., Baro, R.K.R., Nath, V.: Electronic toll collection system using barcode technology. In: Nath, V., Mandal, J.K. (eds.) Nanoelectronics, Circuits and Communication Systems. LNEE, vol. 511, pp. 549–556. Springer, Singapore (2019). https://doi.org/10.1007/978-981-13-0776-8_51
4. Fournier-Viger, P., Lin, J.C.W., Kiran, R.U., Koh, Y.S., Thomas, R.: A survey of sequential pattern mining. Data Sci. Pattern Recogn. **1**(1), 54–77 (2017)
5. Ibrahim, A., Sastry, S., Sastry, P.S.: Discovering compressing serial episodes from event sequences. Knowl. Inf. Syst. **47**(2), 405–432 (2015). https://doi.org/10.1007/s10115-015-0854-3
6. Iwanuma, K., Takano, Y., Nabeshima, H.: On anti-monotone frequency measures for extracting sequential patterns from a single very-long data sequence. In: IEEE Conference on Cybernetics and Intelligent Systems, 2004, vol. 1, pp. 213–217. IEEE (2004)
7. Laxman, S., Sastry, P., Unnikrishnan, K.: Discovering frequent episodes and learning hidden Markov models: a formal connection. IEEE Trans. Knowl. Data Eng. **17**(11), 1505–1517 (2005)
8. Laxman, S., Tankasali, V., White, R.W.: Stream prediction using a generative model based on frequent episodes in event sequences. In: Proceedings of the 14th ACM SIGKDD International Conference on Knowledge Discovery and Data Mining, pp. 453–461 (2008)
9. Liu, D., Lung, C.H., Lambadaris, I., Seddigh, N.: Network traffic anomaly detection using clustering techniques and performance comparison. In: 2013 26th IEEE Canadian Conference on Electrical and Computer Engineering (CCECE), pp. 1–4. IEEE (2013)
10. Mannila, H., Toivonen, H., Verkamo, A.I.: Discovery of frequent episodes in event sequences. Data Min. Knowl. Disc. **1**(3), 259–289 (1997)
11. Ng, A., Fu, A.W.: Mining frequent episodes for relating financial events and stock trends. In: Whang, K.-Y., Jeon, J., Shim, K., Srivastava, J. (eds.) PAKDD 2003. LNCS (LNAI), vol. 2637, pp. 27–39. Springer, Heidelberg (2003). https://doi.org/10.1007/3-540-36175-8_4
12. Zhu, L., Haobo, H.: Patent survey on electronic toll collection(etc) technologies. China Invention Patent **16**(S2), 95–99 (2019)
13. Zhu, Y., Okita, H., Hanaoka, S.: A practical approach for network fault detection. In: 2016 International Conference on Computing, Networking and Communications (ICNC), pp. 1–5. IEEE (2016)

Parallel Cache Prefetching for LSM-Tree Based Store: From Algorithm to Evaluation

Shuo Zhang[1,2], Guangping Xu[1,2](✉), YuLei Jia[1,2], Yanbing Xue[1,2], and Wenguang Zheng[1,2]

[1] Tianjin Key Lab. of Intelligence Computing and New Software Technology, Tianjin 300384, China

[2] Computer and Communication Engineering School, Tianjin University of Technology, Tianjin 300384, China

xugp@email.tjut.edu.cn

Abstract. The Log-Structured Merge-Tree has efficient writing performance and performs well in big data scenarios. An LSM-tree transforms random writes into batch sequential writes through the design of a multilayer storage structure. However, as the core operation, the compaction inevitably results in degrading periodically in the read performance. Regular but irregular data compaction operations make the cache challenging to track the access information of data blocks. This work studies how to address the cache invalidation problem. We propose a two-phase parallel prefetching approach, which can effectively improve the cache invalidation when the compaction occurs. Our experimental results show our method can effectively improve read performance.

Keywords: LSM-tree · Key-value store · Caches · Minor cache · Cache invalidation

1 Introduction

Nowadays, key-value store engines are widely used in different large-scale storage systems to support various wide-area network applications. The Log-Structured Merge-Tree (LSM-tree) [1,2] is one of the most popular back-end design choices for key-value stores, such as BigTable [4], LevelDB [3], RocksDB [5] and Cassandra [18]. However, one of the primary issues of the LSM-tree is that read operations tend to have high latency.

As we know, an LSM-tree contains multiple levels, and the query of a key-value pair may check data from the lower level to the higher level until target data are found. This process inevitably incurs multiple I/Os, which amplifies the read operation. The hierarchical design is a trade-off between reading and writing performance. Thus, the LSM-tree is a data index structure with high write throughput but low read throughput. It is highly desirable to develop techniques that can improve the query performance of LSM-tree without sacrificing the write performance.

Y. Lai et al. (Eds.): ICA3PP 2021, LNCS 13155, pp. 222–236, 2022.
https://doi.org/10.1007/978-3-030-95384-3_15

In order to meet these challenges, there are two possible ways to improve read performance. As a straightforward way, some research works tend to accelerate the compaction operation [12,14,16]. These solutions try to decrease the frequency of compactions by relaxing the sorted data layout, which may decrease write performance to some extent. The alternative way is to resort the classic data caching approach, which is the focus of our work. Data caching is a widely used technique to improve the read performance of storage systems. By loading and storing frequently accessed data in the cache, it can greatly improve the read performance by reducing the number of I/O operations. However, importing the caching mechanism to LSM-tree is not trivial due to the compaction operation. The compaction operation frequently reorganizes data and merges data into the next level. We observe that the previous statistical information for access frequency of key-value pairs is still valuable to estimate the reorganized blocks after the compaction operation. From such observation, we prefetch a part of compacted blocks into cache directly if they have notable reuse hints. At the same time, for those blocks which have inconspicuous reuse hints, we may carefully choose a subset of them into cache by an extensive method.

This paper proposes a parallel cache prefetching approach to improve the read performance of LSM-tree, which consists of two parallel phases. The first phase tracks the position of compacted key-value pairs during the compaction and prefetches blocks that contain the hot key-value pairs into the cache. The second phase frequently tracks accessed key-value pairs, which aims to prefetch valuable blocks with temporal locality into the cache. The two-phase prefetching approach can be integrated with LSM-tree. It can make the read operation not affected by the compaction and retain all merits of the LSM-tree. We implement the two-phase prefetching approach on an LSM-tree simulator and conduct comprehensive experiments using four mix ratio Yahoo! Cloud Service Benchmark [10] (YCSB) workloads. The experimental results show that the two-phase prefetching approach can significantly increase the cache hit rate after the compaction, ranging from 2.56X to 3.12X.

2 Background

2.1 LSM-tree

The LSM-tree is a storage structure which turns random writes into sequential writes. The LSM-tree contains multiple components C_0, C_1, ..., C_k. Among them, C_0 resides in the memory and the other components reside on the disk. The memory component, C_0, also called MemTable, is usually implemented in data structures such as *SkipList* [15] or *B+tree*. When the memtable is full, it is flushed to the disk in the form of SSTable (Sorted Strings Table). SSTable contains data blocks and index blocks. The index block can be used to locate the position of data blocks for a given key.

In the LSM-tree, the merger operation comprises the flush operation and the compaction operation. When C_0 reaches the capacity limit, data will be transformed into an SSTable and written to C_1 by the flush operation. The compaction operation is responsible for data merging from C_1 to C_k. In order to improve write efficiency, it is often to flush records to C_1 without relocating them. The compaction reorganizes data among SSTables of neighboring components.

In short, the operations of the LSM-tree can be described as follows. When a new key-value pair is written, it is put into the active MemTable. Memtables are later flushed to disk, and the stored file is called an SSTable. SSTables are immutable. The data of the SSTables contained is never updated or deleted in place. Instead of updating the data in place, the LSM-tree writes the changes to a new memtable and then a new SSTable. The compaction operation merges the SSTables together and removes those staled data. If there is an update, the newest key-value pair is kept by compaction and written to the new SSTable, and the staled is discarded.

2.2 Cache in LSM-tree

The cache can enhance the query speed of the key-value store, Since keeping hot data in the cache can effectively decrease read I/Os. The LSM-tree, such as LevelDB and RocksDB, usually adopts the block cache, where the blocks could be data block, index block, or Bloom Filter (BF) block. Blocks can be located in the cache by the SSTable ID and the block offset. The block cache can be beneficial for both point lookup and range query operations. Read I/Os can be saved as long as the target block is cached. In this paper, we focus on such cache organization.

As stated above, the compaction operation reorganizes the data layout and compacts some SSTables into new ones. Thus some blocks will not be accessed in the future. If the block cache contains these blocks, which causes the so called *cache invalidation problem*. The periodical compaction makes a difference to keep hot data blocks in the cache. So the statistics for the access frequency of blocks may not be useful to the cache replacement policies. In addition, those invalid data blocks may be viewed as hot blocks incorrectly and keep the cache space. In particular, when an upper level has no space for new SSTables, the compaction will occur to the lower level in an iterated style. There will be a large number of cache misses for the following queries. It will lead to an increase in the number of invalid cached blocks and cache misses.

Therefore, the compaction will impact the cache performance for queries. So it is necessary for an LSM-tree since it collects garbage and ensures disk access efficiency.

3 Proposed Approach

3.1 Methodology and Architecture

As analyzed in Subsect. 2.2, the root cause of the cache invalidation problem lies in the reorganization of SSTables on the disk. After the compaction, the generated data will be stored to the new SSTable. The data blocks of the compacted

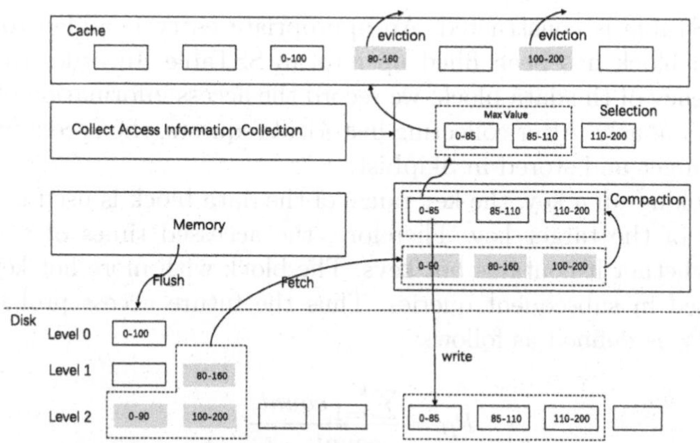

Fig. 1. The diagram of the cache architecture.

SSTables will become invalid and will not be accessed again. The references of those data blocks will also become invalid. If they have been marked as hot data by the cache, it will take some time to evict them from the cache. After compaction, a large number of new blocks will be generated, and their values for future accesses will be re-estimated. In the long-running, the set of new SSTables are becoming larger.

Our *methodology* is to exploit a two-phase parallel approach to estimate the access values of new blocks and then prefetch the most valuable ones into cache. The two-phase parallelism is to balance the requirements of efficiency and accuracy. In the first phase, it is a lightweight process to select the most valuable data blocks among the new blocks of a recent compaction to prefetch into a block cache. The search space is relative small. So we use a greedy strategy to meet the requirement of efficiency. At the same time, the second phase is to search valuable data blocks in a larger search space. We adopt a random-based optimization, the modified binary particle swarm optimization (MBPSO) [19], to select hot data blocks in a larger search space to prefetch. MBPSO improves the position change function of BPSO so that the particles have a larger range of motion and the search becomes more effective. Therefore, the design challenge of both the parallel phases is to prefetch blocks into cache cooperatively.

The diagram of the overall *architecture* is shown in Fig. 1, which has the following main components: the data collection, the compaction, and the selection. It shows the interaction of the prefetching process among cache, memory, and disk when compaction occurs. Our proposed prefetching approach aims to effectively reduce the number of cache misses caused by cache invalidation without extending the execution time of compaction.

3.2 Access Information Collection

Recall that an SSTable includes multiple data blocks, and key-value pairs are arranged orderly in each block. During compaction, the index block is created while the SSTable is constructed. An appropriate entry is added to the index when a data block has been filled up into an SSTable. In order to count the access frequency of the data block, we record the access information of key-value pairs asynchronously. After collecting historical requests, the access frequency of KV pairs can get and stored in SkipList.

When looking up a key, the key range of the data block is used to determine the location of the target key. Therefore, the accessed times of a data block depend on whether it contains hot keys. The block with more hot keys is likely to be accessed in subsequent queries. Thus the future access probability of a data block P_B is defined as follows:

$$P_B = \frac{\sum_{i=1}^{k} count_i}{count_{total}} \tag{1}$$

where $count_i$ represents access times of the ith KV pair of the data block B, k is the number of KV pairs in B, and $count_{total}$ represents the total number of query requests.

As shown in Formula 1, we can obtain P_B by calculating the sum of the access frequency of KV pairs. The larger P_B is, the subsequent access probability is higher.

3.3 Two-phase Approach

In the following presenting our approach, we categorize blocks in a new SSTable into the following two types: If a block includes invalid KV pairs, it called derived block; otherwise, it called non-derived block.

The workflow of the approach is shown in Fig. 2. In the first phase, the data blocks with higher P_B are selected from the derived blocks. And then they are put into the block cache. The second phase is an asynchronous selection phase assisted by a minor cache.

The First Phase. The first phase collects derived blocks and prefetches them into the block cache, see Algorithm 1. Derived blocks contain KV pairs splits from the invalid cache blocks. The challenge of cache invalidation is how to locate the position of the hot key after the compaction.

The probability that the data block is accessed, P_B, can be calculated by Formula 1. Let V_B be the value of the data block B. We have the following formula,

$$V_B = P_B * w = P_B * e^{-(l-1)} \tag{2}$$

where w denotes weight, l means the number of the level where the data block locates.

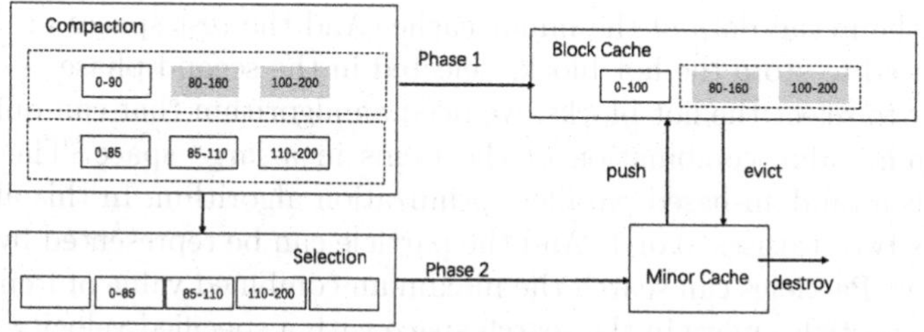

Fig. 2. The two-phase prefetching approach

In Formula 2, we use parameter w to represent the weight of the data block. The data in the LSM-tree is stored hierarchically, and the most popular design currently divides the underlying storage into seven layers. Due to the append log of the LSM-tree, the lower levels tend to contain more valid data. When there are more updates in the workload, the blocks stored in the higher layer may never be accessed. If these data blocks are prefetched into the block cache, it will waste many cache space. So when two data blocks have the same key, we should put the data block of the lower level into the block cache.

We use Formula 2 to determine the value of the data block. Then the priority queue is used to determine which generated data blocks should be prefetched into the block cache. The priority queue here is implemented based on a small top heap and has a capacity limit. The data block left in the queue will be prefetched into the cache.

Algorithm 1. The first phase prefetching algorithm

Input: Derived blocks $D\{B_i, V_i\}$, the size of invalid blocks N
Output: prioryQueue Q

$n \leftarrow 0$
while B_i in D **do**
  $Q.push(B_i)$
  $n \leftarrow n + B_i.size()$
  while $n > N$ **do**
    $n \leftarrow n - Q.pop().size()$
  end while
end while
return Q

The Second Phase. During the second phase, we want to cache the rest part of the derived blocks without evicting the original data of the cache and searching the most valuable blocks in a larger set. In our solution, the minor cache is introduced to store the blocks of the second phase. The rest derived blocks are treated as the initial data of the minor cache. And the rest spaces of the minor cache are used to store the hot blocks selected in the second phase.

In order to select the hot blocks, we need an algorithm that can quickly find the maximum value combination of the items in a large space. The MBPSO algorithm is a random-based parallel optimization algorithm. In this algorithm, an item has two statuses: 0 or 1. And the particle can be represented by a vector with β items. Particles can search the maximum combined value of items in parallel. These particles move in the search space with a specified velocity to search the optimal solution. In the original BPSO algorithm, the sigmoid function is used as the changing probability function of the status bit. The sigmoid function can cause the search space to decrease when the values of velocity are bigger. The MBPSO algorithm improves the location update function of the BPSO algorithm to obtain a more extensive search space. The improved location update function is as follows:

First, let $x_{id} \in \{0,1\}$ be the status bit of the item, indicating whether the current block is selected or not,

$$x_{id} = \begin{cases} 1 & \text{U(0,1)} < p(x_{id}, v_{id}) \\ 0 & \text{otherwise} \end{cases}$$

During the search, the algorithm adopts the changing probability function of the status bit,

$$p(x_{id}, v_{id}) = \frac{x_{id} + v_{id} + Vmax}{1 + 2V_{max}} \tag{3}$$

where $v_{id} \in$ [-V_{max}, V_{max}] is the velocity of particle and $U(0,1)$ is a random number between 0 and 1.

It can be seen from Formula 3 that the changing probability of a status bit is within a rectangular area. Thus, the MBPSO still has an efficient search operation when the velocity is bigger. When there are many elements, efficient search operations can make it find the optimal value faster. The new position update function increases the exploration rate of particles, and it is more effective in large-scale discrete space.

Considered that non-derived blocks contain a large number of blocks, we use the MBPSO algorithm to search the most valuable blocks in a larger set during the second phase. It will search the most valuable blocks from the non-derived blocks and put them into the minor cache. During the second phase, the particle is defined as a vector containing n blocks. The combined value of the particles can be calculated by Formula 2. Algorithm 2 depicts the process of single-particle optimization. Using the algorithm, we can update the position vector containing β items. The α is the number of iterations of the algorithm. During the execution of the algorithm, a particle corresponds to a thread. The current thread will block and wait for the notification at the beginning of a loop. And it will update the status bit when the execution is complete. Firstly, the particle will calculate the fitness value of the items. Secondly, the particle will update velocity by p_{best} and p_{gbest}. And it will calculate the changing probability of the status bit. The p_{best} is the position of the current particle with the historical optimal fitness

Algorithm 2. The single particle optimization algorithm

```
Initialize the particle postion vector x{x_i}_{i=1}^β and velocity vector v{v_i}_{i=1}^β
Initialize the value vector val{val_j}_{j=1}^α
for j from 1 to α do
   Wait the starting notification of the current thread
   for i from 1 to β do
      val[j] ← val[j] + x[i]
      if val[j] is the best value of the particle then
         p_best ← x
      end if
      v[i] ← v[i] + c_1 * rand * (p_best[i] − x[i]) + c_2 * rand * (g_best[i] − x[i])
      p[i] ← (x[i] + v[i] + V_max)/(1 + 2 * V_max)
      if U(0, 1) < p[id] then
         x[i] ← 1
      else
         x[i] ← 0
      end if
   end for
   notify the main thread
end for
```

value. And g_{best} is the position of the particle with the historical optimal fitness value. Last but not least, the particle will update position vector according to the condition of $U(0,1) < p_{id}$.

Formally, Algorithm 3 depicts the procedure of the second phase prefetching algorithm. When the beginning of the selection, the system will start the γ particle thread. And notify the particles to execute when all threads are ready. When the particles have been executed, the main thread will be notified and continue to calculate g_{best}. During the selection, the calculation of the particle is carried out independently. We can choose the number of particles according to the performance of the machine.

The minor cache is created at the beginning of the second phase in our solution and then destroyed as the block cache hit rate stabilizes. If the compaction occurs during the startup stage of the LSM-tree, the hit cache is difficult to reach a stable state. In this case, the minor cache will be destroyed when the start of another compaction.

In addition to improving the recovery speed of the cache hit rate, the minor cache can also delay eviction of the data block. When the data block in the minor cache is accessed, it will be pushed into the block cache. Moreover, the data block evicted from the block cache will be put into the minor cache.

In statistics, the range of data is the difference between the largest and smallest values. We can determine the degree of dispersion of the data through a range. We use a time window to record the block cache hit rate, and then calculates the difference between the current value and the stable value before the cache invalidation. When the difference is small enough, the hit rate of the block cache is considered to be stable.

Algorithm 3. The second phase prefetching algorithm

Initialize $p_{best}\{\{p_{best_i}\}_{i=1}^{\beta}\}_{h=1}^{\gamma}$, $g_{best}\{g_{best_i}\}_{i=1}^{\beta}$
Start γ *particle thread*
notifyall
for j *from* 1 *to* α **do**
    waitall
    for h *from* 1 *to* γ **do**
        if $\{\{p_{best_i}\}_{i=1}^{\beta}\}[h]$ *has the best combined value* **then**
            $g_{best_h} \leftarrow \{\{p_{best_i}\}_{i=1}^{\beta}\}[h]$
        end if
    end for
    notifyall
end for

4 Experimental Evaluation

We implement a caching prototype on an LSM-tree KV store and implement a two-phase prefetching approach based on it. To exam the proposed approach performance, we evaluate the cache hit rate, QPS, read I/Os, and read latency under four workloads with different read and write ratios.

4.1 Experimental Setup

The machine that we use consists of an AMD Ryzen-5 3500U 2.1 GHz processor, 16 GB main memory, and 512 GB SSD. In our experiments, the size of the writing buffer and the block cache is set to 8 MB and 500 MB, respectively. The value of the ratio between levels is 10. The preloaded data is 10^6 KV pairs, and the size of the key-value pair is 1KB. We collect access information in parallel. The length of the time window is variable, a time interval between two compactions.

We conduct a series of experiments to evaluate the performance of the LSM-tree with cache prefetching approachs. In order to evaluate the effect of the two-phase parallel prefetching approach, we test three different prefetching strategies, including Least Recently Used (LRU) algorithm as the baseline, the first-phase prefetching algorithm and the two-phase prefetching approach.

4.2 Workloads

We use four workloads with mixed read and write ratios, generated by Yahoo! Cloud System Benchmark (YCSB) with various configurations, to evaluate the performance of the two-phase prefetching approach. Table 1 shows the parameters of these workloads. YCSB can simulate the workloads of real applications. Today's internet applications commonly receive far more read requests than write requests, making the workloads much more read-heavy. Read, write and update operations exist at the same time in real application scenarios [20]. Thus, we use a Zipfian distribution to generate the workloads.

Table 1. Workloads.

Workloads	Description
A	80.0% reads, 15.0% updates, 5.0% writes
B	80.0% reads, 10.0% updates, 10.0% writes
C	70.0% reads, 7.5% updates, 22.5% writes
D	70.0% reads, 15.0% updates, 15.0% writes

4.3 Algorithm Performance Evaluation

In this section, we conduct a series of experiments to evaluate the performance of the LSM-tree with different prefetching approaches under four workloads with mixed read and write ratios.

Block Cache Hit Rate. Figure 3 shows the hit rates of the baseline and our prefetching strategies. We can see that the performance of the first phase prefetching algorithm is better on workload A than workload C. This is mainly because compactions in workloads with a high percentage of update operations contain more duplicate data. As the number of update operations increases, the cascading compaction decreases. Duplicate data is deleted during compaction. Thus, the first phase prefetching algorithm can easily track the location of the invalidation KV pairs in this situation. Suppose users are sensitive to CPU consumption in the workloads with more update operations. In that case, they can only run the first phase prefetching algorithm. It can also effectively reduce the cache miss.

From Fig. 3, we can see that the cache hit rates with the first phase prefetching algorithm are increased by 1.5X under workload A and workload B. Under the workloads with more write operations such as Workload C and Workload D, the cache hit rates are increased by 0.7X. As the writing ratio increases, cascading compaction also increases. After the cascading compactions, the key-value pairs in the invalidation blocks will be dispersed into multiple blocks. It results in tracking the invalidation KV pairs difficultly. Thus, we need the second phase to enhance the first-phase prefetching algorithm.

We can also see that the cache hit with the two-phase prefetching approach increases by 2.65X after the compaction under four mixed workloads. Because we put the rest derived blocks of the first phase into the minor cache. This ensures that key-value pairs associated with the invalidation block are not lost. And then, the second phase's prefetching works. The non-derived blocks that contain hot KV pairs are put into the minor cache. The impact of the size of the minor cache on the performance will be discussed later.

Read I/Os. We also evaluate read I/Os under four workloads with mixed read and write ratios. The number of reading I/Os is an essential indicator for evaluating our prefetching approach. If the number of reading I/Os is fewer, the number of cache hits is higher. This means read operation does not have to fetch blocks from the disk every time. From Fig. 4, we can intuitively see the effect of the prefetch algorithm on the number of disk accesses.

Figure 4 shows the number of reading I/Os of the LSM-tree with different prefetch strategies. We can see that the number of reading I/Os of the first prefetching algorithm is decreased by 17% under workload A and workload B. And it is decreased by 11% under workload C and workload D. Workload A and workload B have more read operations, so the number of reading I/Os is more. Workload C has more write operations, so the number of reading I/Os can not be effectively reduced with the first phase algorithm. This is mainly because write-intensive workloads may result in a large number of cascading compaction. This will cause the invalidation keys to be split into multiple blocks and decrease the access number of blocks. Thus, a large number of reading requests will still be sent to the disk. So the second phase is essential to the problem of cache invalidation.

The number of reading I/Os of the LSM-tree with the two-phase prefetching approach is decreased by 20% based on the first phase prefetching algorithm. In the original LSM-tree with the cache, read performance may drop sharply after compaction. The root cause is that the number of reading I/Os increased suddenly. The two-phase prefetching approach can effectively decrease the number of reading I/Os by the minor cache and the prefetching algorithm. Thus, it can make key-value stores more stable.

The Size of the Minor Cache. We evaluate how the size of the minor cache impacts the performance of the prefetching algorithm. The baseline is an LRU model without a cache prefetching algorithm. The percentage of the minor cache to the total cache is expressed by the parameter c. Figure 5 shows the cache hit rate with the two-phase prefetching approach over 1500 seconds. We can see that the cache hit rate will gradually increase with the increase of c. It is mainly because the number of blocks selected by the second-phase prefetching algorithm increase. As c increases, the number of the selected blocks increases. In addition, a minor cache is also a victim cache. When a block in a minor cache is accessed, this block will be put into the main cache. Suppose the eviction is caused by the placement of a block of the minor cache. In that case, the blocks which are evicted will be put into the minor cache.

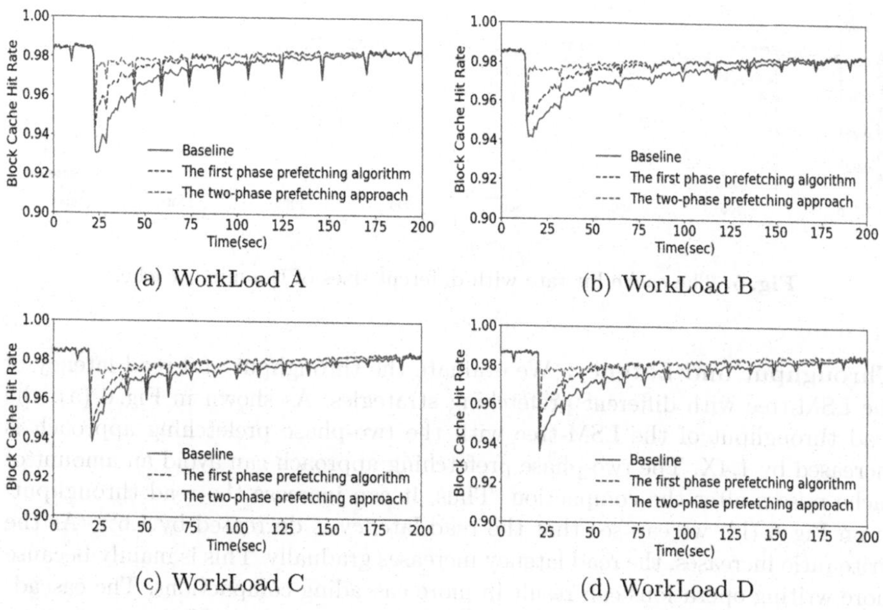

(a) WorkLoad A (b) WorkLoad B

(c) WorkLoad C (d) WorkLoad D

Fig. 3. Cache hit rates with different workloads.

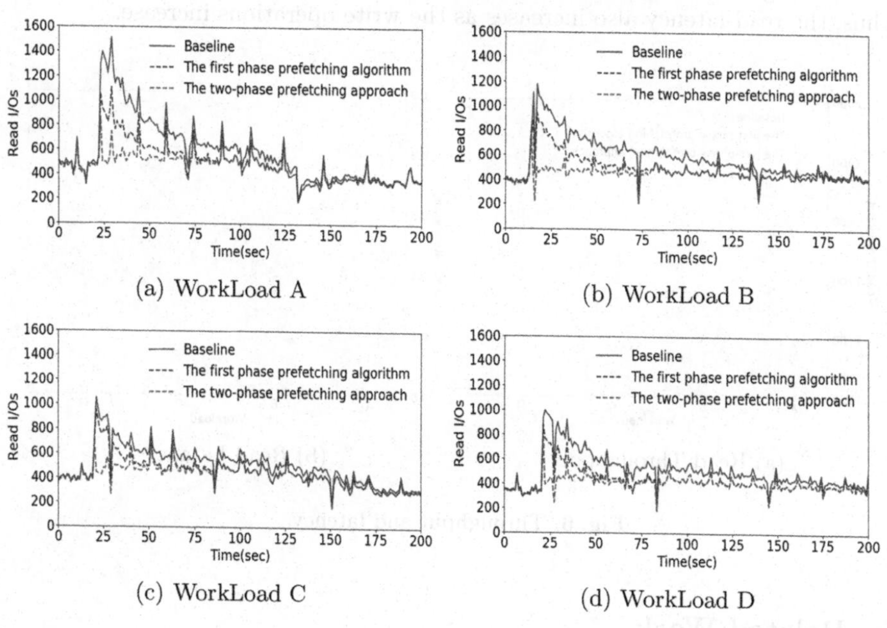

(a) WorkLoad A (b) WorkLoad B

(c) WorkLoad C (d) WorkLoad D

Fig. 4. Read I/Os with different workloads.

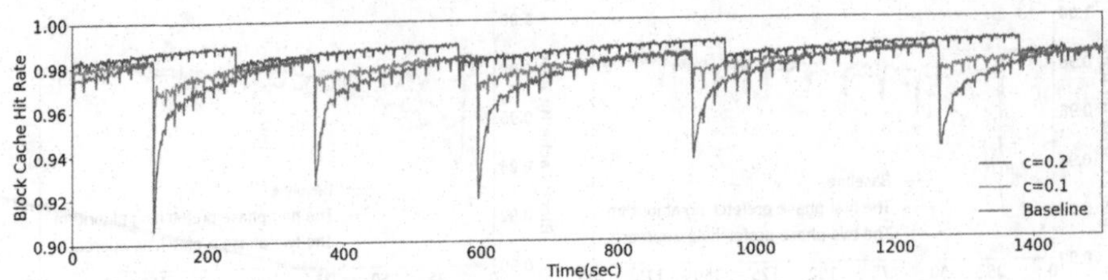

Fig. 5. The cache hit rate with different sizes of the minor cache.

Throughput and Latency. We evaluate the throughput and read latency of the LSM-tree with different prefetching strategies. As shown in Fig. 6 (a), the read throughput of the LSM-tree with the two-phase prefetching approach is increased by 1.4X. The two-phase prefetching approach can avoid an amount of cache misses after the compaction. Thus, it can increase the read throughput. From Fig. 6 (b), we can see that the read latency is decreased by 1.6%. As the write ratio increases, the read latency increases gradually. This is mainly because more writing operations can result in more cascading compactions. The cascading compactions can increase the time of the compactions. If the compaction time is longer, the number of unordered SSTable in the *level* 0 will increase. Thus, the read latency also increases as the write operations increase.

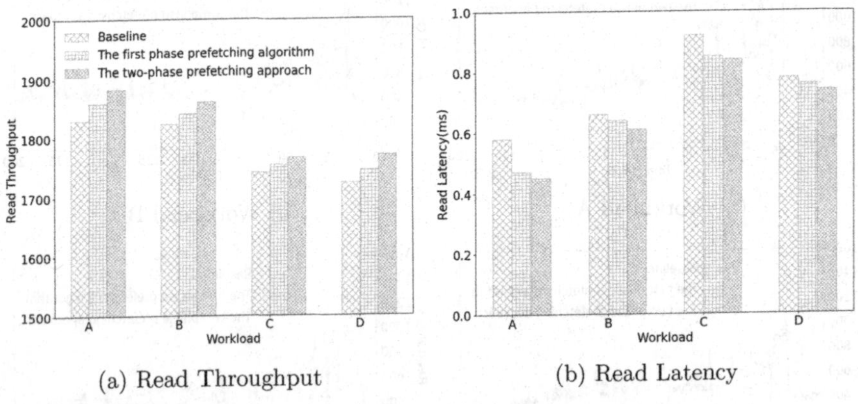

(a) Read Throughput

(b) Read Latency

Fig. 6. Throughput and latency.

5 Related Work

Nowadays, key-value stores have become a fundamental part of the cloud infrastructure [6,7,17]. Cloud based data services ubiquitously use key-value storage systems to store data required with high performance for both read and write operations. For example, Yahoo! has reported that the trend in their typical workloads has changed to have similar proportions of reads and writes [10]. Therefore, it is essential to have optimized key-value stores for both read and write workloads. In the LSM-tree, in order to improve read performance, the LSbM-tree [8] uses the on-disk buffer between the LSM-tree and the cache to delay the cache misses. Inspired by the SkipList, PebblesDB [9] uses a globally ordered but partially un-ordered method at each level to reduce write amplification. But more aggressive compaction will cause the cache to be invalid. Kim and Wu propose LSM-based secondary indexing structures to optimize query performance [25,26]. The tiered LSM-tree [11] achieves very low write amplification and fewer compactions. Thus, the number of compaction and the corresponding cache invalidation can be reduced significantly. TRIAD [24] propose a solution that separates hot keys from cold keys in the memory component to reduce write and read amplification for skewed update workloads. Recently, emerging persistent memory like PCM, MRAM is promised to enhance the performance of the LSM-tree. SLM-DB [13] using persistent memory to maintain a B+-tree index for indexing KVs. The persistent B+-index can accelerate the search of a key. Although the LSM-tree offers a much higher write throughput compared to B+-tree, it often exhibits write stalls since heavy operations such as flushes and merges run in the background [23]. When the compaction between *L0*-*l1* involves a large amount of data, write stalls cause application performance to drop periodically. NoveLSM [22] and MatrixKV [21] use NVM to reduce write stalls and the depth of LSM-trees. SpanDB [27] relocates write-ahead logs (WAL) and the top levels of the LSM-tree to NVMe SSDs for optimize write performance.

6 Conclusion

This paper focuses on the cache invalidation problem in the LSM-tree based store. The read performance suffers from a large number of cache misses due to the compaction operation periodically. In order to solve this problem, we propose a two-phase parallel prefetching approach. We evaluate the cache hit rate, the throughput, read I/Os and latency for four workloads with different read and write ratios. The experimental results show that the two-phase parallel prefetching approach can significantly increase the block cache hit rate after the compaction.

Acknowledgments. This work was supported in part the National Science Foundation of China Projects (61971309) and Tianjin Science Foundation Project (18JCY-BJC85500).

References

1. O'Neil, et al.: The log-structured merge-tree (LSM-tree). Acta Inf. **33**(4), 351–385 (1996)
2. Jagadish, H.V., et al.: Incremental organization for data recording and warehousing. In: Proceedings of the 23rd VLDB (1997)

3. Ghemawat, S., Dean, J.: LevelDB. http://leveldb.org (2011)
4. Chang, F., Dean, J., et al.: Bigtable: a distributed storage system for structured data. ACM Trans. Comput. Syst. (TOCS) **26**(2), 1–26 (2008)
5. Facebook. RocksDB: a persistent key-value store. http://rocksdb.org
6. Basescu, C., et al.: Robust data sharing with key-value stores. In: Proceedings of DSN (2012)
7. DeCandia, G., et al.: Dynamo: amazon's highly available key-value store. ACM SIGOPS Oper. Syst. Rev. **41**(6), 205–220 (2007)
8. Teng, D., et al.: LSbM-tree: re-enabling buffer caching in data management for mixed reads and writes. In: Proceedings of ICDCS (2017)
9. Raju, P., et al.: PebblesDB: building key-value stores using fragmented log-structured merge trees. In: Proceedings of SOSP (2017)
10. Ooper, B.F., et al.: Benchmarking cloud serving systems with YCSB. In: Proceedings of Cloud (2010)
11. Wu, X., et al.: LSM-trie: an LSM-tree-based ultra-large key-value store for small data items. In: Proceedings of USENIX ATC (2015)
12. Dayan, N., et al.: Dostoevsky: Better space-time trade-offs for LSM-tree based key-value stores via adaptive removal of superfluous merging. In: Proceedings of SIGMOD (2018)
13. Kaiyrakhmet, O., et al.: SLM-DB: single-level key-value store with persistent memory. In: Proceedings of USENIX FAST (2019)
14. Athanassoulis, M., et al. MaSM: efficient online updates in data warehouses. In: Proceedings of SIGMOD (2011)
15. Pugh, W.: Skip lists: a probabilistic alternative to balanced trees. Commun. ACM **33**, 668–676 (1990)
16. Pan, F.F., et al.: dCompaction: speeding up compaction of the LSM-tree via delayed compaction. Comput. Sci. Technol. **32**(1), 41–54 (2017)
17. Cockroach Labs. CockroachDB. https://github.com/cockroachdb/cockroach
18. Apache Cassandra. http://cassandra.apache.org
19. Bansal, J.C.: Particle Swarm Optimization. In: Bansal, J.C., Singh, P.K., Pal, N.R. (eds.) Evolutionary and swarm intelligence algorithms. SCI, vol. 779, pp. 11–23. Springer, Cham (2019). https://doi.org/10.1007/978-3-319-91341-4_2
20. Cao, Z., et al.: Characterizing, modeling, and benchmarking RocksDB key-value workloads at Facebook. In: Proceedings of USENIX FAST (2020)
21. Yao, T., Zhang, Y., et al.: MatrixKV: reducing write stalls and write amplification in LSM-tree BasedKV stores with a matrix container in NVM. In: Proceedings of USENIX ATC (2020)
22. Kannan, S., et al. Redesigning LSMs for nonvolatile memory with NoveL SM. In: Proceedings of USENIX ATC (2018)
23. Luo, C., Carey, M.J.: LSM-based storage techniques: a survey. VLDB J. **29**(1), 393–418 (2019). https://doi.org/10.1007/s00778-019-00555-y
24. Balmau, O., et al. TRIAD: creating synergies between memory, disk and log in log structured key-value stores. In: Proceedings of USENIX ATC (2017)
25. Wu, L., et al.: Building efficient key-value stores via a lightweight compaction tree. In: Proceedings of USENIX ICDE (2017)
26. Kim, Y., et al. A comparative study of log-structured merge-tree-based spatial indexes for big data. In: Proceedings of USENIX ICDE (2017)
27. Chen, H., et al.: SpanDB: a fast, cost-effective LSM-tree based KV store on hybrid storage. In: Proceedings of USENIX FAST (2021)

A Hybrid TLBO-TS Algorithm Based Mobile Service Selection for Composite Services

Runbin Xie[1,2], Jianxun Liu[1,2](✉), Guosheng Kang[1,2], Buqing Cao[1,2], Yiping Wen[1,2], and Jiayan Xiang[1,2]

[1] Hunan Provincial Key Lab. for Services Computing and Novel Software Technology, HNUST, Xiangtan, China

[2] School of Computer Science and Engineering, Hunan University of Science and Technology, Xiangtan, China

Abstract. Service selection for composite service has been a hot research issue in service computing field. With the proliferation of mobile devices, service selection confronts new challenges in the mobile environment due to the mobility, unpredictability, and variation of signal strength of mobile networks, since quality of service (QoS) is closely related to these factors. In this work, we aim to address the problem of mobile service selection for composite service in terms of QoS. Specifically, based on the mobility model and mobility-aware QoS computation rule, we propose a hybrid service composition optimization algorithm, named TLBO-TS, by integrating Teaching-Learning-Based Optimization (TLBO) algorithm and Tabu Search (TS) algorithm. Through the optimization of service selection with TLBO-TS algorithm, the global QoS of the generated mobile service composition is approximately optimal. Extensive experiments are conducted and the experimental results show that the proposed approach can derive more optimized mobile service composition with acceptable scalability compared with the traditional approach and other baselines.

Keywords: Mobile service selection · Service composition · Quality of Service · Mobile network · Hybrid algorithm

1 Introduction

Technological advances in fields, such as service computing, cloud computing, and mobile Internet, have led to the exponential increase of online services on the Web [1]. Services with different functionalities can be composed to develop new applications or offer value-added services. In recent years, mobile devices and sensors become prevalent. Due to the easy access and low barrier of Internet, more and more users tend to invoke services to solve their business or personal requirements since the service invocation via mobile systems has provided advantages, for example, convenience and popularity [2]. Under mobile Internet environment, users often are called mobile users, and services are called mobile services. Correspondingly, service selection for composite services is called mobile service selection [3].

Y. Lai et al. (Eds.): ICA3PP 2021, LNCS 13155, pp. 237–256, 2022.
https://doi.org/10.1007/978-3-030-95384-3_16

As there may be many available services with the same functionality, these services are usually differentiated by QoS (Quality of Service) which describes their non-functional characteristics [4]. Thereby, the objective of QoS-aware service selection for composite services is to select the optimal service from the corresponding candidate service set for each task to optimize the QoS criteria of the target composite service. As we know, service selection for composite services is a NP-hard problem in traditional service computing [5]. Similarly, mobile service selection for composite services is also a NP-hard problem in mobile service computing. As the number of mobile service applications increases, mobile service selection becomes more complex. Moreover, due to the mobility, unpredictability, and variation of signal strength in the mobile environment, this problem will be more challenging since quality of service (QoS) is closely related to these factors. Thus, the service selection in the mobile Internet environment is quite different from that in the traditional Internet environment. And service selection algorithms for the traditional Internet environment will not be desirable for the mobile environment since the best composition cannot be generated as the constant mobility makes the performance of service invocation unpredictable and location-based. In this paper, we will focus on the problem of mobile service selection for composition in terms of QoS and propose the resolving method.

Service selection has always been a hot research issue in service computing field. The topic is widely studied in the traditional Internet environment. In order to solve the service selection problem in service composition, traditionally researchers usually consider the comprehensive QoS of the target composite service and transform service selection into an optimization problem. However, there are limited works on mobile service selection. Among these limited works, they can be divided into three scenarios: QoS-driven mobile service selection [6], energy consumption-driven mobile service selection [7], energy-QoS trade-off mobile service selection [8], and QoS-correlation aware mobile service selection [9]. In this paper, we focus on QoS-driven mobile service selection which is first studied in [6]. In this work, the research problem of service selection for mobile service composition is identified, and formal definitions to describe this problem are provided. Besides, a mobility model is introduced to formally model service invocations in mobile environments. Finally, a mobility-aware service selection algorithm is proposed by using the Teaching-Learning-Based Optimization (TLBO) algorithm. And it has been empirically proved that the performance for service selection with TLBO algorithm outperforms the other population-based optimization methods on both optimality and scalability. TLBO is a meta-heuristic successfully applied to a variety of complex optimization problems with only several parameters [10]. However, sometimes it may also fall into the suboptimal solution due to the focus more on the global optimization. Thus, it can be improved further, which is the motivation of this paper.

In this paper, we aim at improving the quality of solution of mobility- and QoS-aware service selection for composite services in the mobile Internet environments. To this end, we propose a novel hybrid heuristic method, named TLBO-TS, by combining TLBO algorithm and Tabu Search (TS) algorithm, in which both the global optimization and local optimization are taken into account during the optimization procedure. Note that in the mobile Internet environment, users usually consider the response time of service invocation, so minimizing the response time is considered as the objective of service selection for composite services in our work. Extensive simulation experiments have been done in this paper. And the experimental results show that our method can generate much better solution than the traditional service selection methods, and also better than the state-of-the-art method in [6] on both optimality and scalability.

The remaining of this paper is organized as follows. Related work is presented in Sect. 2. In Sect. 3, preliminaries related to the service selection problem for the mobile environment are presented, including prerequisite definitions and mobility-aware QoS computation rule, and the corresponding assumptions to our problem. The hybrid TLBO-TS algorithm based mobile service selection for composite services is proposed in Sect. 4. Section 5 evaluates the proposed approach. Finally, we conclude the paper in Sect. 6.

2 Related Work

QoS-aware service selection is a key part of service composition. As for this topic, service selection for composite services is widely studied for non-mobile terminal users' invocation scenario. However, the study of service selection for composite services under the mobile environment is relatively new, only limited works have been done. Thus, we review the related work from the following two aspects: 1) QoS-aware service composition under the traditional Internet environment; 2) QoS-aware service composition under the mobile Internet environment.

2.1 Traditional QoS-Aware Service Composition

As for QoS-aware service selection for composite services under the traditional Internet environment is extensively investigated. By reviewing the existing works, they can be roughly divided into three categories: (1) independent service selection approaches [11, 12]; (2) integer programming based approaches [13–15]; and (3) meta-heuristic based approaches [16–19].

Specifically, independent service selection approaches usually assume that tasks in the target composite service plan are independent. And then services with the best QoS utility from candidate set are selected for each task. This kind of approach mainly focuses on local optimization for each task. Thus, the generated solution is usually far from satisfactory since it cannot solve service selection with global QoS optimization. It is only feasible when tasks from the composite service plan are indeed independent with respect to QoS. However, this situation rarely happens. Therefore, to consider the overall QoS of the composite service, integer programming based approaches are proposed for QoS-aware service composition. The objective of integer programming based approaches is to optimize the overall QoS of the target composite service. However, 0–1 integer programming problem is a NP-hard problem, so it is not easy to find the best solution especially under the large number of Web service candidates. Under the big data environment, to enumerate all the composite solutions is hardly possible. Therefore, to obtain the near-to-optimal solution, meta-heuristic approaches are utilized for service selection, such as Genetic Algorithm (GA) [18], (Particle Swarm Optimization) PSO [5], etc.

Although there are many other variant or hybrid approaches, the main ideas basically belong to the above three categories. Beside, some extended service selection scenarios are also studies, like service selection with multi-user [20], trust [21], QoS correlations [9], etc. These approaches either consider extended service selection scenarios or improve the efficiency of service composition optimization. In summary, these approaches aim at the traditional Internet environment, which cannot fit well the mobile Internet environment since QoS of the composite service is sensitive to the active user's moving location.

2.2 Mobile QoS-Aware Service Composition

With the development of mobile computing and rapid increase of mobile devices, some researchers transfer the topic of service composition from the traditional Internet environment to the mobile Internet environment. Among the works on mobile service composition, they can be divided into three scenarios: QoS-driven mobile service selection [6], energy consumption-driven mobile service selection [7], energy-QoS trade-off mobile service selection [8], and QoS-correlation aware mobile service selection [9]. QoS-driven mobile service selection is to select services for each task to make the whole composition perform the best QoS. Energy consumption-driven mobile service selection aims to reduce the energy consumption of the target composite service in mobile environments as the energy consumption of service can be affected by many factors including users' moving location. Energy-QoS trade-off mobile service selection aims to make a balance between energy consumption and overall QoS of the target composite service. QoS-correlation aware mobile service selection considers QoS correlations between services, because it influences the overall QoS of the composite service.

In our work, we focus on QoS-driven mobile service selection which is first studied in [6]. In [6], a mobility-aware approach for service composition is proposed, which consists of a mobility model, a mobility-aware QoS computation rule, and a mobility-aware service selection algorithm based on TLBO. Thus, this work lays a solid foundation for our work. Although this approach can generate better solution than traditional approaches in a mobile environment, it may also fall into local optimization or the generated solution is not near-optimal enough. Thus, to further improve the service composition optimization, we try to proposed an improved approach based on [6], so that more optimized service composition solution can be generated. Although mobility-aware service composition in mobile communities is also studied in [22] which considers that both service requesters and providers are mobile. This situation would be more challenging. However, in this paper, we only consider that users are mobile while services are not mobile.

3 Preliminaries

In this section, we first present some definitions of the key concepts in the scope of mobile service selection for service composition. Then, the mobility-aware QoS computation rules based on the mobility model will be given. And for the simplicity reason, we also give several assumptions related to the mobile service selection problem. Our approach is proposed based on the given assumptions below.

3.1 Prerequisite Definitions

Definition 1 (Mobile Service). A mobile service s is modeled as a triple $(I(s), O(s), Q(s))$, where:

(1) $I(s) = \{i_1(s), i_2(s)\ldots\}$ is the set of input parameters;
(2) $O(s) = \{o_1(s), o_2(s)\ldots\}$ is the set of output parameters;
(3) $Q(s) =< q_1(s), q_2(s)\ldots q_n(s) > (n \geq 2)$ is an n-tuple where each q_i denotes a QoS property of s such as cost, response time, throughput, or availability.

Assumption 1. Since in the mobile Internet environment, users usually pay much more attention to the response time of service invocations, so we focus on the QoS property, i.e., response time, as the objective of service selection for composite services. And response time of service invocation is closely related with users' mobility since a user's location will determine the wireless signal strength which affects the time spent on data transmission.

Definition 2 (Mobile Service Invocation). A mobile service invocation si is a triple (s, v_i, v_o), where:

(1) s is the invoked service;
(2) v_i is the volume of the input data to s;
(3) v_o is the volume of the output data from s.

Definition 3 (Mobile Service Composition Plan). A mobile service composition plan sp is a 2-tuple (T, R), where:

(1) $T = \{t_1, t_2\ldots t_n\}$ is a set of tasks; each task t_i can be achieved by a set of candidate services C_i;
(2) $R = \{r(t_i, t_j)|t_i \in T, t_j \in T\}$ is a set of relations between tasks in T. $r(t_i, t_j)$ represents the inputs of t_j that depend on the outputs of t_i. R can describe the structure of the service composition plan.

Definition 4 (Mobile Service Composition). A mobile service composition sc is a triple (sp, S, Q), where:

(1) sp is the service composition plan to which sc corresponds;
(2) S is the set of services in sc. Each $s_i \in S$ corresponds to a $t_i \in p.T$;
(3) Q is the global QoS of sc.

Definition 5 (Mobility Model). A user's mobility model is a triple $m = (T, L, DR)$, where:

(1) T is the discrete points during the user's mobility. It is a set of time points;

(2) L is the set of a user's specific location points in the moving path corresponding to time points in T;
(3) DR is a function that maps location points in L to data transmission rates.

Assumption 2. In this paper, we assume that in the service invocation process the movement path of a user is already known. That's because in mobile environment when a user invokes a service composition, the invocation of the previous service will affect the invocation of all the subsequent services, thereby affecting the global response time. With the user's future movement path, we can select suitable services for the user's service composition in advance to obtain optimal global response time. To make our approach more practical, it would be better to predict the mobile user's future moving path, which can be obtained by some mobility prediction methods in [23, 24]. In addition, when the mobile user uses the map navigation function, we can also get the user's movement path.

The mobility of a user will lead to the vary of the mobile network signal. Thus, we get the data transmission rate by a mobility model according to location points from the user's movement path. Based on definition 5, given a time point $t \in T$, we can find a specific location $l \in L$. Then, we can map l to a specific data transmission rate $d = DR(l)$, since the location l determines the signal strength of mobile networks in mobile environment.

3.2 Mobility-Aware QoS Computation Rule

As mentioned before, this paper will focus on the response time for service invocation in the mobile Internet environment. Thus, we define the mobility-aware QoS (MQoS) based on the mobility model as definition 6. And the computation rule of the global mobility-aware QoS (GMQoS) for a composite service is presented as definition 7. Some of definitions are referred to [6].

Definition 6 (Mobility-Aware QoS, MQoS). MQoS describes the performance of the single service in a mobile service composition. In our work, only one QoS property (response time) is considered, so MQoS represents the time cost in the remaining of this paper. The MQoS of a single service s is calculated as follows:

$$MQoS_s = t_{v_i} + Rt_s + t_{v_o} \tag{1}$$

where t_{v_i} is the time for transmitting input data, Rt_s is the response time of s, and t_{v_o} is the time for transmitting output data. t_{v_i} and t_{v_o} can be calculated as Eq. (2), where v_i is the volume of input data and DR_i is the data transmission rate at the location from which the user starts sending the input data. v_o and DR_o are the corresponding variables for the output data.

$$\begin{aligned} t_{v_i} &= \frac{v_i}{DR_i} \\ t_{v_o} &= \frac{v_o}{DR_o} \end{aligned} \tag{2}$$

Assumption 3. The mobile network signal determines the data transmission rate, and the data transmission rate remains constant during the service invocation process including data transmission of input and output. Although the data transmission rate may vary during a service invocation process, we think the variation has little impact on the time spent on data transmission.

Next, we give an example of a user's moving path as shown in Fig. 1 to illustrate the computation process of MQoS. Suppose that a user requests a service at $t_0 = 3\,\mathrm{s}$ which corresponds to the location l_0. Moreover, l_0 corresponds to the corresponding data transmission rate $d_o = 40\,\mathrm{Kb/s}$. Assume that the service's input volume is 1000 Kb, then the transmission time of input data would be $t_{v_i} = 1000/40 = 25\,\mathrm{s}$. Next, the transmission of input data ends at $t_1 = 3 + 25 = 28\,\mathrm{s}$. Assume that the service's response time $Rt_s = 10\,\mathrm{s}$, after the service's execution the user will get to the location l_2 at $t_2 = 28 + 10 = 38\,\mathrm{s}$. Assume that the service's output data volume is 800 Kb and the corresponding data transmission rate at the location l_2 is $d_2 = 50\,\mathrm{Kb/s}$, then the transmission time of output data can be calculated, i.e., $t_{v_o} = 800/50 = 16\,\mathrm{s}$. Finally, the transmission of output data ends at $t_3 = 38 + 16 = 54\,\mathrm{s}$. Thus, the MQoS of this service invocation is $25\,\mathrm{s} + 10\,\mathrm{s} + 16\,\mathrm{s} = 51\,\mathrm{s}$, i.e., $t_3 - t_0 = 54\,\mathrm{s} - 3\,\mathrm{s} = 51\,\mathrm{s}$.

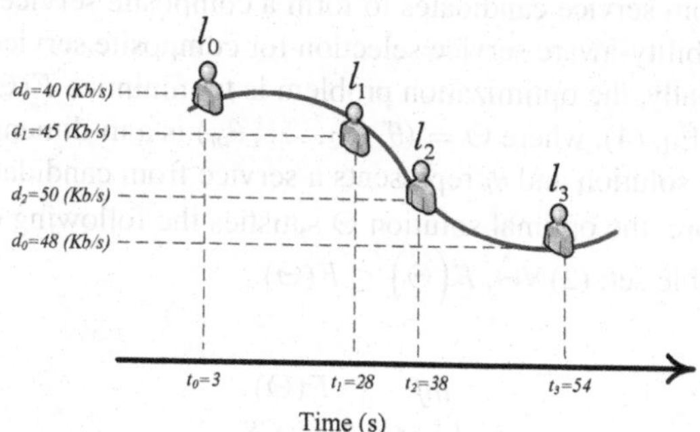

Fig. 1. An example of a user's moving path example for computing MQoS in single service inquest

Definition 7 (Global MQoS, GMQoS). Global MQoS describes the performance of a mobile service composition sc, i.e., the global response time of the composite service. Thus, the global MQoS is calculated as follows:

$$GMQoS(sc) = \underset{s \in sc}{\Phi} MQoS_s \tag{3}$$

where s is a single service, sc represents service composition. Φ is an operator that integrates the local MQoS and what Φ exactly to do depends on the relationship between any two services. The integration rules include summation $\sum$, product $\prod$, the maximum max, and the minimum min. To the ease the calculation, we assume that in this paper the mobile composite services are all in a sequential structure, and we only use the $\sum$ integration rule.

Definition 8 (Mobility-Aware Service Selection for Service Composition). With the path of mobile a user and the covering specific data transmission rate in the path, for a mobile service composition invoked by the user, specific services are selected respective from service candidates to obtain the optimal global QoS (GMQoS).

4 Hybrid TLBO-TS Algorithm for Mobility-Aware Service Selection

In this section, we introduce the details of the hybrid TLBO-TS algorithm for mobility-aware service selection. First, we demonstrate how the mobility-aware service selection problem is transformed into an optimization problem. Then we present the hybrid TLBO-TS algorithm for mobility-aware service selection in detail.

4.1 Mobility-Aware Service Selection as an Optimization Problem

Given a user's path in the mobility model and the data transmission rate obtained from the model within the area covering the path, mobility-aware service selection is to select specific services from service candidates to form a composite service with the optimal GMQoS. Thus, mobility-aware service selection for composite services is an optimization problem. Formally, the optimization problem is to minimize $F(\Theta) = GMQoS(\Theta)$ with constraints as Eq. (4), where $\Theta = (\theta_1, \theta_2, \ldots, \theta_m)$ is a m-dimensional vector representing a feasible solution and θ_i represents a service from candidate service set CS_i for task T_i. Therefore, the optimal solution $\widehat{\Theta}$ satisfies the following conditions: (1) $\widehat{\Theta}$ belongs to the feasible set; (2) $\forall\Theta$, $F\left(\widehat{\Theta}\right) \leq F(\Theta)$.

$$\begin{array}{cc} inf & F(\Theta) \\ subject\ to\ \theta_i \in CS_i & \end{array} \quad (4)$$

4.2 Hybrid TLBO-TS Algorithm

The optimization problem in Eq. (4) is clearly an NP-hard optimization problem. To find the approximately optimal solution, we propose a hybrid heuristic algorithm, named TLBO-TS, by combining Teaching-learning-based optimization (TLBO) algorithm and Tabu Search (TS) algorithm within polynomial time. Next, we present the TLBO algorithm and TS algorithm respectively, and then propose the TLBO-TS algorithm.

4.2.1 Teaching-Learning-Based Optimization Algorithm

Teaching-learning-based optimization (TLBO) algorithm is one of the meta-heuristic optimization methods [25]. The algorithm is inspired by teaching-learning process. The algorithm consists of two basic learning phases: (1) through interaction with the teacher (i.e., teacher phase) and (2) through interaction with the other learners (i.e., the learner phase).

In the mobile service selection optimization problem, the teacher represents the best performance mobile service composition solution obtained so far and learners represent feasible mobile service composition solutions in the group. And subjects provided to the learners are regarded as tasks in a mobile service composition plan. A learner's performance corresponds to the fitness value which represents the GMQoS calculated with Eq. (3) in this paper. There are three steps in TLBO algorithm: population initialization, teacher phase, and learner phase. The TLBO algorithm for our service selection problem is shown in Fig. 2.

1) **Population Initialization.** Common control parameters are required in TLBO, such as the population size P and the maximum iteration number of generations G. So, in our problem we randomly generate P mobile service composition solutions as the learners which can represent like $S^i = \left(s_1^i, s_2^i, \ldots, s_m^i\right), (i = 1, 2, \ldots, P)$, where m is the number of tasks in the mobile service composition plan, and each s_j^i is a concrete service for the ith learner's j^{th} task which selected from the jth task's candidates.
2) **Teacher Phase.** In TLBO, the teacher is considered as the individual with the best fitness value in every generation, so in this phase, each learner enhances themselves from interacting with the teacher. Specifically, $S_{teacher} = (s_1, s_2, \ldots, s_m)$ in our problem is a mobile service composition solution with smallest $GMQoS$ obtained so far, and every feasible mobile composition solution $S^i(i = 1, 2, \ldots, P)$ learns from the $S_{teacher}$ through the difference between the $S_{teacher}$ and the mean value of all learners, where:

$$S_{new}^i = S_{old}^i + difference \tag{5}$$

$$difference = r_i * (S_{teacher} - T_F * S_{mean}) \tag{6}$$

$$T_F = round[1 + rand(0, 1)] \tag{7}$$

where S_{new}^i and S_{old}^i are i^{th} learners before and after teacher phase, r_i is the random number in the range [0, 1] and T_F is the teaching factor which value is 1 or 2, S_{mean} is the average of all learners, calculated as follows.

$$S_{mean} = \frac{1}{P}\sum_{i=1}^{P} S^i \tag{8}$$

where S^i denotes a learner. After learning from the teacher, Eq. (9) is used to update all learners. In Eq. (9), F is the objective function in our mobile service selection optimization problem which used to calculate the fitness value of all leaners according to Eq. (3).

$$\begin{gathered} if\ F\left(S_{new}^i\right) < F\left(S_{old}^i\right): \\ S_{old}^i = S_{new}^i \end{gathered} \tag{9}$$

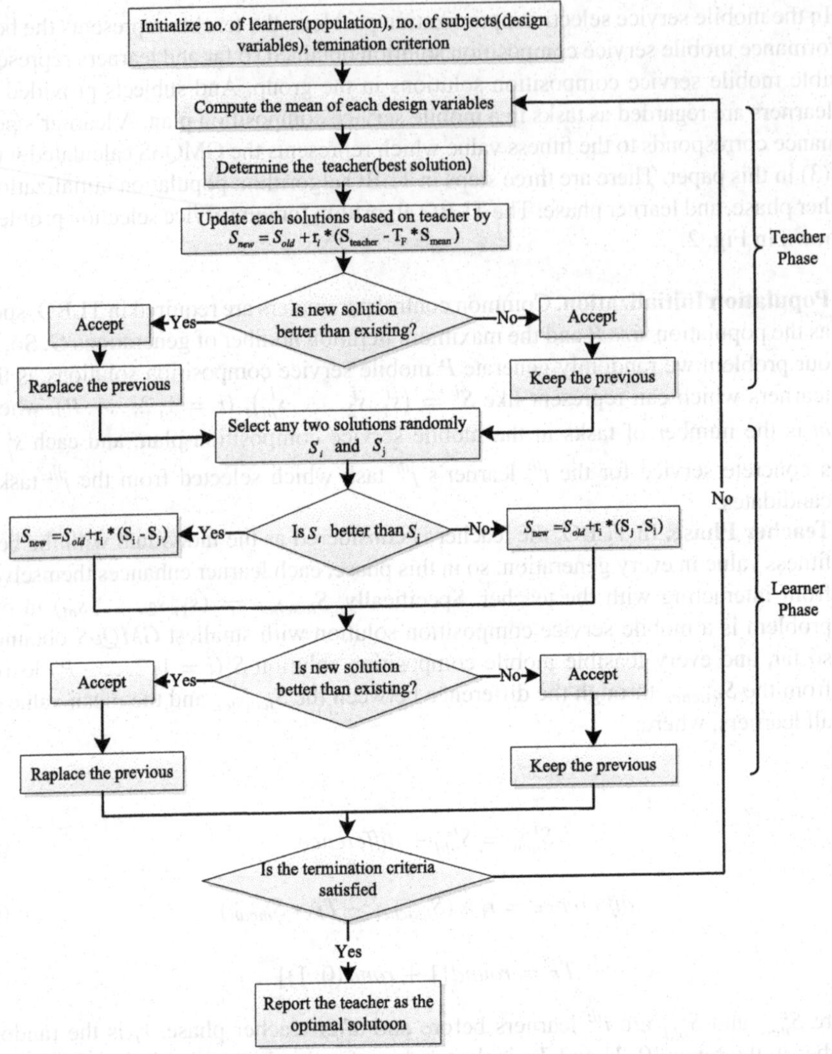

Fig. 2. TLBO flow chart

3) **Learner Phase.** In this phase, learners improve their knowledge by interaction among themselves. This phase can keep the population diverse. Specifically, for each S_i, a learner S_j $(i \neq j)$ is chosen to update S_i with Eq. (10), where r_i is the random number in the range [0, 1]. After this phase, all learners updated themselves with Eq. (9).

$$S^i_{new} = \begin{cases} S^i_{old} + r_i * (S_i - S_j) F(S_i) < F(S_j) \\ S_{old} + r_i * (S_j - S_i) otherwise \end{cases} \tag{10}$$

In this paper, we need to make sure each variable in our service composition solution is an integer and cannot cross the bounds of candidates in every iteration of TLBO algorithm. After the learner phase, the TLBO algorithm ends one iteration, and all learners (mobile service composition solutions) will be updated.

4.2.2 Tabu Search Algorithm

Tabu Search (TS) is a commonly used meta-heuristic algorithm, developed by Glover in [26] and is known for a strong local search base on the neighborhood which prevents its early convergence. The basic idea of TS is searching base on neighborhood solutions which give this algorithm strong local search ability.

In our mobile service selection optimization problem, the solution representation in TS is the same as in TLBO. TS begins from a current mobile service composition solution, then generates its feasible neighborhood solution set by movement which should satisfy the tabu list (i.e., the memory structures in tabu search, more commonly a tabu list is a short-term set of the solutions that have been visited in the recent past) and the aspiration criteria (a simple and commonly used aspiration criteria is to allow solutions which are better than the currently-known best solution). Find the best solution from neighborhood solution set and compare with the current solution according to Eq. (3). If it is better than the current solution, then replace it.

In our problem, the size of tabu list is considered as a function of the scale of the problem and if it offers a better solution than the best performance mobile service composition solution found so far, it is considered as the aspiration criteria. The TS algorithm flowchart is shown in Fig. 3.

Since our service selection is a global service selection, selecting a specific service for a task in the service composition chain may affect the performance of all the subsequent tasks. As for the current solution, we generate its neighborhood solutions by randomly choosing one task. Then the task's concrete service is replaced by its candidates. Meanwhile, the tabu list is taken into consideration in this process. This makes the new solution locate in the vicinity of the current solution. Figure 4 shows an example of how neighborhood solutions are generated from a current solution in TS.

4.2.3 Hybrid TLBO-TS Algorithm

TLBO is a novel population-based algorithm which needs few parameters, easy to control, and has great global exploration capability but lacks ability to explore the solution space locally. Tabu search is a traditional meta-heuristic algorithm which has a good capability for local search. Therefore, we developed a hybrid TLBO and TS optimization algorithm for our mobile service selection problem, which searches the solution space in both global and local, making the generated mobile service composition solution approximately optimal.

The hybrid TLBO-TS algorithm begins with the randomly generated mobile service composition solutions as learners in TLBO. Then after teacher phase and learner phase in TLBO, the best identified solution is input to TS as the current solution. For the current solution, apply TS algorithm and try to improve it from its generated neighborhood solutions. Then after applying TS, the best solution is feedback to the TLBO algorithm

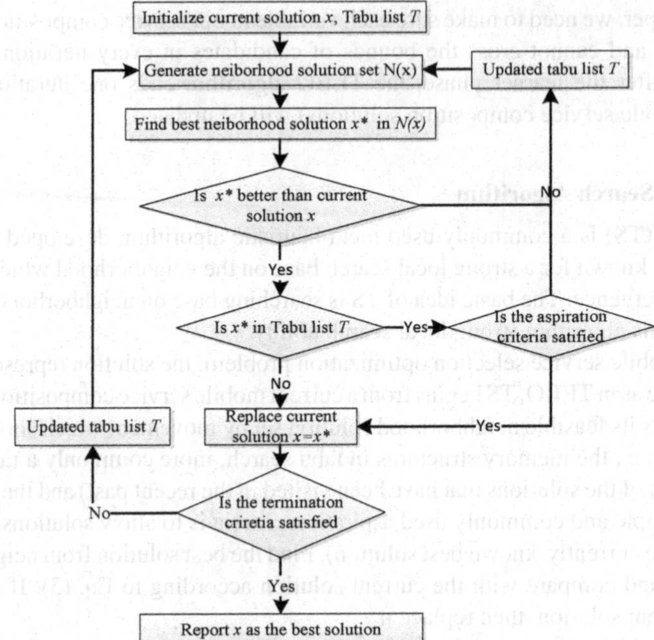

Fig. 3. Tabu search flow chart

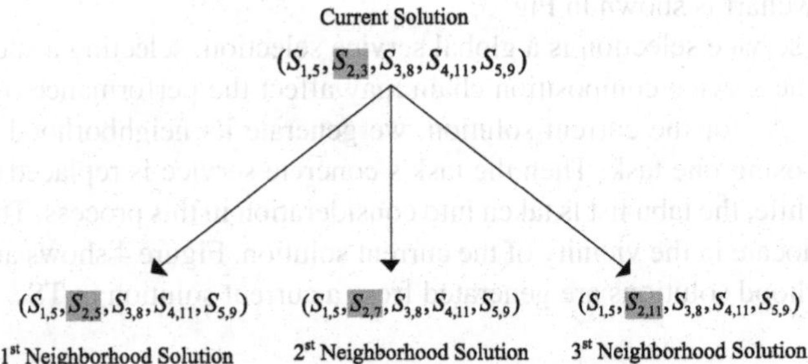

Fig. 4. The generation of neighborhood solutions from a current solution in Tabu Search

as a new teacher. TLBO algorithm is in front while TS algorithm is in back can achieve search for the optimal space as a whole and then locally. The TLBO-TS is iterated until the termination criterion is satisfied. Figure 5 shows the flowchart for the hybrid TLBO-TS optimization algorithm.

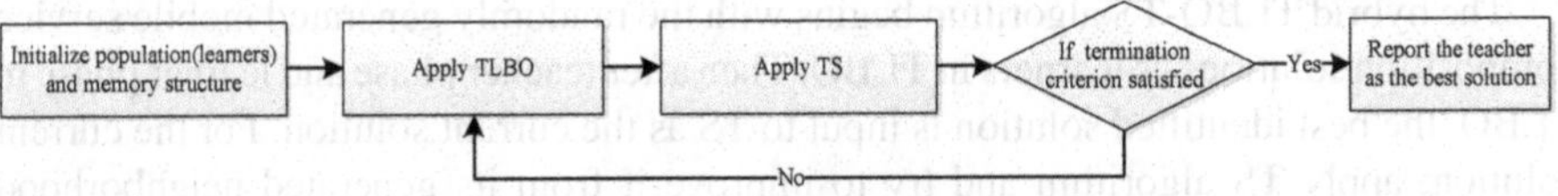

Fig. 5. TLBO-TS flow chart

4.2.4 Algorithm and Analysis

Given a user's mobile service composition invocation, the user's moving path and all the signal strength (data transmission rate) among the path from our mobility model, we present the algorithm for TLBO-TS. Suppose that P is the population size (i.e., the number of mobile service composition solutions), D is the number of tasks in a mobile service composition plan, G is the maximum iteration number, TL is the tabu list, A is the aspiration criteria, and N is the number of neighborhood mobile service composition solutions. The process of the proposed TLBO-TS algorithm can be summarized as Algorithm 1.

Algorithm 1: TLBO-TS

Input: The service invocation; the user's moving path; the signal strength (data transmission rate) among the path; the population size P; the number of tasks D; the maximum iteration generation number G; the tabu list TL; the aspiration criteria A

Output: Teacher T_G (i.e., best solution)

1: Initialization: randomly generate P mobile service compositions, initialize tabu list TL;
2: determine the initial teacher T_0 with the maximum fitness from the generated mobile service compositions;
3: **for** g=1: G
4:     $T_g = T_{g-1}$;
5:     **for** i=1: P
6:         **for** j=1: D
7:             Update solution S_i with Equation (5) and (9); // learning from the teacher
8:             Update solution S_i with Equation (9) and (10); // learning from other learners
9:         **end for**
10:    **end for**
11:    Update T_g with the maximum fitness from the updated mobile service compositions;
12:    **for** k=1: N
13:        Generate N neighborhood solutions of T_i under constraint of TL;
14:        Update TL by adding and expiring the service's movement from the generated neighborhood solutions;
15:    **end for**
16:    Determine the best neighborhood solution S_{neibor} with the maximum fitness from the generated neighborhood solutions under constraint of A;
17:    **if** Fitness(S_{neibor})$<$Fitness(T_g) **then**
18:        $T_g = S_{neibor}$;
19: **end for**
20: **return** T_G;

Given the initial population and parameters, TLBO-TS algorithm executes iteratively to search for the approximate optimal solutions. In each iteration, new solutions are generated by 2 steps: apply TLBO in lines 5–11 and apply TS in 12–18. And the time complexity of the algorithm is clearly polynomial. The overall time complexity with G iterations is $O(G * P * D * N)$.

5 Evaluation

In this section, we evaluate the performance of the proposed meta-heuristic algorithms for mobility service selection. All the algorithms are implemented with Python 3.8 and executed on a laptop computer with Core i5, and Windows 10 using 16 GB RAM. Each experiment repeats 50 times, and we use the average values as the results.

5.1 Dataset Setup and Evaluation Metrics

The experimental data is generated by simulation. We generate service compositions with randomly inserted tasks and structure. As for each task's candidate services, their response times are generated randomly in ascending order from a uniform distribution U.

Table 1. Parameter settings in TLBO-TS

Parameters	Value
Population size	15
Iteration number	100
Input/output data	$U(300, 999)$
Data transmission rate	$120 * (\cos(time) + 1) + 1$
Response time	$U(10, 80)$
Tabu list length	Population size * 0.4
Number of neighborhood solutions	Population size * 0.8

To simplify the problem solving and void other factors affect the results, we assume the candidates for a task require the same input/output data volume, and we generated the input/output data volume also with a uniform distribution U. We use a function $DR = a * (\cos(b * time) + 1) + 1$ to simulate the data transmission rate distribution in user's path, which makes the signal vary between 1 and a+1 with a variation cycle of $2\pi/b$. This function simulates the condition of a user moving with constant velocity and regularly distributed base stations. For tabu list length and neighborhood solution number, we set it as a scale function which depends on the population size. The Table 1 shows the specific parameter setting.

To evaluate the performance of mobility-aware service selection approaches for service composition, we use two evaluation metrics for evaluating the proposed approach, i.e., ***Optimality*** and ***Scalability***. Optimality is measured by the global response time of the derived mobile service composition. And Scalability is measured by the running time of the service selection algorithm. As the algorithm showed in Sect. 4, the number of tasks and the number of candidates per task are the two main factors which affect the scale of the mobile service selection problem. In order to examine the impact of above two parameters on optimality and scalability, we vary the number of tasks and the number of candidates per task in experiments. and a series of experiments are conducted based on the parameter settings in Table 2.

Table 2. Parameter settings

Parameters	Experiment series	
	A	B
Number of tasks	5–25	10
Number of candidates per task	50	20–100

5.2 Optimality Evaluation

In this section, in order to verify the superiority of our algorithm, we conducted a series of experiments, i.e., the optimality evaluation with the default parameters, the optimality evaluation with impact of different task number and the optimality evaluation with impact of different candidate services number to compare the global response time of services selected by TLBO-TS with those of four methods:

- **Standard method**: This method aims to select service with the best performance QoS for each task. Thus, in our scenario, the standard method selects service with the shortest response time.
- **GA for mobility service selection**: Genetic algorithm is a widely used heuristic algorithm for service selection, which has great global search ability. In our scenario, we set the GA's probability of performing crossover *pc* to 0.88 and set the probability of mutation *pm* to 0.2.
- **PSO for mobility service selection**: Particle Swarm optimization algorithm is another widely used heuristic algorithm. In our scenario, we set the PSO's acceleration constants c_1 and c_2 to 2 and set the inertia weight w to 0.8.
- **TLBO for mobility service selection** [6]: TLBO is an emerging meta-heuristic optimization which has been used in mobility service selection problem. This approach has powerful global search ability but lacks local search ability, so that may lead the solution to suboptimal. This is the state-of-the-art method for mobility-aware service selection.

Optimality v.s. Iterations. Figure 6 shows the optimality comparison between TLBO-TS and comparative methods with default number of tasks and candidates per task of above five methods, where x-axis is the iteration number and the y-axis is the global response time of the optimized service composition. As shown in Fig. 6, the standard method can find service compositions with shortest time, but it fails to derive the optimized GMQoS since the varied signal strength. PSO performs better than GA, because PSO has memory structure and optimized solutions are preserved in every iteration, while in GA the previous knowledge may be destroyed as the population changes. Both TLBO and TLBO-TS can obtain a better solution than the other three methods. TLBO as a powerful global algorithm can achieve convergence as the number of iterations increases. TLBO-TS is an improved method of TLBO. Based on the best solution in each iteration, TLBO-TS uses TS to achieve local search by generating neighborhood solutions. As can be seen from Fig. 6, TLBO-TS can find better solutions with the minimum number of iterations than other comparative methods, which shows the effectiveness of our method from the optimality perspective. In the remaining sections, we no longer compare with standard methods due to that it is not applicable to the mobile environment.

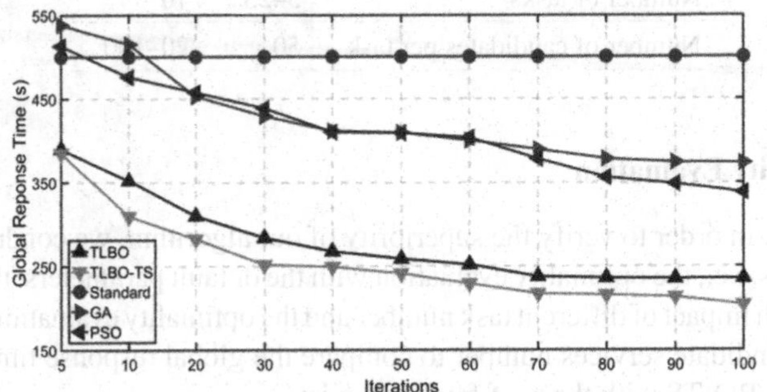

Fig. 6. Convergence curves

Optimality v.s. Number of Tasks. To verify the impact of the number of tasks on the global response time of service requests, we set the parameters to experiment series A as shown in Table 2. The results, shown in Fig. 7, demonstrate that the global response time of four algorithm all rise with the increasing number of tasks. Since other parameters are fixed, with the number of tasks increasing, the dimensionality of the solution increases, more services in the mobile service composition chain will be executed, which results in both the service response time and data transmission time increasing. So, the global response time of service requests increases. PSO performs better than GA when task numbers in low-dimensional. But with the increase of task number, the advantage seems slight. Both TLBO and TLBO-TS perform better than GA and PSO, and the global response time of the optimal mobile service composition returned by TLBO-TS is at least 10% lower than TLBO, the improvement becoming more evident with the increasing numbers of tasks, which verify the superiority of our TLBO-TS method.

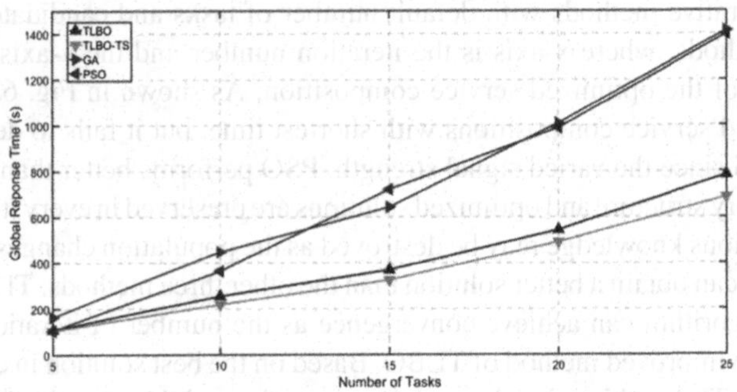

Fig. 7. Impact of number of tasks on global response time

Optimality v.s. Number of Candidates Per Task. Now, we examine the impact of the number of candidates per task on the result of mobile service selection. To this end, we set the experimental parameters to experiment series B as shown in Table 2. As shown in Fig. 8, GA's solution has no obvious change with the increase of number of candidates per task. It shows that even though the search space expands, GA still may lose the optimal solution in iterations due to the randomness of the algorithm and the mobility scenario. All other three algorithms' global response time of service compositions declines with the increasing number of candidates per task. With the increase of candidate services, there are more choices available, which expands search space and improves the selected services. Thus, the global response time of service invokes decreases. Furthermore, TLBO-TS performs better than the other three algorithms regardless of the number of candidate services, and the improvement becomes more evident with the increasing numbers of candidates per task.

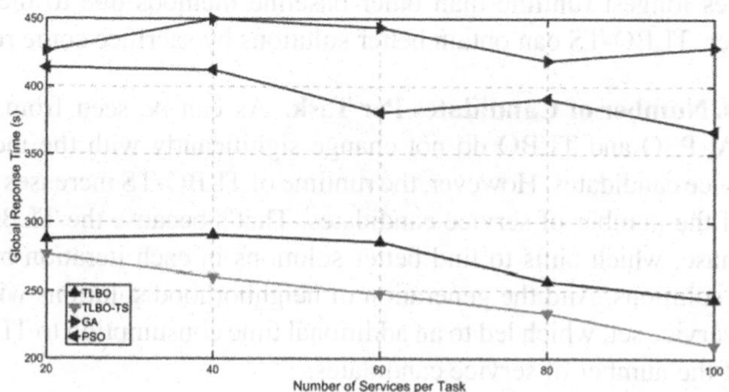

Fig. 8. Impact of number of candidates per task on global response time

Now, we can conclude that our TLBO-TS method maintains better performance with the vary of both the number of tasks or service candidates than other three baseline methods. This is because TLBO-TS not only has powerful global search capability but also local search capability. Therefore, TLBO-TS has best powerful optimality in our mobility scenario.

5.3 Scalability Evaluation

In this subsection, we conduct two sets of experiments to verify the scalability of our method. Analogously, we consider ranges of two parameters, i.e., the number of tasks and the number of candidates per task. To evaluate the impact of the number of tasks and the number of service candidates on the runtime of TLBO-TS, we vary the number of tasks from 5 to 25 and the number of service candidates from 20 to 100. Figure 9 and 10 show the runtime of TLBO-TS and other three baseline methods with the increase of the number of tasks and the number of service candidates respectively.

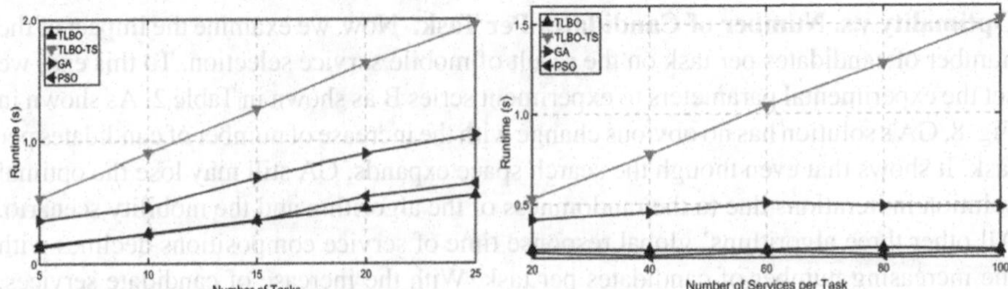

Fig. 9. Impact of number of tasks on runtime

Fig. 10. Impact of number of candidates on runtime

Scalability v.s. Number of Tasks. As can be seen from Fig. 9, the runtime of four methods all increases linearly with the increase of the number of tasks since more tasks increase the dimension of solutions. And TLBO runs faster than other methods. The TLBO-TS takes longest runtime than other baseline methods due to the tabu search phase. However, TLBO-TS can obtain better solutions by sacrifice some runtime.

Scalability v.s. Number of Candidates Per Task. As can be seen from Fig. 10, the runtime of GA, PSO and TLBO do not change significantly with the increase of the number of service candidates. However, the runtime of TLBO-TS increases linearly with the increase of the number of service candidates. That's because the TLBO-TS adds a tabu search phase, which aims to find better solutions in each iteration by generating neighborhood solutions. And the generation of neighborhood solutions will depend on the candidate service set, which led to an additional time consumption to TLBO-TS with the increase of the number of service candidates.

To summarize, TLBO-TS takes longer runtime than other baseline methods with the increase of the number of tasks and the number of tasks. However, the gap of runtime is very close and TLBO-TS can obtain much better solutions than the other baselines, which is acceptable in practice.

6 Conclusion

In this paper, we focus on QoS-aware service selection for composite services in the mobile Internet environment, in which the mobility of users is taken into consideration. To generate an optimized mobile service composition solution, we proposed a hybrid algorithm based on TLBO and TS, named TLBO-TS algorithm, in which global and local MQoS optimizations are both performed. Extensive simulated experiments have been done. And the experimental results show that the proposed algorithm can generate a much better solution than the traditional method, PSO-based method, GA-based method and better than the state-of-the-art methods which only use TLBO algorithm. In future work, we will address the problem of service selection for composite services in mobile Internet environments with complex or uncertain moving paths of users, and consider the condition while service providers are mobile.

Acknowledgment. This work was partially supported by National Key R&D Program of China under grant No: 2020YFB1707602, Educational Commission of Hunan Province of China under Grant No: 20B244, National Natural Science Foundation of China under grant No: 61872139 and 61572187.

References

1. Kang, G., Liu, J., Cao, B., Cao, M.: NAFM: neural and attentional factorization machine for web API recommendation. In: 2020 IEEE International Conference on Web Services (ICWS), pp. 330–337. IEEE (2020)
2. Deng, S., et al.: Toward mobile service computing: opportunities and challenges. IEEE Cloud Comput. **3**(4), 32–41 (2016)
3. Deng, S., Wu, H., Yin, J.: Mobile Service Computing. ATSTC, vol. 58. Springer, Singapore (2020). https://doi.org/10.1007/978-981-15-5921-1
4. Kang, G., Liu, J., Cao, B., Xiao, Y.: Diversified QoS-centric service recommendation for uncertain QoS preferences. In: IEEE International Conference on Services Computing, Beijing, China, pp. 288–295. IEEE (2020)
5. Kang, G., Liu, J., Tang, M., Xu, Y.: An effective dynamic web service selection strategy with global optimal QoS based on particle swarm optimization algorithm. Paper presented at the International Parallel and Distributed Processing Symposium, Shanghai, China (2012)
6. Deng, S., Huang, L., Hu, D., Zhao, J.L., Wu, Z.: Mobility-enabled service selection for composite services. IEEE Trans. Serv. Comput. **9**(3), 394–407 (2014)
7. Deng, S., Wu, H., Tan, W., Xiang, Z., Wu, Z.: Mobile service selection for composition: an energy consumption perspective. IEEE Trans. Autom. Sci. Eng. **14**(3), 1478–1490 (2015)
8. Gelenbe, E., Lent, R.: Energy–QoS trade-offs in mobile service selection. Future Internet **5**(2), 128–139 (2013)
9. Deng, S., Wu, H., Hu, D., Zhao, J.L.: Service selection for composition with QoS correlations. IEEE Trans. Serv. Comput. **9**(2), 291–303 (2014)
10. Rao, R.V.: Teaching Learning Based Optimization Algorithm. Springer, Cham (2016). https://doi.org/10.1007/978-3-319-22732-0
11. Liu, Y., Ngu, A.H., Zeng, L.Z.: QoS computation and policing in dynamic web service selection. Paper presented at the Proceedings of the International World Wide Web Conference (2004)
12. Benatallah, B., Dumas, M., Sheng, Q.Z., Ngu, A.H.H.: Declarative composition and peer-to-peer provisioning of dynamic web services. Paper presented at the Proceedings of the 18th International Conference on Data Engineering (2002)
13. Zeng, L., Benatallah, B., Dumas, M., Kalagnanam, J., Sheng, Q.Z.: Quality driven web services composition. Paper presented at the International World Wide Web Conference (2003)
14. Yu, T., Zhang, Y., Lin, K.: Efficient algorithms for web services selection with end-to-end QoS constraints. ACM Trans. Web (TWEB) **1**(1), 6–32 (2007)
15. Alrifai, M., Risse, T.: Combining global optimization with local selection for efficient QoS-aware service composition. Paper presented at the 18th International Conference on World Wide Web, Madrid, Spain (2009)
16. Kashyap, N., Kumari, A.C., Chhikara, R.: Service composition in IoT using genetic algorithm and particle swarm optimization. Open Comput. Sci. **10**(1), 56–64 (2020)
17. Li, C., Li, J., Chen, H.: A meta-heuristic-based approach for Qos-aware service composition. IEEE Access **8**, 69579–69592 (2020)

18. Liu, S., Liu, Y., Jing, N., Tang, G., Tang, Y.A.: Dynamic web service selection strategy with QoS global optimization based on multi-objective genetic algorithm. In: Zhuge, H., Fox, G.C. (eds.) GCC 2005. LNCS, vol. 3795, pp. 84–89. Springer, Heidelberg (2005). https://doi.org/10.1007/11590354_10
19. Wang, Z., Cheng, B., Zhang, W., Chen, J.: QoS-aware automatic service composition based on service execution timeline with multi-objective optimization. In: 2020 IEEE International Conference on Services Computing (SCC), pp. 296–303. IEEE (2020)
20. Kang, G., Liu, J., Tang, M., Liu, X.F., Fletcher, K.F.: Web service selection for resolving conflicting service requests. Paper presented at the International Conference on Web Services, Washington, DC, USA (2011)
21. Somu, N., Gauthama Raman, M.R., Kirthivasan, K., Shankar Sriram, V.S.: A trust centric optimal service ranking approach for cloud service selection. Future Gener. Comput. Syst. **86**, 234–252 (2018)
22. Deng, S., Huang, L., Taheri, J., Yin, J., Zhou, M., Zomaya, A.Y.: Mobility-aware service composition in mobile communities. IEEE Trans. Syst. Man Cybern. Syst. **47**(3), 555–568 (2016)
23. Yavaş, G., Katsaros, D., Ulusoy, Ö., Manolopoulos, Y.: A data mining approach for location prediction in mobile environments. Data Knowl. Eng. **54**(2), 121–146 (2005)
24. Jain, C.C.R., van den Berg, E.: Location prediction algorithms for mobile wireless systems (2002)
25. Rao, R.V., Savsani, V.J., Vakharia, D.: Teaching–learning-based optimization: a novel method for constrained mechanical design optimization problems. Comput. Aided Des. **43**(3), 303–315 (2011)
26. Glover, F., Laguna, M.: Tabu search. In: Handbook of Combinatorial Optimization, pp. 2093–2229. Springer, Cham (1998)

UPM-DMA: An Efficient Userspace DMA-Pinned Memory Management Strategy for NVMe SSDs

Jinbin Zhu[1,2], Limin Xiao[1,2(✉)], Liang Wang[1,2(✉)], Guangjun Qin[3], Rui Zhang[1,2], Yuting Liu[1,2], and Zhonglin Liu[4]

[1] State Key Laboratory of Software Development Environment, Beihang University, Beijing, China
{jinbinzhu,xiaolm,lwang20,ruizhang1230,null233}@buaa.edu.cn
[2] School of Computer Science and Engineering, Beihang University, Beijing, China
[3] Smart City College, Beijing Union University, Beijing, China
qinguangjun@buaa.edu.cn
[4] North China Institute of Computing Technology, Beijing, China

Abstract. Fast storage devices (e.g., NVMe SSDs), which integrate new storage medium and parallel architecture, provide promising storage solutions for cloud data center loads. However, they make the lengthy IO stack that is negligible in the past become a new performance bottleneck. Current user-mode solutions can eliminate context switches between kernel and user space by moving partial kernel IO stack into userspace. Unfortunately, they introduce additional overhead for pinning memory used by DMA in userspace which is quite time-consuming for IO-intensive workloads. To address this issue, we present UPM-DMA, an alternative and efficient memory management strategy for NVMe SSDs to lighten the IO stack by amortizing per-request latency. UPM-DMA can dynamically select the appropriate pinned memory mode for different IO requests, especially the pinned memory pool for medium data IO requests. We further explore different parameter settings to optimize system performance and memory space efficiency. Finally, we implement the overall strategy as a memory library named UPM libs and integrate it into the SPDK framework. The official benchmarks, SPDK perf, are adopted to evaluate our solution. The experimental results show that compared with the default processing method in SPDK, there is an improvement of at least 17% under various test data sizes.

Keywords: NVMe SSDs · User space · Pinned memory · IO-intensive

1 Introduction

Data-intensive applications (such as RocksDB, Hadoop) exert increasingly intense pressure on the IO system, posing forward higher performance requirements for storage devices [1,2]. At present, there has been much research on

Y. Lai et al. (Eds.): ICA3PP 2021, LNCS 13155, pp. 257–270, 2022.
https://doi.org/10.1007/978-3-030-95384-3_17

the performance of storage system [3,4]. Solid-state drive (SSD) is an popular fast storage device made up purely of flash memory in NAND modules and governed by a controller chip. Non-volatile memory express (NVMe) is a communication standard developed specially for SSDs by a consortium of vendors, including Intel, Samsung, Sandisk, Dell, and Seagate [5]. As a replacement for the traditional hard disk drives (HDDs), NVMe SSDs reduce the latency of IO processing from 2 milliseconds to less than 100 microseconds. Up to now, many cloud providers (Google, Amazon, Alibaba cloud, and Tencent cloud) have configured NVMe SSDs to build high-performance storage systems.

While NVMe SSDs provide low latency IO processing and high system throughput, they make the lengthy IO stack which is negligible in the past, become a new performance bottleneck [6]. The latest research from Intel shows that the Linux kernel takes up too much execution time on the IO stack in NVMe SSDs oriented storage system [6–8]. The main reason is that context switching and interrupt produce a lot of time waste, which make the existing native IO software stack can not fully explore the abilities of NVMe SSDs. For example, when performing DMA operations, context switching needs to be made between user mode and the kernel. Frequent processing of IO requests will cause frequent switching, resulting in high switching overhead.

To address this issue, recent studies of fast storage devices on the storage IO stack tend to access NVMe SSDs directly from user mode [6,7,9,10]. User-mode solutions can eliminate context switches between kernel and userspace by moving partial kernel IO stack into userspace. SPDK is a high-performance user-mode storage tool library developed by Intel [6]. Userspace, polled mode, asynchronous, lockless NVMe driver provides highly parallel access to an SSD from a userspace application.

However, directly access hardware devices in user mode also brings new problems, such as the introduction of pinned memory overhead. We need to execute a specific command to ensure that the mapping of the virtual address to the physical address of the user process cannot be changed for some time. When the operating system marks a page so that its address translation relationship cannot be modified, it is a pinned memory page. There are many reasons for the change in the mapping of virtual addresses to physical addresses. One of the more common reasons is that some physical memory pages are swapped to disk space. It is because physical memory is limited after all. Still, the scope of the virtual address is much larger than the physical memory address space, so when a part of the physical address space is insufficient, the operating system will select some physical page data to save to the disk and then map the new virtual address to the physical address again.

For the issue that the pinning memory overhead introduced when the user-mode applications directly access the fast storage device affects the efficiency of IO request processing, a user-mode pinned memory buffer pool algorithm based on dynamic adjustment is designed and implemented to allocate the pinning memory overhead to multiple IO requests and improve the execution efficiency of IO request. Different pinned memory pool allocation algorithms and free algorithms are designed and implemented for different IO data sizes to improve memory space performance effectively. The internal organization is in the form of memory blocks. The pinned memory and the free memory are managed uniformly. Different parameter settings are explored to optimize memory efficiency for improving system performance and reducing memory space occupation.

Our key contributions are as follows,

- We propose an efficient User-mode Pinned Memory pool that can lighten the IO stack for NVMe by amortizing per-request latency;
- We propose an effective pool management strategy dynamically supporting different IO data sizes, and explore different parameter settings to optimize system performance and memory space efficiency;
- We implemente the overall strategy as a memory library named UPM libs and integrate it into the SPDK framework;
- We evaluatie UPM by SPDK perf and show that it significantly reduces IO latency.

The remainder of this paper is organized as follows. Section 2 presents the related works. Then, we explain our scheme, including UPM design, management strategy, parameter optimization, and integration with SPDK in Sect. 3. In Sect. 4, we evaluate our scheme by SPDK perf. Finally, we conclude our paper in Sect. 5.

2 Related Work

The main problem to solve in the storage IO stack is to reduce the context switching and data copying overhead between the userspace and the kernel, the interrupt processing overhead of IO requests, and the competition overhead of shared resources on the kernel IO stack, etc. Several optimization schemes have been proposed to reduce the time cost of the IO software stack [11,12]. Traditional methods to solve these problems mainly include the polling method, merging IO, and double buffering. These performance improvements are far from enough for NVMe SSDs, a fast storage device.

To fully utilize the performance of NVMe SSDs, the user-mode IO frameworks have been proposed in recent years, which significantly reduces the time overhead of the IO software stack by avoiding context switching. Researchers in [6,7,9,10] designed SPDK, nvmedirect, and unvme, all of which are based on user-mode IO frameworks, where SPDK is released and continuously maintained by Intel. Although this will significantly improve the performance, it will also introduce some new problems. For example, when the user directly accesses the fast storage device, the virtual and physical addresses mapping relationship may be modified. Therefore, pinning the memory during DMA operation is used to keep the mapping relationship from virtual address to physical address unchanged during DMA operation.

This pinned memory operation has a non-negligible overhead. In the face of IO-intensive applications, frequent pinned memory will seriously degrade the execution efficiency of IO requests. At present, the research on DMA buffer operation mainly focuses on the network. For example, there are many pieces of research on RDMA (remote direct memory access) buffer [13], but there are few studies on DMA buffer when users directly access fast storage devices. Although researchers in [14] study the issue of DMA caching, they main consider it from a security perspective. UPM-DMA is a complementary solution to this problem. First, UPM-DMA is a user-mode-based strategy, which maintains the high-performance characteristics of the user-mode IO framework. Second, it amortizes per-request latency by dynamically selecting the appropriate pinned memory mode for different sizes of IO requests.

3 Proposed Method

In this section, we introduce the basic ideas of the proposed UPM, the design of the framework, and the design of specific modules.

3.1 Motivation

To motivate the design of UPM, we first quantify overheads associated with memory pinning and copying under various data sizes. First, we use SPDK's *spdk_dma_malloc* each time to allocate and pin memory for different data sizes. After that, the pinned memory is released immediately. We perform this operation ten times and count the average time cost. At the same time, for the same testing data size, we copy the data to the pinned memory space, calculate the cost of copying data 10 times, and then take the average. The statistical results are shown in Fig. 1.

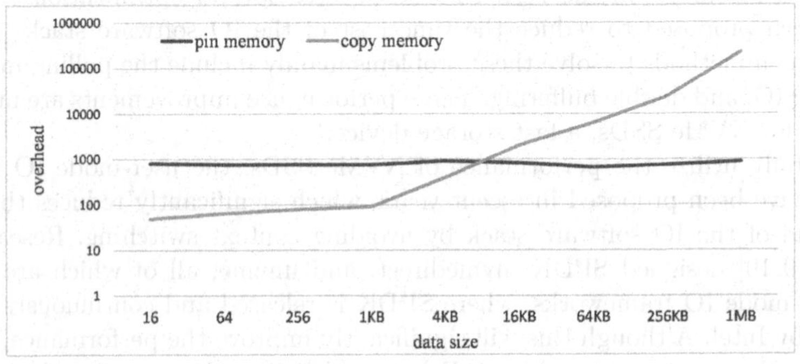

Fig. 1. Time consumption of pin memory and memory copy.

We can make the following observation from Fig. 1, (1) the overhead of pinning memory is almost constant. And it is evident that the total overhead will be substantial when there are a large number of IO requests, which is a severe performance bottleneck; (2) When the 4KB data is below, the memory copy overhead is much smaller than the memory pinned overhead. However, with the increase of data, the cost of memory copying becomes increasingly expensive. Notably, for small data IO, memory copy is adopted, and user-mode storage buffer is used for DMA operation will be an effective way to lighten the IO stack. We make the following conclusions from the observation:

- For small data requests, copying data from pageable to pinned memory is shorter than pin memory. At the same time, because of the small amount of data, only a tiny amount of memory needs to be pinned to meet the requirements, which will not have too much negative impact on the system memory. It means that we can pin a small piece of memory during the initialization phase and keep it pinned all the time.
- For the IO request of big data, the data copy time increases with the data size. In other words, the time consumption of pinned memory is less than the data copy time. It means we can pin a large piece of memory and form a memory pool for dynamic use by IO requests. Nevertheless, at the same time, we should also be aware that if the data is too large, we cannot use the method of the pinned memory pool. Because pinning a large piece of memory for a long time will lead to inefficient use and bring the lack of memory to other system applications.

This analysis motivates us to design the proposed UPM-DMA solution.

3.2 Design of UPM

Based on the above analysis, we design the scheme framework as shown in Fig. 2. UPM consists of three parts, which are selectively applied for small, medium, and big workloads.

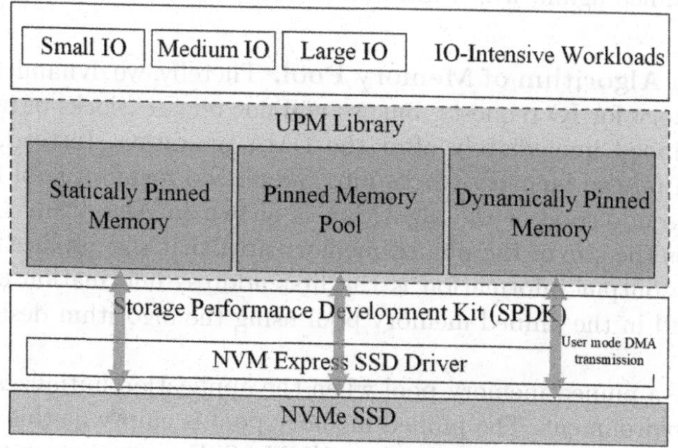

Fig. 2. The framework of UPM-DMA

3.3 Statically and Dynamically Pinned Memory

We adopt the statically pinned memory strategy for small IO requests, equal to or less than 4 KB. We allocate and pin three memory blocks during the library initialization process, and Users can set the size of memory blocks according to their experience. These three memory blocks remain pinned throughout the life cycle of the application. Because of the fast reading and slow writing characteristics of SSD, we use one piece of memory for reading requests and the remaining two blocks for writing requests. Our design is because small data IO requests do not need to take up too much memory. If we maintain a pinned memory state throughout the application's life cycle, we will not waste memory, and we only need to spend one-time pin memory.

For big data's IO request, which is greater than or equal to 4MB, we adopt the strategy of dynamically pinned memory. This strategy dynamically allocates and pins memory when the application reads and writes NVMe SSD through DMA. We unpin and release the pinned memory block immediately after the DMA operation. We cannot use the static pinned memory strategy for IO requests for big data because these requests will take up more memory, and if we occupy memory for a long time, it will have a destructive impact on other processes. Another reason is that the pinned memory time is less than big data's copy time.

3.4 Pinned Memory Pool

We design a pinned memory pool for medium data IO requests and propose the allocation and release algorithms. Unlike small IO requests and big IO requests, the copy and pinned memory costs for medium data are relatively high. The more common way to process these is to pin the memory of the corresponding size for DMA operation for each IO request. In the face of IO-intensive applications, the pinned memory area can not be reused, and then the processed IO requests need to be pinned again, which introduces a lot of pinned memory overhead.

Application Algorithm of Memory Pool. Thereby, we dynamically allocate and pin memory for IO requests, but we will not release blocks of memory that have been pinned immediately after the DMA operation. Instead, we recycle the used and pinned memory blocks into the pinned memory pool for management. The pseudo-code of the algorithm is shown in Algorithm 1. The input information is the size of the pinned memory area that the application needs to allocate. The output information is the first address information of the memory area found in the pinned memory pool using the algorithm designed in this section.

We create a pinned memory pool when the application initializes the NVMe SSD usage environment. The pinned memory pool is empty at this stage, causing the application to initiate the first NVMe SSD access request. When the applicant applies for memory allocation for the first time, we allocate and pin a memory block according to the memory block size configured by the user. After that, we put the memory block into the index for the next memory IO request.

We mark the memory area as used, update the memory block's meta information, and return the first address of the memory area to the application. When this pinned memory is used up, we just change its state to unused and manage it using a linked list, rather than unpinning and releasing it directly.

Algorithm 1. Pinned memory pool allocation algorithm for medium IO data

```
Input: memory_size
Output: memory_address
if memory_pool is null then
    memory_chunk ← alloc_and_pin(pin_size)
    add_memory_pool(memory_pool, memory_chunk)
    memory_address ← find_memory_region(memory_chunk)
    update(memory_chunk)
    return memory_address
else
    for memory_chunk in memory_pool do
        if mem_free_size(memory_chunk) > memory_size then
            memory_address ← find_memory_region(memory_chunk)
            update(memory_chunk)
            return memory_address
        end if
    end for
    if total_memory_size + pin_size > total_pin_size then
        return null
    end if
    memory_chunk ← alloc_and_pin(pin_size)
    memory_address ← find_memory_region(memory_chunk)
    add_memory_pool(memory_pool, memory_chunk)
    update(memory_chunk)
    return memory_address
end if
```

For subsequent IO requests, we will not immediately apply for and fix new blocks of memory. Alternatively, we will first find the fixed memory pool and get the information through the index of the linked list. The core idea is to find out whether there is a pinned memory block that meets the memory pool requirements. If so, we directly extract the corresponding memory from the memory pool instead of applying for and fixing a new memory block. At this time, the time cost of fixed memory is saved. However, if the difference cannot find fixed memory that meets the requirements, we will apply for and fix a new memory block.

Design of Pinned Memory Pool Release Algorithm. We design a fixed memory block release algorithm, and the pseudo-code is shown in Algorithm 2. The input is the first address of the pinned memory area to be released by the application. The output is the memory block's status, representing the status of the execution result of the release process so that the upper application can judge the execution result. The upper layer application may mistakenly pass in a memory address that does not exist in the pinned memory cache pool.

Three different scenarios need to be considered for medium data size memory area release, and different strategies will be adopted for different situations. The accessible memory areas before and after will are merged to prevent fragmented memory space when releasing the memory area. To quickly respond to IO requests, the fragmented integration will not be done when releasing.

Algorithm 2. Pinned memory pool release algorithm for medium IO data

Input: memory_address
Output: status

```
memory_chunk ← find_memory_chunk(memory_pool, memory_address)
if memory_chunk is null then
    return status::error
end if
pin_memory_free(memory_chunk, memory_address)
update(memory_chunk)
mem_pre_region ← chunk_pre_region(memory_chunk, memory_address)
mem_cur_region ← memory_address
while memory_pre_region is free do
    merge_free(memory_chunk, mem_pre_region, mem_cur_region)
    mem_cur_region ← mem_pre_region
    mem_pre_region ← chunk_pre_region(memory_chunk, mem_cur_region)
end while
mem_next_region ← chunk_pre_region(memory_chunk, memory_address)
mem_cur_region ← memory_address
while memory_next_region is free do
    merge_free(memory_chunk, mem_next_region, mem_cur_region)
    mem_cur_region ← mem_next_region
    mem_next_region ← chunk_next_region(memory_chunk, mem_cur_region)
end while
if memory_chunk is free and memory_chunk_time_interval > time then
    delete_memory_chunk(memory_pool, memory_chunk)
    free_and_unpin(memory_chunk)
end if
return status::ok
```

We will traverse the linked list of fixed memory blocks, finding out if there are contiguous, mergeable blocks of memory. If such memory blocks are not found, we will directly mark them as free; if we find blocks that can be merged, we will first merge them to get a larger contiguous fixed memory space. And then mark it as free. After completing the marking of the memory blocks that can be released, we will detect whether the unused time of these memory blocks exceeds the threshold set by the user. Only the memory blocks that exceed the unused time set by the user are released.

3.5 Parameters Setting

Parameters include the size of the pinned memory block for each application, the total size of the memory cache area, and the time threshold when the memory block is not accessed.

Memory block size can be set according to the application load characteristics or empirical values, and the default value is 4MB. We can also adjust the pinned memory block size according to the memory block usage of the application during a period, that is, after the application is executed for some time, judge the memory usage of the current application, and then adjust the memory block size appropriately to improve the performance.

There is no specific policy for setting the total size of the memory block area because the more significant the value is set, the better the concurrency of IO request processing. However, it needs to be set reasonably according to different situations. For example, there is no other application to run on a server that only does disk IO. You can set a considerable value to improve the processing efficiency of IO requests. The setting principle is to set a significant threshold without affecting the operation of other applications.

The primary consideration of the unreachable memory block time threshold is to release some completely free memory blocks and reduce the occupation of memory space when IO processing is less in a period.

4 Experiments and Results

In the experiments, we evaluate the efficiency of the proposed UPM by both simulations and actual environment. And UPM is compared with the default process in SPDK. Various comparative experiments were conducted by SPDK Perf [15], which is the storage benchmark tool. The experimental platform was constructed based on the environment shown in Table 1.

Table 1. The configuration of experiment

Environment	Configuration
CPU	Intel(R) Xeon(R) Gold 5115 CPU @ 2.40 GHz
Memory	128 G
Operating system	CentOS Linux release 7.4.1708 (Core)
SPDK	20.01
GCC	4.8.5

The first set of experiments is designed to test the proposed UPM under different data sizes. With perf, 50000 random IO requests were generated for data less than or equal to 4 KB, 10000 random requests for the medium data with a size of 4 KB to 128 KB, and 1000 random requests for the big data larger than 128 KB. Memory block size is also set to three cases. The blocks used for small IO requests are set to 32 MB, the pinned memory block size is 4 MB, and the memory block size used for data more significant than 128 KB is set to 32 MB.

Figure 3 shows the average response time of IO requests for different data size. The x-axis and the y-axis represent the size of IO requests and the average response time in seconds.

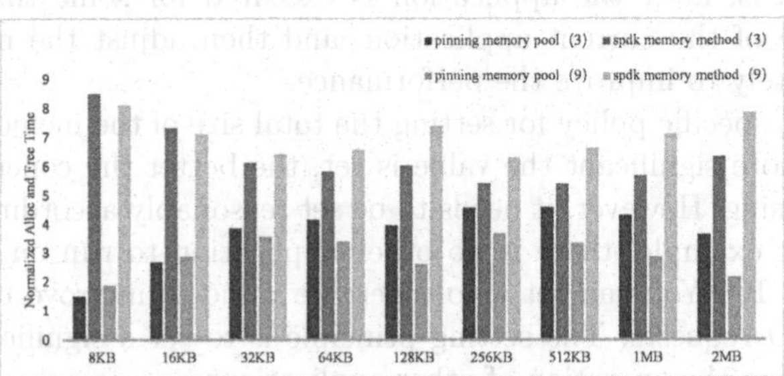

Fig. 3. Effect of IO size on memory algorithm.

It can be seen from Fig. 3 that performance of NVMe SSD is improved under all memory sizes tested, and the performance improvement will be more evident under different data sizes. The main reason is that the algorithm's key idea is to allocate the pinned memory overhead to more requests and reduce the overhead of memory-related operations. It also optimizes small data requests and uses memory copy to reduce the pinned memory overhead effectively. The pinned memory pool is constructed in memory blocks for medium data to allocate the pinned memory overhead to multiple IO requests effectively. It can also be seen that under the same data size, with the increase of program running time, the longer the memory usage time, the more pronounced the improvement effect of the algorithm. Because with the increase of memory usage time, each memory block is used by more requests, the pinned memory overhead of each memory block is allocated to more requests, which reduces the proportion of the overall memory operation-related overhead.

Figure 4 shows the performance improvement effect of the algorithm under various test data sizes. It can be seen from the figure that under most data sizes, with the increase of program running time, the improvement effect of the algorithm becomes more prominent. The main reason is that more requests use the same memory block, and its pinned memory overhead is allocated to more requests accordingly.

The second experiment verifies the efficiency of the algorithm under different program running times. The maximum size of selected memory test data is 64 KB and 128 KB. The test results are shown in Fig. 5. It can be seen from the figure that with the increase of the running time of the simulation program, the effect of the algorithm is more prominent. The main reason is that with the

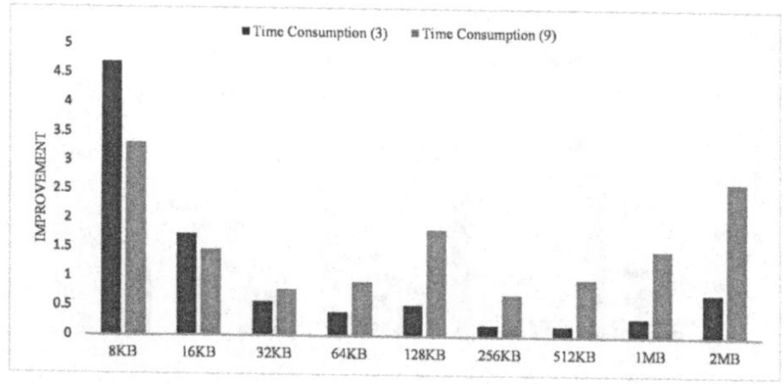

Fig. 4. Effect of multiple data sizes on memory algorithm.

increase of program running time, the memory library in SPDK will apply for and pin memory every time and release memory after use, which introduces a significant overhead. The algorithm allocates the pinned memory overhead to more IO requests, reducing the total memory operation overhead. The results show the effectiveness of the algorithm. Allocating the pinned allocated memory overhead to more requests can effectively improve the performance.

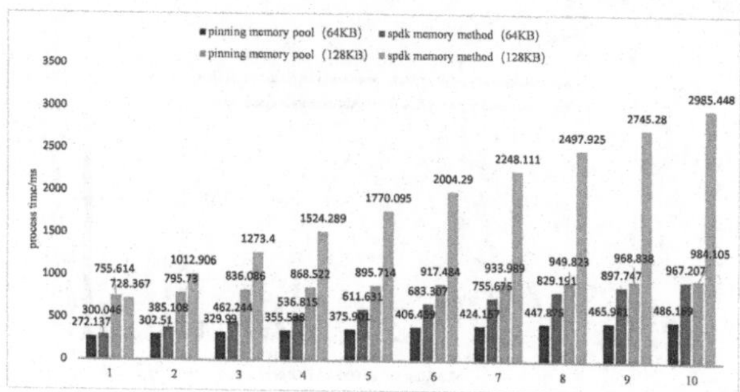

Fig. 5. Influence of different running time on memory Algorithm 1.

We integrate UPM library into SPDK, and generate IO write requests of various data sizes to test the performance improvement by our solutino. The main experimental steps refer to the conventional test steps for the storage system. A certain number of IO write requests with different data sizes are randomly generated each time, memory is allocated through two different memory libraries, and then the IO request is submitted to the asynchronous request queue in the background SPDK for processing. After processing, the asynchronous interface will be called to release the memory area. Finally, count the completion time of all IO requests.

Multiple IO requests of different sizes can put great pressure on NVMe SSD. In Fig. 6, he abscissa represents the maximum data size of the randomly allocated IO requests, i.e., 8 KB, representing the random generation of 8000 or 16000 IO

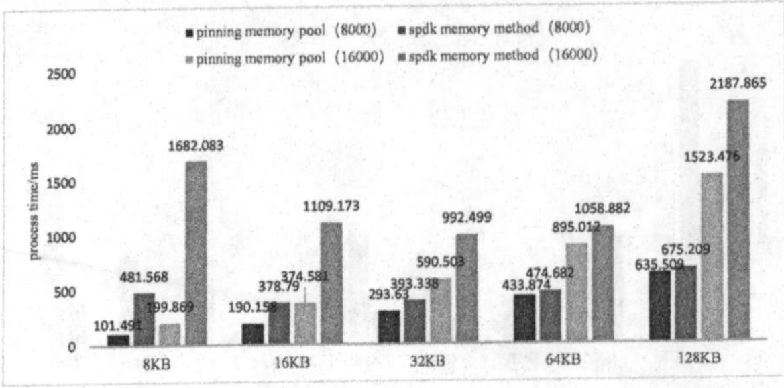

Fig. 6. Influence of different running time on memory Algorithm 2.

writes requests not greater than 8 KB. Then, the two memory libraries directly submit IO requests to the disk using the existing user state and interact directly with the disk. It can be seen from the figure that the algorithm can effectively improve its processing performance after being integrated into SPDK under all data sizes. The main reason is that the key idea of the algorithm is to allocate the pinned memory overhead to more requests and reduce the overhead of memory-related operations.

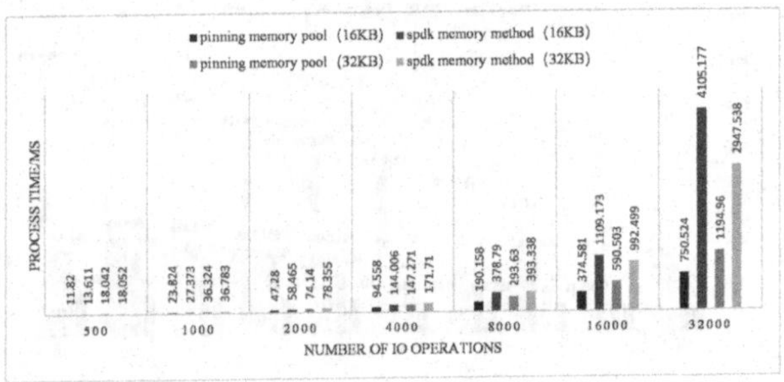

Fig. 7. Effect of integration testing under multiple IO data sizes.

Figure 7 shows how the number of IO requests impact performace of the SPDK integrating these two algorithms. In the figure, two types of IO requests are selected for processing. The maximum sizes are 16 kb and 32 KB, respectively. It can be seen from the figure that as the number of IO requests increases, the SPDK integrating the algorithms improves the performance of IO request processing because the overhead of some memory operations is allocated to more IO requests. In particular, the algorithms are aimed at IO-intensive applications. With the increase in the number of IO requests, the algorithms will improve the processing performance of SPDK, which verifies the effectiveness of the algorithms.

5 Conclusion

The user-mode IO frameworks dramatically improve the access performance of NVMe SSD by avoiding kernel intervention. Nevertheless, they introduce additional overhead, such as pinning memory is a very time-consuming operation for IO-intensive applications. In order to solve this problem, we propose an alternative and efficient memory management strategy named UPM-DMA in this paper. UPM-DMA can amortize per-request latency by dynamically selecting the appropriate pinned memory mode for different sizes of IO requests. We integrate UPM-DMA into SPDK and do experiments based on SPDK's official testing tools, and the experimental results verify the effectiveness of our work. In the future, we will continue to explore efficient memory usage to improve further the performance of the user-mode IO framework accessing NVMe SSD.

Acknowledgment. This work was supported by the National Natural Science Foundation of China under Grant NO.61772053.

References

1. Dong, S.Y., Kryczka, A., Jin, Y.Q., Stumm, M., et al.: RocksDB: evolution of development priorities in a key-value store serving large-scale applications. ACM Trans. Storage **17**(4), 1–32 (2021)
2. Kc, K., Hsu, C.J., Freeh, V.W.: Evaluation of mapreduce in a large cluster. In: Proceedings of the 8th International Conference on Cloud Computing (2015)
3. Huo, Z.S., Xiao, L.M., Guo, M.Y., et al.: Incremental throughput allocation of heterogeneous storage with no disruptions in dynamic setting. IEEE Trans. Comput. **69**(5), 679–698 (2020)
4. Huo, Z.S., Guo, M.Y., Xiao, L.M., et al.: TACD: a throughput allocation method based on variant of cobb-douglas for hybrid storage system. J. Parallel Distrib. Comput. **128**, 43–56 (2019)
5. Intel. White paper: NUC Tested Components and Peripherals list (2016)
6. Yang, Z.Y., Harris, J.R., Walker, B., et al.: SPDK: a development kit to build high performance storage applications. In: Proceedings of the International Conference on Cloud Computing Technology and Science (CloudCom), 11–14 December 2017
7. Yang, Z.Y., Liu, C.P., Zhou, Y.B., et al.: SPDK vhost-NVMe: accelerating IOs in virtual machines on NVMe SSDs via user space vhost target. In: Proceedings of the 8th International Symposium on Cloud and Service Computing (SC2), 18–21 November 2018
8. Kim, H.J., Kim, J.S.: A user-space storage IO framework for NVMe SSDs in mobile smart devices. IEEE Trans. Consum. Electron. **63**(1), 28–35 (2017)
9. Kim, H.J., Yee, Y.S., Kim, J.S.: NVMeDirect: a user-space IO framework for application-specific optimization on NVMe SSDs. In: Proceedings of the 8th USENIX Workshop on Hot Topics in Storage and File Systems (HotStorage), 20–21 June 2016
10. Yang, Z.Y., Wan, Q., Cao, G., et al.: uNVMe-TCP: a user space approach to optimizing NVMe over fabrics TCP transport. In: Proceedings of the Internet of Vehicles. Technologies and Services Toward Smart Cities, 19 January 2020

11. Yu, Y.J., Shin, D.I., Shin, W., Song, N.Y., et al.: Optimizing the block I/O subsystem for fast storage devices. ACM Trans. Comput. Syst. **32**(2), 1–48 (2014)
12. Song, N.Y., Song, Y.S., Han, H., Yeom, H.Y.: Efficient memory-mapped I/O on fast storage device. ACM Trans. Storage **12**(4), 1–27 (2016)
13. Guz, Z., Li, H., Shayesteh, A., Balakrishnan, V.: Performance characterization of NVMe-over-fabrics storage disaggregation. ACM Trans. Storage **14**(4), 1–18 (2018)
14. Tian, K., Zhang, Y., Kang, L.W., et al.: coIOMMU: a virtual IOMMU with cooperative DMA buffer tracking for efficient memory management in direct IO. In: Proceedings of the 2020 USENIX Annual Technical Conference (ATC), 15–17 July 2020
15. https://spdk.io/

AHOA: Adaptively Hybrid Optimization Algorithm for Flexible Job-shop Scheduling Problem

Jiaxin Ye[1,2], Dejun Xu[1,2], Haokai Hong[1,2], Yongxuan Lai[1,2], and Min Jiang[1,2(✉)]

[1] School of Informatics, Xiamen University, Xiamen 361005, Fujian, China
minjiang@xmu.edu.cn

[2] Key Laboratory of Digital Protection and Intelligent Processing of Intangible Cultural Heritage of Fujian and Taiwan, Ministry of Culture and Tourism, Fujian, China

Abstract. The Flexible Job-shop Scheduling Problem (FJSP) is a typical scheduling problem in industrial production that is proven to be NP-hard. The Genetic Algorithm (GA) is currently one of the most widely used algorithms to address the FJSP task. The major difficulty of using GA to solve FJSP lies in how to set the key hyperparameters and improve the convergence speed. In this paper, we propose a hybrid optimization method based on Reinforcement Learning (RL) called the Adaptively Hybrid Optimization Algorithm (AHOA) to overcome these difficulties. The proposed algorithm first merges GA and improved Variable Neighborhood Search (VNS), which aims to integrate the advantages of global and local search ability into the optimization process. Then the double Q-learning offers crossover and mutation rates according to the feedback from the hybrid algorithm environment. The innovation of this work lies in that our method can adaptively modify the key hyperparameters in the genetic algorithm. Furthermore, the proposed method can avoid the large overestimations of action values in RL. The experiment is evaluated on the most widely studied FJSP instances and compared with some hybrid and self-learning algorithms including dragonfly algorithm (DA), hybrid gray wolf weed algorithm (GIWO), and self-learning GA (SLGA), etc. The results show that the proposed method outperforms the latest related algorithms by more than 12% on average.

Keywords: Flexible Job-shop Scheduling Problem · Hybrid algorithm · Reinforcement learning

1 Introduction

Production scheduling [15] plays an essential role in the planning and scheduling of the manufacturing system. There are several well-known production scheduling problems including Job shop Scheduling Problem (JSP) and Flexible Job shop Scheduling Problem (FJSP) [15]. JSP is one of the most challenging parts

Y. Lai et al. (Eds.): ICA3PP 2021, LNCS 13155, pp. 271–287, 2022.
https://doi.org/10.1007/978-3-030-95384-3_18

of these kinds of problems and had been proved to be an NP-hard (Non-deterministic Polynomial-time Hard) problem [13]. It can be described as a set of independent jobs to be processed on multiple available machines, and each job contains a series of operations with a specified order. However, each operation must be processed on a specified machine in JSP. It is incapable of meet the flexible scheduling requirements in the real world.

The FJSP is an essential extension of JSP [4] that can address flexible requirements. It allows each operation to be processed on different machines. This means FJSP can be decomposed into two sub-problems including operations sequencing and machines selection [13]. Where, JSP is only composed of the operations sequencing problem to find the best sequence of operations and machines selection problem is a extention of JSP task to assign suitable machines to each operation.

In recent years, a great deal of algorithms have been proposed to address FJSP like Evolutionary Algorithm (EA), Local Search (LS) Algorithm, and Hybrid Optimization Algorithm (HA). For instance, Ding *et al.* [10] proposed an improved Particle Swarm Optimization (PSO) method based on novel encoding and decoding schemes for the FJSP; Amiri *et al.* [1] presented a Variable Neighborhood Search(VNS) for the FJSP; Li *et al.* [26] proposed a HA based on the Genetic Algorithm (GA) and Tabu Search (TS).

These methods are limited by hyperparameters settings. There is evidence [8] showing that hyperparameters like crossover and mutation operator play a crucial role in the evolution of the population. This means that relying on experience to select parameters will affect the efficiency and performance of the algorithm. For instance, the adverse configuration may cause premature convergence to local extreme rather than the globally optimal. To overcome this defect, many researchers often fix or update in a predetermined way [11]. But it still lacks generalization. Moreover, some methods [5] based on RL can also adjust the hyperparameters adaptively. However, the problem of overestimation is not avoided [17].

Based on these motivations, this work proposes an Adaptively Hybrid Optimization Algorithm named AHOA to search the minimum makespan of FJSP. AHOA first utilizes GA to optimize globally and then introduces improved VNS based on the critical path to optimize locally. Finally, the double Q-learning offers crossover and mutation rates according to the feedback from the hybrid algorithm environment. The contributions are as follow:

1) Our algorithm can adaptively modify the hyperparameters in the HA based on double Q-learning. It can also avoid the overestimations problem of action values compared to other works based on RL.
2) We deploy a hybrid algorithm to solve this problem effectively. It shows an outstanding performance by combining the exploration ability of GA and the exploitation ability of VNS.

The remainder of this work is organized as follows. Section 2 introduces the related work. The problem formulation is presented in Sect. 3, while the proposed algorithm is given in Sect. 4. The experimental result is proposed in Sect. 5. Finally, Sect. 6 describes the conclusion and future work.

2 Related Work

Since Brucker and Schlie [3] first proposed the FJSP in 1990, numerous methods have been suggested to solve the FJSP task in recent years, and most existing algorithms can be classified into two categories: GA based methods and RL based methods. The GA based methods include simple GA and HA. The simple GA is a typically global optimization method [31]. It can scale robustly and easily to complex problems that are difficult to solve by traditional optimization algorithms. For instance, De *et al.* [7] presented an improved GA for the FJSP, in which a new operator based on LS is combined. Lu *et al.* [28] proposed a GA embedded with a concise chromosome representation to solve the distributed FJSP. However, the drawbacks of the simple GA are the poor local optimization ability, and its excessively low convergence speed. For solving these problems, HA based on the GA and LS is proposed. This hybrid method can greatly improve the quality of the solution and the efficiency of optimization. Specifically, the GA optimizer always yields an approximate optimal set or population. The best individual of the set is utilized as the starting point for the local search optimizer to run with. So that, the global searching ability of GA and the local searching ability of LS are combined to solve the FJSP effectively. Wang *et al.* [33] proposed a hybrid algorithm that combines GA and TS. This approach also calls for some improvements, it limited by GA that the hyperparameters cannot be adjusted adaptively during the optimization process.

Therefore, the RL based methods are proposed to overcome this defect. The characteristic of the RL method is self-learning [30]. In other words, it can adaptively adjust the hyperparameters of GA based method. For example, Chen *et al.* [5] proposed a algorithm combined with the SARSA algorithm and Q-learning algorithm to adjust the hyperparameters. However, the problem of overestimation in RL and sparse Q-table are not avoided.

3 Problem Formulation

3.1 Problem Model

The FJSP can be classified into two types of problems including total FJSP and partial FJSP [18]. The total FJSP means each operation can be processed on every machines and the partial FJSP means each operation can be processed only on one or more machines. In this paper, our major research problem is partial FJSP which is described as follows. There are a set $J = \{J_1, J_2, \cdots, J_i, \cdots, J_n\}$ of n jobs and a set $M = \{M_1, M_2, \cdots, M_k, \cdots, M_m\}$ of m machines. The total number of operations is defined as $o = \sum_{i=1}^{n} H_i$. Each job J_i contains a series of operations $O_i = \{O_{i,1}, O_{i,2}, \cdots, O_{i,H_i}\}$ with a specified order. The jth operation $O_{i,j}$ of the ith job J_i can be processed on a machine M_k selected from a set of available machines and $t_{i,j,k}$ is its processing time. The objectives in the FJSP are to find the best sequence of all operations and the most suitable machine for each operation to optimize the makespan, workload, etc. For an intuitive illustration, a specific partial FJSP instance is shown in Table 1. The numbers in the table refers to the processing time $t_{i,j,k}$ and the symbol '–' indicates that operation $O_{i,j}$ can not be processed on a machine M_k. The typical assumptions [27] of FJSP are given as follows:

(1) All jobs can not be processed and all machines are available at the initial time;
(2) The order of precedence between operations of each job must be obeyed;
(3) Each machine M_k can process only one operation in any time;
(4) Each operation $O_{i,j}$ owns at least one machine to be processed;
(5) The processing cannot be interrupted until is finished;
(6) The transportation time of operations and depreciation of machines are ignored.

Table 1. An instance of 3 × 3 partial FJSP.

Jobs	Operation	M_1	M_2	M_3
J_1	$O_{1,1}$	3	2	–
	$O_{1,2}$	2	5	1
J_2	$O_{2,1}$	2	3	2
	$O_{2,2}$	4	–	2
	$O_{2,3}$	–	2	3
J_3	$O_{3,1}$	4	2	1
	$O_{3,2}$	2	–	6

3.2 Optimization Object

In this paper, the optimization object is to obtain a scheduling with the lowest makespan C_{max}. The object function can be presented by Eq. (1) and the constraints are listed in Eq. (2)–Eq. (4). Among these equations, $s_{i,j,k}$ represents the work start time of operation $O_{i,j}$ on M_k machine. Equation (2) shows that the processing time of all operations is positive. Equation (3) represents the order of precedence between operations of each job must be performed. Equation (4) represents that each machine M_k can process only one operation in any time. Equation (5) shows that each operation $O_{i,j}$ owns at least one machine to be processed.

Object:

$$min(C_{\max}) = min(max(C_i)) \tag{1}$$

Subject to:

$$t_{i,j,k} > 0 \text{ , } i = 1, 2, \cdots, N; j = 1, 2, \cdots, H_i; k = 1, 2, \cdots, M \tag{2}$$

$$s_{i,j,k} + t_{i,j,k} \leq s_{i,j+1,k} \text{ , } j = 1, 2, \cdots, (H_i - 1) \tag{3}$$

$$\sum_{i=1}^{N}\sum_{j=1}^{H_i} X_{i,j,k} = 1, X_{i,j,k} = \begin{cases} 1 & \text{If } O_{i,j} \text{ is assigned to } k^{th} \text{ machine} \\ 0 & \text{Otherwise} \end{cases} \tag{4}$$

$$\sum_{k=1}^{M} X_{i,j,k} \geq 1 \tag{5}$$

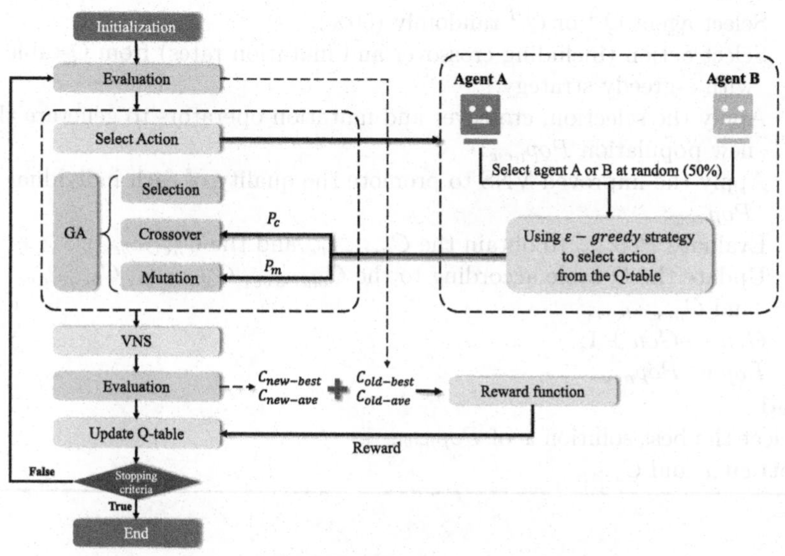

Fig. 1. The workflow of the proposed algorithm

4 Proposed Algorithm

4.1 Workflow of the Proposed AHOA

As Fig. 1 shows, AHOA is designed by merging the hybrid algorithm and double Q-learning algorithm. It accepts an instance of the FJSP as the inputs and uses HA that combines GA and VNS to optimize the makespan. To avoid the performance being severely affected by the preset hyperparameters, the double Q-learning algorithm is introduced to intelligently adjust them. The overall workflow of the proposed algorithm is described as Algorithm 1, and the details are described in the following sub-sections.

Algorithm 1: Adaptively Hybrid Optimization Algorithm

Input: Initial parameters and an FJSP instance.
Output: The best solution x and its makespan C_x.

Initialize two agents Q^A and Q^B, population Pop with P individuals and set $Gen = 1$;
for $Gen = 1 \rightarrow loop_{max}$ **do**
  Evaluate Pop to obtain the average makespan $C_{old-ave}$ and the minimal makespan $C_{old-best}$;
  if *population has stagnated in evolution for 20 iterations* **then**
    Reinitialize the population;
  end
  Select agent Q^A or Q^B randomly (50%);
  Select action (including crossover and mutation rates) from Q-table with ε-greedy strategy;
  Apply the selection, crossover and mutation operators to generate the new population Pop_{new};
  Apply the improved VNS to promote the quality of each individual in Pop_{new};
  Evaluate Pop_{new} to obtain the $C_{new-ave}$ and the $C_{new-best}$;
  Update the Q-table according to the $C_{old-ave}$, $C_{old-best}$, $C_{new-ave}$ and $C_{new-best}$;
  $Gen \leftarrow Gen + 1$;
  $Pop \leftarrow Pop_{new}$;
end
Select the best solution x of Pop_{new};
return x and C_x.

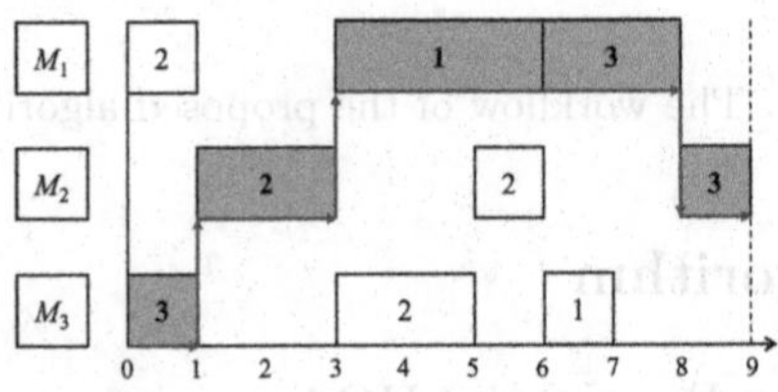

Fig. 2. The example of the critical path.

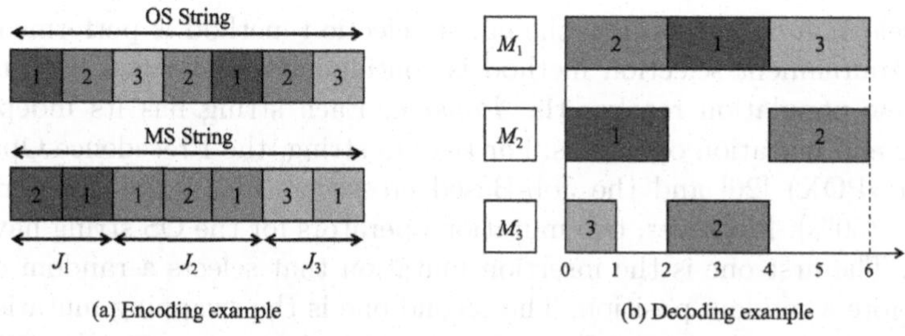

Fig. 3. The encoding and decoding example based on the Table 1

4.2 Hybrid Algorithm

4.2.1 Neighborhood Structure Based on Critical Path

The critical path is the longest path from the start to the end of all the operations in the Gantt chart. There is no interval between any adjacent operations, and the length of the path is equal to the makespan of the current solution. For instance, the combination of the blue blocks including $O_{3,1} \to O_{2,1} \to O_{1,1} \to O_{3,2} \to O_{3,3}$ is the critical path in Fig. 2. The length of this critical path is 9 which is also the end of the Gantt chart. In the case that the critical path length remains the same, the makespan cannot be reduced [35]. Therefore, the operation based on the critical path is introduced to design a problem-specific neighborhood structure for the VNS and the mutation operators for the GA.

The neighbourhood structure can be decomposed into the following steps. First, all the critical paths in the current Gantt chart are found. Among the critical paths, one path is chosen randomly, and the available machine is selected for each operation on this critical path. For VNS, This neighborhood structure modifies the individuals from the GA to generate new neighborhood solutions. Moreover, this structure is also used as a mutation operator to generate new individuals for the new population in the GA.

4.2.2 Genetic Algorithm

In the proposed algorithm, the encoding method proposed by Gao *et al.* [12] is adopted. As shown in Fig. 3, the code composes of two strings. One is called OS (Operation Sequence), and the other is called MS (Machine Sequence). As mentioned in Sect. 1, two sub-problems are needed to solve FJSP. The first is to find the best sequence of operations and the second assigns suitable machines to each operation. The two strings correspond to these two sub-problems respectively. For OS string, the number i that appears j^{th} times represents it is the operation $O_{i,j}$ of job J_i. For MS string, it presents the selected machines for the corresponding operations of each job. Moreover, the decoding method proposed by Gong *et al.* [14] is used in this work to minimize the makespan. It can minimize makespan while obeying the constraints during operations and machines.

For selection operators [34], the elitist selection method is performed firstly and the tournament selection method is continuously performed until the size of the new population reaches the *Popsize*. Each string has its independent crossover and mutation operators. For the OS string, the Precedence Operation crossover (POX) [26] and the Job-Based crossover (JBX) [26] are performed randomly (50%). Moreover, two mutation operators for the OS string have been adopted. The first one is the insertion mutation that selects a random digit to insert before a random position. The second one is the swapping mutation that swaps two random positions. These two mutation operators for the OS string are also performed randomly (50%). For the MS string, the two-point crossover [26] is adopted as the crossover operator. The mutation operator based on the critical path is introduced in this paper. This operator only changes the machine of operation in the critical path which is explicitly described in Sect. 4.2.1.

4.2.3 Variable Neighborhood Search

The LS algorithm is introduced to improve GA in terms of the convergence speed, the quality of individuals, and the local search ability. The VNS is a meta-heuristic LS method proposed by Mladenovi *et al.* [29]. The core idea is neighborhood transformation which has been successfully applied to numerous combinatorial optimization problems.

The VNS contains two procedures called shaking and local search. These procedures only employ a neighborhood structure based on the critical path. The main structure of the improved VNS is composed of external and internal loops. The external loop performs the shaking procedure, and the internal loop performs a local search. The definition of related symbols is as follows. The symbol k_{max} is the max iterations of the external loop ($k_{max} = 2$), and l is the max iterations of the internal loop ($l_{max} = 2$). The symbol $k_{popsize}$ is delineated as the size of the shaking neighborhood ($k_{popsize} = 3$), and $l_{popsize}$ is the size of the local search neighborhood ($l_{popsize} = 3$). Besides, $N(x)$ is presented as the neighborhood structure related to the critical path which is generated from a solution x.

As shown in Algorithm 2, in the first step, the initial solution x generated from the GA is denoted as the global optimal solution and the x' generated from the neighborhood structure $N_k(x)$ is denoted as the local optimal solution. The shaking procedure firstly selects a random solution x' from the kth neighborhood $N_k(x)$ generated base on x. Then the local search procedure selects a random solution x'' from the lth neighborhood $N_l(x')$ generated base on x'. If x'' is better than x', then set x' to x'' and l to 1. The local search procedure continues to be performed until the end of the internal loop iteration. When the end of the local search procedure. If the local optimal solution x' is better than x, then set x to x' and k to 1. Finally, the procedure repeats until the end of the external loop iteration.

Algorithm 2: Variable Neighborhood Search Algorithm

```
Input: The old population P_old, and neighborhood parameters
       k_max, k_popsize, l_max, l_popsize.
Output: The new population P_new.
foreach individual x in P_old do
    Evaluate the global makespan C_x of x;
    for k = 1 → k_size do
        Shaking procedure: pick a random solution x' from k^th
          neighbothood N_k(x) of x;
        for l = 1 → l_size do
            Local search procedure: pick a random solution x'' in
              neighbothood N_l(x') of x';
            if C_x'' < C_x' then
                Update the local optimal solution: x' → x'',and l → 1;
            else
                l → l + 1;
            end
        end
        if C_x' < C_x then
            Update the global optimal solution: x → x',and k → 1;
        else
            k → k + 1;
        end
    end
    P_new = P_new ∪ {x};
end
return P_new.
```

4.3 Double Q-learning Procedure

The double Q-learning is a typical reinforcement learning method proposed by Hasselt *et al.* [17]. It aims to solve the overestimations problem of Q-value defined as $Q(s, a)$. The double Q-learning obtains two intelligent agents Q^A and Q^B. The agents interact continuously with the HA environment to find the most cumulative Q-value based on past experience and update the Q-value from the other agent for the next state. Moreover, each agent can obtain a reward or penalty while performing a selected action and learn to maximize the long-term reward after multi iterations.

The agent will take different actions to get crossover and mutation rates combined as the action in the Q-table. Specifically, the crossover rates group G_c is set to $\{0.9, 0.8, 0.7, 0.6, 0.5\}$ and the mutation rates group G_m is establish to $\{0.3, 0.2, 0.1\}$ in the action list A_{list}. The A_{list} is the Cartesian product of

G_c and G_m. Besides, the symbol $\alpha \in [0, 1]$ represents the learning rate. The $\gamma \in [0, 1]$ represents the attenuation rate for the future reward. The R is defined as the reward shown in Eq. (8).

In the AHOA, we assume that only one static state in the HA environment prevents the sparse Q-table. The specific algorithm is described as Algorithm 3 shows. Firstly, step in the double Q-learning is to initialize agent Q^A and agent Q^B by initializing each Q-table to a zero-value matrix. Then select agent Q^A or agent Q^B to use at random (50%). Assuming that agent Q^A is selected, the ε-greedy strategy as shown in Sect. 4.3.1 is next used to select the action a^* to be performed from the action list A_{list}. The HA obtains the crossover rate and mutation rate from a^* and executes to get an evaluation. Finally, the Q-table is updated according to Eq. (6). In Eq. (6), the above equation is used to update if the agent Q^A is selected. Otherwise, the following equation is used to update the table.

$$\begin{cases} Q^A(s,a) \leftarrow Q^A(s,a) + \alpha \times \left(R + \gamma \times Q^B\left(s,a^*\right) - Q^A(s,a)\right) & (1) \\ Q^B(s,a) \leftarrow Q^B(s,a) + \alpha \times \left(R + \gamma \times Q^A\left(s,a^*\right) - Q^B(s,a)\right) & (2) \end{cases} \tag{6}$$

Algorithm 3: Double Q-Learning Algorithm

Input: The learning rate α, attenuation rate γ, ε-greedy rate ε, crossover rates group G_c, mutation rates group G_m, and makespans $C_{old-ave}$, $C_{old-best}$, $C_{new-ave}$ and $C_{new-best}$.

Initialize Q^A and Q^B;
Select agent Q^A or Q^B randomly (50%);
Select action a^* from Q-table with ε-greedy strategy;
Apply the a^*(the crossover rate P_c and mutation rate P_m) to HA;
Calculate reward R according to the $C_{old-ave}$, $C_{old-best}$, $C_{new-ave}$ and $C_{new-best}$ from HA;
if *Select* Q^A **then**
  /* Update the table of Q^A */;
  $Q^A(s,a) \leftarrow Q^A(s,a) + \alpha \times \left(R + \gamma \times Q^B\left(s,a^*\right) - Q^A(s,a)\right)$;
else if *Select* Q^B **then**
  /* Update the table of Q^B */;
  $Q^B(s,a) \leftarrow Q^B(s,a) + \alpha \times \left(R + \gamma \times Q^A\left(s,a^*\right) - Q^B(s,a)\right)$;
end

4.3.1 Selection Strategy

The selection strategy of double Q-learning is the ε-greedy strategy. In this strategy, an action a^* will be selected randomly from A_{list} if a random number $rand_{[0-1]}$ is higher than the agreed value ε, otherwise the action with the most $Q(s,a)$ will be selected. The definition is shown in Eq. (7).

$$a^* = \begin{cases} \arg max_a Q\left(s,a\right) & rand_{[0-1]} < \varepsilon \\ A_{list}[random] & rand_{[0-1]} \geq \varepsilon \end{cases} \tag{7}$$

4.3.2 Reward Function

The reward function is designed according to the best and the average individual fitness as shown in Eq. (8). In the reward method, $C_{old-ave}$ and $C_{old-best}$ are delineated as the average and the best makespan in the previous population. $C_{new-ave}$ and $C_{new-best}$ are delineated as the average and the best makespan in the current population.

$$R = \begin{cases} 30 \text{ If } C_{new-ave} < C_{old-ave} \text{ and } C_{new-best} < C_{old-best} \\ 15 \text{ If } C_{new-ave} \geq C_{old-ave} \text{ and } C_{new-best} < C_{old-best} \\ -0.5 \text{ If } C_{new-ave} < C_{old-ave} \text{ and } C_{new-best} \geq C_{old-best} \\ -5 \text{ If } C_{new-ave} \geq C_{old-ave} \text{ and } C_{new-best} \geq C_{old-best} \end{cases} \tag{8}$$

4.4 Computation Complexity Analysis

For each generation of the AHOA, the computational complexity can be analyzed as follows. First it is with the computational complexity $\mathcal{O}(P \log P)$ by using quick sorting method to evaluate the population. Then, it is with the computational complexity $\mathcal{O}(P)$ to perform selection operator, with the computational complexity $\mathcal{O}(P \times o)$ to perform crossover operator and with the computational complexity $\mathcal{O}(P \times m \times o)$ to perform mutation operator. Finally, the AHOA applies the VNS based on the critical path with the computational complexity $\mathcal{O}(P \times k_{size} \times k_{pop} \times l_{size} \times l_{pop} \times m^2 \times o)$. Thus, the computational complexity for each generation of the AHOA is $\mathcal{O}[P \times (\log P + o \times k_{size} \times k_{pop} \times l_{size} \times l_{pop} \times m^2)]$.

5 Experiment and Discussions

5.1 Experimental Setup

To demonstrate the performance of the proposed algorithm, ten benchmark instances of the BRdata [2] are tested. The BRdata is a data set consists of 10 test instances (Mk01-Mk10), which are randomly generated using a uniform distribution between given limits. The parameters in the proposed algorithm for these instances are listed in Table 2. The proposed AHOA is implemented in C++ on the AMD Ryzen 5 5600X machine running at 3.7 GHz and the optimization objective of it is makespan. Besides, we select three categories represented algorithms with to compare, including EA based methods, HA based methods, and RL based methods. For instance, Wang *et al.* [32] proposed a hybrid algorithm named GIWO based on gray wolf and invasive weeds, and Chen *et al.* [5] and Han *et al.* [16] both proposed methods based on RL to solve the FJSP.

Table 2. The AHOA parameters.

Parameters	Descriptions	
α	Learning rate	0.6
γ	Attenuation rate	0.8
ε	ε-greedy rate	0.85
G_c	Crossover rates group in the action list A_{list}	$\{0.9, 0.8, 0.7, 0.6, 0.5\}$
G_m	Mutation rates group in the action list A_{list}	$\{0.3, 0.2, 0.1\}$
k_{max}	Max iterations of external loop in VNS	2
l_{max}	Max iterations of internal loop in VNS	2
$k_{popsize}$	Size of the shaking neighborhood	3
$l_{popsize}$	Size of the local search neighborhood	3
P	Size of the population	80
$loop_{max}$	Max iterations of the proposed algorithm	200

5.2 Experimental Results

The results of these comparisons are shown in the Table 3. For the EA based methods, the average improving performance of AHOA is 11%, of which the three testing instances Mk02, Mk04, and Mk06 improved by more than 22.8%; for the HA based methods, the average improving performance of AHOA is 6.7%, of which the two testing instances Mk02 and Mk06 improved by more than 16.7%. The most important difference between the AHOA and the above two methods is the ability to adjust the hyperparameters adaptively. The improved PSO [10] proposes some tuning schemes of parameters with exact mathematical methods. However, the accurate mathematical expression of the parameters tuning schemes may be unavailable as the schemes are affected by various factors, such as dynamic conditions. The result shows that the AHOA is more effective in large-scale complex environments. Besides, it can deal with situations in which the mathematical expression of the parameters tuning schemes is not available. These observations align with the findings in [6].

For the RL based methods, the average improving performance of AHOA is 22%, including the three testing instances Mk06, Mk07, and Mk10 increasing by more than 25%. However, the AHOA performs slightly worse than the self-learning GA (SLGA) [5] on Mk05. Because the setting range of the action list in the SLGA is more refined compared with the AHOA. It can find more suitable parameters for complex environments with enough iterations. The SLGA proposes a different way to combine two RL algorithms to enhance the performance of GA. Nevertheless, it still can not avoid the large overestimations of action values in the RL [17]. The result shows that AHOA performs better in terms of solution quality than other RL based methods. Besides, the Gantt chart of the solution obtained by AHOA for the Mk02 and Mk05 instance are presented in Fig. 4 and Fig. 5.

Table 3. The comparison of the makespan in BRdata

Methods	Year	Mk01	Mk02	Mk03	Mk04	Mk05	Mk06	Mk07	Mk08	Mk09	Mk10
		10 × 6	10 × 6	15 × 8	15 × 8	15 × 4	10 × 15	20 × 5	20 × 10	20 × 10	20 × 15
Deep RL [16]	2020	44	28	245	74	193	123	216	**523**	386	337
Dragonfly Algorithm [19]	2020	52	46	210	88	175	87	–	–	–	–
HLO-PSO [9]	2020	**40**	28	**204**	63	175	71	**144**	**523**	326	238
Improved PSO [10]	2020	**40**	29	**204**	66	175	77	145	**523**	320	239
SLGA [5]	2020	**40**	27	**204**	**60**	**172**	69	**144**	**523**	320	254
GIWO [32]	2021	**40**	32	**204**	65	177	84	156	**523**	331	242
GA	–	42	33	**204**	67	183	86	181	**523**	325	270
VNS	–	42	33	**204**	67	183	86	181	**523**	338	277
HA(GA+VNS)	–	**40**	31	**204**	64	179	78	150	**523**	315	258
Our proposed method	2021	**40**	**26**	**204**	**60**	173	**65**	**144**	**523**	**307**	**233**

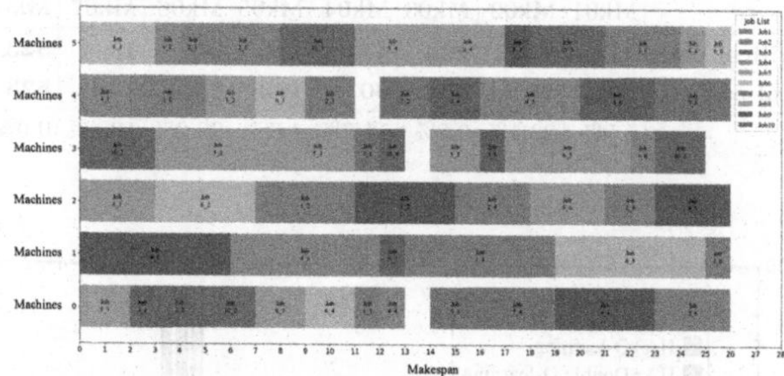

Fig. 4. Gantt chart of the solution of the Mk02 instance

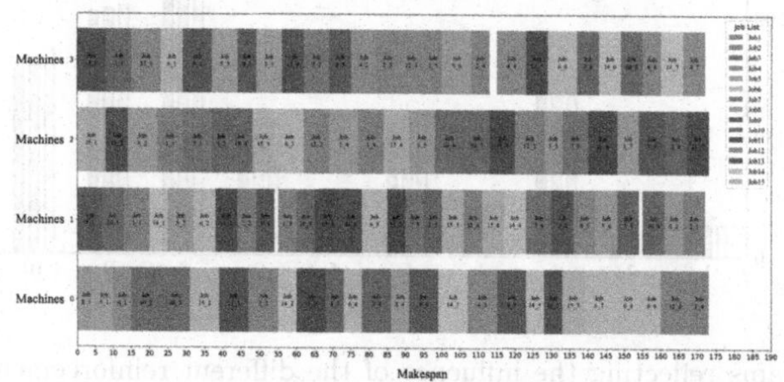

Fig. 5. Gantt chart of the solution of the Mk05 instance

5.3 Ablation Experiment

As mentioned in Sect. 4, we introduce the critical path and reinforcement learning to solve the FJSP task. In order to illustrate the improvement in the proposed algorithm, we make ablation experiments on these two parts.

5.3.1 The Discussion of the Critical Path

As shown in Table 4, an algorithm with critical path can achieve better performance than an algorithm without critical path. The experiment results show that the makespan of the AHOA is 10.7% more than the method without critical path on the ten instances on average. It indicates that the critical path can effectively improve search quality. As the complexity analysis of AHOA in Sect. 4.4, searching the critical path will increase the time cost of the algorithm, which means that the AHOA will spend more time in a single iteration. However, it can reach convergence with fewer iterations.

Table 4. Experiments observing the influence of the critical path on 10 instances.

Method	Mk01	Mk02	Mk03	Mk04	Mk05	Mk06	Mk07	Mk08	Mk09	Mk10
AHOA without critical path	42	32	**204**	67	181	84	159	**523**	326	272
AHOA with critical path	**40**	**26**	**204**	**60**	**173**	**65**	**144**	**523**	**307**	**233**
Improvement	5.0%	23.1%	0.0%	11.7%	4.6%	29.2%	10.4%	0.0%	6.2%	16.7%

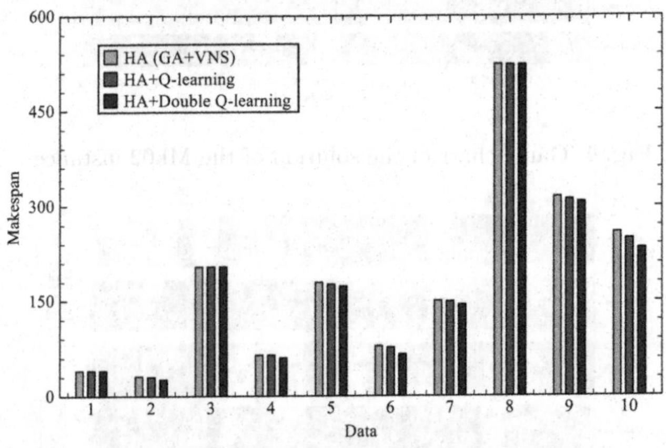

Fig. 6. Histograms reflecting the influence of the different reinforcement learning on 10 instances.

5.3.2 The Discussion of the Reinforcement Learning

We also conduct an ablation experiment on reinforcement learning. It aims to explore the influence of different reinforcement learning methods on HA. As shown in Fig. 6, the makespan of the histogram performs best on the double Q-learning algorithm, which outperforms the HA and the Q-learning algorithm by more than 5.9% and 4.6% respectively. Compared with the HA, the algorithm introduced with reinforcement learning has the characteristics of adaptation which can better adapt to different environments. However, the Q-learning overestimate attempted actions, which will affect the convergence results to a certain extent. The double Q-learning can avoid the positive bias in estimating the action values and achieves better performance. This result shows that the strategy with the double Q-learning has achieved the best results on the ten instances. It indicates that the double Q-learning can improve performance by enhancing generalization and preventing overestimation problems.

6 Conclusion

This paper discusses the FJSP task by proposing a novel algorithm AHOA, combining the hybrid and double Q-learning algorithms. Compared with the widely deployed methods for FJSP, our algorithm can modify the key hyperparameters adaptively in the HA and avoid the overestimations problem of action values compared to other methods based on RL. Moreover, the proposed algorithm is more efficient due to the neighborhood structure based on the critical path. Finally, the experiment results verify that AHOA is suitable for FJSP instances with the high flexibility.

For the future work, we notice that the existing algorithms have achieved remarkable results in static scheduling, but there are few studies on the dynamic scheduling. We would like to investigate the dynamic multi-objective FJSP based on the transfer learning [21,23–25] and domain adaptation learning [20,22]. Besides, the performance of AHOA in practical applications still needs further testing, our future work will also focus on enhancing the generalization capabilities of the algorithm on the larger FJSP data sets.

Acknowledgement. This work was supported in part by the National Natural Science Foundation of China under Grant 61673328, and in part by the Collaborative Project Foundation of Fuzhou-Xiamen-Quanzhou Innovation Demonstration Zone under Grant 3502ZCQXT202001.

References

1. Amiri, M., Zandieh, M., Yazdani, M., Bagheri, A.: A variable neighbourhood search algorithm for the flexible job-shop scheduling problem. Int. J. Prod. Res. **48**(19), 5671–5689 (2010)
2. Brandimarte, P.: Routing and scheduling in a flexible job shop by tabu search. Ann. Oper. Res. **41**(3), 157–183 (1993)
3. Brucker, P., Schlie, R.: Job-shop scheduling with multi-purpose machines. Computing **45**(4), 369–375 (1990)
4. Chaudhry, I.A., Khan, A.A.: A research survey: review of flexible job shop scheduling techniques. Int. Trans. Oper. Res. **23**(3), 551–591 (2016)
5. Chen, R., Yang, B., Li, S., Wang, S.: A self-learning genetic algorithm based on reinforcement learning for flexible job-shop scheduling problem. Comput. Ind. Eng. **149**, 106778 (2020)
6. Dai, P., Yu, W., Wen, G., Baldi, S.: Distributed reinforcement learning algorithm for dynamic economic dispatch with unknown generation cost functions. IEEE Trans. Ind. Inf. **16**(4), 2258–2267 (2019)

7. De Giovanni, L., Pezzella, F.: An improved genetic algorithm for the distributed and flexible job-shop scheduling problem. Eur. J. Oper. Res. **200**(2), 395–408 (2010)
8. Deb, K., Agrawal, S.: Understanding interactions among genetic algorithm parameters. Found. Genet. Algorithms **5**(5), 265–286 (1999)
9. Ding, H., Gu, X.: Hybrid of human learning optimization algorithm and particle swarm optimization algorithm with scheduling strategies for the flexible job-shop scheduling problem. Neurocomputing **414**, 313–332 (2020)
10. Ding, H., Gu, X.: Improved particle swarm optimization algorithm based novel encoding and decoding schemes for flexible job shop scheduling problem. Comput. Oper. Res. **121**, 104951 (2020)
11. Du, Y., Fang, J., Miao, C.: Frequency-domain system identification of an unmanned helicopter based on an adaptive genetic algorithm. IEEE Trans. Ind. Electron. **61**(2), 870–881 (2013)
12. Gao, L., Peng, C., Zhou, C., Li, P.: Solving flexible job shop scheduling problem using general particle swarm optimization. In: Proceedings of the 36th CIE Conference on Computers & Industrial Engineering, pp. 3018–3027. Citeseer (2006)
13. Garey, M.R., Johnson, D.S., Sethi, R.: The complexity of flowshop and jobshop scheduling. Math. Oper. Res. **1**(2), 117–129 (1976)
14. Gong, X., Deng, Q., Gong, G., Liu, W., Ren, Q.: A memetic algorithm for multi-objective flexible job-shop problem with worker flexibility. Int. J. Prod. Res. **56**(7), 2506–2522 (2018)
15. Graves, S.C.: A review of production scheduling. Oper. Res. **29**(4), 646–675 (1981)
16. Han, B., Yang, J.: A deep reinforcement learning based solution for flexible job shop scheduling problem. Int. J. Simul. Model. (IJSIMM) **20**(2), 375–386 (2021)
17. Hasselt, H.: Double q-learning. Adv. Neural Inf. Process. Syst. **23**, 2613–2621 (2010)
18. Ho, N.B., Tay, J.C.: Genace: an efficient cultural algorithm for solving the flexible job-shop problem. In: Proceedings of the 2004 Congress on Evolutionary Computation (IEEE Cat. No. 04TH8753), vol. 2, pp. 1759–1766. IEEE (2004)
19. Hu, S., Zhou, L.: Research on flexible job-shop scheduling problem based on the dragonfly algorithm. In: 2020 International Conference on Artificial Intelligence and Electromechanical Automation (AIEA), pp. 241–245. IEEE (2020)
20. Jiang, M., Huang, W., Huang, Z., Yen, G.G.: Integration of global and local metrics for domain adaptation learning via dimensionality reduction. IEEE Trans. Cybern. **47**(1), 38–51 (2015)
21. Jiang, M., Huang, Z., Qiu, L., Huang, W., Yen, G.G.: Transfer learning-based dynamic multiobjective optimization algorithms. IEEE Trans. Evol. Comput. **22**(4), 501–514 (2017)
22. Jiang, M., Qiu, L., Huang, Z., Yen, G.G.: Dynamic multi-objective estimation of distribution algorithm based on domain adaptation and nonparametric estimation. Inf. Sci. **435**, 203–223 (2018)
23. Jiang, M., Wang, Z., Guo, S., Gao, X., Tan, K.C.: Individual-based transfer learning for dynamic multiobjective optimization. IEEE Trans. Cybern. **51**(10), 4968–4981 (2021)
24. Jiang, M., Wang, Z., Hong, H., Yen, G.G.: Knee point-based imbalanced transfer learning for dynamic multiobjective optimization. IEEE Trans. Evol. Comput. **25**(1), 117–129 (2020)
25. Jiang, M., Wang, Z., Qiu, L., Guo, S., Gao, X., Tan, K.C.: A fast dynamic evolutionary multiobjective algorithm via manifold transfer learning. IEEE Trans. Cybern. **51**(7), 3417–3428 (2021)

26. Li, X., Gao, L.: An effective hybrid genetic algorithm and tabu search for flexible job shop scheduling problem. Int. J. Prod. Econ. **174**, 93–110 (2016)
27. Liu, B., Fan, Y., Liu, Y.: A fast estimation of distribution algorithm for dynamic fuzzy flexible job-shop scheduling problem. Comput. Ind. Eng. **87**, 193–201 (2015)
28. Lu, P.-H., Wu, M.-C., Tan, H., Peng, Y.-H., Chen, C.-F.: A genetic algorithm embedded with a concise chromosome representation for distributed and flexible job-shop scheduling problems. J. Intell. Manuf. **29**(1), 19–34 (2018). https://doi.org/10.1007/s10845-015-1083-z
29. Mladenović, N., Hansen, P.: Variable neighborhood search. Comput. Oper. Res. **24**(11), 1097–1100 (1997)
30. Qi, X., Luo, Y., Wu, G., Boriboonsomsin, K., Barth, M.: Deep reinforcement learning enabled self-learning control for energy efficient driving. Transp. Res. Part C Emerg. Technol. **99**, 67–81 (2019)
31. Rangel-Merino, A., López-Bonilla, J., y Miranda, R.L.: Optimization method based on genetic algorithms. Apeiron **12**(4), 393–408 (2005)
32. Wang, Y., Song, Y.C., Zou, Y.J., Lei, Q., Wang, X.K.: A hybrid gray wolf weed algorithm for flexible job-shop scheduling problem. In: Journal of Physics: Conference Series, vol. 1828, p. 012162. IOP Publishing (2021)
33. Wang, Y., Zhu, Q.: A hybrid genetic algorithm for flexible job shop scheduling problem with sequence-dependent setup times and job lag times. IEEE Access **9**, 104864–104873 (2021)
34. Whitley, D.: A genetic algorithm tutorial. Stat. Comput. **4**(2), 65–85 (1994)
35. Zhang, G., Shao, X., Li, P., Gao, L.: An effective hybrid particle swarm optimization algorithm for multi-objective flexible job-shop scheduling problem. Comput. Ind. Eng. **56**(4), 1309–1318 (2009)

Trace-Navi: A High-Accuracy Indoor Navigation System Based on Real-Time Activity Recognition and Discrete Trajectory Calibration

Yu Wang[1,2], Zhipeng Yu[1,2], Licai Zhu[1,2], and Hao Yang[1,2,3,4](✉)

[1] Nanjing Tech University, Nanjing 211816, Jiangsu, China
[2] Yancheng Teachers University, Yancheng 224001, Jiangsu, China
[3] Suzhou Research Institute, University of Science and Technology of China, Suzhou 215123, Jiangsu, China
[4] Jiangsu Provincial Key Constructive Laboratory for Big Data of Psychology and Cognitive Science, Yancheng Teachers University, Yancheng 224002, Jiangsu, China

Abstract. Pedestrian dead reckoning based on mobile smart devices has become one of the most attractive solutions as the miniaturization of sensors. However, the ambient noise would be inevitable in the process of sampling by sensors of the mobile device, which probably degrades the performance of the dead reckoning system sharply in the wake of accumulative errors. In this paper, we present an indoor navigation system based on real-time activity recognition and discrete trajectory calibration, Trace-Navi. It consists of three modules: activity recognition module, reference trace collection module and navigation module. In the phase of trace construction, a volunteer holds the device to collect trace data by embedded sensors. To establish a credible reference trajectory, our system firstly adopts the activity recognition model based on multi-head CNN and attention mechanism to discriminate activities and estimate corresponding distances. Furthermore, various collected data from different sensors are mingled to confine the indefinable sampling error in a long-term activity. In the phase of real-time navigation, we propose a discrete trajectory matching algorithm to calibrate dead reckoning using the discrete-time difference sequence. In complex scenes with multiple floors, we compare with the current mainstream methods, which employ the fusion of accelerometer and magnetometer (FAM), and gyroscope angular velocity integration (GAVI). Experimental results demonstrate that the performance of Trace-Navi is far superior to that of them. Its accuracy respectively increases 74.33% and 73.8%.

Keywords: Pedestrian dead reckoning · Activity recognition · Signal fusion · Trajectory matching

Y. Lai et al. (Eds.): ICA3PP 2021, LNCS 13155, pp. 288–303, 2022.
https://doi.org/10.1007/978-3-030-95384-3_19

1 Introduction

The widespread use of mobile smart devices has brought development opportunities for smart device-based location services [1,2]. Although global positioning systems (GPS) can achieve sub-meter position estimation in outdoor environments, they cannot achieve precise positioning and navigation in indoor scenes due to refraction and diffraction in signal propagation. Researchers have proposed a variety of indoor positioning and navigation methods, including Wi-Fi [3], Bluetooth [4], RFID [5], UWB [6] and Computer Vision [7]. These technologies have achieved high positioning accuracy. However, most of these technologies require the deployment of infrastructure and have high requirements for indoor scenarios. For instance, Wi-Fi or Bluetooth-based positioning requires the deployment of multiple access points, and the positioning accuracy depends on the density of the collected fingerprint map. Computer vision-based positioning requires high stability in indoor scenes, and a large number of images are needed for positioning and matching. Signal-based ranging positioning methods, such as RFID, UWB, are easily affected by the environment, and it is difficult to ensure positioning accuracy in complex indoor scenes. All of the above mentioned positioning technologies require additional equipment costs, which are not conducive to widespread application.

To date, indoor positioning technology using MEMS sensor components embedded in commercial smart devices for dead reckoning has attracted wide attention. This method does not require pre-deployment of auxiliary equipment. The miniaturization of sensor has enabled more sensing elements to be embedded in smart devices, and their sensing capabilities have also been continuously improved. However, MEMS sensors will inevitably include a large amount of noise data when collecting data, most of the pedestrian dead reckoning methods are facing the challenge of excessive accumulation error, which results in the estimated pedestrian position far away from the true path trajectory.

In this paper, we propose a high-accuracy indoor navigation system based on real-time activity recognition and discrete trajectory calibration: Trace-Navi. The proposed system employs a real-time activity recognition model transferred to the mobile smart device to recognize pedestrian activities and estimates the walking distance of pedestrian based on the recognition result. Afterwards, a fusion filter is adopted to estimate the heading direction of pedestrian, and the pedestrian position is located based on the estimated walking distance and heading direction. Finally, Trace-Navi adopts a discrete magnetic field data sequence matching method to calibrate the pedestrian position and displays the calibrated estimated position coordinates on the system user interface. The main contributions of this paper are as follows:

- We propose a real-time activity recognition method based on attention mechanism and multi-head convolutional neural network, and transfer the activity recognition model to mobile smart devices, realize the real-time activity recognition and achieved a classification accuracy of 98.65%.
- We propose a trajectory matching algorithm based on the discrete-time sequence difference sequence. The proposed algorithm can match the magnetic field data sequences collected by different devices in real time.

- We implement Trace-Navi on smart devices and evaluate the system performance in a complex interior scene with three floors. The two mainstream dead reckoning methods using gyroscope angular velocity integration and acceleration and magnetic field fusion were compared. The performance of Trace-Navi increased by 74.33% and 73.8%.

The remainder of this paper is organized as follows: in Sect. 2, related works about activity recognition and pedestrian dead reckoning are discussed. Section 3 presents the system design of Trace-Navi, the activity recognition module, reference trace collection module and navigation module are introduced. The experimental results of activity recognition and system performance are shown in Sect. 4. Finally, conclusions and future work are discussed in Sect. 5.

2 Related Work

Activity recognition is widely used in the fields of health care, interactive entertainment and assisted living. With the rapid development of electronic technology, the use of sensors embedded in smart devices for activity recognition has attracted wide attention. Song Mi Lee et al. [8] used one-dimensional convolutional neural network to analyze the triaxial data of accelerometer. This method converts the triaxial data into vector amplitude data to achieve high classification accuracy. H. Zhang et al. [9] adopted multi-head convolution neural network combined with attention mechanism to recognize various activities, including walking, standing, sitting, jogging, going upstairs and going downstairs. R. Mutegeki et al. [10] combined convolutional neural network with long and short memory neural network, and the classification accuracy on the public dataset reached 92%.

The widespread use of mobile devices also brings development opportunities for pedestrian dead reckoning algorithms in indoor positioning scenarios. Pedestrian dead reckoning algorithm is mainly composed of three parts: step detection, step length estimation and heading direction estimation. The traditional dead reckoning algorithm requires special equipment to be installed on the user, such as the strap-down inertial navigation system (SINS) [11] by deploying inertial measurement units (IMU) at various positions of the human body to obtain data, and integrating acceleration data and gyroscope data. After obtaining the pedestrian's walking distance and heading direction, this method then adopts the zero-speed update algorithm [12] to limit the influence of acceleration noise on the walking distance calculation. Although the positioning accuracy of this method is better, the special equipment deployed on the body makes the deployment cost of this method high, which is not conducive to wide application. Y. Shu et al. [13] employ the built-in accelerometer, gyroscope, magnetometer and barometer implement dead reckoning algorithm based on smart device sensors, and achieve a positioning accuracy of less than 2 m within 95% of the time in a multi-floor indoor scene. Wonho Kang et al. [14] proposed SmartPDR, the step length estimation and heading direction estimation are optimized, and the positioning accuracy of the maximum positioning error of 1.62 m is achieved in the actual scene. Yuanqing Zheng et al. [15] proposed the Travi-Navi, which adopts the traditional dead reckoning algorithm combined with Wi-Fi fingerprint positioning and the camera on the smart device, the captured pictures and RSSI fingerprint are used to assist in correcting the positioning results given by the dead reckoning, and the final error is limited to less than 4 steps. Erqun Dong et al. [16] transferred the traditional real-time positioning and picture composition technology to mobile smart devices, only the camera of the smart device was used to complete the map construction, the navigation success rate was 98.6%.

However, the sensors embedded in the smart devices have not been professionally calibrated. Incorrect data measurement results will cause huge accumulative error of the dead reckoning algorithm with the increase of walking distance and running time. In order to cope with these challenges, Qu Wang et al. [17] integrated activity recognition into the dead reckoning algorithm, and combined it with magnetic field fingerprint data to calibrate the position, but they did not consider the smart devices that are not equipped with barometer sensors in multi-floor scenarios, and in complex multi-floor scenarios, this method is not applicable. Bang Wu et al. [18] employ a variety of neural network models to recognize different activities and combine them with dead reckoning algorithms. The average positioning error in a multi-story building is only 1.79 m. However, this model has high complexity and is not suitable for running on smart devices.

In this paper, we employ activity recognition model to recognize pedestrian activity, and estimate the step length based on the activity recognition result. Furthermore, we propose a fusion filter based on multiple MEMS sensors to estimate the heading direction, which improves the positioning accuracy.

3 System Design

The proposed Trace-Navi consists of three modules: human activity recognition module, reference trace collection module and navigation module. Human activity recognition module mainly includes three parts: data collection, data processing, and activity recognition. Reference trace collection module collects the following four information: step distance, heading direction, floor detection and magnetic trajectory. The navigation module includes walking progress estimation function and trajectory calibration function. The overall architecture of Trace-Navi is shown in Fig. 1.

3.1 Data Processing

MEMS sensor data is affected by diverse factors, such as environmental noise, equipment coordinate system and navigation coordinate system inconsistency. Original data cannot be directly applied to activity recognition and dead reckoning. In this paper, we adopt coordinate system transformation and noise reduction to process the original data, ensure the processed data meet the requirements of activity recognition and dead reckoning algorithm.

Coordinate System Transformation. The original sensor data is collected based on the coordinate system of the device itself, and the posture of the device

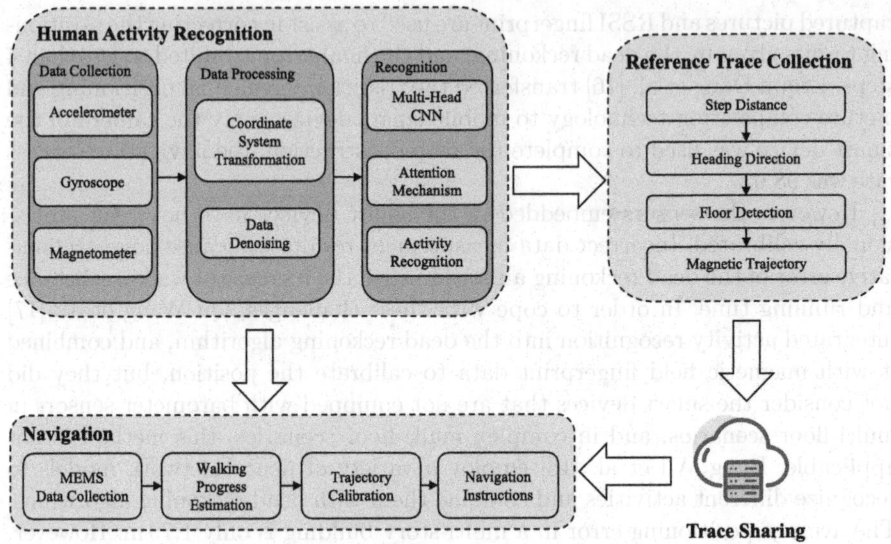

Fig. 1. Architecture of Trace-Navi.

will not always be consistent with the data collector when pedestrians are actually walking. We call the coordinate system based on the device itself Device Coordinate System (DCS), and the coordinate system used for the activity recognition and navigation is called the Navigation Coordinate System (NCS), as shown in Fig. 2.

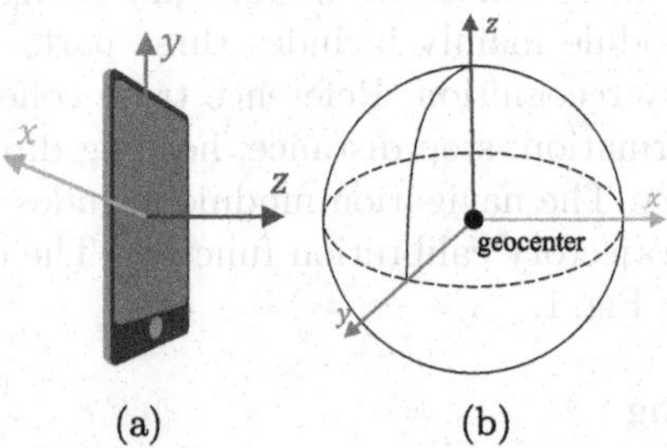

Fig. 2. Coordinate system comparison. (a) Device coordinate system; (b) Navigation coordinate system.

Considering the collected sensor data in device coordinate system as a vector:

$$Acc_d = \left[a_d^x, a_d^y, a_d^z\right]^T \tag{1}$$

$$Mag_d = \left[m_d^x, m_d^y, m_d^z\right]^T \tag{2}$$

where a_d^x, a_d^y and a_d^z are the measurements from 3-axis accelerometer, m_d^x, m_d^y and m_d^z measurements from 3-axis magnetometer. To perform positioning and navigation services in the actual scene, it is necessary to convert the collected sensor data from DCS to NCS. This paper adopts the following coordinate transformation method: assuming that at time t, the three-axis rotation angles of the smart device in DCS are ω, φ, γ, the transformation matrix T that transforms the sensor data vector from the DCS to NCS can be expressed as:

$$T = \begin{bmatrix} cos\varphi cos\gamma & -cos\omega cos\gamma + sin\varphi sin\omega cos\gamma & sin\omega sin\gamma + sin\varphi cos\omega cos\gamma \\ -cos\varphi sin\gamma & -cos\omega cos\gamma - sin\varphi sin\omega sin\gamma & sin\omega cos\gamma - sin\varphi cos\omega sin\gamma \\ -sin\varphi & cos\varphi sin\omega & cos\varphi cos\omega \end{bmatrix} \tag{3}$$

Multiply the transformation matrix T with the collected sensor data to get the sensor data in NCS:

$$Acc' = \left[a_n^x, a_n^y, a_n^z\right]^T = T * Acc = \left[a_d^x, a_d^y, a_d^z\right]^T \tag{4}$$

$$Mag' = \left[m_n^x, m_n^y, m_n^z\right]^T = T * Mag = \left[m_d^x, m_d^y, m_d^z\right]^T \tag{5}$$

Noise Reduction. When the accelerometer is employed to collect acceleration data, its sampling value will be disturbed by environmental noise. Among them, the gravity acceleration has the most serious interference to the collected acceleration data. Assuming that the device is placed on a horizontal desktop, the data collected by the accelerometer is actually the acceleration of gravity G; when the device is in free fall, the collected data is zero. To measure the real acceleration value, it is necessary to filter out the gravitational acceleration contained in the collected data. Trace-Navi adopts a high-pass filter to solve this problem, which is implemented by Eq. 6 and 7:

$$g_t = \alpha * g_{t-1} + (1 - \alpha) * acc_t^{origin} \tag{6}$$

$$acc_t^{real} = acc_t^{origin} - g_t \tag{7}$$

where g_t is the gravity acceleration at time t, α is the high-pass filter parameter, acc_t^{origin} is the acceleration data in the navigation coordinate system, and acc_t^{real} is the final acceleration data after the high-pass filter. In this paper, α is set to 0.8.

The processed data is divided into multiple active instance windows, and a single active data instance window is composed of a series of sensor data. The data of each instance $Data_i$ in the NCS is expressed as:

$$sample_t = \{acc_t^x, acc_t^y, acc_t^z, g_t^x, g_t^y, g_t^z, time\} \tag{8}$$

$$Data_i = \{sample_t, sample_{t+1}, ..., sample_{t+K+1}\} \tag{9}$$

where $sample_t$ represents the sensor data collected at time t, $Data_i$ represents each input activity data segment, K represents the window size of each activity data segment. In this paper, each window size has 50% data overlap with the previous window to learn the data characteristics of the same activity in different time periods. acc_t^x, acc_t^y, acc_t^z represent the three-axis acceleration data in NCS, g_t^x, g_t^y, g_t^z represent the three-axis gyroscope data in NCS. We define the dataset as $Data$=$\{Data_i$,$Label_i\}$.

3.2 Activity Recognition Based on Attention Mechanism and Convolutional Neural Network

Step detection and step length estimation in actual dead reckoning scenes are facing great challenges. This paper proposes a real-time activity recognition model based on multi-head convolution and attention mechanism. The attention mechanism focuses on finding high-value information from a large amount of information in short time, and is suitable for real-time activity recognition model that operates the navigation and position estimation method. Aiming at the actual walking state of pedestrians using smart devices for navigation while walking indoors, and some smart devices are not equipped with barometer sensors, it is impossible to directly use the sensors to determine the data to recognize the activity of upstairs and downstairs. This paper considers simple activities, such as {A1: downstairs}, {A2: fast walk}, {A3: normal walk}, {A4: slow walk} and {A5: upstairs}.

Artificial neural networks have been widely used in many fields [19–22]. In recent years, Convolutional Neural Network (CNN) has been proven to extract high-dimensional data features from original sensor data, while attention mechanism can ignore irrelevant data features and focus on a subset of related features [23]. Since the window size of pedestrian activity data in each walking event is not always be the same, this paper employs Multi-Head CNN to extract the high-dimensional data features of more data length windows. Multi-Head CNN can set CNN parameters in each individual CNN head, and apply different data feature extraction strategies on input data with different window sizes. Trace-Navi combines the attention mechanism on the basis of the CNN in each head, thus the activity recognition model can quickly extract important high-dimensional data features. The structure diagram of the activity recognition model employed in this paper is shown in Fig. 3.

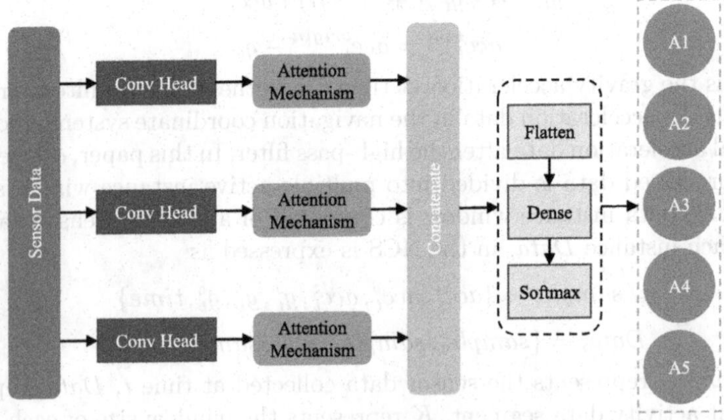

Fig. 3. Activity recognition based on multi-head CNN and Attention Mechanism.

To extract more data features, this paper sets up a multi-head convolutional neural network of 3 heads to process the sensor data vector in NCS, the filter size of CNN in each head is set to 64, and the kernel parameter is 3, 5, and 11 respectively. In order to prevent over-fitting, the dropout layer is adopted after data feature extraction, and the investment rate is set to 0.5. High-dimensional data features are extracted at the core layer of each convolutional head, the attention mechanism employs $Softmax$ as the activation function of 64 neuron weights to generate weighted parameters W_i, and multiplies it with the data feature set element by element. The higher the feature importance of the output of the CNN layer, the closer the output of the attention mechanism is to 1, and vice versa. After that, the data features extracted by the attention mechanism in each head are connected in the concatenate layer. Then the Flatten layer is employed to convert the spliced high-dimensional data features into one-dimensional. Finally, the fully connected layer (Dense) is used as the output layer, and the $Softmax$ function is adopted as the activation function to output the classification result. This paper employs the cross-entropy function as the loss function to minimize the result of the loss function. The loss function is defined as Eq. 10:

$$loss = -\sum_i (y_i log y_p + (1 - y_i) log(1 - y_p)) \tag{10}$$

where i is the serial number of the currently predicted activity instance, y_i represents the real activity of the current sequence, and y_p represents the predicted activity result of the classifier. The trained activity recognition model transferred to smart devices, thus Trace-Navi can recognize pedestrian activity on the mobile smart device.

3.3 Improved Pedestrian Dead Reckoning

The accumulative error of the traditional pedestrian dead reckoning is mainly caused by the heading direction estimation error and the step length estimation error. In this paper, we propose a fusion filter to estimate the heading direction of pedestrians, and a dynamic step length estimation method based on the result of activity recognition is employed to estimate the step length.

Heading Direction Estimation. Traditional dead reckoning methods adopt gyroscope data to calculate the heading direction, as shown in Eq. 11:

$$\theta_{Gyr}(tf) = \theta_{Gyr}(ts) + \int_{ts}^{tf} \omega(t)dt \tag{11}$$

where $\theta_{Gyr}(tf)$ is the heading direction at time tf, $\theta_{Gyr}(ts)$ is the heading direction at time ts, $\omega(t)$ is the output data of the gyroscope (angular velocity) from time ts to time tf. This method has obtained high dead reckoning accuracy on calibrated special equipment. However, the gyroscope sensor embedded in the mobile smart device has not been calibrated, and the response speed is very fast in actual use. Therefore, the collected noise data will be included in the integral calculation, resulting in a short-term direction estimation error. Some other researchers use acceleration data combined with magnetic field strength data to estimate heading direction [14]. The estimated heading direction changes greatly in a short time, but it can accurately reflect the change of the heading direction during a long walking time. In this paper, we propose a fusion filter to fuse the data of accelerometer, magnetometer and gyroscope to estimate the accurate heading direction during walking. The specific structure of the proposed fusion filter is shown in Fig. 4.

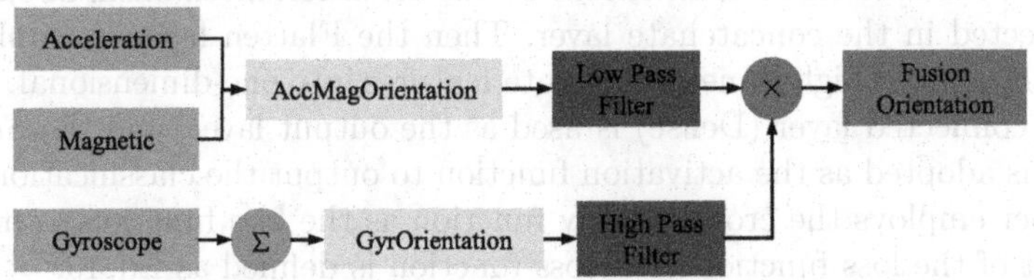

Fig. 4. Structure of the proposed fusion filter.

The fusion filter combines the acceleration data with the magnetic data, and estimates the heading direction obtained from the acceleration and the magnetic data through low pass filter. After that, a high pass filter is employed to remove the low-frequency components in the gyroscope data and reduce the integral of the environmental noise. At the end, the filtered gyroscope data is merged with the filtered output of acceleration and magnetic data. The fusion heading direction is estimated by Eq. 12.

$$FusedOri = f * Gyr + (1 - f) * AccMag \tag{12}$$

where *FusedOri* is the heading direction estimated by the fusion filter, *Gyr* is the heading direction estimated by the gyroscope, *AccMag* is the heading direction estimated by fusion of accelerometer and magnetometer, and f is the filter parameter. The experimental result in Sect. 4 verifies the accuracy of the fusion filtering method in estimating the heading direction of pedestrians during long-time walking.

Step Length Estimation. Trace-Navi estimates the step length based on recognized pedestrian activity. The traditional pedestrian dead reckoning methods employ acceleration data integration to calculate the pedestrian speed and distance. As the system running time increases, the accumulative error increase sharply, and the estimated walking distance is different from real value. This paper adopts the Weinberg [24] model to estimate the step length. We employ the relationship between the pedestrian's dynamic step length and the vertical acceleration reading to estimate the pedestrian's step length:

$$S = L * \sqrt[4]{a_{max} - a_{min}} \tag{13}$$

where S represents the step length, a_{max} and a_{min} are the maximum and minimum values on the Z axis of acceleration, L is a constant related to the step frequency. In general, the value of L increases with the increase of walking speed.

Position Estimation. Trace-Navi estimates the pedestrian location based on heading direction and step length. The parameters of each pedestrian movement consist of the step length and the heading direction. Therefore, the position coordinates (x_n, y_n, z_n) at n^{th} step can be determined by the previous position coordinates $(x_{(n-1)}, y_{(n-1)}, z_{(n-1)})$ at step $n-1^{th}$, the heading direction θ_t and step length l_t at n^{th} step. The position is estimated as Eq. 14:

$$Loc_{user} = (x_n, y_n, z_n) = \begin{cases} x_{n-1} + l_n * sin\theta_n \\ y_{n-1} + l_n * cos\theta_n \\ z_{n-1} + h_{stair} * \Delta \end{cases} \tag{14}$$

where h_{stair} is the height of stair. Δ is 1 or -1 when the recognized activity is going upstairs or downstairs. Otherwise, Δ is *zero*.

3.4 Discrete Trajectory Calibration

Trace-Navi provides the navigation instructions based on walking progress estimation. Traditional pedestrian dead reckoning methods estimate the walking progress based on the walking distance of the pedestrian, and the accumulative error leads to a complete failure of navigation instructions. In this paper, we adopt an improved DTW (Dynamic Time Warping) algorithm to estimate the walking progress and calibrate the trajectory. The algorithm matches magnetic field data collected by the pedestrian's smart device to the data of reference trace. DTW distance is expressed as Eq. 15:

$$D(i,j) = Dist(i,j) + min([D(i-1,j), D(i,j-1), D(i-1,j-1)]) \tag{15}$$

When the same smart device is used for magnetic field data sequence matching, this method has achieved excellent result. However, the magnetic field data collected by different smart devices in the same trajectory are not consistent. In this paper, we improve the traditional DTW algorithm. The core concept of the improved DTW algorithm is to convert the input sequence data into discrete-time difference sequence. The sequence data conversion process is shown as Eq. 16:

$$m'(i) = \frac{(m_i - m_{i-1}) + (\frac{m_{i+1} - m_{i-1}}{2})}{2} \tag{16}$$

where m_i is the ith value of the input magnetic data sequence. The distance between two data sequences is defined as the DTW distance between two changing distribution sequences $m'(i)$ and $m'(j)$.

4 Experiments and Evaluation

4.1 Experimental Setup

We implement Trace-Navi on android mobile smart devices. To ensure Trace-Navi can run stably on the most of mobile smart devices, the sensor data sampling frequency of the data collection module and the navigation module are both set 20 Hz. Figure 5 shows the graphical user interface of Trace-Navi. Meanwhile, according to the experimental results of volunteers and combined with the activity recognition model, the appropriate parameter L is selected, the dynamic step length estimation is adapted to estimate walking distance under mixed activity. After a large number of experiments conducted by many volunteers, the value of L is set to 0.46 in the slow walk stating, in the normal walking state is 0.48, and the value in fast walking state is 0.53. The filter parameter f in the fusion process is set to 0.98.

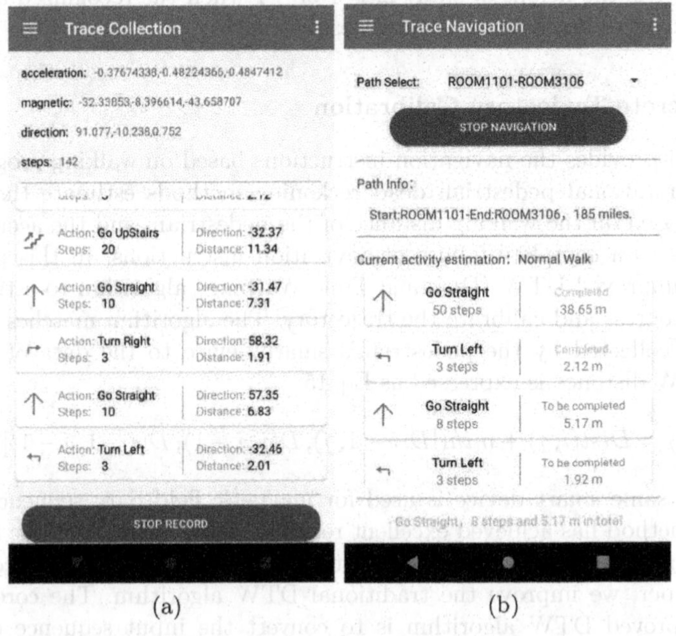

Fig. 5. GUI of Trace-Navi. (a) Collection module user interface. (b) Navigation module user interface.

The experimental environment is a commercial center with three floors, each floor area is about 1000 m^2. The reference trace consists of walking trace on three floors, with a total length of 280 m. Several volunteers constructed our activity dataset in the above scene. The indoor layout is shown in Fig. 6.

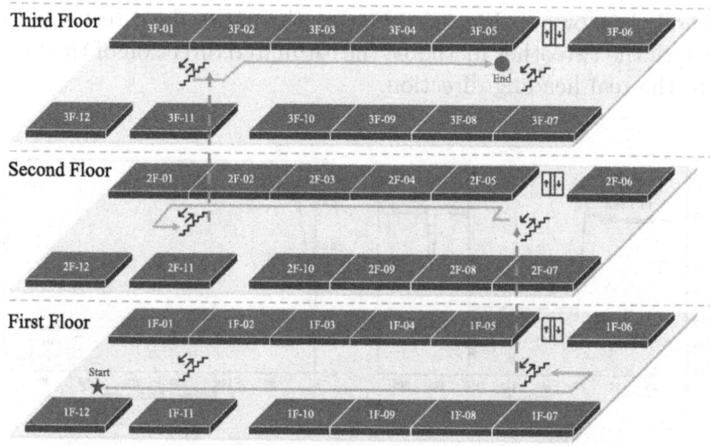

Fig. 6. Experimental environment with walking trace.

4.2 Experimental Results

Classification Accuracy of Activity Recognition Model. We compare various neural networks on several public datasets: Unimib-SHAR(ADL) [25], WISDM [26] and the self-built activity dataset of this paper. Among them, Unimib-SHAR (ADL) dataset contains 9 kinds of daily life activities, including a total of 7579 examples of similar activity data, collected by 30 volunteers between the ages of 18 and 60. The WISDM dataset contains 6 daily activities, a total of 10299 examples of human activities, and is composed of 30 volunteers collecting data of daily life activities on a waist smart device with embedded inertial sensors. We compared the performance of several typical algorithms on three data sets, as shown in Table 1.

Table 1. Comparison of accelerometer inherent errors of common smart devices.

Method	Unimib	WISDM	Our dataset
CNN	92.56%	93.25%	94.37%
LSTM	92.34%	94.71%	93.91%
CNN-LSTM	95.38%	96.81%	96.97%
Proposed	97.23%	98.17%	98.65%

Performance of Trace-Navi. Trace-Navi estimates the pedestrian location based on the data collected by the sensors embedded in the smart device. Figure 7(a) shows the estimated heading direction of pedestrian from gyroscope, fusion estimation of accelerometer and magnetometer, and the proposed algorithm. The Cumulative Distribution Function (CDF) comparison of the three above methods is shown in Fig. 7(b). During the entire dead reckoning process, compared with the two other methods, the estimated direction of the fusion filter is closest to the real heading direction.

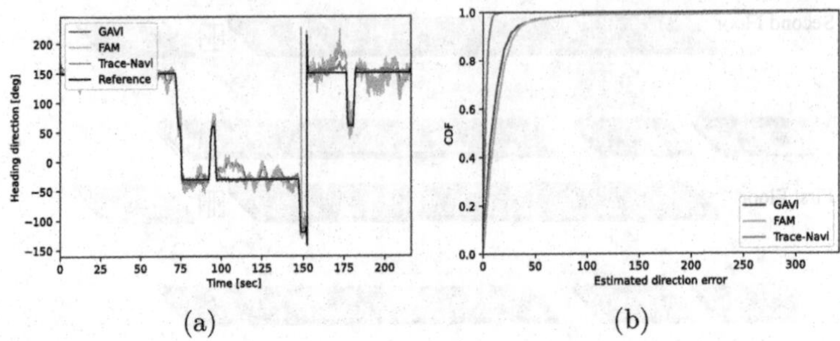

(a) (b)

Fig. 7. Comparison of heading direcion deviations of the three methods. (a) Heading direction estimation with respect to the ground-truth; (b) CDF comparison chart.

Figure 8 depicts the pedestrian trajectory drawn based on three different dead reckoning methods before trajectory calibration.

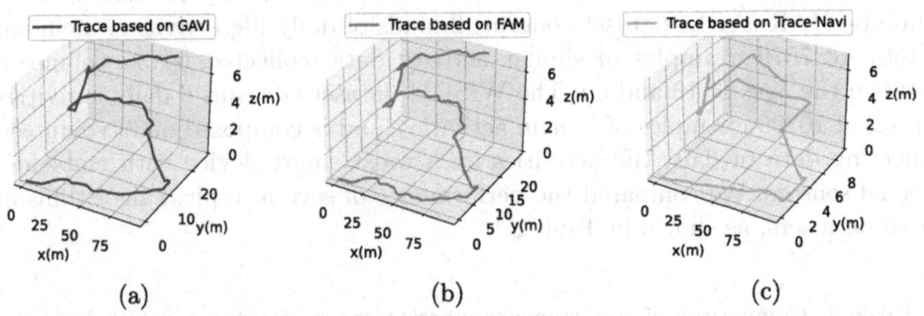

(a) (b) (c)

Fig. 8. The walking path in the experimental scene before trajectory calibration. (a) GAVI trace; (b) FAM trace; (c) Trace-Navi trace.

Table 2. Comparison of three dead reckoning methods before calibration.

Method	EPS	MES	RMSE	MAE
GAVI	9.56	111.13	10.54	11.94
FAM	7.87	83.39	9.13	10.05
Trace-Navi	1.18	1.99	1.41	1.31

Table 3. Comparison of three dead reckoning methods after calibration.

Method	EPS	MES	RMSE	MAE
GAVI	1.98	6.45	2.54	2.44
FAM	1.95	6.30	2.51	2.39
Trace-Navi	0.51	0.50	0.71	0.56

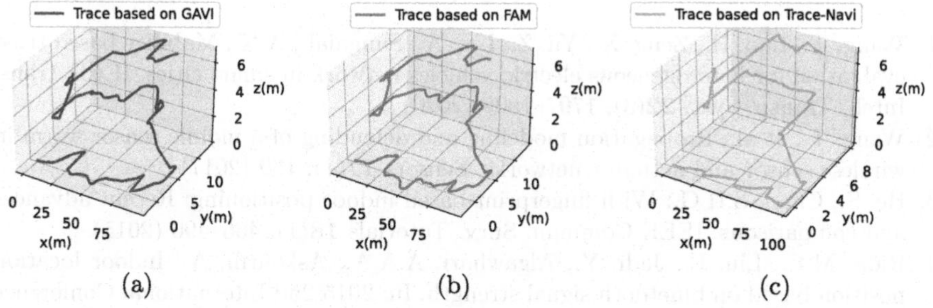

Fig. 9. The Estimation trace in the experimental scene after trajectory calibration. (a)GAVI trace; (b) FAM trace; (c) Trace-Navi trace.

Table 2 summarizes the error of per step (EPS), mean square error (MSE), root mean square error (RMSE) and mean absolute error (MAE) of the three dead reckoning methods. The performance comparison of the three dead reckoning methods after trajectory calibration is detailed in Fig. 9 and Table 3.

5 Conclusions

In this paper, we propose an indoor navigation system, Trace-Navi. The proposed system adopts a real-time activity recognition model transferred to the mobile smart devices to recognize pedestrian activities, and estimates the pedestrian walking distance according to the recognition result. In the future, we will consider using the activity recognition model to recognition more indoor activities. We intend to allow users to improve the performance of the activity recognition model according to their own activity data, so as to further improve the positioning accuracy of Trace-Navi.

Acknowledgment. This work is supported by National Natural Science Foundation of China under Grant No. 61772448, Natural Science Foundation of Jiangsu Province under Grant No. BK20191481, Natural Science Major Project of the Higher Education Institutions of Jiangsu Province of China under Grant No. 17KJA520006, China Postdoctoral Science Foundation funded project under Grant No. 2019M660132, Jiangsu Province Postdoctoral Research Subsidy Program under Grant No. 2019K123, Project on the Industry-University-Research Cooperation of Jiangsu Province under Grant No. BY2021326, Scientific Research Subsidy of Jiangsu Province under Grant No.

BRA2019288, Future Network Scientific Research Fund Project Grant No. FNSRFP-2021-YB-45, Jiangsu Provincial Key Constructive Laboratory for Big Data of Psychology under Grant No. 72591962002G and 72592162003G, Open Foundation of Jiangsu Provincial Key Laboratory of Coastal Wetland Bioresources and Environmental Protection under Grant No. JLCBE14008.

References

1. Wang, T., Luo, H., Zeng, X., Yu, Z., Liu, A., Sangaiah, A.K.: Mobility based trust evaluation for heterogeneous electric vehicles network in smart cities. IEEE Trans. Intell. Transp. Syst. **22**(3), 1797–1806 (2020)
2. Wang, T., et al.: Propagation modeling and defending of a mobile sensor worm in wireless sensor and actuator networks. Sensors **17**(1), 139 (2017)
3. He, S., Chan, S.H.G.: Wi-fi fingerprint-based indoor positioning: Recent advances and comparisons. IEEE Commun. Surv. Tutorials **18**(1), 466–490 (2015)
4. Rida, M.E., Liu, F., Jadi, Y., Algawhari, A.A.A., Askourih, A.: Indoor location position based on bluetooth signal strength. In: 2015 2nd International Conference on Information Science and Control Engineering, pp. 769–773. IEEE (2015)
5. Bouet, M., Dos Santos, A.L.: Rfid tags: positioning principles and localization techniques. In: 2008 1st IFIP Wireless Days, pp. 1–5. IEEE (2008)
6. Ren, A., Zhou, F., Rahman, A., Wang, X., Zhao, N., Yang, X.: A study of indoor positioning based on UWB base-station configurations. In: 2017 IEEE 2nd Advanced Information Technology, Electronic and Automation Control Conference (IAEAC), pp. 1939–1943. IEEE (2017)
7. Tsai, T.H., Chang, C.H., Chen, S.W.: Vision based indoor positioning for intelligent buildings. In: 2016 2nd International Conference on Intelligent Green Building and Smart Grid (IGBSG), pp. 1–4. IEEE (2016)
8. Lee, S.M., Yoon, S.M., Cho, H.: Human activity recognition from accelerometer data using convolutional neural network. In: 2017 IEEE international conference on big data and smart computing (bigcomp), pp. 131–134. IEEE (2017)
9. Zhang, H., Xiao, Z., Wang, J., Li, F., Szczerbicki, E.: A novel IoT-perceptive human activity recognition (HAR) approach using multihead convolutional attention. IEEE Internet Things J. **7**(2), 1072–1080 (2019)
10. Mutegeki, R., Han, D.S.: A CNN-LSTM approach to human activity recognition. In: 2020 International Conference on Artificial Intelligence in Information and Communication (ICAIIC), pp. 362–366. IEEE (2020)
11. Guo, H., Uradzinski, M.: The usability of MTI IMU sensor data in PDR indoor positioning. In: 2018 25th Saint Petersburg International Conference on Integrated Navigation Systems (ICINS), pp. 1–4. IEEE (2018)
12. Skog, I., Handel, P., Nilsson, J.O., Rantakokko, J.: Zero-velocity detection-an algorithm evaluation. IEEE Trans. Biomed. Eng. **57**(11), 2657–2666 (2010)
13. Shu, Y., Shin, K.G., He, T., Chen, J.: Last-mile navigation using smartphones. In: Proceedings of the 21st Annual International Conference on Mobile Computing and Networking, pp. 512–524 (2015)
14. Kang, W., Han, Y.: Smartpdr: smartphone-based pedestrian dead reckoning for indoor localization. IEEE Sens. J. **15**(5), 2906–2916 (2014)
15. Zheng, Y., Shen, G., Li, L., Zhao, C., Li, M., Zhao, F.: Travi-navi: self-deployable indoor navigation system. IEEE/ACM Trans. Networking **25**(5), 2655–2669 (2017)

16. Dong, E., Xu, J., Wu, C., Liu, Y., Yang, Z.: Pair-navi: peer-to-peer indoor navigation with mobile visual slam. In: IEEE INFOCOM 2019-IEEE Conference on Computer Communications, pp. 1189–1197. IEEE (2019)
17. Wang, Q., Luo, H., Xiong, H., Men, A., Zhao, F., Xia, M., Ou, C.: Pedestrian dead reckoning based on walking pattern recognition and online magnetic fingerprint trajectory calibration. IEEE Internet Things J. **8**(3), 2011–2026 (2020)
18. Wu, B., Ma, C., Poslad, S., Selviah, D.R.: An adaptive human activity-aided handheld smartphone-based pedestrian dead reckoning positioning system. Remote Sens. **13**(11), 2137 (2021)
19. Zhao, L., et al.: Novel online sequential learning-based adaptive routing for edge software-defined vehicular networks. IEEE Trans. Wireless Commun. **20**(5), 2991–3004 (2020)
20. Wang, T., Liu, Y., Zheng, X., Dai, H.N., Jia, W., Xie, M.: Edge-based communication optimization for distributed federated learning. IEEE Trans. Netw. Sci. Eng. (2021)
21. Zhao, L., Li, H., Lin, N., Lin, M., Fan, C., Shi, J.: Intelligent content caching strategy in autonomous driving toward 6g. IEEE Trans. Intell. Transp. Syst. 1–11 (2021)
22. Zhao, L., Han, G., Li, Z., Shu, L.: Intelligent digital twin-based software-defined vehicular networks. IEEE Network **34**(5), 178–184 (2020)
23. Vaswani, A., et al.: Attention is all you need. In: Advances in Neural Information Processing Systems, pp. 5998–6008 (2017)
24. Weinberg, H.: Using the adxl202 in pedometer and personal navigation applications. Analog Devices AN-602 Appl. Note **2**(2), 1–6 (2002)
25. Micucci, D., Mobilio, M., Napoletano, P.: Unimib shar: a new dataset for human activity recognition using acceleration data from smartphones. Appl. Sci. **7**(10), 1101 (2017)
26. Kwapisz, J.R., Weiss, G.M., Moore, S.A.: Activity recognition using cell phone accelerometers. ACM SigKDD Explor. Newsl. **12**(2), 74–82 (2011)

Iterative Filling Incomplete Fingerprint Map Based on Multi-directional Signal Propagation in Large-Scale Scene

Yonghao Zhao[1,2], Yanhui Ji[1,2], Licai Zhu[1,2], and Hao Yang[1,2,3,4](✉)

[1] Nanjing Tech University, Nanjing 211816, Jiangsu, China
[2] Yancheng Teachers University, Yancheng 224001, Jiangsu, China
[3] Suzhou Research Institute, University of Science and Technology of China, Suzhou 215123, Jiangsu, China
[4] Jiangsu Provincial Key Constructive Laboratory for Big Data of Psychology and Cognitive Science, Yancheng Teachers University, Yancheng 224002, Jiangsu, China

Abstract. Indoor localization using RSS fingerprint database has become one of the most influential and practical solutions for various types of location service systems, which circumvents proprietary requirements, such as infrastructures, signal transceivers. In order to provide accurate location information, how-ever, a full-scale fingerprint map with high density has to be construct in advance. Furthermore, the established map is obliged to be updated regularly due to instability of RSS or damage of access points. And yet, sampling RSS of abundant reference points (RPs) is undoubtedly an extreme time-consuming and labor-intensive. The interpolation is naturally introduced to lessen the number of samples through collecting partial RPs. This paradigm generally needs to sample adequate RPs for the desirable accuracy in actual surroundings, especially in large-scale scenes. To reduce sampling with along with high-precision, we propose an iterative filling approach for incomplete fingerprint map based on multi-directional signal propagation, called FMS to find missing samplings. Different from the current interpolating concept, FMS do not potentially assume that the missing samples depend on their annular near neighbors, but considers the character of signal propagation to better fitting with the reality. For restraining unfavorable effects caused by reflection, diffraction and re-fraction of signals, FMS mingles multiply direction chains of missing RPs through estimating the RSS value and the weight of each chain. In our experiments, the sampling rates are from 40% to 80%. Extensive experiments demonstrate that FMS has a good stability in various of conditions, including sampling rates and devices. It exceeds 11.35% ~ 24.43% compared to the whole sampling mode, and 14.48% ~ 24.75% compared to the best accuracy of current mainstream methods.

Keywords: Iterative filling · Directional chains · Signal propagation · Missing samplings

Y. Lai et al. (Eds.): ICA3PP 2021, LNCS 13155, pp. 304–318, 2022.
https://doi.org/10.1007/978-3-030-95384-3_20

1 Introduction

Location-based service has gradually become an indispensable service in daily life, which is generally divided into outdoor localization and indoor localization according to the implementation scenario. Due to the deployment of navigation satellite systems (such as Beidou, GPS, etc.), outdoor localization and navigation has been well applied [1–3] and the localization errors can reach sub-meter level. Compared with the outdoor environment, the interference of various obstacles to the satellite signal in the indoor environment leads to the navigation satellite systems are unable to locate. With the widespread deployment and application of wireless network environment [4], researchers generally use the method of wireless signal matching to locate [5]. The basic thought is to make use of the spatial differences of Access Points (APs) in different positions, and take the wireless signal characteristics [6] of reference points (RPs) as the characteristics of physical locations to determine the location. Such characteristics are called Fingerprints, and Received Signal Strength (RSS) is a typical fingerprint characteristic.

It is necessary to collect signal data [7] from many reference points in the interest areas to build a signal fingerprint map, which is extremely time-consuming and labor-intensive [8]. To meet the positioning needs of large-scale scenes, such as business centers, parking lots, conference and administrative centers, etc., huge measurement costs are required [9]. According to the fingerprint localization, the localization accuracy will generally improve when the number of reference points increases. Therefore, it is required to collect as many signal data of reference points as possible to improve the localization accuracy. Because of the inherent characteristics of wireless signals, the signals of the reference points will change over time, resulting in inconsistent fingerprints in the same position when the training phase and the testing phase, thus failing to match [10]. Hence fingerprint databases need to be maintained and updated at regular intervals, which further increases localization costs. Therefore, it is very important to reduce the workload of building and updating fingerprint databases while ensuring the localization errors.

The paper designs a map construction method to find missing sampling (FMS), which iterative filling incomplete and irregular large-scale map from multiple directions through constructing signal propagation model. It aims to reduce the cost of building fingerprint maps. Under the condition of incomplete map, it uses multiple directional chains to estimate the RSS of each unsampled point according to the signal propagation characteristics in the actual environment, so as to complete the whole fingerprint map. Specifically, random and irregular sampling is carried out in the interest areas to judge the distribution rules of sampled points located in different directions of each unsampled point. Different models are used to estimate the signals in different directions of each unsampled point. And the RSS of each unsampled point is obtained from the estimates in these directions. Unsampled points will be iteratively completed with reasonable values, and the interpolated locations are regarded as virtual points. Finally, a fingerprint database composed of sampled points and virtual points is formed, which effectively reduces the cost of building and updating the fingerprint database. Experimental results show that this algorithm reduces the measurement workload while improving the localization accuracy.

In summary, the core contributions are as follows.

1) In this paper, a fingerprint map is built through random sampling, which only needs a few sampled points. Compared to the whole sampling method, we have only collected 40% to 80% of the data.
2) Based on in-depth understanding of the signal propagation, we propose a new method of fitting fingerprint map signals based on the multi-directional chains to make the interpolated values are more suitable for the indoor localization.
3) We have proved that the proposed FMS algorithm has better localization accuracy and algorithm stability under the condition of using less sampled points through extensive experiments. The localization accuracy is improved by 11.35% ~ 24.43% compared with whole sampling, and is improved by 14.48% ~ 24.75% compared with the best results in other methods when the sampling rates are different.

2 Related Work

Massive time and labor costs are required when establishing or updating a database for fingerprint localization methods. It is obviously not practical to sample the perceptual environment intensively when the localization scene is large in the actual sampling process. Researchers always build the fingerprint database by adding virtual points to solve this problem. The points in perceptual environment are not fully sampled, but only some signals are collected when creating the fingerprint map. Then these RSS values are used to simulate the signals at the unsampled positions to expand to obtain a complete fingerprint map, which can greatly reduce the amount of sampling and save cost. While the area of the localization scene is larger, the method of inserting the virtual points into the fingerprint map saves more resources, and the more important it is. Nowadays, the fingerprint map building for WiFi and radio frequency are more representative methods.

In the establishment of WiFi fingerprint map, the method based on mobile crowdsensing [11, 12] is the most common method to alleviate this problem. Ledlie J et al. [13] encourage more users to participate in the process of crowdsourcing by means of incentives, so that users can actively collect information. However, Chenshu Wu et al. [14] use an implicit method to allow users to collect data with unconscious participation which does not need to motivate users to achieve the purpose of explicitly collecting data. Chao G et al. [15] collected a small amount of labeled fingerprint data (i.e., data set containing RSS and location information) and a large number of unlabeled fingerprint data (i.e., data set containing RSS information but no location information), so as to reduce the workload and compensate for the localization accuracy. On the other hand, many scholars build maps based on models [16, 17], which mainly use a small amount of sampled data to fit signals on other unsampled points according to the model in order to fill the entire interest area. Wang B et al. [18] proposed an algorithm for region segmentation and curve fitting through a polynomial model. In each subregion, a fitting function was obtained based on a small number of reference points to get the signals of the unsampled points. Velazquez et al. [19] estimated the position of APs by weighted centroid of fingerprints of sampled points, and evaluated the parameters of signal transmitting power and path loss model by Bayesian method. Feng Y et al.

[20] fitted few existing sampling data through Gaussian Process Regression(GPR) to generate the mean and standard deviation surfaces of local areas, and combined with the particle filter to generate particles for weight assignment. Zhao J et al. [21] combined GPR with K-means to update fingerprint map better.

In the construction of the radio frequency fingerprint map, the LANDMARC algorithm [22] is based on the RFID localization technology to assist in updating the fingerprint database online by adding landmarks. The dynamic environment can be handled without obtaining the map in advance. Many scholars have proposed many improved methods on this basis. Among them, the VIRE algorithm [23] is a localization algorithm of virtual tag based on the LANDMARC algorithm, and its localization accuracy has increased by 13% to 73% in different indoor environments. It has introduced virtual reference tags. By adding virtual tags between the fixed reference tags, the virtual tags can achieve the same effect as the fixed reference tags, and excessive numbers will not cause signal interaction. Therefore, the use of spatial interpolation algorithms [24] in the training phase to reduce the workload of sampling is gradually accepted by more researchers. Hu K et al. [25] used linear interpolation to generate virtual sampling fingerprints in order to increase the density of fingerprints and estimated the accurate position through Bayesian algorithm. Compared with the traditional Bayesian algorithm, the time complexity of O(4n/5) is reduced while its localization accuracy is improved by 6%. Mo L et al. [26] introduced the variable coefficients into the traditional Laiyite criterion to eliminate multipath errors, then used natural neighborhood interpolation for virtual reference marking to improve the stability and accuracy of the similarity measurement of virtual tags. So as to ensure that 80% of the testing tags errors are accumulated within 4cm, and 90% are within 10 cm. Haniz A et al. [27] compared the Kriging interpolation algorithm and its several deformation algorithms on the accuracy of map localization. Kubota R et al. [28] estimated wall penetration loss and distance power loss, and inserted RSS values based on these loss models when building a fingerprint database. They used spatial interpolation to refine the fingerprint map and improved the average localization accuracy by about 25%. T. Guan et al. [29] proposed to use dynamic particles to replace static reference tags in traditional RFID virtual tag localization. They used the Particle Swarm Optimization (PSO) algorithm to update the Monte Carlo sample particle swarm, and gave different weights based on the signal strength difference between the sampled particles and the undetermined tags in order to realize the interpolation effect of the unknown tags. Wu Y H et al. [30] proposed an area-based repair method for the construction of virtual maps, which mainly minimizes the estimation error by limiting the path loss index used to construct the virtual RP and limiting the area. These methods are more or less involved in the idea of interpolation and virtual points, which optimizes the workload of constructing fingerprint maps to a certain extent. However, the above methods have certain limitations, and it is difficult to directly apply to WiFi fingerprint localization.

In this paper, the FMS algorithm is proposed for the map construction of WiFi fingerprint localization to reduce costs. It fully considers the signal propagation characteristics and multiple directions chains with different situations to build an accurate fingerprint map, which improves the localization accuracy and stability.

3 Our Design of FMS

3.1 Basic Idea

Based on the idea of spatial interpolation algorithm, we explore ways to re-duce the workload of establishing fingerprint maps. Since the signal propagation characteristics are not consider in the existing spatial interpolation algorithms, the fingerprint map cannot be filled by simply applying the existing interpolation method. Wireless signal propagation in the medium faces many severe challenges, including the effects of noise, interference, blocking, multipaths and other influences, which changed with time and distance. The signal change shows an overall attenuation trend, that is, path loss. Based on the loss process to summarize the signal propagation, researchers proposed a path loss model. However, there is still a large error in fitting RSSs with path loss model. Therefore, we have improved the method on this basis.

In order to avoid the influence of the above-mentioned problems, we comprehensively consider the neighboring points and the far points in the interpolation process. There are several sampled points in each direction of the interpolated point. We connect the sampled points in the same direction into a chain, which is called directional chain. There is a signal value determined by the directional chain at the position of the interpolated point in each direction. And all the directional chains together determine the RSS of the interpolated point. Different from the interpolation methods that only consider neighboring points, the directional chains make the interpolation algorithm no longer limited to a single or a few neighboring points. It is determined by the multi-level sampled points, which greatly avoids the influence of signal ambiguity.

The details are shown in Fig. 1. The RSSs of surrounding neighboring points are relied on in most traditional interpolation methods. However, multiple directional chains are used to obtain the RSSs of the interpolated points in the FMS algorithm.

Since the correlation is based on the signal propagation direction, the chain model in each direction can be summarized into the following three characteristics.

1) The RSSs of the interpolated points are the most similar to the RSSs of the neighboring sampled points on the directional chain.
2) The RSSs of the interpolated points are related to the RSSs of all sampled points on the directional chain.
3) The chain model in each direction is different, that is, the RSSs of the interpolated points are affected differently in each direction.

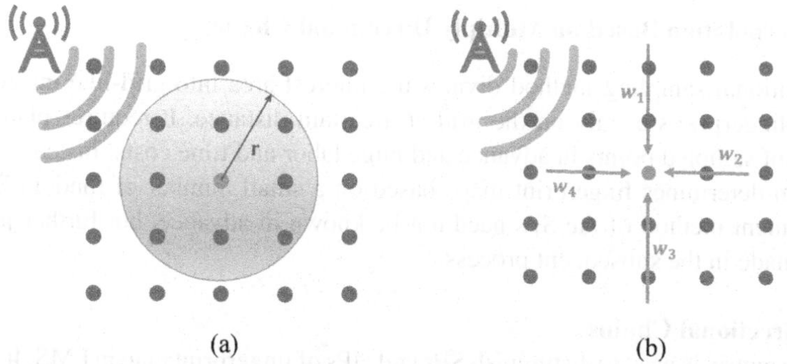

Fig. 1. Schematic of interpolation. (a) Most traditional interpolation methods. (b) FMS

3.2 Framework

The framework of FMS is shown in Fig. 2. To simplify, we use some abbreviations as follows:

- ***RP*** indicates Reference Point. It means that the potential location which would be sampled in the sensing area.
- ***SP*** indicates Sampled Point. It means that the location has been sampled for RSS.
- ***UP*** indicates Unsampled Point. It means that RP which have been not sampled for RSS.
- ***VP*** indicates Virtual Point. It means that the location has been interpolated among UP.

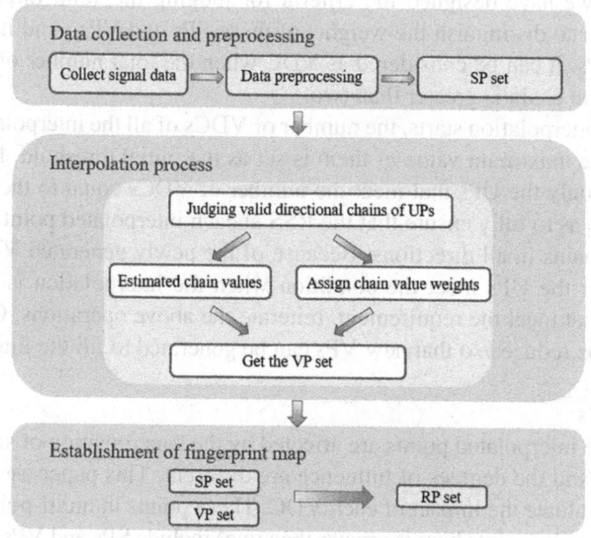

Fig. 2. Framework of FMS

3.3 Interpolation Based on Multiply Directional Chains

The traditional sampling method divides the interest area into grid-like intervals, and collects fingerprints evenly on the grid at a certain distance. It requires planning the location of sampled points in advance and huge labor and time costs. Instead, the FMS algorithm determines fingerprint maps based on a small number of random SPs. The measurement method of the SPs need not be known in advance, but further judgment will be made in the subsequent process.

Valid Directional Chains

The state matrix is used to distinguish SPs and UPs of fingerprint map in FMS. It does not need to be interpolated when the target points are SPs. When the target points are UPs, their interpolations are comprehensively considered from multiple directions. And all SPs in each directional chain are recorded. Five neighboring layers on each directional chain are selected to record in this paper. It should be noted that recording all SPs on each directional chain is to pick out the valid direction number of each interpolated point. And the directional chains in the valid directions are considered. In other words, we only consider the directional chains with the number of SPs that meets the requirements. Theoretically, the SPs should have the greatest impact on the interpolated points in the process of creating the fingerprint map. And the impact of all newly generated VPs on the interpolated points should be smaller as the number of rounds of its generation increases. In other words, the effects of the actual SPs and the generated VPs on the interpolated points need to be considered. Meanwhile, it should also be considered that the VPs generated in different rounds have different effects on the interpolated points. The VPs obtained from multiple SPs should have a greater impact than the VPs generated from multiple VPs.

Therefore, we have designed the criteria for judging the valid directional chains (VDCs) in order to distinguish the weights of these SPs and VPs. And it considers the influence of VPs. It can be considered as VDC when the total number of SPs and VPs on the directional chain is greater than two.

Before the interpolation starts, the number of VDCs of all the interpolated points are counted, and the maximum value of them is set as the initial threshold. In the interpolation process, only the UPs that meet the number of VDCs equal to the threshold are interpolated. So as to fully ensure that the RSS at each interpolated point is determined by the signal chains in all directions. Because of the newly generated VPs, we should rejudge whether the UPs meet the condition when the interpolation is completed. If there are UPs that meet the requirement, reiterate the above operations. Otherwise, the threshold will be reduced so that new VPs can be generated to fill the fingerprint map.

Values of VDCs

The RSSs of the interpolated points are affected by the superposition of signal chains in each direction, and the degrees of influence are different. This paper uses multi-points estimation to evaluate the impact of each VDC. These points in multi-points estimation (the number of multi-points here is greater than two) include SPs and VPs. Thus, we can get the estimated values in each direction of the point.

If a VDC comprises more than two points including the total of RPs and VPs, we fully consider the signal propagation and the model relationship in each directional chain, and apply the path loss logarithmic model. The parameters of the signal attenuation model are fitted by several points, and the estimations of the signals in each direction of the interpolation points are obtained.

Assuming that the interpolated point is $P(x, y)$, and the SPs on the d_k directional chain ($1 \leq k \leq n$) are $P_1^k(x, y), P_2^k(x, y), \cdots\cdots, P_{l(k)}^k(x, y)$ respectively. Similarly, point $P_1^k(x, y)$ is the neighbor of point $P(x, y)$, and point $P_i^k(x, y)$ is the neighbor of point $P_{i+1}^k(x, y)$ when $1 \leq i < l(k)$. We use these points to fit the signal attenuation model

$$P(d) = P(d_0) - 10\eta lg(d/d_0) + \sigma$$

According to the signal attenuation model, the predicted signal on the directional chain is calculated as RSS_P.

$$RSS_P = RSS_{P_1^k} - 10\eta lg\left(\frac{dis(P - P_1^k)}{dis(P_1^k - P_{l(k)}^k)}\right) + \sigma \tag{1}$$

where $dis(P - P_1^k)$ is the distance between point $P(x, y)$ and point $P_1^k(x, y)$, $dis(P_1^k - P_{l(k)}^k)$ is the distance between point $P_1^k(x, y)$ and $P_{l(k)}^k(x, y)$.

Weights of VDCs

There are multiple VDCs at each target point, and each VDC determines the effect of this direction on the interpolated point. And the influence of the estimated value in each direction is different.

The number of points in each direction chain is different in the multi-points estimation method, which leads to uncertainty in the impact of the interpolated point. Meanwhile, there are newly generated VPs on the directional chains. These VPs are obtained by the different rounds of interpolation, and their credibility are also different. It is assumed that the credibility of the SPs is higher than that of the VPs, and the credibility of the VPs generated first is higher than that of the VPs generated later. The impact of the rule on the algorithm is the difference in weight. In addition, the actual RSS values of the SPs on the directional chains are known. According to the signal attenuation model, these points can infer the theoretically fitted RSS values. Thus, all SPs on the chains have two RSS values, which are the actual sampled RSSs and the fitted RSSs. We determine the influence of each directional chain on the interpolated point according to the deviation between the two RSSs of all points on each chain. Specifically, there are v directional chains using multi-points estimation, and the SPs on the jth directional chain ($1 \leq j \leq v$) are $P_1^j(x, y), P_2^j(x, y), \cdots\cdots, P_{l(k)}^j(x, y)$ respectively. Their actual sampled RSSs are $RSS_{P_1^j}, RSS_{P_2^j}, \cdots\cdots, RSS_{P_{l(k)}^j}$, their fitted RSSs are $RSS'_{P_1^j}, RSS'_{P_2^j}, \cdots\cdots, RSS'_{P_{l(k)}^j}$, and the deviation of the directional chain can be expressed as

$$e_j = \sum\nolimits_{i=1}^{v}(RSS_{P_i^j} - RSS'_{P_i^j}) \tag{2}$$

The deviation value of each directional chain based on multi-points estimation can be obtained according to formula (4). And the weight of the corresponding directional chain to the interpolated point can be obtained by these deviations. The greater the deviation of the directional chain, the smaller the impact on the interpolated point.

If the overall weight based on the multi-points estimation method is w_m. The deviations on the v directional chains are $e_1, e_2, \cdots\cdots, e_v$ respectively. The influence factor of the $j^{th}(1 \leq j \leq v)$ directional chain on the interpolated point is

$$w_j = w_m * (\frac{1}{e_j}/\sum_{i=1}^{v}\frac{1}{e_i}) \tag{3}$$

Interpolated Value

Through the above steps, the estimated interpolation value on each directional chain can be calculated according to formula (1), and the weight of the corresponding directional chain can be calculated according to formula (3). Finally, the RSSs of the interpolated points can also be calculated.

$$RSS_p = \sum_{i=1}^{n}(RSS_p^i * w_i) \tag{4}$$

where n is the number of VDCs, RSS_p^i is the estimated interpolation value on the ith directional chain of the interpolated point $P(x, y)$, w_i is the weight corresponding to this VDC.

3.4 Updating VP Set

It should be noted that it is difficult to fill the entire map with only one round of interpolation because of random sampling. Therefore, it is highly probable that continuous iteration is required to achieve dense and complete fingerprint maps through multiple rounds of interpolation. Some new VPs will be generated and add them to the fingerprint map after each round of interpolation algorithm. Then the calculated point P will be added to the VP set and merged with the SP set.

$$Vitural\ Set = Vitural\ Set \cup \{P\}$$

$$Sensing\ Set = Real\ Set \cup Vitural\ Set$$

After each round of interpolation is completed, the threshold will be reduced and the above steps will be repeated until a complete fingerprint map is obtained. In order to make the fingerprint connection deeper and present more regular features, it can be further smoothed. So as to alleviate or eliminate some noise and abnormal values in the process of constructing the fingerprint map.

4 Experiment and Evaluation

4.1 Experiment Settings

This paper executes our experiments in a large factory, whose area covers approximately 1500 m^2 and shape is nearly rectangular. And the details of pre-sampling are in Table 1, including the testing site area, the number and height of APs, sampling interval, sampling equipment, amount of sampling data and measurement time and other attributes.

Our experiments apply six phones that are embedded different types of wireless chips because of individual design patterns for RF collection, including one HUAWEI mate7, one YAAO and one VIVO x20 which is specially designed for the elderly. In addition, the typical RF collection frequency 1 Hz is made used of in the process of sampling.

Table 1. Experiment settings

Parameter	Value
Size	1500 m^2
Number of APs	50
Height of AP	2 m
Sampling density	1 m × 1 m
Devices	HUAWEI mate7, YAAO, VIVO x20
Samples	1.8×10^6

4.2 Comparative Approaches

To adequately evaluate the performance of FMS, this paper compares three different start-of-the-art interpolation algorithms which have been verified to be promoted localization accuracy.

1) Nearest neighbor (NN): The signal values of the interpolated points are equal to that of the nearest SPs.
2) Inverse Distance Weighted (IDW): Determine the weights of the interpolated signals according to the distances from the SPs to the interpolated points.
3) Linear: The interpolation function is a polynomial interpolation method.

The detailed results are shown as follows, including the method with no interpolation (NI). There is no special case, we perform the same preprocessing for all methods (such as SP selection). And all of these methods are implemented using WKNN with K = 5. In addition, we employ the localization error when the whole RPs are sampled in experiment area as a reference one.

4.3 Performance Evaluation

Overall Performance

As is known, localization error is related to the number of RPs in general. That is, the more RPs and the smaller the localization error. Therefore, we need to sample a certain number of RPs.

This section verified the influence of the number of SPs on the localization accuracy. The testing sets of the localization process are SPs, UPs and All RPs. Table 2 shows the overall localization error of FMS, four interpolation methods and NI.

According to the results of Table 2, the localization performances of FMS on different numbers of testing sets are outperform ones of other methods. In particular, its localization accuracy for the three test sets all exceeds whole sampling. In contrast, other interpolation methods are almost inferior or slightly better than the whole sampling case.

Table 2. Overall localization error

Testing set	Method	Sampling rate		
		40%	60%	80%
SP	NI	4.79	4.57	4.48
	FMS	**3.40**	**3.18**	**3.50**
	NN	5.35	4.93	4.62
	IDW	4.37	4.56	4.61
	Linear	4.10	4.36	4.62
UP	NI	4.89	4.55	4.52
	FMS	**4.26**	**3.57**	**3.70**
	NN	5.08	4.82	4.62
	IDW	5.26	5.04	4.94
	Linear	4.90	4.54	4.52
All RP	NI	4.85	4.57	4.48
	FMS	**3.92**	**3.34**	**3.54**
	NN	5.11	4.87	4.62
	IDW	4.90	4.75	4.68
	Linear	4.58	4.44	4.44
Whole sampling		4.42		

According to the above, FMS could have an evident advantage when the sampling rate is about 60%. Therefore, we could sample 60% of the whole sampling in order to obtain the minimum error with as few samples as possible.

Influence of Devices

Figure 3 compares the best results among FMS, NI and other interpolation methods. The best performance among the contrasted approaches is labeled as BP. As the sampling rates of most mobile phones increase, the localization errors become smaller. But the localization error of FMS is still 14.5%-24.7% smaller than the localization error of BP, which fully proves that FMS algorithm has good robustness on different devices. And Fig. 4 shows that the accuracy of these methods using the fingerprint maps built by different devices.

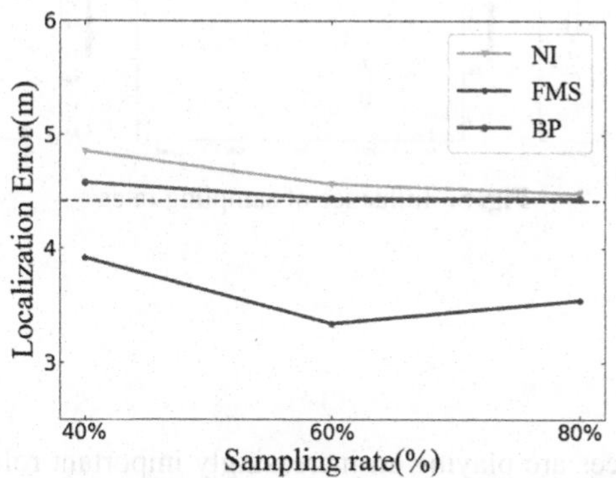

Fig. 3. Comparison of FMS, NI and other algorithms

(Red dotted line is the ground truth, which denotes the result in the case of whole sampling).

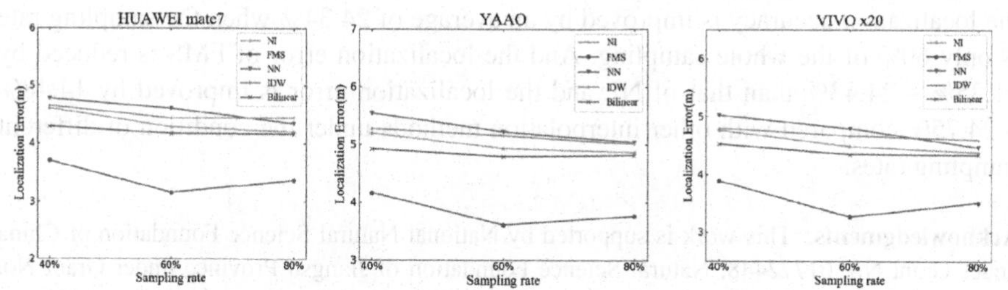

Fig. 4. Comparison of methods under different devices

Influence of Sampling Rates

Figure 5 shows the stability of each method at different sampling rates. As can be seen from the figure, the localization errors of FMS are all minimal for the fingerprint maps built by different devices. The dispersion of the localization errors of these devices is the smallest, which shows that FMS has better robustness under different sampling rates. In contrast, the dispersions of the localization errors have large differences after other methods interpolate the sampling maps of different devices, and the dispersion changes irregularly when the sampling rate is different.

According to the above experiments, FMS has good stability at different sampling rates. And the localization errors when facing different devices is within a small range. Therefore, this method can ensure a better localization effect in the case of limited sampling environment, and alleviate the impact of device heterogeneity.

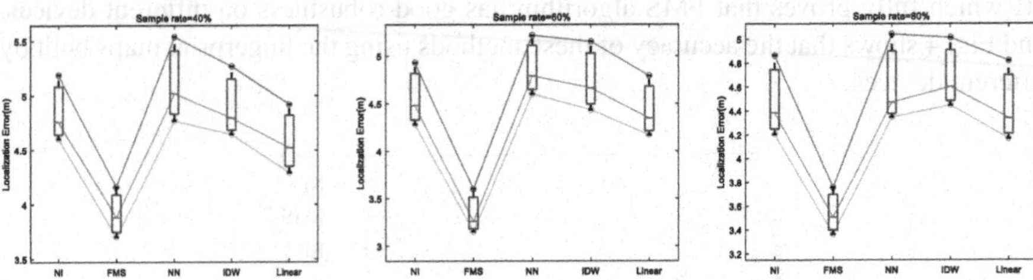

Fig. 5. Influence of sampling rates

5 Conclusions

Location-based services are playing an increasingly important role in daily life. However, the fingerprint location method requires the establishment of a regional map in advance. It requires collecting RSSs at a large number of RP positions. In particular, the fingerprint map needs to be updated regularly due to the time-shift of signal strength. These reasons make the workload huge. This paper proposes a method of finding missing sampling which filling incomplete and irregular large-scale map from multiple directions through constructing signal propagation model. According to experimental verification, the localization accuracy is improved by an average of 24.34% when the sampling rate is only 60% of the whole sampling. And the localization error of FMS is reduced by 11.35% ~ 24.43% than that of NI, and the localization error is improved by 14.48% ~ 24.75% compared with other interpolation methods under the condition of different sampling rates.

Acknowledgments. This work is supported by National Natural Science Foundation of China under Grant No. 61772448, Natural Science Foundation of Jiangsu Province under Grant No. BK20191481, Natural Science Major Project of the Higher Education Institutions of Jiangsu Province of China under Grant No. 17KJA520006, China Postdoctoral Science Foundation funded project under Grant No. 2019M660132, Jiangsu Province Postdoctoral Research Subsidy Program under Grant No. 2019K123, Project on the Industry-University-Research Cooperation of Jiangsu Province under Grant No. BY2021326, Scientific Research Subsidy of Jiangsu Province under Grant No. BRA2019288, Future Network Scientific Research Fund Project Grant No. FNSRFP-2021-YB-45, Jiangsu Provincial Key Constructive Laboratory for Big Data of Psychology under Grant No. 72591962002G and 72592162003G, Open Foundation of Jiangsu Provincial Key Laboratory of Coastal Wetland Bioresources and Environmental Protection under Grant No. JLCBE14008.

References

1. Zhao, L., Yang, K., Tan, Z., et al.: A novel cost optimization strategy for SDN-enabled UAV-assisted vehicular computation offloading. IEEE Trans. Intell. Transp. Syst. (T-ITS) (2020). https://doi.org/10.1109/TITS.2020.3024186
2. Zhao, L., Han, G., Li, Z., Shu, L.: Intelligent digital twin-based software-defined vehicular networks. IEEE Netw. (2020). https://doi.org/10.1109/MNET.011.1900587
3. Zhao, L., Li, H., Lin, N., Lin, M., Fan, C., Shi, J.: Intelligent content caching strategy in autonomous driving towards 6G. IEEE Trans. Intell. Transp. Syst. (T-ITS) (2021). https://doi.org/10.1109/TITS.2021.3114199
4. Wang, T., Luo, H., Zeng, X., Yu, Z., Liu, A., Sangaiah, A.K.: Mobility based trust evaluation for heterogeneous electric vehicles network in smart cities. IEEE Trans. Intell. Transp. Syst. **22**(3), 1797–1806 (2020)
5. Rempel, P., Borisov, A., Siemens, E., et al.: Local system of localization using a WiFi network. Matec Web Conf. **155**, 01014 (2018)
6. Wang, T., et al.: Propagation modeling and defending of a mobile sensor worm in wireless sensor and actuator networks. Sensors **17**(1), 139 (2017)
7. Tian Wang, M., Bhuiyan, Z.A., Wang, G., Qi, L., Jie, W., Hayajneh, T.: Preserving balance between privacy and data integrity in edge-assisted Internet of Things. IEEE Internet Things J. **7**(4), 2679–2689 (2019)
8. Liu, H., Darabi, H., Banerjee, P., et al.: Survey of wireless indoor localization techniques and systems. IEEE Trans. Syst. Man Cybern. Part C **37**(6), 1067–1080 (2007)
9. Alam, F., Chew, M.T., Wenge, T., et al.: An accurate visible light localization system using regenerated fingerprint database based on calibrated propagation model. IEEE Trans. Instrum. Meas. **68**(8), 2714–2723 (2019)
10. Wu, C., Xu, J., Zheng, Y., et al.: Gain without pain: accurate WiFi-based localization using fingerprint spatial gradient. Proc. ACM Interact. Mobile Wearable Ubiquit. Technol. **1**(2), 1–19 (2017)
11. He, S., Chan, S.: Towards crowdsourced signal map construction via implicit interaction of IoT devices. In: IEEE International Conference on Sensing, IEEE (2017)
12. Li, Z., Nika, A., Zhang, X., et al.: Identifying value in crowdsourced wireless signal measurements. In: Proceedings of the International Conference on World Wide Web, New York, NY, USA, ACM, pp. 607–616 (2017)
13. Ledlie, J., Park, J.G., Curtis, D., et al.: Mole: a scalable, user-generated WiFi localization engine. In: International Conference on Indoor Localization & Indoor Navigation, IEEE (2011)
14. Wu, C., Yang, Z., Liu, Y.: Smartphones based crowdsourcing for indoor localization. IEEE Trans. Mob. Comput. **14**(2), 444–457 (2015)
15. Chao, G., Harle, R.: Easing the survey burden: quantitative assessment of low-cost signal surveys for indoor localization. In: 2016 International Conference on Indoor Localization and Indoor Navigation (IPIN), IEEE (2016)
16. Jung, S.H., Moon, B.C., Han, D.: Unsupervised learning for crowdsourced indoor localization in wireless networks. IEEE Trans. Mob. Comput. **15**(11), 2892–2906 (2016)
17. Liu, C., Fang, D., Yang, Z., et al.: RSS distribution-based passive localization and its application in sensor networks. IEEE Trans. Wireless Commun. **15**(4), 2883–2895 (2016)
18. Wang, B., Zhou, S., Liu, W., et al.: Indoor localization based on curve fitting and location search using received signal strength. Ind Electron. IEEE Trans. **62**(1), 572–582 (2015)
19. Velazquez, I.S., Murillo-Fuentes, J.J., Djuric, P.M.: Recursive estimation of dynamic RSS fields based on crowdsourcing and Gaussian processes. IEEE Trans. Signal Process. **67**, 1152–1162 (2019)

20. Feng, Y., Gunnarsson, F.: Distributed recursive Gaussian processes for RSS map applied to target tracking. IEEE J. Sel. Top. Signal Process. **11**, 492–503 (2017)
21. Zhao, J., Gao, X., Wang, X., et al.: An efficient radio map updating algorithm based on K-means and Gaussian process regression. J. Navig. **71**(5), 1055–1068 (2018)
22. Ni, L.M., et al.: LANDMARC: indoor location sensing using active RFID. Wireless Netw. **10**, 407–415 (2004)
23. Zhao, Y., Liu, Y., Ni, L.M.: VIRE: active RFID-based localization using virtual reference elimination. In: 2007 International Conference on Parallel Processing (ICPP 2007), Xi'an, China, p. 56 (2007). https://doi.org/10.1109/ICPP.2007.84
24. Qi, J., Chen, B., Zhang, D.: A calibration method for enhancing robot accuracy through integration of kinematic model and spatial interpolation algorithm. J. Mech. Robot. **13**, 061013 (2021)
25. Hu, K., Yu, M., Liao, X.Y.: Research on improvement to WiFi fingerprint location algorithm. In: International Conference on Wireless Communications, IET (2015)
26. Mo, L., Li, C.: Passive UHF-RFID localization based on the similarity measurement of virtual reference tags. IEEE Trans. Instrum. Meas. **68**, 2926–2933 (2018)
27. Haniz, A., Kim, M., Takada, J.I., et al.: Application of geostatistical techniques for spatial interpolation of location fingerprints. In: 2014 XXXIth URSI General Assembly and Scientific Symposium (URSI GASS), IEEE (2014)
28. Kubota, R., Tagashira, S., Arakawa, Y., et al.: Efficient survey database construction using location fingerprinting interpolation. In: IEEE International Conference on Advanced Information Networking & Applications, IEEE (2013)
29. Guan, T., Wang, D., Su, Y.: Research on RFID virtual tag location algorithm based on Monte Carlo. In: 2021 IEEE 13th International Conference on Computer Research and Development (ICCRD), pp. 68–72 (2021)
30. Wu, Y.H., Chen, Y.L., Sheu, S.T.: Indoor location estimation using virtual fingerprint construction and zone-based remedy algorithm. In: International Conference on Communication Problem-solving, IEEE (2016)

Dynamic Adjustment Policy of Search Driver Matching Distance via Markov Decision Process

Suiming Guo, Pengcheng Zhang(✉), and Qianrong Shen

College of Information Science and Technology, Jinan University, Guangzhou 510000, China
guosuiming@email.jnu.edu.cn

Abstract. The past decade has witnessed the shift from calling taxis within sight to using applications to send location information which enables e-hailing platforms to match passengers to drivers out of sight. After receiving orders, e-hailing platforms search for a suitable driver to accept the order to complete the matching. At present, platforms mainly set a fixed matching distance according to passengers' locations for finding available drivers in this distance, but they do not consider if there are enough drivers available for their passengers. In order to find the properest driver which enables the decrease of passengers' waiting time (i.e., from the time the passenger places the order to the time the driver accepts the order), we propose a model to dynamically change matching distance. The purpose is to find the best matching distance in different supply (i.e., the number of drivers) and demand (i.e., the number of passengers) circumstances. We obtain the required probability information by extracting data from multiple datasets in different supply and demand regions. Later, a Markov Decision Process (MDP) model is established to solve the optimal dynamically changing matching distance policy. At the same time, we propose a reward function to measure the waiting time of passengers and the distance between the driver and the passenger boarding point. Finally, we compared MDP model with three benchmark models, the model of fixed matching distance, dynamic programming matching distance, and linear matching distance. Through comparison, it is concluded that the policy of dynamically changing matching distance based on the MDP model we proposed is best applied. The policy we proposed balance the waiting time of passengers and the distance between the driver and the passenger boarding point, improve platform efficiency.

Keywords: Markov decision process · Dynamic matching distance · Order dispatch · Urban computing · E-hailing

S. Guo and P. Zhang—Contribute equally to this work.

Y. Lai et al. (Eds.): ICA3PP 2021, LNCS 13155, pp. 319–335, 2022.
https://doi.org/10.1007/978-3-030-95384-3_21

1 Introduction

In recent years, online ride-hailing platforms, e.g., Didi and Uber have rapidly emerged, providing new ways for passengers to travel. The way that passengers hailing a taxi has changed from waving to taxis in view to sending a request to the e-hailing platform, and waiting for its assignment for a driver. However, for e-hailing platforms, how to determine the maximum matching distance when helping passengers search for nearby drivers is needed to be solved.

Currently, the platform mainly searches for drivers within a fixed maximum matching distance. Former researchers mainly focused on the determination of maximum fixed matching distance [12], adjustment for the maximum fixed matching distance in consideration of supply and demand [17] and adjustment for fixed matching distance when ride-pooling [9]. However, a fixed matching distance may lead to some problems. If the distance is too small and there are few nearby drivers, it is difficult to match the driver and the waiting time for passengers is prolonged. When the distance is too long, the order dispatch algorithm [10,15,20] costs much time and performance, meanwhile the matched driver is too far from the passenger. A fixed matching distance limits adjustment for the matching distance according to real-time supply and demand, resulting in the inability to reduce the waiting time of passengers in different spatio-temporal. In order to solve the problems caused by its limitations, some scholars proposed that dynamically changing distance may be helpful in searching for drivers [5], but they did not consider the supply and demand.

We considered the supply and demand, and adopted the MDP model to solve a problem, i.e. how to dynamically change the matching distance. Nowadays, MDP has been well applied in many fields, for example, applying MDP in road recommendation for taxis drivers [11,18]. We took two key factors of this problem into consideration: passenger waiting time and matching distance. A dynamic adjustment for the matching distance is necessary to balance these two factors. When there are many drivers around the passenger, the matching distance increases slowly, and the passenger can quickly match a driver who is not far from him/her. When there are few drivers around the passenger, the matching distance increases quickly, and the driver can be quickly matched, which reduces the probability of cancellation [13,14]. We evaluated the MDP-based dynamic matching distance policy, fixed matching distance policy and other matching distance policies, and founded that the policy based on the MDP model is more effective.

Our contributions are as follows:

- A dynamic growth method of matching distance is proposed to balance the waiting time of passengers and the distance from the driver to the boarding point, which bring more benefits to the platform.
- An MDP model is designed to be a solution to the problem about dynamically changing matching distance. Hence a dynamically changing optimal matching distance can be made under different supply and demand conditions.

The later part of this paper is structured as follows, Sect. 2 explains the process of data analysis. Section 3 provides a detailed description of the decision process of MDP model and the way to work out the optimal policy. Section 4 is an evaluation of different policies. Section 5 we summarized the entire experimental process and gave future research directions.

2 Data Analysis

This paper selected the order data of local areas in Chengdu from Didi Chuxing GAIA Initiative dataset [4] and made an analysis. The order data includes order cancellation rate, empty car transfer rate and hexagonal grid data. In this experiment, the order data of one day was selected as the experimental data set, because orders of one day were more than 200,000, which was enough to fulfill the experimental requirements. We analyzed the data that were needed to establish the MDP model of Sect. 3, and determine the probability of transition from the current state to next state under different spatio-temporal conditions.

2.1 Hot Spots and Not-hot Spots

We need to find out hot spots and non-hot spots, and use data from a time period with a large number of orders, so that stability can be ensured with enough data support. A fixed matching distance cannot be changed in different regions after it has been set. In contrast, dynamic matching distance is more flexible. Adopting dynamic matching distance means there exists variable policies in line with different regions.

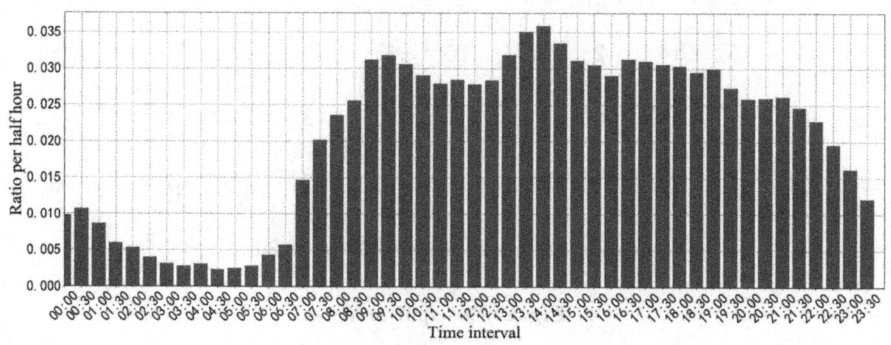

Fig. 1. Histogram of percentage of orders per half hour.

The order data set includes *order_id*, *timestamp* at the beginning and the end, *longitude* and *latitude* of boarding point and destination. We divide the data size every half hour into a group to clearly show the proportion of orders from each time period to total orders. It can be concluded from Fig. 1 that the peak order volume occurs in three time periods: (09:00∼10:30), (13:00∼15:30), (16:30∼18:30).

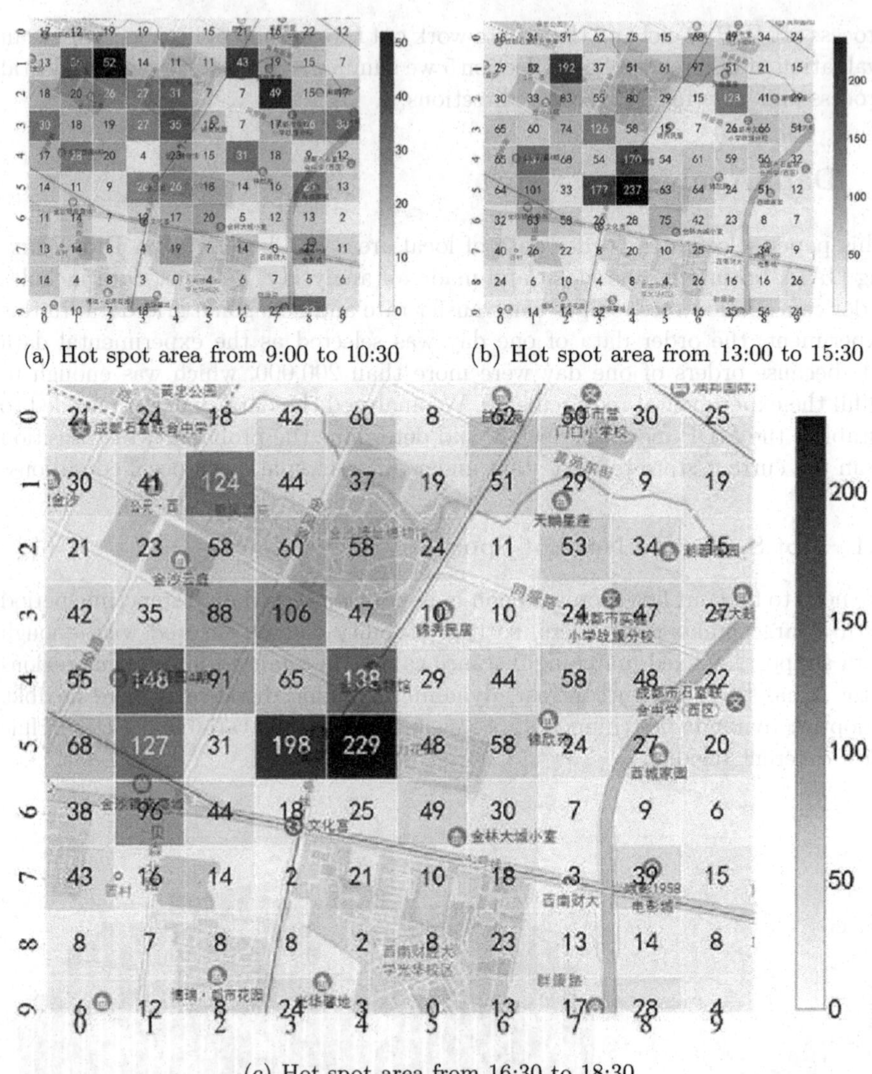

(a) Hot spot area from 9:00 to 10:30

(b) Hot spot area from 13:00 to 15:30

(c) Hot spot area from 16:30 to 18:30

Fig. 2. Hot spot area

In order to accurately locate hot spots and find the order information in the hot spots and non-hot area, we delineate the map area as a rectangular area on the basis of the location of all users when they place an order: longitude (103.597825064∼104.484519464), latitude (30.3196650639∼31.0151659839). In order to correspond to the cancellation rate of the cancellation rate data set of Sect 2.2 with 200 m intervals, the rectangle is divided into 320 × 251 small grids. Each grid spacing is 0.00277092°, approximately equals to 200 m in reality, The calculation result of the order quantity in each grid is shown in the heat map Fig. 2. The graph shows the volume of orders in some areas in different time periods, and the volume of orders increases as the color darkens.

Since the hot area represents the area with a large number of orders, we set the top 100 small grid areas as hot spots. At the same time, the non-hot area corresponds to the place where the number of orders is low. We set the area where the order volume is lower than the average of the total order volume as the non-hot area.

2.2 Cancel Rate

Passengers may cancel the order due to waiting time, dynamic price changes [6–8], psychological cost and other reasons within the waiting time. The existing order cancellation rate data set contains *order_id* and the corresponding cancellation probability when the distance between the passenger and the driver is 200 m, 400 m, 600 m ... 2000 m. Since the order cancellation rate data set and the order data set contain the same *order_id*, it is possible to figure the average cancellation probability (i.e., P_{canc}) of hot spots and non-hot spots in different time periods, which equals the order cancellation rate P_{canc} of passengers in the two regions. Figure 3 shows the order cancellation rate, which increases as the matching distance increases when passengers are in two spots in different time

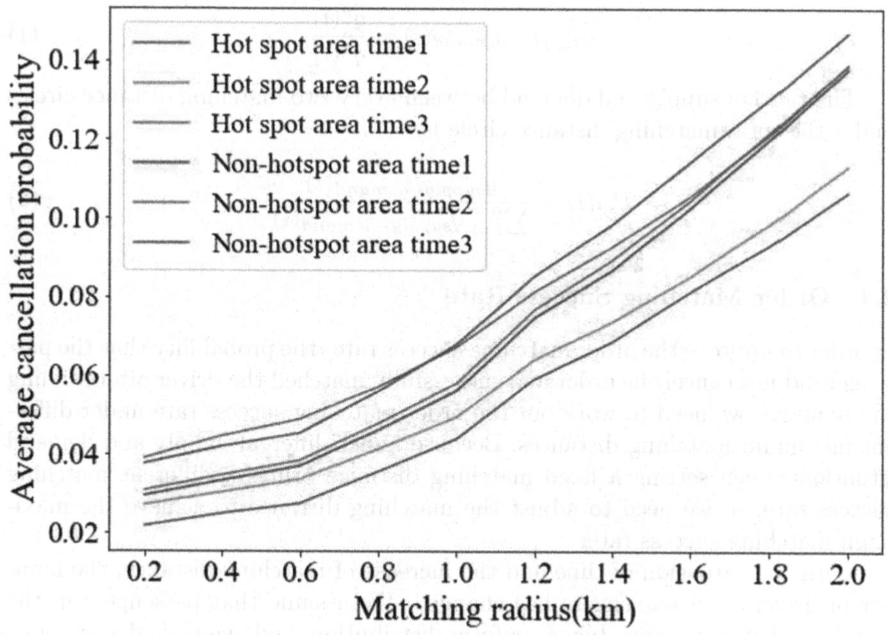

Fig. 3. Average cancellation probability.

periods. This proves that when adopting a fixed matching distance, passengers waiting too long and matching drivers are too far away have a negative impact on the order cancellation rate.

2.3 Supply and Demand Ratio

A fixed matching distance does not allow the platform to adjust the matching distance according to supply and demand. However, the use of dynamic matching distance can make up for this deficiency and reduce the matching time. We need to command the supply and demand around a certain passenger, so the GPS trajectory data set is chosen, which includes *driver_id*, *order_id*, *timestamp*, *longitude*, *latitude*. These data reflect every path from the boarding point to the destination. First, we filtered out the first track data of the same order within a period of time in the GPS track. It refers to the number of pick up points and approximates to passenger demand. Secondly, based on the trajectory data, we determined the number of drivers appearing in each grid over a period of time, which is similar to the supply. We assume that 200 m is a unit when changing matching distance and calculate the quantity of drivers (i.e., q^d) between two matching distance circles (i.e., the passenger is the center of the circle and the matching distance is the radius) n × 200 m and (n + 1) × 200 m and passenger order quantity (i.e., q^p).

Hence the proportion of demand to supply between every two matching distance circles is shown as:

$$q_{supply-demand}(i) = \frac{q^p(i)}{q^d(i)} \tag{1}$$

The ratio of supply and demand between every two matching distance circles under the total matching distance circle is:

$$P_{sd}(i) = \frac{q_{supply-demand}(i)}{\sum_{i=1}^{n} q_{supply-demand}(i)} \tag{2}$$

2.4 Order Matching Success Rate

In order to improve the order matching success rate (the probability that the passenger did not cancel the order and successfully matched the driver after sending the request), we need to work out the order matching success rate under different maximum matching distances. Because under different supply and demand situations, even setting a fixed matching distance brings a different matching success rate, so we need to adjust the matching distance to achieve the maximum matching success rate.

With the extension of time and the increase of matching distance, the number of drivers and passengers has changed. We assume that passengers in the matching distance circle obey a uniform distribution, and available drivers obey a Spatial Poisson Point Process [3]. We assume that the area of the matching distance circle is S, the densities of demand and supply denoted by $\hat{\rho}^d$ and $\hat{\rho}^s$, can be expressed as follows:

$$\hat{\rho}^d = \frac{q^d}{S} \tag{3}$$

$$\hat{\rho}^s = \frac{q^s}{S} \tag{4}$$

The area of each passenger's dominant zone equals$(\hat{\rho}^d)^{-1} = S/q^d$. When the matching distance is d, the range S_{match} of the nearby drivers the passenger can match is expressed as:

$$S_{match} = min((\hat{\rho}^d)^{-1}, \pi d^2) \tag{5}$$

Therefore, the probability that n idle drivers are within the matching area of each waiting passenger can be written as follows [1,16]:

$$P\{n\} = \frac{1}{n!} exp(-S_{match}\hat{\rho}^s)(S_{match}\hat{\rho}^s)^n \tag{6}$$

Yang et al. [17] proposed the calculation method of the matching success rate P_{match} for drivers and passengers belonging to different matching time intervals and different matching distances as follows:

$$P_{match} = 1 - P\{0\} = 1 - exp(-S_{match}\hat{\rho}^s) = 1 - exp(-\hat{\rho}^d min((\hat{\rho}^d)^{-1}, \pi d^2)) \tag{7}$$

3 Markov Decision Process

We have established an MDP model to solve the problem about how to dynamically change matching distance. Table. 1 shows all the parameters required by the MDP model.

3.1 Status and Action

We set the state in the MDP model as s = (I,d,t), containing I, d, t:

- $I \in \{- -1, 0, 1\}$
- d ∈ { 0 : 200 m, 1 : 400 m, 2 : 600 m, 3 : 800 m, 4 : 1000 m, 5 : 1200 m, 6 : 1400 m, 7 : 1600 m, 8 : 1800 m, 9 : 2000 m }
- t ∈ { 0 : 0 s, 1 : 20 s, 2 : 40 s, 3 : 60 s, 4 : 80 s, 5 : 100 s, 6 : 120 s, 7 : 140 s, 8 : 160 s, 9 : 180 s }

Use I to represent the state of the passenger at next time node. $I = -1$ means that the passenger canceled the order, for waiting too long or other reasons. $I = 0$ means that the passenger continues to wait for a driver, and $I = 1$ means that the passenger has matched a driver.

Table 1. Key notations.

Variable	Explanation
I	Indicator, Indicates the current state
D	Distances, The distance of the current passenger
Δd	Distances delta default is 200 m
D	Maximum matching distance
T	Maximum waiting time is 180 s
T	Waiting time
Δt	Time delta default is 20 s
S	State
S	State space
A	Action
A	Action space
P_{match}	The probability of passenger match the driver
P_{canc}	The probability of passenger cancel the order request
P_{wait}	The probability of passenger waiting for the order match
P_{jump}	The probability of passenger jump from one match state to another
α	A factor that balances waiting time and matching distance

Use d to represent the matching distance. For example, {0 : 200 m} means d = 0, showing that the linear distance between the passenger and nearby drivers is 200 m. According to the maximum matching distance mentioned in many articles and the result of data analysis in Chapter two, we assume that the maximum straight line matching distance is 2000 m, hence when the matching distance reaches 2000 m, the searching range will stop expanding. We assume a maximum expansion range, not only to prevent the extension of waiting time and the increased cost of the driver to the boarding point due to the excessive distance, but also because the matching result will become irrelevant to the matching distance when the expansion range reaches a threshold [17]. The matching distance ranges from 200 m to 2000 m. We divide it into 10 different matching distances. The minimum interval Δd of each matching distance is 200 m.

Research has shown that the average time spent on order delivery is 89 s [19]. In order to avoid endless waiting of passengers, we assume that the longest waiting time is T = 180 s, and the platform must match both sides or cancel the order within T. This parameter is acceptable for passengers. We divide the waiting time into 10 equal parts, and each time interval Δt is 20 s. Passenger waiting time is signified as t, e.g., {3 : 60 s} means t = 3 to indicate that the passenger has been waiting for 60 s.

The initial state set t = 0, which means that the matching distance is at the default state d = 0. For example, s = {0, 2, 1} means that the passenger has spent 20 s waiting for the order to match, and the matching distance is 600 m.

We define Action space as A $\in$ { 1 : 200 m, 2 : 400 m, 3 : 600 m, 4 : 800 m, 5 : 1000 m, 6 : 1200 m, 7 : 1400 m, 8 : 1600 m, 9 : 1800 m }. The action { 1 : 200 m } means that when a = 1, the executing action is to extend the matching distance by 200 m.

3.2 State Transition

Table 2. Status transfer matrix.

State/Action	1	2	3	4	5	6	7	8	9
0	1	2	3	4	5	6	7	8	9
1	2	3	4	5	6	7	8	9	
2	3	4	5	6	7	8	9		
3	4	5	6	7	8	9			
4	5	6	7	8	9				
5	6	7	8	9					
6	7	8	9						
7	8	9							
8	9								
9									

When a passenger sends a request, there are three states that can be jumped to: cancel the order, wait for the order, and match successfully. The sum of these three probabilities is 1. According to the P_{canc} and P_{match} obtained from the previous data analysis, the probability of waiting for an order can be calculated as:

$$P_{wait}(j) = 1 - P_{canc}(j) - P_{match}(j) \tag{8}$$

When $I = 1$ or $I = -1$, it will stop jumping, but it will continue to jump to the next state when $I = 0$. The probability of jumping from waiting state to the next state is closely related to the ratio of supply and demand under different matching distances. If the supply exceeds the demand in the current matching distance circle, the probability of matching the driver in the current matching distance circle is relatively large, and the probability of jumping to the next matching distance circle is relatively small. Thus, there is no need to jump too fast, and the right driver can be matched at close range. On the contrary, if passengers' demands are more than the supply of drivers, the speed of jumping to the next matching distance should be increased, because sufficient supply is needed to fulfill the demands of passengers. Hence it is necessary to jump quickly to the matching distance that have sufficient supply. We assume that the transition probability has a linear relationship with the supply-demand

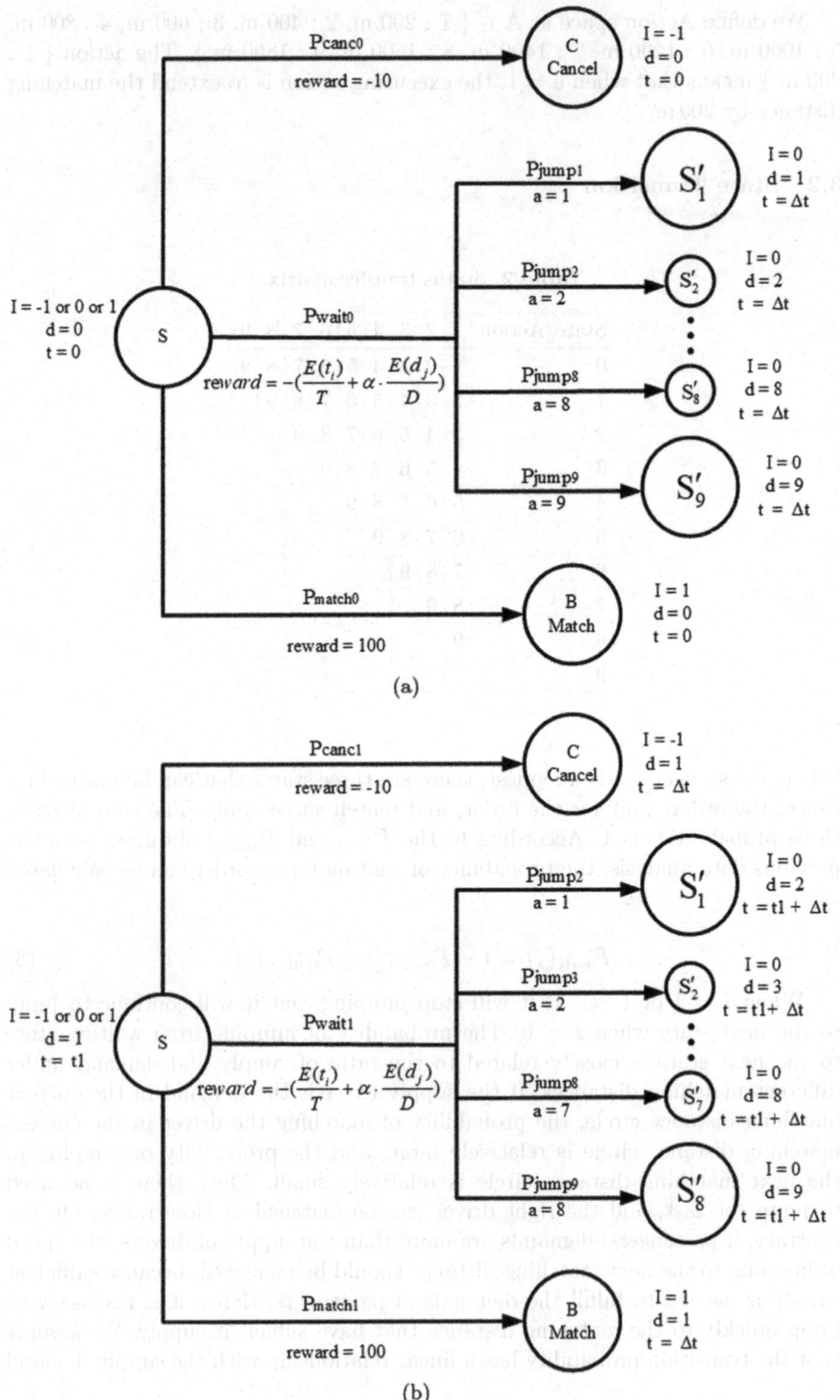

Fig. 4. State transfer diagram

ratio in different matching distance circles, The jumping probability from the state s = (0,0,0) to the next state $s = (I, 1, t), s = (I, 2, t) \ldots$ s = (I,8,t) ($I \in \{-1, 0, 1\}, t \in T$) or s = (I,9,t) ($I \in \{-1, 1\}, t \in T$) is:

$$P_{jump}(i) = P_{wait}(j) * P_{sd}(i) = (1 - P_{canc}(j) - P_{match}(j)) * P_{sd}(i) \tag{9}$$

During state transition, different states correspond to different action ranges. The state transition Table 2 shows all the specified states, which may help readers better understand the state transition process. Figure 4 provides the specific state transition process. Figure 4(a) shows that when the initial test state $s = (I, 0, 0)$, d = 0, it means that the initial matching distance range is 200 m, and the probability of canceling the order is P_{canc0}. If the order is cancelled, the reward will be −10, meanwhile set the state indicator $I = -1$ to indicate the cancellation. Nonetheless the test also has the probability of P_{match0} to make a successful match, hence the reward will be 100, and setting the state as $I = 1$ to indicate a successful match. Besides, another kind of probability is to continue to wait for a match between the passenger and the driver, then the action that can be performed is a = (1,2,3,4,5,6,7,8,9). The time it takes to jump from the state $s = (I, 0, 0)$ to other states is t = 1. When a = 1 is executed, d = 1 occurs in the next state, meanwhile the state can be reflected in $s_1' = (0, 1, 1)$.

When jumping to the state of $s = (I, 1, t)$ as shown in Fig. 4(b), the matching distance is d = 1, and t = 1 is the time taken to jump from the previous state. There are three types of jumping. The order may have a possibility of P_{canc1} to be cancelled, and the state indicator is set to $I = -1$; it may also have a possibility of P_{match1} to be finished, so the state indicator is $I = 1$, the matching distance is d = 1; there is still another kind of possibility P_{wait1} that the passenger needs to wait for more time. When probability of continuing to wait is P_{wait1}, after time t = t1 + Δt, there is a possibility of P_{jump2} to jump from action a = 1 to the state $s_1' = (0, 2, t)$, at this very moment the matching distance is d = 2. The jumping process in other states is similar to the above process.

It should be noted that the available states for each state to jump to are limited. For example, when $s = (I, 6, t)$, only three actions a = (1,2,3) can be executed. When a=3 is executed, the original d = 6, d = d + a is updated. d = 9 reaches the limited maximum matching distance. When reaching the state $s = (I, 9, t)$, considering that endlessly waiting for the driver to take orders is nearly impossible in reality, jumping is limited to two types of $I = -1 or I = 1$ in the state of $s = (I, 9, t)$, so $P_{wait} = 0$, and $P_{match} + P_{canc} = 1$.

3.3 Reward

From the data analysis Sect. 2.2 and article [14], it is concluded that the cancellation rate largely depends on two key factors, namely the time that passengers wait for orders and the matching distance. There are papers [17] that prove the optimization effect of these two factors on the matching system. In order to balance these two factors to reduce the cancellation rate of passengers, so that passengers can be matched to suitable drivers around more quickly, We assume a functional function:

$$reward = -(\frac{E(t_i)}{T} + \alpha \cdot \frac{E(d_j)}{D}) \tag{10}$$

$E(t_i)$ is the expectation of the time required for the current passenger's state. For example, when the passenger is in the state s = (0,5,2), the jump process is $s = (0,0,0) \longrightarrow (0,1,1) \longrightarrow (0,5,2)$. The corresponding time expectation is $E(t_2) = P_{jump1} \times 20 + P_{jump5} \times 20$, which is the expectation of the corresponding time spent in the jump process. Expectation increases as the waiting time prolongs, thereupon $E(t_i)/T$ will be larger. The reward value is negative while continuously waiting. The greater is the expectation corresponding to the time, the larger is the negative value of the reward. As for $E(d_j)$, it represents the expectation of the distance of the current passenger's state. Multiply the corresponding distance by the probability of jumping. The equation is $E(d_j) = 200 \times P_{jump1} + 200 \times P_{jump5}$. If the distance is longer, the expectation will be higher, and $E(d_j)/D$ will be greater. Correspondingly, the negative reward value will be greater, which means the punishment is greater.

At the same time, an α influence factor is also set to adjust the degree of influence on the waiting time and matching distance. It is found that the properest range for setting α is (0.5, 1.5). The reward function is aimed at working out the reward value when the passenger is still waiting for a successful match. When a passenger cancels his/her order, we set $reward = -10$ to indicate a punishment. In order to match passengers and drivers as quickly as possible, we set $reward = 100$ to indicate the reward for a successful order match.

3.4 Solving MDP

Solve the MDP to get the best policy to match the dynamic change of the distance. However, it is difficult to solve various problems (such as complex state space and difficulty in comprehensive display of probabilistic models) in practical applications by directly solving Bellman optimal equation [2]. Therefore, we can find the approximate optimal solution of this problem through policy iteration.

The state value function is represented by $v_\pi(s)$, which means the expected reward obtained through the policy π in the state s. The action value function is represented by $q_\pi(s,a)$, meaning that the expected return obtained through the policy π after executing the action a in the state s. When continuing to wait, $I = 0$. executing the jump action a at present, we can get:

$$q_\pi(s,a) = \sum_{s'} P_{wait} \times [-(\frac{E(t)}{T} + \alpha \cdot \frac{E(d)}{D}) + P_{jump} \times V^*(s')], \tag{11}$$

$$s' \in S, a \in A, t \in T, d \in D$$

When $I = -1$ or 1, it means that the action has been executed, the match is cancelled or succeeds. Hence the state is expressed as $C = (-1,d,t), B = (1,d,t)$.

$$q_\pi(s,.) = \sum_{d \in D} [-10 \times P_{canc} \times V^*(C) + 100 \times P_{match} \times V^*(B)] \tag{12}$$

Choose an optimal policy:

$$\pi(s) = argmax_a[q_\pi(s,a)] \tag{13}$$

Then the optimal state value function can be expressed as:

$$V^*(s) = \begin{cases} max_a q_\pi(s,a) & \text{if s is a wait state} \\ q_\pi(s,.) & \text{if s is a non-wait state} \end{cases} \tag{14}$$

Under certain dynamic space conditions, namely, when MDP model is used in a finite state, the policies obtained are limited. At this time, the action value function and the state value function can be updated to iterate, and the final convergence can be obtained by continuous iteration [21].

$$\pi_0 \xrightarrow{E} v_{\pi 0}, q_{\pi 0} \xrightarrow{I} \pi_1 \xrightarrow{E} v_{\pi 1}, q_{\pi 1} \xrightarrow{I} \pi_2 \xrightarrow{E} \cdots \tag{15}$$

Here, $\xrightarrow{E}$ means policy evaluation, and $\xrightarrow{I}$ means policy improvement.

4 Experiment

4.1 Benchmark

In order to prove that the performance of the MDP model is better than others, we designed three other comparison policies, namely, the fixed matching distance policy, the linear growth policy and the dynamic programming growth policy.

When adopting the fixed matching distance policy, the matching distance is set to 2 km in the initial state and remains unchanged.

When adopting the linear growth policy, once the time interval Δt is passed, it will jump to the next adjacent state, and continue to the state $s = (I, 9, 9)$. The state transition process is expressed as follows:

$$s = (0,0,0) \rightarrow (0,1,1) \rightarrow (0,2,2) \rightarrow \ldots \rightarrow (0,7,7) \rightarrow (0,8,8) \rightarrow (I,9,9) \tag{16}$$

At present, the latest relevant model uses a dynamic programming policy [5] to determine the probability of a jump. The current state of each step is determined by the state of the previous step. The probability of the current jump is $dp[i][j]$, the functional function is $f(i,j) = \frac{E(t_i)}{T} + \alpha \cdot \frac{E(d_j)}{D}$

$$dp[i][j] = min\{dp[i-1][j-k] + f(j-k,j), dp[i][j]\} \tag{17}$$

4.2 Adjusting the α Value

The optimal policy in different spatio-temporal can be obtained by modifying the different α in the reward. Figure 5 reflects the reward value when the α value is adjusted between 0.5 and 1.5 in hotspots and non-hot spots.

Figure 5(a) represents the reward value in time period 1 (09:00~10:30), and Fig. 5(b) represents the reward value in time period 2 (13:00~15:30), Fig. 5(c)

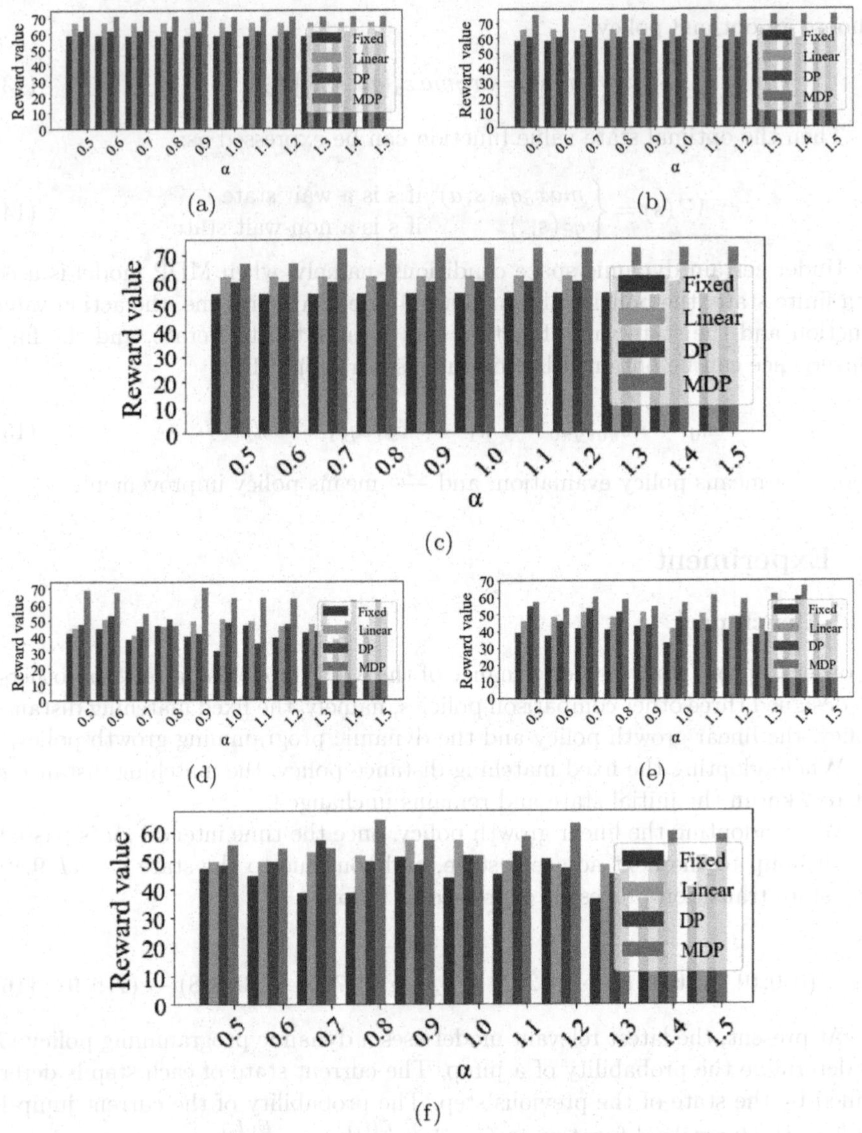

Fig. 5. Reward values for different α's in areas

represents the reward value in time period 3 (16:30∼18:30). As it is shown in the figure, although taking different α every time, reward values are slightly different from each other. The result proves that the change of α have a negligible impact on the policy in hotspots.

Figure 5(d) represents the reward value in the time period 1 (09:00∼10:30). It is clear that when $\alpha = 0.9$, the policy obtained by MDP model has the best effect. Figure 5(e) represents the reward value in time period 2 (13:00∼15:30), and the effect is best when $\alpha = 1.4$. Figure 5(f) represents the reward value in time period 3 (16:30∼18:30), and the best effect is when $\alpha = 0.8$. According to these results, we can draw a conclusion that when using MDP model, the optimal policy in accordance with different time period requires different α. The difference between tests in non-hot spots and hot spots is that the adjustment of the α value in different time periods will have a greater impact on the reward value in non-hot spots. When adjusting the α in the reward in the future, it is necessary to predict an optimal α value based on historical data, which can maximize the reward value under different spatio-temporal conditions.

4.3 Evaluation

It can be seen from Fig. 5 that in hotspots, the dynamic matching distance policy based on the MDP model is much better than other three benchmark policies. In non-hot spots, the optimal policy under different spatio-temporal conditions can be obtained by adjusting α, and the result is also better than other three benchmark policies. The effect of fixed matching distance policy is very poor, the effect of linear growth policy is average, and the effect of dynamic programming policy is not very stable. However, the dynamic matching distance policy obtained by the MDP model has better effect and stability. The best policy we find can better meet the two key factors, namely passenger waiting time and matching distance.

5 Conclusion

We established an MDP model to make the best decision on the dynamic change matching distance under different spatio-temporal conditions. It balances passengers waiting time and matching distance. By comparing the three benchmark policies, adjusting the α value shows that the matching distance dynamic change policy based on the MDP model is better than other policies. In the future, this model can be combined with specific passenger-driver pair matching algorithms (e.g., the best bipartite graph and deep learning). In this way, when operating policy of dynamically changing matching distance, it is possible to calculate the optimal one among multiple drivers that passengers can match at the same time. We can also combine more detailed historical data to predict changes in supply and demand in a short period of time in the future, so as to make the best decision for the dynamic changes of each passenger's matching distance.

Acknowledgement. This work was supported by the National Natural Science Foundation of China (62002135), the Fundamental Research Funds for the Central Universities (11619310) and Science and Technology Projects in Guangzhou (202102021164).

References

1. Arnott, R.: Taxi travel should be subsidized. J. Urban Econ. **40**(3), 316–333 (1996)
2. Bellman, R.: A Markovian decision process. Indiana Univ. Math. J. **6**(4), 679–684 (1957)
3. Stoyan, D., Kendall, W.S., Chiu, S.N., Mecke, J.: Stochastic Geometry and its Applications. Wiley, Hoboken (2013)
4. Data source: Didi Chuxing GAIA Initiative.: Data source: Didi Chuxing GAIA Initiative. https://outreach.didichuxing.com/research/opendata/en/
5. Duan, Y., et al.: Optimizing order dispatch for ride-sharing systems. In: Proceedings of International Conference on Computer Communication and Networks, ICCCN. 2019-July (2019)
6. Guo, S., et al.: A simple but quantifiable approach to dynamic price prediction in ride-on-demand services leveraging multi-source urban data. In: Proceedings of ACM Interactive, Mobile, Wearable Ubiquitous Technologies, vol. 2, no. 3, pp. 1–24 (2018)
7. Guo, S., et al.: Modelling passengers reaction to dynamic prices in ride-on-demand services. In: Proceedings of ACM Interactive, Mobile, Wearable Ubiquitous Technologies, vol. 1, no. 4, pp. 1–23 (2017)
8. Guo, S., et al.: ROD-revenue: seeking strategies analysis and revenue prediction in ride-on-demand service using multi-source urban data. IEEE Trans. Mob. Comput. **19**(9), 2202–2220 (2019)
9. Ke, J., et al.: Data-driven analysis on matching probability, routing distance and detour distance in ride-pooling services. Transp. Res. Part C Emerg. Technol. **124**, 102922 (2021)
10. Qin, Z., et al.: Ride-hailing order dispatching at DiDi via reinforcement learning. Interfaces (Providence) **50**(5), 272–286 (2020)
11. Shou, Z., et al.: Optimal passenger-seeking policies on e-hailing platforms using Markov decision process and imitation learning. Transp. Res. Part C Emerg. Technol. **111**, 91–113 (2019)
12. Sun, L., et al.: Taxi-hailing platforms: inform or assign drivers? Transp. Res. Part B Methodol. **142**(3), 197–212 (2020)
13. Wang, G., et al.: On-demand ride-matching in a spatial model with abandonment and cancellation. SSRN Electron. J. **3414716**, 1–40 (2019)
14. Wang, X. et al.: Customer behavioural modelling of order cancellation in coupled ride-sourcing and taxi markets. Transp. Res. Part B Methodol. **132**, 358–378 (2019)
15. Xu, Z., et al.: Large-scale order dispatch in on-demand ride-hailing platforms: a learning and planning approach. In: Proceedings of the ACM SIGKDD International Conference on Knowledge Discovery Data Mining, pp. 905–913 (2018)
16. Xu, Z., et al.: Optimal parking provision for ride-sourcing services. Transp. Res. Part B Methodol. **105**, 559–578 (2017)
17. Yang, H., et al.: Optimizing matching time interval and matching radius in on-demand ride-sourcing markets. Transp. Res. Part B Methodol. **131**, 84–105 (2019)
18. Yu, X., et al.: A Markov decision process approach to vacant taxi routing with e-hailing. Transp. Res. Part B Methodol. **121**, 114–134 (2019)
19. Zhang, L. et al.: A taxi order dispatch model based on combinatorial optimization. In: Proceedings of the ACM SIGKDD International Conference on Knowledge Discovery and Data Mining, Part F1296, pp. 2151–2159 (2017)

20. Zheng, L., et al.: Order dispatch in priceaware ridesharing. Proc. VLDB Endow. **11**(8), 853–865 (2018)
21. Xiao, Z.: Reinforcement Learning: Theory and Python Implementation. China Machine Press, Beijing (2019)

A Multi-precision Quantized Super-Resolution Model Framework

Jingyu Liu, Dunbo Zhang, Qiong Wang, and Li Shen(✉)

School of Computer, National University of Defense Technology, Changsha 410073, Hunan, China
lishen@nudt.edu.cn

Abstract. Equipment's computing capability has been greatly enhanced at present, which helps deep learning achieve excellent results in various applications, such as super-resolution. However, for higher performance, lower model size and faster computing speed, model compression is widely applied to accomplish the goal. For instance, model quantization is a typical compression method, such as quantization aware training and etc. Quantization aware training can take more quantization loss due to data mapping in model training into account, clamping and approximating the data representation range when updating parameters, which introduces quantization errors into loss function. In the quantization process, we used a quantization strategy that we quantized the model in different stages of combination, and found that some stages of the two super-resolution models' generators based on SRGAN and ESRGAN showed sensitivity to quantization during the process, which greatly reduced the performance. Therefore, according to the quantization sensitivity, we use higher bits integer quantization for the sensitive stage, and get the multi-precision quantized model. For quantizing the SR model automatically, we propose a multi-precision quantization framework in this paper according to the ratio of input and output channels in every stage in the model. We also have our work tested on eight classical data sets of super-resolution. Generally speaking, both the two models' PI values approach the original model's respectively.

Keywords: Framework · Super-resolution model quantization · Quantization aware training · Quantization sensitivity

1 Introduction

Deep learning [1] has been proven to be powerful on tasks including image classification, objection detection, natural language processing and so on. Super-resolution [2–8] is one of the important applications of deep learning in computer vision. Its main function is to improve the clarity of enlarged images and reduce the degradation of image quality caused by image enlargement. From simple mathematical methods to methods based on deep learning, such as SRCNN proposed by Dong et al. [3], SRGAN proposed by Ledig et al. [2] and ESRGAN

Y. Lai et al. (Eds.): ICA3PP 2021, LNCS 13155, pp. 336–353, 2022.
https://doi.org/10.1007/978-3-030-95384-3_22

proposed by Wang et al. [4], the performance of super-resolution reconstruction is constantly improving.

With the fast development of neural network research and application, people want more and more accurate predication, and networks grows deeper and deeper. The memory size of network becomes a problem. The model size is not only a memory usage problem, it's also a memory bandwidth problem. Therefore, the size of the network model becomes one of the main concerns of researchers, especially on especially in an environment with limited resource or power consumption.

Model quantization [9–27], as a means of compressing model, can be applied to model deployment, so that both the model size and the inference delay can be reduced. At present, the sizes of SR models become larger and larger. For instance, a common SRGAN model is about 16 MB in size and has 259G MACs (Multiply-Accumulate Operations), while a common ESRGAN model is about 32 MB in size and 1804G MACs. Figure 1 lists some models and their model size and MACs. Therefore, many researches focus on the methods to reduce model sizes. Quantization is one of the most effective approaches at present, so it has attracted extensive attentions of researchers. Its main idea is to map a data type with a wider representation range and a larger storage space to another data type with a more narrow representation range and a smaller storage space, and therefore reduce the model size and the time overheads. For example, a mapping from high-precision floating point data type to a low-precision one, or from floating point to integer, etc. When a model is quantized, the mapping process inevitably introduces some information loss, and the accuracy of the result model will be reduced accordingly. Therefore, quantization technology will generally be used with other methods together to ensure that the loss of accuracy is as small as possible while effectively reducing the size of model and MACs.

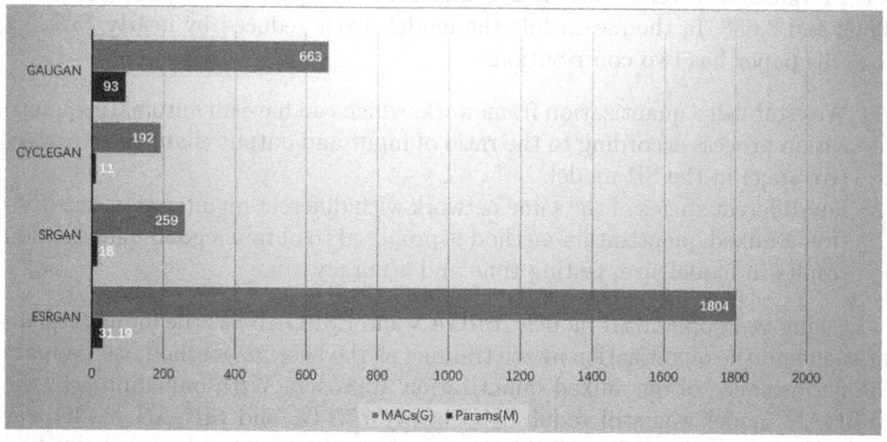

Fig. 1. The model size and calculation amount of super-resolution using deep learning.

However, current model quantization approaches usually have an important drawback, which limits its effectiveness greatly. Existing quantization methods mainly focus on the reduction of model size while ignoring its impact on the model performance (i.e., accuracy). At many cases, quantization effectively compresses the model and shrinks the inference time, but the accuracy of the result also decreased a lot. For example, if simply employing 8-bit integer to replace single-precision floating-point weights, EDSR [7] model will save 73% capacity and get a 43% performance acceleration, but the accuracy will be decreased by 53%. There are some reasons for so much accuracy loss, such as too much information are lost during quantizing. However, for most cases, this is caused because an existing method usually uses a unique quantization strategy to process all stages in the network, ignoring the sensitivity of different stages to data types and results accuracy. For example, for SRGAN and ESRGAN model's basic blocks and upsampling stage, quantization can reduce the size of the model with little effect on accuracy. However, for other stages, the accuracy will decreases rapidly as the size of the model decreases. The detailed statistical results are shown in Fig. 10.

Aiming at the above problems, this paper takes SR model as an example to evaluate the sensitivity of SRGAN and ESRGAN models with quantization aware training at each stage, and identifies the stage with the highest sensitivity. In addition, a mixed quantization method is proposed to obtain a comprehensive quantization results with small model size, short test time and almost unchanged accuracy. In this paper, quantization aware training is selected as the baseline approach when quantizing SR models. The two quantization methods, static post-training quantization and quantization aware training, are used to test the generator part of the SRGAN and ESRGAN models, which ensures the quantization method from the PI value. It is found that the effect of post-training static quantization is far inferior to quantization aware training. Among them, the PI values of SRGAN are 4.6278 and 2.4731 respectively, and ESRGAN is 4.562 and 2.688. In the meanwhile the model size is reduced by nearly 75%.

This paper has two contributions:

(1) We establish a quantization framework, which can have an automatic quantization process according to the ratio of input and output channels of respective stage in the SR model.
(2) For different stages of the same network with different quantization sensitivities, a mixed quantization method is proposed to obtain a good quantization results in model size, testing time and accuracy.

Taking two popular SR models (SRGAN and ESRGAN) as the input of quantization and the quantization aware training as the baseline method, we evaluate the performance of our mixed quantization approach. With our approach, the ESRGAN model was still reduced by nearly 67.14% and SRGAN model was reduced by nearly 68.48%, and the inference time was reduced by nearly 30.48% and 39.85% respectively. What's more, the PI values of SRGAN and ESRGAN are 2.1049 and 2.2075 respectively.

The rest of this article is organized as follows. Section 2 introduces model quantization and super-resolution in brief, and lists some related works. Section 3 introduces the paper's central work, which is the mixed quantization on the basis of quantization sensitivity found in the experiment, and the automatic quantization framework. Section 4 lists the experimental environment and evaluates the performance our approach. Section 5 is the conclusion of our work.

2 Related Work

2.1 Super-Resolution

Super-resolution [2–8] is one of the hottest research areas of low-level image problems in computer vision. Super-resolution technology is mainly to reconstruct images and videos with low-resolution into high-resolution images and videos.

This paper only studies the problem of image super-resolution. The problem of image super-resolution can be divided into multi-image-based super-resolution reconstruction and single-image-based super-resolution reconstruction. Among them, multi-image-based super-resolution reconstruction refers to a method of obtaining multiple low-resolution images that reflect different positions or pixel shifts that need to be obtained in the same scene, and use multiple low-resolution images to obtain high-resolution images. Such as the continuous motion of the object to capture images, etc. And single-image-based super-resolution reconstruction refers to the method of obtaining high-resolution images from a single low-resolution image. This method only uses the information of a low-resolution image to obtain high-resolution images. In many practical applications, due to hardware devices with limited storage capacity, time-sensitive shooting information and other factors, users often cannot obtain multiple low-resolution images reflecting different angles, such as cameras. Time sequence information of captured objects, satellite imaging images, etc. So the super-resolution technology based on single image has a wide range of applications.

This paper mainly studies the super-resolution technology based on single image. In real life, due to hardware constraints or data transmission bottlenecks, the directly obtained images are often small in size and difficult to meet users' needs. Therefore, the size of the image needs to be enlarged to meet the users' processing requirements. In the process of image enlargement, users hope to reduce the quality loss of the image as much as possible, and the goal of super-resolution technology is just to reduce the quality loss during the image enlargement process. Therefore, super-resolution has a wide range of application scenarios and is of great significance to medical imaging, traffic management and other fields.

The core idea of super-resolution reconstruction based on single images is to predict the enlarged images' information based on the information of the low-resolution image and improve the resolution of the enlarged image. Before the emergence of deep learning methods, super-resolution reconstruction based on single images mainly relied on traditional mathematical methods, such as coding methods using sparse dictionaries, interpolation methods such as bilinear interpolation. However, these simple super-resolution reconstruction images obtained by using traditional mathematical methods are still unsatisfactory, because these methods mainly rely on simple mathematical calculations to predict the high-resolution images' RGB pixel values from low-resolution, and the reconstructed images obtained are often blurry, besides, the sensory effect in human eyes is poor, too.

SRCNN [3] applies the model structure of deep learning to the field of super-resolution, and has achieved good results. After SRCNN, the super-resolution reconstruction technology mainly relies on deep learning methods. SRCNN began to use an end-to-end mapping method to directly map low-resolution images to high-resolution images after bicubic interpolation. It fits the nonlinear mapping through a three-layer convolutional network, and finally outputs a high-resolution image result. The structure of the three-layer convolution is explained into three steps: image block extraction and feature representation, feature non-linear mapping and final reconstruction. Then, in order to solve the problem that the bicubic interpolation destroys the rich image information in the low-resolution image, the model cannot be directly used. Figure 2 is the process of solving super-resolution with the deep learning method.

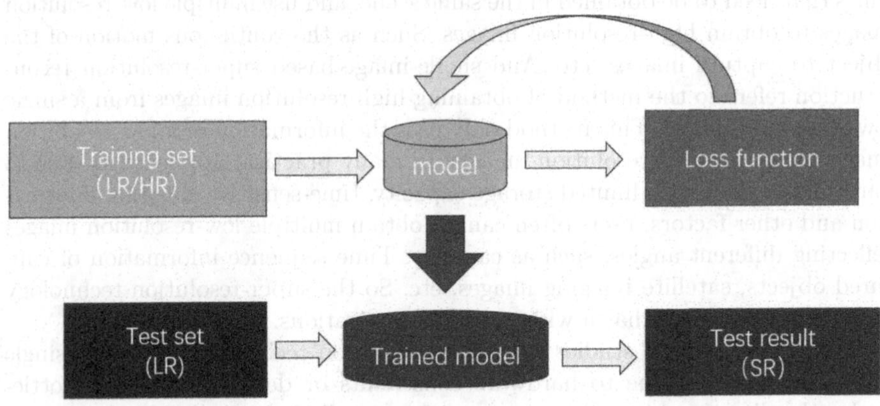

Fig. 2. The process of super-resolution using deep learning.

VDSR [6] takes the low-resolution image in target size obtained by interpolation as the input of the network, and then adds this image and the residual error learned by the network to obtain the final output of the network. Making full use of the idea of residual network, the input low-resolution image and the output high-resolution image are similar to a large extent, that is, the low-frequency information carried by the low-resolution image and the low-frequency information of the high-resolution image Similar, it will take a lot of time to bring this part during training. In fact, we only need to learn the high-frequency part residuals between the high-resolution image and the low-resolution image.

The generative adversarial network [28] is used to solve the super-resolution problem. In addition to the traditional method of using the mean square error as the loss function to obtain a high peak signal-to-noise ratio (PSNR) when training the network, it also uses perceptual loss and adversarial loss to improve the recovery effect of the picture, such as reality, more texture details. Perceptual loss uses the features extracted by the network to compare the differences between the features of the generated image and the target image after passing through the convolutional neural network, so that the generated image and the target image are more similar in semantics and style.

2.2 Model Quantization

Quantization [9–27], as the name implies, is to let the weight and activation of the forward propagation calculation in the neural network and the 32-bit or 64-bit floating point number of the gradient value of the back propagation calculation are represented by low-bit floating point or fixed-point number, and can even be directly calculated. Figure 3 shows the basic idea of converting floating-point numbers into signed 8-bit fixed-point numbers.

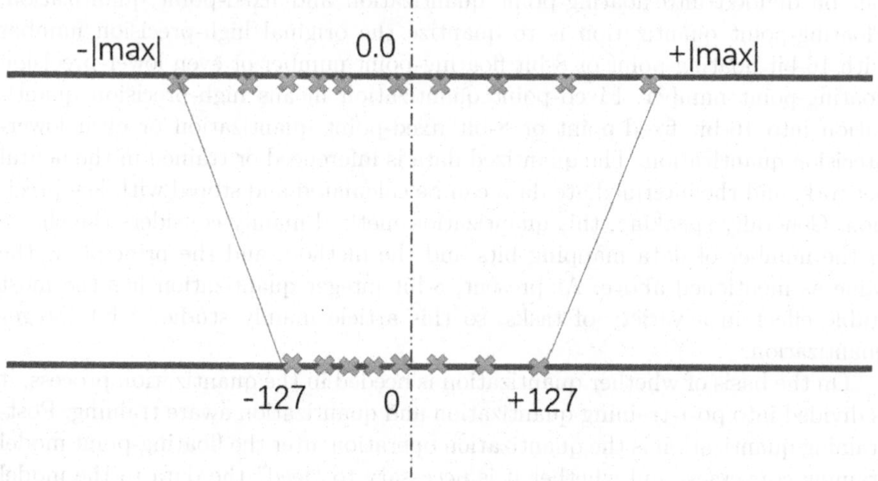

Fig. 3. The process of quantization.

Quantization itself can be divided into linear quantization and non-linear quantization. The steps of non-linear quantization are not fixed, and the method is not fixed, too. Basically, it only reduces the storage size of the model. There is no acceleration and even time complexity in model inference and data processing. So the main discussion is linear quantization. The basic principle of linear quantization is relatively clear. Take the 32-bit floating point to 8-bit integer as an example. Establish the data mapping relationship between the two, from the original data accuracy value to the corresponding quantized value. Its general form can be expressed as:

$$q = round(s \times x + z) \tag{1}$$

Among them, x and q are the numbers before and after quantization, s is called the scaling factor, and z is called the zero point. The zero point is the quantized value of "0" in the original value range. There will be a lot of 0 in the weight or activation (such as zero padding, or through the ReLU function), so we need to make "0" accurately represented after quantization when we quantize. In order to quantize in the range of n-bit integers, then:

$$s = \frac{2^n - 1}{max^x - min^x} \tag{2}$$

Among them, the denominators are the lower (min in above equation or $-|max|$ in Fig. 3) and upper bounds of the value range of the mapping value (such as weight or activation) respectively.

Quantization Method. According to the number of quantization bits, it can be divided into floating-point quantization and fixed-point quantization. Floating-point quantization is to quantize the original high-precision number with 16-bit floating-point or 8-bit floating-point number or even lower-precision floating-point number. Fixed-point quantization means high-precision quantization into 16-bit fixed-point or 8-bit fixed-point quantization or even lower-precision quantization. The quantized data is inferenced or trained in the neural network, and the intermediate data can be calculated and stored with low precision. Generally speaking, this quantization method mainly considers the choice of the number of data mapping bits and the method, and the principle is the same as mentioned above. At present, 8-bit integer quantization has the most stable effect in a variety of tasks, so this article mainly studies 8-bit integer quantization.

On the basis of whether quantization is needed in the quantization process, it is divided into post-training quantization and quantization aware training. Post-training quantization is the quantization operation after the floating-point model training converges, and whether it is necessary to "feed" the data to the model for calibration, it is divided into static quantization and dynamic quantization. The calibration in static quantization is to "feed" data to the model in advance (usually about a few hundred data), and determine the zero point and the scaling factor. For the determination of quantization parameters, sometimes the training environment and training data cannot be obtained, so calibration cannot be used. In dynamic quantization, the weight is quantized in advance, and the activation is dynamically quantized in the forward inference process, that is, the scale value and zero point are calculated once for each layer of the actual floating-point data range, and then quantized. Therefore, calibration is not used, but the maximum and minimum values are determined directly and dynamically based on the input tensor value, and then other parameters are determined. Static quantization is shown in Fig. 4.

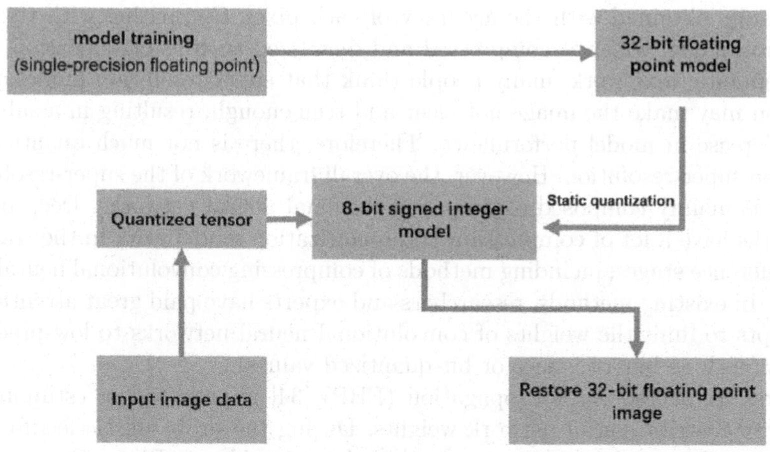

Fig. 4. The static quantization process of the model.

Quantization aware training is model quantization in the process of network training. By using fake quantization in training to simulate the process of training 8-bit integers which use clamping and approximating, so that the weights are able to simulate 8-bit integers to inference and train, but the entire model training is still carried out under the original precision. Quantization aware training is shown in Fig. 5.

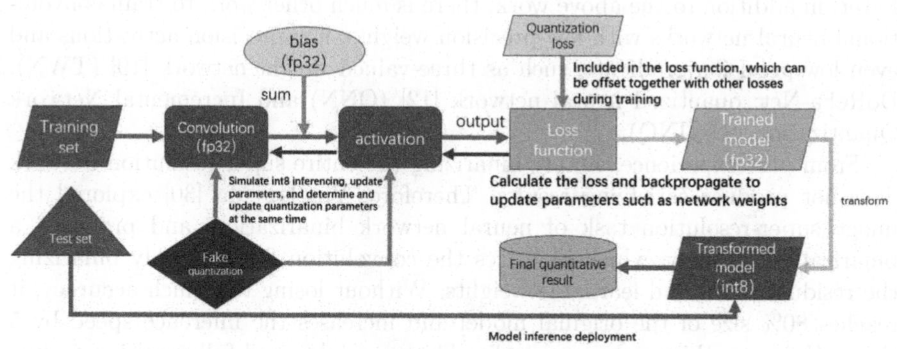

Fig. 5. Quantization aware training process.

2.3 Super-Resolution Model in Quantization

Super-resolution quantization technology [9,29–33] has recently started to do a lot of work. At the beginning, super-resolution reconstruction focused on the results obtained with the accuracy of each pixel. Comparing with the low-precision model which is compressed and quantized such as classification tasks and semantic text work, many people think that super-resolution model quantization may make the image not clear and true enough, resulting in a substantial decrease in model performance. Therefore, there is not much quantization work on super-resolution. However, the overall framework of the super-resolution model is mainly composed of deep convolutional neural networks. Deep neural networks have a lot of compression and quantization model work in the training and inference stages, including methods of compressing convolutional neural networks. In existing methods, researchers and experts have paid great attention to attempts to limit the weights of convolutional neural networks to low-precision values (such as binary values or bit-quantized values).

First, Expected Back Propagation (EBP) [34] is proposed to estimate the posterior distribution of network weights. During the probabilistic feedforward inference, the weights are constrained to be $+1$ and -1. BinaryConnect [10] extends the former idea. First, it proposes to directly binarize the network weights in the training phase, and update the weights which consist of original values according to the gradient of the binarized weights in backforward propagation process. The state-of-the-art classification performance is achieved on a small data set, indicating that binary CNNs may have performance very close to the true value network. Based on BinaryConnect, the binary network binarizes the weight and activation. XNOR-net [13] further expands beyond binary networks and binary connections by combining binary convolution operations and binary inputs in the forward inference process. Although accuracy is sacrificed to some extent, it reduces memory usage and greatly increases computing speed. Later, in addition to the above work, there is much other work to train convolutional neural networks with low-precision weights, low-precision activations and even low-precision gradients, such as three-valued weight network [19] (TWN), DoReFa-Net, quantized neural network [12] (QNN) and Incremental Network Quantization [20] (INQ).

From past experience, simply binarizing the entire super-resolution network does not produce satisfying results. Therefore, Yinglan Ma [30] explored the image super-resolution task of neural network binarization, and proposed a binarization strategy, which binarizes the convolution filter by only binarizing the residual block and learnable weights. Without losing too much accuracy, it reaches 80% size of the original model and increases the inference speed by 5 times. However, this work only studies binary weights and full-precision activation models, and convolution calculations are not bit operations, so the inference speed of the model is not simplified enough. On the basis of predecessors, Jingwei Xin [31] once again used binary weights and even intermediate activations in order to find the best approximation of convolution using bit operations and perform image super-resolution tasks on devices with limited computing resources. Among them, the convolution can be estimated through the bit operation, and the speed is about 58 times faster than the equivalent network of single-precision weights (the model size is also about 32 times compressed). The inference can be done efficiently on the CPU. At the same time, a bit accumulation mechanism is proposed, which approximates the full-precision convolution through an iterative scheme of cascading binary weighting and activation. And it only relies on the existing basic model, without introducing any other additional inference modules, to achieve high-precision one-bit estimation of weights and activation values. Experiment results show that this method can achieve better super-resolution performance with a lighter network structure and fewer operations.

3 Our Work

3.1 Quantize Model in Different Stages in Quantization Aware Learning

The main idea of mixed quantization is to select different mapping to combine them to get model or network higher accuracy. We all know that the two super-resolution models of SRGAN and ESRGAN are divided into several stages: feature extraction block, residual and dense blocks (basic blocks), up-sampling block and high-resolution reconstruction block. As shown in Fig. 6:

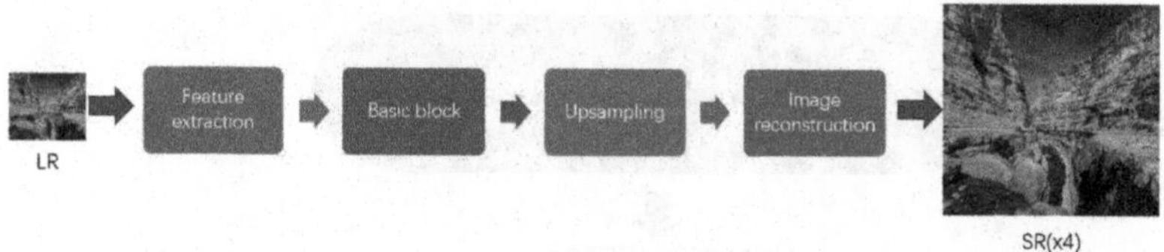

Fig. 6. Super-resolution network framework (generator of GAN network)

SRGAN [2] and ESRGAN [4] are classical super-resolution models in deep learning methods. SRGAN's job is to fool the discriminator network to determine whether the image obtained was generated by the generator network or the original image in the database so as to let the generator network generate high-resolution images from low-resolution images.

ESRGAN [4] is an improved version of SRGAN. First of all, like EDSR [7], the BN layer is removed to reduce artifacts, which can reduce the amount of calculation and generate richer texture details. Secondly, the GAN loss function is upgraded to RaGAN [8], allowing the relative discriminator to predict the authenticity of the image rather than whether the image is "fake image", and network interpolation is also added to generate a richer detailed image than the image interpolation. At the same time, the perceptual loss is performed before the activation function, and the structure of the dense network and the residual network is also added. The combination of residual network and dense connection is called RRDB block.

To get higher quality super-resolution images with less cost, and less training and inference time, Ninghui Yuan [35–37] proposed a multi-model super-resolution framework (MMSR). In the framework, all input images are classified by an approach called TVAT (Total Variance above the Threshold). And the framework prunes the training set according to the complexity of the images in the training set, which significantly reduces the size of the training set. At the same time, the framework can select the specific depth according to the image features of the images to recover the images, which helps to improve the SR-reconstruction effect.

We use quantization aware learning to quantize two SR models, SRGAN and ESRGAN, in different stages, which shows sensitivity in certain stages from PI value. Therefore, we use higher-bits to quantize the sensitive parts, and get the mixed precision quantization SR model. The basic process is below in Fig. 7.

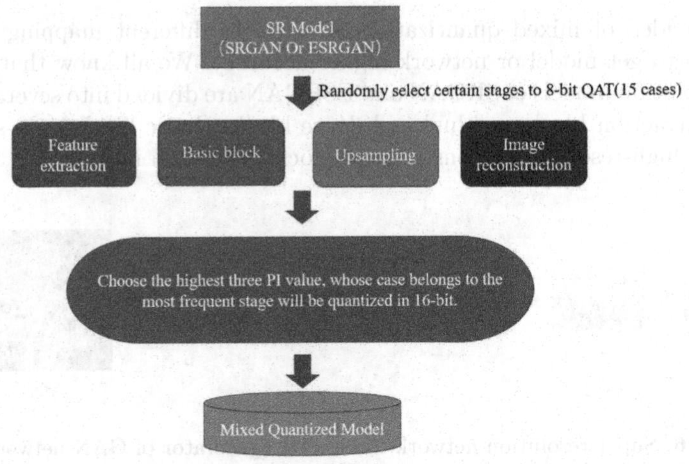

Fig. 7. The basic quantization strategy process

3.2 An Automatic Quantization Framework

From the sensitive parts, we found that the sensitive's input and output. One is to convert the low-resolution image pixel features of the RGB channel into a 64-channel feature map, and the other is to convert the 64-channel feature map back to a super-resolution image of the RGB channel. From the experiment results, quantizing the two parts will get relatively high PI value, and it will be more sensitive if the two parts quantized simultaneously, which will severely affect the effect of the generated image. The detailed operation of the two part is showed in Fig. 8.

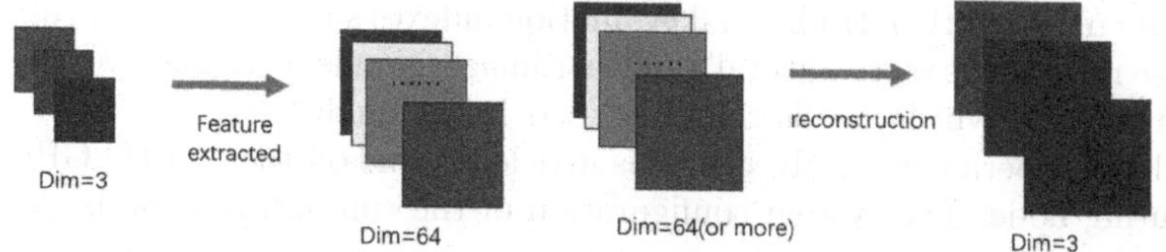

Fig. 8. Two parts' operations in detail.

We found that the two parts with greater quantization sensitivity are the feature extraction part and the image reconstruction part. The stages in the two parts are mainly convolution kernels. The middle two parts of SRGAN and ESRGAN are mainly residual blocks or dense blocks containing convolution. Basically, the channel dimensions have not changed much, and 64 channels are the main ones (although there is channel concatenating, the data shows that quantization is not particularly sensitive to it). So we establish an automatic quantization framework to quantize the two model. First, we traverse all the stages in the model to compute every stage's ratio of input and output channels as a threshold to judge whether to quantize this stage in 8-bit or 16-bit, then we ensure a mixed precision quantization model and use quantization aware learning to get our goal model. The specific process is in Fig. 9.

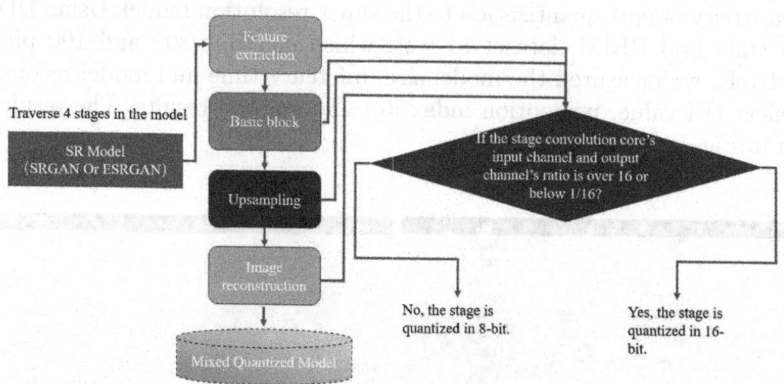

Fig. 9. Two parts' operations in detail.

4 Experiment

4.1 Environment Setup

At present, pytorch [38] supports eight-bit integer quantization, and supports many model quantization methods. The paper uses the pytorch's method to experiment. The evaluation index uses the PI value, which is the perception index, as the criterion. The PI value and the direct perception of human vision are more matched than traditional evaluation indexes such as PSNR (peak signal to noise ratio). The more natural and clear images human eyes observe, the lower the PI value is, which means the higher the image quality.

All the experiment involved in this article is done on the "CPU+GPU" [39] computing node. The system configuration of the computing node is shown in Table 1:

Table 1. System parameters of computing nodes.

HW/SW module	Description
CPU	Intel® Xeon® E5-2660 v3 @2.6 GHz x 2
GPU	NVIDIA Tesla K80 × 4
Memory	64 GB
OS	Linux CentOS 7.4
Development Environment	Anaconda 3, CUDA 9.2, Pytorch 1.7.1

4.2 Environment Result and Analysis

We use quantization aware training to quantize different parts of SRGAN and ESRGAN during the training, and measure the PI values respectively to study the sensitivity of part quantization to the super-resolution model. Using DIV_2K set to train and PRIM dataset to test, which contains 800 and 100 pictures respectively, we measured the model size, inference time and model reconstruction effect (PI value, perception index [5]) and got the results. The results are shown in Fig. 10 below:

Network	1	2	3	4	PI↓
SRGAN	√				2.4266
		√			2.4079
			√		2.3601
				√	2.4666↑
	√	√			2.4363
	√		√		2.8454
	√			√	3.6964↑
		√	√		2.4647
		√		√	2.617
			√	√	2.3603
	√	√	√		2.4071
	√	√		√	3.4352↑
	√		√	√	3.0874↑
		√	√	√	2.39
	√	√	√	√	2.4731
Baseline					2.0817

Network	1	2	3	4	PI↓
ESRGAN	√				2.5215↑
		√			2.452
			√		2.4532
				√	2.463
	√	√			2.6184
	√		√		3.1013↑
	√			√	2.9365↑
		√	√		2.6102
		√		√	2.6986
			√	√	2.4418
	√	√	√		2.5781
	√	√		√	2.7098
	√		√	√	3.0483↑
		√	√	√	2.5903
	√	√	√	√	2.688
Baseline					2.2061

Fig. 10. The results of quantization sensitivity test

From the above results, it can be seen that when concerning about only one part quantization, the image reconstruction part has the highest sensitivity, followed by the feature extraction part, then the upsampling part, and finally the residual basic block part. The image reconstruction module is to directly convert the features obtained through a series of convolution and residual connections into RGB three-channel images through convolution operations. The experimental results show that the feature maps obtained by this part of the module are more sensitively quantized, which has a greater impact on model performance. Therefore, we set an automatic quantization process framework on the basis of the ratio of input and output channels in every stage in the model. The results are as follows in Table 2:

Table 2. Quantized results of ESRGAN and SRGAN models.

Model name	Size-B (MB)	Size-A (MB)	Size-A-M (MB)	PI-O	PI-Q-M	Inf time-B	Inf time-A-M
SRGAN	5.941	1.163	1.952	2.0817	2.1049	82 s	57 s
ESRGAN	65.361	17.4	20.6	2.2061	2.2075	138 s	83 s

Above Table 2, -B represents before quantization, -A represents after quantization. -O represents original model, -M represents mixed precision model, -Q represents quantization aware training. Inf time represents model's inferencing time.

It can be seen that after using our mixed precision quantization framework, the accuracy of the model is better than that of directly using quantization aware training to entire model. Although some model size is sacrificed under the premise of compressing it, it is still optimized by nearly 67.15% and 68.48%, and the inference time is reduced from 82 s and 138 s to 57 s and 83 s. Although model size is compressed worse than straight quantizing all model to some extent, it gets better super-resolution images comparing with the original from PI value. We select two images in the data set as an example. As is shown in Fig. 11. From the figure, we can see that the images quantized in mixed precision have the approximate performance compared to the images without mixed quantization.

To further prove the model quantization's performance, we use another eight image datasets to represent the two models' quantization effect. The results are shown in Fig. 12 and Fig. 13.

According to the eight data sets, we find the performance is improved a lot after we use mixed quantization methods compared to the quantization on the entire models. Even in some data sets, the performance is better than the original precision model.

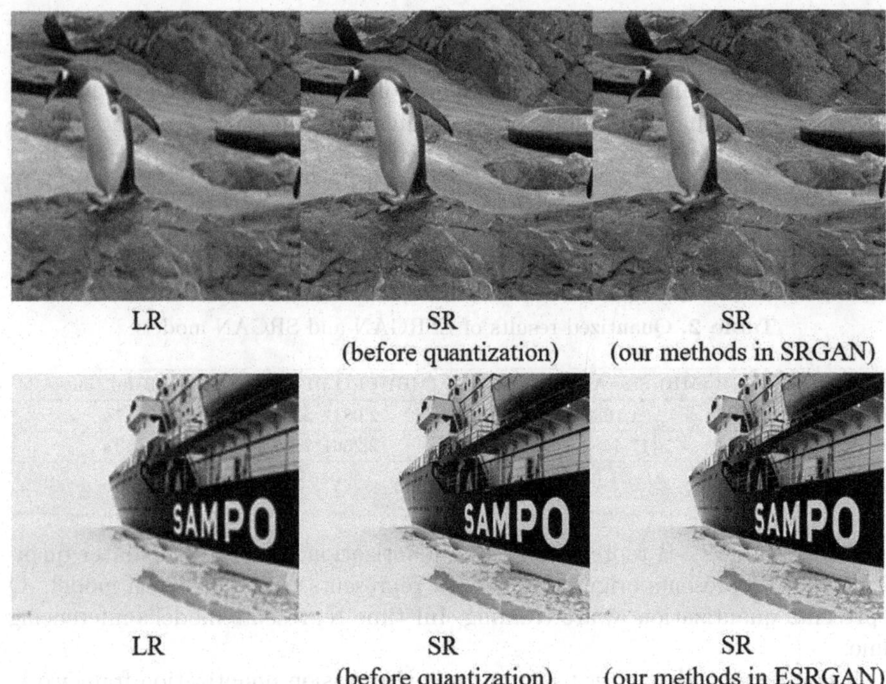

Fig. 11. Two comparison of our method and the original one

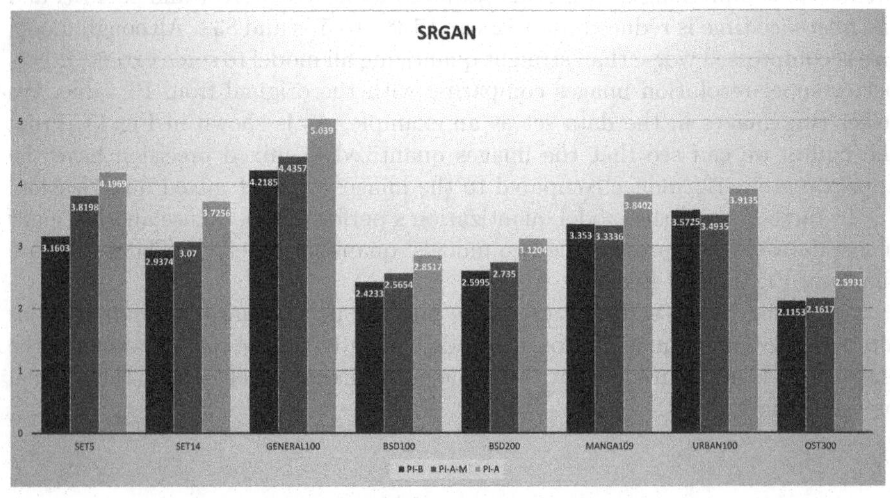

Fig. 12. The results of SRGAN

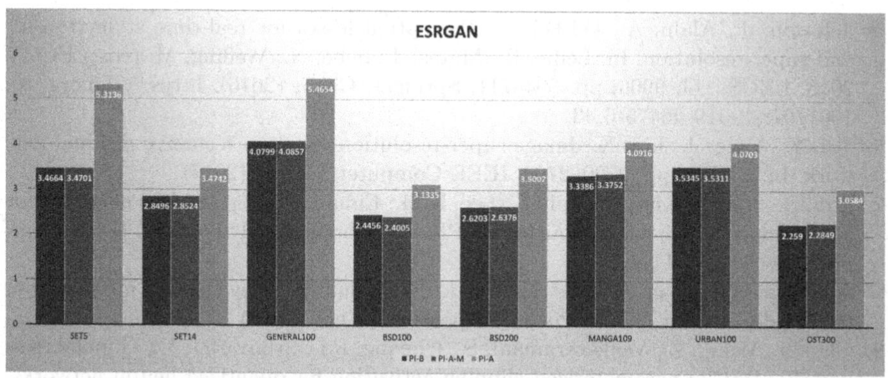

Fig. 13. The results of ESRGAN

5 Conclusion

This paper proposes a new concept (i.e., quantization sensitivity) to describe the degree to which a certain stage of a network model is affected by a specific quantization method. Then, based on the experimental results that the quantization sensitivities usually change in different stages, this paper proposes a mixed quantization method to obtain a better comprehensive results, which is evaluated from three aspects: model size, test time, and accuracy. Evaluation results indicate that with our mixed quantization strategy, the accuracies of two typical SR models are kept almost unchanged while the model size decrease greatly. This combination of multiple quantization methods makes the performance of the model be greatly improved, i.e., the PI value of the image getting inferenced is reduced from 2.4731 to 2.1049 when using SRGAN model, and from 2.688 to 2.2075 when using ESRGAN model. Last but not least, we implement an automatic sensitivity evaluation and mixed quantization method selection framework, and evaluate its performance with more classical SR datasets.

Acknowledgment. This work is supported by National Nature Science Foundation of China (Grant No. 62032001 and 61972407) and Key Laboratory Open Projects Grant No. SZU-GDPHPCL201903.

References

1. Fan, S., Fei, J., Shen, L.: Accelerating deep learning with a parallel mechanism using CPU + MIC. Int. J. Parallel Program. **46**(4), 660–673 (2018)
2. Ledig, C., et al.: Photo-realistic single image super-resolution using a generative adversarial network. In: CVPR, pp. 105–114. IEEE Computer Society (2017)
3. Dong, C., Loy, C.C., He, C.C., Tang, X.: Image super-resolution using deep convolutional networks. CoRR, vol. abs/1501.00092 (2015)
4. Wang, X., et al.: ESRGAN: enhanced super-resolution generative adversarial networks. In: Leal-Taixé, L., Roth, S. (eds.) ECCV 2018. LNCS, vol. 11133, pp. 63–79. Springer, Cham (2019). https://doi.org/10.1007/978-3-030-11021-5_5

5. Johnson, J., Alahi, A., Fei-Fei, L.: Perceptual losses for real-time style transfer and super-resolution. In: Leibe, B., Matas, J., Sebe, N., Welling, M. (eds.) ECCV 2016. LNCS, vol. 9906, pp. 694–711. Springer, Cham (2016). https://doi.org/10.1007/978-3-319-46475-6_43
6. Tai, Y., Yang, J., Liu, X.: Image super-resolution via deep recursive residual network. In: CVPR, pp. 2790–2798. IEEE Computer Society (2017)
7. Lim, B., Son, S., Kim, H., Nah, S., Lee, K.M.: Enhanced deep residual networks for single image super-resolution. In: CVPR Workshops, pp. 1132–1140. IEEE Computer Society (2017)
8. Jolicoeur-Martineau, A.: The relativistic discriminator: a key element missing from standard GAN. In: ICLR (Poster). OpenReview.net (2019)
9. Choi, J., Wang, Z., Venkataramani, S., Chuang, P.I., Srinivasan, V., Gopalakrishnan, K.: PACT: parameterized clipping activation for quantized neural networks. CoRR, vol. abs/1805.06085 (2018)
10. Courbariaux, M., Bengio, Y., David, J.: BinaryConnect: training deep neural networks with binary weights during propagations. In: NIPS, pp. 3123–3131 (2015)
11. Courbariaux, M., Bengio, Y.: BinaryNet: training deep neural networks with weights and activations constrained to +1 or −1. CoRR, vol. abs/1602.02830 (2016)
12. Wu, J., Leng, C., Wang, Y., Hu, Q., Cheng, J.: Quantized convolutional neural networks for mobile devices. In: CVPR, pp. 4820–4828. IEEE Computer Society (2016)
13. Rastegari, M., Ordonez, V., Redmon, J., Farhadi, A.: XNOR-Net: ImageNet classification using binary convolutional neural networks. In: Leibe, B., Matas, J., Sebe, N., Welling, M. (eds.) ECCV 2016. LNCS, vol. 9908, pp. 525–542. Springer, Cham (2016). https://doi.org/10.1007/978-3-319-46493-0_32
14. Howard, A.G., et al.: MobileNets: efficient convolutional neural networks for mobile vision applications. CoRR, vol. abs/1704.04861 (2017)
15. Sa, C.D., et al.: High-accuracy low-precision training. CoRR, vol. abs/1803.03383 (2018)
16. Chu, T., Luo, Q., Yang, J., Huang, X.: Mixed-precision quantized neural networks with progressively decreasing bitwidth. Pattern Recogn. **111**, 107647 (2021)
17. Mishra, A.K., Nurvitadhi, E., Cook, J.J., Marr, D.: WRPN: wide reduced-precision networks. In: ICLR (Poster). OpenReview.net (2018)
18. Zhuang, B., Liu, L., Tan, M., Shen, C., Reid, I.D.: Training quantized neural networks with a full-precision auxiliary module. In: CVPR, pp. 1485–1494. Computer Vision Foundation/IEEE (2020)
19. Li, F., Liu, B.: Ternary weight networks. CoRR, vol. abs/1605.04711 (2016)
20. Zhou, A., Yao, A., Guo, Y., Xu, L., Chen, Y.: Incremental network quantization: towards lossless CNNs with low-precision weights. In: ICLR (Poster). OpenReview.net (2017)
21. Hubara, I., Courbariaux, M., Soudry, D., El-Yaniv, R., Bengio, Y.: Quantized neural networks: training neural networks with low precision weights and activations. CoRR, vol. abs/1609.07061 (2016)
22. Kim, N., Shin, D., Choi, W., Kim, G., Park, J.: Exploiting retraining-based mixed-precision quantization for low-cost DNN accelerator design. IEEE Trans. Neural Netw. Learn. Syst. **32**(7), 2925–2938 (2021)
23. Li, M., Lin, J., Ding, Y., Liu, Z., Zhu, J., Han, S.: GAN compression: Efficient architectures for interactive conditional GANs. In: CVPR, pp. 5283–5293. Computer Vision Foundation/IEEE (2020)

24. Zhuang, B., Liu, J., Tan, M., Liu, L., Reid, I.D., Shen, C.: Effective training of convolutional neural networks with low-bitwidth weights and activations. CoRR, vol. abs/1908.04680 (2019)
25. Cai, H., Gan, C., Wang, T., Zhang, Z., Han, S.: Once-for-all: train one network and specialize it for efficient deployment. In: ICLR. OpenReview.net (2020)
26. Chang, S., et al.: MSP: an FPGA-specific mixed-scheme, multi-precision deep neural network quantization framework. CoRR, vol. abs/2009.07460 (2020)
27. Vasquez, K., Venkatesha, Y., Bhattacharjee, A., Moitra, A., Panda, P.: Activation density based mixed-precision quantization for energy efficient neural networks. CoRR, vol. abs/2101.04354 (2021)
28. Goodfellow, I.J., et al.: Generative adversarial nets. In: NIPS, pp. 2672–2680 (2014)
29. Lee, R., et al.: Journey towards tiny perceptual super-resolution. In: Vedaldi, A., Bischof, H., Brox, T., Frahm, J.-M. (eds.) ECCV 2020. LNCS, vol. 12371, pp. 85–102. Springer, Cham (2020). https://doi.org/10.1007/978-3-030-58574-7_6
30. Ma, Y., Xiong, H., Hu, Z., Ma, L.: Efficient super resolution using binarized neural network. In: CVPR Workshops, pp. 694–703. Computer Vision Foundation/IEEE (2019)
31. Xin, J., Wang, N., Jiang, X., Li, J., Huang, H., Gao, X.: Binarized neural network for single image super resolution. In: Vedaldi, A., Bischof, H., Brox, T., Frahm, J.-M. (eds.) ECCV 2020. LNCS, vol. 12349, pp. 91–107. Springer, Cham (2020). https://doi.org/10.1007/978-3-030-58548-8_6
32. Li, H., et al.: PAMS: quantized super-resolution via parameterized max scale. In: Vedaldi, A., Bischof, H., Brox, T., Frahm, J.-M. (eds.) ECCV 2020. LNCS, vol. 12370, pp. 564–580. Springer, Cham (2020). https://doi.org/10.1007/978-3-030-58595-2_34
33. Jacob, B., et al.: Quantization and training of neural networks for efficient integer-arithmetic-only inference. CoRR, vol. abs/1712.05877 (2017)
34. Soudry, D., Hubara, I., Meir, R.: Expectation backpropagation: parameter-free training of multilayer neural networks with continuous or discrete weights. In: NIPS, pp. 963–971 (2014)
35. Yuan, N., Zhu, Z., Wu, X., Shen, L.: MMSR: a multi-model super resolution framework. In: Tang, X., Chen, Q., Bose, P., Zheng, W., Gaudiot, J.-L. (eds.) NPC 2019. LNCS, vol. 11783, pp. 197–208. Springer, Cham (2019). https://doi.org/10.1007/978-3-030-30709-7_16
36. Yuan, N., Liu, J., Wang, Q., Shen, L.: Customizing super-resolution framework according to image features. In: ISPA/BDCloud/SocialCom/SustainCom, pp. 1189–1196. IEEE (2020)
37. Yuan, N., Zhang, D., Wang, Q., Shen, L.: A multi-model super-resolution training and reconstruction framework. In: He, X., Shao, E., Tan, G. (eds.) NPC 2020. LNCS, vol. 12639, pp. 105–116. Springer, Cham (2021). https://doi.org/10.1007/978-3-030-79478-1_9
38. Imambi, S., Prakash, K.B., Kanagachidambaresan, G.R.: PyTorch. Programming with TensorFlow (2021)
39. Zhang, S., Qin, Z., Yang, Y., Shen, L., Wang, Z.: Transparent partial page migration between CPU and GPU. Front. Comput. Sci. **14**(3), 1–13 (2019). https://doi.org/10.1007/s11704-018-7386-4

An Optimized GPU Implementation of Weakly-Compressible SPH Using CUDA-Based Strategies

Yuejin Cai[1], Jianguo Wei[1], Qingzhi Hou[2(✉)], and Ruixue Gao[1]

[1] College of Intelligence and Computing, Tianjin University, Tianjin 300350, China
[2] School of Civil Engineering, Tianjin University, Tianjin 300350, China
qhou@tju.edu.cn

Abstract. SPH (Smoothed Particle Hydrodynamics) is a meshless method that is widely used to simulate computational fluid problems but is very time-consuming. Although many GPU-based solutions have been widely applied to accelerate the SPH method, the potential bottlenecks of optimizing the GPU implementation are less investigated. This study puts forward a fully optimized GPU-based implementation to accelerate SPH simulations. To this end, different aspects of GPU optimization, including splitting CUDA kernels, using fast instructions, simplifying data arrays, using texture memory and unrolling loops, are developed to speed up the GPU implementation. Our parallel-optimized GPU framework is able to be further applied to implement other SPH-based algorithms. The performance results show that these GPU optimization strategies can greatly improve the computational performance on graphics devices, and the fully optimized GPU implementation is on average 3 times faster than basic GPU implementation without optimizations. Besides, the parallel computing power of an advanced GPU is introduced to accelerate SPH codes with a speedup of 3338x in comparison to its serial version. The same simulation run on a rather outdated GPU is also 693 times as efficient as the implementation achieved on a mainstream single CPU.

Keywords: SPH · GPU · Optimization strategies · Performance improvements · CUDA

1 Introduction

Smoothed particle hydrodynamics (SPH) firstly proposed by Lucy [1] and Gingold and Monaghan [2] is a meshless method in which fluid volume is discretized into a series of Lagrangian particles. The SPH method is Lagrangian in nature, which makes it distinctly different from grid-based methods [3]. It is particularly useful for computational fluid problems with complex shapes and geometries, such as free-surface flows [4], fluid-structure interactions [5], in fields of heat conduction [6] and multi-physics [7]. Compared with the methods based grid, the SPH method is easy to solve these numerical problems with particle interpolation. It can avoid some numerical difficulties such as unphysical oscillations and uncontrolled instabilities and get accurate results. Recently,

Y. Lai et al. (Eds.): ICA3PP 2021, LNCS 13155, pp. 354–369, 2022.
https://doi.org/10.1007/978-3-030-95384-3_23

the realistic and real-time simulation of SPH-based fluid animation is starting to be used to the application of virtual reality (VR) such as VR game and augmented reality (AR) [8]. The physical computations in virtual applications become more robust when using the SPH method. However, high computational requirements are needed when particle interactions in SPH are calculated by interpolating the neighboring particles, especially for large-scale scenarios. This is mainly caused by its frequent neighbor computation and storage, which is very time-consuming. To reduce the neighbor computation costs per simulation step, an improved sorting method for parallel neighborhood search was proposed by Sun et al. [9]. They used a parallel all-prefix-sum algorithm to improve cell-particle sorting on GPUs. Their sorting method performed well when the scale of particles is not very high and the cell number is moderate. And yet, it has no disadvantage when the scale is high. Winkler et al. [10] compared different implementations of Verlet neighbor list with the cell linked list approach and showed the speedup achieved on the two methods differed greatly for different hardware and dimensionality. Their work found that the actual performance of Verlet-list method over cell-linked-list approach decreased with increasing number of particles due to the much higher memory consumption required to store references to all neighbors. In the paper of Band et al. [11], they put forward a compression scheme based on the cell linked list approach to store the neighbor particles by reducing memory consumption. Their implementation of compression scheme on GPUs yielded an average memory saving of 87% for all computations within a simulation step compared to the uncompressed case. Regretfully, it had little effect on the reduction of the total simulation time of the SPH method.

To minimize the computational cost for entire simulation sequences in SPH, many parallel solutions have been presented to take advantage of GPU parallel computing to simulate the SPH fluids that are computationally intensive. DualSPHysics [12] and GPUSPH [13] were the two most common GPU solvers for simulating fluid dynamic problems based on the open-source SPHysics Fortran code. The weakly compressible Navier-Stokes equations [14] were used in the design and coding theory of these two GPU implementations. The main processes in the numerical model included neighbor list, particle interaction and property update were parallelized on GPUs using CUDA invented by NVIDIA [15]. Compared to their CPU versions, the computing efficiency obtained by these two sets of GPU codes was greatly increased on the targeted NVIDIA graphics devices. Different from depending on CUDA-capable devices, AQUAgpusph [16] was developed by the CEHINAV group to accelerate SPH simulations with OpenCL, in which accelerator cards from vendors such as IBM, AMD and Intel can also be used. This GPU-accelerated SPH solver broke the platform dependence and used different graphics accelerator cards, which was easily moldable and extensible with Python. Similarly, SISPH [17] presented by Muta et al. was implemented by PyOpenCL and can work together seamlessly with OpenMP and MPI on massively parallel computers. On the other hand, the SPH-based real-time simulation and rendering of dynamic fluids with GPU computing is being deeply developed and applied for virtual reality research [18]. As expected, such SPH implementations with GPUs markedly decrease computational complexity and save much running time compared to serial and parallel implementations achieved on CPU.

Besides focusing on simple acceleration of SPH method on modern GPUs, however, only a few researches are concerned about parallel optimizations for further speeding up the particle-based simulations. For example, Domínguez et al. [19] developed optimization strategies for the GPU implementation of the SPH method, such as reducing cell size and accessing cells by row in the neighbor search. Their implementation of GPU optimizations for SPH simulation on CUDA gained nearly 56.2x speedups compared with the serial SPH codes. However, these improvements on performance came at the cost of large memory. In the paper of Winkler et al., a shared memory caching algorithm was presented to accelerate 2D hydrodynamic simulations [20]. The shared memory on a GPU chip was used to store and fetch neighbor lists quickly when particle interactions were computed. A major challenge of this method was that it needed to check whether the data to be accessed was in the shared memory. Once the accesses to shared memory of the GPU failed, the computational performance would decrease significantly. Wang et al. [21] introduced a series of kernel optimization strategies of reducing particle properties, utilizing particle reusability, and maximizing the GPU occupancy into the GPU-accelerated SPH codes, which got 20x speedups on their peer2peer model.

Up to now, when implementing the weakly compressible SPH method on the GPU cards, the problems from the computational complexity of CUDA kernels, the latency of the access to global memory on GPU devices, data representation, etc. are not solved well and greatly limit the performance of the GPU code. Thus, this study further focuses on parallel-optimizing these performance problems and proposes different GPU optimization strategies to improve the ability of the whole parallel execution. Based on these improvements, we develop a fully optimized GPU implementation for simulating large-scale fluid flow phenomena with the SPH method.

The rest of this paper is organized as follows. Section 2 gives a brief overview of the weakly compressible SPH (WCSPH) method. Our GPU implementation of this numerical method and optimization techniques are described in Sect. 3, followed by the performance results in Sect. 4. Finally, this paper is concluded in Sect. 5.

2 Numerical Method

In SPH [22], the particle properties at an arbitrary point are interpolated from neighbors. For the physical property A_i at particle i, the particle-average based approximation is given as

$$A_i = \sum_j \frac{m_j}{\rho_j} A_j W_{ij}\left(\boldsymbol{r}_{ij}, h\right) \tag{1}$$

where m_j, ρ_j, A_j represent mass, density and physical property at particle j. $\boldsymbol{r}_{ij} = \boldsymbol{r}_i - \boldsymbol{r}_j$ is the relative position between particle i and j. $W_{ij}\left(\boldsymbol{r}_{ij}, h\right)$ is compact-support kernel function and h is the smoothing length with respect to $\boldsymbol{r}_i$, position of particle i.

Similarly, the original SPH approximation of the derivative of the physical property at particle i is formulated as

$$\nabla A_i = -\sum_j \frac{m_j}{\rho_j} A_j \nabla_i W_{ij} \tag{2}$$

where $\nabla_i W_{ij}$ is the gradient of the kernel function $W_{ij}(\boldsymbol{r}_{ij}, h)$.

Following [23, 24], the discretization of the WCSPH method based on a low-dissipation Riemann solution applied for the continuity and momentum equations is

$$\begin{cases} \frac{d\rho_i}{dt} = 2\rho_i \sum_j \frac{m_j}{\rho_j}(\boldsymbol{v}_i - \boldsymbol{v}^*) \cdot \nabla_i W_{ij} \\ \frac{d\boldsymbol{v}_i}{dt} = -2\sum_j m_j \left(\frac{p^*}{\rho_i \rho_j}\right) \nabla_i W_{ij} \end{cases} \tag{3}$$

where $\boldsymbol{v}$ is the velocity. $\boldsymbol{v}^*$ and p^* are the solutions of the Riemann problem [23].

To meet the weakly compressible assumption, the artificial state equation is applied to relate pressure to density [25]. According to [23, 26], the wall boundary condition is imposed by fixed dummy particles and the interaction between fluid particles and wall particles is computed by solving a one-sided Riemann problem along the wall normal direction. Finally, a kick-drift-kick [24, 27] time integration with time-step size criteria is used to advance the densities, velocities and positions of all fluid particles using the WCSPH scheme.

3 GPU Implementation Framework

SPH-based fluid simulator uses neighbor search and particle summation as its key procedures. Both these two processes are very time-consuming because the neighbor search is performed several times at each time step and the particle summation needs intensive force computations by using (3), which is over all the neighboring particles located in the support domain of the particle of interest. To eliminate this performance bottleneck, it is significant to design a fully efficient parallel implementation to map the WCSPH simulation scheme onto the GPU.

In this section, our work describes a GPU-based data structure and a parallel cell-linked-list method for neighbor search on the GPU. The full GPU-accelerated implementation of SPH simulation is presented. Moreover, different CUDA-based optimization strategies are developed to speed up the GPU-based implementation.

3.1 Data Layout and Neighbor Search

Data Structure. The SPH fluid particles generally require many physical properties, such as mass, volume, density, pressure, velocity and position. Besides, other particle data may be needed when the SPH equations in (3) are calculated on the basis of time-stepping scheme, such as particle type and index. Therefore, it is necessary to use an efficient data representation to store particles.

In general, structure of array (SoA) and array of structure (AoS) are two common data structures. As described in [28, 29], it is better to store particle properties data in SoA instead of AoS to obtain a higher GPU memory bandwidth. This is because that SoA results in a better coalesced memory access while the AoS-based scattered memory access often occurs in GPU implementation. For CUDA-enabled GPU devices, the memory coalescing and successive data manner can satisfy the coalesced rules of global memory access, leading to higher global memory bandwidth and cache-hit rate.

Thus, SoA is used to store particle data in our GPU simulation. For example, a standard GPU data layout using SoA style is shown below.

```
Struct
{
    float3 velocity[];
    float3 position[];
    float density[];
    float pressure[];
    float type[];
    float index[];
} particle;
```

Neighbor Search. In our GPU implementation, the parallel cell-linked-list (CLL) method [11, 30] is performed on the CUDA-capable devices to query neighboring particles. The background grid structure with the same cell size as support radius of parti-cles is used for the searching procedure in the CLL method.

In CLL-based grid system, every particle is assigned to one grid cell through its address. To get the neighbors of particle of interest, the grid-based CLL method only needs to search for particles which located in the same cell or in one of the other neighboring cells. The number of neighboring grid cells is 8 in 2D and 26 in 3D. To quickly query all neighboring particles in a grid cell, the fast radix sorting algorithm [31] is used to reorder particles on the CUDA architecture. Then, we can find the start and end of any given cell by aid of GPU shared memory technique. This approach makes particles in the same cell also close in GPU memory, which achieves memory coherence on the GPU.

3.2 Parallel SPH on GPU

In a SPH simulation, every Lagrangian particle computation is independent of each other. In this way, it is very suitable to be fully implemented in parallel on the GPU using CUDA programming language [32]. Figure 1 illustrates the entire computational procedure of the GPU implementation for the weakly compressible SPH simulation.

According to Fig. 1, we map all the main simulation steps on the modern GPU so as to fully parallelize the SPH method, which are abstracted as CUDA kernels. At the beginning of each physics update, we create the searching CUDA kernel to obtain neighboring particles by using the CLL method discussed in the previous section. Then, in the first computational stage two different CUDA kernels are created to compute density and velocity for fluid particles at the half time step $k + 1/2$ according to (3). With the effort of position kernel, the particle position is updated for the new time step by using the newly updated velocity according to the symplectic time scheme mentioned above. The next calculations will be performed on the new-time-step position data.

This present GPU simulation in the second stage of the continuity computation uses the corresponding CUDA kernel to accumulate density for fluid particles at the next time step.

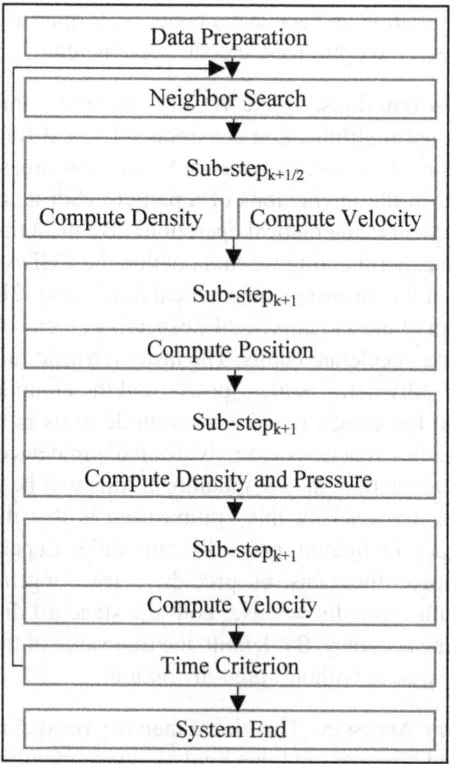

Fig. 1. Flow chart of the GPU implementation for WCSPH.

$k+1$ using the previously updated particle information. The pressure should be also calculated on the basis of the new corrected density in the same CUDA kernel. Similarly, the new-time-step velocity is accumulated for each fluid particle according to momentum calculation CUDA kernel using the new-time-step density and pressure value. Finally, the time-step march CUDA kernel is executed by using different numerical stability conditions in order to get the new time-step size to advance the simulation loop.

In the context of CUDA kernel computation, a lot of CUDA threads are configured and used to compute physical attributes for all fluid particles, which using GPU memory and registers. In this way, the parallel implementation of the SPH model on the GPU can achieve a fast simulation.

3.3 Optimization Strategies for GPU Implementation

When a CUDA kernel is launched for particle calculations, all the CUDA thread blocks work simultaneously on the CUDA kernel by performing thread-level allocation. However, the computational efficiency of the CUDA threads in the CUDA kernel is affected by various factors, such as random and unaligned global memory access, limited on-ship registers and complex cycle calculations. These influential aspects impose restrictions on code execution and reduce parallel performance to some extent. To improve the ability of the CUDA thread execution and achieve a better performance, we develop different GPU-accelerated strategies to optimize the code implementation.

Using Fast Intrinsic Instructions. In the WCSPH method, some mathematical functions such as division and multiplication are frequently used for calculating the mass and momentum equations. For example, at least 80 and 100 times of division and multiplication are involved in the interactions of a particle with its neighboring particles, respectively. These frequent mathematical operations take most time and influence code execution. However, in experimenting we find out that the CUDA programming library has the faster versions of the intrinsic mathematical functions [32], which are supported in device code. They can be used to improve the executive efficiency of the CUDA-based codes by using hardware-accelerated units. The faster intrinsic functions have the same name prefix __ (e.g., __fdividef, __sinf, __powf), and the compiler has to add a directive, *-use_fast_math,* to force each function to compile to its intrinsic counterpart. As mentioned above, these fast functions are only available in device code and when they are used in GPU implementation, the computational time will be greatly decreased.

However, the main drawback of this optimization is that they are less accurate. Particularly, the accuracy of floating-point division varies depending on whether the CUDA compiler uses -prec-div = false or -prec-div = true. In general, when the CUDA codes are compiled with -prec-div = false, both the standard division / operator and __fdividef have the same accuracy. By default, the true value of the code compilation is given with -prec-div = true or without -prec-div option.

Using Texture Memory Accesses. The global memory resided in CUDA-capable devices requires coalesced accesses, and this kind of device memory is accessed via 32-, 64-, or 128-byte memory transactions that must be naturally aligned. If this requirement is not met, fully coalesced accesses could not be achieved when a wrap reads data from global memory. Unfortunately, the memory accesses to fetch neighboring particles are typically non-coalesced in the WCSPH scheme. To improve the access efficiency, the texture memory is used in this study to cache the data of particles using random memory access. In this optimization strategy, we bind the particle arrays to textures and use texture lookups *tex1Dfetch* instead. For example, a texture reference *texVel* is defined to bind the velocity array p*Vel* and use *cudaBindTexture* to inject relevance into *texVel* and *pVel.* Then, the texture data are fetched via *texVel* by *tex1Dfetch* for particle calcu-lation. After the calculations are completed by all threads, the main process calls *cudaUnbindTexture* to unbind *texVel* and *pVel.* Thus, the data are first read from the texture cache, and in case of cache miss they will be accessed from the global memory.

In this optimization, the read-only texture cache can reduce DRAM bandwidth demand if the cache is hit, which avoids accesses to global memory and has a shorter access time. This advantage is apparent when the scale of particles is not very large. But it has no advantage when the scale is big. Besides, the performance improvements.

differ for different CUDA-capable devices. Note that the texture reference is replaced by texture object API in the latest CUDA version.

Unrolling Loop Calculations. The CUDA codes need to cope with different flow control instructions in implementing the WCSPH method, such as *for* loops. For CUDA-enabled architectures, the loop calculations take up most of the computation time in a force calculation and affect the computational efficiency of the CUDA kernel. Therefore, the *#pragma unroll* directive is used to unroll any given loop to increase instruction throughput. This pragma must be placed immediately before the loop and only applies to that loop. However, given that complete unrolling may not be needed in some calculations, unrolling is optionally followed by an integral constant expression (ICE), such as *#pragma unroll C*. In general, if *C* is absent, the loop will be completely un-rolled if its trip count is constant. If *C* evaluates to 1, the compiler will not unroll the loop. If *C* is a non-positive integer or an integer greater than the maximum count, the pragma will be broken. Otherwise, the loop will be unrolled by *C* times.

When this optimization strategy is applied for GPU implementation, the cost of iterative loops can be reduced and the Instruction-level Parallelism (ILP) is automatically inspired in a CUDA kernel. The ILP approach may avoid address calculation in each loop iteration, reducing the consumption of CUDA cores. However, loop unrolling often requires more registers, resulting in low GPU occupancy of the resident CUDA blocks and performance degradation.

Simplifying Data Representation. When implementing the WCSPH method on GPUs, the arrays of particle properties are required as input and output to a CUDA function. The common arrays are used as described in section **Data Structure**. For simplicity, the initial CUDA codes also need row and column arrays for grid cells and create several assistant arrays to store the half-step results. In our new optimization, the array list is simplified by reassembling the arrays that have to be transferred and dis-carding unnecessary arrays. For example, the new array *posp* consists of position and pressure and the new *velrho* array consists of velocity and density. The remaining un-necessary variables such as line and column numbers are calculated from these new creations. This optimization strategy uses float4 vectors to combine four single precision floating point values into a single one.

In this way, the number of arrays involved in CUDA kernel is reduced by at least a half, reducing the overheads of launching CUDA functions. On the other hand, this approach allows CUDA threads to read four values of an array at once instead of retrieving them from different particle arrays, which can reduce the time needed to load data from the global memory.

Splitting Complex CUDA Kernels into Simple Ones. In the beginning, the renewal of the position and density of each fluid particle is performed together with the half-time-step velocity computation, which are included in the first kick step. To improve the complex CUDA kernel computations, we add two CUDA kernels to compute density and position, respectively. In this way, the complex kernels are split into simpler ones and the special CUDA kernel computation is independent of each other. The optimization gains some benefits that speed up the GPU code. First, it improves the ability of the code execution. Besides, it reduces the register usage and improves the overall efficiency of the accesses to global memory of the GPU. This is because the number of physical attributes needed to be computed in a complex CUDA kernel is reduced. But this method has no advantage when the number of computed attributes becomes more and more. The performance improvement also depends on the data layout and CUDA-code programming.

4 Results

In this section, our GPU framework evaluates the SPH simulation on an Nvidia Titan V and Quadro K2200. We use CUDA Toolkit 9.2/GCC 7.5.0/CLion as the development environment. The benchmark of dam break problem [33] is used to test the performance of the present parallel implementation on the GPU. For all simulations, the variable time-step size is obtained by following the previous time-step criteria. The smoothing length h is set to $1.3r$, where r is radius of particles, and the support radius of particles is set to $2h$. The reference density of fluid particles is given by 1000 kg/m^3. The density fluctuation threshold is required to be less than 1% according to the weakly compressible assumption.

Figure 2 sketches the evaluated scene with the fluid dimensions of $H = 1.0$ m and $L = 2.0H$. The tank boundaries are outlined as $L_t = 5.366H$ and $H_t = 3.5H$, depending on the particle sampling distance the actual lengths are rounded.

4.1 GPU Optimization Evaluation

The optimization results on GPUs show the speedup obtained when comparing the performance of the optimized code with the basic version without any optimization.

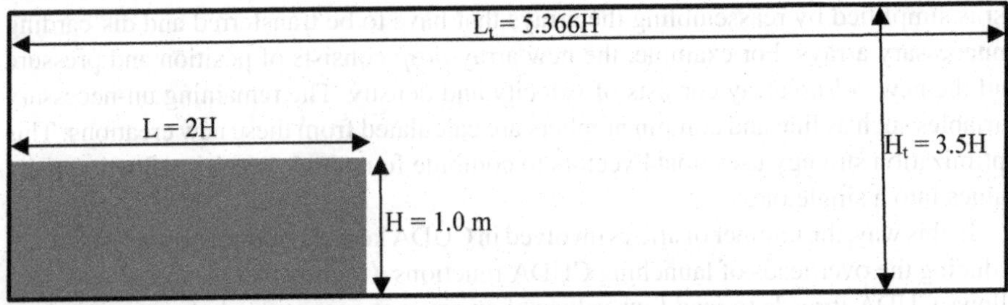

Fig. 2. Sketch of the dam break problem.

Figures 3 and 4 show the achieved speedup of different GPU optimization strategies in comparison to GPU implementation without optimizations on an Nvidia Titan V and Quadro K2200, respectively. Note that each line of the figures denotes the speedup of each GPU optimization including all previous ones. Thus, the line that corresponds to the optimization of loop unrolling includes the previous optimizations: splitting complex kernels into special ones, simplifying data arrays, using fast instructions and texture memory. It can be found that the performance increases rapidly at first and then levels off with the increase of particle number. The peak performance of Volta- and Maxwell-based GPU is 4.1 and 2.8 times that of non-optimized implementations, respectively. When one million particles are involved, the speedup obtained by using optimization of loop unrolling on the Nvidia Titan V and K2200 is 3.3 and 2.8, respectively. Another important differentiator texture memory is applied on K2200, and better performance is obtained when the number of particles is less than 600,000. As the number of particles increases, the speedup decreases because the texture memory is not enough to cache all particle data and may cause a decrease in cache hit ratio. But it has no contribution to Titan V. These comparisons show that the performance improvement achieved by these different optimizations differs significantly for different GPU cards.

Figure 5 shows the contribution of each GPU optimization strategy to performance improvement on the Titan V and K2200 simulating 500,000 particles of the testcase. According to this figure, it is obviously found that the speedup of fully optimized GPU code over the basic GPU version without any optimization is 2.8 for K2200 and 2.9 for Titan V, respectively when 500,000 particles are involved in this case. It is clear that the use of fast instructions with compiler directive *use_fast_math* contributes most significantly to the improvement of execution efficiency, and the efficiency has almost doubled for both the two GPU devices. Simplifying particle arrays to reduce global memory accesses can reduce the computational time by at least 20%. One of the other optimization strategies including separating complex CUDA kernels, unrolling loop and using texture memory at least gains 10% improvement in performance whether the hardware is Titan V or K2200. Besides, this figure also displays that the same GPU optimization also has different effects on performance improvement for different GPU cards.

4.2 Simulation Performance

The section firstly evaluates the simulation of the dambreak test case and compares our present implementation against the existing GPU-SPH solvers dualSPHysics and gpuSPHASE [20] on the GPU. The latest available dualSPHysics is a versatile GPU-SPH solver with all compiler optimization enabled. The gpuSPHASE solver is capable of handling a large number of neighboring particles in two-dimensional SPH simulations. The present GPU implementation in this study uses a fully optimized version with different optimizations described in Sect. 3.3 to test the simulation performance.

The average computational time for a single iteration for a CUDA core at different particle sampling distances is visualized in Fig. 6. In this comparison the values are calculated by dividing the overall computational time by the number of iterations by the number of CUDA cores and tested on the GPU card Titan V. For high particle numbers, our GPU implementation is more than 4.6 times faster than gpuSPHASE and 5 times faster than dualSPHysics. The runtime per iteration per core is 0.0000028 s when simulating over one million particles in our GPU implementation. But the time is more than 0.000014 s whether the GPU implementation is gpuSPHASE or dualSPHysics. Moreover, it is clearly seen that the computational time got by gpuSPHASE and dualSPHysics increases sharply with higher particle numbers while our solver changes very little. These computational results show that the proposed fully optimized GPU implementation in this study has a good scalability on the SPH problem, especially on simulating large-scale particles.

We also compare the performance results of the optimized GPU implementations on Titan V and K2200 with the single-threaded CPU implementation and its multithreaded version. It is noted that the CPU results are not implemented with any of optimizations.

Figure 7 exhibits the speedup achieved with the most efficient GPU implementations in comparison to the single-core and multi-core implementations on CPU. Thus, for example, when simulating for a run involving 200,000 particles, the speedups obtained

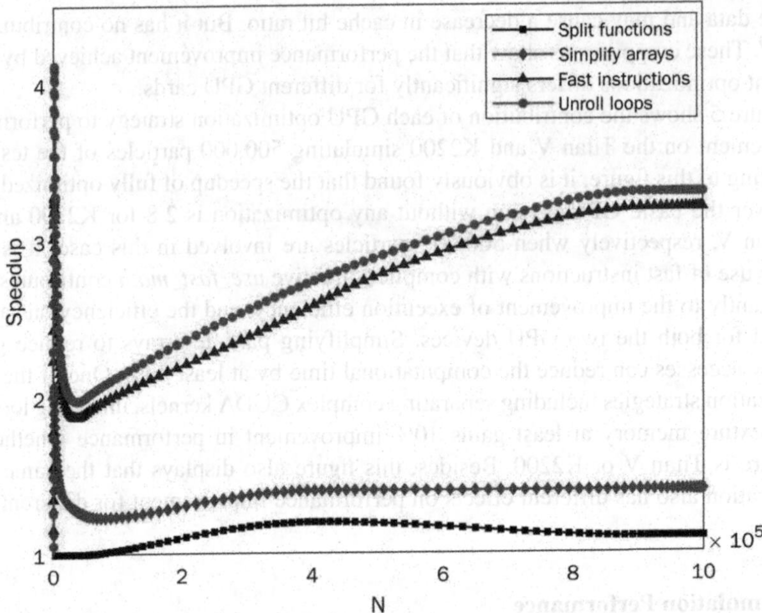

Fig. 3. Speedup of different GPU optimization strategies for different numbers of particles (N) on Titan V.

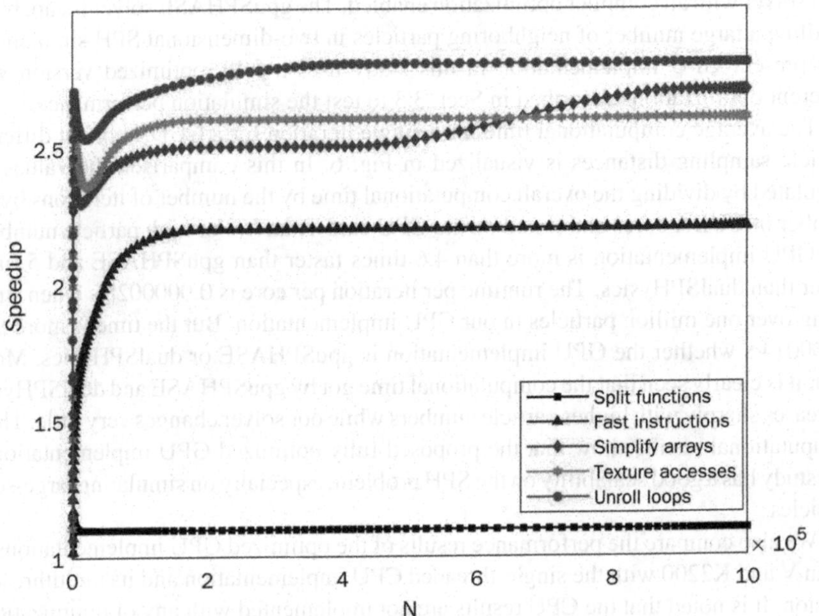

Fig. 4. Speedup of different GPU optimization strategies for different numbers of particles (N) on K2200.

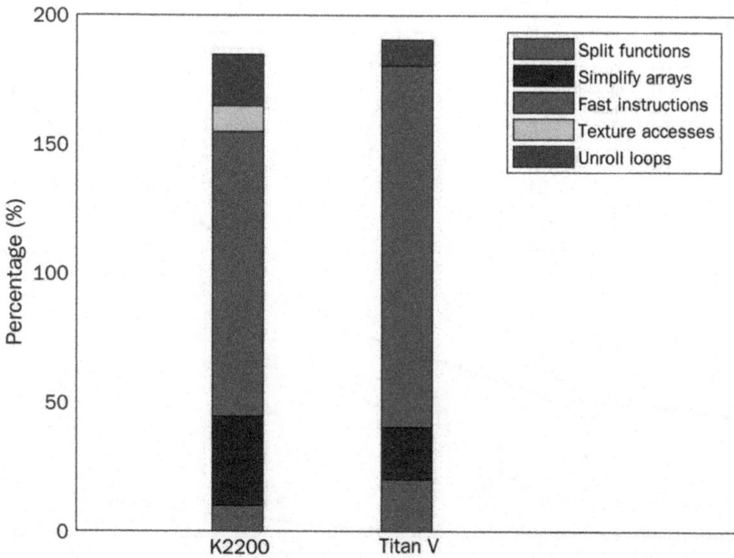

Fig. 5. Contributions of different GPU optimization strategies to performance improvement on K2200 and Titan V in simulating 500,000 particles.

with Titan V and K2200 compared with the single-threaded CPU implementation are about 3300 and 690, respectively. When the multi-core implementation on CPU are compared, the speedup obtained with these two GPUs is about 40% of that of single-core CPU version. Compared with the serial implementations, the total achieved speedup on Titan V is about 5 times that of K2200 when the SPH method is implemented by the fully optimized GPU code. In other words, the speedup obtained on the advanced GPUs with higher computed cores increases at a much greater than the CUDA-enabled GPUs with low computed cores.

Table 1 summarizes the simulation times for the test case run on the GPU and CPU. It again proves that the new GPU-based WCSPH implementation works well on both Volta and Maxwell architectures and achieves a higher performance when compared to the basic CPU implementations. For example, when running a simulation involving 200,000 particles, the whole simulation takes 293.3 h on the single-core CPU and only five minutes on the Titan V, leading to a speedup of 3338.6x. Even on the outdated K2200, the computational time is just 0.42 h. From 25,000 particles to 220,000 particles, the simulation time of the single-threaded CPU code is increased by 29 times and the 4-threaded CPU version is 20 times. However, the GPU simulation run on the advanced Titan V is only increased by 9 times. According to this table, it is seen that the simulation times for CPU codes increase sharply with the increase of the number of particles while this trend is smaller on the GPU. Moreover, since our optimized GPU framework will benefit from more CUDA cores, we can expect higher computational performance on current Ampere and future architectures.

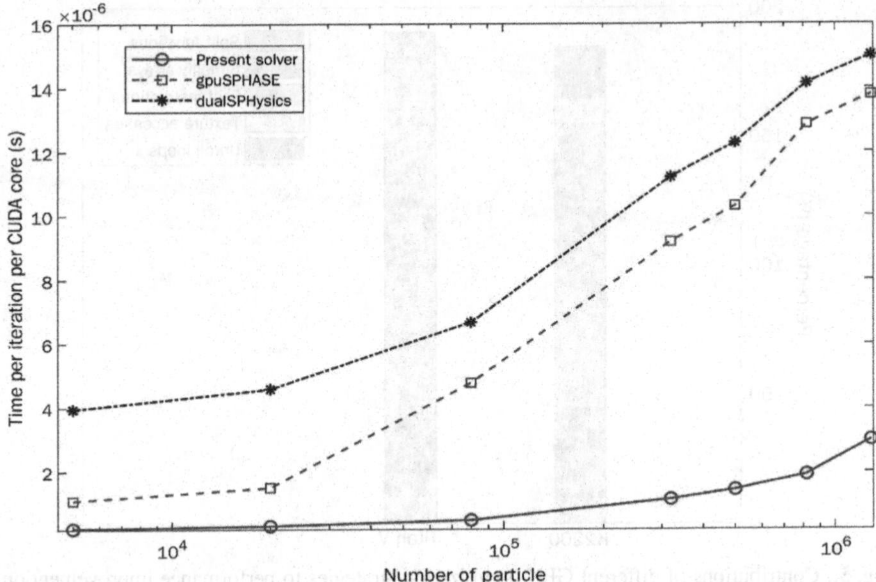

Fig. 6. Simulation performance achieved by the present GPU implementation in comparison to dualSPHysics and gpuSPHASE [20]. The number of particles are involved from the resolutions 0.02, 0.01, 0.005, 0.0025, 0.002, 0.0015625, 0.00125.

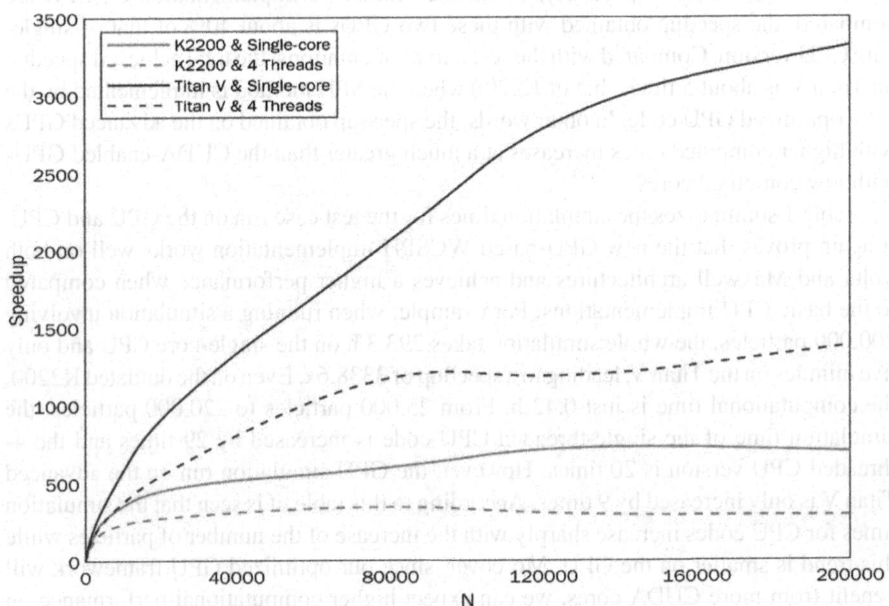

Fig. 7. Speedup for different numbers of particles (N) obtained by the most efficient GPU implementation on Titan V and K2200 in comparison to the CPU implementations (single thread and 4 threads).

Table 1. Runtimes of the GPU and CPU simulations.

Number of particles	Total simulation time			
	GPU		CPU	
	K2200	Titan V	Single core	4 threads
25097	0.02 h	0.01 h	10.55 h	6.18 h
89794	0.12 h	0.03 h	69.90 h	43.32 h
137201	0.21 h	0.05 h	159.58 h	61.90 h
220412	0.42 h	0.09 h	293.28 h	122.39 h

5 Conclusion and Future Work

This paper puts forward a fully optimized GPU-accelerated solver to simulate fluid flows using the WCSPH method based on a low-dissipation Riemann solution. To further speed up the calculation of the mass and momentum equations, optimization strategies are proposed to improve the performance of GPU implementation by splitting CUDA kernels, using fast instructions, simplifying data arrays, using texture memory and unrolling loops. First, the complex CUDA kernels in the GPU-accelerated simulation are split into small and specific CUDA kernels. Then, particle arrays are simplified to reduce the global memory accesses. Only required arrays are created for the calculation of the WCSPH scheme by using the float4 type, and fast instructions compiled by *-use_fast_math* are used to speed up the mathematical arithmetic operations. To avoid non-coalesced global memory access in CUDA calculations, a random memory access method through texture cache is used. Loop unrolling is controlled by using *#pragma unroll* to increase the throughput of the execution instructions computed by CUDA threads. In summary, the GPU optimization strategies are integrated with the basic GPU code to obtain a fully optimized GPU solver for the implementation of the WCSPH method, resulting in a speedup of 3.3x obtained on a Titan V. In the case of K2200, the achieved speedup is 2.8x. Besides, the optimized GPU version outperforms the single-threaded CPU implementation by 693.6x on the Maxwell card while the achieved speedup is 3338.6x using the Volta card.

It is well concluded that our parallel framework on GPU can exploit the large-scale computing capabilities of the GPUs to accelerate serial SPH codes to achieve a higher computational performance. Moreover, it is also well adapted for parallelizing other SPH-based algorithms using CUDA.

We will extend our GPU simulator to support multi-GPU system in the future work.

Acknowledgment. This work is supported by the National Key Research and Development Program of China [grant no. 2020YFC1807905], Tianjin Municipal Science and Technology Program for New Generation of Artificial Intelligence [grant no. 19ZXZNGX00030] and National Natural Science Foundation of China [grant no. 52079090].

References

1. Lucy, L.B.: A numerical approach to the testing of the fission hypothesis. Astrophys. J. **82**, 1013–1024 (1977)
2. Monaghan, J.J.: Smoothed particle hydrodynamics. Annu. Rev. Astron. Astrophys. **30**, 543–574 (1992)
3. Liu, G.R., Liu, M.B.: Smoothed Particle Hydrodynamics: A Meshfree Particle Method. World Scientific, Singapore (2003)
4. Xu, X., Ouyang, J., Yang, B., Liu, Z.J.: SPH simulations of three-dimensional non-Newtonian free surface flows. Comput. Methods Appl. Mech. Eng. **256**, 101–116 (2013)
5. Zhang, A.-M., Sun, P.-N., Ming, F.-R., Colagrossi, A.: Smoothed particle hydrodynamics and its applications in fluid-structure interactions. J. Hydrodyn. **29**, 187–216 (2017)
6. Hosain, M.L., Domínguez, J.M., Bel Fdhila, R., Kyprianidis, K.: Smoothed particle hydrodynamics modeling of industrial processes involving heat transfer. Appl. Energy **252**, 113441 (2019)
7. Domínguez, J.M., Fourtakas, G., Altomare, C., Canelas, R.B., Tafuni, A., García-Feal, O., et al.: DualSPHysics: from fluid dynamics to multiphysics problems. Comput. Part. Mech. (2021). https://doi.org/10.1007/s40571-021-00404-2
8. Zhang, F., Wei, Q., Xu, L.: An fast simulation tool for fluid animation in VR application based on GPUs. Multimedia Tools Appl. **79**, 16683–16706 (2019). https://doi.org/10.1007/s11042-019-08002-4
9. Sun, H.Y., Tian, Y.S., Zhang, Y.L., Wu, J., Wang, S., Yang, Q., et al.: A special sorting method for neighbor search procedure in smoothed particle hydrodynamics on GPUs. In: 44th International Conference on Parallel Processing Workshops, Beijing, pp. 81–85 (2015)
10. Winkler, D., Rezavand, M., Rauch, W.: Neighbour lists for smoothed particle hydrodynamics on GPUs. Comput. Phys. Commun. **225**, 140–148 (2018)
11. Band, S., Gissler, C., Teschner, M.: Compressed neighbour lists for SPH. Comput. Graph. Forum **39**, 531–542 (2020)
12. Crespo, A., Domínguez, J., Rogers, B., Gómez-Gesteira, M., Longshaw, S., Canelas, R., et al.: DualSPHysics: open-source parallel CFD solver based on smoothed particle hydrodynamics (SPH). Comput. Phys. Commun. **187**, 204–216 (2015)
13. Hérault, A., Bilotta, G., Dalrymple, R.A.: SPH on GPU with CUDA. J. Hydraul. Res. **48**, 74–79 (2010)
14. Antuono, M.: Numerical diffusive terms in weakly-compressible SPH schemes. Comput. Phys. Commun. **183**, 2570–2580 (2012)
15. CUDA toolkit. https://developer.nvidia.com/about-cuda. Accessed 29 Sept 2021
16. Cercos-Pita, J.L.: AQUAgpusph, a new free 3D SPH solver accelerated with OpenCL. Comput. Phys. Commun. **192**, 295–312 (2015)
17. Muta, A., Ramachandran, P., Negi, P.: An efficient, open source, iterative ISPH scheme. Comput. Phys. Commun. **255**, 107283 (2020)
18. Nie, X., Chen, L.C., Xiang, T.: Real-time incompressible fluid simulation on the GPU. Int. J. Comput. Games Technol. **2015**, 2 (2015)
19. Domínguez, J.M., Crespo, A.J.C., Gesteira, M.G.: Optimization strategies for CPU and GPU implementations of a smoothed particle hydrodynamics method. Comput. Phys. Commun. **184**, 617–627 (2013)
20. Winkler, D., Meister, M., Rezavand, M., Rauch, W.: gpuSPHASE—a shared memory caching implementation for 2D SPH using CUDA. Comput. Phys. Commun. **235**, 514–516 (2017)
21. Wang, Y.R., Li, L.S., Wang, J.T., Tian, R.: Acceleration of smoothed particle hydrodynamics method on CPU-GPU heterogeneous platform. J. Comput. **40**, 2040–2056 (2017)

22. Liu, M.B., Liu, G.R.: Smoothed particle hydrodynamics (SPH): an overview and recent developments. Arch. Comput. Method Eng. **17**, 25–76 (2010)
23. Zhang, C., Hu, X.Y., Adams, N.A.: A weakly compressible SPH method based on a low-dissipation Riemann solver. J. Comput. Phys. **335**, 605–620 (2017)
24. Rezavand, M., Zhang, C., Hu, X.Y.: A weakly compressible SPH method for violent multi-phase flows with high density ratio. J. Comput. Phys. **402**, 092–109 (2020)
25. Monaghan, J.J.: Simulating free surface flows with SPH. J. Comput. Phys. **110**, 399–406 (1994)
26. Adami, S., Hu, X., Adams, N.: A generalized wall boundary condition for smoothed particle hydrodynamics. J. Comput. Phys. **231**, 7057–7075 (2012)
27. Monaghan, J.J.: Smoothed particle hydrodynamics. Rep. Prog. Phys. **68**, 1703 (2005)
28. Wei, F., Jin, L., Liu, J., Ding, F., Zheng, X.: GPU acceleration of a 2D compressible Euler solver on CUDA-based block-structured Cartesian meshes. J. Braz. Soc. Mech. Sci. Eng. **42**, 250 (2020)
29. Wang, X.L., Qiu, Y.X., Slattery, S.R., Fang, Y., Li, M.C., Zhu, S.C., et al.: A massively parallel and scalable multi-GPU material point method. ACM Trans. Graph. **39**, 1–15 (2020)
30. Green, S.: Particle simulation using CUDA. NVIDIA (2010)
31. Satish, N., Harris, M., Garland, M.: Designing efficient sorting algorithms for manycore GPUs. In: IEEE International Symposium on Parallel and Distributed Processing, pp. 1–10. IEEE Press, Rome Italy (2009)
32. CUDA Toolkit Documentation (v11.4.1). https://docs.nvidia.com/cuda/cuda-toolkit-release-notes/index.html. Accessed 29 Sept 2021
33. Zhou, Z.Q., De Kat, J.O., Buchner, B.: A nonlinear 3D approach to simulate green water dynamics on deck. In: 7th International Conference on Numerical Ship Hydrodynamics, Nantes France, pp. 1–15 (1999)

A Heterogeneous Multi-core Network-on-Chip Mapping Optimization Algorithm

Juan Fang(✉), Haoyan Zhao, Jiayue Zhang, and Jiamei shi

Beijing University of Technology, Beijing 100124, China
{fangjuan,zhangjiayue}@bjut.edu.cn

Abstract. Heterogeneous multi-core Network-on-Chip provides high-speed execution and performance to meet the heavy communication demands of the system. However, the system power consumption and delay increase as the number of cores rises. In this paper, we present KL_SA, a mapping scheme based on the Kernighan-Lin (KL) and Simulated Annealing (SA) algorithms, aiming at the mapping optimization problem on heterogeneous multi-core Network-on-Chip. This scheme combines the advantages of the KL algorithm's efficient partitioning and the SA algorithm's global search for optimal solution and improves the latter in turn. Firstly, we divide the task using KL algorithm, with which result to initial the mapping of the SA algorithm, solving the random initialization problem in the SA algorithm, also increasing the likelihood of getting the optimal solution. Secondly, we do the mapping using the SA algorithm with memory function added in the iterative process. In this way, the current best state is memorized without losing the current optimal solution when escaping from the local optimum, ensuring the global optimal approximate solution obtained. Experiments show large savings in the aspects of system power consumption and delay on the system with the proposed mapping scheme compared to the existing mapping schemes.

Keywords: Network on chip · KL_SA algorithm · Mapping optimization · Global optimal approximate solution

1 Introduction

The emergence of Network-on-Chip (NoC) enables numerous processing units (also known as IP cores) being integrated so as to provide a higher system performance. While NoC contributes to a higher system performance with its speed-up execution, it also brings the problem of energy consumption growth and delay increment. This problem may reduce the life span of chip and affect the stability of chip's operation due to the limitation of chip area and heat dissipation capacity, which limits further improvement of system performance. Thus, it is crucial to improve the energy efficiency in NoC design.

In previous works, low-power, high-performance topology and routing algorithms [1–3] have been designed to lower the NoC energy consumption. In recent years, the study on NoC mapping algorithm with power consumption or delay as the optimization

Y. Lai et al. (Eds.): ICA3PP 2021, LNCS 13155, pp. 370–384, 2022.
https://doi.org/10.1007/978-3-030-95384-3_24

target has gained extensive attention from scholars and designers globally. Since the communication cost among the NoC nodes is largely affected by the communication distances between the nodes of NoC, it is desirable to design an effective NoC mapping algorithm to reduce the average distance between nodes.

The task mapping problem is generally known to be a NP-hard problem [17]. The widely-used method to solve this kind of problem in a reasonable amount of time is to obtain near-optimal solutions using meta-heuristic algorithms, e.g. simulated annealing [13–15]. Studies have shown that implementing an effective NoC mapping algorithm can significantly reduce the overall power consumption of the system, and improve system performance [4–6].

In addition, various analysis and heuristic algorithms for mapping optimization [7–12] have been proposed. Alagarsamy et al. [13] proposed a mapping scheme with the meta-heuristic search algorithm of SA and Tabu search (SAT) to analyze and optimize the power consumption based on NoC system. Jiang et al. [16] devised a new fault model based on the standby core technology, and designed a fault-recognition low-power dynamic task mapping algorithm targeting on NoC system fault, power consumption and system performance. Taassori et al. [17] proposed a linearized Quadratic Allocation Problem (QAP) model, mapping tasks to the kernel, which not only minimized the power consumption, but also improved the performance of NoC. Le et al. [18] designed a mapping optimization algorithm based on distributed search (SS) and improved multi-objective optimization of standard SS, which helped to obtain high-performance mapping layout.

Inspired by the aforementioned studies of NoC mapping algorithm, we discover a pattern in NoC mapping for effectively reducing communication cost among NoC nodes. We then analyze the pattern and explore the principle behind it. Instead of mapping IP cores to NoC nodes directly, we break it into two phases. The first phase is to assign the tasks to IP cores according to the relatedness between the tasks. The second phase is the mapping between the IP cores and the NoC routing nodes. In this way, the more related tasks are assigned to one IP core, then, the relatedness between IP cores is reduced, thereby, the interactions between IP cores are cut down. This means that after the IP cores are mapped to NoC nodes in the second phase, the locality of data communication among NoC nodes becomes higher, which leads to a shorter average communication distance and therefore a lower energy consumption.

Hence, how to divide the tasks according to their relatedness and then how to integrate the two phases leaves us the space to innovate. The key is to find an appropriate algorithm for each phase and combine the two with taking the advantage of them. The KL algorithm [20] has the advantage of efficient partitioning, which can quickly divide the whole space into multiple subspaces. On the other hand, the SA algorithm [21] can avoid the premature phenomenon better by introducing the Metropolis criterion, and has the advantage of global search optimal solution, which can find the solution that is nearest to the optimal solution in the whole solution space.

The contribution of this paper is three-fold. Firstly, we discover a pattern for NoC mapping to reduce energy consumption and analyze the principle behind the pattern. Secondly, we propose a new mapping scheme to reduce power consumption and delay for the heterogeneous multi-core NoC, and the proposed scheme can achieve global optimal approximate solution with less complexity than the existing mapping schemes proved by abundant experimental results. Finally, our proposed scheme provides a new perspective for heterogeneous multi-objective optimization problem by utilizing the partitioning of KL algorithm to remedy the drawback of SA algorithm's random initialization.

In the following sections, we first describe the NoC mapping problem and the mapping optimization model in Sect. 2, then we present the implementation of the proposed scheme in Sect. 3, finally we show the evaluation and analysis in Sect. 4, and conclusion is in Sect. 5.

2 NoC Mapping Problem Description and Mapping Optimization Model

2.1 Overview

In a heterogeneous environment, this paper divides the mapping into two phases, as shown in Fig. 1. In the first phase, the task nodes in the task graph are assigned to the appropriate IP cores. In the second phase, the IP cores that have been bound to the task are mapped to the NoC platform to settle their specific locations in the NoC. T represents the subtask in the application, IP represents the IP core, and R represents the NoC routing node.

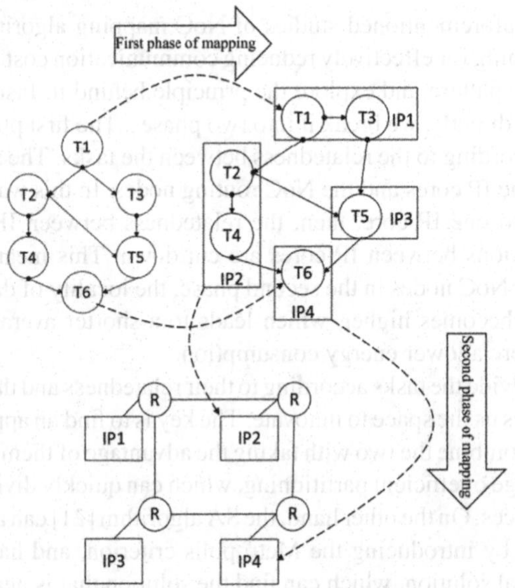

Fig. 1. Two phases of NoC mapping

The NoC used in the study of the mapping scheme in this paper is a 2D Mesh topology, as shown in Fig. 1, where each resource node consists of a processing module (IP) and a routing node (R), which perform data interaction through a network interface (NI). As for the route assignment in NoC mapping, the routing algorithm of static shortest path is applied in our NoC platform. It is because it is simple to understand and implement, and the conclusion can be easily extended to other more complicated NoC architecture.

2.2 Problem Definition

In this section, the mathematical description of the NoC mapping problem is presented. First, we give the following three basic definitions.

Definition 1: The task graph is a directed graph, denoted as $\mathrm{TG}(T, R, V)$, where each vertex $T_i \in T$ represents a subtask, and each directed edge $R_{i,j} \in R$ represents a communication relationship from subtask T_i to T_j, and the edge weight $V_{i,j}$ represents the traffic of subtasks T_i to T_j.

Definition 2: The given core communication graph is a directed graph, denoted as $\mathrm{CG}(C, R, V)$, where each vertex $C_i \in C$ represents an IP core, and each directed edge $R_{i,j} \in R$ represents a communication relationship from the IP core C_i to C_j, and the edge weight $V_{i,j}$ represents the traffic of IP core C_i to C_j.

Definition 3: The given NoC platform topology graph is a directed graph, denoted as $\mathrm{NG}(N, P, E)$, where each vertex $N_i \in N$ represents a resource node in the NoC, $P_{i,j} \in P$ represents the routing path from network node N_i to N_j, and $E_{i,j}$ represents the communication cost by transmitting 1 bit of data from network node N_i to N_j.

Based on the above definitions, the mathematical description of the mapping problem for power consumption and delay is as follows.

2.3 NoC Power Consumption Model

The total energy consumed in the entire NoC system consists of the energy consumed by the IP core E_c in the processing task and the energy consumed during the data communication process E_{net}, as shown in the following formula.

$$\mathrm{Energy} = E_c + E_{\mathrm{net}} \tag{1}$$

The energy consumed by the IP core in processing the task is only related to the type of the core, so the processing power consumption of the v bits data on a specific core is:

$$E_{\mathrm{ip}} = \rho_{\mathrm{e}} * v \tag{2}$$

where ρ_{e} is a constant associated with the type of IP core.
So, the total power consumption on the IP core is:

$$E_c = \sum_{\mathrm{ip}=1}^{m} E_{\mathrm{ip}} \tag{3}$$

where m represents the number of IP core.

For the power consumption in data communication, the power consumption during single bit data transmission is:

$$E_b = E_s + E_a + E_w + E_l \tag{4}$$

where E_b represents the power consumption generated by transferring unit data from one network node to another network node, E_s represents the power consumption generated by the crossbar switch in the router, E_a represents the power consumption generated by the unit data in the internal buffer area of the routing node, E_w represents the power consumption generated by the internal wiring of the router, and E_l represents the power consumption generated by the unit data through the NoC communication interconnection link. The values of E_s, E_a and E_w are mainly related to the internal design of the routing node, and will not change with the communication status in the network. It can be approximated as a constant, so we use E_r to represent it uniformly. Then the power consumption generated during unit data transmission can be expressed by Formula (5).

$$E_b^{i,j} = n_{i,j} * E_r + (n_{i,j} - 1) * E_l \tag{5}$$

Where $E_b^{i,j}$ is the power consumption generated by transmitting unit data from the network node N_i to the network node N_j, and $n_{i,j}$ is the number of network nodes through which the unit data passes during transmission. Therefore, the power consumption generated by communication between node T_i and node T_j can be calculated by the Formula (6).

$$E_{i,j} = v_{i,j} * E_b^{i,j} \tag{6}$$

Where $v_{i,j}$ represents the amount of communication between network nodes N_i and N_j.

Therefore, the total power consumption in data communication can be calculated by the Formula (7).

$$E_{\text{net}} = \sum_{i=1}^{n} \sum_{j=1}^{n} E_{i,j} \tag{7}$$

Where n represents the number of NoC node.

Thus, the power consumption in the entire NoC system can be obtained from the Formula (8).

$$\text{Energy} = \sum_{\text{ip}=1}^{m} E_{\text{ip}} + \sum_{i=1}^{n} \sum_{j=1}^{n} E_{i,j} \tag{8}$$

2.4 NoC Delay Model

The delay in the entire NoC system mainly comes from the time consumption when the IP core processes the data D_c and the time consumption of the data during the NoC transmission D_{net}.

The time consumption of the IP core in processing the data is only related to the type of the core, so the time consumption of the v bits data on a specific core is:

$$D_c = \rho_d * v \tag{9}$$

where ρ_d is a constant associated with the type of IP core.

The time consumption of data in the NoC transmission mainly includes three parts: the delay on the source and sink nodes NI, the delay generated by the routing node during transmission, and the delay of the transmission path interconnection line. Therefore, the time consumption during unit bit data transmission can be expressed as formula (10).

$$D_b = 2*\rho_N + \rho_R + \rho_L \tag{10}$$

Where ρ_N, ρ_R, and ρ_L are the constants of the transmission time on the NI, router, and interconnect lines.

With the above formula, we can get the time consumption of unit data transmission from network node N_i to network node N_j:

$$D_b^{i,j} = 2 * n_{i,j} * \rho_N + n_{i,j} * \rho_R + (n_{i,j} - 1) * \rho_L \tag{11}$$

where $n_{i,j}$ is the number of network nodes through which the unit data passes during the transmission. So, the time consumption of communication between node N_i and node N_j can be calculated by the following formula:

$$D_{\text{net}} = v_{i,j} * D_b^{i,j} \tag{12}$$

where $v_{i,j}$ represents the amount of communication between network nodes N_i and N_j.

Therefore, we can calculate the delay in the entire NoC system based on the time consumption when the IP core processes the data D_c, the time consumption of the data during the NoC transmission D_{net}, and the dependencies between the tasks. The formula is as follows:

$$Delay = D_c + D_{net} \tag{13}$$

2.5 NoC Multi-objective Optimization Model

As the size of the NoC and the complexity of the application increases, the optimization goal of the mapping is no longer limited to a single target such as power consumption, delay or heat balance. The multi-objective optimization is becoming more and more needed. This paper establishes an evaluation function that weighs two optimization goals for the two targets of power consumption and delay simultaneously.

The existing evaluation functions designed for multi-objective optimization mainly include linear weighted summation method, square weighted summation method, ideal point method and other common methods [20]. In this paper, the linear weighted summation method is used to establish the multi-objective optimization evaluation function. For the multi-objective optimization problem of power consumption and delay, the weights

α_e and α_d are determined according to the importance of power consumption and delay, and subjected to the following conditions: $\alpha_e \geq 0$, $\alpha_d \geq 0$, $\alpha_e+\alpha_d = 1$. Therefore, the evaluation function is defined as follows:

$$\text{Fevaluation} = \alpha_e * \text{Energy}(M)+\alpha_d * \text{Delay}(M) \tag{14}$$

where $M \in$ MAP, MAP is the set of all mapping schemes. After the weights α_e and α_d are set, different evaluation values Fevaluation are obtained by changing the mapping scheme M, and the smaller the value of Fevaluation is, the lower the power consumption and delay obtained by the current mapping can be.

3 Implementation of Mapping Scheme Based on KL_SA Algorithm

3.1 NoC Multi-objective Optimization Model

This paper proposes the KL_SA algorithm, which combines the advantages of the KL algorithm and the SA algorithm and improves the latter. The KL_SA algorithm has the advantage of efficient and global search for optimal solution. By using the mapping scheme generated by the KLc_SA algorithm, the communication distance between nodes with large traffic can be effectively reduced, and the solution closest to the optimal solution can be quickly obtained, thereby it can reduce the power and delay of the system.

3.2 NoC Mapping Process Based on KL_SA Algorithm

The mapping of task to IP core: Because the power consumption and delay estimated at this phase does not involve the communication distance, it is only related to the IP core type. Therefore, the basic SA algorithm is adopted as the mapping scheme in this phase. The specific steps are as follows.

1. Randomly assign the subtasks in the task communication graph to the IP core, and initialize the algorithm parameters.
2. Calculate the evaluation value generated by the current allocation result (the result obtained by the evaluation function is called the evaluation value), and then perform the random disturbance, the disturbance mode is to exchange the subtasks of the two different positions with each other.
3. Calculate the evaluation value of the new allocation result. At the same time, a flag bit is added as the current optimal solution to record the optimal mapping result.
4. Calculate the difference between the old and new evaluation value. If the new evaluation value is better than the old evaluation value, replace the old evaluation value with the new evaluation value, otherwise accept the new mapping result with a certain probability. The purpose of this is that because continuing the search on the poorer mapping results may get a better solution, thereby jumping out of the local optimal solution and finding a more ideal solution. The smaller the difference between the current evaluation value and the new evaluation value, the lower the temperature will be and the closer the solution obtained is to the optimal solution. So the acceptance probability will decrease as the difference and temperature decrease.

5. Repeat steps 2 to 4 according to the number of iterations at initialization until pre-specified the number of iterations is reached, and then execute 6.
6. Check the temperature as it is dropping, if it reaches the stop temperature, stop the iteration and record the final distribution result, otherwise, return to step 2 again.

When the algorithm converges as the temperature finally meets the stop criteria, the mapping of the task to the IP core is completed.

The Mapping of IP Core to NoC Platform: The allocation result of the first stage is taken as the initial input, and the KL_SA algorithm is adopted as the NoC mapping scheme in this phase.

The specific steps are as follows.

1. The IP core set in the core communication graph is regarded as a set of nodes, which is recorded as set V.
2. The partitioning operation is performed on the set V, and the steps of dividing the operation are as follows.
   (1) Randomly divide the set into two subsets A and B containing the same number of nodes.
   (2) Calculate the internal communication cost and external communication cost of the subsets A and B respectively. The calculation formula is:

$$E_a = \sum_{v_i \in A, v_j \in B} w(v_i, v_j)$$

$$E_b = \sum_{v_i \in B, v_j \in A} w(v_i, v_j)$$

$$I_a = \sum_{v_i \in A, v_j \in A} w(v_i, v_j)$$

$$I_b = \sum_{v_i \in B, v_j \in B} w(v_i, v_j) \tag{15}$$

where I_a, I_b represents the internal communication cost of the set A, B, and E_a, E_b represents the external communication cost of the set A, B, v_i, v_j represents an IP core node, $w(v_i, v_j)$ represents the communication cost of nodes v_i to v_j.

(3) The nodes in subsets A and B are exchanged until a partition is found, maximizing the difference between internal and external communication costs. The difference D of the communication cost is calculated by the Formula (16).

$$D = I_a + I_b - (E_a + E_b) \tag{16}$$

3. The generated new set is recursively divided until there are only 2 subtasks in each divided subset. Finally get a set of sets, each of which has two subtasks.

4. We use the final set as the IP core to the NoC node initialization mapping method, and then initialize the algorithm parameters. The IP cores are randomly allocated to the NoC of the unit of subset, and 2 nodes in each subset are to be allocated to adjacent positions of NoC, where each node uniquely corresponds to a NoC resource node.
5. Calculate the current mapping evaluation value E_o, and set a flag E_f to be the current optimal solution, initialized to E_o.
6. Perform random perturbation by interchanging the IP core mappings of two different locations.
7. Calculate the evaluation value of the new mapping E_n.
8. If the new evaluation value is better than the old evaluation value $E_n < E_o$, replace the old evaluation value with a new evaluation value $E_o = E_n$, if the new evaluation value is better than the current optimal solution $E_n < E_f$, update the current optimal solution to the new evaluation value $E_f = E_n$.
9. If the conditions in step 8 is not satisfied, accept the new mapping result with a certain probability. This is because if the new result is accepted, it may get a poor solution while the new mapping result may bring the capacity to jump out of the local optimal solution and obtain the global optimal solution. Between accepting and rejecting, the probability value for accepting the new solution is:

$$P = \exp(\Delta E / T) \tag{17}$$

where ΔE represents the difference between the two mapping results, and T represents the current temperature.
10. Repeat steps 6 to 9 according to the number of iterations at initialization until the number of iterations is reached, and then execute 11.
11. Check the temperature as it keeps dropping, if it reaches the stop temperature, stop the iteration and record the final mapping result, otherwise return to step 6.

When the algorithm converges as the temperature finally meets the stop criteria, the mapping of the IP core to the NoC platform is completed.

4 Evaluation

4.1 Experimental Environment and Algorithm Parameter

In this section, we explain how the experiment is designed and how the evaluation of the algorithm performance is done. Our experiments were performed in a 4 × 4 2D Mesh NoC structure, in which a static shortest path routing algorithm [21] is used as the routing algorithm. In order to get more accurate experimental simulation data, we use the BookSim simulation simulator. BookSim [22] is a widely used cyclic precision interconnect network simulator designed by the Stanford University nocs group.

In order to verify the performance of the mapping scheme proposed in this paper, three applications are used to analyze the scheme. The first example is the MPEG-4 decoding system, the second example is the VOPD application, and the third example is the MWD application. We assign the tasks of the three application instances to different IP cores such as DSP, graphics processor, RISCCPU, etc. The mapped NoC platform uses a 4 × 4 2D Mesh structure. The SA algorithm is used as the baseline to compare with the KL_SA algorithm proposed in this paper. The experiments and analysis are carried out on power consumption optimization model, delay optimization model, and multi-objective optimization model.

4.2 Comparative Experiment and Result Analysis

Figure 2 shows the power consumption statistics of three applications using SA algorithm and the KL_SA algorithm with power consumption optimization as a single target. It can be seen from the figure that the mapping scheme generated by the KL_SA algorithm proposed in this paper has a large optimization in power consumption compared with the SA algorithm. For the MPEG-4 decoding system, the mapping scheme generated by the KL_SA algorithm is reduced by 15.9% compared with the SA algorithm. For the VOPD application, the mapping scheme generated by the KL_SA algorithm is reduced by 11.7% compared with the SA algorithm. For MWD applications, the mapping scheme generated by the KL_SA algorithm is reduced by 17.5% compared to the SA algorithm.

Figure 3 shows the delay statistics of three applications using SA algorithm and the KL_SA algorithm with delay optimization as a single target. We use the delay of the SA algorithm as the benchmark for analysis. It can be seen from the figure that the mapping scheme generated by the KL_SA algorithm proposed in this paper has a large optimization in delay than the SA algorithm. For the MPEG-4 decoding system, the mapping scheme generated by the KL_SA algorithm is reduced by 10.5% compared with the SA algorithm. For the VOPD application, the mapping scheme generated by the KL_SA algorithm is reduced by 9.6% compared with the SA algorithm. For MWD application, the mapping scheme generated by the KL_SA algorithm is 13.3% lower than the SA algorithm.

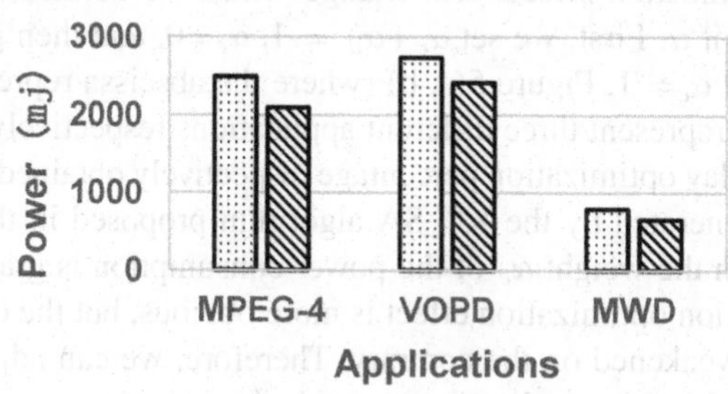

Fig. 2. Power consumption obtained by SA algorithm and KL_SA algorithm

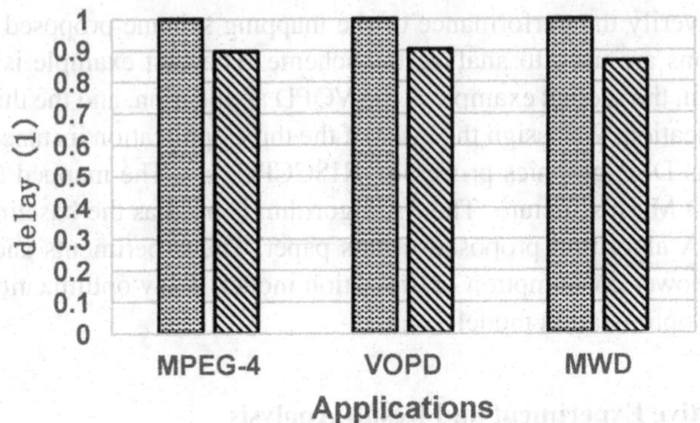

Fig. 3. Delay obtained by SA algorithm and KL_SA algorithm

For the multi-objective optimization of NoC, we performed set of experiments. we set the power consumption and delay weights to $\alpha_e = \alpha_d = 0.5$, Fig. 4(a, b) show the power consumption and delay statistics of the three applications using the SA algorithm and the KL_SA algorithm. We use the power consumption and delay of the SA algorithm as the benchmark. It can be seen from the figure that the mapping scheme generated by the KL_SA algorithm proposed in this paper has a large optimization in power consumption and delay compared with the SA algorithm. For the MPEG-4 decoding system, the mapping scheme generated by the KL_SA algorithm is 10.8% lower in power consumption and 8.5% lower in delay than the simulated annealing algorithm. For VOPD application, the mapping scheme generated by the KL_SA algorithm is better than the SA algorithm. As a result, the power consumption is reduced by 8.7% and the delay is reduced by 6.6%. For the MWD application, the mapping scheme generated by the KL_SA algorithm is 13.7% lower than the SA algorithm in power consumption, and the delay is reduced by 10.3%.

When adjusting the power consumption and delay weight α, the power consumption and delay optimization effects also change. Thus, we do more experiments with adjusting the value of α. First, we set $\alpha_e + \alpha_d = 1$, $\alpha_e = 0$, and then gradually increase the value of α_e until $\alpha_e = 1$. Figure 5(a, b) (where the abscissa represents the value of α_e, the three curves represent three different applications respectively) show the power consumption and delay optimization percentage respectively obtained by comparing the mapping scheme generated by the KL_SA algorithm proposed in this paper with the SA algorithm. When the weight α_e of the power consumption is gradually increasing, the power consumption optimization effect is more obvious, but the delay optimization effect is gradually weakened on the contrary. Therefore, we can adjust the value of α according to the needs of the application to get the best mapping results.

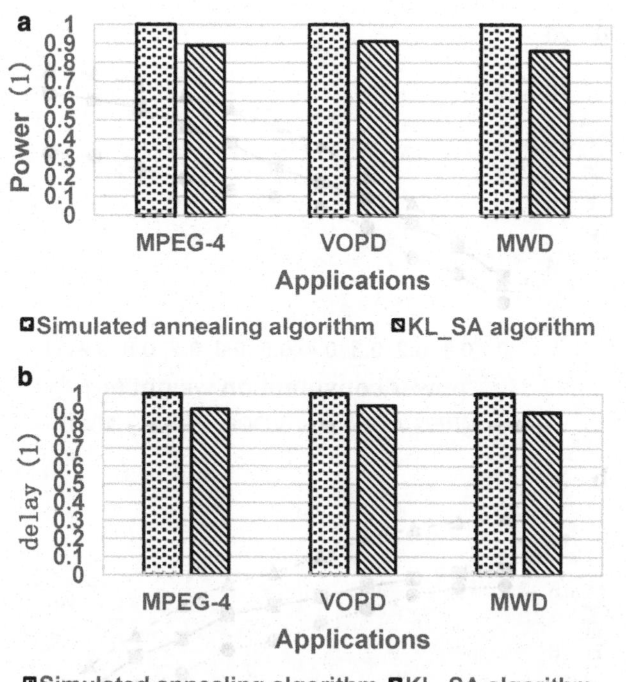

Fig. 4. (a) Power consumption obtained by SA algorithm and KL_SA algorithm (Multi-objective optimization, $\alpha_e : \alpha_d = 1 : 1$). (b) Delay obtained by SA algorithm and KL_SA algorithm (Multi-objective optimization, $\alpha_e : \alpha_d = 1 : 1$).

In order to better illustrate the effectiveness of the proposed scheme, we compare it with the methods in Refs. [23, 24] respectively. For the MPEG-4 decoding system, compared with the SA algorithm, the power consumption of the mapping scheme generated by the KL_SA algorithm proposed in this paper is reduced by 15.9%, and the power consumption of the mapping scheme proposed in the Ref. [23] is reduced by 6.7%. For the VOPD application, the power consumption of the mapping scheme generated by the KL_SA algorithm proposed in this paper is reduced by 11.3%, and the power consumption of the mapping scheme proposed in the Ref. [24] is reduced by 6.2%. Therefore, it can be seen that the mapping scheme proposed in this paper has better optimization effect.

Through experiments and results analysis, under the single-objective optimization condition, the power consumption and delay of the system are optimized by using the NoC mapping scheme generated by the KL_SA algorithm. Under the multi-objective optimization condition, by setting different parameters, the optimization effect of power and delay can be adjusted to better meet the needs of different applications and improve

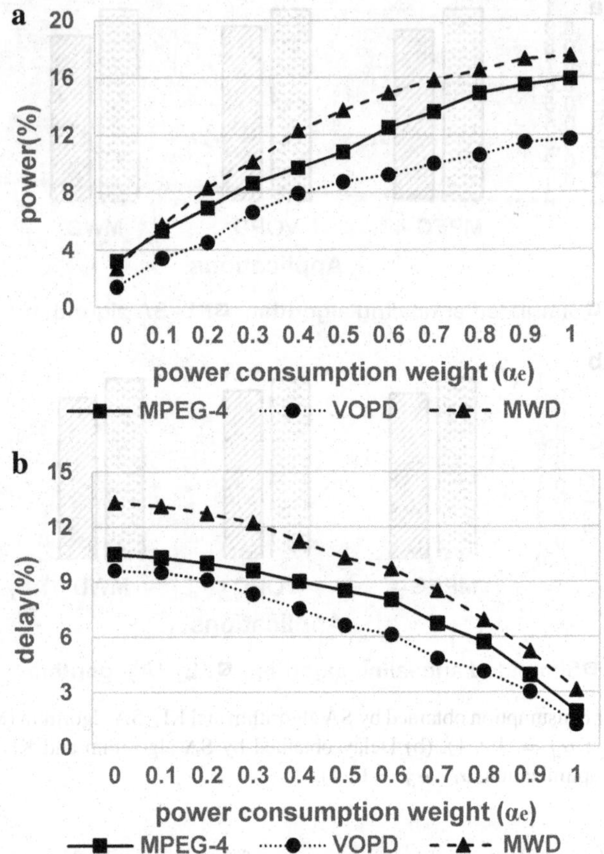

Fig. 5. (a) The power consumption optimization effect. (b) The delay optimization effect.

the flexibility of the mapping scheme. In terms of time complexity, this scheme is determined by the number of application task nodes *n*, the simulated annealing initial temperature *T*, the number of iterations of the simulated annealing algorithm *Iter*, the simulated annealing temperature drop rate *Temp_down*, and the simulated annealing minimum temperature *T_min*. Through calculation, the time complexity can be obtained as: $O^n * \text{Iter} * log_{\text{Temp\_ down}} \frac{\text{T\_min}}{T}$. It can be seen that the time complexity of this scheme has not increased compared to the simulated annealing algorithm. But NoC power consumption and delay have been greatly optimized, which can better illustrate the effectiveness of this scheme.

5 Conclusions

This paper first analyzes the NoC mapping problem in the heterogeneous multi-core circumstances. Then the power optimization model, delay optimization model and multi-objective optimization model are established, on which basis, a NoC mapping scheme based on KL_SA algorithm is proposed. The proposed scheme significantly improves the mapping optimization of heterogeneous multi-core NoC. The simulation experiments are carried out on three general applications. The results show that the mapping scheme proposed in this paper brings significant savings on power consumption and delay compared to the original simulated annealing algorithm.

In the future work, we will deeply study the topology of NoC which design a suitable architecture for the mapping scheme to further optimize the system performance.

Acknowledgement. This work is supported by the National Natural Science Foundation of China (Grant No. 61202076), Beijing Municipal Postdoctoral foundation, Chaoyang District Postdoctoral foundation, along with other government sponsors. The authors would like to thank the reviewers for their efforts and for providing helpful suggestions that have led to several important improvements in our work. We would also like to thank all teachers and students in our laboratory for helpful discussions.

References

1. Dai, H., Qu, H., Zhao, J.: QoS routing algorithm with multi-dimensions for overlay networks. Chisna Commun. **10**(10), 167–176 (2013)
2. Houtian, W., Xiangjun, Z.Q.X., et al.: Cross-layer design and ant-colony optimization based routing algorithm for low earth orbit satellite networks. China Commun. **10**(10), 37–46 (2013)
3. Juan, F., Zhenyu, L., Sitong, L., et al.: Exploring heterogeneous NoC design space in heterogeneous GPU-CPU architectures. J. Comput. Sci. Technol. **30**(1), 74–83 (2015)
4. Ren, X., An, J., Gao, Y., et al.: Fusion strategy with genetic and ant algorithms for low-power NoC mapping. J. Xian Jiaotong Univ. **46**(8), 65–70 (2012)
5. Zang Mingxiang, Z., et al.: Improved shuffled frog-leaping algorithm for low-power network-on-chip mapping. J. Xidian Univ. **42**(1), 118–123 (2015)
6. Qihua, D., et al.: Modified genetic algorithm based method on low-power mapping in network-on-chip[C]. In: International Conference on Applied Science & Engineering Innovation (2015)
7. Cui, H., Zhang, D., Song, G.: A novel mapping algorithm for three-dimensional network on chip based on quantum-behaved particle swarm optimization. Front. Comput. Sci. **11**(4), 622–631 (2017)
8. Srinivasan, K., Chatha, K.S., Konjevod, G.: Linear-programming-based techniques for synthesis of network-on-chip architectures. IEEE Trans. Very Large Scale Integr. (VLSI) Syst. **14**(4), 0–420 (2006)
9. Leary, G., Srinivasan, K., Mehta, K., Chata, K.S.: Design of network-on-chip architectures with a genetic algorithm-based technique. IEEE Trans. Very Large Scale Integr. (VLSI) Syst. **17**(5), 674–687 (2009)
10. Chatha, K.S., Srinivasan, K., Konjevod, G.: Automated Techniques for Synthesis of Application-Specific Network-on-Chip Architectures. IEEE Press, Piscataway, NJ 2008
11. Ost, L., Mandelli, M., Almeida, G.M., et al.: Power-aware dynamic mapping heuristics for NoC-based MPSoCs using a unified model-based approach. ACM Trans. Embed. Comput. Syst. **12**(3), 1–22 (2013)
12. Nadia, N., Carvalho, M.V., da Silva, L., Mourelle, D.M.: Preference-based multi-objective evolutionary algorithms for power-aware application mapping on NoC platforms. Expert Syst. Appl. **39**(3), 2771–2782 (2012)

13. Alagarsamy, A., Gopalakrishnan, L.: SAT: a new application mapping method for power optimization in 2D — NoC. In: International Symposium on Vlsi Design and Test, pp. 1–6 (2016)
14. Lang, H., Liu, A., Xie, M., Wang, T.: UAVs joint vehicles as data mules for fast codes dissemination for edge networking in smart city. Peer-to-Peer Networking Appl. **12**(6), 1550–1574 (2019)
15. Wang, T., Wang, P., Cai, S., Ma, Y., Liu, A., Xie, M.: A unified trustworthy environment establishment based on edge computing in industrial IoT. IEEE Trans. Ind. Inf. **16**(9), 6083–6091 (2019)
16. Jiang, S., Zhou, J., Lu, Z., et al.: A fault-aware low-power-dissipation dynamic mapping algorithm based on NoC1. In: 2017 20th International Conference on Electrical Machines and Systems (ICEMS), pp. 1–5 (2017)
17. Taassori, M., Niroomand, S., Uysal, S., Vizvari, B., Hadi-Vencheh, A.: Optimization approaches for core mapping on networks on chip. IETE J. Res. **64**(3), 394–405 (2018)
18. Qianqi, L., et al.: A multiobjective scatter search algorithm for fault-tolerant NoC mapping optimisation. Int. J. Electron. **101**(8), 1056–1073 (2014)
19. Rabindra, K., Srivastava, P., Sharma, G.K.: Network-on-chip: on-chip communication solution. Int. Rev. Comput. Softw. **5**(1), 22–33 (2010)
20. Niu, G.: The optimization of the search scheme by the simulated annealing algorithm. In: 2016 6th International Conference on Machinery, Materials, Environment, Biotechnology and Computer (2016)
21. Antunes, C.H., Henriques, C.O.: Multi-objective optimization and multi-criteria analysis models and methods for problems in the energy sector. In: Greco, S., Ehrgott, M., Figueira, J.R. (eds.) Multiple Criteria Decision Analysis. ISORMS, vol. 233, pp. 1067–1165. Springer, New York (2016). https://doi.org/10.1007/978-1-4939-3094-4_25
22. Medhi, D., Ramasamy, K.: Network routing, pp. 30–63 (2018)
23. Jiang, N., Becker, D.U., Michelogiannakis, G., et al.: A detailed and flexible cycle-accurate network-on-chip simulator. In: 2013 IEEE International Symposium on Performance Analysis of Systems and Software (ISPASS), IEEE (2013)

A Novel 3D Intelligent Cluster Method for Malicious Traffic Fine-Grained Classification

Baokang Zhao[1], Murao Lin[1], Ziling Wei[1(✉)], Qin Xin[2], and Jinshu Su[1]

[1] College of Computer, National University of Defense and Technology, Changsha 410073, China
{bkzhao,linmurao18,weiziling,sjs}@nudt.edu.cn

[2] Faculty of Science and Technology, University of the Faroe Islands, Torshavn, Faroe Islands, Denmark
qinx@setur.fo

Abstract. Distributed denial of service (DDoS) attacks have become one of the most serious threats to cloud network. Currently, most of the research on DDoS attack mitigation focuses on DDoS traffic detection without considering further analysis (e.g., fine-grained classification of mixed attack traffic). By further analysis, we can provide more support for attack interception and traceback. This paper proposes a new abnormal traffic classification method A3DC (Autoencoder-based Three-Dimensional Linear Cluster) to overcome the difficulty of fine-grained distinguishing DDoS attacks under a small amount of labelled training data in cloud network environment. Based on a novel proposed 3D cluster algorithm, A3DC method consisting of three stages, which are data normalization preprocessing, autoencoder downscaling, and data clustering, is designed. The experimental results on the public data set show that A3DC is significantly superior to existing methods in terms of mixed attack traffic classification and is able to obtain higher detection rate and lower false alarm rate in DDoS attack detection.

Keywords: Auto encoder · DDos detection · Attack classification

1 Introduction

Cloud servers have played a significant role in the advancement of intelligent and connected systems, like Internet of things, smart city, etc. However, as high-performance cloud services continue to deploy diversified technologies, security issues with regards to Cloud server network still prevail. Numerous intrusions and information theft attacks against cloud servers have caused huge losses to society and corporate groups. In recent years, DDoS attacks are one of the biggest

This work was supported by the National Natural Science Foundation of China (Grant No. 61972412) and the Science and Technology Innovation Program of Hunan Province (2020RC2047).

Y. Lai et al. (Eds.): ICA3PP 2021, LNCS 13155, pp. 385–401, 2022.
https://doi.org/10.1007/978-3-030-95384-3_25

threats to cloud servers [1]. The primary purpose of these attacks is to overload the victims by sending large amount of exception request and make its services unavailable for legitimate users. In October 2016, a major DDoS attack on Dyn, a major DNS provider, caused disruption to many major websites, including airbnb, Netflix, PayPal, visa, Amazon, New York Times, reddit, and GitHub. Top game production company Blizzard Entertainment has suffered more than ten DDoS attacks in the past few years, and many games on the platform have been seriously affected. According to neustar's data, the total number of DDoS attacks in the first half of 2020 increased by more than 2.5 times compared with the same period in 2019.

To deal with the threat of DDoS attacks, DDoS attack detection, which is the first step towards attack mitigation and traceback has gained the attention of researchers. The main methods of DDoS attack detection technology can be divided into attack detection based on traffic characteristics and attack detection based on traffic anomalies. The former is mainly based on various characteristic information of known DDoS, combined with expert domain knowledge, to establish a characteristic database of DDoS attacks. By comparing the data information of current network packets with the feature database, we can find whether this system is under DDoS attack. The main implementation methods are feature matching, model inference, expert system and so on. The latter mainly establishes the traffic model by learning the normal and abnormal traffic states to analyze the abnormal changes of the traffic. We can determine whether the traffic is abnormal to judge whether the server is attacked.

With the development of machine learning, increasing research focus on applying machine learning algorithms to attack traffic detection [2]. A considerable amount of attack detection methods are realized through supervised machine learning algorithms. However, these methods suffer from a high reliance on labelled data. In addition, they only focus on the classification of attack traffic with normal traffic and are not able to provide more information for further steps of attack mitigation.

It is concluded that the current DDoS detection algorithm needs further improvement. We need to extract more complete features by conducting further research on the characteristics of DDoS attack traffic. This allows us to implement the training of DDoS attack traffic models using a very small amount of data and reduce the manual work.

From the observation of DDoS attack module, it is found that the traffic from the same attack source presents some similar features in a short time (such as similar frame size, etc.), while the traffic from different attack sources performs different. We want to learn this feature of DDoS and observe whether we can extract attack traffic from the same attack source from a large number of mixed traffic. Figure 1 shows the mixed traffic in a network under DDoS attack. Homologous attack traffic shows the same color, and different sources of attack traffic shows different color. Traffic analysis software (e.g., wireshark) clearly displays that attack traffic is mixed with normal traffic. We can distinguish these attack traffic more carefully by learning the characteristics of homologous DDoS.

This paper proposes A3DC (Autoencoder-based Three-Dimensional Linear Cluster), based on a three-dimensional linear cluster algorithm, to overcome the

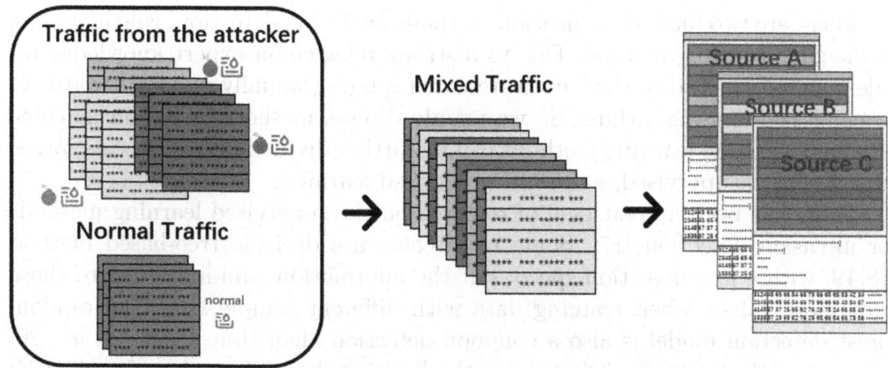

Fig. 1. DDoS mixed traffic

difficulty of fine-grained distinguishing abnormal traffic such as DDoS attacks in cloud network environment. The contributions of this paper can be summarized as follows.

- We exploit the characteristics of DDoS homologous attacks so that the attack traffic can be obtained fine-grained analysis. Thus, in this paper, we realize DDoS heterogeneous attack analysis for the first time. In addition, it also reduce the reliance on labelled data sets for training models.
- We creatively propose an A3DC method, which consists of three stages (i.e., data normalization preprocessing, autoencoder downscaling and data clustering). Besides attack detection, A3DC can achieve a fine-grained classification for the attack traffic under a small amount of labelled training data.
- Experimental results on public datasets show that A3DC significantly outperforms existing methods in classifying mixed attack traffic, and is able to obtain higher detection rates and lower false alarm rates in DDoS attack detection.

Following this introduction, the paper is organized as follows. Section 2 first reviews the related work of DDos attack detection. Section 3 introduces the novel method A3DC, and describes its structure in detail. In Sect. 4 we investigate the dataset CIC-IDS-2017 and show experimental results. The last section summarizes the whole paper.

2 Related Work

As the threat of DDoS attack becoming more and more serious, the focus on DDos detection has been growing in recent years. Some of researchers are concerned about software define network (SDN) [3–5], online social networks (OSNs) [6–8], backbone web traffic [9,10] and the cloud center [11–15]. The detection of DDoS can be the first and foremost part of DDos mitigation [16].

There are two main classification methods for DDoS detection, rule matching or machine learning method. Due to a strong reliance on expert knowledge for rule matching methods, the current research is now gradually moving towards to machine learning algorithms. So we mainly discuss methods based on machine learning. Machine learning methods can be further divided into three categories: supervised, unsupervised, and semi-supervised learning.

There has been a great deal of research on the supervised learning methods for intrusion detection [17]. Some researchers use decision tree-based method [18,19] with a fast detection speed, but the information gain may tend to those with more values when training data with different sample size. The random forest detection model is also a common detection algorithm. Cheng et al. [20] propose a DDoS attack detection method with enhanced random forest (RF) optimized by genetic algorithm based on flow correlation degree (FCD) feature. They use a genetic algorithm based on the FCD feature sequences to optimize two key parameters of the decision tree in the RF. Idhammad et al. [21] present a detection system of HTTP DDoS attacks based on the Information Theoretic Entropy and Random Forest ensemble learning algorithm with a time-based sliding window algorithm.

However, a lot of labeled data is generally required for a supervised learning method. Unsupervised learning does not have this limitation [22]. Zanero et al. [23] introduce a two-tier architecture to overcome the problem of the sheer size of the input. Amini et al. [24] present a real-time solution using unsupervised neural nets to detect known and new attacks in network traffic and observed 97% precision using ART nets, and 95% precision using SOM nets. Almalawi et al. [25] proposes an add-on anomaly threshold technique to identify the observations whose anomaly scores are extreme and significantly deviate from others, and then such observations are assumed to be abnormal. The most common method is the clustering algorithm [26]. Leung et al. present a new density-based and grid-based clustering algorithm that is suitable for unsupervised anomaly detection. Jiang et al. [27] present a clustering-based method for the unsupervised intrusion detection (CBUID). They consider the outlier factor of clusters for measuring the deviation degree of a cluster. Jha et al. [28] propose an immune system based real-time intrusion detection system using unsupervised clustering with a combination of a probabilistic model based T-cell algorithm and a decision tree based B-cell model. Casas et al. [29] propose a novel unsupervised outliers detection approach UNIDS, based on Sub-Space Clustering and Multiple Evidence Accumulation techniques. Wu et al. [30] test and experiment with different clustering algorithms.

Some methods based on clustering and outliers detection [31,32] also achieve good results but with a lack of robustness. Besides, most of the unsupervised learning method can only detect anomalies but not give some further information. Our research will focus on whether unsupervised algorithms can provide more attack information.

Considering these situations, some scholars decide to use semi-supervised method to train the model [33–36]. It can not only reduce the dependence on label data but also improve the detection accuracy. Idhammad et al. [33] present an online sequential semi-supervised ML approach for DDoS detection based on network Entropy estimation, Co-clustering, Information Gain Ratio and Extra-Trees algorithm. Gu et al. [34] propose a semi-supervised weighted k-means detection method. They present a Hadoop-based hybrid feature selection algorithm to find the most effective feature sets and propose an improved density-based initial cluster centers selection algorithm to solve the problem of outliers and local optimal. However, this study still uses 6,0000 labeled data to select the cluster center. Srihari et al. [35] use semi-supervised learning to extract and classify wavelet based features to detect DDoS attacks by using real-time dataset. Casas et al. [37] devise a novel attacks detection and classification technique based on semi-supervised Machine Learning (ML) algorithms to automatically detect and diagnose network attacks with minimal training, and compare its performance to that achieved by other well-known supervised learning detectors. Kaur et al. [38] use a distributed framework to apply semi-supervised machine learning techniques of k-means clustering for labeling a large dataset. Lysenko et al. [36] use the semi-supervised fuzzy c-means clustering and can ensure DDoS botnet detection at the rate at about 95%. It can be seen that the clustering algorithm is widely used in network intrusion detection, and we also introduce the idea of clustering to solve the problem of attack classification.

In summary, existing methods suffer from a high reliance on labelled data. In addition, they only focus on the classification of attack traffic with normal traffic and are not able to provide more information for further steps of attack mitigation.

3 The Proposed Method

In this section, we will introduce the structure of the proposed A3DC detection method. As shown in Fig. 2, the A3DC method is mainly composed of three stages: data initialization module, data dimensionality reduction module and clustering module. The data initialization module is responsible for processing the original data with missing and infinite values. The data downscaling module is responsible for mining the state characteristics of abnormal traffic in low-dimensional space based on AE algorithm. The clustering module is based on the three-dimensional linear clustering algorithm 3DC, which introduces vector distance evaluation index and three-dimensional linear regression fitting algorithm to achieve effective clustering of traffic data in low-dimensional space.

As shown in Fig. 2, in the training phase, the AE-based data dimensionality reduction module will keep the three-dimensional data from hidden layer and transfer the data into the clustering module. The clustering module will maintain and update the line cluster table during the training process, and these line equations will be stored with their category information. In the testing phase, the test data is compressed by the encoder into three-dimensional data and then matched with each line in the table to achieve detection.

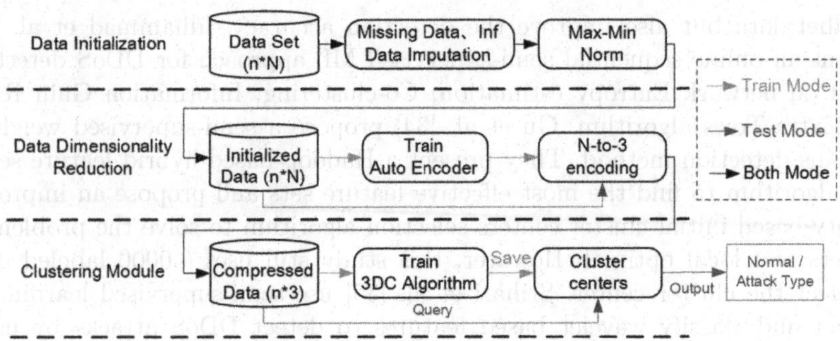

Fig. 2. The entire structure of A3DC

3.1 Data Initialization

In practice, due to the differences in research objects and research environments, the acquired raw data usually have the characteristics of missing data, unbalanced quantity, repeated redundancy, heterogeneous and containing non-numerical data. This makes it difficult to obtain meaningful information if these raw data are directly used for training, and it is difficult to reflect the powerful learning ability of machine learning.

In order to improve the training effect and solve the problems of the raw data mentioned before, the data reconstruction module is very important, which usually contains two parts, i.e., data imputation and data normalization.

Data Imputation. Most of the current publicly available traffic datasets (e.g., KDD, NSL-KDD, etc.) usually suffer from the situation of missing data and infinity(Inf) values. If the relevant data containing these values are directly deleted, it will lead to the loss of data information. Conversely, if all the data are directly retained, not only is it impossible to train the machine learning models, but it may also lead to higher false alarm rates.

Data containing missing values and infinity values are usually processed by direct deletion, mean fill or zero fill. Direct deletion means that data containing missing or infinite values are directly deleted. The disadvantage of this method is that it may make an impact on the overall distribution of the data. It will result in the deletion of a lot of valuable training data and affect the final training results. The mean fill method uses the overall mean of the attribute data in the training data set to fill the data, which makes the overall data relatively average. The zero-value filling method is to fill these values directly with zeros, which is one of the global constant filling methods. Numerous studies have shown that the zero-padding method retains most useful data information and has little effect on the experimental results. Thus, we choose the zero fill method in A3DC.

Data Normalization. The data normalization part is mainly used to solve the problem of excessive differences in data values. The individual features present in the raw data usually have different orders of magnitude and different distributions. If the data is used directly for calculation, features with smaller values will have less impact on the model, while features with larger values will have more impact on the model. In Data Normalization part, we can use standardization methods to unify the impact of different orders of magnitude of data on the model by unifying the distribution range of data in each dimension while preserving the size differences between data in the same dimension. The two commonly used normalization methods are maximum-minimum normalization method and L2 normalization method. After the experiments of two initialization methods, it shows that the Min-max normalization method works better. Thus, in this work, we also use the Min-max normalization method. The idea of this method is given as follows.

- Min-max normalization
  Suppose all the data of the m_{th} dimension is $\{x_1, x_2, ..., x_n\}$. Min-max normalization use the equation blow to make the value fall in the interval [0,1],

$$x_i' = \frac{x_i - \min\limits_{1 \le j \le n} \{x_j\}}{\max\limits_{1 \le j \le n} \{x_j\} - \min\limits_{1 \le j \le n} \{x_j\}} \quad (1)$$

  in which $\min\limits_{1 \le j \le n} \{x_j\}$ and $\max\limits_{1 \le j \le n} \{x_j\}$ respectively represent the maximum and minimum value of the m_{th} dimension.

For the dataset, the process of Min-max normalization is described as follows.

1) Screen out useful features of the dataset,
2) Get the maximum and minimum values of each dimension in the training dataset,
3) Update each data according to the maximum and minimum values.

Now we get valuable data that can be learned directly with machine learning algorithms. Next, we need to compress the data dimension.

3.2 AE-Based Data Dimensionality Reduction

We use AE to achieve data dimensionality reduction. In this section, we will first introduce the working principle of AE, and then its application in A3DC.

An auto-encoder (AE) [39] is one of the simplest neural networks. In order to making the reconstruction output close to the original input, the AE will learn data characteristics of low dimensional space. Therefore, AE provides a way to realize dimension reduction of data.

Suppose the j_{th} data is $D^{(n)} = \{d_1, d_2, ..., d_n\}$, in which d_i means the i_{th} dimension of this piece of data $D^{(n)}$, and n means the dimension of the data is n. The process of AE training is shown in the following equation.

$$h^{(k)} = f(W \cdot D^{(n)} + b) \quad (2)$$

$$z^{(n)} = S(W' \cdot h^{(k)} + b') \tag{3}$$

Equation (2) means AE maps the input data such as $D^{(n)}$ to k-dimensional hidden layer $h^{(k)}$. Equation (3) means AE maps the k-dimensional hidden layer $h^{(k)}$ to n-dimensional output layer $z^{(n)}$. b is k-dimensional biases and b' is n-dimensional biases. W and W' are the weight matrix of $k * n$ fully connected with input layer nodes and hidden layer nodes. We initialize $W' = W^T$. f is a nonlinear transformation function. In this paper, f is sigmoid function, $f(x) = \frac{1}{1-e^x}$, S is the activation function of the decoder, and $(\cdot)$ means a multiplication sign.

AE uses the reconstruction error to monitor whether the model has been trained well. The following loss function can calculate the reconstruction error.

$$\begin{aligned} L &= \|D^{(n)} - z^{(n)}\| \\ &= \sum_{i=1}^{n} [d_i \cdot log z_i + (1 - d_i) \cdot log(1 - z_i)] \end{aligned} \tag{4}$$

In this paper, we use cross entropy to train the module. Parameters W, W', b, b' can be updated by minimizing the loss function. This AE reconstruct features and encode the input as k-dimensional data. When the reconstruction error is tiny, it means that normal traffic can be reconstructed very well, and the network has learned the pattern of normal traffic.

Different from other related research, we focus on the exploration of hidden space. Similarly, Gong et al. [40] propose a scheme to add the attention mechanism in hidden space to improve the accuracy of monitoring. We need to explore the characteristics of hidden space data, which will be introduced in detail in the next sub-section.

After finish training the AE network, only the encoding layer is retained. The training data are transformed into three dimensional arrays by the encoding layer of the network. As shown in Fig. 3, each three dimentional data d_i is transformed into point coordinates x_i, y_i, z_i form. The set of coordinate data points will be transferred into the next module. The acquisition of hidden layer data can be achieved by Algorithm 1.

Algorithm 1

Input: *Dataset*;
Output: Hidden layer, $h^{(k)}$;

Train(Repeat)
for each $D^{(n)} = \{d_1, d_2, ..., d_n\}$ in $Dataset - train$ **do**
  # Calculate the output
  $z^{(n)} = S(W' \cdot h^{(n)} + b') = S(W' \cdot f(W \cdot D^{(n)} + b) + b')$
  # Calculate the loss
  $L_i = \|z^{(n)} - D^{(n)}\|$
  # Update Parameters by Stochastic Gradient Descent Until Convergence
end for
Run # Reduce the dimension

for each $D^{(n)} = \{d_1, d_2, ..., d_n\}$ in $Dataset - test$ **do**
    # Load Parameters $\{W, b\}$
    # Calculate the hidden layer
    $h^{(k)} = f(W \cdot D^{(n)} + b)$
end forreturn $\{h_1^{(k)}, h_2^{(k)}, ..., h_m^{(k)}\}$;
Transfer the processed data to the next module

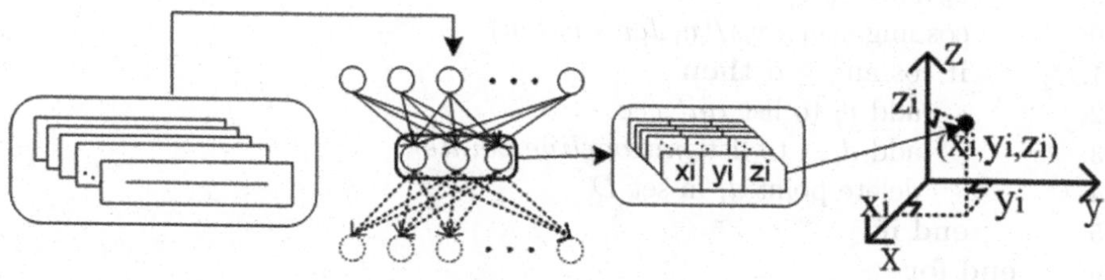

Fig. 3. 3D data transform into 3D points

3.3 3DC-Based Clustering Module

In the previous step, each piece of data is compressed to a three-dimensional vector. Each vector will be considered as a point in space. The task of classifying data is then transformed into the problem of clustering spatial data points. We will apply the proposed 3DC algorithm to the linear clustering module. The main flow process is shown in Fig. 4.

First, we select three points A, B, C from the training point set, and set A to be the initial point. Then we calculate vector $\overrightarrow{AB}$ and $\overrightarrow{AC}$, and set $\overrightarrow{AB}$ to be the initial vector. After completing the initial setup, the angle α between $\overrightarrow{AB}$ and $\overrightarrow{AC}$ will be computed.

If α is small enough, the point C will be considered in the same line as points A and B, and the point C will be added to the point set of the initial vector AB. On the contrary, C will be dropped in this round and a new unclassified point (e.g. the point D in Fig. 4) is extracted to re-execute the operation. After several calculations and judgments, the points in the training point set that may be on the same line are found. If the number of these points is too small to determine a straight line, a new point B will be chosen to form a new initial vector. The above process will be repeated until the set of selected points can be used to calculate the equation of a straight line.

Algorithm 2. Three-Dimensional Linear Clustering Algorithm (3DC)

Input: Compressed 3D dataset: $D = \{d_1, d_2, ..., d_n\}$
Output: Collection of straight lines: $lines = \{l_1, l_2, ..., l_m\}$
$l_i = (pointcoordinate, directionvector)$
$lines = \{\}$
if $len(D)$ >num_min **then**

Initially selected point d_1, and calculate the vectors between d_1 and other data points in D,
all vectors $V = \{d_1d_2, d_1d_3, ..., d_1d_n\}$, initially selected vector $v_1 = d_1d_2$,
set $v_1\_len = \sqrt{v_t^2}$, $init\_vec = \{\}$, $init\_oneline\_points = \{\}$
for v_i in V **do**:
  $\# \; v_i \leftrightarrow d_1d_{i+1} \quad \& \quad 2 \leq i \leq len(D)$
  $v_i\_len = \sqrt{v_i^2}$
  cos_ang=$(v_1 * v_i)/(v_1\_len * v_i\_len)$
  if cos_ang $\geq \alpha$ **then**
    add v_i to list *init_vec*
    add d_{i+1} to list *init_oneline_points*
    delete point d_i in set D
  end if
end for
if len(*init_oneline_points*) $\geq$ *num_min_points* **then**
  Use *init_oneline_points* to calculate the linear equation l
  add l to list *lines*
else
  Set *init_oneline_points* empty
  Put d_1 in the last position of set D
end if
end if

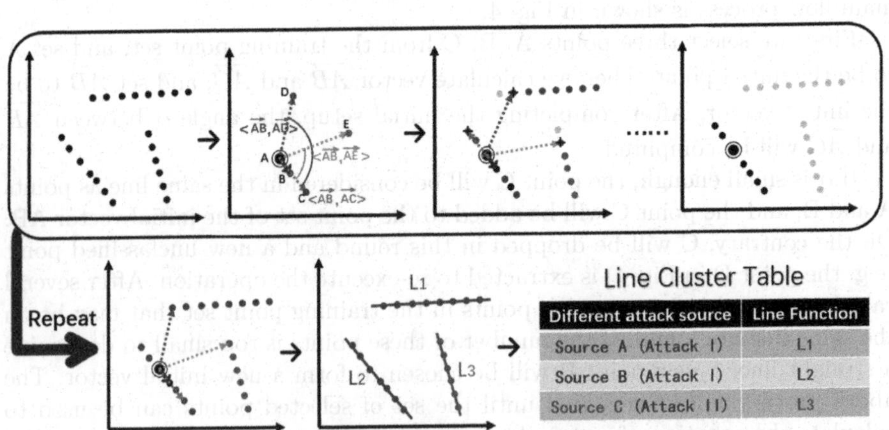

Fig. 4. Get line cluster table

Then we will calculate the line function by these chosen points and remove them from the original training point set. Next, new initial points and initial vectors are reselected, and a new round of computation starts again.

This process will be repeated continuously until the remaining points in the set cannot form a straight line. The training phase divides the data points into different linear sets, and then calculates the central linear equation of the set based on the coordinates of the points in each set, which is stored into the line cluster table. The complete linear clustering process of spatial points will be described in detail in Algorithm 2, where α is a decimal very close to 1.

4 Experimental Results and Analysis

4.1 Public Dataset CIC-IDS-2017

In recent years, the security of the cloud data center has become a hot research point in network security. However, some of the public datasets are out of date and unable to reflect the characteristics of cloud network traffic. We choose "`CICIDS2017 dataset`" [41], which is published by the research team *Institute for Cybersecurity (CIC) and University of Brunswick (UNB)* in 2017, to verify our approach. This dataset contains a variety of attacks, which have high research value [42,43]. There have been many studies on intrusion detection used this public dataset [44–48]. Researchers of CIC extract the 80 traffic features, saved as CSV files, from the dataset using *CICFlowMeter*. These CSV files include source IP, destination IP, source port, destination port, and traffic statistics data, like the Flow Inter arrival time (IAT) related features, the Flow Duration feature, etc.

Our experiment mainly used the dataset of *Benign(normal) in Monday, DDoS in Friday* and some other attack data. We firstly remove the labels of 900 attack data and 3000 normal data as labeled data for model training. The attack data contains three types of attacks, and there is no specific category label, all of which are marked as attacks. Secondly, we exploit about 5000 abnormal data as unlabeled attack data, and 10000 normal data as unlabeled normal data. Finally, we mixed the unlabeled data as test data to verify the detection ability of our method. We also use different combinations of data to show experimental methods from a perspective, which will be discussed in the next section.

4.2 Experimental Evaluation Index

We first give the definition of several indicators.

- True Positive (TP): Positive samples predicted to be positive
- True Negative (TN): Negative samples predicted to be negative
- False Positive (FP): Negative samples predicted to be positive
- False Negative (FN): Positive samples predicted to be negative
- True Positive Rate (TPR): $TPR = TP/(TP + FN)$
- False Positive Rate (FPR): $FPR = FP/(FP + TN)$
- Accuracy (ACC): $ACC = (TP + TN)/(TP + TN + FP + FN)$
- Precision (P): $PRE(N) = TN/(TN + FN); PRE(P) = TP/(TP + FP)$
- F1 measure (F1): $F1 = (2 * PRE * TPR)/(PRE + TPR)$

As known, usually supervised or unsupervised multi-class directly use the original labels to match the detection labels. Nevertheless, for A3DC, we need to add some steps.

In unsupervised attack classification experiment, the one-class attack has more than one *line-cluster*, so we use the following two tips to determine the label of each data in the test (α means a type of attack).

- Determine the class of each line: When we have a linear equation l_k, we will get a mount of points belongs to this cluster. Among these points, most of them belong to attack α, so we mark this line to belong to α. For example, $\alpha\_lines\_set : \{l_3, ..., l_k, ...\}$.
- Determine the class of each data: After calculating the distances between the data d_p to each line, we think that the point belongs to line l_k. By querying, l_k belongs to α_lines_set. So the label of d_p is α.

4.3 Experimental Results

The working principle of A3DC allows the method to achieve unsupervised traffic classification in mixed attack traffic. The total number of attack traffic of Bot in the CIC-IDS-2017 dataset is 1956. In order to make the data balanced and make the algorithm more uniform for the experiments of the three attacks, the experiments will extract the DDoS and DoS Hulk attack data of the same order of magnitude according to the data size of Bot. Meanwhile, the most commonly used clustering algorithm, k-means algorithm, is chosen to compare the effectiveness of 3DC in linear clustering.

As shown in Fig. 5(a), each color indicates the linear cluster classes obtained by 3DC, and the linear aggregation property is very obvious in the space, A3DC is able to realize the descending and clustering analysis of high-dimensional raw data. Figure 5(b) lists the specific metric values of the two clustering algorithms.

The DDoS attack traffic detection results are shown in Fig. 6. Figure 6(a) shows the DDoS attack detection confusion matrix, and Fig. 6(b) shows the detection index of DDoS attack detection. Among them, the detection rate of DDoS can reach 0.973, the accuracy can reach 0.921, and the F1-score can reach 0.946. Compared with the traditional methods, A3DC can achieve effective detection of DDoS attacks by a very small amount of training data.

Figure 7 shows the confusion matrix of detection results for each method in the comparison experiments. The experiments show that the proposed A3DC method has a lower false alarm rate compared to LIBSVM, MLP and Naive Bayes algorithm. And it can also show better detection performance for DDoS attacks, and the accuracy can reach 0.921.

Then, we evaluate the performance of multi-type detection for our proposed scheme. In this experiment, a total of over fifteen thousand data sets consisting of a mixture of normal traffic and multiple attack traffic are used. The experimental results compared with LIBSVM, MLP and Maive Bayes algorithm are shown in Table 1.

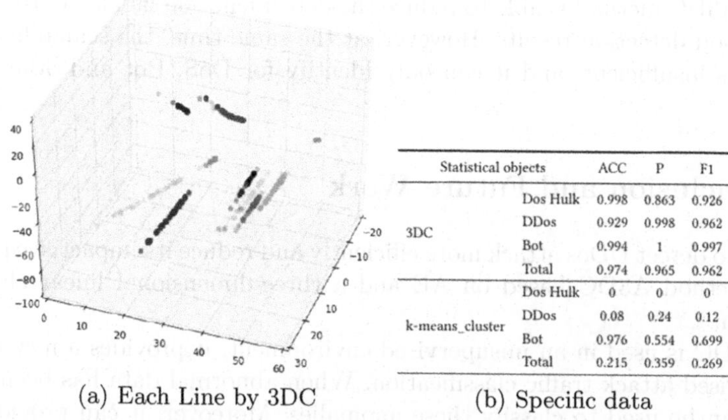

Statistical objects		ACC	P	F1
3DC	Dos Hulk	0.998	0.863	0.926
	DDos	0.929	0.998	0.962
	Bot	0.994	1	0.997
	Total	0.974	0.965	0.962
k-means_cluster	Dos Hulk	0	0	0
	DDos	0.08	0.24	0.12
	Bot	0.976	0.554	0.699
	Total	0.215	0.359	0.269

(a) Each Line by 3DC (b) Specific data

Fig. 5. Experimental result for attack classification by A3DC (Color figure online)

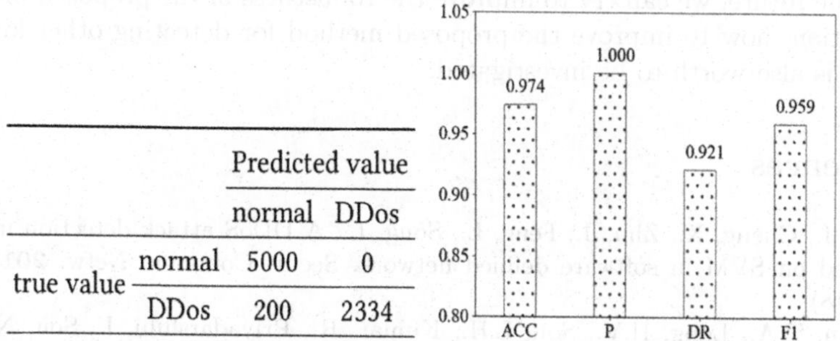

		Predicted value	
		normal	DDos
true value	normal	5000	0
	DDos	200	2334

(a) DDoS detection confusion matrix (b) DDoS detection related metrics

Fig. 6. DDoS detection results

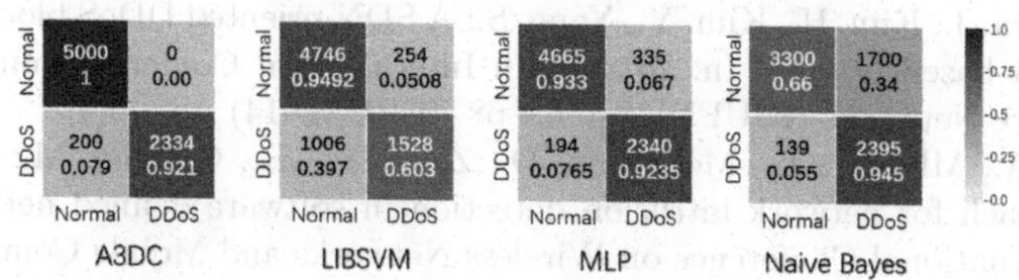

Fig. 7. Confusion matrix of detection results for each algorithm

Table 1. Multi-type attack detection accuracy P

	Our model	LIBSVM [49]	MLP	Naive Bayes
DoS Hulk	**95.3%**	0%	50.67%	89.21%
DDoS	**92.1%**	34.49%	81.14%	63.46
Bot	**99.4%**	12.22%	6.08%	68.6%

The A3DC method is able to reduce the dependence on labeled data sets and obtain good detection results. However, at the same time, the scalability of the method is insufficient, and it can only identify for DoS, Bot and other attack types.

5 Conclusion and Future Work

In order to detect DDos attack more efficiently and reduce its impact, we presents a new method A3DC based on AE and a three-dimensional linear clustering algorithm.

If A3DC is used in an unsupervised environment, it provides a new idea for unsupervised attack traffic classification. When abnormal data has been identified, it can be used to classify these anomalies. Moreover, it can provide more information for unsupervised test results. It can still be an excellent supervised method to detect DDos attack with very little labeled data.

In the future, we can try to improve the robustness of the proposed method. In addition, how to improve the proposed method for detecting other kinds of attacks is also worth to be investigated.

References

1. Ye, J., Cheng, X., Zhu, J., Feng, L., Song, L.: A DDoS attack detection method based on SVM in software defined network. Secur. Commun. Netw. **2018**, 1–8 (2018)
2. Tuan, T.A., Long, H.V., Son, L.H., Kumar, R., Priyadarshini, I., Son, N.T.K.: Performance evaluation of Botnet DDoS attack detection using machine learning. Evol. Intell. **13**(2), 283–294 (2019). https://doi.org/10.1007/s12065-019-00310-w
3. Mousavi, S.M., St-Hilaire, M.: Early detection of DDoS attacks against SDN controllers. In: 2015 International Conference on Computing, Networking and Communications (ICNC), pp. 77–81. IEEE (2015)
4. Lim, S., Ha, J., Kim, H., Kim, Y., Yang, S.: A SDN-oriented DDoS blocking scheme for botnet-based attacks. In: 2014 Sixth International Conference on Ubiquitous and Future Networks (ICUFN), pp. 63–68. IEEE (2014)
5. Tang, T.A., Mhamdi, L., McLernon, D., Zaidi, S.A.R., Ghogho, M.: Deep learning approach for network intrusion detection in software defined networking. In: 2016 International Conference on Wireless Networks and Mobile Communications (WINCOM), pp. 258–263. IEEE (2016)
6. Boshmaf, Y., Muslukhov, I., Beznosov, K., Ripeanu, M.: Design and analysis of a social botnet. Comput. Netw. **57**(2), 556–578 (2013)
7. Fire, M., Katz, G., Elovici, Y.: Strangers intrusion detection-detecting spammers and fake profiles in social networks based on topology anomalies. HFSP J. **1**(1), 26–39 (2012)
8. Savage, D., Zhang, X., Xinghuo, Yu., Chou, P., Wang, Q.: Anomaly detection in online social networks. Soc. Netw. **39**, 62–70 (2014)
9. Zhou, W., Jia, W., Wen, S., Xiang, Y., Zhou, W.: Detection and defense of application-layer DDoS attacks in backbone web traffic. Future Gener. Comput. Syst. **38**, 36–46 (2014)

10. Scholl, T.B.: Methods and apparatus for distributed backbone internet DDoS mitigation via transit providers. US Patent 8,949,459, 3 February 2015
11. Osanaiye, O., Choo, K.-K.R., Dlodlo, M.: Distributed denial of service (DDoS) resilience in cloud: review and conceptual cloud DDoS mitigation framework. J. Netw. Comput. Appl. **67**, 147–165 (2016)
12. Lee, J.-H., Park, M.-W., Eom, J.-H., Chung, T.-M.: Multi-level intrusion detection system and log management in cloud computing. In: 13th International Conference on Advanced Communication Technology (ICACT 2011), pp. 552–555. IEEE (2011)
13. Iqbal, S., et al.: On cloud security attacks: a taxonomy and intrusion detection and prevention as a service. J. Netw. Comput. Appl. **74**, 98–120 (2016)
14. Bakshi, A., Dujodwala, Y.B.: Securing cloud from DDoS attacks using intrusion detection system in virtual machine. In: 2010 Second International Conference on Communication Software and Networks, pp. 260–264. IEEE (2010)
15. Chung, C.-J., Khatkar, P., Xing, T., Lee, J., Huang, D.: NICE: network intrusion detection and countermeasure selection in virtual network systems. IEEE Trans. Dependable Secure Comput. **10**(4), 198–211 (2013)
16. Yu, S., Tian, Y., Guo, S., Wu, D.O.: Can we beat DDoS attacks in clouds? IEEE Trans. Parallel Distrib. Syst. **25**(9), 2245–2254 (2013)
17. Balkanli, E., Alves, J., Zincir-Heywood, A.N.: Supervised learning to detect DDoS attacks. In: 2014 IEEE Symposium on Computational Intelligence in Cyber Security (CICS), pp. 1–8. IEEE (2014)
18. Yi-Chi, W., Tseng, H.-R., Yang, W., Jan, R.-H.: DDoS detection and traceback with decision tree and grey relational analysis. Int. J. Ad Hoc Ubiquit. Comput. **7**(2), 121–136 (2011)
19. Lakshminarasimman, S., Ruswin, S., Sundarakantham, K.: Detecting DDoS attacks using decision tree algorithm. In: 2017 Fourth International Conference on Signal Processing, Communication and Networking (ICSCN), pp. 1–6. IEEE (2017)
20. Cheng, J., Li, M., Tang, X., Sheng, V.S., Liu, Y., Guo, W.: Flow correlation degree optimization driven random forest for detecting DDoS attacks in cloud computing. Secur. Commun. Netw. **2018**, 1–14 (2018)
21. Idhammad, M., Afdel, K., Belouch, M.: Detection system of HTTP DDoS attacks in a cloud environment based on information theoretic entropy and random forest. Secur. Commun. Netw. **2018**, 1–12 (2018)
22. Laskov, P., Düssel, P., Schäfer, C., Rieck, K.: Learning intrusion detection: supervised or unsupervised? In: Roli, F., Vitulano, S. (eds.) ICIAP 2005. LNCS, vol. 3617, pp. 50–57. Springer, Heidelberg (2005). https://doi.org/10.1007/11553595_6
23. Zanero, S., Savaresi, S.M.: Unsupervised learning techniques for an intrusion detection system. In: Proceedings of the 2004 ACM Symposium on Applied Computing, pp. 412–419 (2004)
24. Amini, M., Jalili, R., Shahriari, H.R.: RT-UNNID: a practical solution to real-time network-based intrusion detection using unsupervised neural networks. Comput. Secur. **25**(6), 459–468 (2006)
25. Almalawi, A., et al.: Add-on anomaly threshold technique for improving unsupervised intrusion detection on SCADA data. Electronics **9**(6), 1017 (2020)
26. Leung, K., Leckie, C.: Unsupervised anomaly detection in network intrusion detection using clusters. In: Proceedings of the Twenty-Eighth Australasian Conference on Computer Science, vol. 38, pp. 333–342 (2005)
27. Jiang, S.Y., Song, X., Wang, H., Han, J.-J., Li, Q.-H.: A clustering-based method for unsupervised intrusion detections. Pattern Recogn. Lett. **27**(7), 802–810 (2006)

28. Jha, M., Acharya, R.: An immune inspired unsupervised intrusion detection system for detection of novel attacks. In: 2016 IEEE Conference on Intelligence and Security Informatics (ISI), pp. 292–297. IEEE (2016)
29. Casas, P., Mazel, J., Owezarski, P.: Unsupervised network intrusion detection systems: detecting the unknown without knowledge. Comput. Commun. **35**(7), 772–783 (2012)
30. Wu, W., Alvarez, J., Liu, C., Sun, H.-M.: Bot detection using unsupervised machine learning. Microsyst. Technol. **24**(1), 209–217 (2016). https://doi.org/10.1007/s00542-016-3237-0
31. Portnoy, L.: Intrusion detection with unlabeled data using clustering. Ph.D. thesis, Columbia University (2000)
32. Eskin, E., Arnold, A., Prerau, M., Portnoy, L., Stolfo, S.: A geometric framework for unsupervised anomaly detection. In: Barbará, D., Jajodia, S. (eds.) Applications of Data Mining in Computer Security. ADIS, vol. 6, pp. 77–101. Springer, Boston (2002). https://doi.org/10.1007/978-1-4615-0953-0_4
33. Idhammad, M., Afdel, K., Belouch, M.: Semi-supervised machine learning approach for DDoS detection. Appl. Intell. **48**(10), 3193–3208 (2018)
34. Yonghao, G., Li, K., Guo, Z., Wang, Y.: Semi-supervised k-means DDoS detection method using hybrid feature selection algorithm. IEEE Access **7**, 64351–64365 (2019)
35. Srihari, V., Anitha, R.: DDoS detection system using wavelet features and semi-supervised learning. In: Mauri, J.L., Thampi, S.M., Rawat, D.B., Jin, D. (eds.) SSCC 2014. CCIS, vol. 467, pp. 291–303. Springer, Heidelberg (2014). https://doi.org/10.1007/978-3-662-44966-0_28
36. Lysenko, S., Savenko, O., Bobrovnikova, K.: DDoS botnet detection technique based on the use of the semi-supervised fuzzy c-means clustering. In: ICTERI Workshops, pp. 688–695 (2018)
37. Casas, P., D'Alconzo, A., Settanni, G., Fiadino, P., Skopik, F.: Poster: (semi)-supervised machine learning approaches for network security in high-dimensional network data. In: Proceedings of the 2016 ACM SIGSAC Conference on Computer and Communications Security, pp. 1805–1807 (2016)
38. Kaur, G.: A novel distributed machine learning framework for semi-supervised detection of botnet attacks. In: 2018 Eleventh International Conference on Contemporary Computing (IC3), pp. 1–7. IEEE (2018)
39. Vincent, P., Larochelle, H., Lajoie, I., Bengio, Y., Manzagol, P.-A., Bottou, L.: Stacked denoising autoencoders: learning useful representations in a deep network with a local denoising criterion. J. Mach. Learn. Res. **11**(12), 3371–3408 (2010)
40. Gong, D., et al.: Memorizing normality to detect anomaly: memory-augmented deep autoencoder for unsupervised anomaly detection. In: Proceedings of the IEEE International Conference on Computer Vision, pp. 1705–1714 (2019)
41. Sharafaldin, I., Lashkari, A.H., Ghorbani, A.A.: Toward generating a new intrusion detection dataset and intrusion traffic characterization. In: ICISSP, pp. 108–116 (2018)
42. Ring, M., Wunderlich, S., Scheuring, D., Landes, D., Hotho, A.: A survey of network-based intrusion detection data sets. Comput. Secur. **86**, 147–167 (2019)
43. Khraisat, A., Gondal, I., Vamplew, P., Kamruzzaman, J.: Survey of intrusion detection systems: techniques, datasets and challenges. Cybersecurity **2**(1), 1–22 (2019). https://doi.org/10.1186/s42400-019-0038-7
44. Koroniotis, N., Moustafa, N., Sitnikova, E., Turnbull, B.: Towards the development of realistic botnet dataset in the internet of things for network forensic analytics: Bot-IoT dataset. Future Gener. Comput. Syst. **100**, 779–796 (2019)

45. Vinayakumar, R., Alazab, M., Soman, K.P., Poornachandran, P., Al-Nemrat, A., Venkatraman, S.: Deep learning approach for intelligent intrusion detection system. IEEE Access **7**, 41525–41550 (2019)
46. Vijayanand, R., Devaraj, D., Kannapiran, B.: Intrusion detection system for wireless mesh network using multiple support vector machine classifiers with genetic-algorithm-based feature selection. Comput. Secur. **77**, 304–314 (2018)
47. Chaabouni, N., Mosbah, M., Zemmari, A., Sauvignac, C., Faruki, P.: Network intrusion detection for IoT security based on learning techniques. IEEE Commun. Surv. Tutor. **21**(3), 2671–2701 (2019)
48. Moustafa, N., Jiankun, H., Slay, J.: A holistic review of network anomaly detection systems: a comprehensive survey. J. Netw. Comput. Appl. **128**, 33–55 (2019)
49. Chang, C.-C., Lin, C.-J.: LIBSVM: a library for support vector machines. ACM Trans. Intell. Syst. Technol. (TIST) **2**(3), 1–27 (2011)

Predicting Students' Academic Performance Based on Improved PSO-Xgboost: A Campus Behavior Perspective

Zhongyu Liang[1,2], Xiaoqiang Di[1,2,3](✉), Zhen Liu[1,2], Xu Liu[1,2], Xingxu Zhang[1,2], and Zhi Yu[1,2]

[1] School of Computer Science and Technology, Changchun University of Science and Technology, Changchun, China
dixiaoqiang@cust.edu.cn

[2] Jilin Province Key Laboratory of Network and Information Security, Changchun, China

[3] Information Center, Changchun University of Science and Technology, Changchun 130022, China

Abstract. Performance prediction research has become an important research task in the education field. Previous studies mainly used questionnaires and specific learning systems to collect data. Due to the richness of data sources and the diversity of features, the model's low prediction accuracy is still challenging. Therefore, extracting features related to academic performance and improving the prediction accuracy of the model are the keys to advance research and development. In this paper, we use multi-source campus data to mine student behavior characteristics from different angles. The results of Pearson's correlation coefficient show that there is a certain correlation between the features we extracted and academic performance. In particular, we designed a model based on the combination of XGBoost and an improved PSO algorithm to solve the multi-classification problem of academic performance. Finally, we verify on the real campus data set, and the results prove that the model can predict students' academic performance with high accuracy.

Keywords: Campus big data · Academic performance prediction · PSO-Xgboost · Particle swarm optimization

1 Introduction

Academic performance is one of the key indicators to evaluate the level of education. In recent years, with the widespread application of Educational Data

This work was supported in part by the Science and Technology Development Plan Project, Jilin, China, under Grant 20190302070GX and Grant ZD18027, and in part by the Education Department of Jilin Province under Grant JJKH20200795KJ, and Grant GH180148 and in part by the Next Generation Internet Technology Innovation Project of Cernet under Grant NGIICS20190503.

Y. Lai et al. (Eds.): ICA3PP 2021, LNCS 13155, pp. 402–421, 2022.
https://doi.org/10.1007/978-3-030-95384-3_26

Mining (EDM) technology, the study of academic performance prediction has become an important research direction and task in the field of education [1]. This not only helps educators design in-time interventions but also promotes personalized education. Specifically, it is of great use to the education of students with poor academic performance. If educators can predict their academic performance in advance, this will not only provide students with effective help and cultivate good study habits, but also enable relevant personnel to provide better services for education and teaching [2]. On the other hand, the development and application of EDM is conducive to the education system stakeholders, such as educators, administrators, and researchers to provide feedback, whose purpose is to help education managers understand the reasons for students' abnormal behaviors and help researchers discover hidden knowledge and patterns from a large amount of student behavior data to better understand the educational structure and evaluation of learning effect [3].

In the past ten years, the challenge we faced are the insufficient data sources. First of all, the source of data involves the privacy of the research object. The acquisition of data often requires a lot of human intervention, material resources and economic costs, which will hinder the development of educational data mining research. Therefore, more and more researchers collect data through specific online learning systems. In recent years, with the development of information technology, the above shortcomings can be overcome by the information system, and various behavioral data of students can be collected in a more convenient and hidden way. Another important challenge is how to extract features related to students' academic performance from massive campus data, and use these features to develop a predictive model with higher accuracy. The data acquired in the campus information system has problems such as multiple semantics, multiple noises, and missing data. The data preprocessing and feature selection methods are relatively rough. Therefore, the forecast results are not ideal. How to dig out valuable information from rich data and use it to build predictive models with better performance is also a key step in the research.

Our research first obtains multi-source campus data from the school information system, including student basic information data, consumption data, course performance data, and Internet data. Then the data is effectively cleaned, fused and classified. Secondly, valuable characteristics are extracted from different types of data. We adopt the method of self-defined formula to construct the characteristics of students' daily behaviors and analyzes the correlation between these characteristic information and students' academic performance. In addition, we propose a new hybrid model based on particle swarm optimization (PSO) and XGBoost, which uses joint inertia weights to improve the PSO algorithm. This makes the convergence speed and optimization ability of the model significantly improved. Finally, a large number of experiments are conducted on real data sets, and the proposed method is compared with a variety of machine learning methods. The results show that the algorithm we proposed effectively improves the performance of the classification results.

The main contributions of this paper are as follows:

1. We used and classified multi-source campus data, and discovered some new characteristics related to students' academic performance from different perspectives.
2. In order to improve the convergence speed of the algorithm, we have improved the inertia weight of the particle swarm optimization (PSO). Experiments have shown that our algorithm has better performance.
3. We propose a joint model based on PSO and Xgboost, which not only effectively optimizes the parameters of the model, but also improves the accuracy of classification.

2 Related Work

The process of converting raw educational data into useful information that could have a further great impact on educational research and practices is called Educational Data Mining (EDM) [4]. EDM technology comprehensively applies theories and technologies in other fields to solve problems in education research and teaching practice. By analyzing and mining education-related data, EDM technology can discover and solve various problems in education [5]. For example, it provides decision support for managers, helps students understand their own learning efficiency, and provides technical means for teachers to predict students' academic performance. Online learning systems [6], social networks [7], and smart mobile devices [8] provide a lot of applications and data for EDM research. For example, literature [9] uses text mining technology to analyze questions and discussion records in online courses, revealing the strong correlation between dropouts in different courses and dropout behaviors of friends. Bardh Prenkaj builds a model based on historical student course data on MOOCs [10]. Using this model, it is possible to predict whether a certain student can successfully complete the course. The literature [11] uses social analysis technology and random graph model to analyze the e-mails sent between students and uses graphs to represent the social relationships of students sending and receiving e-mails. Rui Wang and Gabriella Harari [12] used passive sensor data and self-reports from student smartphones to build prediction models.

Predicting students' academic performance is one of the earliest and most popular applications of EDM to predict students' future academic performance based on existing data, such as using machine learning algorithms to predict final scores or academic performance based on students' performance records. In addition, academic performance will also lead to different prediction results based on data sources, research methods, and types of EDM applications. EDM research usually uses machine learning or statistical techniques to analyze the data collected by traditional educational research methods in the early stage. Questionnaire surveys [13], classroom observations [14] and other methods are included, but these methods often have more interference with personal wishes and are not universal. Specifically, literature [15–17] mainly uses students' self-reports and questionnaires to reveal that physical activity, intelligence level and sleep habits are important factors affecting academic performance. Driven by the rapid development of the Internet, more open and visual data are used, including data from student forums, online learning platforms [18], etc., to help scholars build models of students' behavior, motivation and learning strategies. Furthermore, it can predict student performance and establish a curriculum and learning recommendation system. It can also provide students with an adaptive and personalized learning system. However, the data provided by the online learning system can no longer meet the needs of contemporary researchers. In recent years, many colleges and universities have established various advanced information management systems to collect student activity data on campus. Therefore, more and more scholars use campus data to predict students' academic performance, help students improve their learning efficiency or provide decision support for education managers [19]. For example, Liang Zhao et al. [20] proposed a model named Augmented Education to predict student performance. Ashwin S and Gayathri R used multiple classifiers to classify and cluster student learning behavior data [21].

3 Methodology

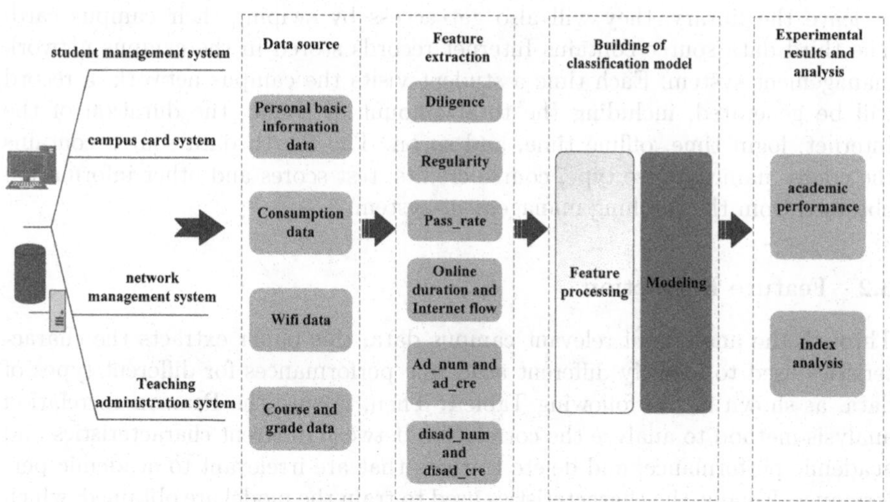

Fig. 1. Campus data analysis framework.

The overall framework is shown in Fig. 1. It contains four modules: data source, data feature extraction, modeling of student performance classification, and analysis and comparison of the results through experiments. Firstly, the data source module collects relevant data of students' campus life based on the student management system, campus card system, campus network management system and teaching management system, including basic personal information and consumption, Wifi data, course and grade results. Secondly, the feature extraction module extracts and selects effective features related to academic performance from different types of data. Thirdly, the PSO optimized Xgboost algorithm is used to train the academic performance classification model. In addition, we improve the inertia weight of the PSO, which effectively improved the performance of the algorithm. Finally, the model is applied to the real data set for verification. Each module is described in detail as follows:

3.1 Data Source

The data collected in this paper comes from four data sources. The first data source comes from a student management system that contains personal and basic information data. When a student enrolls in the college, his/her information will be entered into the system. It contains the following items: student ID, name, gender, student type (undergraduate, master, doctor), registration date, home address, ID number, etc. The second data source involves the campus card system. The campus card system records card swiping data related to each student's daily life activities. When a student swipes a card in a cafeteria or a store, the system will record the consumption amount, consumption time, balance in the card and the ID of the POS machines. When students enter or leave the library, they will also get access by swiping their campus card. The third data source contains Internet records stored in the campus network management system. Each time a student visits the campus network, a record will be generated, including the total amount of traffic, the duration of the Internet, login time, offline time, and so on. The fourth data source contains the course name, course type, course credits, test scores and other information obtained from the teaching management system.

3.2 Feature Extraction

Through the analysis of relevant campus data, this paper extracts the characteristics used to identify different academic performances for different types of data, as shown in the following Table 1. Then, we use the Pearson correlation analysis method to analyze the correlation between different characteristics and academic performance, and delete features that are irrelevant to academic performance. Finally, the characteristics used to train the model are obtained, which include regularity of life, diligence, online duration and traffic, the number and credits of dominant subjects, the situation of failing course, etc. These characteristics are selected for the following reasons: (1) They are all high-frequency behaviors, so we have a large number of records; (2) These characteristics can reflect students' lifestyles [22], learning behaviors and Internet usage [23], and are related to students' performance; (3) These characteristics help us quantify college students' campus behavior and further analyze the relationship with students' performance.

Table 1. Feature description.

Feature	Description
Diligence	The first smart card record every day
Life regularity	Regularity of life measured based on the concept of entropy
Ad_num	Number of courses with a score greater than 90
Disad_num	Number of failed courses
Ad_cre	Credits for courses with a score greater than 90
Disad_cre	Credits for failing courses
Pass_rate	The status of each student passing the exam
Online_duration	Average weekly online time
Internet_flow	Average weekly Internet flow

Life Regularity. Whether a student eats on time is an indicator that reflects whether the student's living condition is orderly within a period of time, and reflects the regularity of the student's life to a certain extent. This article uses information entropy and student consumption data to calculate the regularity of life. First of all, we divide meal times of each day into three periods, breakfast time is 6:00–10:00, lunch time is 11:00–15:00, and dinner time is 16:00–19:00. For each period, the time interval spans 30 min and is coded as 1,..., N. Then we count the distribution of students' meals in various time periods. Inspired by the concept of information entropy, we define life entropy to quantify the regularity of students' lives. Firstly, we define the formula for calculating the student's dining probability p_i in each time period:

$$p_i = \frac{T_i}{\sum_{i=1}^{N} T_i}, i = 1, 2, ..., N \tag{1}$$

Where T_i represents the number of meals in the ith time interval, and p_i represents the probability of meals in the ith time interval.

The formula for calculating life entropy R is as follows:

$$R = -\sum_{i=1}^{N} p_i log p_i \tag{2}$$

When $\sum_{i=1}^{N} T_i = 0$, it means that the student does not have any consumption records within the statistical time range. At this time, we stipulate that R=-1. According to the definition of entropy, the smaller the value of R, the higher the orderliness of the students' three meals and the higher the regularity of life.

However, only from the consumption data of students, the regularity of students' lives has the disadvantage of uneven distribution. The R value of students who spend less days in the cafeteria but relatively concentrated in the consumption range is relatively small; Students who spend more days in the cafeteria but whose consumption range is slightly less concentrated have a higher R value (the smaller the regularity index, the more regular the behavior). Therefore, this paper proposes a penalty factor β to balance the regularity of students' lives, defined as follows:

$$\beta = \frac{TotalD}{d} \tag{3}$$

Where $TotalD$ represents the total number of days in the statistical time range, and d represents the number of days the student spends in the statistical time range, The updated life regularity R is calculated as follows:

$$R = -\sum_{i=1}^{N} p_i logp_i \times \beta \tag{4}$$

Diligence. The earliest consumption time of a student each day can be approximated as the student's wake-up time. This reflects the diligence level of the students to a certain extent. Specifically, when calculating the diligence of students, we first obtain a record of the time when the student uses the campus card for the first time each day. We convert the raw date-time format into Unix timestamp. Then calculate the average value of the student's first swiping time in the semester. Finally, it is linearly transformed and mapped to [0,1]. The specific formula of student diligence D is as follows:

$$D = \frac{T_t - T_0}{f} \tag{5}$$

Where T_t represents the average time that the student uses the campus card for the first time in a semester. Since the time period we define for breakfast time is 6:00–10:00, f represents the timestamp of this time period. The value of T_0 is the timestamp at 6 o'clock in the morning. The smaller the value of D, the earlier the first time the student swipes the card, that is, the better the student's diligence.

Online Duration and Internet Flow. The formula of online duration and Internet flow is as follows:

$$IT = \frac{\sum_{i=1}^{K} t_i}{w} \tag{6}$$

$$IF = \frac{\sum_{i=1}^{K} F_i}{w} \tag{7}$$

Where K represents the frequency of students surfing the Internet in a semester; t_i represents the ith online duration of the student(measured by timestamp); F_i represents the ith Internet flow of the student; w represents the number of weeks we count.

The Learning Situation of the Dominant Subject and the Failure of the Course. We count students' course data for a semester, and take the number and credits of dominant subjects ($score \geq 90$), the number and credits of failed courses ($score < 60$), and the course pass rate as performance characteristics. The specific calculation formulae are as follows:

$$ad\_num/disad\_num = \sum course_i \tag{8}$$

$$ad\_cre/disad\_cre = \sum credit_i \tag{9}$$

$$pass\_rate = \frac{C}{M} \tag{10}$$

Where $course_i$ and $credit_i$ represent the number of courses and course credits respectively. $pass\_rate$ represents the student's course pass rate in a semester, C represents the number of courses that passed the exam, and M represents the total number of courses in a semester.

Academic Performance. We aggregate the student performance data to calculate the students' final average grade. However, the difficulty of courses in different majors is different. It is biased to only use the average grade of students to measure the academic performance of students. Therefore, we introduced the percentage of professional rankings and used it as an indicator to measure students' academic performance. In our study, performance prediction is regarded as a classification problem. According to the university's scholarship evaluation criteria, the students distributed in the interval [0-0.2] are marked as GPA_A, the students distributed in the interval (0.2-0.5] are marked as GPA_B, and the remaining students are marked as GPA_C.

$$grade_{ave} = \frac{\sum_{i=1}^{m} grade_i \times credit_i}{\sum_{i=1}^{m} credit_i} \tag{11}$$

$$GPA\_percentage_i = \frac{avg\_rank_i}{total\_num_i} \tag{12}$$

Where $grade_{avg}$ represents the average score of the course, $grade_i$ represents the test score of the ith course, m represents the total number of courses studied in this semester, avg_rank_i represents the ranking based on the average score of students, and $total\_num_i$ represents the total number of students, $GPA\_percentage_i$ represents the percentage of the student's ranking in the major. The lower the $GPA\_percentage_i$, the better the student's academic performance.

3.3 Building of Classification Model

This module includes two steps: feature processing and modeling. The values of features and academic performance are distributed in different ranges. Therefore, in order to avoid the impact of measurement units, the data needs to be standardized. We use the maximum and minimum normalization method, which is given by

$$X_{ij} = \frac{X_{ij} - X_{min}}{X_{max} - X_{min}} \tag{13}$$

where X_{max} and X_{min} respectively represent the maximum and minimum values of the jth feature. In addition, we introduced in detail the Xgboost algorithm, the PSO algorithm, the improvement of the inertia weight in the PSO algorithm, and the steps to optimize the Xgboost model using the improved PSO algorithm.

XGBOOST. Chen et al. proposed a scalable end-to-end tree boosting system called Xgboost [24], which is widely used in Kaggle competitions. In fact, the Xgboost algorithm is based on GBDT, which is composed of many decision trees and is usually used in the field of classification and regression. Compared with GBDT, Xgboost has the advantage of supporting linear classifiers. The Taylor expansion of the objective function is carried out by introducing the second derivative to make the result more accurate. Given a data set, there are n samples and m features

$$D = \{(X_1, y_1), (X_2, y_2), ...(X_i, y_i)\} \; (|D| = n, X_i \in R^m, y_i \in R) \tag{14}$$

The Xgboost model uses an additive training method to optimize the objective function, that is, the optimization process of the latter step depends on the result of the previous step. The objective function of Xgboost is defined as

$$obj^{(t)} = \sum_{i=1}^{n} l\left(y_i, \widehat{y}_i^{(t-1)} + f_t(x_i)\right) + \Omega(f_t) + constant \tag{15}$$

Where the first term $l(y_i, \widehat{y}_i)$ is the training loss function. The second term Ω is the regularization term of model, shown as

$$\Omega(f) = \gamma \cdot T + \frac{1}{2}\lambda \sum_{j=1}^{T} {w_j}^2 \tag{16}$$

both γ and λ are customization parameters. T and w denote the number of leaf node and score, respectively.

Obviously, our next step is to find a f_t that minimizes the objective function. We carry out a second-order Taylor expansion of formula (15). The objective function is approximately:

$$obj^{(t)} = \sum_{i=1}^{n} \left[g_i f_t(x_i) + \frac{1}{2} h_i f_t(x_i)^2 \right] + \gamma \cdot T_t + \lambda \frac{1}{2} \sum_{j=1}^{T} w_j^2 \tag{17}$$

Where g_i is the first derivative and h_i is the second derivative, they can be described as

$$g_i = \partial_{\widehat{y}_i^{(t-1)}} l\left(y_i, \widehat{y}_i^{(t-1)}\right) \tag{18}$$

$$h_i = \partial^2_{\widehat{y}_i^{(t-1)}} l\left(y_i, \widehat{y}_i^{(t-1)}\right) \tag{19}$$

Finally, the objective function is optimized, and the optimal solution is:

$$w_j^* = -\frac{\sum_{i \in I_j} g_i}{\sum_{i \in I_j} h_i + \lambda} \tag{20}$$

$$obj^* = -\frac{1}{2}\sum_{j=1}^{T}\frac{\left(\sum_{i \in I_j} g_i\right)^2}{\sum_{i \in I_j} h_i + \lambda} + \gamma \cdot T \tag{21}$$

PSO. The particle swarm optimization algorithm is an evolutionary computing technology proposed by two scholars named Kennedy and Eberhart in 1995 [25]. The two scholars were inspired by the cooperative and information-sharing behaviors among individuals in biological groups by observing the foraging behavior of bird swarm. Particles are the simulation of birds. Figure 2 describes the process of determining the optimal solution for particles.

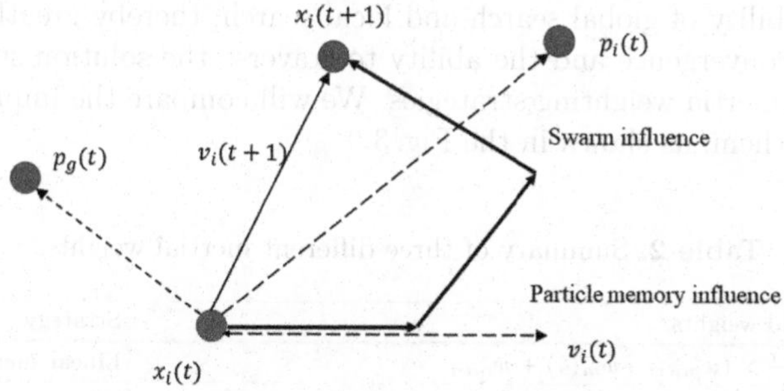

Fig. 2. Graphical representation of the particle evolution.

Assuming that there are N particles in the D-dimensional search space, and each particle represents a solution. The position of the $i-$th particle is $X_{id} = (x_{i1}, x_{i2}, ..., x_{iD})$, and the velocity of the $i-$th particle is $V_{id} = (v_{i1}, v_{i2}, ..., v_{iD})$. The historical optimal position found by the $i-$th particle is $P_i = (p_{i1}, p_{i2}, ..., p_{iD})$, which is called p_{best}, and the global optimal position found by all particles is expressed as $P_g = (p_{g1}, p_{g2}, ..., p_{gD})$, which is also called g_{best}.

During each iteration, the velocity and position of each particle updated by the following formulae

$$v_{id}^{(t+1)} = w \times v_{id}^{t} + c_1 \times r_1 \times \left(p_{id}^{t} - x_{id}^{t}\right) + c_2 \times r_2 \times \left(p_{gd}^{t} - x_{id}^{t}\right) \tag{22}$$

$$x_{id}^{t+1} = x_{id}^{t} + v_{id}^{t+1} \tag{23}$$

w is the inertia weight; $i = 1, 2, ..., N$; $d = 1, 2, ..., D$; t is the current number of iterations; v_{id} is the velocity of the particle; c_1 and c_2 are learning factors; the range of r_1 and r_2 is (0, 1).

The particle speed update formula consists of three parts: the first part is the speed before the particle, and the second part is the "cognition" part, which expresses the particle's own thinking. This can be interpreted as the distance between the particle's current position and its optimal position. The third part is the "society" part, which represents information sharing and cooperation between particles. It is the distance between the current position of the particle and the best position of the swarm.

Improved PSO XGBoost. As we all know, inertia weight plays an important role in balancing the global search and local search of particles. From the point of view of statistical analysis, the overall performance of PSO is strongly affected by the inertia weight. We also observed that a relatively larger inertia weight value helps to improve the global search ability, and the smaller inertial weight is conductive to accurate local search of the search space. This paper combines the nonlinear inertia weight with the sigmoid function (y = 1/(1 + exp(−x)) in the neural network to construct a new joint inertial weight. Its purpose is to balance the ability of global search and local search, thereby greatly improving the speed of convergence and the ability to traverse the solution space. Table 2 lists common inertia weighting strategies. We will compare the improved inertia weights with them, as shown in the Fig. 3.

Table 2. Summary of three different inertial weights.

Different inertial weights	Strategy
$w_0(t) = \frac{t_{max}-t}{t_{max}} \times (w_{max} - w_{min}) + w_{min}$	Linear inertia weight
$w_1(t) = -(w_{max} - w_{min}) \times \left(\frac{t}{t_{max}}\right)^2 + w_{max}$	Nonlinear inertia weight
$w_2(t) = \begin{cases} w_{max} - (w_{max} - w_{min}) \times (\frac{t}{t_{max}})^2, & (t \le \alpha t_{max}, \alpha = 0.21) \\ \frac{1}{1+e^{\frac{10t-2t_{max}}{t_{max}}}} + 0.4, & otherwise \end{cases}$	Joint inertia weight

Among them, t and t_{max} in Table 2 respectively represent the current number of iterations and the maximum allowable number of iterations. w(t) is used to represent the inertial weight of the $t-$th iteration. In order to show the changing trend of inertia weight, their iterative curves are shown in Fig. 3.

For linear inertia weights, the value of w decreases linearly from the initial value w_{max} to the final value w_{min}. However, because the slope is constant,

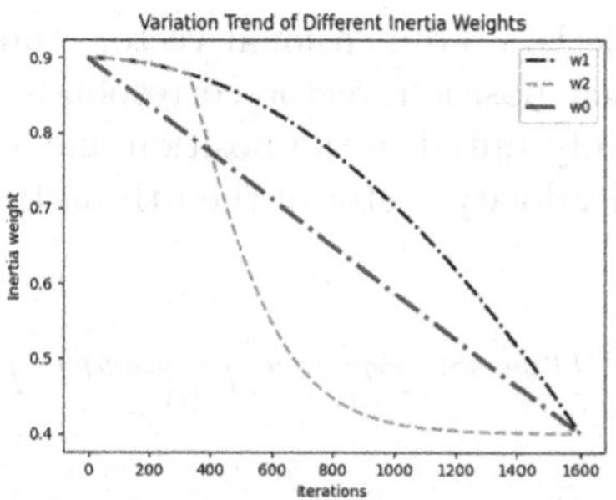

Fig. 3. Curves of different inertia weights

the speed change always maintains the same level. If a good initial value is not produced at the beginning of the iteration, then as the number of iterations increases and the speed decreases rapidly, it is likely that it will eventually fall into a local optimum. For the nonlinear inertia weight, although its slope is constantly changing, as the number of iterations increases, the value of the nonlinear inertia weight is always greater than the value of the linear inertia weight. It is not conducive to the local search of particles. For the joint inertia weight, in the early stage of the iteration, its weight changes within a larger value interval, which is beneficial to explore the entire search space. As the number of iterations increases, the weight gradually decreases, which is beneficial to the local search of particles. In addition, the slope of the joint inertia weight is constantly changing, and the value of its weight is dynamically adjusted according to the evolution speed of the particles, which can effectively avoid the situation where the particles fall into the local optimal solution. All in all, the joint inertia weights we proposed can effectively balance the global search and local search capabilities of particles, and can obtain better optimization results in a shorter time.

The Xgboost model includes general parameters, booster parameters and learning task functions. Table 3 shows the information about the Xgboost model parameters: eta, min child weight, gamma, subsample and colsample by tree.

Table 3. Information about parameters of the Xgboost model.

Parameter	Default	Range	Explain
Eta	0.3	[0, 1]	Learning rate
Min_child_weight	1	[0, ∞]	Minimum leaf weight
Gamma	0	[0, ∞]	Related to loss function
Subsample	1	(0, 1]	The proportion of sub-samples
Colsample_bytree	1	(0, 1]	The ratio of feature sampling

In this study, we used an improved particle swarm algorithm to optimize five parameters that have a greater impact on the model. Therefore, the position attribute of each particle is a 5-dimensional vector, and its range refers to the entire search space. Each position vector corresponds to a different Xgboost parameter. Then randomly initialize the position and velocity of the particles. The position vector and velocity vector of the i-th particle in the $t-$th iteration are expressed as

$$P_{i(t)} = \left[P_{i(t)}^{eta}, P_{i(t)}^{min\_child\_weight}, P_{i(t)}^{gamma}, P_{i(t)}^{subsample}, P_{i(t)}^{subsample\_bytree}\right] \tag{24}$$

$$V_{i(t)} = \left[V_{i(t)}^{eta}, V_{i(t)}^{min\_child\_weight}, V_{i(t)}^{gamma}, V_{i(t)}^{subsample}, V_{i(t)}^{subsample\_bytree}\right] \tag{25}$$

Then, we assign the position vector to the corresponding parameters of the model, and use the performance of the training set as the fitness value. False positive (FP) denotes the negative cases mistakenly detected as positive. False negative (FN) denotes the positive cases mistakenly detected as negative. The smaller the fitness value, the better the performance. The fitness value is defined as follows

$$Fitness = F_i^{(t)} = FN + FP \tag{26}$$

For the $i-$th particle, its individual optimal value at the $t-$th iteration is expressed as

$$p_{id}^{(t)} = min\left(F_i^{(j)}\right), 0 \leq j \leq t \tag{27}$$

Suppose there are m particles, the global optimum of the $i-$th particle at the $t-$th iteration is

$$p_{gd}^{(t)} = min\left(P_{kd}^{(t)}\right), 0 \leq k \leq m \tag{28}$$

We provide a concise and complete pseudo-code to help us better understand the optimization process of the algorithm. The pseudo-code algorithm is explained as follows:

Algorithm 1. Improved PSO-XGBoost Algorithm

Input: Number of Iteration *n_iterations*, Training data X, Initialize Parameters,x_{id}, v_{id}, P_i, P_i_*value*, P_g, P_g_*value*, *fitness_cadidate*
Output: Global optimal solution P_g after iteration

```
for iteration t = 1 to n_iterations do
    for i ∈ N do
        Calculate fitness_cadidate according to the position vector x_id
        if P_i_value >fitness_cadidate then
            Update the parameters P_i and P_i_value
        if P_g_value >fitness_cadidate then
            Update the parameters P_g and P_g_value
    for i ∈ N do
        Calculate the "new position" by Eq.(23)
```

Our goal is to use the improved PSO to find the optimal solution to the parameters of the Xgboost model. First, initialize the particle swarm parameters, including $n\_iterations$, x_{id}, v_{id}, P_i, P_i_value, P_g, P_g_value. Each particle traveled circularly through the search space at an initial speed. As described in line 3, in each iteration, the particle calculates its current fitness value (fitness_cadidate) according to its own position vector and training data. The fitness value is derived from the False positive and False negative of the Xgboost model and is used to evaluate the performance of PSO. The smaller the fitness value, the better the algorithm performance. According to the description in lines 4–8, if the fitness value of the historical optimal position of the particle is greater than the current fitness value, we should update the historical optimal position of the current particle and its fitness value. If the fitness value of the particle's global optimal position is greater than the current fitness value, we should update the current particle's global optimal position and its fitness value. Finally, as described in line 12, we continuously update the velocity of the particles through Eq. 22. Then calculate the particle's "new position" according to its velocity vector and the global optimal position, until the particle reaches the maximum number of iterations or its fitness value reaches the best, then the iteration process is terminated and the optimal parameters of the Xgboost model are obtained. Otherwise, continue the iterative calculation.

4 Experiments

4.1 Datasets and Evaluation Index

The data set of this study is derived from a data management system of a university, and it mainly contains the students' consumption records, grade records and Internet records of students, and all records have been anonymized. Due to the incomplete data of some students, a total of 2199 students were used in this experiment after screening. The statistical data is shown in Table 4. We divide the data set into training set (70%) and test set (30%). The classification model is trained with the training set and the test data set is applied. Finally, the verification index is extracted to calculate the robustness of the model.

Table 4. Basic statistics of our dataset.

Data type	Data size
Number of consumptions	820080
Number of Internet data	677053
Number of grades	80783

We use the following indicators to compare the prediction effect of the model:

$$precision = \frac{TP}{TP + FP} \tag{29}$$

$$recall = \frac{TP}{TP + FN} \tag{30}$$

$$F1 - score = \frac{2 \times precision \times recall}{recall + precision} \tag{31}$$

Precision represents the probability of the sample that is actually positive among all the samples that are predicted to be positive. The recall rate is also called sensitivity, which can reflect the proportion of a certain type of data that can be correctly identified by the model. F1 score is to measure the comprehensive performance of the model on two indicators. It is the harmonic mean of precision and recall. Its value ranges from 0 to 1. The larger the value, the better the effect of the model.

4.2 Correlation Analysis

After extracting features from multi-source data, we use Pearson coefficients for correlation analysis, and select features that are relevant to academic performance for follow-up experiments. As shown in the Table 5, the Pearson correlation coefficient falls within the range of $[-1, 1]$. The greater the absolute value, the higher the correlation. It can be seen from the Table 5 that the Pearson coefficients of the number of dominant disciplines, credits of dominant subjects, and the course pass rate are negative, indicating that they are negatively correlated with academic performance; similarly, diligence, regularity, and course failure Status and Internet access are positively correlated with academic performance.

Taking diligence and regularity as examples, we analyzed the distribution of students with different academic performance. Specifically, it can be seen from Fig. 4(a) that students with good academic performance usually have better eating habits. Figure 4(b) shows that diligence and academic performance are linearly correlated, and students with poor academic performance have their first activities later in the day. It can be seen that the features extracted from different dimensions have a certain correlation with academic performance, and it is feasible to use them as the features of predicting students' academic performance.

4.3 Results Analysis

In order to reveal the feasibility of these methods, the curve of fitness value with the number of iterations can be obtained. It can be easily seen from Fig. 5 that the evolution speed of $w_2 - PSO$ is faster, and compared with the corresponding particle swarm algorithm, it can quickly converge to the optimal solution. We can also clearly observe that in the early stage of the iteration, the convergence speed of $w_0 - PSO$ is the fastest, but the final fitness value is poor. The evolution speed of $w_1 - PSO$ and $w_2 - PSO$ in the first few iterations is slightly slower, but after that, $w_2 - PSO$ will converge to the global optimal value very significantly and has the best fitness value. Therefore, the PSO algorithm with joint inertia

Table 5. Correlation coefficient and P-value.

feature	Correlation coefficient	P-value
Diligence	0.57	$2.67e^{-19}$
Regularity	0.66	$3.81e^{-28}$
Ad_num	−0.6	$1.97e^{-22}$
Disad_num	0.73	0.001
Ad_cre	−0.61	$6.77e^{-22}$
Disad_cre	0.71	0.001
Internet_time	0.51	$1.94e^{-15}$
Internet_traffic	0.54	$3.09e^{-16}$
Pass_rate	−0.87	0.001

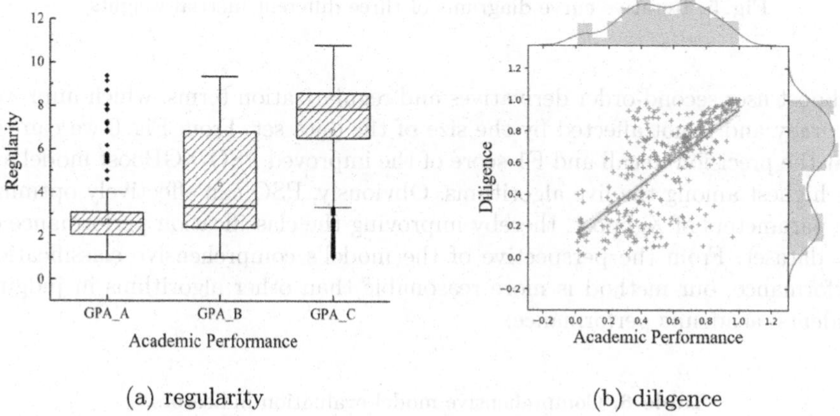

(a) regularity (b) diligence

Fig. 4. Correlation between student behavior and academic performance

weights proposed in this paper can achieve the best performance in most cases by better balancing the capabilities of global search and local search in the solution space.

In order to verify the effectiveness of the model, we compare the improved PSO-Xgboost model with the traditional classifier. The experimental results are shown in Table 6. LR, SVM, KNN, and RF perform poorly. The principle is that we cannot use logistic regression to solve nonlinear problems, and it is difficult to fit the true distribution of the data. The classic SVM only gives two classification algorithms, but in data mining, it is difficult to use SVM to solve multi-classification problems. KNN is very dependent on training data, and its fault tolerance is too poor, which will directly lead to inaccurate predicted data. RF may not produce good classification for small data or low-dimensional data. In addition, the classification model constructed with Xgboost is superior to the other four algorithms in terms of precision, recall and F1-score, because

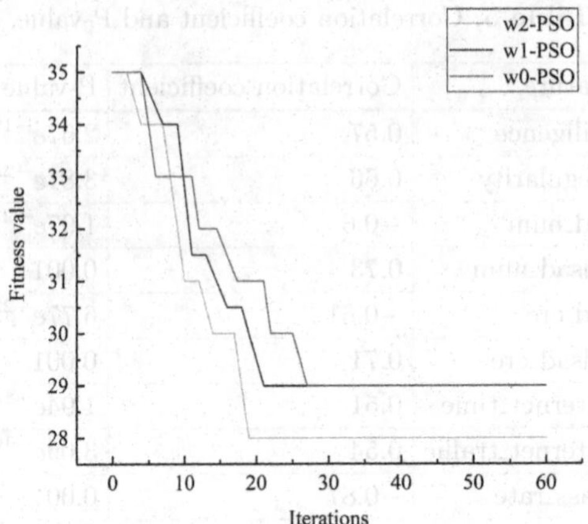

Fig. 5. Iterative curve diagrams of three different inertial weights.

Xgboost uses second-order derivatives and regularization terms, which improves accuracy and is not affected by the size of the data set. From Fig. 6 we can see that the precision, recall and F1-score of the improved PSO-XGBoost model are the highest among the five algorithms. Obviously, PSO can effectively optimize the parameters of Xgboost, thereby improving the classification performance of the dataset. From the perspective of the model's comprehensive classification performance, our method is more reasonable than other algorithms in judging students' academic performance.

Table 6. Comprehensive model evaluation indicators.

Method	Precision	Recall	F1-score
Logistic Regression	0.780	0.783	0.781
RandomForest	0.857	0.851	0.853
KNN	0.820	0.807	0.811
XGBoost	0.862	0.854	0.856
SVM	0.806	0.801	0.803
Improved PSO-XGBoost	0.894	0.891	0.892

Next, we compared our model with Xgboost optimized using three algorithms of GA, BCO and traditional PSO, applied the model on the test data set, and calculated indicators to verify its effectiveness and highlight the superiority of our proposed algorithm over other algorithms. Table 7 shows the comparison of verification indicators. We can see that all four optimizers are able to achieve promising results, which are better than traditional Xgboost. Moreover, the optimization effect of the PSO on Xgboost is higher than the other two,

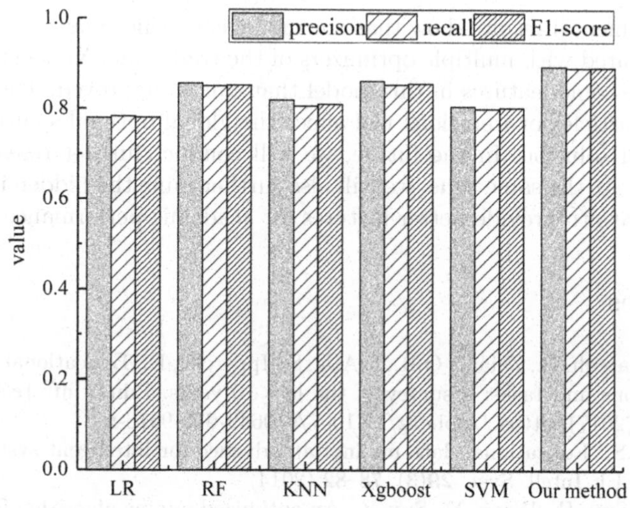

Fig. 6. Performance comparison of six algorithms.

indicating that the use of the PSO is effective for Xgboost optimization. In addition, comparing our algorithm and the PSO algorithm, the former is better than the latter in recall and F1-score. Therefore, the Xgboost classifier optimized by our improved PSO algorithm has high classification performance and achieves promising verification indicators.

Table 7. Comprehensive model evaluation indicators.

Optimization	Precision	Recall	F1-score
BCO -Xgboost	0.882	0.869	0.872
PSO-Xgboost	0.891	0.885	0.887
GA-Xgboost	0.877	0.869	0.871
Improved PSO-Xgboost	0.894	0.891	0.892

5 Conclusion

Improving the accuracy of academic performance prediction has always been an important research task in the field of educational data mining. The Xgboost model can be effectively applied to campus data to improve the prediction performance of multi-classification problems. In this article, we not only dig out new valuable features from a large number of data sets, but also use the PSO algorithm to search for the optimal parameters of the Xgboost model to find the best solution faster. In addition, we also discussed how to optimize Xgboost with

PSO. Experimental results show that our model can achieve better classification results compared with multiple optimizers of the traditional Xgboost algorithm.

There are some features in this model that can be improved. The size of the number of particles or iterations will affect the global optimal solution and the local optimal solution. In the future, we will conduct further research on the above issues. At the same time, we will also further mine the hidden information in campus data to provide better features for academic performance prediction.

References

1. Bakhshinategh, B., Zaiane, O.R., ElAtia, S., Ipperciel, D.: Educational data mining applications and tasks: a survey of the last 10 years. Educ. Inf. Technol. **23**(1), 537–553 (2017). https://doi.org/10.1007/s10639-017-9616-z
2. Baker, R.S.: Educational data mining: an advance for intelligent systems in education. IEEE Intell. Syst. **29**(3), 78–82 (2014)
3. Shao, Z., Sun, H., Wang, X., Sun, Z.: An optimized mining algorithm for analyzing students' learning degree based on dynamic data. IEEE Access **8**, 113543–113556 (2020)
4. Hooshyar, D., Pedaste, M., Yang, Y.: Mining educational data to predict students' performance through procrastination behavior. Entropy **22**(1), 12 (2020)
5. Dutt, A., Ismail, M.A., Herawan, T.: A systematic review on educational data mining. IEEE Access **5**, 15991–16005 (2017)
6. Liang, K., Zhang, Y., He, Y., Zhou, Y., Tan, W., Li, X.: Online behavior analysis-based student profile for intelligent e-learning. J. Electr. Comput. Eng. **2017** (2017)
7. Friedland, L., Jensen, D., Lavine, M.: Copy or coincidence? A model for detecting social influence and duplication events. In: International Conference on Machine Learning, pp. 1175–1183. PMLR (2013)
8. Gong, J., et al.: Understanding behavioral dynamics of social anxiety among college students through smartphone sensors. Inf. Fusion **49**, 57–68 (2019)
9. Feng, W., Tang, J., Liu, T.X.: Understanding dropouts in MOOCs. In: Proceedings of the AAAI Conference on Artificial Intelligence, vol. 33, pp. 517–524 (2019)
10. Prenkaj, B., Velardi, P., Stilo, G., Distante, D., Faralli, S.: A survey of machine learning approaches for student dropout prediction in online courses. ACM Comput. Surv. (CSUR) **53**(3), 1–34 (2020)
11. Uddin, S., Thompson, K., Schwendimann, B., Piraveenan, M.: The impact of study load on the dynamics of longitudinal email communications among students. Comput. Educ. **72**, 209–219 (2014)
12. Wang, R., Harari, G., Hao, P., Zhou, X., Campbell, A.T.: Smartgpa: how smartphones can assess and predict academic performance of college students. In: Proceedings of the 2015 ACM International Joint Conference on Pervasive and Ubiquitous Computing, pp. 295–306 (2015)
13. Zollanvari, A., Kizilirmak, R.C., Kho, Y.H., Hernández-Torrano, D.: Predicting students' GPA and developing intervention strategies based on self-regulatory learning behaviors. IEEE Access **5**, 23792–23802 (2017)
14. Godwin, K., Almeda, V., Petroccia, M., Baker, R., Fisher, A.: Classroom activities and off-task behavior in elementary school children. In: Proceedings of the Annual Meeting of the Cognitive Science Society, vol. 35 (2013)

15. Muñoz-Bullón, F., Sanchez-Bueno, M.J., Vos-Saz, A.: The influence of sports participation on academic performance among students in higher education. Sport Manag. Rev. **20**(4), 365–378 (2017)
16. Laidra, K., Pullmann, H., Allik, J.: Personality and intelligence as predictors of academic achievement: a cross-sectional study from elementary to secondary school. Pers. Individ. Differ. **42**(3), 441–451 (2007)
17. Dewald, J.F., Meijer, A.M., Oort, F.J., Kerkhof, G.A., Bögels, S.M.: The influence of sleep quality, sleep duration and sleepiness on school performance in children and adolescents: a meta-analytic review. Sleep Med. Rev. **14**(3), 179–189 (2010)
18. Liu, Z., Yang, C., Rüdian, S., Liu, S., Zhao, L., Wang, T.: Temporal emotion-aspect modeling for discovering what students are concerned about in online course forums. Interact. Learn. Environ. **27**(5–6), 598–627 (2019)
19. Francis, B.K., Babu, S.S.: Predicting academic performance of students using a hybrid data mining approach. J. Med. Syst. **43**(6), 1–15 (2019)
20. Zhao, L., et al.: Academic performance prediction based on multisource, multifeature behavioral data. IEEE Access **9**, 5453–5465 (2020)
21. Satyanarayana, A., Ravichandran, G.: Mining student data by ensemble classification and clustering for profiling and prediction of student academic performance. In: American Society for Engineering Education (2016)
22. Cao, Y., et al.: Orderliness predicts academic performance: behavioural analysis on campus lifestyle. J. R. Soc. Interface **15**(146), 20180210 (2018)
23. Shaojie, Q., Li, K., Zhang, S., Wang, Y.: Predicting achievement of students in smart campus. IEEE Access **6**, 60264–60273 (2018)
24. Chen, T., Guestrin, C.: XGBoost: a scalable tree boosting system. In: Proceedings of the 22nd ACM SIGKDD International Conference on Knowledge Discovery and Data Mining, pp. 785–794 (2016)
25. Júnior, D.A.D., et al.: Automatic method for classifying COVID-19 patients based on chest X-ray images, using deep features and PSO-optimized XGBoost. Expert Syst. Appl. **183**, 115452 (2021)

Motion-Sequence Authentication System: Guard for Smart Phones

Yuzheng Dong, Yanfeng Zhao, Ziyue Wang, Juan He(✉), and Liubin Zhu

School of Information Science and Technology, Northwest University, Xi'an, Shaanxi, China
hejuan@stumail.nwu.edu.cn

Abstract. In recent years, the mobile privacy protection is becoming increasingly critical due to the popularity of smartphones. The owner needs a "second line of guard" (user behavior authentication) when unlocking methods are attacked. The traditional user authentication approaches either face smudge attacks or can only work on dedicated devices. In this paper, we present a Motion-sequence Authentication System (MAS), an accurate and robust security authentication system that is not limited to expensive phones. MAS (Motion-sequence Authentication System) distinguishes user categories according to the unique characteristics of different user motion sequences. It is a rapid, non-contact and unobtrusive method of user authentication without predefined motions. MAS exploits Markov model to track the behavior of smartphone users, it can achieve real-time user authentication by utilizing the user's short-term interaction with the smartphone. Our experiments in multiple real environments show that MAS can achieve higher than 94% accuracy for authenticating user motion sequences, which fills the gap with motion sequence recognition and provides a way of thinking for the development of human-computer interaction and information security.

Keywords: Human-computer interaction · Behavior recognition · Multi-scenario security authentication · Markov model

1 Introduction

With the rich functionalities and enhanced computing capabilities available on smart phones, users not only store sensitive information, but also use privacy-sensitive applications on smart phones [19–21]. Consider that the smart phone can be accessed by many people other than the owner, such as be shared with families and friends, be stolen by a thief, or be peeped by an attacker, smart phone owners face increasing risk of privacy leakage [1]. If the smartphone can distinguish the owner and attacker during their accessing, it would be the "second line of guard" for owners.

The existing authentication methods fall short on one or more of the following aspects. The widely adopted PIN/pattern password is inherently vulnerable to shoulder surfing attacks and smudge attacks [2, 10]. Methods of the highest security are using the physiological biometrics of different persons (e.g., fingerprint and face) [14, 17, 18]. However, these solutions require dedicate hardware, such as fingerprinting scanner.

Y. Lai et al. (Eds.): ICA3PP 2021, LNCS 13155, pp. 422–441, 2022.
https://doi.org/10.1007/978-3-030-95384-3_27

A recent alternative is to use the behavior biometrics (e.g., hand waving, gait, touch-based biometrics) [3, 5], which however either incurs high delay (e.g., gait and touch behavior-based methods have to collect more than 10s motion data for accurate authentication [16]) or impose additional burden to users (i.e., perform a defined motion [3]). Moreover, the smart phone owners may not be willing to take distrust action to reduce permission deliberately before sharing. Besides, the above solutions cannot distinguish the attackers and the friendly users, providing them with the same access permission [4]. However, different from the attacker, the owners may allow the friendly user to access the insensitive applications (such as the mobile games, browser, YouTube, etc.) on the smart phone, but do not want them to access sensitive applications.

Under this circumstance, it would be good for a smart phone to identify who is the current user (the owner, a friendly user or an attacker) instantly and inconspicuously, as well as provide specific and targeted privacy protection and access control automatically for each role [7, 11, 15]. We meet many challenges to this version. First, without relying on a pre-defined distinguishable behavior, we have to identify three kinds of users only based on users interacting with the smart phone. Consider that different people (especially the friendly user and the attacker) exhibit neither consistent nor distinguishing behavior when they interact with the smart phone, it's hard to find features that can constantly identify the users with high accuracy. Second, a great challenge comes from the low-latency requirement. This means that we have to identify the user instantly based on their short-term behaviors, which lasts for short time.

In this paper, we for the first time propose a multi-user authentication system for smart phones, named MAS. MAS is able to instantly identify whether the current user is the owner, a friendly user or an attacker based on only the 2s key behavior, i.e., pick up and unlock the smart phone of the user. The idea behind MAS comes from the following intuitive observations:

i) Different users have the distinctive habit (or motion sequence) when he/she performs the key behavior;
ii) A person has distinctive features when he/she performs each independent motion in the key behavior. Obtaining the abovementioned fine-grained information in real time, however, is a very challenging task. Based on a joint consideration of the user's motion patterns and the order of motions, we provide a model based on Markov chain, which continuously track motions of the smart phone user, and instantly estimate how likely the motions are performed by the smart phone owner, a friendly user, or an attacker.

This paper also provides details of designing and implementing such a system. Specifically, we design a motion segmentation algorithm to detect the transition between two motions from the noisy sensing data. Then we leverage the distinct feature contained in each sub-segment of the unlocking motion, instead of the entire motion, to estimate the probability that the unlocking motion is performed by the smart phone owner himself/herself. At the same time, under the influence of Covid-19, wearing masks have become a necessary means of protection, which limits the application range of face recognition to a certain extent [23–25]. MAS can identify users independently of facial features, and does not need to take off the mask during detection, so it has great application potential in air defense work of Covid-19 epidemic situation.

Finally, we summarize the following contributions:

i) We observe the distinctive accessing habit of different types of smartphone users, and propose a model based on Markov chain to continuously track motions of the smartphone user, which can identify different users instantly and accurately.
ii) Based on the proposed model, we introduce MAS, the first system which can distinguish whether the current user is the owner or a friendly user or an attacker in real time, and then provides necessary privacy protection and access control based on the identification result.
iii) We implement MAS on several platforms (including Samsung S4, m1 mental, MI 2s, and OPPO r9s). The extensive experiments demonstrate the accuracy and robustness of MAS.

2 Related Work

Parallel work on identification can be classified into three categories: passwords/PINs/patterns, physiological biometrics and behavioral biometrics. However, passwords/PINs/patterns are inherently not security, since malicious users can unlock the smartphone by peeping attacks or smudge attacks. Physiological biometrics need extra hardware and also can be spoofed. Moreover, these solutions fail to achieve a goal of distinguishing multi-users.

The two key technologies in behavioral biometrics domain are feature based and similarity based.

Feature based: Feature based recognition approaches mainly exploit the distinctive features among different behaviors, extracting features of a behavior to establish a classifier [22–24]. For example, GEAT [3] is a gesture-based user authentication system, which extracts behavioral-related feature of predesigned 10 sliding gestures and uses a SVDE classifier. It achieves an average equal error rate of 0.5% to identify legitimate and illegitimate for each gesture. Touchalytics [5] presents a continuous authentication scheme that leverages features of a swipe. An evaluation of Touchalytics shows that with the SVM or the KNN classifier it provides an EER of 4% to distinguish owner and other people. LXG [6] devices following 9 features of users' finger movements during a swipe gesture and uses an SVM classifier to check the current behaviors of user against the owner's, which achieve an equal error rate (EER) of 8%.

They perform a high accuracy under the assumption that different behaviors have distinguishable features, which is not hold for the multi-user identification scenario, where the motions performed by different people are very similar, e.g., take out the smartphone. A pre-designed behavior to capture distinctive feature may impose additional burden to users and show distrust to another user.

Similarity based: Similarity based behavior recognition methods primarily maintain a group of well-defined behavior profiles. They distinguish behaviors based on similarity metrics (e.g., DTW and EMD) to evaluate the similarity between the sampled signals and the pre-constructed behavior. For example, GTGF [9] proposes a continuous mobile authentication that converts the touch traces to images and computes the score between two users' image, it achieves an EER of 2.62%. Sae-Bae et al. [13] provides a user authentication system based on 22 designed touch gestures that compute DTW distance and Frechet distance between users' traces to authenticate legal and illegal user. Luca et al. [8] proposes a user authentication system that directly computes the distance between touch traces using DTW algorithm.

However, it may perform low accuracy in such an identification scenario due to the variability of both the user and smartphone scenario, even the same motion (e.g., take-out smartphone) performed by the same user rarely have a fixed pattern [12].

Different from all past work, MAS combines users' order of using smartphone motions with independent motion patterns to authenticate owner, friendly user or attacker implicitly and instantly.

3 Preliminary

It may perform low accuracy in such an authentication scenario due to the variability of both the user and smart phone scenario, even the same motion (e.g., take-out smart phone) performed by the same user rarely have a fixed pattern [12]. The user behavior is short in duration and has little information to refer to. Unlike previous work, MAS combines a user's smart phone action sequence with independent action patterns to implicitly and instantly verifies an owner, user-friendly or attacker.

3.1 User Classification This Style for Level Two

The users are divided into three categories:

i) Owner: The owner of the smart phone device with using his/her smart phone freely.
ii) Friendly users: Smart phone owners share their device to them, but hiding the sensitive information unobtrusively.
iii) Attackers: The users who use the smart phone without owner's permission, the smart phone system does not provide any access permission with them.

In the first experiment, we observe how owners use their smart phones. The experiment lasts for 10 days and collects 489 samples for the key behavior of owner. The experiments are conducted in two scenarios: static scenario and dynamic scenario. In the second experiment, we firstly ask the owner to share their smart phones to 31 volunteers and then ask 20 volunteers to borrow a smart phone from 4 owners. We observe the key behavior of owners and borrowers respectively, collecting 489 samples. In the third experiment, we ask 20 volunteers to pretend to be attackers to 'peep' privacy without being noticed. We collect 40 samples for the key behavior of attackers when 'stealing' sensitive information. Some samples, for example, user picks up the smart phone, but does not unlock or use the smart phone, are invalid samples, and we throw them away.

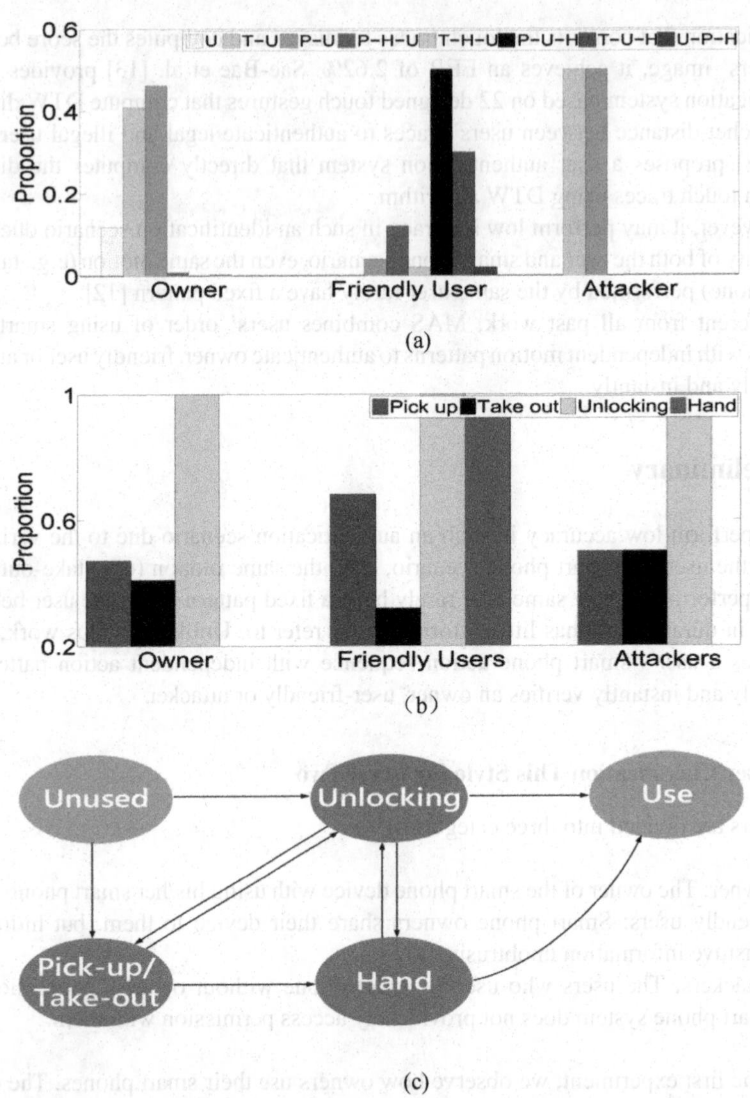

Fig. 1. Motion sequence of key behavior performed by different user: (a) probability of motion sequence in different users; U: Unlocking motion, T: Take-out motion; P: Pick-up motion, H: Hand motion (b) probability of each independent motion during a key behavior; (c) Motion sequence recognition process.

3.2 Data Analysis

In this subsection, we will further analyze the smart phone use behavior of the owner, friendly users and attackers. We abstract the user's key behavior into a motion sequence composed of several independent motions. We found that users' independent motion mainly include:

i) Pick-up: picking up the phone from the desktop or other places;
ii) Take-out: taking the phone out of the bag or pocket;
iii) Unlocking: unlocking the smart phone;
iv) Hand: handing the phone to others. They will form different motion sequences in different usage scenarios.

Motion Sequence of the Key Behavior Performed by Different User. In Fig. 1(a), we plot the statistical results of different users' habits (or motion sequence) during the key behavior. We observed that, when the owner uses the smart phone, the sampled motion sequences are {unlocking} or {pick up/take out, unlocking}. However, when the friendly users use the smart phone, the sampled motion sequences are {pick up/take out, unlocking, hand}, {take out/pick up, hand, unlocking}. When the attackers use the smart phone, the sampled motion sequences are {pick up/take out, unlocking}, which a hand motion will not appear in the key behavior of attacker. Also, we show the transitions between each motion in Fig. 1(c), including eight motion sequences which is performed by owner, friendly users or attackers. The main cause behind such a difference is:

i) Smart phone owners share their device to friendly users, usually accompanied by a handing motion.
ii) Attackers peep private information of smart phone sneakily, they may steal the smart phone from owner's pocket, desk, etc., with a take-out or pick-up motion.

In Fig. 1(b), we then show the statistical results of independent motion in the habit of users. With the above observations, we can find that different users have distinctive habit when performing the key behavior.

Analysis of Independent Motion Performed by Different User. In this subsection, we analyze the differentiation and consistency features of different people when performing an unlocking motion. As shown in Fig. 2(a) and Fig. 2(b), we plot the time series of velocity and acceleration sampled of unlocking motion performed by different users. Specifically, at the top of the figures show the velocity and acceleration data when two unlocking motions performed by the same owner, at the bottom of the figures show the velocity and acceleration data during two unlocking motions performed by another person. We can see that the pattern of unlocking motion is very similar when performed by the same owner, while different from that performed by another non-owner.

Figure 2(c) shows the distribution of normalized duration time when unlocking motions performed by the same owner and other people, which shows that the duration time of owner is smaller than other people and overlap is very small.

The main reason behind this difference is:

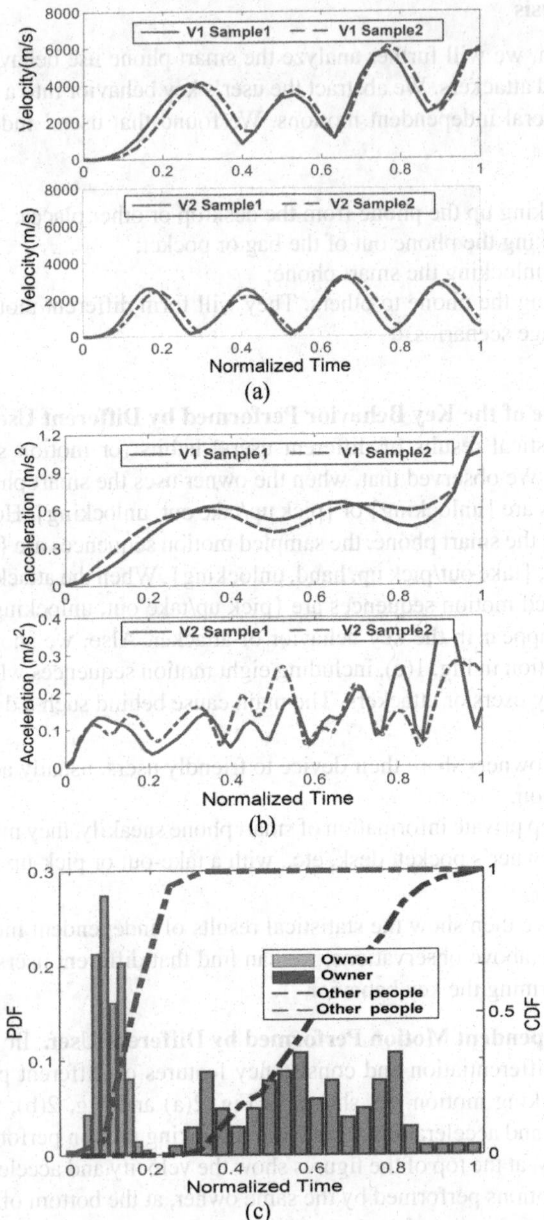

Fig. 2. The macro button chooses the correct format automatically. The unlocking motion performed by different people: (a) time series of velocity sampled during unlocking motion; (b) time series of acceleration sampled during unlocking motion; (c) distribution of normalized time distance between different users (e.g., owner and other people). PDF: Probability Density Function.

i) The owner is more familiar with their password. As a consequence, showing a faster velocity, shorter time, more stable and greater acceleration;
ii) The duration time of other people is more decentralized.

The above observation implies that a person has consistent and distinguishing features when she/he performs unlocking motion in the key behavior.

4 Overview

Based on the observation in Sect. 3, we propose MAS, a multi-user identification detection system that is able to instantly and inconspicuously identifying who is accessing the smartphone (the owner, a friendly user or an attacker) using only the inertial sensors. Figure 4 shows the overview of MAS. To distinguish different users, we go through the following four steps:

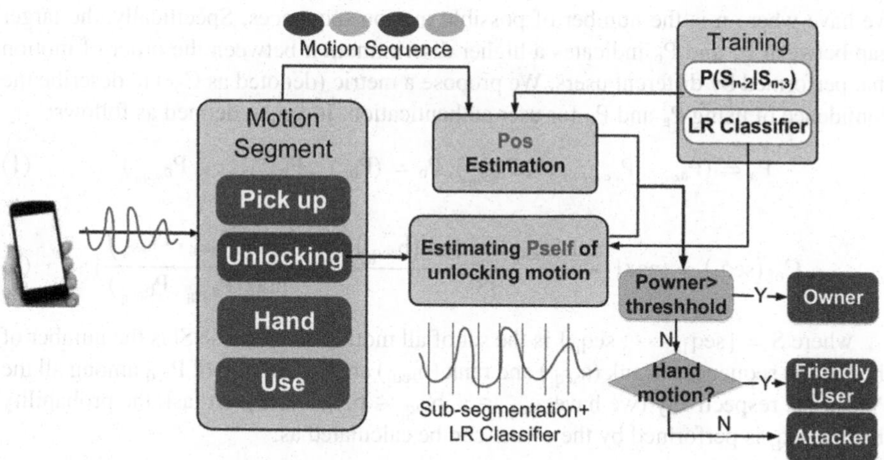

Fig. 3. Overview of MAS (Motion-sequence Authentication System)

- Pre-processing: the sampled sensor data are processed via a low-pass filter and a sliding mean filter to remove high frequency noises.
- Motion sequence detection and segmentation: in this component, MAS segments the smoothed sensor data into a sequence of sub-segments, where each segment contains an independent motion S_i. The output of this component is a motion sequence S_1, S_w.
- P_{self} estimation: after obtaining the motion sequence, we take the unlocking segment as the input of P_{self} estimation component, which evaluate how likely the unlocking motion is performed by owner.
- User authentication: We take a joint consideration of both the order of the motions and the P_{self} of the unlocking motion as input to distinguish the owner, the friendly user and the attacker.

5 System Design

5.1 A Model for User Authentication

In this paper, we present a Markov-based model to address this problem. When the user performs a key behavior (denoted as S_n). After segment, we get a motion sequence $S_n^{(p)} = \{S_{n-1}, \ldots, S_{n-w}\}$,where w is the number of independent motions in the key behavior. Assume that the transition probability between S_i and S_{i-1} is $P_{(S_i|S_{i-1})}$. The probability that user perform the key behavior in the order of $\{S_{n-1}, \ldots, S_{n-w}\}$ can be calculated as:

$$P_{seq} = \prod\nolimits_{i=n-w}^{n} P(S_i|S_{i-1}, S_{i-2})$$

We use two metrics, i.e., P_a and P_b to evaluate the probabilities that the 8 motion sequences (which is shown in Fig. 1(c)) performed by the owner.

Take-out motion detection: The difference in magnitude of take-out motion, step on a stair and hand motion. Himself/herself and by other people, respectively. Specifically, we have where n is the number of possible motion sequences. Specifically, the larger gap between P_a and P_b indicates a higher discrimination between the order of motion that performed by different users. We propose a metric (denoted as C_{of}) to describe the confidence of using P_a and P_b for user authentication. It can be defined as follows:

$$P_a = (P_{a_{seq_1}}, P_{a_{seq_2}}, \ldots, P_{a_{seq_n}}), P_b = (P_{b_{seq_1}}, P_{b_{seq_2}}, \ldots, P_{b_{seq_n}}) \tag{1}$$

$$C_{of}(seq_i) = \max\{\frac{rank(a_{seq_i}) - rank(b_{seq_i})}{|S|}, 1 - \frac{\min(P_{a_{seq_i}}, P_{b_{seq_i}})}{\max(P_{a_{seq_i}}, P_{b_{seq_i}})}\} \tag{2}$$

where $S = \{seq_1, \cdots, seq_n\}$ is the set of all motion sequences, |S| is the number of the motion sequences. Rank (a_{seq_i}) and rank (b_{seq_i}) are the ranking of P_{seq} among all the P_a and P_b, respectively (we have $a_{seq_i} \in a$, $b_{seq_i} \in b$, $a, b \in S$). At last, the probability that the seq_i is performed by the owner can be calculated as:

$$P_{os}(seq_i) = \alpha P_{ai} + (1 - \alpha) C_{of}(seq_i) \tag{3}$$

Where α is the weighing coefficient, which is set as 0.7 in our implementation.

We denote the probability that the unlocking motion performed by the owner as P_{self}. Based on $P_{os}(seq_i)$ and P_{self}, we propose a normalized metric to evaluate how likely the detected motion sequence $S_{current}$ is performed by the owner as follows:

$$P_{owner} = P_{os}(S_{current}) \times P_{self}. \tag{4}$$

Clearly, a high P_{owner}, namely a high probability that the smart phone is using by the user himself/herself, is achievable under the following two conditions:

i) The motion sequence $S_{current}$ comparisons with the habit of the owner, leading to a high P_{os};
ii) The unlocking motion exhibits high similarity with that performed by the owner himself/herself, leading to a high P_{self}.

5.2 Motion Detection and Segmentation

The goal of the motion detection and segmentation component is to extract the pick up, take out, hand or unlocking motion, from the smoothed sensor data, and output a motion sequence $\{S_{n-1}, \cdots, S_{n-w}\}$.

Pick-Up/Take-Out Detection. The pick-up motion refers to the motion that the user picks up a smart phone from a desk. The take-out motion refers to the motion that the user takes out a smart phone from his/her pocket or handbag.

We consider the pick-up and take-out motions as the same motion, since the only difference between them is the initial attitude of the smart phone. One naive solution to detect the pick-up/take-out motions is to detect the sudden change of the acceleration data with a predefined threshold. However, some other motions, such as step on a stair, might have a similar impact on the acceleration data, as shown in Fig. 4(a).

Our idea to solve this problem is to use the gyroscope. Specifically, comparing with pick-up/take-out motion, Fig. 4(b) shows that the motion of step on a stair has a marginal impact on the gyroscope data. However, we meet another problem that the handing motion has a similar impact on gyroscope data. Fortunately, we find that different from the handing motion, the pick-up/take-out motion will lead to the changes in the altitude of a smart phone.

Unlocking Detection. To unlock a smart phone, the user usually first lights the touch screen, then enters a password/pattern to unlock the smart phone. We use the built-in API of the smart phone to detect this motion.

Handing Detection. To detect the handing motion, one potential solution is to capture the sudden change in gyroscope data. However, pick-up/take-out motion would incur similar change in the gyroscope data, as shown in Fig. 4. Fortunately, when we observe the components of the gyroscope data on different axis, we find that different from the pick-up/take-out motion which incurs significant changes in all the three axes, the handing motion only incurs changes in the Z-axis.

In MAS, we take both the changes in Y-axis and Z-axis of gyroscope sensor data into account, which detect the hand motion with two predefined thresholds (threshold 3 and threshold 4). When a handing motion is performed at the time of t with coming the new data (gyroscope sensor data on y-axis w_y^t and gyroscope sensor data on z-axis w_z^t), the algorithm collects the samples w_y^m and w_z^m within last T second to compare with w_y^t and w_z^t find the maximal difference. If $\left|w_z^t - w_z^m\right| <$ threshold3 and $\left|w_y^t - w_y^m\right| <$ threshold4, we identify the t as the start of handing motion.

After identifying individual motion during performing using behavior, MAS obtains a motion sequence which might be {take-out, unlocking, hand}. In addition, we take independent motion as the input of P_{self} estimation component for further analysis.

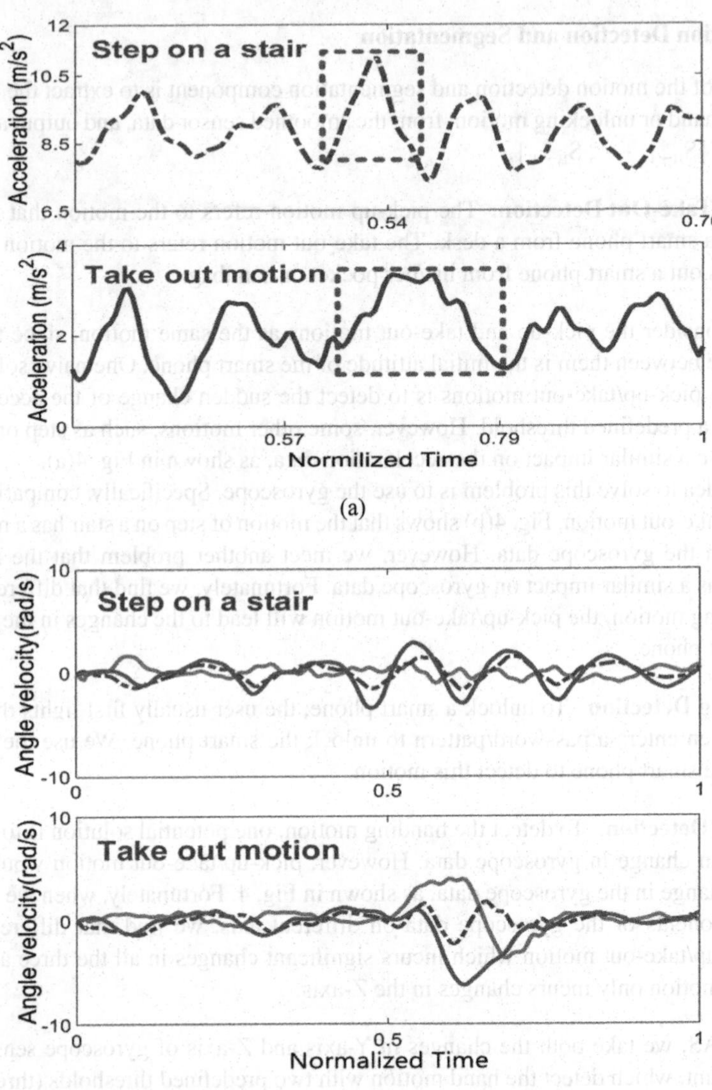

Fig. 4. The difference between take-out motion and step on a stair: (a) accelerometer data; (b) gyroscope data.

5.3 Pself estimation

The target of P_{self} estimation component is to evaluate how likely the independent motion is performed by the owner. With the analysis in Sect. 3, we find that a person has distinctive features (e.g., velocity, acceleration and time duration) when performing unlocking motion.

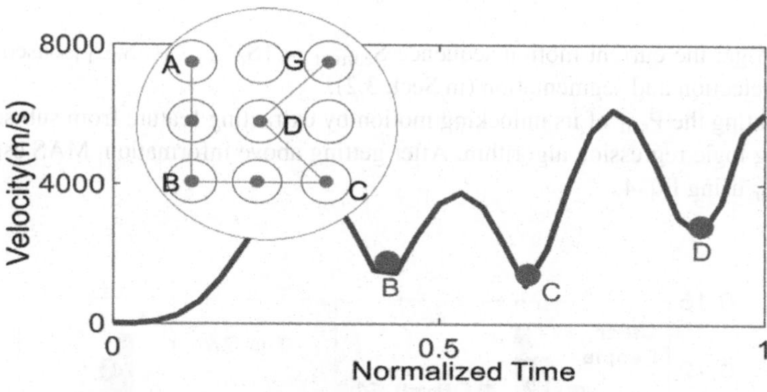

Fig. 5. P_{self} estimation

To obtain above fine-grained information, we find that, the distinguishable features may show at different sub-segment of an unlocking gesture. Thus, compare with extracting feature from the entire motion, we extract features from sub-segment of unlocking motion have higher differentiation degree. However, how many numbers of sub-segments should we segment an unlocking motion is a challenging task. First, too short time duration of a sub-segment may hide the consistent behavior. Second, too large time duration of a sub-segment may average out the distinctive information from the feature, failing to identify different users.

As shown in Fig. 5, we show an example of lock pattern and its corresponding time series of velocity sampled when performing such a gesture. At the key point, such as B, C and D point, Fig. 5 shows that they cause obvious changes in speed and direction. We propose an algorithm that we first leverage the key points to segment unlocking motion into sub-segment, such as {AB, BC, CD, CE}, then calculate P_{self} of entire unlocking motion based on the feature from each sub-stroke and use a logistic regression classifier.

To detect the key point, we leverage the minimum speed which typically indicate a key point. At the time of t, we assume that the speed of P point is v_t, the algorithm traces back to get the mean speed v_m within last T second, and trace forward to colorredget the mean speed vn within T seconds. With a given threshold, if $v_t - v_m$ < Threshold and $v_t - v_n$ < Threshold, we identify the P point as the key point.

We segment the unlocking motion based on the detected key point. We then extract features from sub- segment. Then, we exploit the logic regression algorithm to calculate P_{self}. The main reason for choosing such an algorithm is that:

i) its output ranges from 0 to 1 which is meet with the normalized result of P_{self};

ii) Logic regression algorithm with computational simplicity (O(n)) helps us detect the motion instantly.

5.4 A Model for User Authentication

When the new data comes, we first remove high frequency noise in data using a low-pass filter. Then,

i) we can get the current motion sequence $S_{current} = \{S_{n-w}, \cdots, S_{n-1}\}$ based on the motion detection and segmentation (in Sect. 3.2);
ii) Estimating the P_{self} of its unlocking motion by extracting feature from sub-segment and using logic regression algorithm. After getting above information, MAS estimates the P_{owner} using Eq. 4.

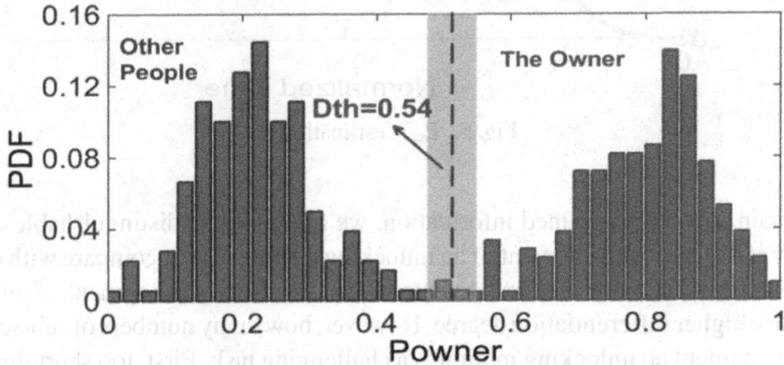

Fig. 6. Distributions of P_{owner} for the motion sequences of key behavior performed by the user and other people.

Figure 6 shows the distribution of P_{owner} for the motion sequences of using smart phone performed by the owner himself/herself and other people. The result shows that the overlap is very small. As a consequence, we set the threshold $D_{th} = 0.54$, which $P_{owner} > D_{th}$ indicates the owner. Further, if $P_{owner} < D_{th}$ and a "and motion" is detected, MAS identifies it as friendly user.

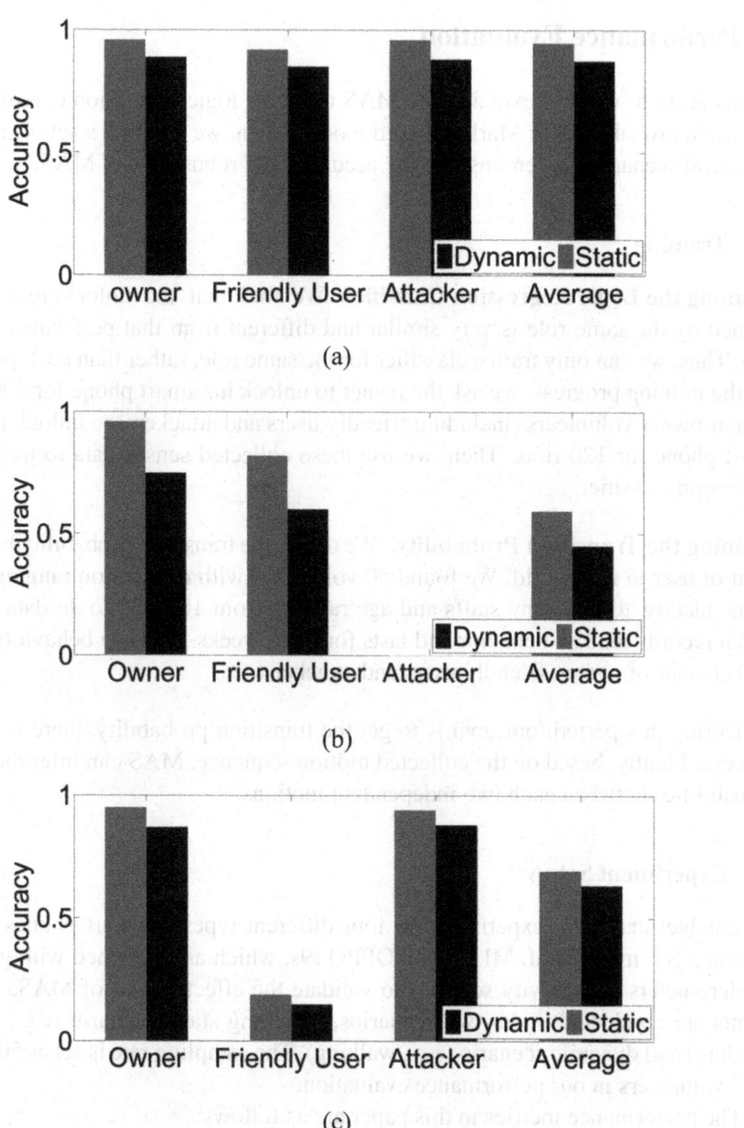

Fig. 7. Performance under different condition: (a) Accuracy of MAS; (b) Accuracy of P_{OS}; (c) Accuracy of P_{self}.

6 Performance Evaluation

In this section, we first explain how MAS trains its logic regression classifier and the transition probability for Markov-based model. Then, we conduct a set of experiments in several scenarios to demonstrate the accuracy and robustness of MAS.

6.1 Training

Training the Logic Regression Classifier. We find that the unlocking motion performed by the same role is very similar and different from that performed by another role. Thus, we can only train a classifier for the same role, rather than each person. During the training progress, we ask the owner to unlock his smart phone for 100 runs, and 30 non-owner volunteers (including friendly users and attackers) to unlock the owner's smart phone for 120 runs. Then, we use these collected sensor data to train the logic regression classifier.

Training the Transition Probability. We obtain the transition probability based on the habit of user in real world. We found 50 volunteers with occupation ranging from students, faculty, to company staffs and age ranging from 18 to 55 to do data collection, which records the key behavior and lasts for three weeks. The key behaviors including the behavior of owner, friendly users and attackers.

During this period, our goal is to get the transition probability, there is no privacy concern. Finally, based on the collected motion sequence, MAS can infer the transition probabilities between each two independent motions.

6.2 Experiment Setup

We conduct extensive experiment on four different types of smart phones (including Samsung S4, m1 mental, MI 2s, and OPPO r9s, which are equipped with gyroscopes, accelerometers and gravity sensors) to validate the effectiveness of MAS. The experiments are conducted under two scenarios, including static scenario (e.g., sitting and standing) and dynamic scenario (e.g., walking). The sampling rate is set as 50 Hz. There are 7 volunteers in our performance evaluation.

The performance metrics in this paper are as follows:

i) True Positive Rate (TPR): the fraction of cases where MAS correctly recognizes the other people;
ii) False Positive Rate (FPR): the fraction of cases where MAS mistakenly recognizes the owner of other people;
iii) Accuracy: the fraction of cases where MAS correctly recognizes owner, friendly user or attacker respectively.

In order to evaluate the performance of MAS, we implement the following approaches for comparison:

- MAS: the system in this paper.
- P_{os}: identify the user by exploiting only the P_{os} of motion sequence.
- P_{self}: identify the user by exploiting only the P_{self} of independent unlocking motion performed by the user.

6.3 Accuracy of MAS

To evaluate the accuracy of MAS, we conduct three experiments in an office, in which users are sitting or standing (a static scenario). In the first experiment, we ask an owner to use his/her smart phone himself/herself for 36 runs whenever he/she wanted; In the second experiment, three of volunteers is required to borrow the smart phone from the owner for 36 runs. To imitate motions of the real attacker, three volunteers pretend to be attackers and try to use the owner's smart phone when the owner leaves or does not pay attention to. The results are shown in Fig. 9.

The results have shown that in the relatively ideal condition (a static scenario), when MAS distinguishes between owner and other people, the TPR of MAS can be as high as 95%, and the FPR is lower than 4%. Compared to LXG (identify the user based on the features of users' finger movements when people interact with touch screen.) only identifies owner and attacker, MAS identifies multi-users (e.g., owner, friendly users and attackers) with high accuracy, as shown in Fig. 7. Meanwhile, the FPR of P_{self} or P_{os} can be as high as 57% and 40% respectively, which means that MAS has a good ability to identify user (friendly user and attacker).

6.4 Robustness of MAS

Performance in Identifying Each Independent Motion. Our results show a high accuracy in identifying pick-up motion, take-out motion and hand motion. As we know, if MAS mistakenly detects an independent motion in the key behavior (e.g., treating a hand motion as pick-up motion), which is likely to identify users incorrectly. Thus, a low motion detection accuracy leads to a low accuracy of MAS.

To study the effectiveness of MAS in identifying each independent motion, we ask 7 volunteers to perform the key behavior (including pick-up motion, take-out motion and hand motion) with random order in a static scenario and dynamic scenario respectively. Also, there is no apparent performance degrade of MAS is observed in different scenarios. The accuracy of distinguishing pick-up motion is as high as 98% in statics scenario. When people are in a dynamic scenario, the accuracy is more than 85%.

Performance in Different Working Conditions. In this section, we evaluate MAS under different conditions (a static scenario and dynamic scenario) to show its robustness. In this section, we have done another experiment. The result is shown in Fig. 7. According to the results in Fig. 7, no matter the user is in a static scenario or dynamic scenario, MAS outperforms all other methods. Also, the results show that all the three methods perform worse in the dynamic scenario, with 7%, 9%, and 20% performance degradation for MAS, P_{os} and P_{self} method respectively, since the inertial sensor data exhibits larger variance when the user is walking.

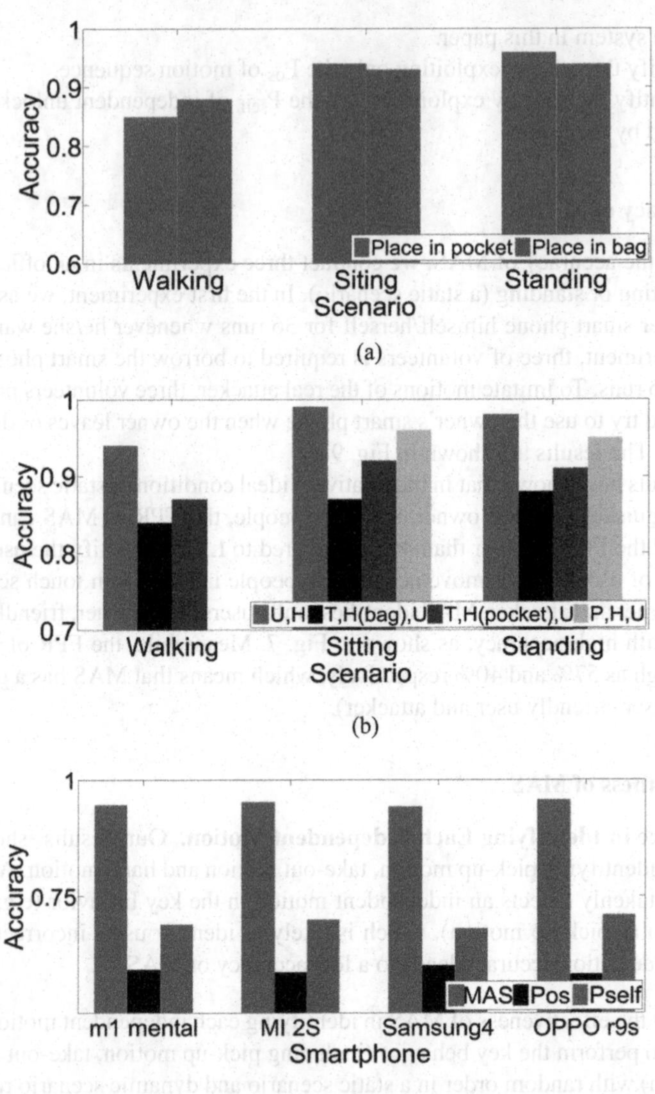

Fig. 8. (a) Accuracy of MAS in identifying take-out motion; (b) Accuracy of MAS in identifying hand motion; (c) Performance in different smart phone. "U, H": hand motion is performed after unlocking; "T, H(bag), U"/"T, H(pocket), U": take out the smart phone from bag/pocket, hand, unlock; "P, H, U": pick up, hand, unlocking.

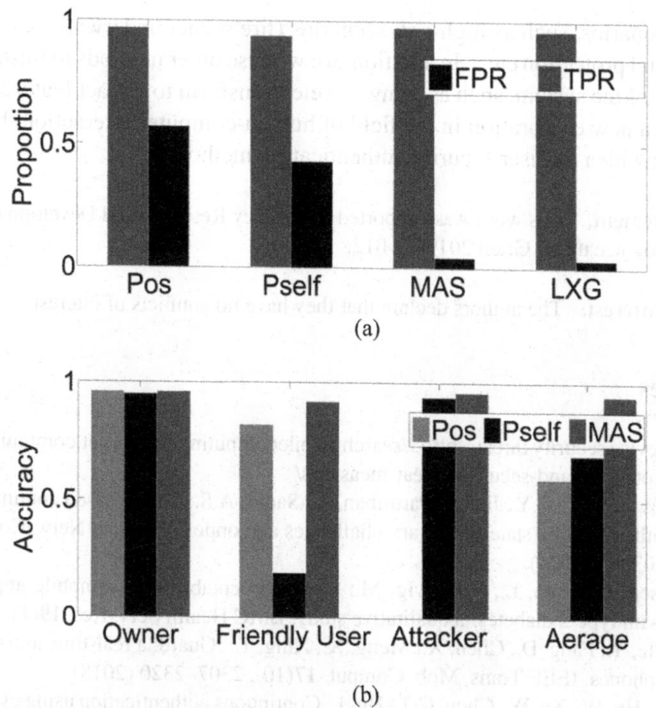

Fig. 9. (a) TPR and FPR comparison of different methods in distinguishing owner and non-owner; (b) Overall accuracy comparison of different methods in identifying multi-user.

Performance in Different Smart Phone. We conduct experiments on different types of smart phones, including Samsung S4, M1 mental, MI 2s, and OPPO R9s. The results in Fig. 8(c) have shown that MAS is applicable on different scenarios, and is better than P_{os} and P_{self} methods.

7 Conclusion

In this paper, we present a novel smart phone authentication approach, MAS, for non-contract multi-user authentication. In short, different users have different behavioral habits (or sequences of motions) when performing critical motions, and MAS uses a sequence of motions as a unique biometric. The system is easy to use, unobtrusive, and hard to counterfeit. It can be applied to multiple scenarios. We tested MAS in static scenarios (such as sitting and standing) and dynamic scenarios (walking), and evaluated the performance of seven volunteers. Experimental results showed that the system has certain accuracy and robustness. In particular, MAS was the most stable (only 7% less) in dynamic scenario, compared to P_{os} (9% less) and P_{self} (20% less). Therefore, it is reasonable to believe that MAS provides a novel, high-precision and high-stability viable approach to existing user identification technologies. At present, our work is focused on common scenarios. In the future, we plan to improve MAS's user recognition capability in special scenarios, such as high-risk scenarios (fire scene, underwater, Covid-19 care unit with facial protection etc.). In addition, we will use other methods to further improve the accuracy of the system, such as using wavelet transform to extract feature parameter. This will be a new exploration in the field of human-computer interaction, but also put forward a new idea for user security authentication method.

Funding Statement. This work was supported by the Key Research and Development Program of Shaanxi Province under Grant 2019GY-012.

Conflicts of Interest. The authors declare that they have no conflicts of interest.

References

1. Mobile device security threats. http://searchmobilecomputing.techtarget.com/guides/Mobile-device-protection-and-security-threat-measures/
2. Zaidi, A.Z., Chong, C.Y., Jin, Z., Parthiban, R., Sadiq, A.S.: Touch-based continuous mobile device authentication: state-of-the-art, challenges and opportunities. J. Netw. Comput. Appl. **191**(1), 103162 (2021)
3. Torbjørnsen, A., Ribu, L., Rønnevig, M.: Users' acceptability of a mobile application for persons with type 2 diabetes: a qualitative study. BMC Health Serv. Res. **19**(1), 641 (2019)
4. Jin, M., He, Y., Fang, D., Chen, X., Meng, X., Xing, T.: iGuard: a real-time anti-theft system for smartphones. IEEE Trans. Mob. Comput. **17**(10), 2307–2320 (2018)
5. Zhang, Y., Hu, W., Xu, W., Chou, C.T., Hu, J.: Continuous authentication using eye movement response of implicit visual stimuli. Proc. ACM Interact. Mob. Wearable Ubiquit. Technol. **1**(1), 1–15 (2018)
6. Bevan, C., Fraser, D.S.: Different strokes for different folks? Revealing the physical characteristics of smartphone users from their swipe gestures. Int. J. Hum. Comput. Stud. **88**(12), 51–61 (2016)
7. Wang, X., Yu, T., Mengshoel, O., Tague, P.: Towards continuous and passive authentication across mobile devices: an empirical study. In: 10th ACM Conference on Security and Privacy in Wireless and Mobile Networks (WiSec 2017), pp. 35–45. Association for Computing Machinery, New York (2017)
8. Alghamdi, S.J., Elrefaei, L.A.: Dynamic authentication of smartphone users based on touchscreen gestures. Arab. J. Sci. Eng. **43**(2), 789–810 (2017). https://doi.org/10.1007/s13369-017-2758-x
9. Ferrag, M.A., Maglaras, L., Derhab, A., Janicke, H.: Authentication schemes for smart mobile devices: threat models, countermeasures, and open research issues. Telecommun. Syst. **73**(2), 317–348 (2019). https://doi.org/10.1007/s11235-019-00612-5
10. Meng, W., Li, W., Wong, D.S.: Enhancing touch behavioral authentication via cost-based intelligent mechanism on smartphones. Multimed. Tools Appl. **77**(2), 30167–30185 (2018)
11. Cao, H., Chang, K.: Nonintrusive smartphone user verification using anonymized multimodal data. IEEE Trans. Knowl. Data Eng. **31**, 479–492 (2018)
12. Nguyen, T., Roy, A., Memon, N.: Kid on the phone! Toward automatic detection of children on mobile devices. Comput. Secur. **8**(4), 334–348 (2019)
13. Alzubaidi, A., Kalita, J.: Authentication of smartphone users using behavioral biometrics. Hum.-Comput. Interact. **46**(12), 256 (2019)
14. Song, C., Wang, A., Ren, K., Xu, W.: EyeVeri: a secure and usable approach for smartphone user authentication. In: 35th Annual IEEE International Conference on Computer Communications, IEEE INFOCOM 2016, San Francisco, CA, USA, pp. 1–9 (2016)

15. Guo, Y., Yang, L., Ding, X., Han, J., Liu, Y.: OpenSesame: unlocking smart phone through handshaking biometrics. In: IEEE INFOCOM, Turin, Italy, pp. 365–369. IEEE (2013)
16. Zou, Q., Wang, Y., Wang, Q., et al.: Deep learning-based gait recognition using smartphones in the wild. IEEE Trans. Inf. Forensics Secur. **1**(1), 99 (2020)
17. Raghavendra, R., Busch, C., Yang, B.: Scaling-robust fingerprint verification with smartphone camera in real-life scenarios. In: IEEE 6th International Conference on Biometrics: Theory, Applications and Systems, pp. 1–8. IEEE (2013)
18. Viebke, A., Memeti, S., Pllana, S., et al.: CHAOS: a parallelization scheme for training convolutional neural networks on Intel Xeon Phi. J. Supercomput. **75**(1), 197–227 (2019)
19. Xu, X., Yu, J., Chen, Y., et al.: TouchPass: towards behavior-irrelevant ontouch user authentication on smartphones leveraging vibrations. In: (MobiCom 2020) The 26th Annual International Conference on Mobile Computing and Networking (2020)
20. Izumoto, D., Yamazaki, Y.: Security enhancement for touch panel based user authentication on smartphones. In: 2019 Asia-Pacific Signal and Information Processing Association Annual Summit and Conference (APSIPA ASC). IEEE (2020)
21. Saini, B.S., et al.: A three-step authentication model for mobile phone user using keystroke dynamics. IEEE Access **8**(1), 125909–125922 (2020)
22. Kang, T., Ji, S., Jeong, H., Zhu, B., Kim, J.: WearAuth: wristwear-assisted user authentication for smartphones using wavelet-based multi-resolution analysis. IEICE Trans. Inf. Syst. **12**(1), 1976–1992 (2019)
23. Kim, S.-J., Kim, J.-M., Jo, I.-J.: Multimedia image data processing on smartphone for authentication. Multimedia Tools Appl. **78**(5), 5287–5303 (2018). https://doi.org/10.1007/s11042-017-5600-2
24. Liu, X., Shen, C., Chen, Y.: Multi-source interactive behavior analysis for continuous user authentication on smartphones. In: Zhou, J., et al. (eds.) CCBR 2018. LNCS, vol. 10996, pp. 669–677. Springer, Cham (2018). https://doi.org/10.1007/978-3-319-97909-0_71
25. Acien, A., Morales, A., Vera-Rodriguez, R., Fierrez, J.: Smartphone sensors for modeling human-computer interaction: general outlook and research datasets for user authentication. In: 2020 IEEE 44th Annual Computers, Software, and Applications Conference (COMPSAC), Madrid, Spain, pp. 1273–1278. IEEE (2020)

Edge Computing and Edge Intelligence

Joint Optimization Scheme of Multi-service Replication and Request Offloading in Mobile Edge Computing

Chenxi Li, Guanghui Li(✉), Shihong Hu, Chenglong Dai, and Dong Li

School of Artificial Intelligence and Computer Science, Jiangnan University, Wuxi 214122, Jiangsu, China
ghli@jiangnan.edu.cn

Abstract. To meet the ever-increasing service quality requirements of end-users and enable delay-sensitive applications to be completed within a tolerable time, Mobile Edge Computing (MEC) offloads the request of users to the edge servers that are closer to the end equipment. However, deploying a single service replication in an appropriate edge node is difficult to deal with all requests of users for multiple services. In addition, after the service replication is deployed at the edge node, a corresponding user request offloading scheme is also required. Considering the heterogeneity of edge servers, this paper studies the joint optimization problem of multi-service replication and request offloading. Firstly, we present an edge computing architecture with multi-service replication, and define the multi-service replication and request offloading as a joint optimization problem. Secondly, a multi-service replication algorithm called Multireplicas Greedy Best (MGB) is proposed to solve the joint optimization problem. Finally, the simulation experiments are carried out. The experimental results show that the proposed algorithm can effectively reduce the overall delay compared with the random strategy, the nearest node offloading strategy, the particle swarm algorithm, and the greedy algorithm.

Keywords: Mobile Edge Computing · Service replication · Request offloading · Multi-objective optimization

1 Introduction

In recent years, Internet of Things (IoT) technology has developed rapidly [1,2]. Statistics show that there are about 30 billion IoT devices in the world by 2021. Furthermore, it is estimated that by 2025, mobile IoT devices will grow to 75 billion [3]. Many time-sensitive applications on IoT devices require fast and safe services, such as video surveillance, traffic flow mapping, personalized multimedia, and data sharing. These requirements usually need interactive behavior with high Quality of Service (QoS) [4], but most of the embedded processors and storage capacity of IoT devices have limited resources and cannot be used for

Y. Lai et al. (Eds.): ICA3PP 2021, LNCS 13155, pp. 445–459, 2022.
https://doi.org/10.1007/978-3-030-95384-3_28

large-scale computing, storage, and communication [5]. The traditional method is to upload all the requirements to the cloud server data center for processing. As a typical example, Mobile Cloud Computing (MCC) was introduced as a solution. MCC is a resource scheduling framework using a resource-sharing model, which can solve the problem of insufficient resources used by IoT devices, especially those applications that require a lot of computing resources.

Supporting mobile applications through public Wi-Fi networks has received significant research attention. The 5th Generation (5G) Mobile Communication Technology and mobile IoT devices (i.e., mobile phones, tablets) have promoted delay-sensitive services such as AR/VR, healthcare, intelligent transportation, and location positioning [4]. However, MCC is not enough to support the communication and computing of IoT devices in 5G. Therefore, Mobile Edge Computing (MEC) has been attached to great importance in recent years. The core idea of MEC is to offload some requests to edge servers instead of offloading them to the cloud server for processing. In the MEC architecture, computing and storage resources are usually deployed at the edge of the network, and edge servers are deployed in the wireless access network. The edge nodes can be any computing resource of the network to offload the user requests. In this case, the offloading request from the user equipment to each service replication can be served by the nearest edge server first, thereby avoiding long-distance network transmission.

A challenging problem in the MEC system is the scarcity of resources. Compared with cloud servers, the resources of edge nodes are limited. Therefore, it is an unworkable scheme to run large-scale applications on a single edge node in MEC. Moubayed et al. [6] proposed an effective way to solve this problem, allowing users to run their application requests on multiple edge nodes. The essential issues that need to be resolved are to find the optimal replication location of the service on multiple nodes and find the location where users request to offload.

In the field of cloud computing, Thai et al. [7] proposed some service replication strategies, which tended to place service replications on the cloud. Compared with cloud servers, edge nodes are more widely distributed in the network topology, and the storage capacity of edge nodes is more limited. Therefore, the strategies proposed by Thai cannot be applied directly. Many studies have discussed service replication placement strategies. Naas et al. [8] proposed some service placement strategies in the edge computing environment. However, these strategies only focused on placing a single service replication to the appropriate edge node. When there are multiple data consumers from different locations requesting the same service, a single service replication cannot meet the latency requirements of all consumers. Multi-service replication was mentioned by the work in [9], the authors investigated multi-service replication as a stochastic game, but they ignored the user-requested offloading problem.

To reduce user request latency under the framework of MEC, researchers have done a lot of work. It can be roughly divided into two categories: One category is to optimize the scheme of service replications. For example, Naas constructed the replication problem as a Generalized Assignment Problem (GAP) [8], which used edge node location information to reduce the total delay. The second category is to find the optimal edge nodes to offload user requests.

There is a big difference between a single-user scenario and a multi-user scenario in reality. Making an optimal offloading decision under the condition of considering the resource allocation of edge nodes is a complex problem. However, they ignored the impact of edge node resource allocation on computation offloading [10].

The multi-service placement problem has been proven to be NP-hard, and use CPLEX MILP solver to find out the best replication placement will take a lot of time [8]. Therefore, to effectively solve the problem, we propose a joint optimization scheme of multi-service replication and request offloading mechanism based on an enhanced greedy algorithm. This mechanism realizes the optimal allocation of edge node resources, and makes an offloading calculation decision.

Our main contributions of this study are highlighted as follows:

(1) Considering the diversity of user requests and the heterogeneity of edge nodes in MEC, we model the multi-replication service placement problem as a joint optimization problem of edge node service replication and user request offloading.
(2) With the primary goal of minimizing total delay and algorithm execution time, we propose a heuristic algorithm Multireplicas Greedy Best (MGB) based on an enhanced greedy strategy.
(3) We evaluate the performance of the proposed algorithm by simulation experiments. The results show that MGB outperforms the baselines in terms of the reduction of total delay and execution time.

2 Multi-edge Computing System Model

The edge computing architecture proposed in this paper is shown in Fig. 1. The system consists of a remote cloud server, multiple edge nodes, and IoT devices.

Multi-replica edge computing architecture is a hierarchical structure. From the bottom to the top, IoT devices encompass wireless end-users objects such as sensors, robots, smartphones, and cameras, which are in the bottom layer. These devices generate continuous and periodic service requests, which need to be processed and stored in edge nodes or cloud servers. The middle layer above the IoT device layer is the edge network layer. This layer is divided into two parts: signal receiving and request processing. Edge nodes maintain computation services for users via various virtualization techniques (e.g., containers and virtual machines). Due to the heterogeneity of edge nodes, the network conditions and resources of edge nodes may be different. The request processing part represents the physical infrastructure for processing the request, which contains the physical machines called MEC nodes. Finally, there is a cloud layer composed of cloud servers on top of the edge network layer. The cloud servers store heterogeneous services and cover the geographic area where the edge nodes are located. The edge nodes and IoT devices are connected via wireless networks. Generally, services can be replicated at any edge node, and when an IoT device generates a request, the request can be processed by the edge node.

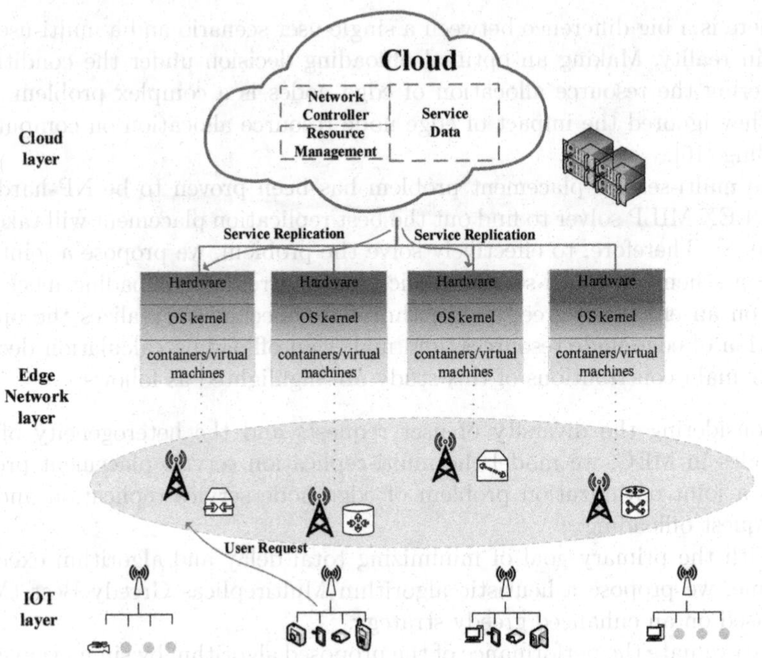

Fig. 1. Multi-replica edge computing architecture

Heterogeneous edge nodes may include gateways (GW), Region Point of Presences (RPOPs), and Local Point of Presences (LPOPs), etc., denoted by $C = \{C_1, ..., C_k, ..., C_m\}$. In the rest of this paper, when referring to edge nodes, k is used interchangeably with C_k to simplify the representation. The services in the cloud server can be replicated on the edge node, which is marked as $S = \{S_1, ..., S_i, ..., S_n\}$, and i will be used interchangeably with S_i later when referring to the service replication. IoT devices generate service requests, and these services will be processed on edge nodes. The generated user requests are denoted as $R = \{R_1, ..., R_j, ...R_q\}$, and we use j interchangeably when referring to request R_j. The service data block for service replication and user request is denoted by $data_i$, which is the unit block for data transmission.

The overall delay consists of the following three parts:

(1) V_{ik} : The delay required for the service replication i to be replicated on the edge node k. The edge nodes are usually heterogeneous, making the speed at which they download the service replication from the cloud server different. Therefore, the heterogeneity of edge nodes must be considered to calculate the service replication time. It can be formulated as follows:

$$V_{ik} = \frac{data_i}{\delta} \cdot td_{ik} \tag{1}$$

Here, td_{ik} is the delay required for the edge node k to download the unit data block of type i request from the cloud server, $data_i$ is the size of the replication i transferred between cloud and edge, δ is the unit data block to be transmitted.

(2) X_{ik}: The delay required to run the service i on the edge node k. Because the edge node is heterogeneous, the delay of processing unit data on the edge node is different, represented by tu_k. λ is the impact ratio of data block size to the service operation, which can be a given positive constant between [0,1] according to the actual situation. The relationship between them is given as follows:

$$X_{ik} = \frac{\lambda data_i}{\delta} \cdot tu_k \quad (2)$$

(3) T_{ijk}: The delay required of the service i to offload the user request j at the edge node k through wireless communication. The delay is proportional to the total amount of data requested $data_i$, and the delay is inversely proportional to the wireless communication transmission rate w_{jk} from the user request j to the edge node k.

$$T_{ijk} = \frac{\lambda data_i}{w_{jk}} \quad (3)$$

The wireless communication transmission rate w_{jk} can be represented by a wireless transmission model based on Shannon-Hartley theorem [11], which is shown in formula (4), where d represents the distance from the terminal device to the edge, v indicates the path loss exponent, and P represents the current device transmit power, h denotes the channel fading coefficient and ω_0 represents the Gàussian white noise power. Based on Shannon's theorem, when a request is offloaded to an edge node with bandwidth B, the transmission rate can be expressed as:

$$w_{jk} = B \log_2 \left(1 + \frac{P|h|^2}{\omega_0 d_{jk}^{\vartheta}}\right) \quad (4)$$

Therefore, the overall delay cost consists of the time required for the service replication i to be replicated on the edge node k, the time required to run the request i on the edge node k, and the user j to request service i at the edge node k through the wireless communication link. It is worth noting that the more service copies that are replicated on edge nodes, the more time it takes to store services, and the time for users to request services may decrease; on the contrary, if fewer service copies are replicated on edge nodes, the storage service time will be reduced, but at the same time the user request service time may increase.

There are multiple replication edge nodes in the edge computing architecture, and there may be multiple options when the service is replicated. We set a decision variable p_{ik}. Assuming that service i is stored in the cloud server, and the data block is transferred to the edge node k during service replication. In order to indicate the replicate state of service i on edge node k, the decision variable p_{ik} is assigned binary value: when the service i is replicated on the edge node k, $p_{ik} = 1$; otherwise, it is 0.

Furthermore, to indicate whether the user request j is offload to service i, the decision variable q_{ij} is defined as follows: when $q_{ij} = 1$, it means that the user j is to request the service i, and 0 otherwise.

Similarly, the decision variable r_{jk} represents the situation that the request j is processed on the edge node k. $r_{jk} = 1$ means that the request j is processed on the edge node k; otherwise, it is 0. Only when there is a service replication on the edge node k, the request j can be processed on the edge node k. In addition, for each user, the request only needs to flow to one edge node, so the following condition should be satisfied:

$$\sum_{k=1}^{K} r_{jk} = 1 \tag{5}$$

Based on the above discussion, we model the joint optimization problem of service replication and user request offloading as follows:

$$\text{Minimize} \sum_{k=1}^{K} p_{ik} \cdot V_{ik} + \sum_{j=1}^{J} \sum_{i=1}^{I} \sum_{k=1}^{K} q_{ij} \cdot r_{jk} \cdot (X_{ik} + T_{ijk}) \tag{6}$$

s.t.

$$\sum_{i=1}^{I} data_i p_{ik} \leq \text{stor}_k, \forall k = 1, 2, 3... \tag{7}$$

$$\sum_{k=1}^{K} r_{jk} = 1 \tag{8}$$

$$p_{ik} = 0, \forall i, \forall k : data_i > stor_k, i = 1, 2, 3 \ldots, k = 1, 2, 3 \ldots \tag{9}$$

The objective of this function is to find the optimal service replication and user request offloading plan to minimize overall delay. The constraint (7) ensures that the service replication stored on the edge node must be smaller than the storage capacity $stor_k$ of the edge node. Constraint (8) can guarantee that each request is completely offloaded to an edge node. Constraint (9) indicates that when the storage capacity of the service replication is greater than the storage capacity of the edge node, no replication will perform at the current node.

3 Joint Optimization Algorithm of Service Replication and User Request Offloading

Service replication and user request offloading is a complex multi-objective optimization problem, which involves a trade-off between two conflicting objectives. This paper attempts to find an effective solution for this problem based on an enhanced greedy strategy. This strategy is more efficient and can reduce the total delay cost of service replication significantly.

The data transmission process of service replication and user request is shown in Fig. 1. Before services are replicated, the cloud server stores services which can handle various user requests. In order to ensure that each user request can be offloaded to edge nodes, the number of service replications should exceed the number of user requests. At the same time, due to the limited storage capacity of the edge node, it is necessary to determine whether the remaining storage capacity of the edge node can accommodate a new one before replicating. We traverse all edge nodes and create the array of ready-to-use service replication solutions as a_i.

After determining the location of the service replication at the edge node, the user request needs to be offloaded to the edge node. The request generated by IoT devices can be offloaded to the edge node through wireless communication. In order to find the optimal edge node to offload the user request, an original idea is offloaded the request to the closest edge node of this user. However, there are two problems. One is that the nearest edge node may not replicate the service; the other is that the computing capability of the edge node closest to the user is uncertain. So it is necessary for each edge node to determine whether the service can handle the offloading request and consider each edge node computing capability.

We design a joint optimization algorithm for service replication and user request offloading based on an enhanced greedy strategy in order to improve efficiency while maintaining the quality of the solution. Algorithm 1 is the proposed service replication algorithm. The first line of the algorithm traverses all replication schemes and records all replication schemes as $AL = \{a_1, a_2, ..., a_n\}$. Lines 2–12 iteratively calculate the delay time of each service replication scheme. Before all the replicate schemes are traversed, the service delay time calculation will be repeated. In each iteration, the user request offloading time is updated to minimize total user request time according to Algorithm 2. Lines 6–8 determine whether the current replicate plan exceeds the upper limit of the edge node storage capacity. If it exceeds, no service replication will be performed. Finally, we find the optimal replicate edge nodes.

After the algorithm determines the replicate location of all services, it also needs to determine user request offloads to which edge node. This paper sets a heuristic rule: all IoT users need to consider the current requesting users and the remaining unallocated edge node users. We design an algorithm based on an enhanced greedy strategy called Multireplicas Greedy Best (MGB). The algorithm runs in two stages: In the first stage, it determines the best replication location among the edge nodes and finds the minimum replicate cost of the service on the edge node. In the second stage, the algorithm considers which edge node to offload the users request.

Algorithm 1. Optimal Multi-service Replication

Require: Set of edge nodes C, Service data size $DATA$, Edge node storage capacity $stor$
Ensure: Optimal strategy of replication and overall delay
Enumerate all replication strategies $AL = \{a_1, a_2, ..., a_n\}$
for all $a_i \in AL$ **do**
  for all $k \in C$ **do**
    Calculate the replicate delay V_{ik} through Formula (1)
    Call the Optimal Request Offloading algorithm
    if $\sum_i^I data_i p_{ik} >$ $stor_k$ **then**
      The service can not replicate on edge nodes, so offload the user request to the cloud server
    end if
  end for
  Calculate the replication delay for all edge nodes $V = \sum_k^K V_{ik}$
  According to Formula (6) calculate the minimum overall delay
end for
Find the optimal multi-service replication strategy with the minimum overall delay for all replications.

Algorithm 2 is the user request offloading algorithm. The input is the service replication strategy of Algorithm 1, and the overall delay of all user requests is output. Line 1 of the algorithm initializes the minimum offloading delay. Lines 2–7 describe the request offloading that takes into account the cost of communication between users and edges. According to formulas (2) and (3), the edge nodes are arranged according to the offloading delay and the service running delay. Line 8 indicates the offloading delay and the service running delay in an arrangement. To choose among candidate replications the best node to offload the request, lines 9–21 describe a feasible offloading solution for all user requests. To find these solution, we first use greedy algorithm. However, the smallest offloading delay of the user request for offloading may not be globally optimal in overall delay. So we proposed MGB, which takes into account the smallest offloading delay node and the second smallest offloading delay node. In MGB, we calculate the overall delay of the smallest offloading delay node and the second smallest offloading delay node. If the overall delay offloading at the second smallest offloading delay node which is less than overall delay of the smallest offloading delay node, the edge node with the second smallest delay will be offloaded.

Algorithm 2. Optimal Request Offloading

```
Require: Replication strategies a_i, Replication delay V, IoT user request R
Ensure: minimum offloading delay of all user request
Initialize minDelay = MaxValue
Traverse all user requests R = {R_1, ..., R_j, ..., R_q}
for all R_j ∈ R do
    Determine whether there is a replication store in node, and offloading the service
    when the edge node has a replication
    According to Formulas (2), (3) calculate the offloading delay and the service
    running delay X_ik and T_ijk
    Calculate retOffloadingDelay[j] = Σ_{k=1}^{K} Σ_{i=1}^{I} X_ik + T_ijk
end for
Arrange the retOffloadingDelay in order
for all R_j ∈ R do
    Offloading the request R_j on edge node with the smallest delay in arrangement
    retOffloadingDelay
    Calculate OverallDelay = X + V + K
    if OverallDealy < minDelay then
        Take the current delay minDelay = OverallDealy
        Remove the edge node from the replicated edge node list
    else
        Offloading the request on edge node with the second smallest delay in arrange-
        ment retOffloadingDelay
        Calculate OverallDelay = X + V + K
        Take the delay minDelay = OverallDealy
        Remove the edge node from the replicated edge node list
    end if
end for
Find the optimal request offloading strategy with the minimum overall delay for
all requests
```

4 Simulation and Performance Evaluation

To verify the effectiveness of the MGB strategy, we conducted simulation experiments. We compare the performance of the proposed algorithm with the random algorithm, greedy algorithm, nearest edge node offloading algorithm, and particle swarm algorithm.

4.1 Experimental Settings

The simulation experiments are conducted on an Intel 3.7 GHz Core i5 system with 8 GB RAM. The simulation scenario includes a set of IoT devices, edge nodes, and a cloud server infrastructure (as shown in Fig. 1). The edge node is composed of heterogeneous devices such as GW, RPOP, and LPOP. Like literature [12]. This article sets the storage capacity of GW to 16 GB, LPOP storage capacity is set to 32 GB, the storage capacity of RPOP is set to 128 GB, and the size of the cloud service center (Cloud Service, CS) is set to 128 TB. As shown in Table 1:

Table 1. The storage capacity of Nodes.

DataHost	GW	LPOP	RPOP	CS
Storage capacity	16 GB	32 GB	128 GB	128 TB

To verify the effectiveness of the multi-service replication strategy, a wireless transmission model based on Shannon-Hartley theorem is used. To meet the needs of real users, this article uses 5G wireless transmission parameters as the wireless communication link. The channel bandwidth is denoted by B, assuming that there is no significant change in an edge-computing architecture, which is 3 GHZ. The transmit power is denoted by P, which is uniformly distributed in the range of 100 mW–320 mW. Two different fading coefficients h are considered, which are 1.0 and 1.1, respectively. In the same channel, the Gaussian white noise power w conforms to the normal distribution. The consumption coefficient v is fixed to 1 in the experiment. As shown in Table 2:

Table 2. Parameters of wireless communication link.

Notation	Description	Value
B	Wireless bandwidth	2 MHz
P	Transmission power	1 mW–3.2 mW
h	Channel fading coefficient	1.0 and 1.1
w	White Gaussian noise power	Normal distribution $X - N(0, 50)$
v	Path loss exponent	1

At the same time, we set the communication delay among different edge nodes as shown in Table 3.

Table 3. Delay of different wireless communication links.

Channel type	IOT-GW	IOT-LPOP	IOT-RPOP	IOT-CS	GW-CS	LPOP-CS	RPOP-CS
Latency (ms)	5	10	20	100	5	10	20

4.2 Performance Index

The purpose of service replication and user request offloading in the edge computing architecture is to provide users a better experience when they request. Therefore, this experiment mainly evaluates two important indices for time-sensitive applications, namely, the overall delay of the edge replication and the execution time of MGB.

(1) Overall delay: This article defines the delay cost as the sum of the time required for the service copy to be replicated on the edge node, the time required to run the request on the edge node, and the time required for the user to request service at the edge node through the wireless communication link.
(2) Execution time: The execution time refers to the time from the simulation data generation to the completion of the service replication and user request offloading.

We compare the MGB algorithm with the following four baselines:

- Random algorithm: The replications are placed randomly in all edge nodes, and the user requests are offloaded randomly to all edge nods [13].
- The nearest edge node offload: This strategy offloads user requests to the edge node which is closest to the user, regardless of the service replication delay and edge node computing power [11].
- Greedy algorithm: Greedy algorithm is a classic heuristic algorithm, which can effectively solve the NP-hard problem [14]. This strategy finds the best offloading edge node for each user request as far as possible unless there is no available edge node.
- Particle swarm algorithm: Particle swarm algorithm (PSO) uses particle swarm optimization to perform user request offloading [15]. The algorithm finds the optimal solution by simulating the migration activities of natural bird groups. Each bird group represents a candidate solution to the optimization problem. The parameter settings in PSO have a great influence on the performance of the algorithm. Considering the execution time of the algorithm and the quality of the solution, we set 100 as the number of iterations and 0.5 as the maximum particle speed later in the experiment.

4.3 Experimental Results and Baselines

The total delay of various algorithms are shown in Fig. 2 when the number of service replication is 6, 8, 10, and 20, respectively. The x-axis is the number of offloading requests on the replicate service node, and the y-axis is the total delay between the service replication and the request offloading.

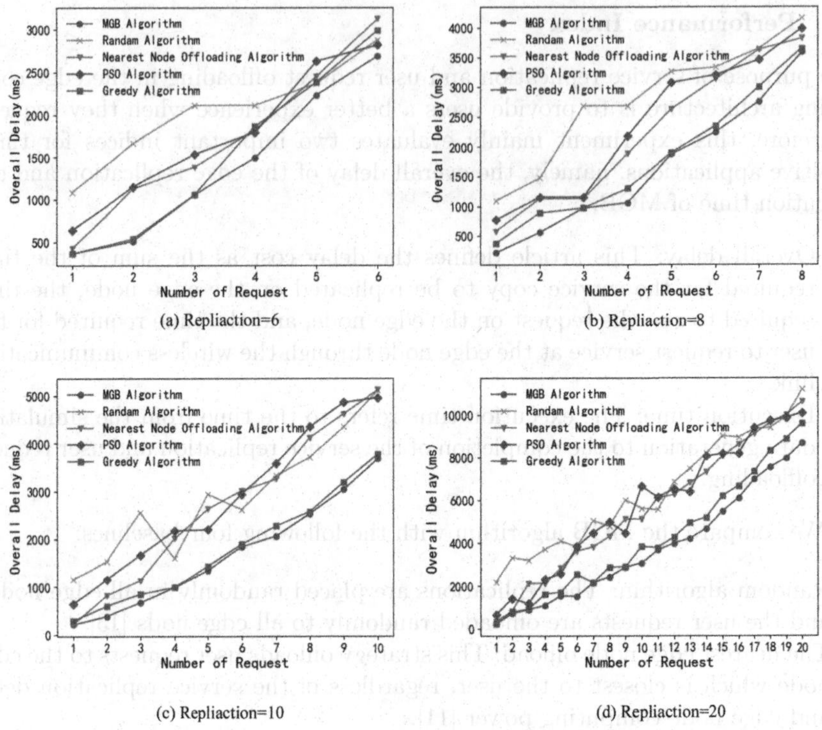

Fig. 2. Overall delay under different number of replications

For each algorithm, we run ten times to get the average value of the overall delay time. It was found that when replication is 20, the MGB algorithm reduced the delay by 41.6% compared with random replication for service offloading. In addition, the MGB algorithm is reduced by 11.45%, which is 6.5% less than the particle swarm algorithm, and 1.9% less than the naive greedy algorithm. This is because the random algorithm does not consider the geographical distribution of user requests, the delay of offloading, and the processing capabilities of different edge nodes for services replication, so the delay is relatively high, and the nearest edge node offloading algorithm only considers the distance between the user request from the edge node. The PSO algorithm is a better solution, but the PSO algorithm has too many parameters, which makes it difficult to adjust those parameters quickly and effectively. The greedy algorithm can reach the optimal solution in some cases, but in more cases, it may easy to fall into a local optimal solution. In contrast, the MGB algorithm attempts to replicate fewer services to edge nodes and uses an enhanced greedy algorithm to offload user requests so as to avoid falling into a local optimum and greatly reduce the delay time. Table 4 shows the overall delay reduction ratio of MGB vs. baselines. Where the second line represents the number of replications, and columns 2–5 give the delay reduction percentages compared with the baselines.

Table 4. Overall delay reduction ratio MGB vs baselines

Baselines	Number of replications			
	6	8	10	20
Random algorithm	14.06%	23.84%	28.04%	41.60%
Nearest node offloading	10.13%	10.22%	13.64%	11.45%
PSO	6.68%	6.17%	7.43%	6.50%
Greedy algorithm	4.54%	2.95%	2.53%	1.90%

Figure 3 shows the overall delay variation with the number of service replications. The x-axis is the number of user requests, and the y-axis is the overall delay between service replication and request offloading. We use random algorithm to replicate services and MGB algorithm to offload user requests. In Fig. 3(a), we plot the overall delay for user request J = 6. As expected, the overall delay is gradually decreasing until 9 service replications. However, after that point, the overall delay increase gradually. This is because more service replications may result in smaller user requests to offload, but too many service replications will cause the replicate time to be too long, thereby increasing the overall delay. And in Fig. 3(b), the lowest latency is 14 service replications, when the user request is 10. Therefore, it is necessary to comprehensively consider service replication and user request offloading to reduce the overall delay time.

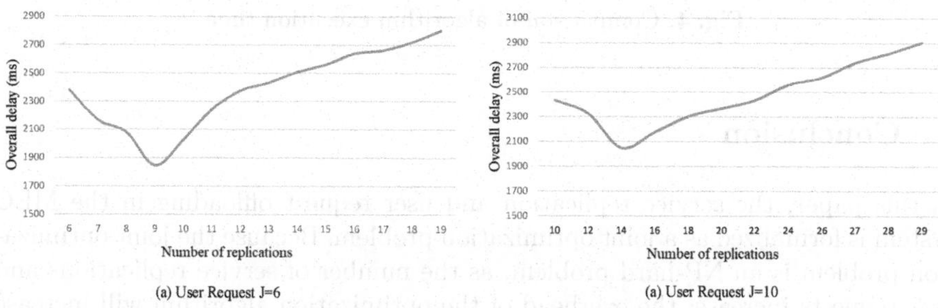

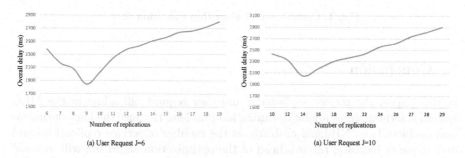

Fig. 3. The overall delay variation with the number of replications

In Fig. 4, the random algorithm, greedy algorithm and particle swarm algorithm are selected as the baseline algorithm to compare the execution time with the MGB algorithm. The x-axis is the number of service replications, and the y-axis is the algorithm execution time. In the experimental setting, a total of 20 edge nodes have performed service replications. The results show that the execution time of the random algorithm is the shortest. And the execution time of the MGB algorithm is shorter, basically similar to the time of the random algorithm and greedy algorithm. However, the particle swarm algorithm has the longest execution time, especially when user requests is 20, the execution time is one order of magnitude higher than MGB.

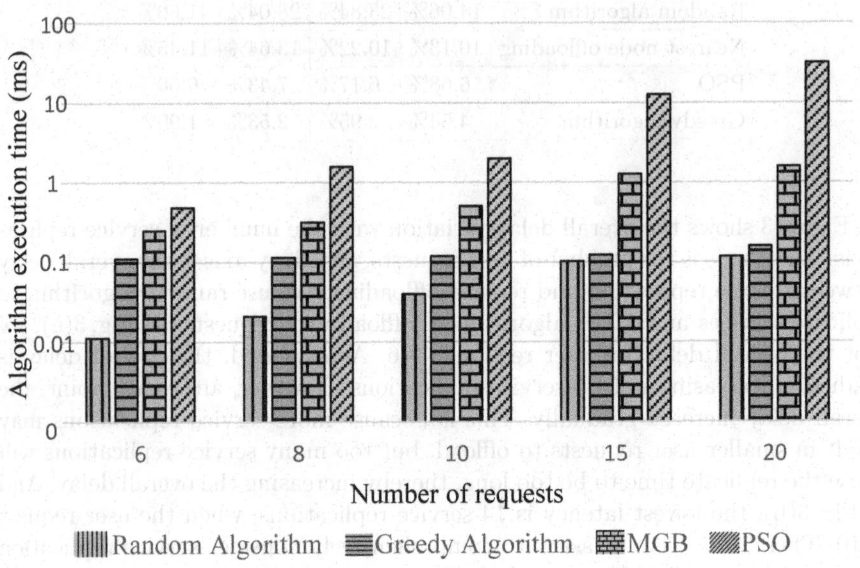

Fig. 4. Comparison of algorithm execution time

5 Conclusion

In this paper, the service replication and user request offloading in the MEC system is formalized as a joint optimization problem. Because the joint optimization problem is an NP-hard problem, as the number of service replications and user requests increase, the overhead of the optimization algorithm will increase dramatically. In order to solve the problem efficiently, we proposed the MGB algorithm and analyzed its performance. Our experiment compares the MGB algorithm with baselines algorithm. The result shows that the MGB algorithm has higher efficiency and can find a solution with lower overall delay for the above mentioned joint optimization problem. In the future, we intend to evaluate our algorithm in real scenarios. Also, we plan to develop multi-service replication based mechanisms for capacity provisioning and resource planning in edge computing systems.

Acknowledgement. This work was supported by the National Natural Science Foundation of China (No. 62072216).

References

1. Hu, S., Li, G.: Fault-tolerant clustering topology evolution mechanism of wireless sensor networks. IEEE Access **6**, 28085–28096 (2018)
2. Chen, M., Wang, T., Ota, K., Dong, M., Liu, A.: Intelligent resource allocation management for vehicles network: an A3C learning approach. Comput. Commun. **151**, 485–494 (2020)
3. Cisco Systems. https://www.cisco.com/c/zh_cn.html. Accessed 8 Aug 2021
4. Wang, T., Cao, Z., Wang, S., et al.: Privacy-enhanced data collection based on deep learning for Internet of vehicles. IEEE Trans. Industr. Inf. **16**(10), 6663–6672 (2019)
5. Wang, T., Jia, W., Xing, G., et al.: Exploiting statistical mobility models for efficient Wi-Fi deployment. IEEE Trans. Veh. Technol. **62**(1), 360–373 (2012)
6. Moubayed, A., Shami, A., Heidari, P., Larabi, A., Brunner, R.: Edge-enabled V2X service placement for intelligent transportation systems. IEEE Trans. Mob. Comput. **20**(4), 1380–1392 (2021)
7. Thai, M., Lin, Y., Lai, Y., Chien, H.: Workload and capacity optimization for cloud-edge computing systems with vertical and horizontal offloading. IEEE Trans. Netw. Serv. Manage. **17**(1), 227–238 (2020)
8. Naas, M.I., Parvedy, P.R., Boukhobza, J., Lemarchand, L.: iFogStor: an IoT data placement strategy for fog infrastructure. In: Fog and Edge Computing (ICFEC), pp. 97–104 (2017)
9. Liu, X., Yu, J., Feng, Z., Gao, Y.: Multi-agent reinforcement learning for resource allocation in IoT networks with edge computing. China Commun. **17**(9), 220–236 (2020)
10. Yu, X., Tang, L.: Competition and cooperation between edge and remote clouds: a Stackelberg game approach. In: IEEE 4th International Conference on Computer and Communications (ICCC), pp. 1919–1923 (2018)
11. Wang, Y., Sheng, M., Wang, X., Wang, L., Li, J.: Mobile-edge computing: partial computation offloading using dynamic voltage scaling. IEEE Trans. Commun. **64**(10), 4268–4282 (2016)
12. Meye, P., Raipin, P., Tronel, F., Anceaume, E.: Toward a distributed storage system leveraging the DSL infrastructure of an ISP. In: 11th Consumer Communications and Networking Conference (CCNC), pp. 533–534 (2014)
13. Chang, W., Wang, P.: An adaptable replication scheme in mobile online system for mobile-edge cloud computing. In: IEEE Conference on Computer Communications Workshops (INFOCOM WKSHPS), pp. 109–114 (2017)
14. Kiani, A., Ansari, N., Khreishah, A.: Hierarchical capacity provisioning for fog computing. IEEE Trans. Netw. **27**(3), 962–971 (2019)
15. Lin, B., et al.: A time-driven data placement strategy for a scientific workflow combining edge computing and cloud computing. IEEE Trans. Industr. Inf. **15**(7), 4254–4265 (2019)

Flying MEC: Online Task Offloading, Trajectory Planning and Charging Scheduling for UAV-Assisted MEC

Qian Wei, Tao Ouyang, Zhi Zhou(✉), and Xu Chen

School of Computer Science and Engineering, Sun Yat-Sen University, Guangzhou, China
{weiq22,ouyt9}@mail2.sysu.edu.cn, {zhouzhi9,chenxu35}@mail.sysu.edu.cn

Abstract. Unmanned aerial vehicle (UAV) with moderate computing resources has been deployed in current mobile edge computing (MEC) system to enhance the service coverage and computing capabilities of MEC servers. In such a UAV-assisted MEC system, UAV flies to the users with poor communication condition and executes the computation tasks in cooperation with the terrestrial MEC server, then the task latency perceived by users can be greatly reduced. However, frequent flight and computation offloading may drain its limited on-board battery and further deteriorate the service performance. Thus, we investigate the problem of joint online task offloading, flying trajectory and charging scheduling for UAV-assisted MEC system, with the goal of maximizing the system-wide energy efficiency in the long-term. Deriving such complicated but efficient strategies is non-trivial, since the long-term problem is NP-hard and requires future uncertain information (e.g., stochastic arrival of user tasks). To address these dual challenges, we carefully combine an online optimization technique with an approximate optimization method into a joint optimization framework, through (1) decomposing the long-term problem into a series of one-shot problems with Lyapunov optimization technique, and (2) solving the resulted NP-hard and non-linear one-shot problem with block coordinate descent and successive convex approximation methods. The efficacy of proposed online algorithm is verified by both rigorous theoretical analysis and extensive trace-driven evaluations.

Keywords: Mobile edge computing · Unmanned aerial vehicle · Lyapunov optimization · Task offloading · Trajectory planning

1 Introduction

Over the past decade, the booming of intelligent mobile and IoT devices has significantly spurred the development of Internet-of-Things (IoT) and driven

This work was supported in part by the National Science Foundation of China (No. U1711265, No. 61972432); the Program for Guangdong Introducing Innovative and Entrepreneurial Teams (No. 2017ZT07X355); the Pearl River Talent Recruitment Program (No. 2017GC010465).

Y. Lai et al. (Eds.): ICA3PP 2021, LNCS 13155, pp. 460–475, 2022.
https://doi.org/10.1007/978-3-030-95384-3_29

the emergence of intelligent applications such as video surveillance, augmented reality and smart home. These smart applications generally demand abundant computation resources and high energy quotas for real-time task processing, which incurs a conflict between the expected quality-of-service and limited computation resources of the mobile and IoT devices.

To address these issues, mobile edge computing (MEC) [1] has been proposed to provide service environment and cloud-computing capabilities at the network edges that are in closer proximity to user devices, leading to greatly reduced service latency [2,3].

While recognizing the superiority of MEC in reducing the service latency, we should also note that a MEC node is typically a tiny server cluster with limited resource capacity. This resource scarcity makes it vulnerable to request surges when the number of user requests bursts in service peak times. Specifically, for request surges that may overwhelm the limited resource capacity of the MEC node, the execution latency for each user request would increase dramatically due to the resource sharing among the huge number of user requests. To address this issue, a natural idea is known as cloud-edge cooperation [4], i.e., offloading the extra user requests to the powerful cloud to reduce the execution latency. However, the approach inevitably increases the communication latency when offloading requests to the remote cloud. In addition to cloud-edge cooperation, cross-edge service migration has been also explored to attack the above issue [5]. However, when migrating a service from a highly-occupied MEC node to a nearby idle MEC node, additional communication latency and service switching cost would be incurred.

Along a different line, in this paper, we advocate an unmanned aerial vehicle (UAV)-assisted approach to alleviating the above performance issue incurred by limited resource capacity of the MEC node [6]. As an emerging complement to the current MEC ecosystem, UAV assists the MEC node to improve communication connectivity and expand service coverage, by taking advantage of their moderate computing capability, high maneuverability and flexibility for on-demand deployment [7–9]. Attracted by such salient features, UAV-assisted communication has quickly draw much attention. For example, Google has recently launched a project named *Skybender*, the aim of this pilot project is to bring 5G network connections to users through UAVs. Besides, for the upcoming 6G era (6G wireless) [10], UAV is also envisioned as a key role in the 6G network architecture that integrates space, air and ground. In a UAV-assisted MEC system as illustrated in Fig. 1, when the user requests burst at service peak times, the UAV can fly to a subset of users with poor communication condition and executes the computation tasks in cooperation with the terrestrial MEC server, then the task latency experienced by user devices can be greatly reduced.

However, with the limitation of on-board battery capacity and service coverage, fully amplifying the benefit of UAV against dynamical user workload is by no means trivial. Specifically, with dynamical user workload that shows both temporal and spatial characteristics, the flight trajectory of UAV should be dynamically planned to follow the user workload and maximally reduce the

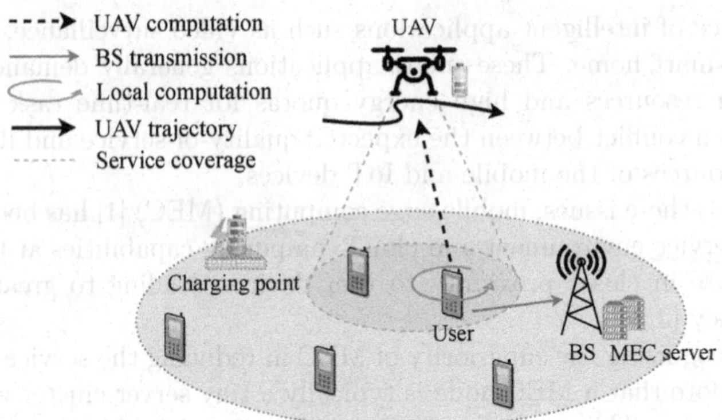

Fig. 1. An illustration of rechargeable UAV-assisted MEC system.

request latency. Besides, with the limited on-board battery capacity of UAV, the charging operation should be periodically scheduled to reduce its service interruption. Finally, with a computation hierarchy across the user device, UAV and MEC server, the task of each user device can be flexibly partitioned into three parts and executed by the user device, UAV and MEC server cooperatively to reduce latency and energy consumption. Keeping the above factors in mind, we advocate an online framework for joint task offloading, trajectory planning and charging scheduling for UAV-assisted MEC, towards the goal of optimizing the system-wide energy-efficiency in the long-term.

By formulating the problem of joint online task offloading, trajectory planning and charging scheduling in the long-term as a mixed-integer non-linear programming, we observe that this problem is rather difficult, due to the following dual challenges. First, it is NP-hard as it can be reduced from the classical Knapsack problem which is known to be NP-hard. Second, the long-term problem is a time-coupling problem that involves future system information. However, in practice, parameters such as user request arrivals typically fluctuate over time and thus cannot be readily predicted. To address these dual challenges, we carefully combine an online optimization technique with an approximate optimization method into a joint optimization framework, through (1) decomposing the long-term problem into a series of one-shot problems with Lyapunov optimization technique, and (2) solving the resulted NP-hard and non-linear one-shot problem with block coordinate descent and successive convex approximation methods.

2 System Model and Problem Formulation

2.1 Task Offloading Model

As shown in Fig. 1, we consider a UAV-assisted MEC system where the UAV acting as a flying edge node and the terrestrial MEC server attached to a base station (BS) are deployed to cooperatively serve a set of ground user devices denoted as $\mathcal{K} \triangleq \{1, 2, \cdots, K\}$[1]. Given such a computation hierarchy across the user device, UAV and MEC server, we assume that the task of each user device can be partitioned into three parts and executed by the user device, UAV and MEC server cooperatively. Note that such a *partial offloading* setup is consistent to the recent literature, and it is practical for tasks such as mobile data streaming and processing [11].

To better capture the system dynamics such as time-varying user device tasks and communication conditions, we first discrete the long time period T into Z equal time slots, indexed by $z \in \mathcal{Z} \triangleq \{1, 2, \cdots, Z\}$. The task generated by user device $k \in \mathcal{K}$ at time slot z is characterized by $A_k[z]$, in terms of the amount of required computing resources. Then, we define variables $\beta_k[z]$ as the ratio of task data offloaded to the terrestrial MEC server and $\mu_k[z]$ as the ratio of task data executed locally at user device k, which satisfy:

$$0 < \beta_k[z] < 1,\ 0 < \mu_k[z] < 1,\ \forall k, z. \tag{1}$$

2.2 Communication Model and UAV Trajectory

Without loss of generality, we construct a 3D Cartesian coordinate system to describe the location of the MEC server, UAV and user devices. Let $\mathbf{u}_0 = (x_0, y_0)$ and $\mathbf{u}_k = (x_k, y_k)$ be the horizontal coordinate of BS and user device k. To avoid the collision with buildings, we assume the UAV flies at the fixed altitude H, and its time-varying horizontal coordinate is $\mathbf{w}[z] = (x[z], y[z])$. For ease of exposition, the elemental slot length $\tau = T/Z$ should be sufficiently small to ensure the location of UAV is approximately unchanged during each time slot. In practice, the speed of UAV is constrained by

$$V_{\min}\tau \le \|\mathbf{w}[z] - \mathbf{w}[z-1]\| \le V_{\max}\tau. \tag{2}$$

As for the task data transmission, the channel power gain from UAV to user device k follows the free-space path loss model, which is expressed as

$$h_{k,u}[z] = \rho_0 d_{k,u}^{-2}[z], \tag{3}$$

where ρ_0 denotes the channel power at the reference distance $d_0 = 1\,m$ and $d_{k,u}[z] = \sqrt{\|\mathbf{w}[z] - \mathbf{u}_k\|^2 + H^2}$ denotes the distance between user device k and UAV at time slot z. To improve the signal-to-interference-plus-noise ratio (SINR), we assume that the user always transmits at its maximum power P. Then the transmission rate of user device k is

$$r_{k,u}[z] = B\log_2(1 + \frac{Ph_{k,u}[z]}{\sigma^2}), \tag{4}$$

where B is the channel bandwidth and σ^2 is the noise power of the additive white Gaussian noise (AWGN) at the UAV.

[1] Due to space limit, we consider the single UAV case for simplicity. We will consider multi-UAV case in a future work.

Similarly, the transmission rate between the MEC node and user device k is given by

$$r_{k,b} = B \log_2(1 + \frac{P\rho_0}{\sigma^2 \|\mathbf{u}_0 - \mathbf{u}_k\|^2}). \tag{5}$$

Due to the limited coverage and communication capability of the UAV, we assume that at each time slot z, the UAV can serve a limited number of user devices at most. By further introducing a binary variable $b_k[z]$ to indicate whether user device k is served by UAV ($b_k[z] = 1$) or not ($b_k[z] = 0$) at time slot z, we have the following UAV capacity constraints,

$$b_k[z] \in \{0, 1\}, \ \forall k, z, \tag{6}$$

$$\sum_{k=1}^{K} b_k[z] \leq C, \ \forall z. \tag{7}$$

It implies the UAV is limited to serve C users at most in one time slot.

2.3 UAV Charging and Battery Model

Since the capacity of UAV's on-board battery is limited, the UAV should be periodically charged to guarantee long-term flight for continuous computing service. Here we use a binary variable $g[z]$ to denote whether the UAV is charged at time slot z ($g[z] = 1$) or not ($g[z] = 0$). Since the mutual influence between the discharging (i.e., serving user devices) and charging process may deteriorate the service performance, it is generally enforced that at each time slot z, the UAV cannot be simultaneously charging or serving user devices, i.e.,

$$g[z]b_k[z] = 0, \ \forall k, z. \tag{8}$$

In order to avoid a sudden outage of the UAV during the flight, we consider a fixed charging station for energy replenishment to guarantee its long-term performance [12]. We denote $e[z]$ as the battery level at time slot z. To guarantee the reliable performance of UAV, the battery level should be no smaller than the current UAV energy consumption $E_{uav}[z]$, i.e.,

$$e[z] \geq E_{uav}[z], \ \forall z. \tag{9}$$

Considering that the charging station is horizontally located at $\mathbf{u}_c = (x_c, y_c)$, the UAV can fly back to the charging station from any point of the considered MEC coverage. Based on the above discussion, the energy overhead for UAV charging is modeled as:

$$E_f[z] = \eta E^{fly}[z] + \lfloor \frac{e[z]}{E_{\max}} \times 10 \rfloor, \tag{10}$$

where the first item is the energy cost of flying from the former location to charging station and we will make the specific definition in Sect. (2.4) later, and the second item is the energy cost of charging operation [13]. It is natural to consider that when the battery level of UAV is low, the MEC system encourages it to supplement energy. Besides, if the UAV makes a charging decision, it will supplement the energy of $E_{\max}$ in one time slot.

Since the current battery level only depends on previous battery level, UAV energy consumption and charging process, the UAV battery model is updated as

$$e[z+1] = e[z] - E_{uav}[z] + g[z]E_{\max}. \tag{11}$$

Considering that the UAV generally operates within a long-term energy cost budget, we advise to optimize the system performance under the long-term energy cost budget. It satisfies:

$$\lim_{Z\to\infty} \frac{1}{Z}\sum_{z=0}^{Z-1} E_{uav}[z] = \lim_{Z\to\infty} \frac{1}{Z}\sum_{z=0}^{Z-1} g[z]E_{\max}. \tag{12}$$

2.4 Energy Consumption Model

We are now ready to formulate the energy consumption of the user device, UAV and MEC server. We first denote $C_{k,l}$ as the computing capability of user device k, i.e., CPU frequency, and $C_{k,u}$ is the computing capability of UAV allocated to user device k. Thus, the energy consumption of local computing in user device k is expressed as

$$E_{k,l}^{cop}[z] = \kappa_1 \mu_k[z] A_k[z] C_{k,l}^2. \tag{13}$$

Similarly, the energy consumed by UAV to process the offloaded tasks is

$$E_{k,u}^{cop}[z] = \kappa_2 (1 - \beta_k[z] - \mu_k[z]) A_k[z] C_{k,u}^2, \tag{14}$$

where κ_1 and κ_2 are effective capacitance coefficients relying on chip architectures [14].

Based on the mentioned communication model for user device k, the energy consumption of UAV transmission and BS transmission are respectively given by

$$E_{k,u}^{tra}[z] = P\frac{(1 - \beta_k[z] - \mu_k[z])A_k[z]}{r_{k,\mathrm{uav}}[z]}, \tag{15}$$

and

$$E_{k,b}^{tra}[z] = P\frac{\beta_k[z]A_k[z]}{r_{k,b}}. \tag{16}$$

In general, due to its powerful computing capability, we ignore the computation energy consumption at BS and only consider the queuing energy consumption of tasks, $E_{k,b}^{que}[z]$, which is related to the amount of task in the queue.

Since energy consumption of UAV trajectory mainly depends on its speed [8], we construct a simplified flying model as

$$E^{fly}[z] = \theta\left(\frac{\|\mathbf{w}[z] - \mathbf{w}[z-1]\|}{\tau}\right)^2, \tag{17}$$

where $\theta = 0.5M\tau$ and M is relevant to the UAV's payload.

Based on the above model, the total energy consumption of the whole system for computing tasks at time slot z can be denoted as $E_{sys}[z] = E_{uav}[z] + E_{user}[z] + E_{bs}[z]$, which is mainly composed of three items.

The first item is UAV energy consumption,

$$E_{uav}[z] = (1 - g[z])(\sum_{k=1}^{K} b_k[z]E_{k,u}^{cop}[z] + \eta E^{fly}[z]) + g[z]E_f[z], \tag{18}$$

where η is a penalty coefficient of the flying energy to alleviate the magnitude difference [14]. The second item is user energy consumption,

$$E_{user}[z] = (1 - g[z])\sum_{k=1}^{K} b_k[z]E_{k,u}^{tra}[z] + \sum_{k=1}^{K}(E_{k,l}^{cop}[z] + E_{k,b}^{tra}[z]). \tag{19}$$

The third item is energy cost at BS,

$$E_{bs}[z] = \sum_{k=1}^{K} E_{k,b}^{que}[z]. \tag{20}$$

Since the data size of computing results is far smaller than the offloaded tasks, we ignore the energy consumption for this part.

2.5 Problem Formulation

The rechargeable UAV-assisted MEC system seeks to maximize the system-wide energy efficiency, by jointly optimizing the task offloading, trajectory planning and charging scheduling. More explicitly, our goal is to achieve the trade-off between the energy consumption of system and UAV and strike the balance between the system-wide energy efficiency and charging cost budget.

$$\min_{b,\beta,\mu,g,\mathbf{w}} \lim_{Z\to\infty} \frac{1}{Z}\sum_{z=0}^{Z-1} E_{sys}[z] \tag{21}$$

$$\text{s.t.} \sum_{k=1}^{K} b_k[z]r_{k,u}[z] \geq R_{\min},\ \forall k, z, \tag{21a}$$

$$1 - \beta_k[z] - \mu_k[z] \leq b_k[z],\ \forall k, z, \tag{21b}$$

$$(1), (2), (6), (7), (8), (9), (12).$$

Constraint (21a) ensures the communication quality for UAV offloading and constraint (21b) implies the ratio requirement of task offloading. We observe that problem (21) is rather difficult due to the following challenges. First, due to binary variables of $\boldsymbol{b}$ and $\boldsymbol{g}$ and non-linear constraints of (2) and (21a), problem (21) is a mixed-integer non-linear programming (MINLP) problem in the long-term. It is NP-hard as it can be reduced from the classical Knapsack problem which is known to be NP-hard. Beside, since the users' task arrival as well as the battery of UAV is time-varying and unpredictable, the lack of future information impedes the derivation of the optimization problem. Moreover, the long-term time-averaged charging budget constraint and the objective involve the time-coupling variable $e[z]$. Thus, these challenges call for an online manner which can efficiently optimize decisions on-the-fly without predicting the future system information.

3 Online Algorithm for Joint Task Offloading, Trajectory Planning and Charging Scheduling

To deal with unknown and unpredictable future information, we first decouple the original problem (21) into a series of queue stability control problems based on the Lyapunov optimization framework [15]. Unfortunately, due to its combinatorial nature, the transformed problem is NP-hard. In this regard, we adopt block coordinate descent and successive convex approximation methods to derive near-optimal policies for current task offloading, trajectory planning and charging scheduling.

3.1 Problem Transformation via Lyapunov Optimization

Queue Stability. When the energy queue is stable **i.e.,** $\lim_{Z\to\infty} \mathbb{E}\{e[z]\}/Z = 0$, the long-term time-averaged energy consumption would not exceed charging cost budget [15]. To stabilize the energy queue, we define a wide-adopted quadratic Lyapunov function

$$L(\Theta[z]) \triangleq \frac{1}{2}e^2[z]. \tag{22}$$

It represents a scalar metric of cost queue congestion. Intuitively, a small value of $L(\Theta[z])$ implies $e[z]$ is small.

In order to keep the stability of queue by persistently pushing the quadratic Lyapunov function towards a bounded level, we introduce a one-shot conditional Lyapunov drift

$$\triangle(\Theta[z]) = \mathbb{E}\left\{L(\Theta[z+1]) - L(\Theta[z])|\mathbf{e}[z]\right\}, \tag{23}$$

Then, we incorporate this draft into the system-wide energy consumption optimization, and further push the quadratic Lyapunov function towards a bounded level to stabilize the queue.

Drift-Plus-Penalty Minimization. Based on the above, the original problem has been decomposed into a series of one-shot real-time optimization problems. Then we have that the one-shot drift-plus-penalty objective, i.e.,

$$D(\Theta[z]) = \triangle(\Theta[z]) + V\mathbb{E}\left\{E_s[z]|\mathbf{e}[z]\right\}, \tag{24}$$

where V is a non-negative constant control parameter. It empowers system to make a flexible policy between the energy consumption and charging scheduling. Since this drift-plus-penalty objective $D(\Theta[z])$ can be bounded as

$$D(\Theta[z]) \leq B + V\mathbb{E}\left\{E_s[z]|\mathbf{e}[z]\right\} + \mathbb{E}\left\{e[z](g[z]E_{\max} - E_{uav}[z])|\mathbf{e}[z]\right\}, \quad (25)$$

where $B \triangleq E_{max}^2$ is a constant value for all time slots, we can consistently minimize the right hand side (R.H.S) of (25) to derive a near-optimal policy. Hence, the long-term problem can be approximately optimized (21) as a queue stability problem at each time slot z.

$$\begin{aligned} \min_{\mathbf{b},\boldsymbol{\beta},\boldsymbol{\mu},\mathbf{g},\mathbf{w}} \quad & e[z](g[z]E_{\max} - E_{uav}[z]) + VE_s[z] \quad (26) \\ \text{s.t.} \quad & (1),\ (2),\ (6),\ (8),\ (21a),\ (21b). \end{aligned}$$

Note that the time-coupling among long-term policies can be tackled by consistently updating the current queue, which implicitly characterize the follow-up service performance (i.e., a trade-off between energy consumption of system and UAV) based on past decision-making. Besides, solving approximated one-shot problem (26) only requires current information of queue backlog and system, rather than global one.

3.2 Online Control Algorithm

Since the charging decision only affects UAV, we first divide the objective into two optimization problems, i.e., whether the UAV is charged or not.

During charging procedure, the UAV dose not serve for user devices. Then the optimization problem is formulated as:

$$\begin{aligned} \min_{\boldsymbol{\beta},\boldsymbol{\mu}} \ & V\Big\{E_f[z] + \sum_{k=1}^{K}(E_{k,l}^{cop}[z] + E_{k,b}^{tra}[z] + E_{k,b}^{que}[z])\Big\} + e[z](E_{\max} - E_f[z]) \quad (27) \\ \text{s.t.} \ & (1),\ \beta_k[z] + \mu_k[z] = 1,\ \forall k, z. \quad (27a) \end{aligned}$$

In this manner, the UAV does not participate in computing and only consumes the charging cost. Each user tasks can be separated into local or BS. Obviously, problem (27) is convex with respect to variables $\boldsymbol{\beta}$ and $\boldsymbol{\mu}$, we thus utilize CVX approach to solve it [16].

When the UAV is not charged, it may provide service for user devices. In order to make it more tractable, we relax the binary discrete variable in (6) into continuous variable. The related optimization problem is

$$\begin{aligned} \min_{\mathbf{b},\boldsymbol{\beta},\boldsymbol{\mu},\mathbf{w}} \quad & -e[z]E_{uav}[z] + VE_s[z] \quad (28) \\ \text{s.t.} \quad & 0 \leq b_k[z] \leq 1,\ \forall k, z, \quad (28a) \\ & (1),\ (2),\ (8),\ (21a),\ (21b). \end{aligned}$$

Such a relaxation implies that the objective value of problem (28) serves as an upper bound of the original problem. Although relaxed, it is still non-convex because of the non-convex constraints of (2) and (21a). In the following, we apply the block coordinate descent and successive convex optimization techniques to solve the relaxed problem.

User Scheduling Optimization. The subproblem of (28) is a user scheduling problem (29), where the task offloading ratio and UAV trajectory $\{\boldsymbol{\beta}, \boldsymbol{\mu}, \mathbf{w}\}$ are fixed. We optimize the following problem:

$$\min_{\boldsymbol{b}} \sum_{k=1}^{K} b_k[z]\Big\{(V - e[z])E_{k,u}^{cop}[z] + VE_{k,u}^{tra}[z]\Big\} \tag{29}$$
$$\text{s.t.} \quad (7),\ (21\text{a}),\ (21\text{b}),\ (28\text{a}).$$

Since problem (29) is convex respect to $\{\boldsymbol{b}\}$, it is can be solved efficiently by utilizing the optimization tool such as CVX.

Task Offloading Optimization. The goal of optimizing task offloading is to perform offloading tasks on the service platform with stronger computing capability and closer to users. For any given user scheduling and UAV trajectory $\{\boldsymbol{b}, \mathbf{w}\}$, the task offloading ratio of problem (28) is

$$\begin{aligned} \min_{\boldsymbol{\beta},\boldsymbol{\mu}} V\sum_{k=1}^{K}(b_k[z]E_{k,u}^{tra}[z] + E_{k,l}^{cop}[z] + E_{k,b}^{tra}[z] + E_{k,b}^{que}[z]) \\ + (V - e[z])\sum_{k=1}^{K} b_k[z]E_{k,u}^{cop}[z] \end{aligned} \tag{30}$$
$$\text{s.t.} \quad (1),\ (21\text{b}).$$

The problem (30) is convex respect to the task offloading ratio variables $\{\boldsymbol{\beta}, \boldsymbol{\mu}\}$, which can be solved directly by CVX tool.

UAV Trajectory Optimization. With determined user scheduling and task offloading ratio $\{\boldsymbol{b}, \boldsymbol{\beta}, \boldsymbol{\mu}\}$, the subproblem about UAV trajectory consists of flying and communication energy consumption which is related to $\{\mathbf{w}\}$. And note that it is still a non-convex problem because the left hand side of (2) and (21a) are non-convex. Thus, we introduce slack variables $\mathbf{S} = \{S_{k,u}[z] = \|\mathbf{w}[z] - \mathbf{u}_k\|^2, \forall k, z\}$. Then, we have following inequalities by applying the first order Taylor expansion at the given point,

$$S_{k,u}[z] \geq \left\|\mathbf{w}^i[z] - \mathbf{u}_k\right\|^2 + 2(\mathbf{w}^i[z] - \mathbf{u}_k)^T(\mathbf{w}[z] - \mathbf{w}^i[z]), \tag{31}$$

$$\begin{aligned} \|\mathbf{w}[z] - \mathbf{w}[z-1]\|^2 \geq \left\|\mathbf{w}^i[z] - \mathbf{w}^i[z-1]\right\|^2 \\ + 2(\mathbf{w}^i[z] - \mathbf{w}^i[z-1])^T(\mathbf{w}[z] - \mathbf{w}[z-1])^2. \end{aligned} \tag{32}$$

To proceed, we adopt the successive convex approximation (SCA) technique to track constraint (21a). Specifically, linearizing the part $\log(H^2 + S_{k,u}[z])$ at a given point, the constraint (21a) can be reformulated as

$$\sum_{k=1}^{K} b_k[z]B(\log_2(\frac{P\rho_0}{\sigma^2} + H^2 + S_{k,u}[z]) - r_{k,u}^{\text{ub}}[z]) \geq R_{\min}, \tag{33}$$

where

$$r_{k,u}^{\text{ub}}[z] = \log_2(S_{k,u}^i[z] + H^2) + \frac{\log_2 e}{S_{k,u}^i[z] + H^2}(S_{k,u}[z] - S_{k,u}^i[z]) \tag{34}$$

It can be easily verified that (31)–(33) are convex. Hence, we obtain the optimal UAV trajectory by analyzing the upper bound of (35).

$$\begin{aligned} &\min_{\mathbf{w}} \ (V - e[z]) \sum_{k=1}^{K} \eta E^{fly}[z] + V \sum_{k=1}^{K} b_k[z] E_{k,u}^{tra}[z] \\ &\text{s.t.} \ \ (31), \ (32), \ (33). \end{aligned} \tag{35}$$

Based on the previous analysis, we summarize our joint optimization of online task offloading, trajectory planning and charging scheduling (JOTTC) algorithm in Algorithm 1.

3.3 Performance Analysis

Next, we briefly illustrate the convergence property of Algorithm 1 at Steps 6–13. For ease of exposition, we define $\Psi(\boldsymbol{b}^i, \boldsymbol{\beta}^i, \boldsymbol{\mu}^i, \mathbf{w}^i)$, which is denoted as the objective function value of problem (28) over i-th iteration. It follows that

$$\begin{aligned} \Psi(\boldsymbol{b}^i, \boldsymbol{\beta}^i, \boldsymbol{\mu}^i, \mathbf{w}^i) &\geq \Psi(\boldsymbol{b}^{i+1}, \boldsymbol{\beta}^i, \boldsymbol{\mu}^i, \mathbf{w}^i) \\ &\geq \Psi(\boldsymbol{b}^{i+1}, \boldsymbol{\beta}^{i+1}, \boldsymbol{\mu}^{i+1}, \mathbf{w}^i) \\ &\geq \Psi(\boldsymbol{b}^{i+1}, \boldsymbol{\beta}^{i+1}, \boldsymbol{\mu}^{i+1}, \mathbf{w}^{i+1}). \end{aligned} \tag{36}$$

It observes that Algorithm 1 at Steps 6–13 is non-increasing over each iteration. Thus, it is guaranteed to converge.

Furthermore, we denote E_s^{opt} as the optimal solution to original problem (21) and $\epsilon > 0$ is a constant to represent the distance between the time-averaged energy consumption of UAV and long-term charging cost budget.

Theorem 1. *For any control parameter V, implementing the JOTTC algorithm can achieve:*

$$\lim_{z \to \infty} \frac{1}{Z} \sum_{z=0}^{Z-1} \mathbb{E}\left\{E_s[z]\right\} \leq \frac{B}{V} + E_s^{opt}, \tag{37}$$

$$\lim_{z \to \infty} \frac{1}{Z} \sum_{z=0}^{Z-1} \mathbb{E}\left\{e[z]\right\} \leq \frac{B + V E_s^{opt}}{\epsilon}. \tag{38}$$

The detailed proof is shown in [17]. In short, our proposed algorithm achieves the $[O(1/V), O(V)]$ performance trade-off.

Algorithm 1. Joint Optimization of Online Task Offloading, Trajectory Planning and Charging Scheduling.

Initialize: $A_k[z]$, $e[0] = E_{\max}$, $\boldsymbol{b}^0$, $\boldsymbol{\beta}^0$, $\boldsymbol{\mu}^0$, $\mathbf{w}^0$.
for each time slot $z \in \mathcal{Z}$ **do**
  if UAV is charged **then**
    Solve problem (27) to obtain $\boldsymbol{\beta}$, $\boldsymbol{\mu}$.
  end if
  if UAV is not charged **then**
    repeat
      Solve problem (29) for given $\{\boldsymbol{\beta}^i, \boldsymbol{\mu}^i, \mathbf{w}^i\}$ and denote the optimal solution as $\{\boldsymbol{b}^{i+1}\}$.
      Solve problem (30) for given $\{\boldsymbol{b}^{i+1}, \boldsymbol{\beta}^i, \boldsymbol{\mu}^i, \mathbf{w}^i\}$ and denote the optimal solution as $\{\boldsymbol{\beta}^{i+1}, \boldsymbol{\mu}^{i+1}\}$.
      Solve problem (35) for given $\{\boldsymbol{b}^{i+1}, \boldsymbol{\beta}^{i+1}, \boldsymbol{\mu}^{i+1}, \mathbf{w}^i\}$ and denote the optimal solution as $\{\mathbf{w}^{i+1}\}$.
      Update $i = i + 1$.
    until Converge to the prescribed accuracy.
  end if
  Compare system energy consumption and obtain the minimum, then select the current system decisions.
  Update the queue backlog $e[z]$.
  Update the time slot $z = z + 1$.
end for

4 Performance Evaluation

In this section, we demonstrate simulation results to evaluate the performance and efficacy of our algorithm. In the simulation setting, we consider that $K = 4$ user devices are randomly distributed in a 2D area of 100×100 m^2. We consider that the computing capability of UAV $C_{k,u}$ is 1200 MHz, while that of each user $C_{k,l}$ is randomly distributed within [300, 400] MHz [14]. Due to the resource limitation, we assume that UAV can serve at most $C = 3$ user devices in one time slot. To better characterize the user behaviors, we use telecommunication dataset to simulate its generated tasks $A_k[z]$. Other parameters [18] are summarized in Table 1.

Table 1. Simulation parameters

Parameter	Value	Parameter	Value	Parameter	Value
σ^2	−100 dBm	ρ_0	−50 dB	B	1 MHz
H	100 m	$V_{\min}$	25 m/s	$V_{\max}$	50 m/s
P	1 W	κ_1, κ_2	10^{-28}	κ_2	10^{-28}
M	9.65 kg	$E_{\max}$	72 kJ	$R_{\min}$	58 bps/Hz

In this paper, we adopt the other five benchmark schemes to verify the efficient performance of our algorithm JOTTC. The first is *Fly Nearest*, which means UAV always provides service for the nearest user devices. Contrarily, *UAV Static* denotes that it always stays at the initial location. UAV using *Random* randomly serves user devices in each time slot. *Fly Circularly* represents UAV flies with circular trajectory [6]. If the system is not configured with UAV, user devices are served locally or by BS in scheme *No UAV*.

We first carry out experiments under different values of control parameter V to show the performance of energy consumption trade-off between the system and UAV. As shown in Fig. 2, these two curves jointly validate the $[O(1/V), O(V)]$ performance trade-off in (37) and (38). In brief, the parameter V is beneficial to achieve better system-wide energy efficiency. Moreover, we can obtain the near-optimal solutions by setting the proper value of V.

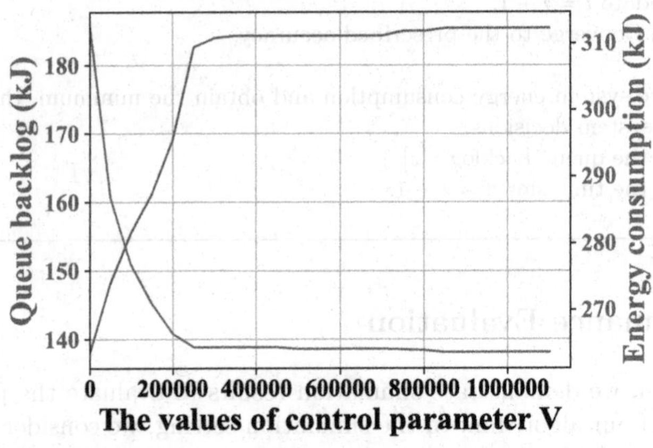

Fig. 2. Performance trade-off under different values of V.

Figure 3 describes the time-averaged system energy consumption under the different algorithms with $Z = 200$. Obviously, the trends of time-averaged system energy consumption are consistent among all algorithm, i.e., rotated S-curve. Evidently, JOTTC outperforms the others, i.e., the energy reduction decreases around from 10.1% to 57.9%. Compared to *No UAV*, JOTTC shows that the UAV considerably enhances the service ability of system and reduces user energy consumption. As for *UAV Static*, limited service coverage deteriorates the system-wide energy efficiency due to inflexible flying trajectory optimization. As for *Fly Circularly* without charging scheme, the battery will be exhausted in the long run, which further terminates subsequent services, and results huge system energy consumption. It implies that the charging process is vital and desirable for the UAV-assisted MEC system.

Figure 4 depicts the optimal flying trajectory of UAV on the horizontal plane. Different colors indicate the number of service provided by UAV for user devices

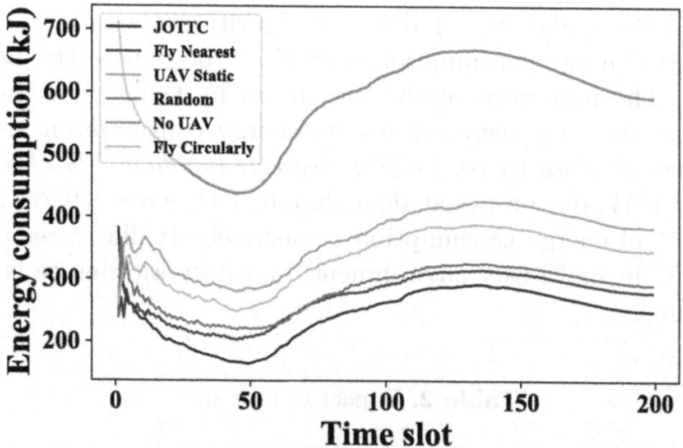

Fig. 3. Time-averaged energy consumption of system under different algorithms.

with $Z = 100$. Figure 5 illustrates the real-time task size of user devices which are correspond to circle points with different radius in Fig. 4. As we can see, the flying trajectory of UAV depends on the random user requests and is related to the position of charging station. Combined with Fig. 4 and Fig. 5, the UAV is more inclined to serve the user device who generates large tasks and is in closer proximity to the charging station. Even though the average task size of User1 and User3 are approximate, the UAV tends to fly to the user device close to charging station, i.e., User1. Besides, we notice that the UAV may fly to unexpected positions, such as up-right position. The reason is that at the beginning, the system was not stable, and UAV was unable to make precise decisions based on a small amount of information. So it may fly to decentralized service points, but this is a rare phenomenon. With the further exploration, the online algorithm which dynamic adjusts the trajectory planning is line with the change of queue and becomes stable.

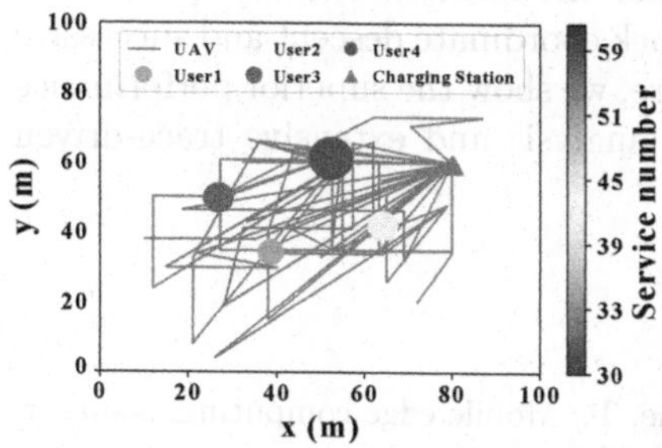

Fig. 4. The optimal real-time trajectory of UAV in the total period.

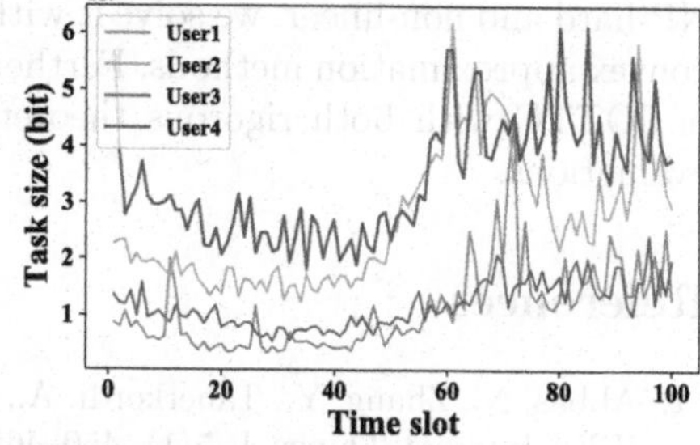

Fig. 5. The real-time task size of all users in the total period.

To explore the scalability of different algorithms, we focus on the time-averaged system energy consumption over $Z = 100$, where the users increase from 4 to 20. The numerical results are shown in Table 2. As the number of users increases, the time-averaged system energy consumption also increases. When the users increase to 10, for *Fly Nearest*, *Random*, *Fly Circularly*, *UAV Static* and *No UAV*, our proposed algorithm JOTTC saves 8.9%, 12.2%, 25.6%, 27.3% and 69% of energy consumption respectively. It illustrates the effectiveness of JOTTC in multi-user environment, and its performance is better than other algorithms.

Table 2. Impact of user size

Algorithm	$K = 4$	$K = 6$	$K = 10$	$K = 20$
JOTTC	263.21	368.35	956.74	2061.51
Fly Nearest	292.09	483.89	1051.26	2231.57
Random	281.69	508.79	1089.77	2242.82
Fly Circularly	360.25	573.80	1285.90	2089.36
UAV Static	394.62	610.40	1315.28	2517.29
No UAV	627.94	1215.16	3103.21	6020.14

5 Conclusion

In this paper, we study the problem of joint online task offloading, flying trajectory and charging scheduling for the UAV-assisted MEC system. We develop a novel online algorithm of JOTTC to achieve a desirable balance between the system-wide energy efficiency and charging cost budget. To tackle the unpredictable future system information, which involves stochastic arrival of user tasks, we decompose the long-term problem into a series of one-shot problems with Lyapunov optimization technique. Since the resulted one-shot problem is NP-hard and non-linear, we solve it with block coordinate descent and successive convex approximation methods. Furthermore, we show the superior performance of JOTTC with both rigorous theoretical analysis and extensive trace-driven evaluations.

References

1. Abbas, N., Zhang, Y., Taherkordi, A., Skeie, T.: Mobile edge computing: a survey. IEEE Internet Things J. **5**(1), 450–465 (2017)
2. Hu, Y.C., Patel, M., Sabella, D., Sprecher, N., Young, V.: Mobile edge computing-a key technology towards 5G. ETSI white paper (2015)
3. Zhou, Z., Chen, X., Li, E., Zeng, L., Luo, K., Zhang, J.: Edge intelligence: paving the last mile of artificial intelligence with edge computing. Proc. IEEE **107**(8), 1738–1762 (2019)

4. Zhou, Z., Yu, S., Chen, W., Chen, X.: CE-IoT: cost-effective cloud-edge resource provisioning for heterogeneous IoT applications. IEEE Internet Things J. **7**(9), 8600–8614 (2020)
5. Ouyang, T., Zhou, Z., Chen, X.: Follow me at the edge: mobility-aware dynamic service placement for mobile edge computing. IEEE J. Sel. Areas Commun. **36**(10), 2333–2345 (2018)
6. Wu, Q., Zeng, Y., Zhang, R.: Joint trajectory and communication design for multi-UAV enabled wireless networks. IEEE Trans. Wireless Commun. **17**(3), 2109–2121 (2018)
7. Hu, X., Wong, K.K., Yang, K., Zheng, Z.: UAV-assisted relaying and edge computing: scheduling and trajectory optimization. arXiv preprint arXiv:1812.02658 (2018)
8. Jeong, S., Simeone, O., Kang, J.: Mobile edge computing via a UAV-mounted cloudlet: optimization of bit allocation and path planning. IEEE Trans. Veh. Technol. **67**(3), 2049–2063 (2017)
9. Wang, Y., Ru, Z.Y., Wang, K., Huang, P.Q.: Joint deployment and task scheduling optimization for large-scale mobile users in multi-UAV-enabled mobile edge computing. IEEE Trans. Cybern. **50**(9), 3984–3997 (2019)
10. Zhang, Z., et al.: 6G wireless networks: vision, requirements, architecture, and key technologies. IEEE Veh. Technol. Mag. **14**(3), 28–41 (2019)
11. Saleem, U., Liu, Y., Jangsher, S., Li, Y.: Performance guaranteed partial offloading for mobile edge computing. In: IEEE Global Communications Conference, pp. 1–6. IEEE (2018)
12. Chen, W., Zhao, S., Shi, Q., Zhang, R.: Resonant beam charging-powered UAV-assisted sensing data collection. IEEE Trans. Veh. Technol. **69**(1), 1086–1090 (2019)
13. Xu, J., Zhu, K., Wang, R.: RF aerially charging scheduling for UAV fleet: a Q-learning approach. In: International Conference on Mobile Ad-Hoc and Sensor Networks, pp. 194–199. IEEE (2019)
14. Zhang, J., Zhou, L., Tang, Q., Ngai, E.C.H., Hu, X., Zhao, H., Wei, J.: Stochastic computation offloading and trajectory scheduling for UAV-assisted mobile edge computing. IEEE Internet Things J. **6**(2), 3688–3699 (2018)
15. Neely, M.J.: Stochastic network optimization with application to communication and queuing systems. Synth. Lect. Commun. Netw. **3**(1), 1–211 (2010)
16. Boyd, S., Boyd, S.P., Vandenberghe, L.: Convex Optimization. Cambridge University Press, Cambridge (2004)
17. Zhou, Z., et al.: Carbon-aware load balancing for geo-distributed cloud services. In: IEEE 21st International Symposium on Modelling, Analysis and Simulation of Computer and Telecommunication Systems, pp. 232–241. IEEE (2013)
18. Hu, Q., Cai, Y., Yu, G., Qin, Z., Zhao, M., Li, G.Y.: Joint offloading and trajectory design for UAV-enabled mobile edge computing systems. IEEE Internet Things J. **6**(2), 1879–1892 (2018)

Multiple Workflows Offloading Based on Deep Reinforcement Learning in Mobile Edge Computing

Yongqiang Gao[1,2,3](✉) and Yanping Wang[2,3]

[1] Engineering Research Center of Ecological Big Data, Ministry of Education, Hohhot 010021, China
gaoyongqiang@imu.edu.cn

[2] Inner Mongolia Engineering Laboratory for Cloud Computing and Service Software, Hohhot 010021, China
csyanping@mail.imu.edu.cn

[3] College of Computer Science, Inner Mongolia University, Hohhot 010021, China

Abstract. With the maturity of 5G technology and the popularization of smart terminal devices, the applications running on mobile terminals are becoming more and more diversified. Most of them are complex, computationally intensive, and time-sensitive applications such as workflow and machine learning tasks. The traditional cloud computing model is far away from the mobile terminal and thus cannot meet the stringent requirements of these applications on delay and energy consumption. As a new computing model, mobile edge computing can better solve the above problems. Mobile edge computing sinks part of the computing and storage resources in the cloud to the edge of the network close to the mobile device. With computational offloading, complex applications are offloaded to nearby edge servers for execution, which leads to low delay and energy consumption. The existing researches mainly focus on independent task offloading in mobile edge computing, and thus they are not suitable for workflow tasks offloading with dependence on mobile edge computing. This paper proposes a multiple workflows offloading strategy based on deep reinforcement learning in mobile edge computing with the goal of minimizing the overall completion time of multiple workflows and the overall energy consumption of multiple user equipments. We evaluate the performance of the proposed strategy by simulation experiments based on real-world parameters. The results show that the proposed strategy performs better than other alternatives in terms of the overall completion time and the overall energy consumption.

Keywords: Multiple workflows offloading · Multi-objective optimization · Multi-agent DDPG

Supported in part by the National Natural Science Foundation of China under Grant 61662052, in part by the Natural Science Foundation of Inner Mongolia Autonomous Region under Grant 2021MS06002, in part by the Science and Technology Planning Project of Inner Mongolia Autonomous Region under Grant 2021GG0155, and in part by the Major Research Plan of Inner Mongolia Natural Science Foundation under Grant 2019ZD15.

Y. Lai et al. (Eds.): ICA3PP 2021, LNCS 13155, pp. 476–493, 2022.
https://doi.org/10.1007/978-3-030-95384-3_30

1 Introduction

With the rapid development of 5G technology and smartphones, there is an increasing trend of employing mobile devices to handle time-sensitive and computationally intensive tasks which require a lot of computing and storage resources [1]. However, mobile devices have limited hardware resources and thus it is difficult for such devices to support time-sensitive and computationally intensive applications. Although offloading computing tasks to remote cloud servers can alleviate the drawbacks of limited resources of mobile devices, cloud computing suffers from very high latency due to long-distance network transmission [2]. To overcome the latency issues in cloud computing, mobile edge computing is proposed to meet the growing demands for low latency.

Mobile edge computing is a new computing paradigm in which computational tasks are performed mainly on servers at the edge of network, which provides users with nearby computing and storage services [3, 4]. The task offloading technology in mobile edge computing provides a promising solution to the limited computing resources in mobile devices. However, it is challenging to find an optimal task offloading strategy in mobile edge computing as tasks can be offloaded to multiple edge servers. In recent years, many efforts have focused on utilizing heuristic and meta-heuristic algorithms to generate approximate optimal task offloading strategy in mobile edge computing environment [5–9]. However, the existing researches mainly consider the offloading of independent tasks, and thus they are not suitable for highly dependent workflow task offloading. In addition, deep reinforcement learning has been proven to be a promising method to solve combinatorial optimization problems. Therefore, this paper focuses on the offloading of dependent tasks and proposes a task offloading strategy based on deep reinforcement learning to shorten the workflow completion time and save mobile devices' energy consumption. Specifically, the main contributions of this paper are as follows:

- We establish a multi-objective optimization model for multiple workflows offloading in mobile edge computing, which minimizes the total completion time of multiple workflows and the total energy consumption of multiple user's equipments.
- We develop a multi-agents deep deterministic policy gradient algorithm (MADDPG) to find the best offloading strategy by adopting centered training with decentralized execution.
- We evaluate the effectiveness of the proposed offloading algorithm through simulation experiments. The results show that our algorithm is better than the current state-of-the-art algorithms.

The remainder of this paper is organized as follows. Section 2 introduces related work. The problem formulation is described in Sect. 3. The proposed algorithm is illustrated in Sect. 4. Section 5 discusses the experimental results. Finally, the conclusion is presented in Sect. 6.

2 Related Work

Many efforts have been devoted to using heuristic algorithms and meta-heuristic algorithms to solve mono-objective task offloading in mobile edge computing (MEC). Jia et al. [10] proposed an online task offloading algorithm that can minimize the completion time of applications on mobile devices. Hu et al. [11] proposed a game-based computational offloading algorithm to solve the task offloading problem. Its goal is to minimize the energy consumption of each mobile device by performing the task on the edge server. Li et al. [12] aimed to minimize the energy consumption of mobile devices in MEC under both power and delay constraints. Different from these studies, we focus on multi-objective task offloading in a MEC environment.

There is also some work focusing on using deep reinforcement learning algorithms to solve multi-objective task offloading problems. Tong et al. [13] proposed an adaptive task offloading and resource allocation algorithm in the MEC environment that uses the deep reinforcement learning method to determine task offloading and allocate computing resource. Lu et al. [14] proposed deep reinforcement learning methods to solve the offloading problem of multiple service nodes with the goal of optimizing energy consumption and load balancing. Tang et al. [15] proposed a model-free deep reinforcement learning-based algorithm where each device determines its offloading decision without knowing the task models and offloading decision of other devices. Different from these studies, we apply a deep reinforcement learning algorithm to solve multiple workflows offloading in the MEC environment.

Recently, there have been several studies on using deep reinforcement learning methods to solve single workflow offloading problems. Zhu et al. [16] proposed a deep Q-networks algorithm (DQN) to reduce the workflow completion time and the user device's energy consumption. Huang et al. [17] designed a reinforcement learning-based workflow scheduling policy to minimize the workflow completion time under the risk probability constraint. Song et al. [18] studied a multi-objective offloading problem of dependent tasks in edge computing and developed an adaptive reinforcement learning algorithm to solve the offloading problem. Different from these studies, we proposed a multiple agents algorithm based on deep reinforcement learning to solve multi-objective offloading problem of multiple workflows in a MEC environment.

3 Problem Definition

3.1 System Model

As shown in Fig. 1, we consider a MEC system with multiple user equipments (UEs), multiple proxy servers (FSs) and multiple edge servers (ESs). UE is responsible for generating the workflow and executing the unoffloaded workflow tasks. FS deploys a multi-objective task offloading algorithm. ES is responsible for executing the offloaded workflow tasks. The specific workflow of the system is described as follows. First, UE sends a scheduling request to FS. Second, FS periodically executes the multi-objective task offloading algorithm to find the optimal offloading for the workflow generated by each UE and then returns them to the corresponding UE. Next, each UE sends the workflow tasks to ESs based on the receipt of offloading strategy. Finally, each ES executes the offloaded workflow tasks and returns the results to the corresponding UE.

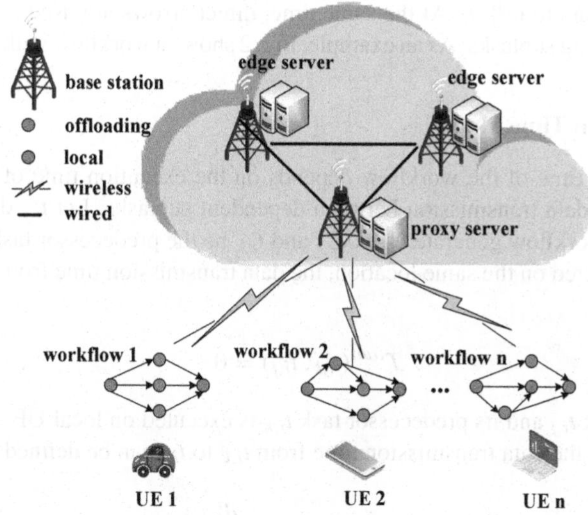

Fig. 1. System model.

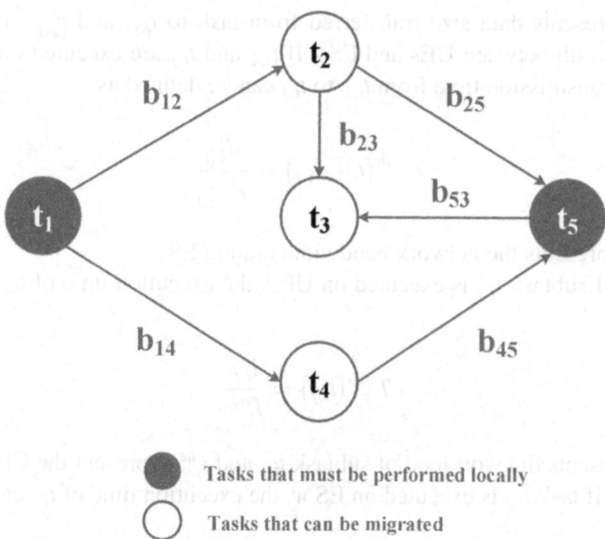

Fig. 2. A workflow application with five subtasks.

3.2 Workflow Model

Generally, the workflow application is represented by a directed acyclic graph $W =$ *G(T,B)* where $T = \{t_1, t_2, \ldots t_n\}$ represents a set of n tasks and B represents the set of dependency between subtasks. Each dependency $b_{ij} \in B$ has a weight $d_{i,j}$ which represents the size of the transmitting data between t_i and t_j. $b_{i,j}$ indicates a precedence constraint that the task t_j cannot begin its execution before t_i finishes and sends all the needed output data to task t_j. At the same time, direct arrows are used to indicate data dependence among subtasks. As an example, Fig. 2 shows a workflow with five subtasks.

3.3 Completion Time Model

The completion time of the workflow depends on the execution time of each subtask and the time of data transmission between dependent subtasks. Let $t_{i,j}$ denote the j-th subtask of the workflow generated by UE i and $t_{i,k}$ be the predecessor task of $t_{i,j}$. If $t_{i,k}$ and $t_{i,j}$ are executed on the same location, the data transmission time from $t_{i,k}$ to $t_{i,j}$ can be defined as

$$T^{tran}(t_{i,k}, t_{i,j}) = 0 \tag{1}$$

If one of subtask $t_{i,j}$ and its predecessor task $t_{i,k}$ is executed on local UE and another is executed on ES, the data transmission time from $t_{i,k}$ to $t_{i,j}$ can be defined as

$$T^{tran}(t_{i,k}, t_{i,j}) = \frac{d^i_{k,j}}{r^i_{edge}} \tag{2}$$

where $d^i_{k,j}$ represents data size transferred from task to $t_{i,j}$, and r^i_{edge} represents the network bandwidth between UEs and ESs. If $t_{i,k}$ and $t_{i,j}$ are executed on the different ESs, the data transmission time from $t_{i,k}$ to $t_{i,j}$ can be defined as

$$T^{tran}(t_{i,k}, t_{i,j}) = \frac{d^i_{k,j}}{r^i_{local}} \tag{3}$$

where r^i_{local} represents the network bandwidth among ESs.

Similarly, if subtask $t_{i,j}$ is executed on UE i, the execution time of $t_{i,j}$ can be given by

$$T^{pro}(t_{i,j}) = \frac{l_{i,j}}{f^{ue}_i} \tag{4}$$

where $l_{i,j}$ represents the workload of subtask $t_{i,j}$ and f^{ue}_i represent the CPU processing speed of UE i. If task $t_{i,j}$ is executed on ES w, the execution time of $t_{i,j}$ can be given by

$$T^{pro}(t_{i,j}) = \frac{l_{i,j}}{f^{edge}_w} \tag{5}$$

where f^{edge}_w represent the CPU processing speed of ES w. Based on the above model, the start time of $t_{i,j}$ can be given by

$$T^{start}(t_{i,j}) = \begin{cases} T^{fst} & pred(t_{i,j}) = \emptyset \\ \max\left\{ \max\limits_{t_{i,k} \in pred(t_{i,j})} \left(T^{tran}(t_{i,k}, t_{i,j}) + T^{fin}(t_{i,k})\right), T^{fst} \right\} & otherwise \end{cases} \tag{6}$$

where T^{fst} represents the earliest start execute time of task $t_{i,j}$ and $pred\left(t_{i,j}\right)$ represents the set of all predecessor subtasks of $t_{i,j}$. $T^{fin}\left(t_{i,k}\right)$ represents the completion time of $t_{i,k}$ which is defined as

$$T^{fin}\left(t_{i,k}\right) = T^{start}\left(t_{i,k}\right) + T^{pro}\left(t_{i,k}\right) \tag{7}$$

Finally, the completion time of the workflow generated by *UE i* can be given by

$$T_i^{ue} = \max_{j \in U_i} T^{fin}\left(t_{i,j}\right) \tag{8}$$

where U_i represents the set of all subtasks of the workflow generated by UE i. Since there are n UEs, the total completion time of workflows generated by all UEs can be defined as

$$T = \sum_{i=1}^{n} T_i^{ue} \tag{9}$$

3.4 Energy Consumption Model

The energy consumption of UE depends on the amount of energy consumed by executing subtasks and transmitting data. If subtask $t_{i,j}$ is executed on local UE, the energy consumption caused by executing $t_{i,j}$ can be defined as

$$Energy^{pro}\left(t_{i,j}\right) = \frac{l_{i,j}}{f_i^{ue}} \times P_i^{act} \tag{10}$$

where P_i^{act} denotes the power consumption of the UE i during executing subtask $t_{i,j}$. If subtask $t_{i,j}$ is offloaded to ES w, the UE energy consumption can be defined as

$$Energy^{pro}\left(t_{i,j}\right) = \left(\frac{l_{i,j}}{r_{local}^{i}} + \frac{l_{i,j}}{f_w^{edge}} + Q_{i,j}^{w}\right) * P_i^{idel} \tag{11}$$

where P_i^{idle} indicates the power consumption when UE i is idle, and $Q_{i,j}^{w}$ indicates the average waiting time of $t_{i,j}$ in ES w. In addition, if one of subtask $t_{i,j}$ and its predecessor task $t_{i,k}$ is executed on local UE i and another is executed on ES w, the UE's energy consumption caused by data transmission between UE and ES can be defined as

$$Energy^{tran}\left(t_{i,k}, t_{i,j}\right) = \frac{d_{k,j}^{i}}{r_{local}^{i}} * P_i^{tran} \tag{12}$$

where P_i^{tran} represents the power consumption of UE i when transmitting data between $t_{i,k}$ and $t_{i,j}$. If subtask $t_{i,j}$ and its predecessor task $t_{i,k}$ is executed on the same location, the UE's energy consumption caused by data transmission between UE and ES can be defined as

$$Energy^{tran}\left(t_{i,k}, t_{i,j}\right) = 0 \tag{13}$$

Therefore, the energy consumption of UE i can be defined as

$$Energy_i^{ue} = \sum_{j \in U_i} Energy^{pro}(t_{i,j}) + \sum_{b_{k,j} \in B} Energy^{tran}(t_{i,k}, t_{i,j}) \quad (14)$$

Finally, the total energy consumed by all UE can be given by

$$E = \sum_{i=1}^{n} Energy_i^{ue} \quad (15)$$

3.5 Problem Formulation

In this paper, we aim to minimize the completion time of multiple workflows and the energy consumption of multiple UEs in the MEC system. Let SU be a set of n UEs, and SP be a set of m ESs. The binary variable $x_{i,j,l}$ indicates whether the j-th subtask of the workflow generated by UE i is allocated to the location $l \in SP \cup \{UE\, i\}$ for execution. Based on the above model, the multiple workflows task offloading problem can be formulated as follows:

$$\textbf{Minimize} \quad T \quad (16)$$

$$\textbf{Minimize} \quad E \quad (17)$$

Subject to

$$T^{fin}(t_{i,k}) + T^{tran}(t_{i,k}, t_{i,j}) \le T^{start}(t_{i,j}) \quad \forall t_{i,j} \in U_i, \forall t_{i,k} \in pred(t_{i,j}) \quad (18)$$

$$\sum_{l \in SP \cup \{UE\, i\}} x_{i,j,l} = 1 \quad \forall i \in SU, \forall j \in U_i \quad (19)$$

$$\sum_{i \in SU} \sum_{j \in U_i} x_{i,j,l} \le TH_l \quad \forall l \in SP \quad (20)$$

$$\sum_{j \in U_i \wedge l = i} x_{i,j,l} \le UH_i \quad \forall i \in SU \quad (21)$$

$$x_{i,j,l} \in \{0, 1\} \quad \forall i \in SU, \forall j \in U_i, \forall l \in SP \cup \{UE\, i\} \quad (22)$$

Constraints (18) indicate that each subtask must be executed after its predecessor subtasks. Constraint (19) ensures that each task can only be allocated to one location. Constraints (20) and (21) ensure that the number of executing tasks on ES l and UE i does not exceed the ES's maximal capacity TH_l and the UE's maximal capacity UHl respectively. Constraint (22) defines the domain of the variables of the problem.

4 Algorithm Design

The optimization problem described above is an NP-hard problem and it takes a long time to get the optimal solution. This paper proposes a multiple agents DDPG algorithm to obtain the optimal solution in an acceptable time. First, we transform the formulated problem into a Markov decision process (MDP) mode with N UE agents. The MDP model can be defined as a triplet $M =< s, o, a >$ where $s = \{s_1, s_2, \ldots, s_N\}$ is a set of states, $o = \{o_1, o_2, \ldots, o_N\}$ is a set of observations and $a = \{a_1, a_2, \ldots, a_N\}$ is a set of actions. The state set describes the possible offloading strategy. For each given states $s \in S$, the UE agent utilizes the policies to select an action from action spaces based on the current observations. The following describes the definition of *s*, *o* and *a*.

- **State space.** For the sake of comprehensively considering the characteristics of UEs and ESs in MEC, the state space of each UE agent is define as $s_i^t = \left(X_i^t, RU_1^t, \ldots, RU_j^t, \ldots, RU_N^t, RE_1^t, \ldots, RE_k^t, \ldots, RE_M^t, T^t, E^t\right)$ at time slot t where X_i^t denotes the subtask offloading decision matrix for the workflow generated by UE i, and RU_j^t and RE_k^t denote the remaining capacity of UE j and ES k respectively.
- **Observation.** At time slot t, the observations of each UE agent i can be defined as $o_i^t = \left(X_i^t, RU_1^t, \ldots, RU_N^t, T_i^{ue^t}, Energy_i^{ue^t}\right)$.
- **Action space.** At time slot t, the action spaces of each UE agent i can be defined as $a_i^t = X_i^t$.
- **Reward function.** The reward function is used to assess the impact of the action taken by an agent at a given state. Inspired by [19], we use the weighted sum of task completion time and energy consumption as the reward function. The reward function of all agents is defined as follows:

$$r = -(\alpha T + (1 - \alpha)E) \tag{23}$$

where α indicates the weight, and T and E represent the total completion time and energy consumption respectively.

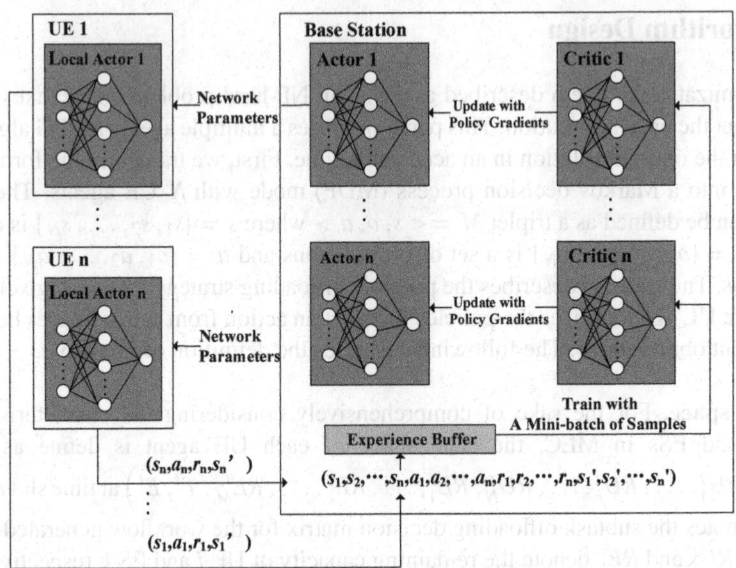

Fig. 3. Network training and parameter updating process of the MADDPG framework.

Next, we describe the proposed MADDPG algorithm. In the traditional single-agent DDPG algorithm, the agent learns to improve its strategy by interacting with the environment and gets the optimal strategy from the environment finally. But, from the perspective of a single agent, the environment is complex and dynamic, which brings great difficulties to the learning process. Thus, we proposed a MADDPG algorithm to solve the above problem. Figure 3 shows the overall framework of MADDPG with N agents. The main steps of the MADDPG algorithm are shown in Algorithm 1.

MADDPG works in a way of centralized training and decentralized execution. Each UE agent inputs local information such as the offloading strategy of workflow subtasks, the completion time of workflow generated by UE i, the remaining resources of UE, and the energy consumption of UE i into the distributed actor network to generate global task offloading strategies. At the same time, the centralized critic network is used to update the actor network's weight.

Each UE agent i firstly observes the current environment and obtains the initial state, and then generates the subtask offloading strategy by inputting the state into the actor network. After that, all UE agents carry out their corresponding offloading actions, and then evaluate the reward and enter the next state. At the same time, the agents store (s_t, a_t, r_t, s_{t+1}) into a centralized replay buffer. The above steps repeat until the replay buffer is full. Finally, H samples are randomly selected from the replay buffer and the actor network and the centralized critic network are trained using H samples.

Algorithm 1 MADDPG-based workflow offloading algorithm for *N* UEs and *M* ESs

```
Randomly initialize the critic network Q(s,a|θ^Q) and the actor network μ(s|θ^μ).
Initialize the replay buffer R, the learning rates of the actor network and the
  critic network, the discount factor γ, the maximum learning episode M, the
  maximum training steps T per episode, and the random process ψ
Initialize the network layout with N UEs and M ESs
for episode = 1 to M
  Each UE agent i generates the initial state o_i^t
  Generates the global state s^t
  for t = 1 to T
    For each UE agent i, generates action a_i^t = μ_{θ_i^μ}(o_i^t) + ψ^t, carry out
      the actions a^t = {a_1^t, ..., a_N^t}, observe the reward r_t, enter the next
      state s_{t+1}
    Store (s_t, a_t, r_t, s_{t+1}) in R
    s^t = s^{t+1}
    for i = 1 to N
      Randomly select H samples from R
      y^t = r^t + γ Q_i^{θ_i^{Q'}}(s^{t+1}, a_1^{t+1}, ..., a_N^{t+1})
      Update the critic network according to the following loss function:
        L(θ_i^Q) = (1/H) Σ_h (y^h − Q_i^{θ_i^Q}(s^h, a_1^h, ..., a_N^h))^2
      Update the actor network using the sampled gradient:
        ∇_{θ_i^μ} ς(μ_i) ≈ (1/H) Σ_h ∇_{θ_i^μ} μ_i(o_i^h) ∇_{a_i} Q_i^{θ_i^Q}(s^h, a_1^h, ..., a_N^h)
    end for
    Update the target critic networks according to the following equation:
      θ_i^{Q'} = τ θ_i^Q + (1 − τ) θ_i^{Q'}
    Update the target actor networks according to the following equation:
      θ_i^{μ'} = τ θ_i^μ + (1 − τ) θ_i^{μ'}
  end for
end for
```

5 Experimental Evaluation

In this section, we first describe the experimental platform and then the experimental results are presented and analyzed.

5.1 Experimental Settings

To verify the effectiveness of the proposed MADDPG algorithm, we build a MEC system with 6 UEs and 20 ESs. The settings for various parameters in the simulate MEC environment are shown in Table 1. We use Google's TensorFlow-CPU 2.2.0 as the deep reinforcement learning framework, and the proposed algorithm is implemented by Python 3.7. Six real-world workflow applications from the Pegasus project [20] are used in the simulation experiment. They are Montage (w1), CyberShake (w2), EpiGenomics (w3), Sipht (w4), LIGO (w5) and Spark (w6). Figure 4 shows their structural frameworks. Each simulation experiment uses a type of workflow application. In the proposed MADDPG algorithm, the actor network has twelve layers and the critic network has six layers. The number of neurons in each layer is fifty. Table 2 shows the training optimizer and the hyperparameters for the MADDPG network. All simulation experiments are conducted on a standard laptop with an AMD Ryzen 5 3500U CPU at 2.10 GHz × 4 and 16 GB RAM.

Table 1. The parameters setting.

Symbol	Definition	Value
p_i^{act}	The power consumption of UE i when CPU is active state	1.5 w
p_i^{idel}	The power consumption of UE i when CPU is idle state	0.2 w
p_i^{tran}	The power consumption of UE i during data transmission	0.6 w
f_i^{ue}	The CPU processing speed of UE i	600 MHz
f_w^{edge}	The CPU processing speed of ES w	2500 MHz
r_{local}^i	The network bandwidth among ESs	10 Gbps
r_{edge}^i	The network bandwidth between UEs and ESs	100 Mbps
TH_w	The maximal capacity of ES w	500 subtasks
UH_i	The maximal capacity of UE i	100 subtasks

Table 2. The hyperparameters of the MADDPG algorithm.

Hyperparameter	Value
Episode M	200
Steps T	1000
Batch size H	64
Discount factor γ	0.96
Learning rate	0.001
Replay buffer size R	3500
Optimizer	Adam

To verify the performance of the proposed MADDPG algorithm, we utilize the following algorithms for comparison.

- Random offloading algorithm (RA): All workflow tasks generated by each UE are randomly allocated to UEs or ESs for execution [21].
- DQN-based offloading algorithm (DQN): All workflow tasks generated by each UE are allocated to UEs or ESs for execution based on DQN algorithm [16].
- Single-agent DDPG-based algorithm (SADDPG) [22]. The SADDPG is a variant of the proposed algorithm and unlike the proposed algorithm it applied a single agent to learn the multiple workflows offloading strategy.

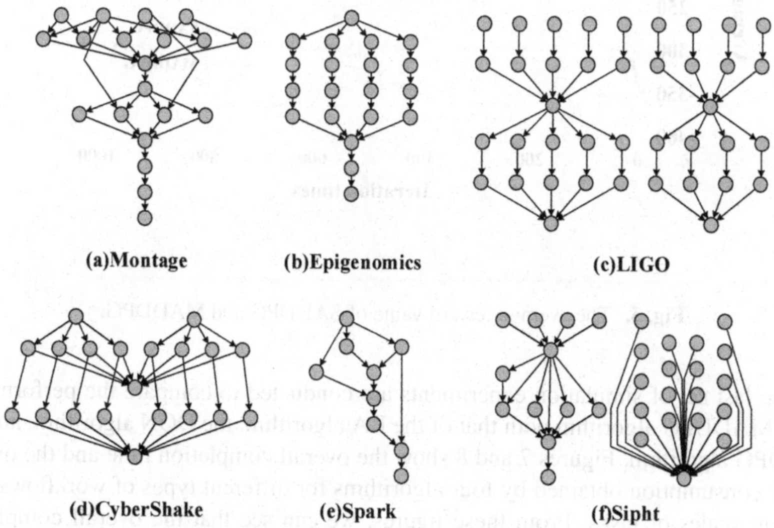

Fig. 4. Structure of six types of workflows.

5.2 Performance Evaluation

In the first set of simulation experiments, we compared the average reward value obtained by the MADDPG algorithm with that obtained by the SADDPG algorithm. Figure 5 shows the corresponding results. It can be seen from the Fig. 5 that the SADDPG algorithm has poorer convergence than the MADDPG algorithm. The reason is that the MADDPG algorithm can effectively learn stable strategies through centralized training and distributed execution which leads to a better reward value.

In the second set of simulation experiments, we evaluate the impact of the parameter α on the workflow completion time and the UE energy consumption. Figure 6 shows the workflow completion time and the UE energy consumption obtained by the MADDPG algorithm under different α values. As can be seen, a small α value leads to long completion time, and a large α value results in high energy consumption. This indicates that the MADDAPG algorithm does not work well when α is too large or too small. Therefore, α is set to 0.5 in this paper because it can lead to a good tradeoff between the workflow completion time and the UE energy consumption.

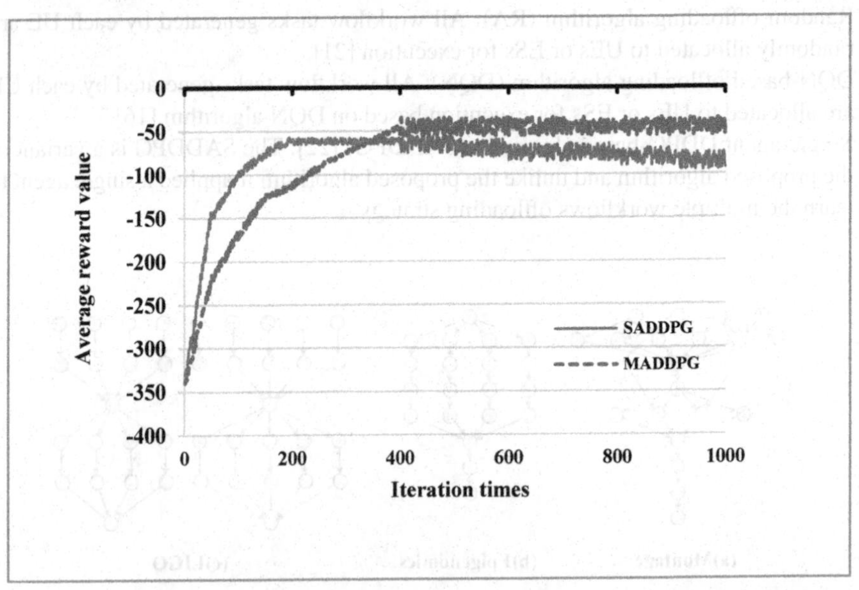

Fig. 5. The average reward value of SADDPG and MADDPG.

The last set of simulation experiments are conducted to compare the performance of the MADDPG algorithm with that of the RA algorithm, the DQN algorithm, and the SADDPG algorithm. Figures 7 and 8 show the overall completion time and the overall energy consumption obtained by four algorithms for different types of workflows with different scales of tasks. From these figures, we can see that the overall completion time and the overall energy consumption obtained by four algorithms increase with the number of tasks. The reason is that the increase in the number of tasks will result in extra resource requirements and long completion time. In addition, the MADDPG

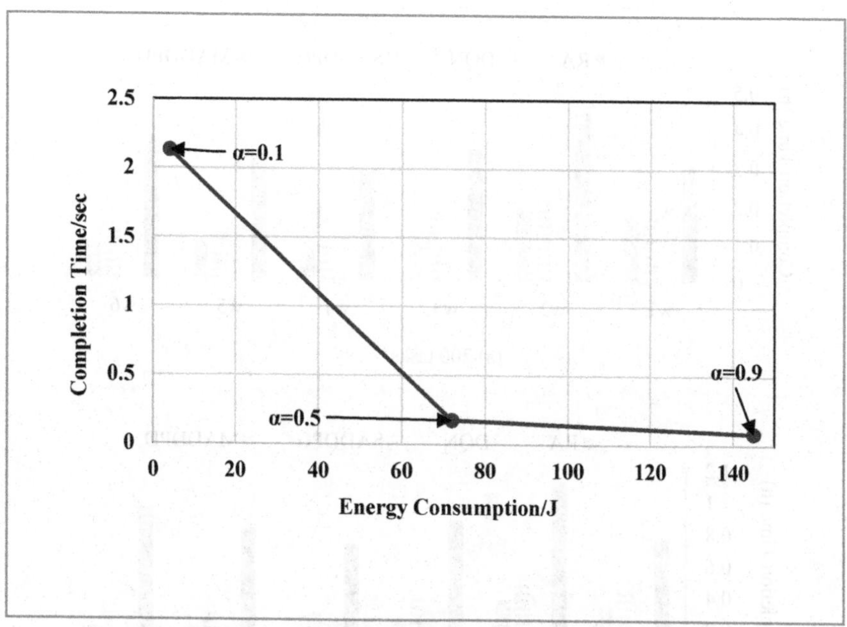

Fig. 6. The tradeoff between the completion time and the energy consumption.

algorithm performances best in terms of the overall completion time and the overall energy consumption. This is because the MADDPG algorithm can search for the solution space more effectively and globally so that it can find the offloading policy with relatively low energy consumption and relatively short completion time. Finally, the RA algorithm has the worst performance. The reason is that random offloading will bring about a large amount of data transmission which requires a lot of time and consumes huge amounts of energy.

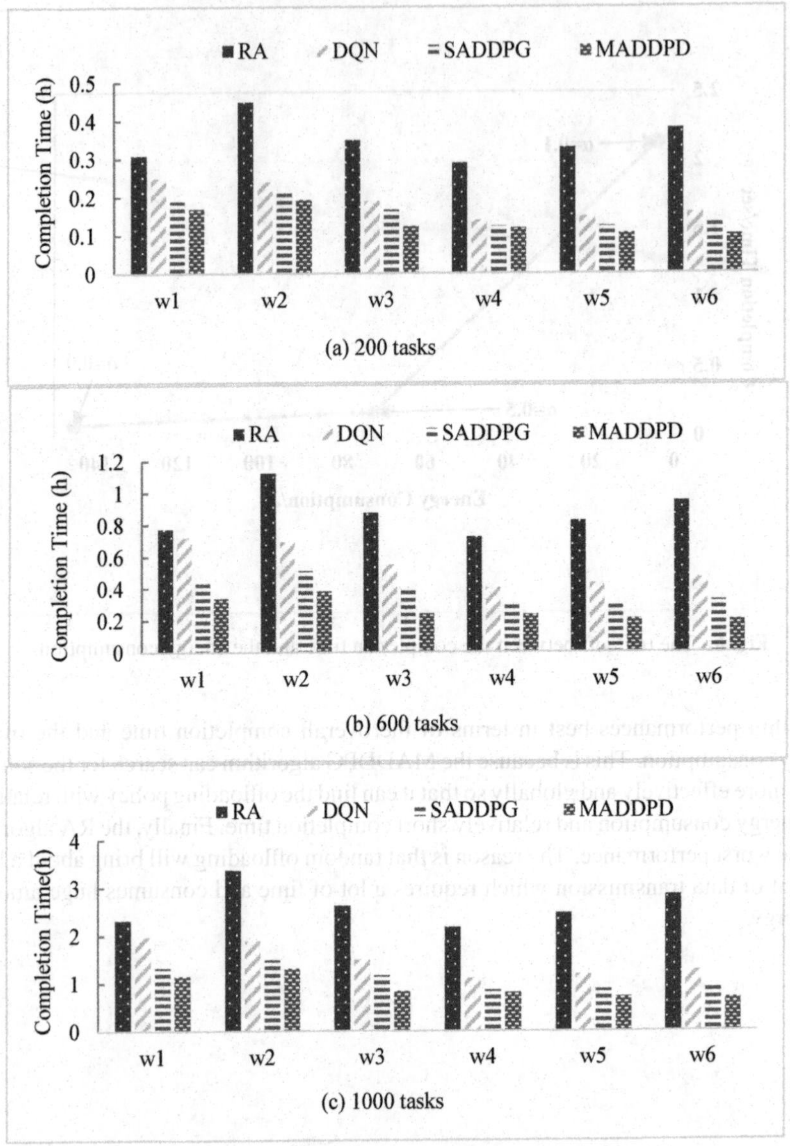

Fig. 7. Completion time for six types of workflows under different number of tasks.

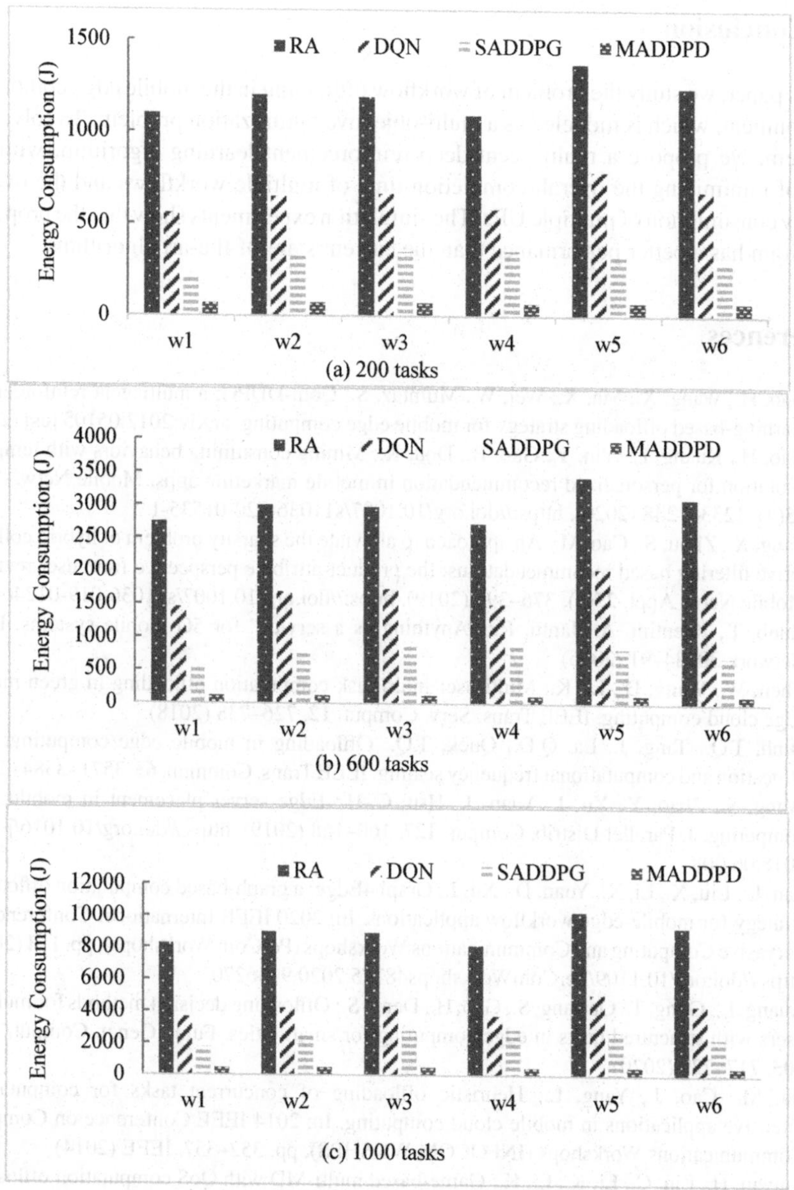

Fig. 8. Energy consumption for six types of workflows under different number of tasks.

6 Conclusion

In this paper, we study the problem of workflow offloading in the mobile edge computing environment, which is modeled as a multi-objective optimization problem. To solve this problem, we propose a multi-agent deep reinforcement learning algorithm, with the goal of minimizing the overall completion time of multiple workflows and the overall energy consumption of multiple UEs. The simulation experiments show that the proposed algorithm has a better performance than the current state-of-the-art algorithms.

References

1. Gao, H., Wang, X., Ma, X., Wei, W., Mumtaz, S.: Com-DDPG: a multiagent reinforcement learning-based offloading strategy for mobile edge computing. arXiv:2012.05105 [cs] (2020)
2. Gao, H., Kuang, L., Yin, Y., Guo, B., Dou, K.: Mining consuming behaviors with temporal evolution for personalized recommendation in mobile marketing apps. Mobile Netw. Appl. **25**(4), 1233–1248 (2020). https://doi.org/10.1007/s11036-020-01535-1
3. Yang, X., Zhou, S., Cao, M.: An approach to alleviate the sparsity problem of hybrid collaborative filtering based recommendations: the product-attribute perspective from user reviews. Mobile Netw. Appl. **25**(2), 376–390 (2019). https://doi.org/10.1007/s11036-019-01246-2
4. Taleb, T., Ksentini, A., Jantti, R.: "Anything as a service" for 5G mobile systems. IEEE Network **30**, 84–91 (2016)
5. Chen, W., Wang, D., Li, K.: Multi-user multi-task computation offloading in green mobile edge cloud computing. IEEE Trans. Serv. Comput. **12**, 726–738 (2018)
6. Dinh, T.Q., Tang, J., La, Q.D., Quek, T.Q.: Offloading in mobile edge computing: task allocation and computational frequency scaling. IEEE Trans. Commun. **65**, 3571–3584 (2017)
7. Wang, S., Zhao, Y., Xu, J., Yuan, J., Hsu, C.-H.: Edge server placement in mobile edge computing. J. Parallel Distrib. Comput. **127**, 160–168 (2019). https://doi.org/10.1016/j.jpdc.2018.06.008
8. Fan, L., Liu, X., Li, X., Yuan, D., Xu, J.: Graph4Edge: a graph-based computation offloading strategy for mobile-edge workflow applications. In: 2020 IEEE International Conference on Pervasive Computing and Communications Workshops (PerCom Workshops), pp. 1–4 (2020). https://doi.org/10.1109/PerComWorkshops48775.2020.9156270
9. Kuang, L., Gong, T., OuYang, S., Gao, H., Deng, S.: Offloading decision methods for multiple users with structured tasks in edge computing for smart cities. Futur. Gener. Comput. Syst. **105**, 717–729 (2020)
10. Jia, M., Cao, J., Yang, L.: Heuristic offloading of concurrent tasks for computation-intensive applications in mobile cloud computing. In: 2014 IEEE Conference on Computer Communications Workshops (INFOCOM WKSHPS), pp. 352–357. IEEE (2014)
11. Junyan, H., Liu, C., Li, K., Li, K.: Game-based multi-MD with QoS computation offloading for mobile edge computing of limited computation capacity. In: Tang, X., Chen, Q., Bose, P., Zheng, W., Gaudiot, J.-L. (eds.) NPC 2019. LNCS, vol. 11783, pp. 16–27. Springer, Cham (2019). https://doi.org/10.1007/978-3-030-30709-7_2
12. Li, S., Tao, Y., Qin, X., Liu, L., Zhang, Z., Zhang, P.: Energy-aware mobile edge computation offloading for IoT over heterogenous networks. IEEE Access **7**, 13092–13105 (2019)
13. Tong, Z., Deng, X., Ye, F., Basodi, S., Xiao, X., Pan, Y.: Adaptive computation offloading and resource allocation strategy in a mobile edge computing environment. Inf. Sci. **537**, 116–131 (2020)

14. Lu, H., Gu, C., Luo, F., Ding, W., Liu, X.: Optimization of lightweight task offloading strategy for mobile edge computing based on deep reinforcement learning. Futur. Gener. Comput. Syst. **102**, 847–861 (2020)
15. Tang, M., Wong, V.W.: Deep reinforcement learning for task offloading in mobile edge computing systems. IEEE Trans. Mobile Comput. (2020)
16. Zhu, A., et al.: Computation offloading for workflow in mobile edge computing based on deep Q-learning. In: 2019 28th Wireless and Optical Communications Conference (WOCC), pp. 1–5. IEEE (2019)
17. Huang, B., Xiang, Y., Yu, D., Wang, J., Li, Z., Wang, S.: Reinforcement learning for security-aware workflow application scheduling in mobile edge computing. Secur. Commun. Netw. **2021**, 1–13 (2021)
18. Song, F., Xing, H., Wang, X., Luo, S., Dai, P., Li, K.: Offloading dependent tasks in multi-access edge computing: A multi-objective reinforcement learning approach. Future Generation Computer Systems. (2021)
19. Cheng, Z., Min, M., Liwang, M., Huang, L., Gao, Z.: Multi-agent DDPG-based joint task partitioning and power control in fog computing networks. IEEE Internet Things J. **9**, 104–116 (2021)
20. Bharathi, S., Chervenak, A., Deelman, E., Mehta, G., Su, M.-H., Vahi, K.: Characterization of scientific workflows. In: 2008 Third Workshop on Workflows in Support of Large-Scale Science, pp. 1–10. IEEE (2008)
21. Yuan, Y., Qian, L., Jia, G., Yu, L., Yu, Z., Zhao, Q.: Efficient computation offloading for service workflow of mobile applications in mobile edge computing. Mobile Inf. Syst. **2021**, 1–11 (2021)
22. Lu, H., Gu, C., Luo, F., Ding, W., Zheng, S., Shen, Y.: Optimization of task offloading strategy for mobile edge computing based on multi-agent deep reinforcement learning. IEEE Access **8**, 202573–202584 (2020)

An Optimized Greedy-Based Task Offloading Method for Mobile Edge Computing

Wei Zhou[1], Chuangwei Lin[1], Jirun Duan[1], Ke Ren[1], Xuyun Zhang[2], and Wanchun Dou[1](✉)

[1] State Key Laboratory for Novel Software Technology, Nanjing University, Nanjing, China
{zw,lcw,duanjirun,rk}@smail.nju.edu.cn, douwc@nju.edu.cn

[2] Department of Computing, Macquarie University, Sydney, Australia
xuyun.zhang@mq.edu.au

Abstract. With the development of smart mobile devices (SMDs), computationally intensive and latency-sensitive applications are emerging. However, Mobile devices have limited processing power by nature. To overcome this problem, mobile edge computing enables users to offload tasks to proximal edge servers for faster task computation. Most studies in task offloading consider stable systems and ignore the number of tasks fluctuating over time. Poor offloading decisions will overload edge servers during peak periods, which leads to significantly high latency. To address this challenge, an optimized greedy-based offloading method (OGOM) is designed to offload tasks. OGOM adopts different offloading strategies depending on the server load factor. When edge servers are highly loaded, OGOM offloads some of the tasks to more idle servers instead of the servers with the lowest theoretical latency to achieve load balancing. Simulation results show that the OGOM is effective in avoiding edge server overload. In addition, OGOM reduces latency by an average of 20% compared to the normal greedy-based offloading method.

Keywords: Mobile edge computing · Greedy-based method · Task offloading · Load factor

1 Introduction

In recent years, the use of mobile devices has become increasingly widespread. A large number of computationally intensive and latency-sensitive applications have emerged, such as autonomous driving, online gaming, and virtual reality. However, mobile devices have limited capacity [1, 2] to meet the ever-expanding computational demands [3]. Therefore, offloading some of the tasks to remote servers is a promising solution. Traditionally, some of the computational tasks are offloaded to cloud with powerful computational capabilities [4]. However, cloud computing uses a centralized processing model with high latency for users located far from the computing center, and this high latency is unavoidable because of the physical distance [5]. To solve this problem, mobile edge computing (MEC) [6] comes to the forefront. In MEC, a large number of edge servers are

Y. Lai et al. (Eds.): ICA3PP 2021, LNCS 13155, pp. 494–508, 2022.
https://doi.org/10.1007/978-3-030-95384-3_31

deployed with certain processing power in the vicinity of users to provide fast services. Combined with the high-speed transmission capability of 5G technology [7], the latency can even be controlled to about 1 ms.

Due to the limited computing power of edge servers, mobile devices cannot offload tasks arbitrarily [8, 9], otherwise it may lead to overloading some edge servers. Moreover, after deciding to offload to an edge server, inappropriate edge server selection choices can also degrade the QoS of users. Therefore, how to make effective task offloading decisions is a key challenge in MEC.

In this paper, average latency is focused as much as possible in a multi-server, multi-user MEC environment. There have been a number of related studies, such as [8, 10]. Greedy-based algorithm has been heavily favored due to their simplicity and impressive performance. However, most of the studies does not take into account that in practice the number of tasks to be processed fluctuates over time. The number of tasks is even several times higher than usual during some time periods. For example, after 8 pm, the number of tasks in online games are much higher than during the day. In most cases, edge servers have ample resources, but when there is a large number of offload requests in a short period of time, greedy offload strategy tends to overload some edge servers. When the server is under high load, some tasks need to wait for the previous tasks to release their resources before they get executed. At this time, if the edge server continues to receive a large number of offload tasks, it can only upload these tasks to the cloud for execution. So, the latency will be significantly increased. To address this problem, this paper designs an optimized greedy-based offloading method. Our main contributions are as follows:

1. An optimized greedy-based offloading method is proposed. The special feature is that the method adopts different offloading policies according to the load factor of servers in the MEC. During busy times, the actual offload policy has a certain probability of adopting suboptimal edge servers. During idle times, the same task may be offloaded to multiple edge servers for redundant computations.
2. Simulation results show that OGOM is less prone to server overload and has the lowest average latency compared to several basic methods.

The remainder of this paper is organized as follows. Section 2 reviews related works. The system model and problem are described in Sect. 3. The online offloading method is proposed in Sect. 4. The simulation results and discussions are given in Sect. 5. Finally, the conclusion remarks are given in Sect. 6.

2 Related Work

There are a lot of researches about computational offloading in MEC. The main optimization metrics are latency, energy, cost. Some researches consider several metrics in combination.

A heuristic algorithm is proposed in [11] to solve allocation problem. The computation resource allocation and the time allocation of the MEC servers are optimized by the bisection search method. The transmit power allocations are obtained in closed-form. MEC servers' allocation and the channel allocation are optimized by the swap matching algorithm. [12] considers joint computation offloading and task caching optimization in a cellular network. By formulating the problem as mixed-integer non-linear programming, they find the optimal solution. [13] notices that tasks are relevant in a single edge server and singer mobile user (MU) system. they formulate the problem as a mixed integer non-linear programming (MINLP) that jointly optimizes the service caching placement, computation offloading decisions, and system resource allocation to minimize the computation latency and energy consumption of the MU.

There are also some researches (e.g., [14, 15]) that combine practical scenarios for optimization. [14] constructs a MEC-enabled 5G health monitoring system and proposes a potential game based decentralized approach for Internet of Medical Things (IoMT), with the target of minimizing the system-wide cost. [15] notices the dependency between tasks, and then express it as a mixed integer optimization problem. A deep reinforcement learning (DRL) framework is proposed to solve the combinatorial optimization problem. [16] considers the offloading problem under Multi-access edge computing. A two-tier hierarchical MEC enabled FiWi enhanced HetNet-based architecture is designed for computation offloading. Experiments show that both local and nonlocal computing resources can achieve low response time and energy consumption for MUs. [17] considers the scenario where multiple MUs offload duplicate tasks to the edge network and share the data required for computing tasks. They design a joint computation offloading and data caching model to minimize the overall execution latency for all mobile users.

However, none of these articles consider the fluctuation of the number of tasks as an important factor. We notice this problem and propose a new method.

3 System Model

In this section, we first introduce the model of MEC. Then, we describe the latency. Finally, the minimization latency problem is defined.

3.1 MEC Model

In this paper, MEC model is a three-tier edge computing architecture, including user layer, edge layer, and cloud layer, as shown in Fig. 1. The connection between the base station and the cloud is not as simple as in Fig. 1. Since this is not what we need to consider, so we simplified it. In this scenario, the layout of the base station is in the form of a cellular. User devices are connected to edge servers through the base station. Considering a system with n users and m edge servers, we use U_i to denote user i, $i = 1, 2, \ldots, n$ and E_j to denote edge server $j, j = 1, 2, \ldots, m$. When offloading tasks, tasks need to occupy the resources of edge servers. Edge servers have computational resources, storage resources, etc. We abstract the resources owned by edge server j as resource S_{E_j}. This abstraction reduces multidimensional resources to one dimension and will simplify the model, but will not affect the conclusion. Both user device and edge server have their own computational capabilities, denoted as C_{U_i} and C_{E_j}. E_j has a task waiting queue of size Q_{E_j} to cache tasks that have not been allocated resources. An edge server overloaded means that the queue is full. Only when a server overloaded, it will choose to offload the task to the cloud. Tasks are created by the user devices. For each task T_k, denote the task data size as D_{T_k}, and the required resources as S_{T_k}. For the offloaded tasks, the user devices upload them to edge servers through the base station. The edge server has three options: execute, forward to another edge, and upload to the cloud. If an edge server receives a forwarded task, it can only execute it or upload it to the cloud. We assume that edge servers in a region continuously exchange their situation with each other.

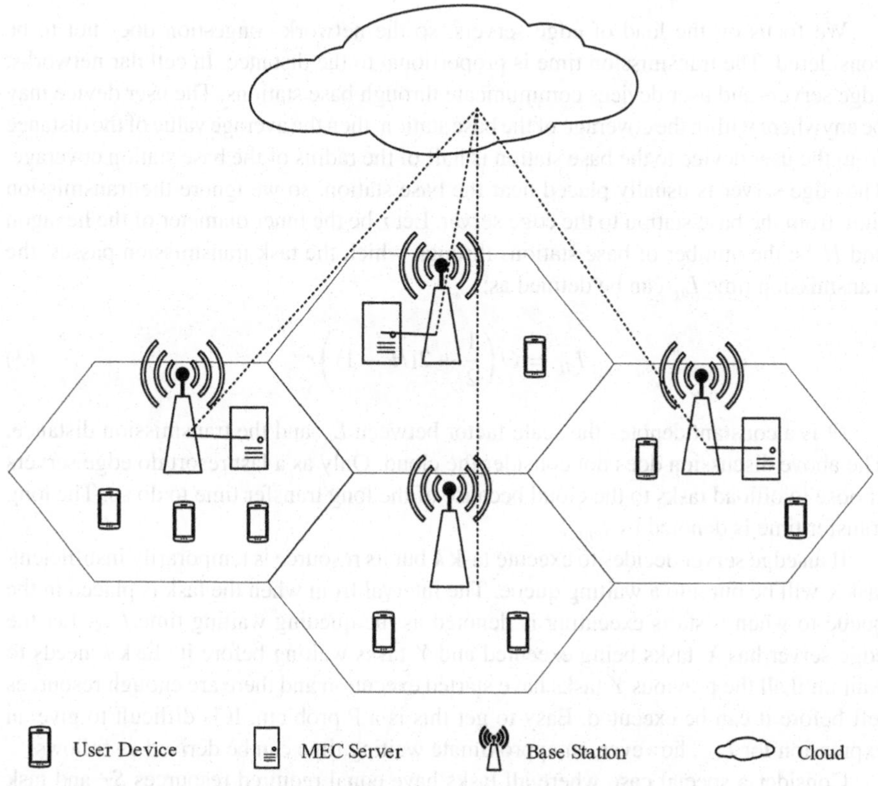

Fig. 1. MEC system model

3.2 Latency Constraints

We refer to the elapsed time from the generation of task k to the acquisition of the result as the latency, denoted as L_k. L_k is the sum of several link time consumption, including computation time L_{c_k}, transmission time L_{t_k} and queuing waiting time L_{w_k}. So, L_k is defined as follows:

$$L_k = L_{c_k} + L_{t_k} + L_{w_k} \tag{1}$$

Next, we will give the calculation of L_{c_k}, L_{t_k}, and L_{w_k}.

Tasks will only be executed on one of the user devices, edge servers, or cloud. We assume that the computational power of the cloud far exceeds that of user devices and edge servers, then L_{c_k} is as follow:

$$L_{c_k} = \begin{cases} \frac{D_{T_k}}{C_{U_i}}, \text{ if executed on user device } i \\ \frac{D_{T_k}}{C_{E_j}}, \text{ if executed on edge server } j \\ 0, \quad \text{ if executed on cloud} \end{cases} \tag{2}$$

We focus on the load of edge servers, so the network congestion does not to be considered. The transmission time is proportional to the distance. In cellular networks, edge servers and user devices communicate through base stations. The user device may be anywhere within the coverage of the base station, then the average value of the distance from the user device to the base station is half of the radius of the base station coverage. The edge server is usually placed near the base station, so we ignore the transmission time from the base station to the edge server. Let r be the inner diameter of the hexagon and H be the number of base stations through which the task transmission passes, the transmission time L_{t_k} can be defined as:

$$L_{t_k} = \Theta\left(\frac{1}{2} + 2(H-1)\right)r \tag{3}$$

Θ is a constant denotes the scale factor between L_{t_k} and the transmission distance. The above discussion does not consider the cloud. Only as a last resort do edge servers choose to offload tasks to the cloud because of the long transfer time to do so. The long transfer time is denoted by l_{max}.

If an edge server decides to execute task k but its resource is temporarily insufficient, task k will be put into a waiting queue. The interval from when the task is placed in the queue to when it starts executing is denoted as the queuing waiting time L_{w_k}. Let the edge server has X tasks being executed and Y tasks waiting before it. Task k needs to wait until all the previous Y tasks have started execution and there are enough resources left before it can be executed. Easy to get this is a P problem. It is difficult to give an expression for L_{w_k}, however, an approximate waiting time can be derived as follows:

Consider a special case where all tasks have equal required resources S_T and task data size D_T, and the progress of ongoing tasks are "uniform", i.e., for X ongoing tasks, their remaining execution time is an equation with tolerance:

$$t = \frac{L_c}{X} \tag{4}$$

In this case, one task will be completed every$t$, so we get L_{w_k}:

$$L_{w_k} = Yt \tag{5}$$

Previously we have made the assumption that the cloud is extremely powerful in terms of computational power. Thus, we ignore the queuing wait time L_{w_k} of the cloud.

3.3 The Minimization Latency Problem

Our goal is to design an offloading method, which can minimize the average latency of the tasks. Based on the analysis in the previous section, L_k needs to be discussed in three cases. If task k is executed on the user device, L_k is:

$$\frac{D_{T_k}}{C_{U_i}} \tag{6}$$

If task k is executed on the edge server, combining (2) (3) (4) (5), L_k is:

$$\frac{D_{T_k}}{C_{E_i}} + \Theta\left(\frac{1}{2} + 2(H-1)\right)r + \frac{YL_c}{X} \tag{7}$$

If task k is executed on the cloud, L_k is a constant l_{max} which is much bigger than the value of (6) and (7). Let us denote (6) (7) by Lu_k and Le_k. We define the problem in terms of vector multiplication. Let.

$$\overrightarrow{(L_k)} = (Lu_k, Le_k, ll)^T \tag{8}$$

T is the matrix transpose symbols. A total of N tasks are generated in the MEC system over a period of time. The offloading strategy is represented as a matrix $\overrightarrow{x}$ with N rows and 3 columns, and the values of the matrix elements can only be 0 or 1. Each row corresponds to an offloading decision for a task. A row with one and only one element having a value of 1 indicates that the method is selected for offloading. For example, (0, 0, 1) indicates the choice of offloading to the cloud. Let.

$$\overrightarrow{avg} = \underbrace{\frac{1}{N}, \ldots, \frac{1}{N}}_{N} \tag{9}$$

The strategy $\overrightarrow{x}$ we require is:

$$\overrightarrow{x} = \underset{\overrightarrow{x}}{argmin}(\overrightarrow{avg}\,\overrightarrow{x}\,\overrightarrow{L_k}) \tag{10}$$

4 Optimized Greedy-Based Offloading Method

In this section, we first introduce the greedy-based offloading method (GOM). Then, we discuss the problems of the method. Finally, we propose the optimized greedy-based offloading method (OGOM).

4.1 Greedy-Based Offloading Method

The greedy-based offloading method has three steps, which occur at the user device, the direct edge server, and the forwarded edge server.

1. **User device decision:** The user device needs to save two elements: the candidate edge servers, and the average latency L_{avg} of the last a tasks which were not performed on the user's device. When making a decision on a task, it first needs to estimate the latency L_k for local execution by Eq. (6). Then compare it with L_{avg}, if L_{avg} is smaller, choose to offload and upload the task to the candidate edge server. If L_{avg} is larger, the execution starts locally. At the same time, user device will send an interrogation message to the edge server containing only the size of the current task. The server will not execute the task, but will return a latency prediction for offloading the task. The purpose of this is to prevent accidental environmental changes (such as network fluctuations, server failures, etc.) from causing L_{avg} to be large and the task will never be offloaded.
2. **Direct edge server decision:** The edge server that receives the task directly called direct edge server. Every edge server will save basic information about other edge servers in the proximity, including computational capacity and whether they are currently overloaded. Every once in a while, this information will be passed to the neighboring servers. If an interrogation message is received, a latency prediction is returned directly. If an offload task is received, the time for direct execution and forwarding to other servers for execution is first estimated. According to the greedy idea, this scheme is used as long as the selected servers with the lowest latency (including itself) are not overloaded. If all servers are overloaded, uploading to the cloud is chosen.
3. **Forward edge server decision:** The edge server that receives the forwarding task called forward edge server. The forward server only determines if it is overloaded itself. if so, task will be uploaded to the cloud. If not, tasks will be executed locally.

The performance of this method is good if the edge server has sufficient resources, but in practice the average latency of this method rises sharply when encountering peak offload periods.

4.2 Problems in GOM

In general, GOM performs well. Offload methods with the lowest theoretical latency are usually the best in practice if the edge server has sufficient resources. However, when the edge server is under-resourced, GOM performance deteriorates dramatically with the number of tasks. As an example, let the time interval for exchanging status information between servers be d. At time t, server A informs the other servers that it is under low load. In the subsequent time d, there is a spike in task generation. According to the GOM, servers close to A are considered to have the lowest latency in forwarding tasks to A when they need to forward them. Therefore, a large number of tasks are forwarded to server A, leading the overload of server A. To solve this problem, it is necessary to introduce a load balancing method that does not allow a large number of tasks to be forwarded to the same server. In addition, when the server is idle, the idle resources can be used to reduce the latency.

4.3 OGOM

To address the above issues, an optimized greedy-based offloading method (OGOM) are proposed. The main steps of the method remain unchanged, but a new parameter μ need to be consider. μ is introduced to indicate the load factor of the edge server, whose value is the ratio of the server's task waiting queue to be occupied.

There are q tasks in the waiting queue of edge server$j$, labeled 1, 2, ..., q, then μ can be calculated by the following equation

$$\mu = \frac{\sum_{i=1}^{q} D_{T_i}}{Q_{E_j}} i = 1, 2, \ldots, q \tag{11}$$

In addition, two values μ_h and μ_l need to be defined with the constraints:

$$0 < \mu_l \leq \mu_h \leq 1 \tag{12}$$

The steps of the OGOM method are as follows:

1. **User device Decision:** The list of candidate edge servers saved by the user device should include each server's μ. When making a decision on a task, it first needs to estimate the latency L_k for local execution by Eq. (6). Then compare it with L_{avg}. If L_{avg} is smaller and the value of μ does not exceed the threshold μ_h, choose to offload and upload the task to the candidate edge server. The task will be labeled as NORMAL; if μ exceeds μ_h, the task is likewise uploaded to the edge server and execution starts locally as well. The task will be labeled as PARALLEL. If L_{avg} is larger and $\mu \leq \mu_l$, the execution starts locally and the task is offloaded simultaneously with probability p:

$$p = \left(\frac{L_k}{L_{avg}}\right)^2 \tag{13}$$

The task will be labeled as PROB. In other cases, the task is performed directly on the user device.

Algorithm 1 Direct Edge Server Decision Algorithm

Input: task data volume $D_T$task resources required S_T, estimated execution time on user device L_U, task label $LABEL$

```
set μ_l,μ_h
accept ← false
Estimate latency of each server,the list of tuple(serverId,latency) is denoted as array L.
Ascending sort L by lantency.
if LABEL = PARALLEL then
    for i = 1,2,...,n do
        if μ_i < μ_h then
            accept ← true
            Send accept message to user device.
            break
        end if
    end for
else if LABEL = PROB then
    for i = 1,2,...,n do
        if μ_i < μ_l and L[i] < L_U then
            accept ← true
            Send accept message to user device.
            break
        end if
    end for
else
    accept ← true
end if

j ← L[0][serverId]
if accept ≠ true then
    Send reject message to user device
else if μ_i < μ_h then
    Forward the task to server j. if j is current server, add the task to the
    waiting queue directly.
else
    each server has a probability of p to be forwarded the task.p is calculated
    by equation (14).
end if
```

2. **Direct edge server decision:** Every edge server will save basic information about other edge servers in the proximity, including computational capacity and load factor μ. Direct edge server adopts different strategies depending on the label of the received task. When a task with label NORMAL is received, the edge server first calculates the estimated latency which is denoted as L_{E_j} of execution on each edge server j in the proximity. The L_E are ranked. If the scenario with the lowest latency is executing on the current server and the current server is not overloaded, the scenario is directly

selected. If the scenario with the lowest latency is executing on another server and that server has $\mu_j \leq \mu_h$, then the scenario is directly selected. Otherwise, for all servers that are not overloaded, each server has a probability of p to be selected. p is expressed as follows:

$$p = \frac{1/\left(L_{E_j}\mu_j\right)}{\sum_{j=1}^{n}\left(1/\left(L_{E_j}\mu_j\right)\right)} \tag{14}$$

When a task with label PARALLEL is received, if there exists a server j with $\mu_j < \mu_h$, the task is processed as NORMAL task and a message is immediately replied to the user device. Then, the user device will stop the local execution on user device. Otherwise, the edge server will reject the task without doing any other operation. When a task with label PROB is received, if there exists a server j with $\mu_j < \mu_l$ and $L_{E_j} < p$, the task is processed as NORMAL task and a message is immediately replied to the user device. Otherwise, the edge server will reject the task. The pseudo-code of the algorithm is given in Algorithm 1.

3. **Forward edge server decision:** The same as GOM.

4.4 Optimization of OGOM

At the end we briefly analyze the optimization of OGOM. The optimized method does not change the frequency of information interaction with the server and hardly changes the amount of transmission in the network. The difference from the normal greedy algorithm is in the first and second decision. The decision on the user device has the probability of performing a redundant computation, which has the purpose of slightly reducing the latency expectation of the task. Regardless of how the edge server's state changes, latency is the minimum value for local execution and offload execution. It obviously reduces latency. Although this will increase the load on the edge server, it does not actually make an impact because the redundant computation is limited by the second decision, which generally only occurs when the server is idle. The change in the second decision is also well understood. Overload increases latency dramatically, so when a server is found to be at risk of overload, offloading tasks to multiple servers with higher theoretical latency can reduce latency instead.

5 Performance Analysis

5.1 Experiment Setup

We use simulated experiments to validate our method. The data are randomly generated within a reasonable range and multiple sets of experiments are performed. Consider a cellular network as shown in the Fig. 2, with a number of users under each base station coverage and a total of four edge servers. Servers are labeled 1–4 in left-to-right top-to-bottom order Since this is a simulation experiment. According to a certain information

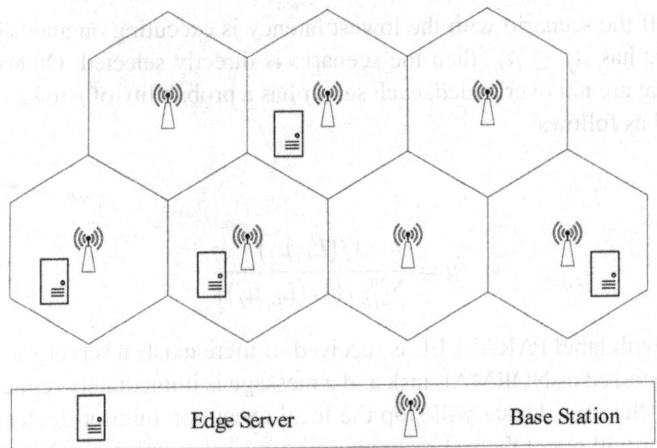

Fig. 2. Cellular network

we found, the latency of each hop is assumed to be 5 ms. the latency of any edge server to the cloud is set to 100 ms.

According to our investigation, the computing power of mobile phones to calculate single-precision floating-point numbers is around a few hundred MOPS, while the 5G edge computing box is about 15 TOPS. TOPS means that the processor can perform 10^{12} operations in a second. A very large value that is not conducive to the next calculation, so we scale it down. Similarly, the task data size will be reduced in the same proportion. The computational capacity of the user device is random between [1, 2], while the computational capacity of the edge server is random between [20, 50]. The value of D_{T_k} was reduced by a factor of 1000 in order to make the latency unit uniform in milliseconds. Table 1 shows all variables' value range.

Table 1. Parameter settings

Variable	Description	Value range
S_{E_j}	Resources of edge server j	[500,1000]
S_{T_k}	Resources required for task k Execution	[1, 5]
C_{U_i}	Computational capabilities of user device i	[1, 2]
C_{E_j}	Computational capabilities of edge server j	[20, 50]
Q_{E_j}	Size of waiting queue	Same as S_{E_j}
D_{T_k}	Task data size	[20, 200]

The experiment will be carried out in the following form: every 1ms, there will be Y tasks generated by Y different user devices need to be processed, the location of the user is random. When the results of all tasks in the first millisecond are returned, count the average latency of the tasks generated in this second. The set of Y is {1, 100, 200, 300, 400, 500}.

There are three methods used to compare with OGOM, which are GOM, always offload and Optimal. A brief description of the four comparative experiments is as follows:

- GOM: normal greedy-based offloading method mentioned in the previous section.
- Always Offload (AO): Tasks are not executed on the user's device and are not allowed to be forwarded.
- Optimal (OPT): For measuring the performance of OGOM. Due to the servers and the users cannot know the status of others in real time, the optimal strategy exists only in theory.

5.2 Comparison and Analysis

Figure 3 and Fig. 4 shows the results of the experiment. Figure 3. shows the average latency and load factor of four edge servers. Figure 4 shows each of the four servers' load factor. Observing Fig. 3(a), as the number of tasks per millisecond increases, the latency of all methods gets higher and improves faster.

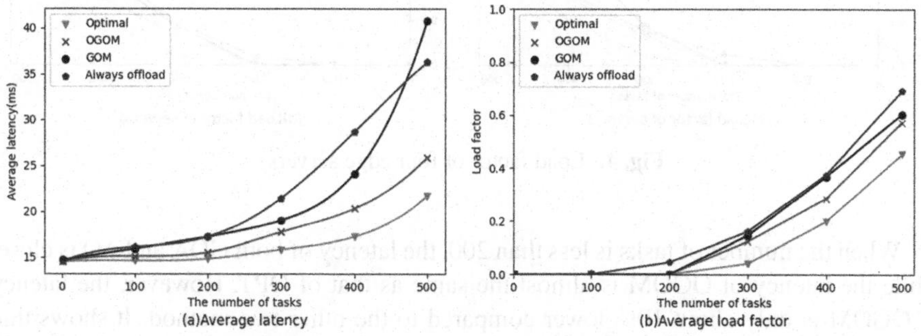

Fig. 3. Average latency and load factor of four edge servers

First, we analyze the performance optimal (OPT) method. The latency of the OPT method remains essentially unchanged until 200 tasks. The reason for this is that when the number of tasks is less than 200, the total resources of the edge server are more than the resources needed for all tasks. So, with perfect scheduling, no tasks need to be queued. Figure 3(b) also shows that when the number of tasks is less than 200, the average load factor of the OPT method is 0. When the number of tasks is between 200 and 400, the latency of the OPT method grows essentially linearly. It means the OPT method does not overload when the number of tasks is less than 400. The growth in latency comes mainly from queuing and forwarding. When the number of tasks exceeds 400, the latency growth of the OPT method starts to become faster. The number of tasks has exceeded the upper limit of the carrying capacity of the edge service at this point, so some tasks have to be uploaded to the cloud for execution.

By analyzing the latency of the OPT method, we get two thresholds: when the number of tasks is greater than 200, the total resources demanded by the tasks will exceed the total resources of the edge server, and when the number of tasks is greater than 400, overload is bound to occur. Next, we compare the performance comparison of OGOM, GOM and (AO) based on these two thresholds.

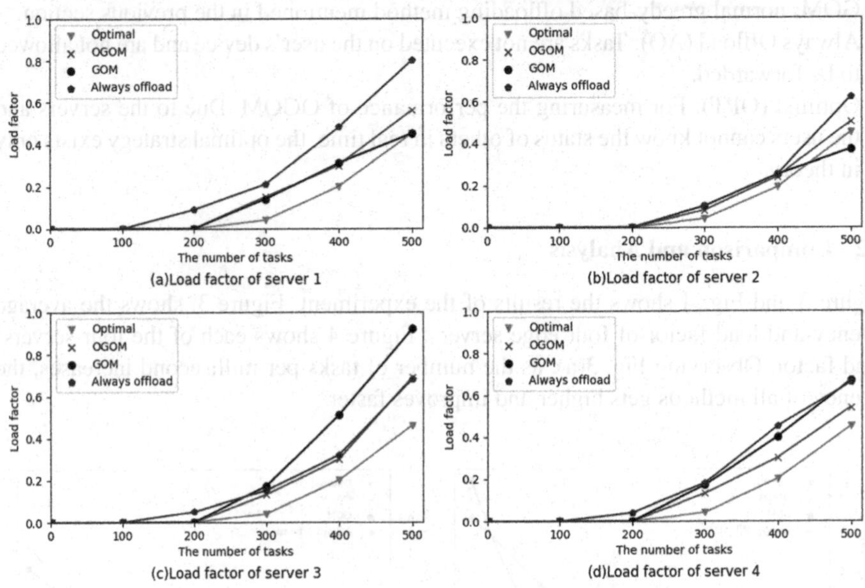

Fig. 4. Load factor of four edge servers

When the number of tasks is less than 200, the latency of both GOA and AO is close, while the latency of OGOM is almost the same as that of OPT. However, the latency of OGOM is only about 10% lower compared to the other two method. It shows that the offloading policy does not matter when all edge servers are idle. Even if all tasks are chosen to be offloaded nearby or randomly, it does not lead to task queuing, so the difference in latency is not significant.

When the number of tasks is between 200 and 400, the latency of all three algorithms is improving, and the one that improves the most is AO and the slowest is OGOM. According to Fig. 3(b), we can see that the average load factor of OGOM is lower, AO and GOM are basically equal. OGOM will probabilistically choose the non-greedy optimal solution when the server is under high load, so it can achieve a certain degree of load balancing. In the case that the average load ratios of AO and GOM are similar, the latency of the AO algorithm is significantly higher than that of GOA. The reason can be found in the Fig. 4(a). On server 1, the load factor with AO method is significantly higher than the other methods. Server1 is one of the nearest servers for the user devices within the service range of 1, 2, 3, 6 four base stations, the tasks received under the AO algorithm are higher than the other three servers. Server 1 overload occurs more frequently, resulting in high latency.

When the number of tasks is greater than 400, overload is already bound to occur, so the latency of all algorithms starts to grow more rapidly. A more specific phenomenon is that the latency of the GOM gradually exceeds that of AO, we find the answer in Fig. 4(c). It shows that the load factor of server3 rises sharply from 300 onwards, even reaching 0.91 at 500. Server 3 is adjacent to both server 1 and 2, so the reason for its high load is not hard to imagine. With their own high load, both server 1 and 2 will give priority to forward tasks to server 3, causing the load on 3 to increase rapidly. Surprisingly, OGOM reduces latency by 50% compared to the other two methods when the number of tasks reaches 500.

In summary, OGOM has better performance compared to OA and GOM. As the load on the edge server increases, OGOM becomes more and more effective. According to the results of several experiments, OGOM can reduce latency by about 20% on average compared to GOM and AO.

6 Conclusion

In this paper, a task offloading method which uses an optimized greedy-based offloading algorithm is proposed. It takes different offloading methods depending on the load of the edge server. When the edge server is idle, OGOM probabilistically performs redundant offloading. When the edge server is busy, OGOM probabilistically selects different edge servers for offloading so as to avoid overloading a particular edge server as much as possible. Simulation results show that the OGOM reduces the probability of overloading the edge server and lowers the average latency of tasks.

Although the experimental part of this paper has done some simplification, the work of this paper is still of great enlightening significance. In a distributed scenario, it takes time for servers to exchange information. Between two information exchanges, the information is not updated, which can lead to a series of problems. In an edge environment, an edge server informs other servers that it has sufficient resources. That may cause it to be overloaded if multiple servers offload a large number of tasks to that server before the next information exchange. We notice the problem and propose a feasible method to solve it. For future research, we will apply the method to practical scenarios, such as industrial Internet, Internet of vehicles, etc.

References

1. Zhou, Z., et al.: When mobile crowd sensing meets UAV: energy-efficient task assignment and route planning. IEEE Trans. Commun. **66**(11), 5526–5538 (2018)
2. Lin, L., Liao, X., Jin, H., Li, P.: Computation offloading toward edge computing. Proc. IEEE **107**(8), 1584–1607 (2019)
3. Wang, S., et al.: Adaptive federated learning in resource constrained edge computing systems. IEEE J. Sel. Areas Commun. **37**(6), 1205–1221 (2019)
4. Yao, D., Yu, C., Yang, L.T., Jin, H.: Using crowdsourcing to provide QoS for mobile cloud computing. IEEE Trans. Cloud Comput. **7**(2), 344–356 (2019)
5. Gedeon, J., Meurisch, C., Bhat, D., Stein, M., Wang, L., Mühlhäuser, M.: Router-based brokering for surrogate discovery in edge computing. In: 37th International Conference on Distributed Computing Systems Workshops (ICDCSW), pp. 145–150. IEEE, Atlanta (2017)

6. Hu, Y.C., Patel, M., Sabella, D., Sprecher, N., Young, V.: Mobile edge computing—a key technology towards 5G. ETSI White Paper **11**(11), 1–16 (2015)
7. Panwar, N., Sharma, S., Singh, A.K.: A survey on 5G: the next generation of mobile communication. Phys. Commun. **18**, 64–84 (2016)
8. Jošilo, S., Dán, G.: Computation offloading scheduling for periodic tasks in mobile edge computing. IEEE/ACM Trans. Netw. **28**(2), 667–680 (2020)
9. Jošilo, S., Dán, G.: Selfish decentralized computation offloading for mobile cloud computing in dense wireless networks. IEEE Trans. Mobile Comput. **18**(1), 207–220 (2019)
10. Sheng, M., Dai, Y., Liu, J., Cheng, N., Shen, X., Yang, Q.: Delay-aware computation offloading in NOMA MEC under differentiated uploading delay. IEEE Trans. Wirel. Commun. **19**(4), 2813–2826 (2020)
11. Wan, Z.L., Xu, D., Xu, D., Ahmad, I.: Joint computation offloading and resource allocation for NOMA-based multi-access mobile edge computing systems. Comput. Netw. **196**, 108256 (2021)
12. Chen, Z.X., Chen, Z., Jia, Y.: Integrated task caching, computation offloading and resource allocation for mobile edge computing. In: 2019 IEEE Global Communications Conference (GLOBECOM), pp. 1–6. IEEE, Waikoloa (2019)
13. Bi, S., Huang, L., Zhang, Y.A.: Joint optimization of service caching placement and computation offloading in mobile edge computing systems. IEEE Trans. Wirel. Commun. **19**(7), 4947–4963 (2020)
14. Ning, Z., et al.: Mobile edge computing enabled 5G health monitoring for internet of medical things: a decentralized game theoretic approach. IEEE J. Sel. Areas Commun. **39**(2), 463–478 (2021)
15. Yan, J., Bi, S., Zhang, Y.J.A.: Offloading and resource allocation with general task graph in mobile edge computing: a deep reinforcement learning approach. IEEE Trans. Wirel. Commun. **19**(8), 5404–5419 (2020)
16. Ebrahimzadeh, A., Maier, M.: Cooperative computation offloading in FiWi enhanced 4G HetNets using self-organizing MEC. IEEE Trans. Wirel. Commun. **19**(7), 4480–4493 (2020)
17. Zhang, N., Guo, S., Dong, Y., Jiang, Q., Jiao, J.: Joint task offloading and data caching in mobile edge computing. In: 15th International Conference on Mobile Ad-Hoc and Sensor Networks (MSN), pp. 234–239. IEEE, Shenzhen (2019)

Location Aware Workflow Migration Based on Deep Reinforcement Learning in Mobile Edge Computing

Yongqiang Gao[1,2,3](✉) and Xiaolei Liu[2,3]

[1] Engineering Research Center of Ecological Big Data, Ministry of Education, Hohhot 010021, China

[2] Inner Mongolia Engineering Laboratory for Cloud Computing and Service Software, Hohhot 010021, China

[3] College of Computer Science, Inner Mongolia University, Hohhot 010021, China

gaoyongqiang@imu.edu.cn, 31909094@mail.imu.edu.cn

Abstract. Prompted by the remarkable progress in mobile edge computing, there is an increasing need for executing complex applications on the edge server. These complex applications can be described using workflows which is a set of interdependent tasks. Existing studies focus on offloading the workflow tasks to the nearby edge servers in order to achieve high quality of service, However, the original edge server with the offloaded workflow tasks may be far away from the users due to the high mobility of users in mobile edge computing (MEC). Therefore, it is a key challenge to make good decisions on where and when the workflow tasks are migrated in the light of user's mobility. In this paper, we propose a workflow task migration algorithm based on deep reinforcement learning with the goal of optimizing the cost of workflow migration under delay-guarantee constraints. The proposed algorithm firstly utilizes the Recurrent Neural Network (RNN) based model to predict the mobile location of users, and then applies a dynamic programming algorithm to calculate the completion time of workflow. Finally, an improved Deep Q Network (DQN) algorithm is adopted to find the optimal workflow migration strategy. In order to assess the performance of the proposed algorithm, extensive simulations are carried out for four well-known scientific workflows. The experimental results show that the proposed algorithm can meet threshold at lower costs in comparison with the state-of-the-art approaches applied to similar problems.

Keywords: Mobile edge computing · Workflow migration · Mobility · Deep reinforcement learning

Supported in part by the National Natural Science Foundation of China under Grant 61662052, in part by the Natural Science Foundation of Inner Mongolia Autonomous Region under Grant 2021MS06002, in part by he Science and Technology Planning Project of Inner Mongolia Autonomous Region under Grant 2021GG0155, and in part by the Major Research Plan of Inner Mongolia Natural Science Foundation under Grant 2019ZD15.

Y. Lai et al. (Eds.): ICA3PP 2021, LNCS 13155, pp. 509–528, 2022.
https://doi.org/10.1007/978-3-030-95384-3_32

1 Introduction

With the rapid development of mobile Internet and Internet of things, various mobile devices such as smart phones and tablet computers are increasingly being used in real life. People are increasingly looking forward to running more computing intensive tasks on mobile device, such as online games, virtual reality (VR), augmented reality (AR), face recognition and so on. These computing intensive tasks usually consume a lot of computing resources, and have strict requirements for quality of service (QoS) such as delay [1]. However, due to the lack of resources, mobile devices can't handle these computing intensive tasks independently and efficiently. To address this issue, mobile cloud computing (MCC) is proposed to provide high-quality services for mobile users. It combines cloud computing and mobile computing to provide rich computing, storage and network resources for applications in the way of centralized cloud. However, the proliferation of mobile devices leads to a great pressure on the uplink and downlink bandwidth in MCC. What is worse, it will bring long response time and large communication overhead. In this context, mobile edge computing (MEC) emerges. MEC is a new edge technology which uses mobile base station to extend cloud computing services to the edge of the network. It provides fast and powerful computing and storage capacity support to mobile devices. Meanwhile, it solves the problems of long delay and too much bandwidth pressure in traditional methods, which greatly meets the requirements of QoS.

The concept of workflow is evolved from the manufacturing process. By defining workflow as different tasks, these tasks are executed and monitored according to certain rules and processes, so as to achieve the purpose of improving work efficiency. With the popularity of computers and the development of network technology, workflow technology has been widely used in various enterprise management systems, such as logistics, e-commerce, medical, administrative office and so on [2]. It separates the business process from the application and manages it independently, which makes it easier for the software to support the changeable business process, and can process tasks in parallel more efficiently, so as to improve the efficiency of work. The core of workflow management system is workflow engine. One of its main tasks is to assign tasks according to the workflow model, that is, to select execution resources for tasks according to certain assignment strategies [3]. Task assignment strategies have a significant impact on the performance of workflow. Because MEC has the technical characteristics of high bandwidth, low latency and location awareness, more and more enterprises submit workflow tasks to MEC environment for execution, so as to quickly get the running results and meet the requirements of QoS.

The main challenge of workflow management in mobile edge computing is how to design effective task offloading and migration strategy to meet the threshold of workflow. Task offloading is an important way to solve the resource constrained problem of mobile devices. By sending all or part of the workflow tasks of mobile devices from local to resource-rich edge servers for execution, task offloading can speed up the calculation and reduce the execution time of tasks with the powerful processing ability of edge servers. Mobility is one of the most important characteristics of MEC, which is different from traditional computing methods. When mobile devices move between different base stations, the unfinished workflow tasks either continue to calculate in source MEC server (MEC server where the task is located before migration), and then transfer the result to mobile devices through the backhaul link, or migrate unfinished workflow tasks to target MEC server (MEC server where the task is migrated) closer to mobile devices for calculation. The former will bring higher data transmission cost, while the latter will result in higher migration cost, and the delay caused by the former and the latter is also different. Therefore, how to effectively deal with the migration of workflow to minimize the migration cost while meeting the user's QoS has become a challenging problem in industry and academia.

In this paper, we proposed a workflow migration strategy based on deep reinforcement learning in MEC, which can meet the user's requirements for delay and minimize the cost of migration. The contributions of this paper are as follows:

(1) Considering the mobility of user, we establish a location prediction model, and use the RNN based method to predict the location movement of user. Based on the location prediction model, we model the workflow task migration problem as a constrained single-objective optimization problem.
(2) We propose a workflow task migration algorithm based on the deep reinforcement learning framework to solve the task migration problem, in which the dependent workflow tasks is considered, and the completion time of the workflow is calculated by using the dynamic programming algorithm.
(3) We conduct a large number of simulation experiments to evaluate the effectiveness of the proposed algorithm. We compare the proposed algorithm with three methods: Always Migration Policy (AMP), No Migration Policy (NMP) and Markov Approximation Based Placement Policy (MAbPP) [4]. The results show our proposed algorithm performs better than other algorithms.

The rest of this paper is structured as follows. A survey on related works is presented in Sect. 2. Section 3 presents the models and problem formulation. Section 4 describes the proposed workflow migration algorithm in detail. The experimental evaluations are presented in Sect. 5. Finally, the conclusion is drawn in Sect. 6.

2 Related Work

So far, many efforts have been devoted to research on independent task offloading in MEC. You et al. [5] studied the problem of task offloading in systems based on TDMA and OFDMA. The goal is to minimize the system energy consumption under the constraint of task delay. Heydari et al. [6] modeled the task offloading problem as a Markov decision process, and designed and implemented a strategy based on deep reinforcement learning to optimize the energy consumption and delay of the system. These studies are different from our work, they only focus on the offloading of a single independent task to the MEC server without considering the complexity of tasks.

There is also some researches on task offloading focusing on constraints between multiple tasks. Houssemeddine mazouzi et al. [7] considered the offloading of multi task applications with dependencies, and proposed an adaptive offloading algorithm to minimize the energy consumption of the terminal under the constraints of threshold and offloading decision. Peng et al. [8] modeled the energy consumption, time consumption and cost of workflow applications using cloudlets and cloud resources in MEC as a multi-objective optimization problem, and proposed a corresponding multi-objective computing unloading method based on non dominated sorting genetic algorithm to find the best offloading strategy for all workflow applications. Different from our work, although these studies consider the complexity of tasks, they only consider the simple offloading situation, ignoring the need for task migration when user move to other regions.

Recently, there have been several studies on task migration in MEC. Wang et al. [9] modeled user mobility as a Markov process and task migration as a sequential decision-making problem, and proposed a heuristic algorithm to optimize the cost of migration. Zhang et al. [10] used deep reinforcement learning technology to predict the migration location of tasks, and proposed a task migration algorithm to maximize the benefits of users. Ouyang et al. [4] designed a mobile aware online service placement framework, proposed a task migration mechanism based on Lyapunov optimization and Markov model, and developed a distributed algorithm to quickly obtain the approximate solution of the migration problem, so as to maximize the quality of service under the constraint of operation cost. Different from the above research, considering the mobility of user and the complexity of workflow tasks, we propose a workflow task migration strategy to minimize the migration cost.

3 System Model and Problem Formulation

3.1 System Overview

We consider a MEC scenario with M edge servers and P base stations. MEC servers is represented by set $M = \{m_1, m_2, m_3, m_4, ..., m_t\}$, base station is represented by set $B = \{b_1, b_2, b_3, b_4, ..., b_p\}$, and mobile device is represented by u. Mobile devices and base stations use wireless communication technology to establish connection. A workflow is divided into n subtasks and offloaded to virtual machines running at edge servers for calculation. Before the completion of workflow, user may move to other regions. By predicting the moving position of user in advance, we decide whether to migrate the tasks and where to migrate, so as to minimize the cost of migration under the requirement of workflow threshold.

Figure 1 shows the workflow tasks migration architecture. Each region has an edge server as the front-end proxy server, and mobile device interact with other servers in the corresponding region through the front-end proxy server. An additional server is used to deploy our workflow migration engine which consists of location information collector, location movement predictor and workflow migration decider. The working steps of the workflow migration engine are as follows: Firstly, the location information collector records the user's location reported by the front-end proxy server in real time, and forward it to the location movement predictor. Next, the location movement predictor uses the Long Short-Term Memory (LSTM) to predict the user's location in the future based on the user's historical location trajectory, and forwards it to the workflow migration decider. Finally, the workflow migration decider builds an optimization model based on the predicted location, and uses the deep reinforcement learning based algorithm to solve the decision scheme of workflow migration.

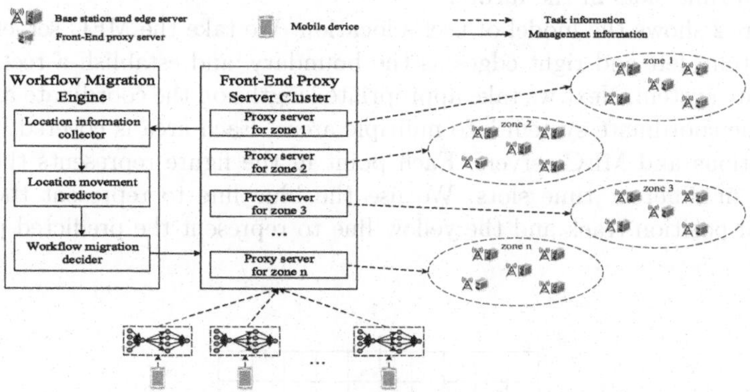

Fig. 1. Workflow tasks migration architecture.

3.2 Workflow Model

In this paper, a workflow is described as a directed acyclic graph (DAG), presented by a two-tuple $G = (V, E)$, where $V = \{v_1, v_2, v_3, v_4, ..., v_n\}$ represents the collection of n subtasks, and each edge $e_{(i,j)} = (v_i, v_j) \in E(i, j = 1, 2, ..., n, i \neq j)$ denotes a dependency between task v_i and task v_j, that is, the task v_j cannot start its execution before v_i finishes and sends the output data to task v_j. Each edge is associated with a weight $d_{i,j}$, which represents the size of data transmission from task v_i to task v_j. Each task $v_i \in V$ is modeled as a quaternion $v_i = (w_i, t_i, l_i, k_i)$, where w_i indicates the workload of the task v_i, t_i is the migration strategy of task v_i, expressed by 0 or 1, $t_i = 0$ means that task v_i will not be migrated and execute on the source MEC server, $t_i = 1$ means that task v_i will be migrated and execute on the target MEC server. l_i is a mapping relationship of task migration from source MEC server m_i to target MEC server m_j, which is expressed by : $l_i : m_j = f(m_i), \forall i, j \in m$. If the current task does not migrate, then m_i and m_j are the same. k_i represents the size of output data after task v_i is executed. The direct precursor set and direct successor set of task v_i are represented as $pred(v_i), v_i \in V$ and $succ(v_i), v_i \in V$ respectively. A task without a precursor task is called an entry task, and a task without a successor task is called an exit task. If a workflow has multiple entry tasks and exit tasks, two virtual tasks are added to workflow to ensure that this workflow has only one entry task and one exit task.

3.3 Location Movement Prediction Model

Because user may move to other regions before workflow is completed, it is necessary to predict the user's next location and migrate tasks in advance. We consider the location coordinates of the user in each time slot, and assume that it does not change in this time slot. Based on the historical trajectory formed by the user's location of multiple time slots in the past, we predict the location of several time slots in the future.

Figure 2 shows the model of user's location. We take the MEC server at the top, bottom, left and right edges as the boundary, and establish a rectangular coordinate system. Next we take appropriate lengths on the coordinate axis, and divide the coordinate system into multiple areas. Each area is covered by some base stations and MEC servers. Each point in the figure represents the user's location in different time slots. We use the blue line to represent the user's historical position track and the yellow line to represent the predicted position track.

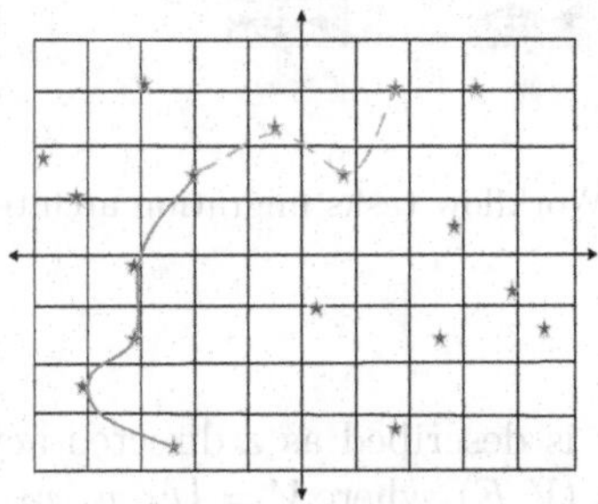

Fig. 2. The model of user's location. (Color figure online)

Based on the time series analysis method, we use the Recurrent Neural Network (RNN) to predict the user's moving trajectory. RNN is a kind of neural network with time sequence characteristics. Compared with ordinary feedforward neural network (FNN), RNN has certain memory storage capacity and can consider the output of the previous time and the input of the current time at the same time. Therefore, RNN is good at dealing with problems with sequence correlation. LSTM is a special RNN, which uses gating mechanism to control input, memory and other information and make prediction in the current time step. LSTM overcomes the gradient dispersion problem of general RNN when dealing with long-term dependence. General RNN has only a hidden state h_t which is sensitive to short-term input. In contrast, LSTM adds a cell state c_t which can keep long-term state, and uses forget gate, input gate and output gate to control c_t.

The model used to predict user's location in this paper is GRU, which is a variant of LSTM. It includes the reset gate and the update gate. The update gate is composed of the forget gate and the input gate. GRU model merged c_t and h_t. Compared with the standard LSTM model, GRU model is simpler, faster in training speed and requires fewer samples. Its structure is shown in Fig. 3.

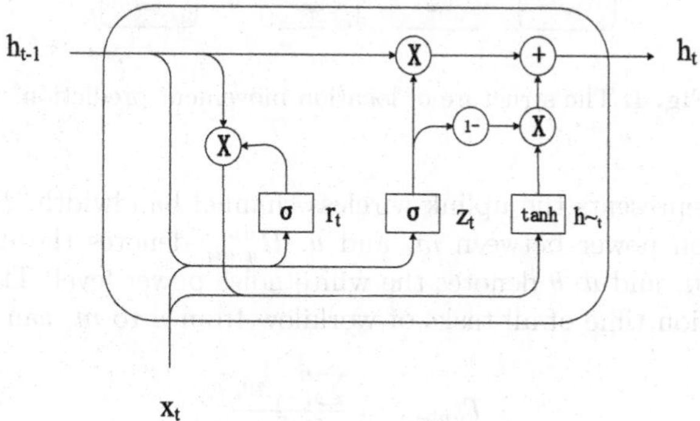

Fig. 3. The architecture of GRU model.

Figure 4 shows the structure of location movement prediction. The structure contains three layers: input layer, RNN layer (GRU) and output layer. GRU takes the sequence $x = \{x_1, x_2, x_3, x_4, ...x_k\}$ as the input and outputs the sequence $y = \{y_1, y_2, y_3, y_4, ...y_k\}$. As far as user's location forecasting be concerned, x can be regarded as prior user's historical trajectory information, and y is the predicted user position.

3.4 Response Delay Model

The response delay consists of three parts:uplink data transmission time, workflow completion time and downlink data transmission time.

Uplink Data Transmission Time. We assume that all tasks of workflow are offloaded to MEC servers for calculation. Referring to [11], the uplink data transmission rate between mobile device u and MEC server m_i close to user is defined as

$$R_{u,m_i}^{up} = W_{u,m_i}^{up} \log_2(1 + \frac{P^{up} H_{u,m_i}^{up}}{\theta}) \tag{1}$$

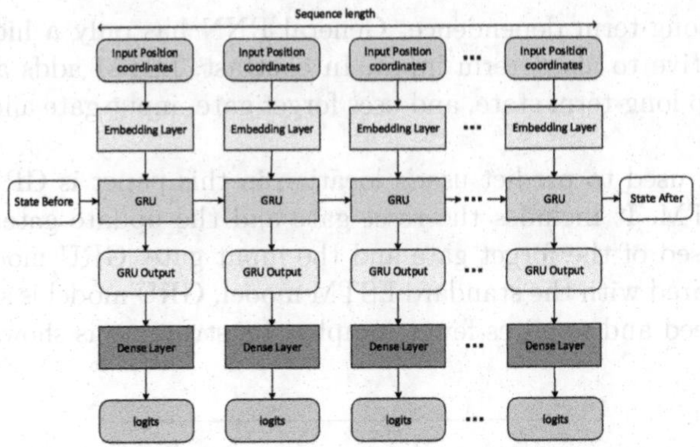

Fig. 4. The structure of location movement prediction.

where W^{up}_{u,m_i} represents the uplink wireless channel bandwidth. P^{up} represents the transmission power between m_i and u. H^{up}_{u,m_i} denotes the uplink channel gain between m_i and u. θ denotes the white noise power level. Thus the uplink data transmission time of all tasks of workflow from u to m_i can be calculated as

$$T_{uplink} = \frac{\sum_{i=1}^{n} w_i}{R^{up}_{u,m_i}} \tag{2}$$

Downlink Data Transmission Time. After the workflow is completed, the output data of exit task is transmitted to u through the downlink. We assume that the downlink data transmission rate and uplink data transmission rate are the same. If the exit task is running on server m_j, downlink data transmission rate $R^{down}_{m_j,u}$ can be expressed as

$$R^{down}_{m_j,u} = W^{down}_{m_j,u} \log_2(1 + \frac{P^{down} H^{down}_{m_j,u}}{\theta}) \tag{3}$$

Correspondingly, the downlink data transmission time of all tasks of workflow from m_j to u can be calculated as

$$T_{downlink} = \frac{k^{exit}_i}{R^{down}_{m_j,u}} \tag{4}$$

where k^{exit}_i is the amount of the output data of the exit task.

CPU Computing Time. We assume that the processing rate of each MEC server is r_m. If task v_i is executed on MEC server, the computing time can be given by

$$T_{com}(v_i) = \frac{w_i}{r_m} \tag{5}$$

Data Transfer Time Between Dependent Tasks. We suppose that the average bandwidth between edge servers is b, and thus the time of data transmission from task v_i to task v_j can be calculated as

$$T_{trans}(v_i, v_j) = \frac{d_{i,j}}{b} \tag{6}$$

Task Migration Time. If task v_i needs to migrate, the migration time can be expressed as

$$T_{mig}(v_i) = \frac{w_i}{b} \tag{7}$$

Workflow Completion Time. Let $ST(v_i)$, $CT(v_i)$, and $CT(v_i^{exit})$ represent the start time of task v_i, the completion time of task v_i, and the completion time of exit task v_i^{exit}, respectively. Therefore, the completion time MS of workflow can be defined as

$$MS = CT(v_i^{exit}) \tag{8}$$

$$CT(v_i) = ST(v_i) + T_{com}(v_i) + T_{trans}(v_i, v_j) \tag{9}$$

$$ST(v_i) = CT(v_t) + T_{mig}(v_i), v_t \in pred(v_i) \tag{10}$$

Response Delay. Based on the above model, the response delay can be expressed as

$$T_{total} = T_{uplink} + MS + T_{downlink} \tag{11}$$

3.5 Migration Cost Model

If the task needs to be migrated, the total cost of workflow migration consists of two parts: virtual machine rental cost and data transmission cost.

Virtual Machine Rental Cost. Suppose that $C_{VM}(v_i)$ denotes the rental cost of virtual machine handling task v_i, and thus it can be calculated as follows

$$C_{VM}(v_i) = T_{com}(v_i) * c(VM) \tag{12}$$

where $c(VM)$ is the virtual machine rental cost per unit time.

Data Transmission Cost. We assume that the link bandwidth is the same and set it to W. The cost $C_W(v_i)$ of data transmission caused by migrating task v_i can be expressed as

$$C_W(v_i) = (T_{trans}(v_i, v_j) + T_{mig}(v_i)) * c(W) \tag{13}$$

where $c(W)$ denotes the data transmission cost per unit time. The total cost can be calculated as

$$C_{total}(v_i, v_j) = \begin{cases} C_{VM}(v_i), & \text{if } t_i = 0 \\ C_{VM}(v_i) + C_W(v_i), & \text{if } t_i = 1 \end{cases} \quad (14)$$

where $t_i = 0$ means that the task v_i will not migrate, and $t_i = 1$ indicates task v_i will migrate.

3.6 Problem Formulation

Our goal is to design an effective workflow migration strategy to minimize the cost of workflow migration while ensuring the threshold. Based on the models defined above, the problem of workflow tasks migration can be formulated as

$$Minimize \quad \sum_{i=1}^{n}(C_{VM}(v_i) + t_i \cdot (C_W(v_i))) \quad (15)$$

$$Subject\,to \quad T_{total} \leq D \quad (16)$$

$$\sum_{i=1}^{n}(x_{i,j} \cdot RE_i^{CPU}) \leq CA_j^{CPU}, \forall j \in M \quad (17)$$

$$\sum_{i=1}^{n}(x_{i,j} \cdot RE_i^{BAND}) \leq CA_j^{BAND}, \forall j \in M \quad (18)$$

$$\sum_{j=1}^{t} x_{i,j} = 1, \forall i \in N \quad (19)$$

$$t_i \in \{0, 1\}, \forall i \in N \quad (20)$$

$$x_{i,j} \in \{0, 1\}, \forall i \in N, j \in M \quad (21)$$

The objective function (15) minimizes the cost of workflow migration, where the decision variable t_i indicates whether task v_i is migrated or not. Constraint (16) ensures that response delay does not exceed the threshold D predefined by the user. Constraint (17) and (18) ensures that the CPU resources and bandwidth resources on each edge server can meet the resource requirements of the task assigned to it. RE_i^{CPU} and RE_i^{BAND} represent the CPU and bandwidth resource requirements of task v_i, respectively. CA_j^{CPU} and CA_j^{BAND} represent the upper limit of CPU and bandwidth resources available on edge server j, respectively. The decision variable $x_{i,j}$ suggests whether task v_i is migrated to edge server j. M is the set of edge servers. Constraint (19) ensures that each subtask can only be migrated to one edge server. N indicates the set of subtasks in the workflow. Constraint (20) and (21) defines the value range of two decision variables, respectively.

4 The Proposed Algorithm

Because the workflow migration problem defined above is NP-hard combinatorial optimization problem, it is difficult to find the best solution in almost acceptable time. This section shows how to apply a workflow migration algorithm based on deep reinforcement learning (HDPDQN) to solve the above problem.

Reinforcement learning refers to a kind of problems that are constantly learning from interaction (with the environment) and the methods to solve such problems. Reinforcement learning problems can be described as an agent learning continuously from the interaction with the environment to obtain the maximum reward. Theoretically, agent explore the overall state and take corresponding actions, and then it can make the optimal decision. However, in the face of some unknown situations, it is difficult for agent to make correct decisions because it lack the ability to summarize previous experience. Therefore, we use DQN based algorithm to solve this problem. Based on Q-learning, DQN uses deep neural network (DNN) to predict Q-value, which solves the problem of "dimensional disaster" of Q-learning and the problem of generalized experience of agent, so that agent can choose actions according to Q-value to make the optimal task migration decision.

4.1 Encoding and Decoding Scheme

In this subsection, we define the basic elements of reinforcement learning. Agent is the front-end proxy server, which contains all the relevant information of users, base stations and MEC servers. And it can make decisions on whether the workflow task is migrated or not.

Before defining the state and action, we first encode the MEC servers in all regions, starting with the letter a. Then we encode the workflow. We turn the directed acyclic graph representing the workflow into a topological sorting to maintain the dependency between workflow subtasks, and the sequence element is the ID of each subtask. We define state s as a migration strategy of workflow, which is represented by a sequence, and each state corresponds to a sequence. Each element of the sequence represents the migration of each subtask of the workflow, the target server to be migrated to represented by the corresponding MEC server code, and number 0 is used to represent the non-migration decision.

Action is defined as whether to perform migration action for each task in the state, and can be expressed as a triplet $a = (V_i, t_i, M_i)$, where V_i is the encoding of task v_i in the current state. If $t_i = 0$, it means that task v_i will not migrate, executed on the source MEC server, and if $t_i = 1$, it means that task v_i will migrate. M_i is the target MEC server number after v_i is migrated.

After performing an action on the current state, a new state, namely a strategy sequence, will be generated. In order to show the dependencies between tasks, and the relationship between the task and its migration, we replace the sequence subscript with the ID of the task in the workflow topology sequence in turn, and reorder it according to the subscript to generate the final strategy sequence. At this time, the element corresponding to the task represented by the subscript is the migration strategy of the task. Figure 5 shows the encoding and decoding scheme.

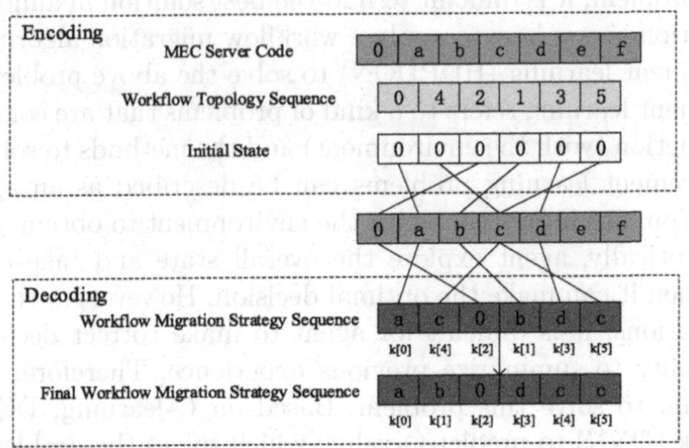

Fig. 5. Encoding and decoding scheme.

Since the objective of the above optimization problem is to minimize the cost of task migration, we define the reward function as follows

$$R_{(s,a)} = \frac{1}{C_{total}} \tag{22}$$

The smaller the cost of migration, the greater the reward.

4.2 Dynamic Programming Based Workflow Completion Time Calculation

Because the completion time of the workflow is the time when the last task is completed, we start from the exit task v_i^{exit} of the workflow, and look forward for its parent task until the entry task is found. We store the set of task at the same level for easy searching. From an exit task to an entry task, we calculate the completion time and store it with a collection. Finally, we find a valid path with the largest completion time and the time corresponding to this path is the completion time of the workflow. The input of the algorithm is a migration strategy of workflow, and the output is the completion time of the workflow. Algorithm 1 gives the pseudo code of the above algorithm.

4.3 The Proposed Workflow Task Migration Algorithm

The proposed HDPDQN algorithm uses an improved DQN based algorithm to solve the single objective optimization problem defined in the model. A large number of migration strategies, that is, states, will appear in the process of selecting the optimal migration strategy. DQN inputs the eigenvector of a state,

Algorithm 1: Dynamic Programming Based Workflow Completion Time Calculation

Input: A workflow task migration strategy
Output: Workflow completion time

Initialize workflow path completion time collection $COMP$, set to blank, and initialize node storage collection $STOR$, set to blank;
if *v_i is the first node of workflow* **then**
  $MS(v_i) = T_{com}(v_i)$;
  put $MS(v_i)$ into set $COMP$;
else
  find the parent node v_t of v_i node in reverse, and store them all in set $STOR$;
for *each node in $STOR$* **do**
  $MS(v_t) = MS(v_i) + T_{com}(v_t) + T_{trans}(v_t, v_i) + T_{mig}(v_t)$;
  put $MS(v_t)$ into set $COMP$;
Output the maximum value in set $COMP$;

and calculates it through the Q-network, and estimates the value function by function approximation, so as to better solve the problem of excessive number of states. The value function described by parameter ω is

$$\tilde{Q}(s, a, \omega) \approx Q_\pi(s, a) \tag{23}$$

In the current state s, when we execute action a, we will get a new state s' corresponding to an eigenvector $\phi(s')$. The improved DQN algorithm DDQN uses two identical Q-networks to reduce the dependence between the calculation of the target Q-value and the parameters of the Q-network to be updated. A current Q-network is used to select actions and update the model parameters, another target Q-network Q' is used to calculate the target Q-value, and the model parameter of Q' is ω'. The network parameters of the target Q-network do not need to be updated iteratively. Instead, they are copied from the current Q-network every other period of time, that is, delayed updating, so as to reduce the correlation between the target Q-value and the current Q-value. On this basis, DDQN decouples the selection of target Q-value action from the calculation of target Q-value. It is no longer to find the maximum Q-value of each action directly in the target Q-network, but first find the action corresponding to the maximum Q-value in the current q-network, so as to eliminate the problem of over estimation of target Q-value. The equation is as follows:

$$a^{max}(s', \omega) = \arg\max_{a'} Q(\phi(s'), a, \omega) \tag{24}$$

Then DDQN use the selected action $a^{max}(s', \omega)$ to calculate the target Q-value in the target Q-network. The calculation equation is as follows:

$$y = R + \gamma Q'(\phi(s'), \arg\max_{a'} Q(\phi(s'), a, \omega), \omega') \tag{25}$$

Algorithm 2: The Proposed Workflow Task Migration Algorithm

Input: Topological sorting sequence of workflow $[A_1, A_2, A_3, ..., A_n]$; threshold D; Number of iteration rounds T; Number of MEC servers m; Attenuation factor γ; Current Q-network Q; Target Q-network Q'; Number of samples with batch gradient decline n; target Q-network parameter update frequency f.

Output: Optimal workflow migration strategy satisfying constraints

Initialize candidate solution set $SPACE = \emptyset$;
Initialize total cost set $TotalCost = \emptyset$;
Initialize all Q-value to 0, randomly initialize the current Q-network parameter ω, and initialize the target Q-network parameter $\omega' = \omega$;
Initialize experience playback set $ER = \emptyset$;
Initialize workflow migration policy sequence $s = [0, 0, 0, ..., 0]$;
for $i = 1$ *to* T **do**
    Take s as the first state;
    For each subtask, select the target MEC server according to the predicted user's next location to generate action a;
    Generate a new workflow migration policy ns;
    Calculate the completion time ct of ns using dynamic programming based workflow completion time calculation algorithm;
    Calculate the response delay T_{total};
    if $T_{total} <= D$ **then**
        Calculate the total cost tc of ns and add it to $TotalCost$;
        Add ns to $SPACE$;
        Store the s, tc, ns triples in ER;
        $s \leftarrow ns$;
    else
        $s \leftarrow ns$;
    Take n samples $s_k, tc_k, ns_k, k = 1, 2, ..., n$ from ER and calculate the target Q-value y_k according to (25);
    Using mean square loss function $\frac{1}{n}\sum_{k=1}^{n}(y_k - Q(\phi(s_k), a_k, \omega)^2$, update the parameter ω of Q-network through gradient reverse transmission of neural network;
    If T % $f = 1$, update the parameter $\omega' = \omega$ of target Q-network;
    Turn to step 8;
Traverse the state of the candidate solution set $SPACE$;
Find the minimum value in the $TotalCost$ set by sorting, and assign its subscript to MIN_TC_INDEX;
Output the state $SPACE[MIN\_TC\_INDEX]$ with minimum total cost ;

We use dynamic programming algorithm to calculate the workflow completion time corresponding to each state. Algorithm 2 gives the pseudo code of the above algorithm.

5 Performance Evaluation

In this part, we verify the accuracy of the RNN based location movement prediction algorithm and the effectiveness of the proposed HDPDQN algorithm through simulation experiments.

5.1 Experiment Setup

In the simulation experiment, we define a MEC environment with 5 regions, and 6 MEC servers in each region, therefore, the maximum number of MEC servers considered in the experiment is 30. The defined threshold range is 20–70 s. As suggested in [12], the simulation parameter settings are shown in Table 1. We build the neural network using TensorFlow 2.5. The number of iterations T is set to 10000. The neural network settings are as follows: the attenuation factor γ is set to 0.9, the learning rate is set to 0.001 and the number of samples with batch gradient descent n is equal to 32.

When predicting the user's location movement, the historical trajectory data of user comes from two public data sets ETH [13] and UCY [14], with five scenes. ETH dataset is divided into Hotel and ETH, including two scenes with 750 different pedestrians respectively. UCY dataset is divided into ZARA-01, ZARA-02 and UCY, including two scenes with 786 people. We use 80% of data as a training set and 20% of data as a test set to predict three future locations through five historical locations.

In order to evaluate the performance of our proposed algorithm in solving the problem of workflow task migration, we use four workflow applications from different scientific fields released by Pegasus project [15], namely Montage for astronomy, SIPHT for bioinformatics, CyberShake for seismic science and LIGO for gravitational wave physics. Figure 6 shows the structure of the scientific workflow we used.

5.2 Simulation Results

In the first group of simulation experiments, we evaluate the accuracy of the proposed location prediction model. We report the prediction accuracy with two metrics:

1. Average Display Error (ADE): the average Euclidean distance difference between each predicted location and each real location.

$$\frac{\sum_{j=1}^{N} \frac{\sum_{i=1}^{n} \sqrt{(\hat{x_i^j}-x_i^j)^2+(\hat{y_i^j}-y_i^j)^2}}{n}}{N} \tag{26}$$

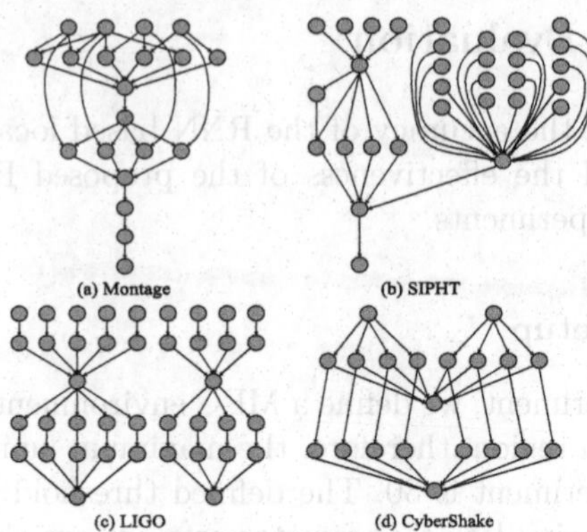

Fig. 6. The structure of the scientific workflow we used.

Table 1. Experiment parameters.

Parameters	Value
Number of MEC servers	30
Threshold range	{20 s, 30 s, 40 s, 50 s, 60 s, 70 s}
Uplink wireless channel bandwidth W_{u,m_i}^{up}	1 Mbps
Downlink wireless channel bandwidth $W_{m_j,u}^{down}$	1 Mbps
Uplink transmission power P^{up}	1 W
Downlink transmission power P^{down}	1 W
Uplink channel gain H_{u,m_i}^{up}	10^{-6}
Downlink channel gain $H_{m_j,u}^{down}$	10^{-6}
White noise power level θ	10^{-9} W
Processing rate of each MEC server r_m	1200 MHz
The average bandwidth between edge servers b	8 Mbps
Virtual machine rental cost per unit time $C(VM)$	1.5
Data transmission cost per unit time $C(W)$	1

Table 2. Predictive accuracy of four location prediction models.

Metric	IGP	LIN	LTA	Our-model
ADE	0.38	0.54	0.43	0.35
FDE	0.69	0.98	0.72	0.62

2. Final Display Error (FDE): the average Euclidean distance difference between the predicted location of the end point and the real location of the end point.

$$\frac{\sum_{j=1}^{N}\sqrt{(\hat{x_i^j}-x_i^j)^2+(\hat{y_i^j}-y_i^j)^2}}{N} \tag{27}$$

In (26) and (27), $\hat{x_i^j}$ and $\hat{y_i^j}$ represent the X-coordinates and Y-coordinates of predicted position, respectively. x_i^j and y_i^j represent the X-coordinates and Y-coordinates of real position, respectively. We compared our model with three commonly used models including a Constant Speed Model (LIN) [16], a Collision Avoidance Model (LTA) [16], and a Iterative Gaussian Process (IGP) [17]. Table 2 shows the values of two metrics for four prediction models. We also show in Fig. 7 the real track and the predicted track of four pedestrians. From the table, we can see that our model has better prediction accuracy compared with other three models. Because the linear model can only simulate the linear paths, the prediction error around the non-linear regions is higher. Our model can simulate more complex trajectories, so its prediction accuracy is better than that of the linear model.

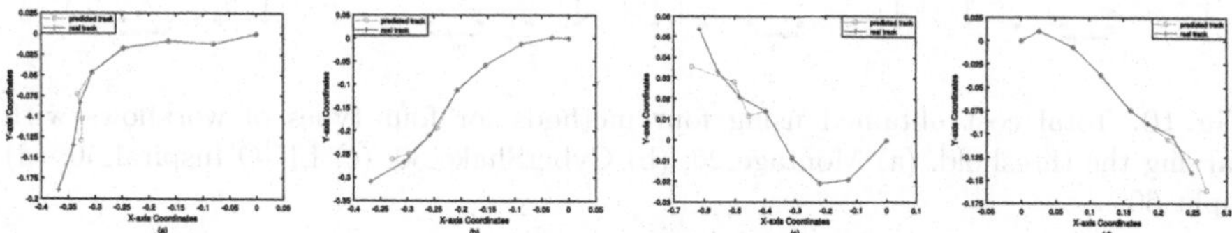

Fig. 7. Predicted and real trajectory for four pedestrians. (a) pedestrian 1; (b) pedestrian 2; (c) pedestrian 3; (d) pedestrian 4.

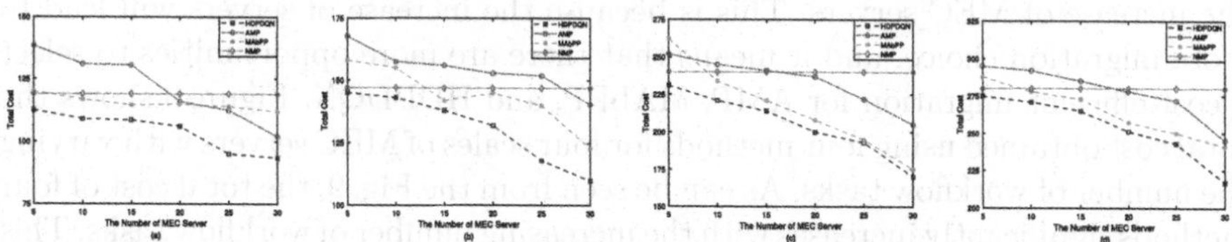

Fig. 8. Total cost obtained using four methods for four types of workflows with varying the number of MEC servers. (a) Montage_25; (b) CyberShake_30; (c) LIGO Inspiral_50; (d) Sipht_60.

In the second group of simulation experiments, we verify the effectiveness of the proposed workflow task migration algorithm by comparing it with the following methods: (1) Always Migration Policy (AMP); (2) No Migration Policy (NMP); (3) Markov Approximation Based Placement Policy (MAbPP) [6]. This simulation experiment uses Montage workflow with 25 tasks, CyberShake workflow with 30 tasks, LIGO workflow with 50 tasks and SIPHT workflow with 60 tasks. The predefined threshold for response delay is 20 s.

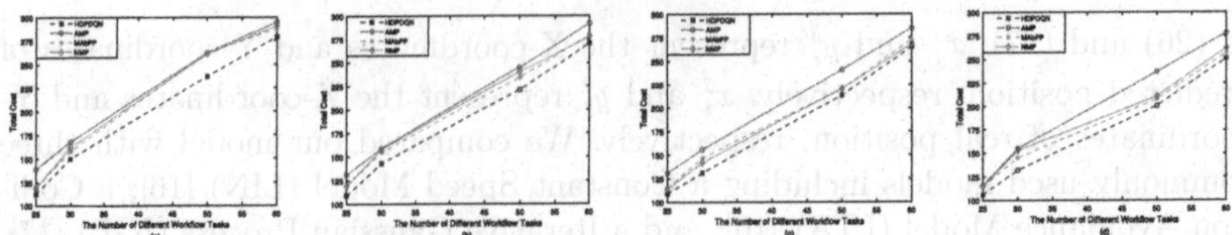

Fig. 9. Total cost obtained using four methods for four scales of MEC servers with varying the number of workflow tasks. (a) 5 MEC servers; (b) 10 MEC servers; (c) 15 MEC servers; (d) 20 MEC servers.

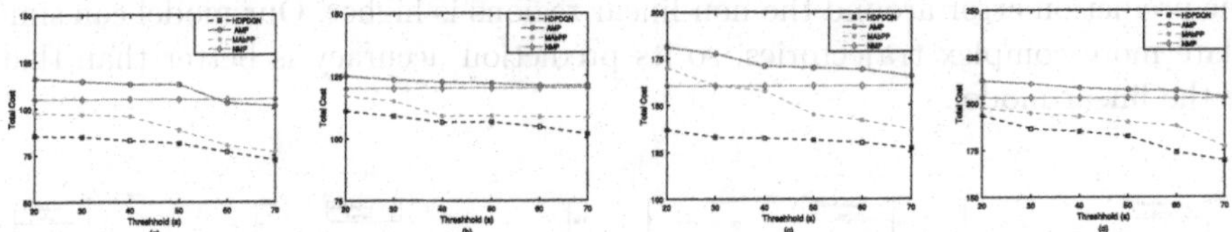

Fig. 10. Total cost obtained using four methods for four types of workflows with varying the threshold. (a) Montage_25; (b) CyberShake_30; (c) LIGO Inspiral_50; (d) Sipht_60.

Figure 8 shows the total cost obtained using four methods for four types of workflows with varying the number of MEC servers. It can be seen from the Fig. 8, the total cost of AMP, MAbPP and HDPDQN decreases gradually with the increase of MEC servers. This is because the increase of servers will lead to more migration choice, and it means that there are more opportunities to select a cost-efficient migration for AMP, MAbPP, and HDPDQN. Figure 9 shows the total cost obtained using four methods for four scales of MEC servers with varying the number of workflow tasks. As can be seen from the Fig. 9, the total cost of four methods significantly increases with the increasing number of workflow tasks. This is because more workflow tasks will introduce additional virtual machine rental costs and data transmission costs. Figure 10 shows the total cost obtained using four methods for four types of workflows with varying the threshold. As can be seen from the Fig. 10, the total cost of four schemes decreases slowly with the extending of threshold. This is because the extended threshold will cause that AMP, MAbPP, and HDPDQN have more opportunities to choose a migration strategy with lower migration costs. In addition, these figures show that our method perform best in terms of minimum migration cost. This is because our method can search the solution space of workflow migration problem more efficiently and globally so that it can find relatively low cost migration.

6 Conclusion

This paper studies the workflow task migration problem in mobile edge computing environment. We formulates it into a single objective optimization problem, and proposes an HDPDQN algorithm based on deep reinforcement learning to solve the optimization problem. The proposed algorithm considers the mobility of user, the dependency between workflow tasks, and the user's requirements for delay. The experimental results show that our workflow task migration strategy can save more costs.

References

1. Mao, Y., You, C., Zhang, J., Huang, K., Letaief, K.B.: A survey on mobile edge computing: the communication perspective. IEEE Commun. Surv. Tut. **19**(4), 2322–2358 (2017)
2. Adhikari, M., Amgoth, T., Srirama, S.N.: A survey on scheduling strategies for workflows in cloud environment and emerging trends. ACM Comput. Surv. (CSUR) **52**(4), 1–36 (2019)
3. Masdari, M., Zangakani, M.: Efficient task and workflow scheduling in inter-cloud environments: challenges and opportunities. J. Supercomput. **76**(1), 499–535 (2019). https://doi.org/10.1007/s11227-019-03038-7
4. Ouyang, T., Zhou, Z., Chen, X.: Follow me at the edge: mobility-aware dynamic service placement for mobile edge computing. IEEE J. Sel. Areas Commun. **36**(10), 2333–2345 (2018)
5. You, C., Huang, K., Chae, H., Kim, B.-H.: Energy-efficient resource allocation for mobile-edge computation offloading. IEEE Trans. Wirel. Commun. **16**(3), 1397–1411 (2016)
6. Heydari, J., Ganapathy, V., Shah, M.: Dynamic task offloading in multi-agent mobile edge computing networks. In: 2019 IEEE Global Communications Conference (GLOBECOM), pp. 1–6. IEEE (2019)
7. Mazouzi, H., Achir, N., Boussetta, K.: Elastic offloading of multitasking applications to mobile edge computing. In: Proceedings of the 22nd International ACM Conference on Modeling, Analysis and Simulation of Wireless and Mobile Systems, pp. 307–314 (2019)
8. Peng, K., et al.: An energy-and cost-aware computation offloading method for workflow applications in mobile edge computing. EURASIP J. Wirel. Commun. Netw. **2019**(1), 1–15 (2019)
9. Wang, S., Urgaonkar, R., Zafer, M., He, T., Chan, K., Leung, K.K.: Dynamic service migration in mobile edge computing based on Markov decision process. IEEE/ACM Trans. Netw. **27**(3), 1272–1288 (2019)
10. Zhang, C., Zheng, Z.: Task migration for mobile edge computing using deep reinforcement learning. Fut. Gener. Comput. Syst. **96**, 111–118 (2019)
11. Dinh, T.Q., Tang, J., La, Q.D., Quek, T.Q.S.: Offloading in mobile edge computing: task allocation and computational frequency scaling. IEEE Trans. Commun. **65**(8), 3571–3584 (2017)
12. Liu, J., Mao, Y., Zhang, J., Letaief, K.B.: Delay-optimal computation task scheduling for mobile-edge computing systems. In: 2016 IEEE International Symposium on Information Theory (ISIT), pp. 1451–1455. IEEE (2016)
13. Pellegrini, S., Ess, A., Schindler, K., Van Gool, L.: You'll never walk alone: modeling social behavior for multi-target tracking. In: 2009 IEEE 12th International Conference on Computer Vision, pp. 261–268. IEEE (2009)
14. Lerner, A., Chrysanthou, Y., Lischinski, D.: Crowds by example. In: Computer Graphics Forum, pp. 655–664. Wiley Online Library (2007)

15. Bharathi, S., Chervenak, A., Deelman, E., Mehta, G., Su, M.-H., Vahi, K.: Characterization of scientific workflows. In: 2008 3rd Workshop on Workflows in Support of Large-Scale Science, pp. 1–10. IEEE (2008)
16. Gupta, A., Johnson, J., Fei-Fei, L., Savarese, S., Alahi, A.: Social GAN: socially acceptable trajectories with generative adversarial networks. In: Proceedings of the IEEE Conference on Computer Vision and Pattern Recognition, pp. 2255–2264 (2018)
17. Trautman, P., Ma, J., Murray, R.M., Krause, A.: Robot navigation in dense human crowds: statistical models and experimental studies of human-robot cooperation. Int. J. Robot. Res. **34**, 335–356 (2015)

Recode-Decode-and-Compare: An Efficient Verification Scheme for Coded Edge Computing Against Collusion Attack

Zhaobo Lu[1], Jin Wang[1(✉)], Jingya Zhou[1], Jianping Wang[2], and Kejie Lu[3]

[1] Computer Science and Technology, Soochow University, Suzhou 215006, China
20195227041@stu.suda.edu.cn, {wjin1985,jy_zhou}@suda.edu.cn

[2] Computer Science, City University of Hong Kong, Hong Kong, China
jianwang@cityu.edu.hk

[3] Computer Science and Engineering, University of Puerto Rico at Mayagüez, Puerto Rico, USA
kejie.lu@upr.edu

Abstract. Recently, coded computing has great breakthroughs in *edge computing*, which can support and optimise many latency-sensitive and computation-intensive applications. In edge computing, a practical issue is that, comparing to computing devices configured in the cloud data center, edge devices may be unreliable and untrustworthy. Since coded computing has potential in speeding up the computation and protecting the confidentiality of computing data, *coded edge computing* (CEC) emerges. For a major security issue: verify the correctness of computation result in CEC, the existing efficient work performs verification by comparing multiple final results. The attackers (malicious edge devices) studied before are independent, while collusion attack is the realistic and important scenario in distributed computing and edge computing. To improve the security of CEC, in this paper, we study the verification issue of collusion attack and design the efficient verification scheme: *Recode-Decode-and-Compare* (RDC). By re-encoding the encoded data, collusive attackers cannot infer the linear relationship between different intermediate results. Therefore, the user can verify the correctness by comparing multiple final results. We also conduct solid theoretical analyses to show the successful verification probability and verification overheads. Finally, extensive simulation experiments demonstrate that the RDC verification scheme can efficiently verify the computation result and outperforms the state-of-the-art works.

Keywords: Edge computing · Coded computing · Verification · Collusion attack

1 Introduction

Recently, coded computing has great breakthroughs in *edge computing*, which can support and optimise many latency-sensitive and computation-intensive

Y. Lai et al. (Eds.): ICA3PP 2021, LNCS 13155, pp. 529–548, 2022.
https://doi.org/10.1007/978-3-030-95384-3_33

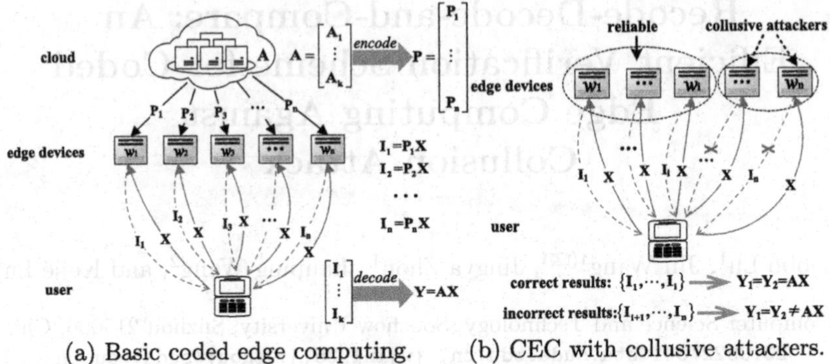

(a) Basic coded edge computing. (b) CEC with collusive attackers.

Fig. 1. Coded edge computing.

applications, such as self-driving, virtual reality, machine learning and big data analytics [1–10].

Firstly, coded computing guarantees that the user can recover the final result as long as enough number of intermediate results are returned by edge devices [8,11]. Therefore, the user can mitigate the impact of the slower edge devices (stragglers) to speed up the computation tasks. Secondly, to protect the data confidentiality, many researchers have studied the secure coded distributed computing so that edge devices cannot obtain the original information of computing data [12–14].

We now take the matrix multiplication as an example to illustrate the basic CEC. In Fig. 1(a), the system consists of the cloud, n edge devices and the user. The user requires the result of matrix multiplication $\mathbf{AX}$, in which $\mathbf{A}$ is in the cloud and $\mathbf{X}$ is in the user. First, the cloud encodes $\mathbf{A} = \left[\mathbf{A}_1^\top, \cdots, \mathbf{A}_k^\top\right]^\top$ into $\mathbf{P} = \left[\mathbf{P}_1^\top, \cdots, \mathbf{P}_n^\top\right]^\top$, in which $n > k$.

For the ith edge device w_i, it will obtain $\mathbf{P}_i$ and $\mathbf{X}$. Then, it computes the intermediate result $\mathbf{I}_i = \mathbf{P}_i\mathbf{X}$ and sends $\mathbf{I}_i$ to the user. Due to the application of coding, the user can recover the final result $\mathbf{Y} = \mathbf{AX}$ by any set of k intermediate results. Therefore, the impact of slower $n - k$ edge devices can be mitigated [8] and edge devices cannot obtain the original information of $\mathbf{A}$ [15].

In addition to the issue of data confidentiality, another major security issue: verify the correctness of the computation result in CEC, is also important.

The intermediate results returned by edge devices may be incorrect due to unexpected hardware failures, software bugs, financial incentives or attacks [5]. When incorrect intermediate results are used in the decoding operation, the final result will be incorrect.

The main traditional schemes for verifiable outsourced computing are usually based on homomorphic encryption and computation replication. For the user device with limited computation and storage capability, the verification complexity of homomorphic encryption schemes is too high [16–18]. For instance, using the latest HElib library, the runtime of the matrix multiplication in homomorphic encryption mode is over 10^3 times higher than the direct multiplication of unencrypted matrices [18]. Actually, homomorphic encryption based schemes are more suitable to be performed on a powerful computation device. For computation replication schemes, computation tasks have to be repeatedly processed on multiple edge devices to tolerate a small number of Byzantine faults (or attackers), which is a huge burden for computation resources in edge computing [19–21]. Therefore, they cannot fulfill the effectiveness requirement of edge computing. To fully utilize the properties of linear coding and efficiently verify the correctness of computation results, the *Decode-and-Compare* (DC) verification scheme has been proposed [22]. The user should decode multiple sets of k intermediate results and compare final results until two same final results are obtained, which are considered to be correct. In the above literatures, the DC verification scheme is much more efficient than other schemes when the scale of computation is large.

A drawback of the DC verification scheme is the need to ensure that attackers are independent. However, in distributed computing and edge computing, collusion attack is the realistic and important scenario [23]. Since any set of k intermediate results can be decoded into the same final result, collusive attackers can infer the linear relationship between them. Based on it, incorrect intermediate results will be modified and can lead to same incorrect final results which may be obtained by the user. For example, in Fig. 1(b), collusive attackers guarantee that any set of k incorrect intermediate results can lead to the same incorrect final result. When the user decodes two same incorrect final results before two correct final results, the incorrect final result will be obtained $\mathbf{Y}_1 = \mathbf{Y}_2 \neq \mathbf{AX}$.

In this paper, we will study the verification issue of collusion attack in CEC. Specifically, we take the matrix multiplication as a representative computation task, because matrix multiplication is one of the most basic components of linear computation and machine learning algorithms, such as gradient descent [24], linear regression [8], neural networks [10] and federated learning [9].

To against the collusion attack, we analyze the probability that the DC verification scheme obtains the incorrect final result and get a conclusion that it is related with the attackers ratio. When the number of collusive attackers is large, the DC verification scheme is impossible to obtain the correct final result. Therefore, we design an efficient verification scheme: *Recode-Decode-and-Compare* (RDC), which mainly contains two parts: *re-encoding encoded data blocks* and *comparing final results*. By re-encoding the encoded data, collusive attackers cannot infer the linear relationship between different intermediate results and design feasible collusion attack method. Therefore, the user can verify the final result by comparing multiple final results. The theoretical analyses and simulation experiments also show that the RDC verification scheme can efficiently verify the computation results when attackers are collusive.

The main contributions of this paper are summarized as follows:

- We analyze the drawback of the DC verification scheme and give a potential collusion attack method. The theoretical analysis shows that (1) any set of k incorrect intermediate results can be decoded into same incorrect final result and (2) the probability that the DC verification scheme will obtain the incorrect final result is related with the attacker ratio.

- We design the *Recode-Decode-and-Compare* (RDC) verification scheme and conduct solid theoretical analyses to give the successful verification probability and verification overheads. The theoretical analyses demonstrate the RDC verification scheme can efficiently verify the computation result of collusion attack with small storage, computation and communication overheads.
- We conduct extensive simulation experiments to evaluate the successful verification probability and verification time consumption of the proposed RDC verification scheme. We also compare it with the DC verification scheme and another baseline scheme, *i.e.*, computing locally at the user device. The results demonstrate that the RDC verification scheme can efficiently verify the computation result and outperforms the state-of-the-art works.

The rest of this paper is organized as follows. First, we introduce the system model and attack model in Sect. 2. Next, in Sect. 3, we analyze the drawback of the DC verification scheme, and then give the main idea and solid theoretical analyses of the proposed RDC verification scheme. Finally, extensive simulation experiments are conducted in Sect. 4 before we conclude this paper in Sect. 5.

2 Problem Modeling

In this section, we will introduce the basic CEC system model and the attack model of this paper.

2.1 System Model of Basic Coded Edge Computing

In this paper, we study the basic CEC system and take the matrix multiplication as the representative computation task [8,10–14,22].

The system consists of the cloud, the user and a set of k' edge devices $w = \{w_1, \dots, w_{k'}\}$. The user requires the result of matrix multiplication $\mathbf{AX}$ in a sufficiently large finite field q. The data matrix $\mathbf{A} \in \mathbb{F}_q^{m \times n}$ is in the cloud and the user matrix $\mathbf{X} \in \mathbb{F}_q^{n \times l}$ is in the user. To utilize the distributed computation power of edge devices, the matrix multiplication $\mathbf{AX}$ will be encoded into multiple subtasks and then performed at k' edge devices.

First, the cloud, which is a trustworthy device, divides the data matrix $\mathbf{A}$ into k blocks with the same size by row[1] $\mathbf{A} = \left[\mathbf{A}_1^\top, \cdots, \mathbf{A}_k^\top\right]^\top$, in which $k < k'$. Then the cloud generates an encoding matrix $\mathbf{B} \in \mathbb{F}_q^{k' \times k}$. Before we give the encoding operations, we define the following operator (Table 1):

[1] If m is not divisible by k, $k\lceil \frac{m}{k} \rceil - m$ zero rows should be padded into $\mathbf{A}$. In most distributed machine learning algorithms, $m \gg k' > k$ and the overheads led by the additional padded rows can be omitted. Without loss of generality, we assume that $\frac{m}{k}$ is always an integer in this paper.

Table 1. Notations

Notations	Meaning
$\mathbf{A}$	The $m \times n$ dimensional data matrix
$\mathbf{B}$	The $k' \times k$ dimensional encoding matrix
$\mathbf{B}_i$	The ith row of $\mathbf{B}$, which is also the encoding vector of $\mathbf{P}_i$
D	The set of recoding matrices $\mathbf{D} = \{\mathbf{D}_1, \ldots, \mathbf{D}_s\}$, in which $\lvert\mathbf{D}\rvert = s$ and $\mathbf{D}_i$ is the $k_i \times k_i$ dimensional matrix
$\mathbf{D}^{-1}$	The set of recovered matrices $\mathbf{D}^{-1} = \{\mathbf{D}_1^{-1}, \ldots, \mathbf{D}_s^{-1}\}$
$\mathbf{I}_i$	The ith intermediate result, $\mathbf{I}_i = \mathbf{P}_i^{\nabla}\mathbf{X}$
$\mathbf{I}_i^{\triangle}$	$\mathbf{I}_i^{\triangle} = \mathbf{D}_j^{-1} \circledast \mathbf{I}_i$, in which $\mathbf{D}_j^{-1}$ is the recovered matrix of $\mathbf{I}_i$
$\mathbf{P}_i^{\nabla}$	$\mathbf{P}_i^{\nabla} = \mathbf{D}_j \circledast \mathbf{P}_i$, in which $\mathbf{D}_j$ is the recoding matrix of $\mathbf{P}_i$
$\mathbf{X}$	The $n \times l$ dimensional user matrix
$\mathbf{Y}$	The final result, $\mathbf{Y} = \mathbf{AX}$
k	The number of blocks divided from data matrix $\mathbf{A}$
k'	The number of edge devices
α	The coding redundancy ratio
β	The attacker ratio, $0 \leq \beta \leq 1$

Definition 1 ***(Block matrix multiplication*** $\circledast$***).*** *For an $m \times n$ dimensional matrix $\boldsymbol{A}$ and a $k' \times k$ dimensional matrix $\boldsymbol{B}$, suppose that $\boldsymbol{A}$ is divided into k blocks with the same size by row $\boldsymbol{A} = \left[\boldsymbol{A}_1^\top, \cdots, \boldsymbol{A}_k^\top\right]^\top$, $\boldsymbol{B}_i$ is the ith row of $\boldsymbol{B}$ and $\boldsymbol{B}_{i,j}$ is the ith row jth column element of $\boldsymbol{B}$, then we will have the $\frac{mk'}{k} \times n$ dimensional matrix $\boldsymbol{P} = \boldsymbol{B} \circledast \boldsymbol{A} = \left[\boldsymbol{P}_1^\top, \cdots, \boldsymbol{P}_{k'}^\top\right]^\top$, in which each block of $\boldsymbol{P}$ is a $\frac{m}{k} \times n$ dimensional matrix:*

$$\boldsymbol{P}_i = \boldsymbol{B}_i \circledast \boldsymbol{A} = \sum_{j=1}^{k} \boldsymbol{B}_{i,j}\boldsymbol{A}_j, \forall i \in \{1, \cdots, k'\}. \tag{1}$$

Then, the cloud uses $\mathbf{B}$ to encode the data matrix $\mathbf{A}$ into the $\frac{mk'}{k} \times n$ dimensional *encoded data matrix* $\mathbf{P} = \mathbf{B} \circledast \mathbf{A} = \left[\mathbf{P}_1^\top, \cdots, \mathbf{P}_{k'}^\top\right]^\top$. In the encoding operation, $\mathbf{B}_i$ is the *encoding vector* of $\mathbf{P}_i$ and $\alpha = \frac{k'}{k}$ is the *coding redundancy ratio*. For the ith block $\mathbf{P}_i$, it should be allocated to the ith edge device w_i. We also note that any k rows of $\mathbf{B}$ are linearly independent to make sure that the final result can be recovered.

Next, when the ith edge device w_i receives the user matrix $\mathbf{X}$, it will calculate the matrix multiplication of $\mathbf{P}_i$ and $\mathbf{X}$, $i \in \{1, \ldots, k'\}$. Then, the edge device should send the *intermediate result* $\mathbf{I}_i = \mathbf{P}_i\mathbf{X}$ to the user. $\mathbf{B}_i$ is also the encoding vector of ith intermediate result $\mathbf{I}_i$.

Finally, when the user receives at least k intermediate results, it selects k intermediate results and forms them into an $m \times l$ dimensional *to-be-decoded result* $\mathbf{I}'$. Their encoding vectors are also formed into a $k \times k$ dimensional matrix $\mathbf{B}'$. $\mathbf{B}'^{-1}$ is the decoding matrix and the user can recover the $m \times l$ dimensional *final result* $\mathbf{Y}$ by a block matrix multiplication $\mathbf{Y} = \mathbf{B}'^{-1} \circledast \mathbf{I}' = \mathbf{AX}$. If the to-be-decoded result $\mathbf{I}'$ contains the incorrect intermediate result, the final result is incorrect $\mathbf{Y} \neq \mathbf{AX}$. Therefore, the correctness of final result should be verified.

2.2 Attack Model

In this subsection, we introduce the attack model considered in this paper. In CEC, each correct intermediate result $\mathbf{I}_i$ may be modified by fault edge devices or attackers. The intermediate result $\mathbf{I}_i$ is modified into the incorrect intermediate result $\mathbf{I}_i^* \neq \mathbf{I}_i$, which is also equivalent to adding a $\frac{m}{k} \times l$ dimensional non-zero matrix $\mathbf{R}_i$ to the correct intermediate result $\mathbf{I}_i$, *i.e.*, $\mathbf{I}_i^* = \mathbf{I}_i + \mathbf{R}_i$ in which $\mathbf{R}_i \neq 0$.

Let β be the attacker ratio, $0 \leq \beta \leq 1$, *i.e.*, there are totally $\beta k'$ attackers and $\beta k'$ incorrect intermediate results.

In the collusion attack, assume that $\beta k'$ attackers can collude with each other and share their encoded data blocks. When attackers control at least two sets of k intermediate results, since any k intermediate results can be decoded into the same final result, attackers can infer the linear relationship between them. Based on it, attackers can modify intermediate results and make sure any k modified intermediate results can be decoded into the same incorrect final result. When attackers control at most one set of k intermediate results, they cannot infer the linear relationship and we assume that attackers randomly modify intermediate results.

3 The Recode-Decode-and-Compare Verification Scheme

In this section, we first analyze the drawback of the DC verification scheme and give a collusion attack method. Then, we give the main idea and algorithm of the RDC verification scheme. Finally, we conduct solid theoretical analyses about the performance of the RDC verification scheme, *e.g.*, the successful verification probability and verification overheads.

3.1 The Drawback of the DC Verification Scheme

In this subsection, we first give a feasible collusion attack method. Then, we analyze the probability that the final result obtained by the DC verification scheme is incorrect when attackers are collusive, which demonstrates that the DC verification scheme cannot against the collusion attack.

In the collusion attack, attackers can infer the linear relationship between different intermediate results and modify intermediate results based on it. We consider the case that the DC verification scheme obtains the incorrect final result. Since only two same final results may be obtained by the DC verification scheme, at least two same incorrect final results are required.

We first analyze the required attacker ratio that attackers can control at least two sets of different k intermediate results and infer the linear relationship between them to design the same incorrect final results.

Lemma 1. *Attackers can control at least two sets of k intermediate results, iff $\beta \geq \frac{k+1}{\alpha k}$.*

Proof. If attackers can control at least two sets of k intermediate results, $\binom{\beta k'}{k} \geq 2$. Since $\beta k'$ must be an integer, $\beta k' \geq k+1 > k$. Since $k' = \alpha k$, we have $\alpha\beta k \geq k+1$, which can also be simplified to $\beta \geq \frac{k+1}{\alpha k}$.

If the attacker ratio $\beta \geq \frac{k+1}{\alpha k}$, the number of incorrect intermediate results $\beta k' \geq k+1$. Therefore, attackers can control $\binom{\beta k'}{k} \geq k+1 \geq 2$ sets of different k intermediate results.

Remark 1. *If the attacker ratio $\beta < \frac{k+1}{\alpha k}$, there are at most one set of k incorrect intermediate results. Therefore, attackers cannot design two same incorrect final results to perform the collusion attack.*

Let $t = \beta k'$ be the number of incorrect intermediate results and $v = (1-\beta)k'$ be the number of correct intermediate results. The required attacker ratio given in Lemma 1 can also be expressed as $t \geq t+1$.

Let two to-be-decoded results, each of which contains k incorrect intermediate results, be $\mathbf{I}_1'^*$ and $\mathbf{I}_2'^*$. Assume that $\mathbf{I}_1'^* = \mathbf{I}_1' + \mathbf{R}_1'$ and $\mathbf{I}_2'^* = \mathbf{I}_2' + \mathbf{R}_2'$, in which $\mathbf{I}_1'$ and $\mathbf{I}_2'$ are the correct to-be-decoded results corresponding to $\mathbf{I}_1'^*$ and $\mathbf{I}_2'^*$, and $\mathbf{R}_1'$ and $\mathbf{R}_2'$ are $m \times l$ dimensional non-zero matrices. Moreover, $\mathbf{B}_1'^{-1}$ and $\mathbf{B}_2'^{-1}$ are their decoding matrices. Let the $k \times k$ dimensional matrix be $\mathbf{C} = \mathbf{B}_2'\mathbf{B}_1'^{-1}$, the two incorrect final results be $\mathbf{Y}_1^* = \mathbf{B}_1'^{-1} \circledast \mathbf{I}_1'^*$ and $\mathbf{Y}_2^* = \mathbf{B}_2'^{-1} \circledast \mathbf{I}_2'^*$. The sufficient and necessary condition that $\mathbf{Y}_1^* = \mathbf{Y}_2^*$ follows:

Lemma 2. $\mathbf{Y}_1^* = \mathbf{Y}_2^*$, *iff* $\mathbf{C} \circledast \mathbf{R}_1' = \mathbf{R}_2'$.

Proof.

$$
\begin{aligned}
&\mathbf{B}_1'^{-1} \circledast \mathbf{I}_1'^* = \mathbf{B}_2'^{-1} \circledast \mathbf{I}_2'^* \\
\Leftrightarrow &\mathbf{B}_1'^{-1} \circledast (\mathbf{I}_1' + \mathbf{R}_1') = \mathbf{B}_2'^{-1} \circledast (\mathbf{I}_2' + \mathbf{R}_2') \\
\Leftrightarrow &\mathbf{B}_1'^{-1} \circledast (\mathbf{B}_1' \circledast \mathbf{AX} + \mathbf{R}_1') = \mathbf{B}_2'^{-1} \circledast (\mathbf{B}_2' \circledast \mathbf{AX} + \mathbf{R}_2') \\
\Leftrightarrow &\mathbf{AX} + \mathbf{B}_1'^{-1} \circledast \mathbf{R}_1' = \mathbf{AX} + \mathbf{B}_2'^{-1} \circledast \mathbf{R}_2' \\
\Leftrightarrow &\mathbf{B}_1'^{-1} \circledast \mathbf{R}_1' = \mathbf{B}_2'^{-1} \circledast \mathbf{R}_2' \\
\Leftrightarrow &(\mathbf{B}_2'\mathbf{B}_1'^{-1}) \circledast \mathbf{R}_1' = \mathbf{R}_2' \\
\Leftrightarrow &\mathbf{C} \circledast \mathbf{R}_1' = \mathbf{R}_2'
\end{aligned}
$$

From the above, we give a simple collusion attack method, which can lead to same incorrect final results by two sets of k incorrect intermediate results.

1. When $t \geq k+1$, attackers first randomly select k intermediate results, which are linearly independent, and compose to an $m \times l$ dimensional to-be-decoded result $\mathbf{I}_1'$. Then, select another set of k intermediate results and also compose to an $m \times l$ dimensional matrix $\mathbf{I}_2'$.
2. Since k intermediate results of $\mathbf{I}_1'$ are linearly independent, attackers can determine a $k \times k$ dimensional matrix $\mathbf{C}$, which satisfies $\mathbf{C} \circledast \mathbf{I}_1' = \mathbf{I}_2'$.

3. Then, attackers randomly generate an $m \times l$ non-zero matrix $\mathbf{R}'_1$ and compute another $m \times l$ matrix $\mathbf{R}'_2 = \mathbf{C} \circledast \mathbf{R}'_1$.
4. Finally, $\mathbf{I}'^*_1 = \mathbf{I}'_1 + \mathbf{R}'_1$ and $\mathbf{I}'^*_2 = \mathbf{I}'_2 + \mathbf{R}'_2$ are divided into k blocks respectively. Each block, as incorrect intermediate results, is returned into the user.

According to Lemma 2, since $\mathbf{R}'_2 = \mathbf{C} \circledast \mathbf{R}'_1$, these two incorrect final results decoded by $\mathbf{I}'^*_1$ and $\mathbf{I}'^*_2$ are same, $\mathbf{Y}^*_1 = \mathbf{B}'^{-1}_1 \circledast \mathbf{I}'^*_1 = \mathbf{B}'^{-1}_2 \circledast \mathbf{I}'^*_2 = \mathbf{Y}^*_2$. When the user decodes $\mathbf{Y}^*_1$ and $\mathbf{Y}^*_2$, the incorrect final result will be obtained.

Remark 2. *Since attackers cannot obtain intermediate results which belong to other v reliable edge devices, they only can design at most $\binom{t}{k}$ same incorrect final results, each of which is decoded by k incorrect intermediate results.*

In the following analysis, we assume that attackers can modify all t intermediate results and make sure any k of them can be decoded into same incorrect final result.

Even if there exist same incorrect final results, the user still can obtain the correct final result. Because when two correct final results are decoded before two same incorrect final results, the user will obtain the correct final result. Then, we will analyze the probability that the final result obtained by the DC verification scheme is incorrect when attackers are collusive. To facilitate the analysis, we make following symbol definitions:

$$T = \binom{k'}{k}, T^{\vee} = \binom{v}{k}, T^{\times} = \binom{t}{k}, T^{*} = T - T^{\vee} - T^{\times}, \tag{2}$$

in which T is the number of total final results, $T^{\vee}$ is the number of correct final results, $T^{\times}$ is the number of incorrect final results which decoded by k incorrect intermediate results and T^* is the number of incorrect final results which decoded by both correct and incorrect intermediate results.

Since attackers cannot obtain intermediate results which belong to v reliable edge devices, they are impossible to infer the linear relationship between them. Therefore, to make the analysis easier, we assume that these T^* incorrect final results are always different[2].

Let the probability that the final result obtained by the DC verification scheme is incorrect be $P(F)$. $P(F)$ is also the probability that the user obtains two same incorrect final results before obtaining two correct final results. Assume that when the user finishes the verification, there are $j (0 \leq j \leq 1)$ correct final results and i incorrect final results. The range of i is from 2 to $2 + T^*$, which contains 2 same incorrect final results and at most T^* incorrect final results which decoded by both correct and incorrect intermediate results. Therefore, the user requires $i + j$ rounds of decoding to obtain the incorrect final result.

Theorem 1. *When $t \geq k + 1$, the probability that the final result obtained by the DC verification scheme is incorrect* $P(F) = \sum_{i=2}^{2+T^*} \sum_{j=0}^{1} \left(\frac{2\binom{T^{\times}}{2}\binom{T^{\vee}}{j}\binom{T^*}{i-2}}{\binom{T}{i+j}(i+j)} \right)$.

[2] The probability that two of these T^* incorrect final results are same is very low and we will give the comprehensive analysis of it in the future work.

Algorithm 1: RDC verification scheme (cloud part).

Input: $\mathbf{A}, k', k, w, m, n$

generate the $k' \times k$ dimensional encoding matrix $\mathbf{B}$;
generate the set of s matrices $\mathbf{D} = \{\mathbf{D}_1, \ldots, \mathbf{D}_s\}$;
generate the set of s parameters $\sigma = \{\sigma_1, \ldots, \sigma_s\}$;
$\mathbf{P} = \mathbf{B} \circledast \mathbf{A} = \left[\mathbf{P}_1^\top, \cdots, \mathbf{P}_{k'}^\top\right]^\top$;
for $i = 1; i \leq s; i = i + 1$ **do**
  for $h = \sum_{j=1}^{i} \sigma_j - \sigma_i + 1; h \leq \sum_{j=1}^{i} \sigma_j; h = h + 1$ **do**
    $\mathbf{P}_h^\nabla = \mathbf{D}_h \circledast \mathbf{P}_h$;
    send $\mathbf{P}_h^\nabla$ to hth edge device w_h;
compute the recovered matrices $\mathbf{D}^{-1} = \{\mathbf{D}_1^{-1}, \ldots, \mathbf{D}_s^{-1}\}$;
send $\mathbf{B}, \mathbf{D}^{-1}$ and σ to the user;

Proof. In $i + j$ rounds of decoding, there are totally $\binom{T}{i+j}$ decoding situations. Therefore, the probability that $i+j$ rounds of decoding contain 2 same incorrect final results and j correct final results is $\frac{\binom{T^\times}{2}\binom{T^\vee}{j}\binom{T^*}{i-2}}{\binom{T}{i+j}}$.

The total number of decoding sequences is $(i + j)!$. For two same incorrect final results, one of them must be the $(i + j)$th round of decoding. Therefore, the number of eligible decoding sequences is $2(i + j - 1)!$. The probability that the decoding sequence is eligible is $\frac{2(i+j-1)!}{(i+j)!} = \frac{2}{i+j}$.

Finally, since i is from 2 to $2 + T^*$, the probability that the obtained final result is incorrect $P(F) = \sum_{i=2}^{2+T^*} \sum_{j=0}^{1} \left(\frac{2\binom{T^\times}{2}\binom{T^\vee}{j}\binom{T^*}{i-2}}{\binom{T}{i+j}(i+j)} \right)$.

Remark 3. *For the probability $P(F)$, there are some conclusions:*

(1) when k' and α are fixed, $P(F)$ is only related with the attacker ratio β;
(2) when $t < k+1$, i.e., $\beta < \frac{k+1}{\alpha k}$, $P(F) = 0$ because there do not exist two same incorrect final results;
(3) when β increases, $P(F)$ will also increase;
(4) when $v < k + 1$, i.e., $\beta > 1 - \frac{k+1}{\alpha k}$, $P(F) = 100\%$ because there do not exist two correct final results.

According to the above analyses, when attackers are collusive and $t \geq k + 1$, the DC verification scheme may obtain the incorrect final result. When β is sufficient large, the user even is impossible to obtain the correct final result. Therefore, the DC verification scheme has drawback and it is necessary to design a verification scheme to against collusion attack.

3.2 The Main Idea of the RDC Verification Scheme

In this subsection, we will introduce the main idea and algorithm of the proposed RDC verification scheme.

Algorithm 2: RDC verification scheme (user part).

Input: $\mathbf{B}, \mathbf{D}^{-1}, \bar{\mathbf{I}}, k', k, s, \sigma$
Output: $\mathbf{Y}$
$\bar{\mathbf{I}}^{\triangle} = \emptyset$;
while *receive* $\mathbf{I}_i$ *and* $|\bar{\mathbf{I}}^{\triangle}| \leq k'$ **do**
  for $j = 1; j \leq s; j = j + 1$ **do**
    if $\sum_{h=1}^{j} \sigma_h - \sigma_j + 1 \leq i \leq \sum_{h=1}^{j} \sigma_h$ **then**
      $\mathbf{I}_i^{\triangle} = \mathbf{D}_j^{-1} \circledast \mathbf{I}_i$;
      $\bar{\mathbf{I}}^{\triangle} = \bar{\mathbf{I}}^{\triangle} \cup \{\mathbf{I}_i^{\triangle}\}$;
      break;
  for $h = \binom{|\bar{\mathbf{I}}^{\triangle}|-1}{k} + 1; h \leq \binom{|\bar{\mathbf{I}}^{\triangle}|}{k}; h = h + 1$ **do**
    select k recovered intermediate results to compose to $\mathbf{I}'_h$ and their encoding vectors to compose to $\mathbf{B}'_h$;
    compute the decoding matrix $\mathbf{B}'^{-1}_h$;
    $\mathbf{Y}_h = \mathbf{B}'^{-1}_h \circledast \mathbf{I}'_h$;
    for $j = 1; j < h; j = j + 1$ **do**
      if $\mathbf{Y}_j = \mathbf{Y}_h$ **then**
        $\mathbf{Y} = \mathbf{Y}_h$;
        return $\mathbf{Y}$;
return false;

We will describe the work of the cloud, edge devices and the user to give key components of the RDC verification scheme. First is the work of the **cloud**, which is also shown in Algorithm 1.

- **Encode the data matrix**: The system model gives this step where the data matrix $\mathbf{A}$ is encoded into k' blocks $\{\mathbf{P}_1, \cdots, \mathbf{P}_{k'}\}$. This is also the work of the cloud in the basic CEC.
- **Generate recoding matrices and recoding parameters**: Then, the cloud generates the set of s matrices $\mathbf{D} = \{\mathbf{D}_1, \ldots, \mathbf{D}_s\}$, in which $\mathbf{D}_i$ is the $k_i \times k_i$ dimensional full rank matrix[3]. The cloud continues to determine a set of s parameters $\sigma = \{\sigma_1, \ldots, \sigma_s\}$, in which $\sum_{j=1}^{s} \sigma_j = k'$ and σ_i is the *recoding parameter* of $\mathbf{D}_i$. The range of σ_i is given in Lemma 6.
- **Re-encode**: Next, the cloud uses $\mathbf{D}_i$ to re-encode the encoded data blocks which range from $\sum_{j=1}^{i} \sigma_j - \sigma_i + 1$ to $\sum_{j=1}^{i} \sigma_j$. $\mathbf{P}_h^{\nabla} = \mathbf{D}_i \circledast \mathbf{P}_h$ is the hth *recoded encoded block*.
- **Communication**: The cloud sends $\mathbf{P}_h^{\nabla}$ to hth edge device w_h, $\forall h \in \{1, \ldots, k'\}$, and computes the *recovered matrices* $\mathbf{D}^{-1} = \{\mathbf{D}_1^{-1}, \ldots, \mathbf{D}_s^{-1}\}$, in which $\mathbf{D}_i^{-1}$ is the inverse matrix of $\mathbf{D}_i$. Finally, the cloud sends the encoding matrix $\mathbf{B}$, recovered matrices $\mathbf{D}^{-1}$ and recoding parameters σ to the user.

[3] Assume that $\frac{ml}{k}$ can be divisible by k_i, *i.e.*, no additional zero row is added into the encoded data block. Therefore, the process of re-encoding will not lead to the differences in the rows of different encoded data blocks.

Noted that, the whole process of the cloud can be pre-processed before the task starts. Next, we give the work of **edge device**.

- **Communication and computation**: For ith edge device w_i, it receives the recoded encoded block $\mathbf{P}_i^{\nabla}$ from the cloud and the user matrix $\mathbf{X}$ from the user. Then compute the multiplication between $\mathbf{P}_i^{\nabla}$ and $\mathbf{X}$ and send the intermediate result $\mathbf{I}_i = \mathbf{P}_i^{\nabla}\mathbf{X}$ to the user.

The work of edge devices has no difference from the basic CEC. If w_i is reliable, $\mathbf{I}_i = \mathbf{P}_i^{\nabla}\mathbf{X}$. Otherwise, $\mathbf{I}_i$ is the incorrect intermediate result. Finally, we show the work of the **user** and describe how to obtain the correct final result. The algorithm of the user is also given in Algorithm 2.

- **Communication**: The user first receives the encoding matrix $\mathbf{B}$, recovered matrices $\mathbf{D}^{-1}$ and recoding parameters σ from the cloud. Then, send $\mathbf{X}$ to all edge devices and receive intermediate results from them.
- **Recover the intermediate results**: For the intermediate result $\mathbf{I}_i$ from w_i, it recovers $\mathbf{I}_i$ with jth recovered matrix, in which $\sum_{h=1}^{j}\sigma_h - \sigma_j + 1 \leq i \leq \sum_{h=1}^{j}\sigma_h$. Let $\mathbf{I}_i^{\triangle} = \mathbf{D}_j^{-1} \circledast \mathbf{I}_i$ be the ith *recovered intermediate result* and the set of recovered intermediate results be $\bar{\mathbf{I}}^{\triangle}$. $|\bar{\mathbf{I}}^{\triangle}| \leq s$.
- **Verification**: When $|\bar{\mathbf{I}}^{\triangle}| \geq k$, the user can decode final results until obtaining two same final results. In the hth round of decoding, select a new set of k recovered intermediate results to compose to $\mathbf{I}_h'$ and their encoding vectors to compose to $\mathbf{B}_h'$. Compute the decoding matrix $\mathbf{B}_h'^{-1}$ and decode the final result $\mathbf{Y}_h = \mathbf{B}_h'^{-1} \circledast \mathbf{I}_h'$. If there exists a final result $\mathbf{Y}_j = \mathbf{Y}_h, (1 \leq j < h)$, $\mathbf{Y}_j$ and $\mathbf{Y}_h$ are correct final results.

3.3 Successful Verification Probability

In this subsection, we will analyze the successful verification probability $P(S)$, *i.e.*, the probability that the proposed RDC verification scheme can verify the correctness of two incorrect final results.

To further analyze whether the RDC verification scheme can against collusion attack, we assume that the incorrect final results are obtained based on two to-be-decoded results $\mathbf{I}_1'^{*}$ and $\mathbf{I}_2'^{*}$, each of which contains k incorrect intermediate results that have been recovered. $\mathbf{B}_1'^{-1}$ and $\mathbf{B}_2'^{-1}$ are their decoding matrices. Let $\mathbf{C} = \mathbf{B}_2'\mathbf{B}_1'^{-1}$. When $\mathbf{B}_1'^{-1} \circledast \mathbf{I}_1'^{*} = \mathbf{B}_2'^{-1} \circledast \mathbf{I}_2'^{*}$, they will lead to the same incorrect final results.

If there are $c(0 \leq c \leq k-1)$ common intermediate results which returned by same edge devices in both $\mathbf{I}_1'^{*}$ and $\mathbf{I}_2'^{*}$, without loss of generality, let the first c intermediate results of $\mathbf{I}_1'^{*}$ and $\mathbf{I}_2'^{*}$ are common intermediate results. Then, we can have the following equation about $\mathbf{I}_i'^{*}$, $\mathbf{I}_i'$ and $\mathbf{R}_i'$ $\forall i \in \{1, 2\}$:

$$\mathbf{I}_i'^{*} = \begin{bmatrix} \mathbf{I}_{i,1}'^{\triangle} \\ \vdots \\ \mathbf{I}_{i,c}'^{\triangle} \\ \mathbf{I}_{i,c+1}'^{\triangle} \\ \vdots \\ \mathbf{I}_{i,k}'^{\triangle} \end{bmatrix}, \mathbf{I}_i' = \begin{bmatrix} \mathbf{I}_{i,1}' \\ \vdots \\ \mathbf{I}_{i,c}' \\ \mathbf{I}_{i,c+1}' \\ \vdots \\ \mathbf{I}_{i,k}' \end{bmatrix}, \mathbf{R}_i' = \begin{bmatrix} \mathbf{R}_{i,1}' \\ \vdots \\ \mathbf{R}_{i,c}' \\ \mathbf{R}_{i,c+1}' \\ \vdots \\ \mathbf{R}_{i,k}' \end{bmatrix}, \tag{3}$$

in which $\mathbf{I}_{i,j}'^{\triangle}$ is the incorrect intermediate result that has been recovered, $\mathbf{I}_{i,j}'$ is the corresponding correct intermediate result that should be returned by attackers and $\mathbf{R}_{i,j}'$ is the block which has been added to $\mathbf{I}_{i,j}'$. Let the recovered matrix of $\mathbf{I}_{i,j}'^{\triangle}$ be $\mathbf{D}_{i,j}^{-1}$, therefore we have $\mathbf{I}_{i,j}'^{\triangle} = \mathbf{D}_{i,j}^{-1} \circledast (\mathbf{I}_{i,j}' + \mathbf{R}_{i,j}')$. When $j \leq c$, we also have $\mathbf{I}_{1,j}'^{*} = \mathbf{I}_{2,j}'^{*}$, $\mathbf{I}_{1,j}' = \mathbf{I}_{2,j}'$ and $\mathbf{R}_{1,j}' = \mathbf{R}_{2,j}'$.

Let $\mathbf{C} = \begin{bmatrix} \boldsymbol{\lambda} \\ \mathbf{C}' \end{bmatrix} \begin{matrix} \}c \\ \}d \end{matrix}$, in which $d = k - c (1 \leq d \leq k)$ means the number of intermediate results which returned by different edge devices in $\mathbf{I}_1'^{*}$ and $\mathbf{I}_2'^{*}$. Next, we analyze the structure of matrix $\mathbf{C}$, *i.e.*, the structure of the $c \times k$ dimensional matrix $\boldsymbol{\lambda}$ and the $d \times k$ dimensional matrix $\mathbf{C}'$. Let $\boldsymbol{\lambda} = [\boldsymbol{\lambda}_1^\top, \ldots, \boldsymbol{\lambda}_c^\top]^\top$ and $\mathbf{C}' = [\mathbf{C}_1'^\top, \ldots, \mathbf{C}_d'^\top]^\top$, in which $\boldsymbol{\lambda}_i$ and $\mathbf{C}_i'$ are the row vectors with length of k.

Let $\mathbf{E}_c$ be the $c \times c$ dimensional identity matrix and $\mathbf{0}_{d \times c}$ be the $d \times c$ dimensional zero matrix. The top c rows of $\mathbf{C}$, *i.e.*, $\boldsymbol{\lambda}$, has the following structure:

Lemma 3. $\boldsymbol{\lambda} = \begin{bmatrix} \mathbf{E}_c \\ \mathbf{0}_{d \times c} \end{bmatrix}^\top$.

Proof. In basic CEC, for the correct intermediate results of $\mathbf{I}_1'$ and $\mathbf{I}_2'$, no matter what $\mathbf{I}_1'$ and $\mathbf{I}_{2,i}'$ are, they must satisfy that $\boldsymbol{\lambda}_i \circledast \mathbf{I}_1' = \mathbf{I}_{2,i}', \forall i \in \{1, \ldots, c\}$. Since $\mathbf{I}_{1,i}'$ and $\mathbf{I}_{2,i}'$ are common intermediate results which from the same edge device, we also have $\mathbf{I}_{1,i}' = \mathbf{I}_{2,i}'$. Therefore, $\boldsymbol{\lambda}$ must be $\begin{bmatrix} \mathbf{E}_c \\ \mathbf{0}_{d \times c} \end{bmatrix}^\top$.

Lemma 4. $\mathbf{C}'$ *is a* $d \times k$ *dimensional matrix without zeros.*

Proof. If there exist some zeros in $\mathbf{C}_i'$, $\mathbf{C}_i'$ has less than k non-zero elements. Since $\mathbf{C}\mathbf{B}_1' = \mathbf{B}_2'$, $(c+i)$th row vector of $\mathbf{B}_2'$ is the linear combinations of less than k row vectors of $\mathbf{B}_1'$. Since any k row vectors of $\mathbf{B}$ are linearly independent, all row vectors of $\mathbf{B}$ are the linear combinations of k other row vectors of $\mathbf{B}$. Since $\mathbf{B}_1'$ and $\mathbf{B}_2'$ is the sub-matrix of $\mathbf{B}$, $(c+i)$th row vector of $\mathbf{B}_2'$ is impossible to be the linear combinations of less than k row vectors of $\mathbf{B}_1'$. Therefore, $\mathbf{C}'$ must be a $d \times k$ dimensional matrix without zeros.

Assume that $\mathbf{C}_{i,j}'$ is the jth element of the row vector $\mathbf{C}_i'$. The sufficient and necessary condition that $\mathbf{B}_1'^{-1} \circledast \mathbf{I}_1'^{*} = \mathbf{B}_2'^{-1} \circledast \mathbf{I}_2'^{*}$ follows:

Lemma 5. *$B_1'^{-1} \circledast I_1'^{*} = B_2'^{-1} \circledast I_2'^{*}$ iff*

$$R'_{2,c+i} = \sum\nolimits_{j=1}^{k} C'_{i,j} D_{2,c+i} \circledast D_{1,j}^{-1} \circledast R'_{1,j}, \forall i \in \{1, \ldots, d\}.$$

Proof. $\mathbf{B}_1'^{-1} \circledast \mathbf{I}_1'^{*} = \mathbf{B}_2'^{-1} \circledast \mathbf{I}_2'^{*}$ equals to $\mathbf{C} \circledast \mathbf{I}_1'^{*} = \mathbf{I}_2'^{*}$, which is also the following two equations:

$$\mathbf{I}_{2,i}'^{\triangle} = \boldsymbol{\lambda}_i \circledast \mathbf{I}_1'^{*}, \quad \forall i \in \{1, \ldots, c\}. \tag{4}$$

$$\mathbf{I}_{2,c+i}'^{\triangle} = \mathbf{C}'_i \circledast \mathbf{I}_1'^{*}, \quad \forall i \in \{1, \ldots, d\}, \tag{5}$$

Since $\boldsymbol{\lambda} = \begin{bmatrix} \mathbf{E}_c \\ \hdashline \mathbf{0}_{d \times c} \end{bmatrix}^{\top}$ and $\mathbf{I}_{1,i}'^{\triangle} = \mathbf{I}_{2,i}'^{\triangle}, \forall i \in \{1, \ldots, c\}$, Eq. (4) is always true.

There must have $\mathbf{D}_{2,c+i}^{-1} \circledast \mathbf{I}'_{2,c+i} = \sum_{j=1}^{k} \mathbf{C}'_{i,j} \mathbf{D}_{1,j}^{-1} \circledast \mathbf{I}'_{1,j}$. Then, Eq. (5) can be simplified to

$$\begin{aligned}
&\mathbf{I}_{2,c+i}'^{\triangle} = \mathbf{C}'_i \circledast \mathbf{I}_1'^{*} \\
\Leftrightarrow &\mathbf{D}_{2,c+i}^{-1} \circledast (\mathbf{I}'_{2,c+i} + \mathbf{R}'_{2,c+i}) = \sum\nolimits_{j=1}^{k} \mathbf{C}'_{i,j} \mathbf{D}_{1,j}^{-1} \circledast \mathbf{I}_{1,j}'^{\triangle} \\
\Leftrightarrow &\mathbf{D}_{2,c+i}^{-1} \circledast (\mathbf{I}'_{2,c+i} + \mathbf{R}'_{2,c+i}) = \sum\nolimits_{j=1}^{k} \mathbf{C}'_{i,j} \mathbf{D}_{1,j}^{-1} \circledast (\mathbf{I}'_{1,j} + \mathbf{R}'_{1,j}) \\
\Leftrightarrow &\mathbf{D}_{2,c+i}^{-1} \circledast \mathbf{R}'_{2,c+i} = \sum\nolimits_{j=1}^{k} \mathbf{C}'_{i,j} \mathbf{D}_{1,j}^{-1} \circledast \mathbf{R}'_{1,j} \\
\Leftrightarrow &\mathbf{R}'_{2,c+i} = \sum\nolimits_{j=1}^{k} \mathbf{C}'_{i,j} \mathbf{D}_{2,c+i} \circledast \mathbf{D}_{1,j}^{-1} \circledast \mathbf{R}'_{1,j}
\end{aligned}$$

The condition that two incorrect final results are same is given in Lemma 5. Next, we will analyze the suitable range of recoding parameter σ_i. To make that attackers cannot design same incorrect final results,

Lemma 6. *$1 \leq \sigma_i \leq k, \forall i \in \{1, \ldots, s\}$.*

Proof. When $\sigma_i \geq k+1$, there exist two to-be-decoded results, each of which obtained based on k incorrect intermediate results that have the same recoding matrices. Therefore, for them, the condition in Lemma 5 becomes

$$\begin{aligned}
&\mathbf{R}'_{2,c+i} = \sum\nolimits_{j=1}^{k} \mathbf{C}'_{i,j} \mathbf{D}_{2,c+i} \circledast \mathbf{D}_{1,j}^{-1} \circledast \mathbf{R}'_{1,j} \\
\Leftrightarrow &\mathbf{R}'_{2,c+i} = \sum\nolimits_{j=1}^{k} \mathbf{C}'_{i,j} (\mathbf{D}_{2,c+i} \circledast \mathbf{D}_{1,j}^{-1}) \circledast \mathbf{R}'_{1,j} \\
\Leftrightarrow &\mathbf{R}'_{2,c+i} = \sum\nolimits_{j=1}^{k} \mathbf{C}'_{i,j} \mathbf{R}'_{1,j}
\end{aligned}$$

In Sect. 3.1, obviously that attackers can make $\mathbf{R}'_{2,c+i} = \sum_{j=1}^{k} \mathbf{C}'_{i,j} \mathbf{R}'_{1,j}$. Therefore, there exist at least two same incorrect final results when $\sigma_i \geq k+1$.

When $1 \leq \sigma_i \leq k$, the analysis will be given in Theorem 2, which shows the probability that two incorrect final results are same. Since the probability is very low, attackers cannot make sure that incorrect final results are same.

Finally, we give the successful verification probability of the RDC verification scheme against collusion attack:

Theorem 2. *When $t \geq k+1$ and $1 \leq \sigma_i \leq k, \forall i \in \{1, \ldots, s\}$, the successful verification probability of the RDC verification scheme $P(S) \geq 1 - \frac{1}{\left(\sum_{j=1}^{\frac{ml}{k}} \binom{\frac{ml}{k}}{j} (q-1)^j\right)^d}$.*

Proof. Since attackers cannot obtain the recovered matrices $\mathbf{D}$, they cannot infer the linear relationship between different intermediate results to satisfy Lemma 5. Therefore, attackers can only randomly modify the intermediate results to make the incorrect final results same.

Assume that $\mathbf{R}'_{2,c+i}$ contains $j (1 \leq j \leq \frac{ml}{k})$ non-zero elements in the finite field $\mathbb{F}_q$. Therefore, $\mathbf{R}'_{2,c+i}$ has totally $\sum_{j=1}^{\frac{ml}{k}} \binom{\frac{ml}{k}}{j} (q-1)^j$ selections. However, $\mathbf{R}'_{2,c+i}$ has at most one situation to equal to $\sum_{j=1}^{k} \mathbf{C}'_{i,j} \mathbf{D}_{2,c+i} \circledast \mathbf{D}_{1,j}^{-1} \circledast \mathbf{R}'_{1,j}$. Therefore, the probability that $\mathbf{R}'_{2,c+i} = \sum_{j=1}^{k} \mathbf{C}'_{i,j} \mathbf{D}_{2,c+i} \circledast \mathbf{D}_{1,j}^{-1} \circledast \mathbf{R}'_{1,j}$ is no more than $\frac{1}{\sum_{j=1}^{\frac{ml}{k}} \binom{\frac{ml}{k}}{j} (q-1)^j}$.

Since i ranges from 1 to d, the probability that $\mathbf{B}_1'^{-1} \circledast \mathbf{I}_1'^{*} = \mathbf{B}_2'^{-1} \circledast \mathbf{I}_2'^{*}$ is no more than $\prod_{i=1}^{d} \frac{1}{\sum_{j=1}^{\frac{ml}{k}} \binom{\frac{ml}{k}}{j} (q-1)^j}$. Finally, we can get the successful verification probability that $P(S) \geq 1 - \prod_{i=1}^{d} \frac{1}{\sum_{j=1}^{\frac{ml}{k}} \binom{\frac{ml}{k}}{j} (q-1)^j} = 1 - \frac{1}{\left(\sum_{j=1}^{\frac{ml}{k}} \binom{\frac{ml}{k}}{j} (q-1)^j\right)^d}$.

Remark 4. *In the DC verification scheme, attackers can infer the linear relationship between different intermediate results because any k of them can be decoded into the same final result. However, the RDC verification scheme re-encodes the encoded data blocks so that attackers cannot design efficient attack method to obtain two same incorrect final results.*

Remark 5. *According to Theorem 2, when the finite filed q and the size of intermediate result $\frac{ml}{k}$ are sufficiently large, $P(S)$ will always approach to 1 and incorrect final results are almost impossible to be same and verified to be correct. Therefore, the RDC verification can efficiently against the collusion attack.*

3.4 Verification Overheads

In this subsection, we will analyze the verification overheads of the RDC verification scheme. Specifically, we focus on the additional storage, computation and communication overheads led by the RDC verification scheme. The verification overheads do not contain the basic CEC and the rounds of decoding and comparisons, because the DC verification scheme has already given them [22].

For the storage, computation and communication overheads, we take one element as the storage unit, one multiplication between two elements as the computation unit and transferring one element as the communication unit. Next, we analyze the additional verification overheads of the RDC verification scheme and summarize them in Table 2.

Table 2. The additional verification overheads of the RDC verification scheme.

	Storage	Computation	Communication
Cloud	$\sum_{i=1}^{s}(k_i)^2 + s$	$\sum_{i=1}^{s}\left(\frac{\sigma_i k_i mn}{k} + (k_i)^3\right)$	$\sum_{i=1}^{s}(k_i)^2 + s$
Edge device	0	0	0
User	$\sum_{i=1}^{s}(k_i)^2 + s$	$\leq \sum_{i=1}^{s} \frac{\sigma_i k_i ml}{k}$	0

The cloud should generate s recoding matrices and recoding parameters before re-encoding k' encoded data blocks. Therefore, s recoding matrices and recoding parameters should be storaged and sent to the user. The storage and communication overheads are all $\sum_{i=1}^{s}(k_i)^2 + s$. The computation overhead of re-encoding k' encoded data blocks and computing the recovered matrices is $\sum_{i=1}^{s}\left(\frac{\sigma_i k_i mn}{k} + (k_i)^3\right)$. Since edge devices only compute the matrix multiplication between $\mathbf{P}_i^{\nabla}$ and $\mathbf{X}$, they will lead to no addition verification overheads.

For the user, it should storage s recovered matrices and recoding parameters, the storage overhead is $\sum_{i=1}^{s}(k_i)^2 + s$. For k' intermediate results, the computation overhead of recovering them is no more than $\sum_{i=1}^{s} \frac{\sigma_i k_i ml}{k}$.

From Table 2, we know that verification overheads are related to the parameters s and k_i. Next, we will give the range of s.

Lemma 7. $\lceil \alpha \rceil \leq s \leq k'$.

Proof. From Lemma 6, we have $1 \leq \sigma_i \leq k$. Since $1 \leq \sigma_i \leq k$, $\sum_{i=1}^{s} = k'$ and s is an integer, we have $\frac{k'}{k} \leq s \leq \frac{k'}{1}$, *i.e.*, $\lceil \alpha \rceil \leq s \leq k'$.

Remark 6. *When k_i is very small, attackers can guess the recoding matrices, which may has the higher successful attack probability than randomly modifying intermediate results. Due to the limit of paper space, this kind of attack method and the range of k_i will be analyzed in the future work.*

Since the whole process of the cloud can be pre-processed and only needs to be done once, additional verification overheads of the cloud can be ignored. Moreover, there is no additional overhead at the edge devices. Therefore, only the storage and computation overheads are required at the user device during the computation task.

The computation overhead that the user locally computes the matrix multiplication between $\mathbf{A}$ and $\mathbf{X}$ is mnl. The storage overhead of $\mathbf{A}$ and $\mathbf{X}$ is $mn+nl$. In most machine learning based algorithms, the parameters about the size of matrix, *i.e.*, m, n and l, may even be tens of thousands, while the parameters about the RDC verification scheme, *i.e.*, s, σ_i and k_i can be at most tens. Therefore, $\sum_{i=1}^{s}(k_i)^2 + s \ll mn + nl$ and $\sum_{i=1}^{s} \frac{\sigma_i k_i ml}{k} \ll mnl$.

Finally, we can summarize that, utilizing the distributed computing power of edge computing and the RDC verification scheme, the computation task can be speed up efficiently.

4 Simulation

In this section, we conduct extensive simulation experiments to evaluate the performance of proposed RDC verification scheme under different system parameters. Specifically, we focus on two evaluation criteria: (1) successful verification probability $P(S)$, *i.e.*, the probability that the verification scheme can obtain the correct final result; (2) verification time consumption, *i.e.*, the time that the user can obtain the final result and finish the verification. For each point shown in the simulation figures, we generate 1000 simulation instances to show the average value. The program is implemented based on Java and carried out on the computer with Intel Core i5-9300H CPU 2.40 GHz.

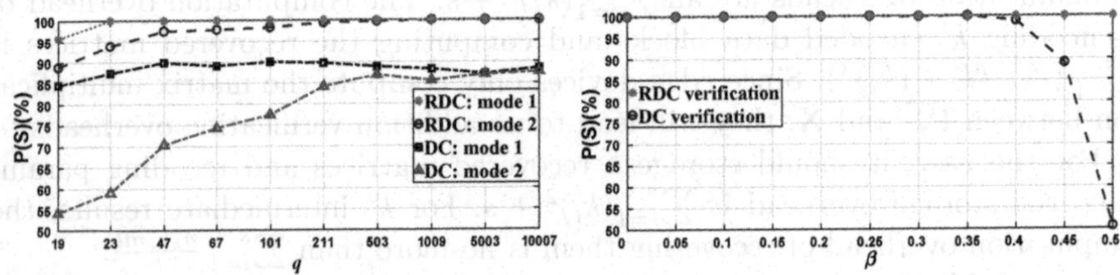

(a) Successful probability when changing q. (b) Successful probability when changing β.

Fig. 2. Successful verification probability when changing q and β.

We mainly compare the RDC verification scheme with two baseline schemes: (1) *Decode-and-Compare* (DC) verification scheme, which is proposed in [22][4]; (2) *Localize Computation* (LC): the user performs the computation task locally, which is also secure and reliable. When $\beta \geq \frac{k+1}{\alpha k}$ (shown in Lemma 1), any k incorrect intermediate results can be decoded into the same incorrect final result under the case without recoding. Otherwise, attackers randomly modify the intermediate results. All elements of incorrect intermediate results are modified unless otherwise specified.

In the simulations, we consider the following system parameters:

- q: the size of finite filed.
- m, n and l: sizes of the data matrix $\mathbf{A}$ and user matrix $\mathbf{X}$.
- α: the coding redundancy ratio, $\alpha = \frac{k'}{k}$.
- β: the attacker ratio, there are $\beta k'$ incorrect intermediate results.
- k': the number of total edge devices.

The default values of them are $q = 1009$, $m = 1000$, $n = 5000$, $l = 10$, $\alpha = 4.0$, $\beta = 0.45$ and $k' = 20$. Moreover, we also set parameters $s = \max\{4, \lceil\alpha\rceil\}$ and $k_i = 2$, which means that the user has little storage overheads to storage the recovered matrices and recoding parameters (only storage $5s$ elements).

[4] The DC verification scheme has the successful verification probability of nearly 100% and the improvement of more than $10^3\times$ than homomorphic encryption on the verification time consumption when attackers are independent.

4.1 Simulation Results

In the first group of simulations, we compare the successful verification probability of the DC verification scheme and RDC verification scheme under different values of q and β. In Fig. 2(a), we set two attack modes:

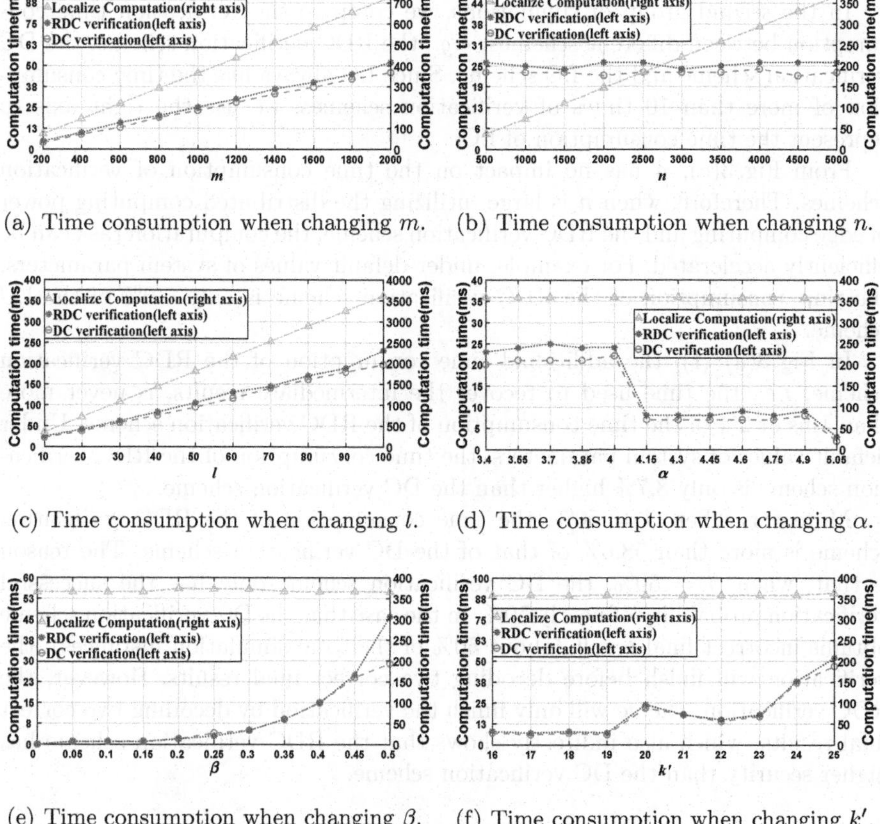

(a) Time consumption when changing m. (b) Time consumption when changing n.

(c) Time consumption when changing l. (d) Time consumption when changing α.

(e) Time consumption when changing β. (f) Time consumption when changing k'.

Fig. 3. Time consumption when changing different system parameters.

- mode 1: attackers modify all elements of intermediate results;
- mode 2: attackers only modify one element of intermediate results.

As shown in Fig. 2(a), it is obvious that the successful verification probability of the RDC verification scheme is always higher than that of the DC verification scheme. The successful verification probability of both verification schemes will increase with the increase of q. Moreover, under attack mode 2, the successful verification probability of both verification schemes will be lower. However, when q is large, the successful verification probability of both two attack modes will fluctuate in a fixed range. For example, when $q \geq 503$, the successful verification probability of the RDC verification scheme is always 100% and that of the DC verification scheme is always about 88.0%.

In Fig. 2(b), when β increases, the successful verification probability of the DC verification scheme will decrease, which is as analyzed in Theorem 1. However, even when $\beta = 50\%$, the successful verification probability of the RDC verification scheme is still 100%.

In the second group of simulations, we compare the verification time consumption between different schemes, *e.g.*, the RDC verification scheme, the DC verification scheme and the LC scheme. Since LC always has the time consumption of more than 10 times of verification schemes, we use the right axis to represent the time consumption of LC.

From Fig. 3(c), n has no impact on the time consumption of verification schemes. Therefore, when n is large, utilizing the distributed computing power of edge computing and the RDC verification scheme, the computation task can be efficiently accelerated. For example, under default values of system parameters, the time consumption of the RDC verification scheme is only 6.9% of the LC scheme.

In Fig. 3(a)–(f), the additional time consumption of the RDC verification scheme, *i.e.*, the time used to recover the intermediate results, is never more than 1ms or 2% of the time consumption of the RDC verification scheme. Under default values of system parameters, the time consumption of the RDC verification scheme is only 8.7% higher than the DC verification scheme.

Moreover, when $\beta = 50\%$, the time consumption of the RDC verification scheme is more than 58.6% of that of the DC verification scheme. The reason is that, when $\beta = 50\%$, the DC verification scheme only has the successful verification probability of nearly 50%. In the case that the DC verification scheme obtains incorrect final result (about 50% of the total simulation instances), the verification will finish before decoding two correct final results. However, the RDC verification scheme will only finish the verification by decoding two correct final results, which also indirectly shows that the RDC verification scheme has higher security than the DC verification scheme.

4.2 Simulation Conclusions

From the simulation results, we have the following conclusions:

- The simulations shown in Fig. 2 demonstrate that the RDC verification scheme has better potential to against the collusion attack than the DC verification scheme. By selecting a large finite filed q, the successful verification probability of the RDC verification scheme can be 100%, no matter what the attacker ratio or attack mode is.
- The simulations in Fig. 3 demonstrate that the RDC verification scheme has no more than 10% additional time consumption than the DC verification scheme. However, the RDC verification scheme has the higher security (successful verification probability) than the DC verification scheme. Therefore, the RDC verification scheme is more suitable for the verification in CEC.

– When n is large, although the user can perform localize computation to make sure the final result is correct, the time consumption is about 14 times higher than the RDC verification scheme. Since the RDC verification scheme can efficiently against the collusion attack and edge computing can reduce the runtime of computation, the RDC verification scheme can be applied in CEC to optimise the computation task.

5 Conclusion

In this paper, we studied the verification issue of collusion attack in coded edge computing. Since the existing scheme cannot efficiently against the collusion attack, we proposed the Recode-Decode-and-Compare (RDC) verification scheme. By re-encoding the encoded data, attackers cannot infer the linear relationship between different intermediate results. Then, the user can obtain the correct final result by comparing different final results. We also conducted solid theoretical analyses to show the successful verification probability and verification overheads. Finally, extensive simulation experiments demonstrate that the proposed RDC verification scheme can efficiently against the collusion attack and outperforms the state-of-the-art works.

Acknowledgment. This work is supported in part by National Natural Science Foundation of China (62072321, 61972272), Six Talent Peak Project of Jiangsu Province (XYDXX-084), China Postdoctoral Science Foundation (2020M671597), Jiangsu Postdoctoral Research Foundation (2020Z100), Suzhou Planning Project of Science and Technology (SNG2020073, SS202023, SYG202024), Tang Scholar of Soochow University, Collaborative Innovation Center of Novel Software Technology and Industrialization, and Soochow University Interdisciplinary Research Project for Young Scholars in the Humanities.

References

1. Qiu, T., Chi, J., Zhou, X., Ning, Z., Atiquzzaman, M., Wu, D.: Edge computing in industrial internet of things: architecture, advances and challenges. IEEE Commun. Surv. Tut. **22**(4), 2462–2488 (2020)
2. Wang, T., et al.: Privacy-enhanced data collection based on deep learning for Internet of vehicles. IEEE Trans. Industr. Inf. **16**(10), 6663–6672 (2020)
3. Liu, Y., Peng, M., Shou, G., Chen, Y., Chen, S.: Toward edge intelligence: multiaccess edge computing for 5G and Internet of things. IEEE Internet Things J. **7**(8), 6722–6747 (2020)
4. Wu, Y., Huang, H., Wu, N., Wang, Y., Bhuiyan, M.Z.A., Wang, T.: An incentive-based protection and recovery strategy for secure big data in social networks. Inf. Sci. **508**, 79–91 (2020)
5. Mao, Y., You, C., Zhang, J., Huang, K., Letaief, K.: A survey on mobile edge computing: the communication perspective. IEEE Commun. Surv. Tut. **19**(4), 2322–2358 (2017)
6. Li, S., Maddah-Ali, M., Avestimehr, A.: Coding for distributed fog computing. IEEE Commun. Mag. **55**(4), 34–40 (2017)

7. Li, S., Maddah-Ali, M., Yu, Q., Avestimehr, A.: A fundamental tradeoff between computation and communication in distributed computing. IEEE Trans. Inf. Theor. **64**(1), 109–128 (2018)
8. Lee, K., Lam, M., Pedarsani, R., Papailiopoulos, D., Ramchandran, K.: Speeding up distributed machine learning using codes. IEEE Trans. Inf. Theor. **64**(3), 1514–1529 (2018)
9. Prakash, S., et al.: Coded computing for low-latency federated learning over wireless edge networks. IEEE J. Sel. Areas Commun. **39**(1), 233–250 (2021)
10. Dutta, S., Cadambe, V., Grover, P.: Short-dot: computing large linear transforms distributedly using coded short dot products. IEEE Trans. Inf. Theor. **65**(10), 6171–6193 (2019)
11. Yu, Q., Maddah-Ali, M., Avestimehr, A.: Polynomial codes: an optimal design for high-dimensional coded matrix multiplication. In: Advances in Neural Information Processing Systems (NIPS), pp. 4406–4416. MIT Press (2017)
12. Yang, H., Lee, J.: Secure distributed computing with straggling servers using polynomial codes. IEEE Trans. Inf. Forensics Secur. **14**(1), 141–150 (2019)
13. Wang, J., Cao, C., Wang, J., Lu, K., Jukan, A., Zhao, W.: Optimal task allocation and coding design for secure edge computing with heterogeneous edge devices. IEEE Trans. Cloud Comput. (2019, early access)
14. Bitar, R., Parag, P., El Rouayheb, S.: Minimizing latency for secure coded computing using secret sharing via staircase codes. IEEE Trans. Commun. **68**(8), 4609–4619 (2020)
15. Bhattad, K., Narayanan, K.: Weakly secure network coding. In: 1st Workshop on Network Coding Theory and Applications (NETCOD), pp. 281–285 (2005)
16. Zhang, L.F., Safavi-Naini, R.: Verifiable delegation of computations with storage-verification trade-off. In: Kutyłowski, M., Vaidya, J. (eds.) ESORICS 2014. LNCS, vol. 8712, pp. 112–129. Springer, Cham (2014). https://doi.org/10.1007/978-3-319-11203-9_7
17. Song, W., Wang, B., Wang, Q., Shi, C., Lou, W., Peng, Z.: Publicly verifiable computation of polynomials over outsourced data with multiple sources. IEEE Trans. Inf. Forensics Secur. **12**(10), 2334–2347 (2017)
18. Halevi, S., Shoup, V.: Faster homomorphic linear transformations in HElib. In: Annual International Cryptology Conference, pp. 93–120 (2018)
19. Zhang, Y., Zheng, Z., Lyu, M.: BFTCloud: a Byzantine fault tolerance framework for voluntary-resource cloud computing. In: IEEE International Conference on Cloud Computing, pp. 444–451. IEEE (2011)
20. Milosevic, Z., Biely, M., Schiper, A.: Bounded delay in Byzantine-tolerant state machine replication. In: IEEE International Symposium on Reliable Distributed Systems, pp. 61–70. IEEE (2013)
21. Rui, G., Rodrigo, R., Nuno, P.: Efficient middleware for Byzantine fault tolerant database replication. In: European Conference on Computer Systems (EuroSys), pp. 107–122. ACM (2011)
22. Fu, M., Wang, J., Wang, J., Lu, K., Jukan, A., Gu, F.: Decode-and compare: an efficient verification scheme for coded edge computing. In: IEEE/ACM International Symposium on Quality of Service (IWQoS), pp. 1–6. IEEE/ACM (2020)
23. Wang, Z., Cheung, S., Luo, Y.: Information-theoretic secure multi-party computation with collusion deterrence. IEEE Trans. Inf. Forensics Secur. **12**(4), 980–995 (2017)
24. Tando, R., Lei, Q., Dimakis, A., Karampatziakis, N.: Gradient coding: avoiding stragglers in distributed learning. In: International Conference on Machine Learning (ICML), pp. 3368–3376. ACM (2017)

MGFL: Multi-granularity Federated Learning in Edge Computing Systems

Shangxuan Cai, Yunfeng Zhao, Zhicheng Liu, Chao Qiu, Xiaofei Wang(✉), and Qinghua Hu

College of Intelligence and Computing, Tianjin University, Tianjin 300350, China
{shangxuancai,yfzhao97,liuzhicheng,chao.qiu,xiaofeiwang, huqinghua}@tju.edu.cn

Abstract. As a promising machine learning framework in the big data era, federated learning (FL) allows multiple mobile devices to collaboratively train a model without transmitting raw data, thus has attracted widespread attention in both academia and industry. Considering that heterogeneous mobile devices with limited resources and data diversity are bound to impact the actual performance of some training nodes. However, conventional FL could not support collaborative training with multi-granularity neural networks. To this end, we propose multi-granularity federated learning (MGFL) that contains two mechanisms serving for same-granularity FL and cross-granularity FL. MGFL customizes a personalized model for each device by designing a divergence-based similarity measurement method in same-granularity FL. Further, it adjusts the empirical risk loss function to break the restriction of cross-granularity FL. Experimental evaluations demonstrate the positive guidance of the fine-granularity model to the coarse-granularity model, which significantly improves the performance of the coarse-granularity model. Besides, our method shows superiority on both independently identically distribution (IID) and non-IID data.

Keywords: Federated learning · Edge computing · Multi-granularity model · Personalized model

1 Introduction

The Internet of Things gets rapidly development since the increased demand for convenience in life. Its network applications, such as mobile heterogeneous electric vehicle networks in smart city [1] and wireless sensor and actuator networks [2], catalyzes enormous amounts of data from edge devices every second. Since these data contain rich and valuable information, they have attracted wide attention from researchers in recent years. Considering that these messages often need more complex calculations to decode the valuable parts of them, some practical technologies are applied to edge computing [3–5]. As an indispensable universal service for big data processing [6], deep learning (DL) promotes the intelligence

Y. Lai et al. (Eds.): ICA3PP 2021, LNCS 13155, pp. 549–563, 2022.
https://doi.org/10.1007/978-3-030-95384-3_34

of mobile devices under the influence of the popularity of deep neural network (DNN) models, which have become an essential part of edge computing systems.

However, there is generally insufficient data on one mobile device to sustain training a high-quality model. Data are often owned by different entities, which means that individual's data privacy is compromised if centralized improperly [7,8]. Thus, privacy and data integrity become essential points for application or data owners [9]. With this in mind, individuals are reluctant to interact and share data to train the model. Federated learning (FL) is proposed to alleviate privacy anxiety and achieve cooperative training among numerous edge devices (a.k.a. clients) [10]. In recent years, FL has drowned extensive discussions for its superior ability on data privacy protection. The general FL system permits all clients to keep their data locally and only upload parameters or gradients to collaboratively train models under the coordination of a central server.

Kairouz *et al.* [11] discussed the existing problems and future development of FL, in addition to which there are still many bottlenecks limiting the applications of FL in edge computing. Conventional FL only allows all clients to share one global model, which is not suitable for heterogeneous clients. One factor of heterogeneity is the diversity of data distribution that makes the aggregated model fail to apply to each participant. Another factor is the constrained resources, including storage, computing power, energy, and other such things. Those heterogeneous clients with constrained resources fail to train DL models ideally, resulting in each client being forced to get a model with low performance.

As coarse-granularity models typically require fewer training resources than fine-granularity models, a sensible approach is to run DL models with coarse granularity if the client does not impose a requirement for granularity. E.g., a fine-granularity model generally runs on resource-rich clients to identify cars, boats, men, women, cats, dogs, etc. In contrast, a coarse-granularity model runs on resource-constrained clients to identify coarse-granularity individuals, such as vehicles, people, animals, etc. The heterogeneous clients in edge computing systems have constrained resources and diverse data, leading to multi-granularity models.

As client diversity is not conducive to the regular cooperation among multiple clients, many researchers have attempted to solve this dilemma in FL. Some studies try to design a personalized model for each client [12] and address the challenge of different data distributions among clients. The works in [13,14] select relevant and irrelevant clients according to the data distributions, and works in [15,16] aggregate similar client models through the idea of clustering. Another work in [17] considers the condition of the heterogeneous resource and non-IID data in clients, aiming to solve the selection problem of clients. However, these works only focus on the differences in data samples and lack consideration of similar samples with different labels. The studies in [18,19] analyze the collaboration among multiple tasks, and the work in [20] further explores the non-IID data effect based on multi-task. However, they all ignore the situation of multi-granularity tasks. Although data labels are different among multi-granularity models, which is contrary to the principle of FL, the similarity among samples is the same as the original intention of FL that aims to make up for the lack of samples among clients.

To address the challenges above, we propose a novel FL method, named multi-granularity federated learning (MGFL), aiming to promote the cooperative training among models with different granularity. MGFL adopts the same-granularity FL to provide a personalized model for each client and adopts cross-granularity FL with adjusting the empirical risk loss function, which further improves the models' performance of coarse-granularity clients.

Our method focuses on the scenario of multi-granularity clients while performing significantly on non-IID data. The main contributions of this paper are summarized as follows:

- We propose a novel FL method named MGFL, which contains two mechanisms, i.e., same-granularity FL and cross-granularity FL, promoting FL among models of different granularity clients in edge computing systems.
- We design a divergence-based similarity measurement method to provide a personalized model for each client in same-granularity FL. Under the guidance of fine-granularity models, we improve models' accuracy of coarse-granularity clients by adjusting the empirical risk loss function in cross-granularity FL.
- We conduct extensive experiments demonstrating that MGFL is more effective for adaptive FL with multi-granularity clients. It shows better performance on both IID and non-IID data than other baselines and achieves great performance by accurately capturing relationships among clients.

The remainder of this paper is organized as follows. Section 2 shows the motivation of this work. Section 3 presents the system model and the proposed MGFL mechanism. Extensive performance evaluations are presented in Sect. 4. Section 5 concludes this paper.

2 Motivation

We compare the model differences between coarse-granularity and fine-granularity clients when performing image classification tasks. Since parameters are basic representations of the models, we Visualize the differences in models' image feature extraction capabilities with different granularity by comparing parameters. We choose two models with different granularity based on pretrained Wide-Resnet [21] as observation objects and compare all layers except the final dense layer.

Let models A and B be trained under the CIFAR-100 dataset [22] with the same samples and different granularity labels, i.e., coarse-granularity and fine-granularity labels, respectively. We measure the distance between the output vectors of the two models at each layer using the Euclidean distance, which indicates the difference in the parameters of each layer. As shown in Fig. 1(a), there are significant differences in the convolutional layer parameters between the coarse-granularity and the fine-granularity model. In particular, as the network layers deepen, the difference in parameters between layers increases. It illustrates that there are significant differences between models with different granularity even trained with the same samples.

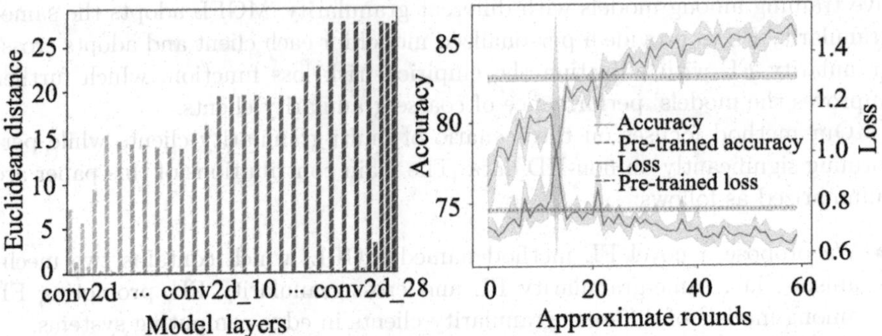

Fig. 1. Two key observations for motivating cross-granularity FL: (a) Euclidean distance of all layers between coarse-granularity model A and fine-granularity model B; (b) Model A's performance during the process of approximating to B's parameters.

Next, we attempt to make the coarse-granularity model A's parameters approximate to those of the fine-granularity model B and test the models' accuracy and loss after several rounds of modifications to the parameters. The results are shown in Fig. 1(b), where the dashed and solid lines represent the performance of the initial pre-trained and the approximating models, respectively. The performance of the coarse-granularity model gradually improves as the parameters continue to approach those of the fine-granularity model and eventually surpass the performance of the pre-trained model. The result show the positive guiding effect of fine-granularity models to coarse-granularity models in feature-extracting of convolution layers and lay the foundation of cross-granularity FL.

Based on the above two observations, we expect to overcome the limitation of conventional FL that only trains the same models and collaborates to train coarse-granularity and fine-granularity models with widely varying model parameters. On the other hand, we aim to give full play to the guidance of the fine-granularity model to the coarse-granularity model during FL to further improve the coarse-granularity model's performance. Therefore, we propose the MGFL method to satisfy the collaborative training among multi-granularity clients in edge computing systems.

3 System Model

In this section, we design MGFL to realize FL among multi-granularity clients containing similar model structures and samples.

3.1 The Multi-granularity Federated Learning Framework

As shown in Fig. 2, extensive clients at the edge are managed by server and divided into coarse-granularity clients $\mathcal{N} = \{1, 2, ..., N\}$ and fine-granularity clients $\mathcal{M} = \{1, 2, ..., M\}$. At the beginning of MGFL, each client sends its model weight as well as a part of its shared data and accuracy of recent rounds to the server. The aggregation phase is divided into two stages dominated by same-granularity FL and cross-granularity FL, respectively. In the early stage, the same granularity models are measured by a similarity method based on the Jensen-Shannon divergence (JS divergence) that urges same-granularity models to aggregate the new personalized models. In the later stage, the fine-granularity clients perform same-granularity FL while coarse-granularity clients perform cross-granularity FL with the guidance of fine-granularity clients. In cross-granularity FL, each coarse-granularity model selects the most relevant fine-granularity model as the target to guide it and then modify parameters with the help of fine-granularity one. The server sends personalized models back to each client, and clients will start the next round of local training. The entire training is iterated until each client receives the highest-performing model. The corresponding algorithm is shown in Algorithm 1. Assume that the system is in a quasi-static state during a round, which means no clients join or leave during a round.

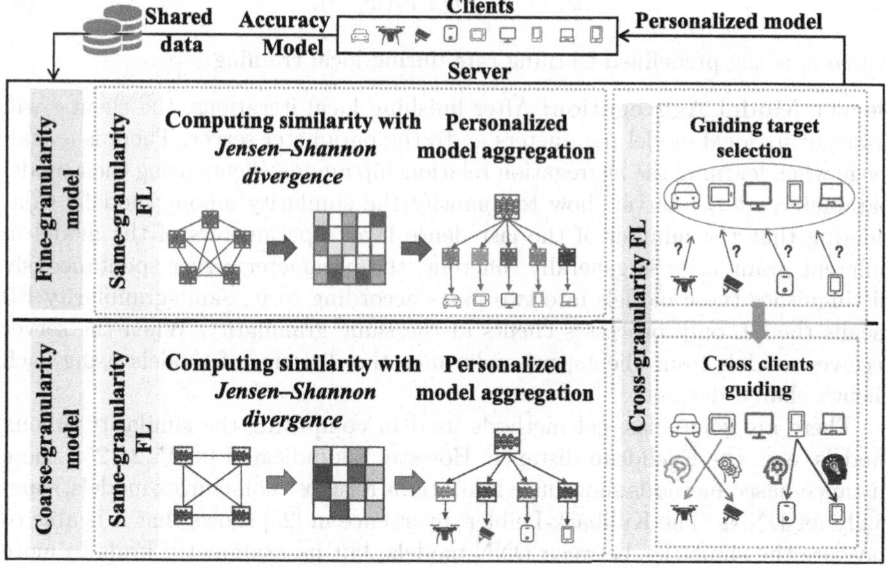

Fig. 2. The framework of MGFL.

3.2 Same-Granularity Federated Learning

One of the challenges of designing MGFL is the non-IID data kept locally by the client, which severely affects the quality of aggregation due to differences in sample classification. To decrease the impact caused by it and design personalized models on non-IID data [23], the early stage of MGFL captures the relationship among clients using a similarity-based method. By taking the coarse-granularity clients as an example, the problem at the early stage is formulated as follows:

$$\min_{W} \sum_{i=1}^{N} \frac{1}{N_i} \sum_{j=1}^{N_i} F_i(w_i, x_{i,j}, y_{i,j}), \tag{1}$$

where N is the number of total involved clients. N_i is the number of training data samples in client i. $(x_{i,j}, y_{i,j}) \in \mathbb{R}^E \times \mathbb{R}$ is the j-th training data pair of client i where E is the number of data features; $W = [w_1, ..., w_N]^{\mathrm{T}}$ is the weight matrix to be estimated consisting of all the local models' weight. F_i is the loss function of the i-th client model.

Local Model Computation: Each client i will parallelly optimize (e.g., using stochastic gradient descent methods) its model weight w_i based on its local data (x_i, y_i). In the t-th local iteration, each client computes its local update, i.e.,

$$w_i^t = w_i^t - \eta_l \nabla F_i(w_i^{t-1}), \tag{2}$$

where η_l is the predefined learning rate during local training.

Server Model Aggregation: After finishing local iterations, the client i will transmit its local model parameters w_i to the parameter server. There is a challenge when learning the aggregation relationship among clients using their model parameters at the server: how to quantify the similarity among models. Considering that the number of the last dense layer's parameters of the model in different granularity is generally different, the parameter server spontaneously distinguishes these models into two parts according to it. Same-granularity FL means the FL only processes clients in the same granularity. When the server receives models from clients, we endurance the distance of models using each client's shared dataset.

There are some classical methods used in computing the similarity among models, e.g., the Euclidean distance. However, as indicated in [15,24,25], these distance-based methods show huge limitations for most non-convex models, especially for DNNs. The Kullback-Leibler divergence in [25] shows that it is able to measure the similarity between DNN models, but its asymmetry leads to more complex optimization. We choose the similarity measurement method based on the vector from the output layer to address these key points to compute the matching relationship between the current model and others. JS divergence is used to measure how one probability distribution is different from the other and is widely used in similarity measurement. The JS divergence of two DNN models (w_i, w_j) can be expressed as:

Algorithm 1: MGFL under T iterations

Input: Initial models of all the clients $w^0_{i\in\mathcal{N}\cup\mathcal{M}}$ with local iteration $t=0$.
Output: Personalized models $w_{i\in\mathcal{N}\cup\mathcal{M}}$.

for *t=1,2,...,T* **do**
  for *each client* $i \in \mathcal{N}\cup\mathcal{M}$ **do**
    Client i solves local problem (2) and derive w_i. (**Local model computation**)
  end
  All the clients transmit their updated $w^t_{i\in\mathcal{N}\cup\mathcal{M}}$ and the local model accuracy $P_{i\in\mathcal{N}}$ to the server.
  After server receiving $w^t_{i\in\mathcal{N}}$, $w_{i\in\mathcal{N}}$ can be obtained by
  if *in the early stage* **then**
    calculating (6). (**Same-granularity FL**)
  else
    if $t\%\theta == 0$ **then**
      calculating (9). (**Same-granularity FL**)
    else
      calculating (9). (**Cross-granularity FL**)
    end
  end
  Server calculates (6) after receiving $w^t_{i\in\mathcal{M}}$ and obtains $w_{i\in\mathcal{M}}$. (**Server model aggregation**)
  The server broadcasts $w_{i\in\mathcal{N}\cup\mathcal{M}}$ to clients such that $w^t_i = w_i$.
end

$$
\begin{aligned}
JSD(w_i||w_j) &= \frac{1}{2}P_i \log\frac{2P_i}{P_i+P_j} + \frac{1}{2}P_j \log\frac{2P_j}{P_i+P_j},\\
P_i = g(w_i, x^S_i),&\quad P_j = g(w_j, x^S_i),
\end{aligned}
\tag{3}
$$

where x^S_i is the shared data of client i. $g(w_i, x^S_i)$ denotes the output of model w_i on the input data x^S_i. A smaller JS divergence indicates the more similar the two models are.

We then obtain a JS divergence matrix $D_{JS} \in \mathbb{R}^{N\times N}$, with $D_{JS}(i,j) = JSD(w_i||w_j)$. In order to construct the similarity matrix D, we first normalize each row of the JS divergence matrix and get the normalized matrix $\hat{D}_{JS}$, i.e.,

$$
\hat{D}_{JS}(i,:) = \frac{(I - D_{JS}(i,:)) - \min\{I - D_{JS}(i,:)\}}{\max\{I - D_{JS}(i,:)\} - \min\{I - D_{JS}(i,:)\}},
\tag{4}
$$

where I is a N-dimensional vector with all elements being 1. The similarity matrix are defined as the dot product of the normalized matrix and its transpose, which is shown as:

$$
D = \sqrt{\hat{D}_{JS}\cdot \hat{D}^{\mathrm{T}}_{JS}}.
\tag{5}
$$

Although the JSD is symmetric in principle, the computed $\hat{D}_{JS}$ is not entirely symmetric, which is not conducive to reaching consensus on the client-side to make the aggregated model more generalized. Therefore, a practical approach is to use the inner product operation to create the ultimate similarity matrix completely symmetric. The server will aggregate the same-granularity models from clients according to the similarity matrix D to get new personalized models and return them to the clients for the next round of training. The aggregated results are shown below.

$$W = D \times W^t. \tag{6}$$

Similarly, the fine-granularity clients also perform same-granularity FL with execution details similar to those described above.

3.3 Cross-Granularity Federated Learning

Compared with the coarse-granularity model, the fine-granularity model is the higher level participant containing mature and complex knowledge, which results in the practical and logical cross direction of different granularity being fine to coarse. During the later stage of training, the server performs cross-granularity FL to facilitate clients' collaboration between different granularity, promoting more extraordinary performance for coarse-granularity models. We formulate the coarse-granularity model problem in the later stage as follows:

$$\begin{gathered}\min_{W} \sum_{i=1}^{N} \frac{1}{N_i} \sum_{j=1}^{N_i} F_i(w_i, x_{i,j}, y_{i,j}) + \lambda \phi_i(w_i, x_{i,j}^S, y_{i,j}^S), \\ \phi_i(w_i, x_{i,j}^S, y_{i,j}^S) = \sum_{r=1}^{|x_i^S|} \frac{\| \sigma(w_i, x_{i,r}^S) - \sigma(w_{h(i)}, x_{i,r}^S) \|^2}{| x_i^S |},\end{gathered} \tag{7}$$

where $w_{h(i)}$ is the model of the fine-granularity client $h(i)$ that is selected to assist the coarse-granularity client i for its model performance improvement, $\sigma(\cdot)$ means the DNN models' output before the last dense layer. The second term $\phi_i(w_i, x_i^S, y_i^S)$ calculates the difference between the output vector before the coarse-granularity and the fine-granularity model's dense layer.

The premise of addressing the problem of Eq. (7) is to find out the correspondence between the coarse-granularity model and the fine-granularity model, i.e., how to ensure the value of $h(i)$ in Eq. (7). The optimal solution can be expressed as follows:

$$\begin{gathered}h(i) = \operatorname*{arg\,max}_{m \in \mathcal{M},\ \epsilon(w_m, x_i^S, y_i^S) - A_i > 0} [\epsilon(w_m, x_i^S, y_i^S) - A_i], \\ \epsilon(w_m, x_i^S, y_i^S) = \frac{\sum_{j=1}^{|x_i^S|} \mathbb{I}_{\{Hg(w_m, x_{i,j}^S) = y_{i,j}^S\}}}{|x_i^S|}, \quad m \in \mathcal{M},\end{gathered} \tag{8}$$

where H is the prior knowledge matrix that contains the correspondence between coarse-granularity classes and fine-granularity classes. It is possible to transform fine-granularity classes into coarse-granularity classes using the matrix H.

$\epsilon(w_m, x_i^S, y_i^S)$ represents the conversion accuracy of the m-th fine-granularity model on the i-th coarse-granularity client's shared dataset. The function $\mathbb{I}(\cdot)$ is the indicator function and A_i is the accuracy of the coarse-granularity client i on the local dataset, which has been sent to the server.

To avoid redundant guidance, the coarse-granularity model only selects one fine-granularity model with the most prominent performance help for federated training at a time. The coarse-granularity model performs iterative weight updates in the server as:

$$w_i = w_i^t - \eta_s \nabla \phi_i(w_i, x_i^S, y_i^S), \tag{9}$$

where η_s is the learning rate in the server update process.

It is worthy to notice that the coarse-granularity model only updates parameters except for the dense layer and retrain the dense layer using shared data before returning it to the clients. The cross-granularity FL process only performs in later stage of the system, whose goal is to ensure that the fine-granularity models are mature enough to guide the coarse-granularity models.

4 Performance Evaluation

In this section, we evaluate the performance of MGFL and compare it with other methods. Our experimental evaluations focus on three aspects: performance on different data settings, improvement by cross-granularity FL, and the relationship between the same and cross-granularity models.

4.1 Dataset and Experiment Settings

We consider the CIFAR-100 dataset, which contains common things in life divided into 100 and 20 classes, corresponding to fine-granularity labels and coarse-granularity labels. Thus, each coarse-granularity class includes five fine-granularity classes. The dataset contains about 50000 training samples and 10000 testing samples uniformly distributed among coarse-granularity classes and fine-granularity classes.

In the evaluations, we set $N = 5$, $M = 5$ and number the clients under each granularity as 0–4. For coarse-granularity clients and fine-granularity clients with the same order (e.g., coarse-granularity client 0 and fine-granularity client 0), we set the same data distribution to facilitate the observation of relationship in cross-granularity FL. We adopt Wide-Resnet as the DNN model with multiple convolutional layers to serve for an image classification task. All evaluation results are implemented on Tensorflow 2.4.1 deployed on a server with Intel(R) Gold 6226R CPU, NVIDIA GeForce RTX 3090, and Windows Server 2019.

To compare the performance of MGFL and other methods on both IID and non-IID data, we apply different data settings. 1) IID data: each client holds the same distribution data; 2) non-IID data: each client holds samples of different 5–6 classes. It is assumed that clients hold the same distribution data belonging to a group, and the number of groups ranges from 2 to 5. Considering that tasks with different granularity in real life are usually deployed on different parties' clients, some parties that have similar data distribution while others have pretty different data distribution. That leads to the emergence of groups, so the data setting 2) is more common in practice. By default, when the confusion degree is 2, clients in $\{0, 1, 2\}$ are a group with the same data distribution, and the other group consists of clients in $\{3, 4\}$. Similarly, the three groups are $\{0, 1\}$, $\{2, 3\}$ and $\{4\}$ at confusion degree 3, the four groups are $\{0, 1\}$, $\{2\}$, $\{3\}$ and $\{4\}$ at confusion degree 4, and each client is a group at confusion degree 5.

4.2 Experiment Results

Performance on Different Data Settings: We first evaluate the performance of MGFL under different data settings and compare it with other methods. Under the non-IID data, we set four confusion degrees *C2*, *C3*, *C4* and *C5* by dividing all clients into 2, 3, 4, and 5 groups. E.g., if all clients with the same granularity are divided into two groups, each contains 5–6 coarse-granularity classes or 25–30 fine-granularity classes. In that case, the confusion degree is two and is denoted by *C2*. We choose the best mean test accuracy (BMTA) [13] as the performance indicator defined as the highest mean test accuracy achieved overall communication rounds of training. Figure 3 shows the performance of coarse-granularity and fine-granularity models using four methods, MGFL, Alone, Cosine, and FedAvg [10], respectively, under training 300 epochs. The Cosine method replaces the JS divergence of MGFL with the cosine distance, and the Alone method only trains the model locally without FL.

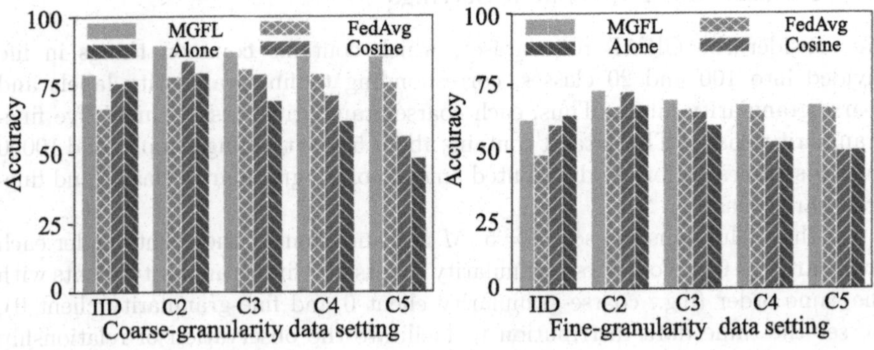

Fig. 3. BMTA of four methods under (a) coarse granularity and (b) fine granularity at different confusion degrees.

As shown in Fig. 3, MGFL exhibits superior performance on both coarse-granularity and fine-granularity clients, especially on the non-IID data. Since FedAvg was originally designed for clients with the same distribution, it is able to achieve high accuracy on IID data and performs poorly on non-IID data. MGFL shows high performance close to FedAvg on IID data and far better than it on non-IID data. Since the Cosine method uses the cosine distance commonly used in distance-based personalized FL [13,15], it captures the similarity among clients in low confusion degrees and has some advantages. However, it gradually loses its impact in the face of high confusion degrees. The Alone method shows the opposite trend to the Cosine method in that it performs the worst at the confusion degree 2. Moreover, it offers better performance as the confusion degree rises, indicating that the more complex the data, the more difficult it is to perform FL.

Performance Improvement by Cross-Granularity FL: Figure 4 shows the performance variation curves of coarse-granularity models in Group 1 and Group 2 at confusion degree 2. As shown in the figure, the clients' performance in both Group 1 and Group 2 using the MGFL method shows a significant increase and convergences to higher accuracy in the later stage than the other methods. The cross-granularity FL causes that in the later stage, where the fine-granularity model guides the coarse-granularity model resulting in a significant performance improvement.

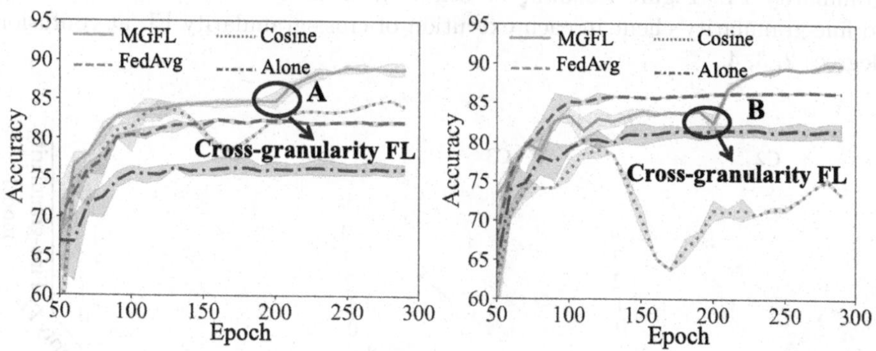

Fig. 4. The performance of coarse-granularity models in (a) Group 1 and (b) Group 2 at confusion degree 2.

It is worth noting that the performance of coarse-granularity models first degrades slightly at the beginning of the cross-granularity FL, as shown in Fig. 4 points A and B, due to improper coordination and incompatibility. That also shows that the parameters of the fine-granularity model are not fully applicable to the coarse-granularity model, so the parameters of the coarse-granularity model do not be directly replaced by those of the fine-granularity model. Even parameter approximation (shown in Fig. 1) requires many rounds of iterations to eliminate the incompatibility and improve performance. In the later stage, fine-granularity models perform the same-granularity FL as in the early stage except for guiding coarse-granularity models. As shown in Fig. 5, the overall performance of MGFL is better than other methods and has minor performance fluctuation in both Group 1 and Group 2.

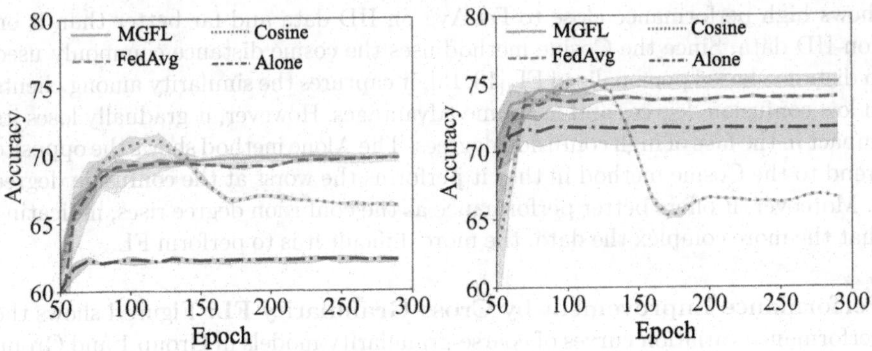

Fig. 5. The performance of fine-granularity models in (a) Group 1 and (b) Group 2 at confusion degree 2.

Relationship in the Cross-Granularity FL: Selecting the most relevant fine-granularity model for the coarse-granularity model is a critical step in the cross-granularity FL. Figure 6 shows the selection of each coarse-granularity client to fine-granularity client in each execution of cross-granularity FL at confusion degrees 2, 3, 4.

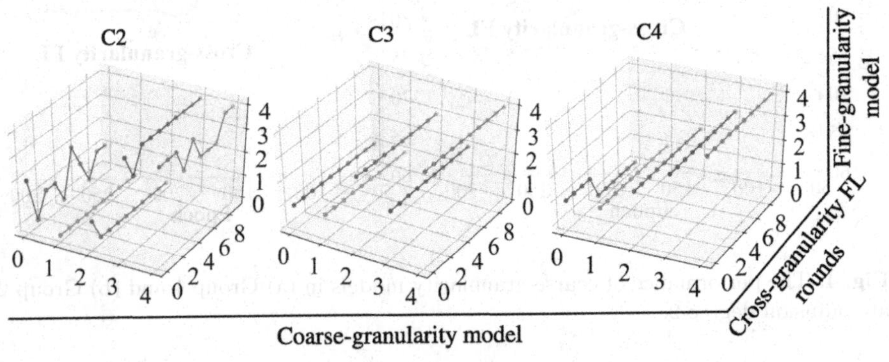

Fig. 6. Relationships in cross-granularity FL at different confusion degrees.

As shown in the figure, MGFL can precisely choose the most similar fine-granularity model to guide fine-granularity one in each cross-granularity FL. E.g., coarse-granularity client 0 at *C2* selects the clients in group {0, 1, 2} due to keeping the same data distribution. Though selection becomes more complex as the number of groups increases, MGFL has the ability to overcome the complexity and make the right choices. Another observation based on the results is that the smaller the confusion degree, the more fluctuating the clients' selection when performing the cross-granularity FL. Compared to the results at confusion degree 4, the coarse-granularity clients at confusion degree 2 fluctuate considerably with different clients selected in each cross-granularity FL. The more the number of groups, the fewer the number of clients in the group and the fewer the number of clients to select from, resulting in more minor fluctuations in selection.

Relationship in the Same-Granularity FL: We aim to leverage the inherent group relationship learned by MGFL to select similar models for aggregation in the same-granularity FL while improving the overall accuracy performance. We visualized the similarity matrix of MGFL and Cosine methods in the early rounds at different confusion degrees, exhibiting the group structure. As shown in Fig. 7, compared with the Cosine method, the group structure in our framework is generally learned precisely in the early rounds, which can be interpreted as having closer similar values within the same group. E.g., the nine values in the upper left corner of MGFL at *C2* are superior to the nine values in the upper left corner of Cosine at *C2*. Besides, our method captures the same-granularity relationship more accurately, especially as the confusion degree increases.

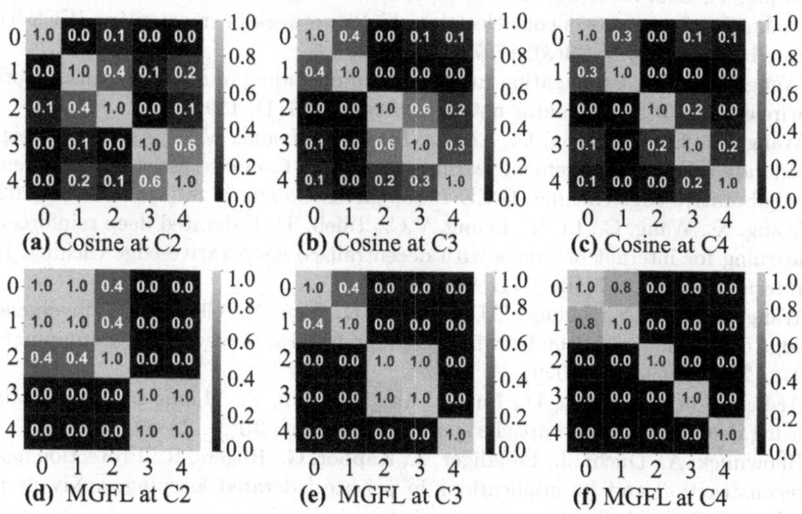

Fig. 7. Relationships in same-granularity FL using (a)–(c) the Cosine and (d)–(f) the MGFL methods at different confusion degrees.

Although learning group relationships become more difficult as the degree of confusion increases, MGFL can overcome this difficulty and learn relationships more accurately than the Cosine method. Due to the limitation of the length of the paper, we only show the results of the coarse-granularity models, and the behaviors of the fine-granularity models are similar to that of the coarse-granularity model.

5 Conclusion

In this paper, we address the challenge of collaborative training among multi-granularity clients. Specifically, we develop a novel method named MGFL that introduces two mechanisms for fine-granularity clients and coarse-granularity clients. MGFL designs personalized models for different-granularity clients using a divergence-based similarity method in same-granularity FL and breaks the restriction cross-granularity FL by adjusting the empirical risk loss function. Extensive experiments show that our method out-performs other baselines on both IID and Non-IID data and proves its superior performance even under confusing data distribution.

Acknowledgement. This work is supported by research on key technologies of electrical cloud-edge-end collaborative AI model sharing in Science and Technology Project of State Grid Headquarters (2021, Power base support technology - 30).

References

1. Wang, T., Luo, H., Zeng, X., Yu, Z., Liu, A., Sangaiah, A.K.: Mobility based trust evaluation for heterogeneous electric vehicles network in smart cities. IEEE Trans. Intell. Transp. Syst. **22**(3), 1797–1806 (2020)
2. Wang, T., et al.: Propagation modeling and defending of a mobile sensor worm in wireless sensor and actuator networks. Sensors **17**(1), 139 (2017)
3. Wang, X., Li, R., Wang, C., Li, X., Taleb, T., Leung, V.C.: Attention-weighted federated deep reinforcement learning for device-to-device assisted heterogeneous collaborative edge caching. IEEE J. Sel. Areas Commun. **39**(1), 154–169 (2020)
4. Wang, X., Wang, C., Li, X., Leung, V.C., Taleb, T.: Federated deep reinforcement learning for internet of things with decentralized cooperative edge caching. IEEE Internet Things J. **7**(10), 9441–9455 (2020)
5. Wang, X., Han, Y., Leung, V.C., Niyato, D., Yan, X., Chen, X.: Convergence of edge computing and deep learning: a comprehensive survey. IEEE Commun. Surv. Tut. **22**(2), 869–904 (2020)
6. Alsheikh, M.A., Niyato, D., Lin, S., Tan, H., Han, Z.: Mobile big data analytics using deep learning and apache spark. IEEE Netw. **30**(3), 22–29 (2016)
7. Bhowmick, A., Duchi, J., Freudiger, J., Kapoor, G., Rogers, R.: Protection against reconstruction and its applications in private federated learning. arXiv preprint arXiv:1812.00984 (2018)
8. Bonawitz, K., et al.: Practical secure aggregation for federated learning on user-held data. arXiv preprint arXiv:1611.04482 (2016)
9. Wang, T., Bhuiyan, M.Z.A., Wang, G., Qi, L., Wu, J., Hayajneh, T.: Preserving balance between privacy and data integrity in edge-assisted internet of things. IEEE Internet Things J. **7**(4), 2679–2689 (2019)
10. Hard, A., et al.: Federated learning for mobile keyboard prediction. arXiv preprint arXiv:1811.03604 (2018)
11. Kairouz, P., et al.: Advances and open problems in federated learning. arXiv preprint arXiv:1912.04977 (2019)
12. Tan, A.Z., Yu, H., Cui, L., Yang, Q.: Towards personalized federated learning. arXiv preprint arXiv:2103.00710 (2021)

13. Huang, Y., et al.: Personalized cross-silo federated learning on non-IID data. In: Proceedings of the AAAI Conference on Artificial Intelligence, vol. 35, pp. 7865–7873 (2021)
14. Nagalapatti, L., Narayanam, R.: Game of gradients: mitigating irrelevant clients in federated learning. In: Proceedings of the AAAI Conference on Artificial Intelligence, vol. 35, pp. 9046–9054 (2021)
15. Sattler, F., Müller, K.R., Samek, W.: Clustered federated learning: model-agnostic distributed multitask optimization under privacy constraints. IEEE Trans. Neural Netw. Learn. Syst. **32**, 3710–3722 (2020)
16. Ouyang, X., Xie, Z., Zhou, J., Huang, J., Xing, G.: ClusterFL: a similarity-aware federated learning system for human activity recognition. In: Proceedings of the 19th Annual International Conference on Mobile Systems, Applications, and Services, pp. 54–66 (2021)
17. Nishio, T., Yonetani, R.: Client selection for federated learning with heterogeneous resources in mobile edge. In: 2019 IEEE International Conference on Communications (ICC), ICC 2019, pp. 1–7 (2019)
18. Smith, V., Chiang, C.K., Sanjabi, M., Talwalkar, A.: Federated multi-task learning. arXiv preprint arXiv:1705.10467 (2017)
19. Liu, F., Wu, X., Ge, S., Fan, W., Zou, Y.: Federated learning for vision-and-language grounding problems. In: Proceedings of the AAAI Conference on Artificial Intelligence, vol. 34, pp. 11572–11579 (2020)
20. Corinzia, L., Beuret, A., Buhmann, J.M.: Variational federated multi-task learning. arXiv preprint arXiv:1906.06268 (2019)
21. Zagoruyko, S., Komodakis, N.: Wide residual networks. arXiv preprint arXiv:1605.07146 (2016)
22. Krizhevsky, A., Hinton, G., et al.: Learning multiple layers of features from tiny images (2009)
23. Zhao, Y., Li, M., Lai, L., Suda, N., Civin, D., Chandra, V.: Federated learning with non-IID data. arXiv preprint arXiv:1806.00582 (2018)
24. Ghosh, A., Hong, J., Yin, D., Ramchandran, K.: Robust federated learning in a heterogeneous environment. arXiv preprint arXiv:1906.06629 (2019)
25. Jiang, B., Pei, J., Tao, Y., Lin, X.: Clustering uncertain data based on probability distribution similarity. IEEE Trans. Knowl. Data Eng. **25**(4), 751–763 (2011)

Energy Efficient Priority-Based Task Scheduling for Computation Offloading in Fog Computing

Jiaying Yin[1], Jing Fu[2], Jingjin Wu[1](✉), and Shiming Zheng[1]

[1] Division of Science and Technology, BNU-HKBU United International College, Zhuhai, Guangdong, People's Republic of China
n830006230@mail.uic.edu.cn, jj.wu@ieee.org, 1630005069@alumni.uic.edu.cn
[2] School of Engineering, RMIT University, Melbourne, VIC, Australia
jing.fu@rmit.edu.au

Abstract. Fog computing offers a flexible solution for computational offloading for Internet of Things (IoT) services at the edge of wireless networks. It serves as a complement to traditional cloud computing, which is not cost-efficient for most offloaded tasks in IoT applications involving small-to-medium levels of computing tasks. Given the heterogeneity of tasks and resources in fog computing, it is vital to offload each task to an appropriate destination to fully utilize the potential benefit of this promising technology. In this paper, we propose a scalable priority-based index policy, referred to as the Prioritized Incremental Energy Rate (PIER), to optimize the energy efficiency of the network. We demonstrate that PIER is asymptotically optimal in a special case applicable for local areas with high volumes of homogeneous offloaded tasks and exponentially distributed task durations. In more general cases with statistically different offloaded tasks, we further demonstrate the improvement of PIER over benchmark policies in terms of energy efficiency and the robustness of PIER to different task duration distributions by extensive simulations. Our results show that PIER can perform better than benchmark policies in more than 78.6% of all simulation runs.

Keywords: Priority-based policies · Energy efficiency · Fog computing · Computational offloading · Internet of Things

1 Introduction

Fog computing, a term originally introduced by Cisco, is considered a refinement to the traditional cloud computing where intensive computational tasks are offloaded to the data centers in the central cloud by migrating part of cloud

The work described in this paper is supported by Student Interdisciplinary Research Fund from BNU-HKBU United International College, College Research Grant from BNU-HKBU United International College R201911, and Zhuhai Basic and Applied Basic Research Foundation Grant ZH22017003200018PWC.

Y. Lai et al. (Eds.): ICA3PP 2021, LNCS 13155, pp. 564–577, 2022.
https://doi.org/10.1007/978-3-030-95384-3_35

computing and storage capabilities to the edge of the network [10]. This concept is particularly suitable for the realization of the Internet of Things (IoT). In IoT applications such as smart cities, remote gaming, and instant healthcare, offloading intensive computational tasks from mobile terminals (MTs) is frequently required because of the limited computing power of the MTs. In the IoT era, *fog nodes* (FNs) at the network edge, including base stations, micro data centers, or other access devices, are usually considered more appropriate offloading destinations as the central cloud is often geographically far from the MTs. The high latency caused by propagation between the MT and the central cloud would be unacceptable for these delay-sensitive applications [12].

Another advantage of fog computing is higher energy efficiency [9]. The MTs are usually operated by batteries with a limited lifespan and thus are not suitable for most computational tasks. At the same time, the data centers in the central cloud are generally much less energy efficient for offloading tasks of small to moderate sizes. Furthermore, additional power consumption is required for long-distance transmissions between the MT and the central cloud. The FNs, closer to the MTs and connected to the electrical grid, offer a reliable platform for energy-efficient computations.

While FNs have higher computing and storage capabilities than MTs, the types of resources and the number of resources for each kind in an FN are still limited compared to the central cloud [9]. Meanwhile, the latency and the energy consumption for a particular task generated by an MT may vary if it is offloaded to different FNs. Two main factors causing the difference are the quality of the wireless connection between the MT and the FN and the availability of required resources in the FN. Therefore, from the perspective of an *Edge Infrastructure Provider* (EIP) who owns and operates the FNs, an effective task scheduling policy is essential to fully exploit the benefits of fog computing.

In this paper, we consider a heterogeneous fog computing architecture with different classes of tasks. The task classes are differentiated by the resources required for each task. At the same time, FNs have distinct energy profiles and different sets of resources that can execute the offloaded computing tasks from MTs. In addition, the allocation of offloading tasks is not only subject to the availability of computing resources in FNs but also constrained by the number of wireless channels between transmitting MT and FN. On the other hand, the central cloud can be considered as the last resort of offloading when all computing resources in FNs are occupied, as additional latency and power consumption would be incurred if a task is offloaded further. Our objective in this paper is to propose a task scheduling policy aiming to optimize the energy efficiency, namely minimizing the ratio of long-run average power consumption of the network to the long-run average throughput, subject to the availability of transmission channels and computing resources.

Existing work on computational offloading policies in fog computing mainly adopts a static optimization approach by considering a single instance of the network or assuming that the arrival pattern of tasks is fully known. It has been shown that, under such conditions, an efficient offloading policy could improve utilization of resources in FNs [2], reduce the energy consumption of the network [9], guarantee fairness among MTs in terms of transmission rate [3], improve the Quality of Experience of mobile users [8], or achieve an appropriate trade-off between the QoS requirement of users and energy-related cost of the EIP [15]. While static optimization approaches can restrict the complexity of the problem to a reasonable level, the main shortcoming of such methods is that the stochastic nature of network traffic is ignored. As most offloaded tasks by personal users have relatively short durations, the availability of computational resources is expected to change frequently over time. As a result, static optimization approaches are likely to miss the potential benefit gained by dynamically reusing the resources released by completed tasks. While more recent studies (e.g. [1,6]) proposed online algorithms that consider the stochastic factor, various important issues, such as the heterogeneity of MTs and FNs, concurrent constraints from a limited number of channels and resources, and non-linear cost functions, are still ignored. Such deviations may affect the practicability of proposed policies.

Motivated by the facts above, we propose a scalable and robust *priority-based index policy* in this paper, which makes real-time decisions on the offloading destination based on the instantaneous state of the network upon the initiation of an offloading task to maximize the energy efficiency of the network. We will demonstrate that the performance gap between the proposed policy and the optimal solution diminishes as the system size increases in a particular case appropriate for a small-area environment where the offloaded tasks are generated from densely distributed mobile devices, while both tasks and FNs are largely homogeneous (e.g., large sports events [16]). Such property is desirable for the problem studied in this paper, as offloading in fog computing generally involves a large-scale system with huge numbers of MTs and FNs. In more general cases, we will also demonstrate that the proposed policy outperforms benchmark policies adopted in existing work.

While similar ideas have been applied in our previous work [4,5,13], we now consider a tailor-made model for the offloading problem in fog computing, where a task may be offloaded to an FN at the network edge or the central cloud, subject to the availability of both computing resources and wireless transmission channels. We will also consider that the power consumption at the network edge is, in fact, a piece-wise function of the load, as an "idle power" would be needed to activate the hardware equipment supporting a resource group when at least one unit of the resources is occupied [9]. To summarize, the model we consider in this paper involves many practical features in task scheduling in fog computing and thus is significantly different from the fundamental models in existing studies.

We will describe the model in Sect. 2 and the stochastic optimization problem in Sect. 3. We then introduce the proposed scheduling policy in Sect. 4, present numerical results in Sect. 5, and give conclusions in Sect. 6.

2 Network Model

Let $\mathbb{R}_0$, $\mathbb{R}_+$ and $\mathbb{N}_+$ represent the sets of non-negative reals, positive reals and positive integers, respectively. Let $[N]$ represent a set $\{1, 2, \ldots, N\}$ for any $N \in \mathbb{N}_+$.

We consider a wireless network with orthogonal transmission channels[1], through which wireless connections between mobile terminals (MTs) and fog nodes (FNs) at the network edge are established. Computing tasks generated by MTs can be appropriately offloaded to FNs through the transmission channels. For the sake of presentation, we refer to all these storage and computing resources in FNs and the cloud as *abstracted resource components* (ARCs), which are classified in *groups* (referred as *ARC groups* thereafter) based on their functionalities and geographical locations. Denote K as the number of ARC groups at the network edge, and let $C_k \in \mathbb{N}_+$ represent the *capacity* of ARC groups $k \in [K]$, which is the total number of ARC units supporting the execution of offloaded tasks in the ARC group.

FNs are connected to the central cloud through wired cables, where tasks can be further offloaded. As the capacity of the cloud is much larger than the capacity of any ARC groups $k \in [K]$ in FNs [3], we assume that there are always sufficient resources in the cloud to support all types of offloaded tasks, that is, the cloud has infinite capacity.

Similarly, consider $J \in \mathbb{N}_+$ classes of offloaded tasks, classified by the differences in their locations of origin, application styles, and relevant requirements on computing resources. As mentioned earlier, an offloaded task can be completed at the network edge or further offloaded to the cloud. We refer to the tasks belonging to class $j \in [J]$ as j-tasks. Tasks of the same type can be those generated by MTs in the same area, with the same application styles, and has the same requirements on ARCs.

If a j-task is accommodated by ARC group $k \in [K]$, $w_{j,k} \in [C_k]$ units of ARCs will be occupied by this task. These occupied units will be released upon the completion of the task (that is, when the computational results are transmitted back to the respective MT) and can be reused by future tasks [7]. While, if a j-task cannot be served by ARC group k, because of mismatch in geographical locations or required functions, we set $w_{j,k} \to +\infty$ to prohibit j-tasks from being assigned there.

We divide the network edge into $L \in \mathbb{N}_+$ *destination areas*, and denote $\mathscr{K}_\ell \subset [K]$ as the set of ARC groups in destination area $\ell \in [L]$. We assume, without loss of generality, that all $\mathscr{K}_\ell$ are non-empty and mutually exclusive for different $\ell \in [L]$. For each sender-receiver pair implied by MTs generating j-tasks and ARC groups in destination area ℓ, the transmission rate is *i.i.d.* with a mean of $\mu_{j,\ell} \in \mathbb{R}_+$. We further denote $\Gamma_{j,\ell} \in \mathbb{N}_+$ as the number of wireless transmission channels for a (j, ℓ) pair (referred as (j, ℓ)-channels hereafter), and assume that each j-task offloaded to destination area ℓ occupies one (j, ℓ)-channel. As transmission

[1] Orthogonal wireless channel allocations eliminate intra-cell interference and utilize frequency spectrum resources more efficiently.

to the cloud must be via the network edge, a j-task offloaded to the cloud may choose an arbitrary destination area $\ell \in [L]$ and occupy one corresponding (j, ℓ)-channel. By this definition, it is impossible for a j-task to be offloaded when all (j, ℓ)-channels for all $\ell \in [L]$ are occupied by existing tasks.

We consider the situation where the computing powers at the network edge or in the cloud are sufficient to complete the tasks in a relatively short period. In this sense, the durations of computational operations of offloaded tasks are generally much shorter than the transmission time; that is, the processing time is dominated by the transmission time between the MT and the FN for offloaded tasks completed at the network edge, and transmission time between the MT and the central cloud via the network edge for tasks offloaded to the central cloud [3,15].

By the arguments above, the expected duration of j-tasks computed by an ARC group $k \in \mathscr{K}_\ell$ at the network edge are assumed to take independently and identically distributed random values, with a mean of $1/\mu_{j,\ell}$.

For offloading to the cloud, two transmission segments, namely from the MT to the network edge and from the network edge to the cloud, are needed. For notation consistency, we define a particular ARC group k_c, which has an infinite capacity $C_{k_c} \to +\infty$. As a cable backbone network usually connects the FNs and the cloud, we can assume that the transmission time between the network edge and the cloud, denoted by D_0, is the same for all tasks via all destination areas. As the number of channels is limited, tasks offloaded to the cloud will always transmit via the area equipped with the fastest available channel. Therefore, the effective mean service rate for tasks offloaded to the cloud via destination area ℓ is

$$\mu'_{j,\ell} = \frac{1}{\left(\frac{1}{\mu_{j,\ell}} + D_0\right)}.$$

We assume that the arrivals of tasks generated from user class $j \in [J]$ form a Poisson process with a mean rate $\lambda_j \in \mathbb{R}_+$. Power consumption of the network edge consists of *idle* and *operational* power. The idle power is the basic consumption incurred when an ARC group is activated (which means the hardware components supporting the ARC group must be powered on). At the same time, the operational power is consumed when ARC units are engaged in processing tasks [11]. The relevant components of ARCs, deployed in FNs such as micro BSs or embedded servers, can be de-activated when not in use for power saving, and activated upon the arrival of a task occupying at least one ARC unit [14][2]. In a dynamic system such as a wireless network, the amount of the operational power consumption is affected by the real-time loads of ARC groups. We let $\varepsilon_k^0 \in \mathbb{R}_+$ and $\varepsilon_k \in \mathbb{R}_+$ represent the amounts of the idle power and operational power consumption per computing unit of ARC group $k \in [K]$. Similar

[2] While the more general cases with non-negligible activation delay and power consumption can be addressed by integrating the vacation queuing model with activation cost and delay as in [14], they will complicate the analysis and we do not consider them in this paper due to the limited space.

power consumption models have been empirically justified and widely applied in existing research, where idle power and operational power values depend on specific hardware [9,14]. Detailed discussion about the power consumption of the network edge will be provided in Sect. 3. Similarly, we denote the power consumption of an ARC unit for computing a j-task in the cloud as $\bar{\varepsilon}_j \in \mathbb{R}_+$. As demonstrated in existing research, a larger amount of power consumption is required for transmission when a task is offloaded further [3]. Therefore, for all $j \in [J]$ and $k \in [K]$, we have $w_{j,k}\varepsilon_k < \bar{\varepsilon}_j$ by definition.

Finally, we introduce a *scaling parameter* $h \in \mathbb{N}_+$ to emphasize the scale of the network. We define $\lambda_j := h\lambda_j^0$ for all $j \in [J]$, $\varepsilon_k^0 := h\tilde{\varepsilon}_k^0$, $C_k := hC_k^0$ for all $k \in [K]$, and $\Gamma_{j,\ell} := h\Gamma_{j,\ell}^0$ for all $j \in [J], \ell \in [L]$ where $\lambda_j^0, \tilde{\varepsilon}_k^0, C_k^0, \Gamma_{j,\ell}^0 \in \mathbb{R}_+$. The assumption that λ_j, C_k, ε_k^0 and $\Gamma_{j,\ell}^0$ all increase proportionally with h is reasonable in practical scenarios in edge computing, that the capacity of wireless channels and ARC groups are designed just to meet the potential demand. Correspondingly, a large h is more appropriate for modelling the network environment in a densely populated area where tasks are generated more frequently.

3 Stochastic Optimization Problem

Define $[K]^* = [K] \cup \{k_c\}$ as the set of ARC groups in the network edge and the cloud. Denote $X_{j,k}(t) \in \mathbb{N}_0$, where $j \in [J]$, $k \in [K]^*$, as the number of j-tasks being served by ARC units in group k at time $t \geq 0$. Also, we let $Y_{j,\ell}(t)$ represents the number of occupied (j,ℓ)-channels at time t. Because of the limited capacities of ARC groups and finite number of transmission channels, these variables should satisfy

$$\sum_{j\in[J]} w_{j,k}X_{j,k}(t) \leq C_k,\ \forall k \in [K],\ t \geq 0, \tag{1}$$

$$\sum_{k\in\mathscr{K}_\ell} X_{j,k}(t) \leq \Gamma_{j,\ell},\ \forall j \in [J],\ \forall \ell \in [L],\ t \geq 0, \tag{2}$$

$$\sum_{k\in[K]^*} X_{j,k}(t) \leq \sum_{\ell\in[L]} \Gamma_{j,\ell},\ \forall j \in [J],\ t \geq 0. \tag{3}$$

Let $\boldsymbol{X}(t) = (X_{j,k}(t) : j \in [J], k \in [K]^*)$, and the state space of process $\{\boldsymbol{X}(t),\ t \geq 0\}$ (the set involves all possible values of $\boldsymbol{X}(t)$ for $t \geq 0$) be

$$\mathscr{X} = \prod_{j\in[J]} \prod_{k\in[K]^*} \{0, 1, \ldots, \mathcal{A}_k\}, \tag{4}$$

where $\prod$ represents Cartesian product and $\mathcal{A}_k$ represents an upper limit of the maximum number of requests that can be concurrently accommodated by ARC group $k \in [K]^*$. Specifically,

$$\mathcal{A}_k = \begin{cases} \min\left\{\lfloor \frac{C_k}{w_{j,k}} \rfloor, \Gamma_{j,\ell(k)}\right\}, & \text{if } k \in [K], \\ \sum_{j\in[J]} \sum_{\ell\in[L]} \Gamma_{j,\ell}, & \text{otherwise.} \end{cases}$$

Note that although $\mathscr{X}$ is larger than the set of possible values of $\boldsymbol{X}(t)$, the process $\boldsymbol{X}(t)$ will be further constrained by (1) to (3) in our optimization problem that will be defined later in this section.

Similarly, we define action variables $a_{j,k}(\boldsymbol{x}) \in \{0,1\}$, $\boldsymbol{x} \in \mathscr{X}$, as a function of the state space for each $j \in [J]$ and $k \in [K]^*$. If $a_{j,k}(\boldsymbol{x}) = 1$, then $w_{j,k}$ units of ARC in group k are selected to serve an incoming j-task when $\boldsymbol{X}(t) = \boldsymbol{x}$; and otherwise, ARC group k is not selected. Specifically, an incoming j-task when $\boldsymbol{X}(t) = \boldsymbol{x}$ means the first j-task coming after and exclude time t; that is, the process $\boldsymbol{X}(t)$ is defined as left continuous in $t \geq 0$.

We define the action space, which is the set involving all possible values of action variables, as

$$\mathscr{A} = \{0,1\}^{J\times(K+1)}. \tag{5}$$

Let $\boldsymbol{a}(\boldsymbol{x}) = (a_{j,k}(\boldsymbol{x}) : j \in [J], k \in [K]^*)$, $\boldsymbol{x} \in \mathscr{X}$. For any $\boldsymbol{a}(\boldsymbol{X}(t)) \in \mathscr{A}$ and $\boldsymbol{X}(t) \in \mathscr{X}$, it should further satisfy

$$\sum_{k\in[K]^*} a_{j,k}(\boldsymbol{X}(t)) \leq 1,\ \forall j \in [J]. \tag{6}$$

Since the probability of tasks simultaneously generated by users in the same class is assumed to be zero, it is reasonable to always select a single ARC group to serve an incoming j-task as described by (6).

A scheduling policy is determined by the action variables for all states $\boldsymbol{x} \in \mathscr{X}$. We add a superscript and rewrite the action variables as $a^{\phi}_{j,k}(\boldsymbol{x})$ representing the action variables under policy ϕ, which can then be considered as a mapping from $\mathscr{X}$ to $\mathscr{A}$. Similarly, since the stochastic process $\{\boldsymbol{X}(t), t \geq 0\}$ is conditioned on the underlying policy, we rewrite previously defined state variable $\boldsymbol{X}(t)$ as $\boldsymbol{X}^{\phi}(t)$. We let $\varPhi$ represents the set of all such policies ϕ.

The long-run average throughput and power consumption of the network system under a policy ϕ are given by (7) and (8), respectively.

$$\begin{aligned}\mathcal{L}^{\phi} := \lim_{T\to\infty} \frac{1}{T}\Bigg[&\sum_{j\in[J]}\sum_{k\in[K]} \int_0^T \mu_{j,l(k)} X^{\phi}_{j,k}(t)\, dt \\ &+ \sum_{j\in[J]}\sum_{\ell\in[L]} \int_0^T \mu'_{j,\ell}\left(Y_{j,\ell}(t) - \sum_{k\in\mathscr{K}_\ell} X^{\phi}_{j,k}(t)\right) dt \Bigg]\end{aligned} \tag{7}$$

$$\begin{aligned}\mathcal{E}^{\phi} := \lim_{T\to\infty} \Bigg[&\sum_{j\in[J]}\sum_{k\in[K]} \left(\frac{1}{T}\int_0^T \varepsilon_k w_{j,k} X^{\phi}_{j,k}(t)\, dt\right) + \sum_{k\in[K]}\left(\frac{1}{T}\int_0^T \varepsilon^0_k \varTheta(X^{\phi}_{j,k}(t))\, dt\right) \\ &+ \frac{1}{T}\sum_{j\in[J]}\int_0^T \left[\sum_{\ell\in[L]}\left(Y_{j,\ell}(t) - \sum_{k\in\mathscr{K}_\ell} X^{\phi}_{j,k}(t)\right) dt\right] \overline{\varepsilon}_j \Bigg]\end{aligned} \tag{8}$$

The two terms in (7) represent the long-run average throughput of the network edge and the central cloud, respectively. For (8), the first and second terms represent the long-run average operational and idle power at the network edge, the last term stands for the long-run average power consumption for computing tasks offloaded to the cloud. In addition, $\ell(k)$ represents the destination area $\ell \in [L]$ for $k \in \mathscr{K}_\ell$, and $\Theta(x)$ for $x \in \mathbb{R}$ is the Heaviside function, defined as

$$\Theta(x) = \begin{cases} 1, & \text{if } x > 0, \\ 0, & \text{otherwise.} \end{cases}$$

4 Scheduling Policy

We aim to optimize the *energy efficiency* of the network, which is equivalent to minimizing the ratio of the long-run average power consumption to the long-run average throughput:

$$\min_{\phi \in \Phi} \mathcal{E}^\phi / \mathcal{L}^\phi, \tag{9}$$

We refer to the optimization problem with the objective function (9) and constraints (1) to (3) and (6) as the *task offloading scheduling problem* (TOSP). The TOSP forms an instance of the *resource allocation problem* discussed in [4], which consists of parallel *restless multi-armed bandit problems* (RMABPs) coupled by capacity constraints. A policy ϕ is considered better if it produces a smaller value of the objective function (9).

4.1 Index Policy

Conventional dynamic optimization techniques are generally not applicable for the TOSP as the computational complexity is prohibitively high for fog computing systems of practical scales due to the large state space of $\boldsymbol{X}(t)$. Therefore, we consider an *index policy*, which only requires a given sequence of *indices* to be assigned to all possible processes $\boldsymbol{X}^\phi(t)$ for a certain policy $\phi \in \Phi$. The indices will then determine the priorities of ARC groups to allocate relevant resources for an incoming task.

We propose an offloading policy, called *Prioritized Incremental Energy Rate* (PIER), and intuitively explain the computation of indices in PIER as follows.

When an incoming j-task is offloaded to the network edge at time t, the EIP needs to decide which ARC group to allocate for handling the task. Note that the power consumption of the network will be increased no matter where the task is computed. We refer such increment as the *incremental power rate* $c_{j,k}(t)$ hereafter. Specifically, suppose the task is computed at the network edge with resources from ARC group $k \in [K]$. In that case, the incremental power rate is $\varepsilon_k^0 + w_{j,k}\varepsilon_k$ if none of the ARC units in the group k are occupied upon the arrival of the task, or $w_{j,k}\varepsilon_k$ otherwise. The expected incremental power rate for computing a j-task in the cloud is $\bar{\varepsilon}_j^c$. We compute the indices of ARC group k

for a j-task, as the ratio of the service rate of ARC group k (determined by the destination area that the group is in) to the incremental power rate of the ARC group if selected, given by $\mu_{j,\ell(k)}/c_{j,k}(t)$. Heuristically, the PIER ranks all ARC groups (including those at the network edge and central cloud) with sufficient resource units by descending order of the indices for a task and then always selects the best ARC group at the moment to optimize the energy efficiency.

By comparing the index values of all feasible offloading destinations for task $j \in [J]$ at time t, when the network state is $\boldsymbol{X}(t) \in \mathscr{X}$, we can obtain the set of available ARC groups with the highest index, denoted as,

$$\tilde{\mathcal{V}}_j(\boldsymbol{X}(t)) := \arg \max_{k \in \mathcal{K}_j(t)} [\mu_{j,\ell(k)}/c_{j,k}(t)], \tag{10}$$

where $\mathcal{K}_j(t)$ is the set of ARC groups that satisfy the channel and capacity constraints (1) to (3) for an incoming j-task at t. Note that the task will be served if $|\tilde{\mathcal{V}}_j(\boldsymbol{X}(t))|$ is no less than one, and will be blocked if $|\tilde{\mathcal{V}}_j(\boldsymbol{X}(t))| = 0$. In cases where $|\tilde{\mathcal{V}}_j(\boldsymbol{X}(t))| > 1$, more than one ARC groups are as efficient as each other, and ties can be broken arbitrarily. We then define that, for any $\boldsymbol{a}(\boldsymbol{X}(t)) \in \mathscr{A}$ and $\boldsymbol{X}(t) \in \mathscr{X}$,

$$a_{j,k}^{\mathrm{PIER}}(\boldsymbol{X}(t)) = \begin{cases} 1, & \text{if } k = \min\left[\tilde{\mathcal{V}}_j(\boldsymbol{X}(t))\right], \\ 0, & \text{otherwise}, \end{cases} \tag{11}$$

where $\min[\tilde{\mathcal{V}}_j(\cdot)]$ returns the minimal element in set $\tilde{\mathcal{V}}_j(\cdot)$ given that $\tilde{\mathcal{V}}_j(\boldsymbol{x}) \neq \varnothing$. When $\tilde{\mathcal{V}}_j(\boldsymbol{x}) = \varnothing$, $a_{j,k}^{\mathrm{PIER}}(\boldsymbol{X}(t))$ is set to 0 for all k, and the incoming j-task is considered blocked. Furthermore, for an incoming j-task, the computational complexity of PIER is linear in the number of ARC groups that the task has access to, which is capped at $K+1$. In addition, only binary information that whether each ARC group has extra capacity to accommodate the incoming task is required for making a decision.

4.2 Asymptotic Optimality

As in [4, Corollary 1], an index policy that prioritizes the process with the highest indices is *asymptotically optimal* as $h \to \infty$. The proposed PIER, while specifically designed for the TOSP, is in the same vein as priority-based policies proposed in [4,5,13], which have been shown to approach the optimal solution in stochastic optimization problems as the scale of the system is sufficiently large. Particularly, when $J = 1$, $K = L$ (that is, there is only one ARC group in every destination area at the network edge), $\Gamma_{j,\ell}^0 = 1$ for all $j \in [J]$, $\ell \in [L]$, $C_k^0 = 1$ for all $k \in [K]$, and the duration of tasks are exponentially distributed, the PIER is equivalent to a special case of a policy that has been proposed and proved to be asymptotic optimal in [5]. We will demonstrate the asymptotic optimality of the PIER with a numerical example in Sect. 5. For more general cases, due to the space constraint, we will also numerically demonstrate and compare the performances of PIER and other benchmark policies in the TOSP in Sect. 5.

5 Numerical Results

In addition to the proposed PIER policy, we consider two benchmark policies for performance comparison: the *Prioritized Transmission Rate* (PTR) policy and the *Prioritized Least Power Consumption* (PLPC) policy. Under the PTR policy, an incoming task always offloads to the ARC group with the fastest transmission rate, given that the capacity constraint is satisfied, to minimize the user's waiting time for task completion. The PLPC policy, on the other hand, always chooses the ARC group with the minimal additional energy consumption rate among all groups with available capacity for an incoming task to reduce the energy bill for EIPs.

For all the simulation results presented in this paper, the 95% confidence intervals based on the Student t-distribution are within 5% of the observed mean.

5.1 Verification of Asymptotic Optimality

We first consider a particular example with a single task class ($J = 1$), five destination areas with a single accessible ARC group in each destination area ($K = L = 5$), each (j, ℓ) pair has the equivalent number of wireless transmission channels ($\Gamma^0_{j,\ell} = 1$), identical capacities across all ARC groups ($C^0_k = 1$ for all $k \in [K]$), negligible idle power ($\varepsilon^0_k = 0$) and relatively high arrival rates such that some tasks will have to be offloaded to the cloud or dropped as the computing and communication resources at the network edge are insufficient to accommodate all requests. This scenario is appropriate for fog computing in densely populated environments, such as significant sports events, where tasks and available resources are largely homogeneous, and the relevant hardware components are always kept active (thus, the idle power consumption under all policies is the same and can be ignored for comparison purposes) to satisfy the enormous offloading demands [16]. Under these conditions, the PIER is a special instance of the policy that has been proved to be asymptotically optimal in [5]. That is, as the scaling parameter $h \to \infty$, the energy efficiency of PIER approaches the optimal solution.

The above conclusion is verified numerically in Fig. 1, where we compare the energy efficiency of PIER and two benchmark policies for different values of h. The durations of tasks served in each destination area are exponentially distributed with mean values 0.587051, 2.65982, 0.547387, 1.1986949, and 4.78274, respectively, and the fixed transmission time from the edge to the cloud is $D_0 = 5$. The power consumption per unit of occupied computing resources ε_k are set to be 1.08316, 10.0584, 1.18651, 8.0544, and 16.0324 for each of the five ARC groups at the network edge, and the power consumption for completing

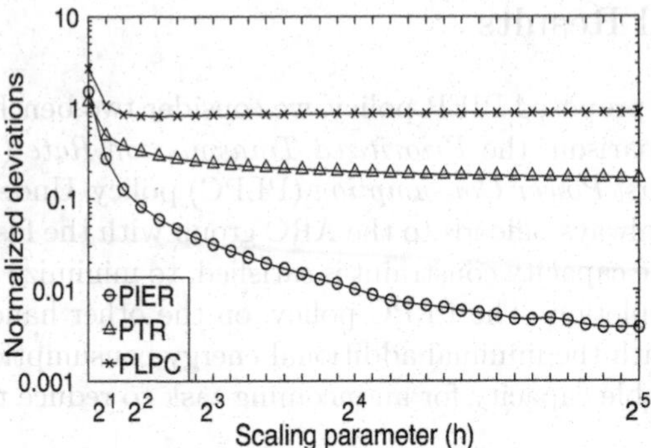

Fig. 1. Normalized deviations of PIER and benchmark policies versus scaling parameter h.

one task at the cloud is $\bar{\varepsilon}_j = 51.4714$. Also, we set the λ_j^0 for the only task class to be 5.182638 and consider $w_{j,k} = 1$ for all $k \in [K]$. The *normalized deviation* in Fig. 1 is defined as

$$\frac{\mathcal{E}^{\phi}/\mathcal{L}^{\phi} - \mathcal{E}^{\phi^O}/\mathcal{L}^{\phi^O}}{\mathcal{E}^{\phi^O}/\mathcal{L}^{\phi^O}}$$

for policy ϕ, where ϕ^O represents the optimal policy. As shown in Fig. 1, the normalized deviation for PIER is below 1% for $h \geq 10$ and converges to 0 as h increases further, while the normalized deviations of PTR and PLPC stay relatively high even for large values of h. Therefore, in this case, PIER is the most appropriate policy, due to its significant superior performance in terms of energy efficiency compared to PTR and PLPC, and advantage in terms of computational complexity over the optimal solution.

5.2 Performance Comparison in General Cases

We now consider more general cases, with two task classes and five ARC groups at the network edge, where the number of computing resources units required for each class in all ARC groups ($w_{1,k}$ or $w_{2,k}$ for all $k \in K$) is randomly generated from $\{1, 2\}$. The five ARC groups are distributed in four destination areas, all of which are accessible for both task classes. Tasks from j-class could be transmitted to destination area ℓ through a (j,ℓ)-channel, where the number of (j,ℓ)-channels for each (j,ℓ) pair is randomly generated from $\{6, 7, 8\}$. For each ARC group, the capacity C_k is randomly generated from $\{3, 4, 5, 6\}$, and the power consumption per unit of occupied resources is randomly generated from $[0.1, 20]$. We also set the idle power consumption $\varepsilon_k^0 = 0.5\varepsilon_k C_k$ according to the power consumption profile of a small scale FN such as a micro cellular gateway [9], and keep $D_0 = 5$.

Note that the computational complexity to obtain the optimal solution, as demonstrated in [5,13], is extremely high even in the special case demonstrated in the previous subsection. It is computationally prohibitive to obtain the optimal solution in the general cases considered in this subsection. Therefore, we will focus on the improvement of PIER over the benchmark policies.

In Fig. 2, we demonstrate the improvement of energy efficiency achieved by PIER as compared to the two benchmark policies in aforementioned scenarios. The cumulative distribution curves of energy efficiency ratios ($\frac{\mathcal{E}^{\phi_1}/\mathcal{L}^{\phi_1}}{\mathcal{E}^{\phi_2}/\mathcal{L}^{\phi_2}}$, where ϕ_1 and ϕ_2 represent two different policies, a ratio less than 1 indicates that ϕ_1 achieves better performance on energy efficiency than ϕ_2) of PIER to PTR and PLPC are presented in three subfigures, where the scaling parameter h is set to 1, 10 and 20, respectively. Each curve is plotted based on the results of 500 independent runs. We observe that, PIER is more efficient compared to PTR and PLPC for over 86.2% and 71% of simulation runs in all cases, respectively.

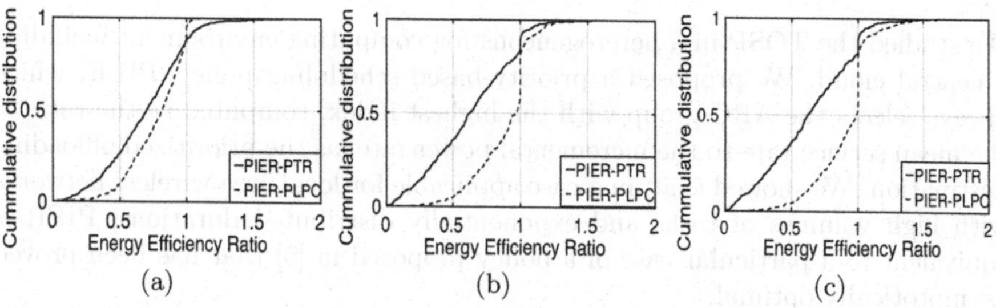

Fig. 2. Cumulative distribution of energy efficiency ratios of PIER to that of the benchmark policies PTR and PLPC with different scaling parameters: (a) $h = 1$; (b) $h = 10$; (c) $h = 20$.

5.3 Robustness to Distributions of Task Durations

We then alter the distribution of task durations to explore the robustness of PIER. We consider three distributions: deterministic distribution, Pareto distribution with finite variance (shape parameter = 2.002, denoted as Pareto-1) and Pareto distribution with infinite variances (shape parameter = 1.981, denoted as Pareto-2). The ranges of all parameters are the same as those in Fig. 2 with $h := 1$. In Fig. 3, we plot the cumulative distribution of the relative difference in power consumption between the one with the exponential distribution and the one with the specified distribution obtained over 500 independent simulation runs under PIER. The results indicate that the mean relative differences in all tested cases are within ±4% for all simulation runs. That is, PIER is robust within the demonstrated cases as its performance is not very sensitive to the distribution of task durations that could be affected by nature of different offloaded tasks.

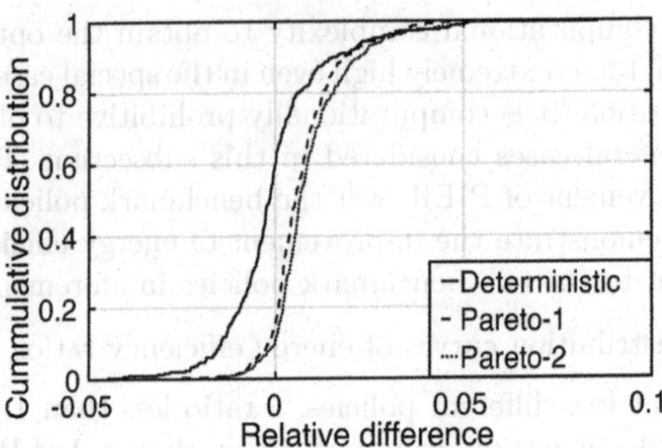

Fig. 3. Robustness of PIER in terms of energy efficiency under different distributions of task durations.

6 Conclusions

We studied the TOSP in a heterogeneous fog computing environment including a central cloud. We proposed a priority-based scheduling policy, PIER, which always selects the ARC group with the highest index, computed as the ratio of the mean service rate to the incremental power rate, as the prioritized offloading destination. We showed that, in a case applicable for local area wireless networks with high volumes of tasks and exponentially distributed durations, PIER is equivalent to a particular case of a policy proposed in [5] that has been proved asymptotically optimal.

In addition, we demonstrated the improvements of the proposed policy in terms of energy efficiency by extensive numerical experiments for more general cases with multiple classes of tasks and ARC groups. Compared with two benchmark policies, PTR and PLPC, that had been commonly adopted in existing work, the PIER acts more effectively in 78.6% of all simulation runs, with significantly better results achieved in most cases. We also demonstrated that, although PIER is derived under the assumption that the task durations are exponentially distributed, its robustness to the different distributions of task durations enables the possibility to apply the policy to a broader range of practical scenarios.

References

1. Chang, Z., Liu, L., Guo, X., Sheng, Q.: Dynamic resource allocation and computation offloading for IoT fog computing system. IEEE Trans. Industr. Inf. **17**(5), 3348–3357 (2020)
2. Deng, R., Lu, R., Lai, C., Luan, T.H., Liang, H.: Optimal workload allocation in fog-cloud computing toward balanced delay and power consumption. IEEE Internet Things J. **3**(6), 1171–1181 (2016)

3. Du, J., Zhao, L., Feng, J., Chu, X.: Computation offloading and resource allocation in mixed fog/cloud computing systems with min-max fairness guarantee. IEEE Trans. Commun. **66**(4), 1594–1608 (2017)
4. Fu, J., Moran, B., Taylor, P.: Restless bandits in action: resource allocation, competition and reservation. Oper. Res. **70**(1), 416–431 (2021)
5. Fu, J., Moran, B.: Energy-efficient job-assignment policy with asymptotically guaranteed performance deviation. IEEE/ACM Trans. Netw. **28**(3), 1325–1338 (2020)
6. Gao, B., Zhou, Z., Liu, F., Xu, F., Li, B.: An online framework for joint network selection and service placement in mobile edge computing. IEEE Trans. Mob. Comput. (2021)
7. Gillam, L., Katsaros, K., Dianati, M., Mouzakitis, A.: Exploring edges for connected and autonomous driving. In: IEEE Conference on Computer Communications Workshops (INFOCOM WKSHPS), IEEE INFOCOM 2018, pp. 148–153. IEEE (2018)
8. He, X., Wang, K., Huang, H., Miyazaki, T., Wang, Y., Guo, S.: Green resource allocation based on deep reinforcement learning in content-centric IoT. IEEE Trans. Emerg. Top. Comput. **8**(3), 781–796 (2018)
9. Jalali, F., Hinton, K., Ayre, R., Alpcan, T., Tucker, R.S.: Fog computing may help to save energy in cloud computing. IEEE J. Sel. Areas Commun. **34**(5), 1728–1739 (2016)
10. Kim, H.S.: Fog computing and the internet of things: extend the cloud to where the things are. Int. J. Cisco (2016)
11. Krishnasamy, S., Akhil, P., Arapostathis, A., Sundaresan, R., Shakkottai, S.: Augmenting max-weight with explicit learning for wireless scheduling with switching costs. IEEE/ACM Trans. Netw. **26**(6), 2501–2514 (2018)
12. Mukherjee, M., Shu, L., Wang, D.: Survey of fog computing: fundamental, network applications, and research challenges. IEEE Commun. Surv. Tut. **20**(3), 1826–1857 (2018)
13. Wang, Q., Fu, J., Wu, J., Moran, B., Zukerman, M.: Energy-efficient priority-based scheduling for wireless network slicing. In: 2018 IEEE Global Communications Conference (GLOBECOM), pp. 1–6. IEEE (2018)
14. Wu, J., Wong, E.W., Chan, Y.C., Zukerman, M.: Power consumption and GoS tradeoff in cellular mobile networks with base station sleeping and related performance studies. IEEE Trans. Green Commun. Netw. **4**(4), 1024–1036 (2020)
15. You, C., Huang, K., Chae, H., Kim, B.H.: Energy-efficient resource allocation for mobile-edge computation offloading. IEEE Trans. Wirel. Commun. **16**(3), 1397–1411 (2016)
16. Zhang, B., Chen, D.: Resource scheduling of green communication network for large sports events based on edge computing. Comput. Commun. **159**, 299–309 (2020)

Space-Heuristic Navigation and Occupancy Map Prediction for Robot Autonomous Exploration

Ping Zhong[1], Bolei Chen[1], Yongzheng Cui[1], Hanchen Song[2], and Yu Sheng[1(✉)]

[1] School of Computer Science and Engineering, Central South University, Changsha, China
{ping.zhong,boleichen,214712191,shengyu}@csu.edu.cn

[2] Changsha Intelligent Driving Institute Ltd., Changsha, China
song.hc@cidi.ai

Abstract. Efficient autonomous exploration in unknown environments is a challenging and basic problem in the field of robotics. Due to the lack of priori information of the environment, the robot cannot clearly select the region with high information gain for exploration. In addition, the existing greedy exploration strategy leads to repeated back-and-forth movements of the robot and inefficient exploration. Intelligent bodies like humans often rely on past experience to infer the structural characteristics of the environment which are used to assist exploration. To give the robot spatial awareness, we propose a method of predicting map occupancy by using Generative Adversarial Networks (GANs). The adversarial loss and the feature extraction loss used in the training of GANs are introduced in detail. Further, we propose a novel utility function for the evaluation of exploration goals based on path length and structural characteristics of the environment. The utility function is used to guide the robot to efficiently explore step by step. We also propose a space-heuristic path planning method named CI-RRT* for robot navigation. We demonstrate the superiority of the proposed methods through comparison and ablation experiments in simulation environments. The experimental results prove that our method is superior to the existing methods.

Keywords: Autonomous exploration · Path planning · Generative adversarial networks · Navigation

1 Introduction

Autonomous exploration is an important research topic in the field of robotics, which is widely used in edge computing [1], smart city [2], 2D/3D reconstruction [3] and mining [4]. The so-called autonomous exploration means that the robot selects suitable exploration goals in an unknown environment for autonomous navigation until the construction of the environment map is completed. Due to the lack of priori information of the environment, the robot needs to continuously

Y. Lai et al. (Eds.): ICA3PP 2021, LNCS 13155, pp. 578–594, 2022.
https://doi.org/10.1007/978-3-030-95384-3_36

detect the frontier points in the current map as candidate exploration goals. The frontiers refer to the critical area of the known area and the unknown area in the current map. Subsequently, the robot uses a certain utility function to evaluate the candidate goals, and selects the best frontier point as the exploration goal. For example [5,6], the frontier points that are close to the robot and have more information gain are favored by the robot.

However, constructing the utility function by only considering the trade-off between distance and information gain is inherently greedy. Such a utility function encourages the robot to get the most rewards at the least cost. Therefore, the robot inevitably abandons the area currently explored, and goes to the frontier of more information gain. As a result, the robot keeps moving back and forth, reducing the efficiency of exploration. Another problem faced by autonomous exploration is the inability to reasonably evaluate the potential information gain of frontier points. The main reason for this problem is that the robot lacks the space occupancy information of the unknown environment. To solve the above problems, we use Generative Adversarial Networks (GANs) [7] to predict and complete the current incomplete map. Based on the predicted map, we can use the spatial structure information of the map to reasonably calculate the information gain for the frontier points. Further mining the spatial information of the prediction map, we construct a novel utility function using distance, information gain and the structural characteristics of the map. This novel utility function guides the robot to efficiently explore step by step, rather than greedily.

GANs are widely used for high-resolution image generation in computer vision [8]. GANs generally consist of a generator (generative network) and a discriminator (discriminative network). The two networks are trained alternately, and their capabilities are simultaneously improved until the data generated by the generator can pass the review of the discriminator. In our work, we use GANs to generate the occupancy of an unknown environment. We use discriminators of different scales to discriminate the maps generated by the generative network. We calculate feature loss based on the features extracted from different layers of the discriminators, and combine it with the adversarial loss to train the GANs. The combination of the two losses helps to improve the performance of the generator and ensure the stability of the training process.

In terms of robot path planning, based on the Informed-RRT* [9] algorithm and the Motion Planning Corridor (MPC) [10], we propose a **C**orridor-**I**nformed algorithm for space-heuristic navigation named CI-RRT*. We construct the MPC through a series of geometric transformations, and introduce the algorithm of uniform sampling in the MPC. By limiting the fine-tuning of Informed-RRT* to the MPC, the CI-RRT* algorithm takes less time to sample the shortest path. We evaluate the proposed methods in simulation environments and compare our autonomous exploration methods with the advanced methods.

2 Related Work

2.1 Autonomous Exploration

Robot autonomous exploration methods are generally divided into two categories: frontier-based [5,6,11–13] and sampling-based [14,15] autonomous exploration strategies. H. Umari et al. [5] pioneers the use of Rapidly-Exploring RandomTree (RRT) for frontiers detection, and the frontier with the largest information gain is provided to the robot for exploration. In contrast, the method in [6] incrementally constructs a Probability Road Map (PRM) for frontiers detection and robot path planning. The exploration strategies in [5] and [6] both encourage robots to explore greedily, that is, always choose exploration goals that are close in distance and have large information gains. Such exploration strategies will cause the robot to constantly change the exploration region and explore inefficiently. Liu et al. [11] proposes a method of using reference objects to guide the robot to efficiently explore step by step. However, it is expensive to use additional neural networks to only identify and locate reference objects. For the autonomous exploration of UAVs, [12] proposes a different frontier selection scheme to achieve the purpose of increasing the flight speed and the exploration rate. Zhou et al. [13] models the autonomous exploration process as a variant of the traveling salesman problem and propose an incremental frontier information structure to maintain the frontiers. As representatives of sampling-based methods, Bircher et al. [14] first uses the concept of next best view (NBV) in 3D exploration in a receding horizon fashion. To achieve global coverage and maximize the utility of the exploration path, [15] continuously expands and refines a single tree utilizing a rewiring scheme inspired by RRT*.

2.2 Path Planning

The robot path planning problems are usually solved by graph searching or stochastic incremental sampling. Graph searching methods such as A*, can often find the shortest paths quickly if the solution exists. However, the paths generated by graph searching methods are difficult to guarantee safety for robots because the paths may be too close to the obstacles. Among stochastic sampling schemes, though RRT-based schemes are probabilistically complete, they are demonstrated almost surely suboptimal [16]. Therefore, several optimized variants of RRT such as RRT*, Direct-DRRT* [17], MOD-RRT* [18], and Informed-RRT* [9] are proposed to search a set of waypoints to approximate the optimal path. However, the paths searched by Direct-DRRT* have poor smoothness and the length optimality can not be guaranteed. MOD-RRT* does not fundamentally improve the efficiency of RRT*. Similarly, Informed-RRT* improves the smoothness of the path and shortens the length of the path at the expense of more computation time. In our work, we propose CI-RRT* based on MPC [10] and Informed-RRT*. By constraining the fine-tuning process in a MPC, CI-RRT* can reduce the redundant sampling space.

2.3 Learning-Based Exploration

In recent years, more and more work uses deep learning methods to assist robots in autonomous exploration and navigation [19–22]. [19] employs a generative neural network to predict unknown regions of a partially explored map, and use the prediction to enhance the exploration in an information-theoretic manner. Based on generative networks, [20] proposes an uncertainty-aware occupancy map prediction method for robot navigation. M. Saroya et al. [21] uses convolutional neural networks to predict the topological features of unknown environments to guide robot exploration. [22] presents a new methodology on learning-based path planning for autonomous exploration of subterranean environments using aerial robots. Autonomous exploration strategies based on reinforcement learning are gradually being proposed [23–25]. For search and rescue applications, [23] combines traditional frontier-based exploration methods with deep reinforcement learning to enable robots to autonomously explore unknown cluttered environments. For office scenarios, [24] proposes to use deep reinforcement learning to predict the long-term exploration sequence of unexplored sub-regions. Using a deep reinforcement learning model to replace the goal selection decision-making module, Li et al. [25] proposes a complete autonomous exploration framework.

3 Space-Aware Occupancy Map Prediction and Space-Heuristic Navigation

In this section, we will introduce how to use generative adversarial networks for occupancy map prediction. We will introduce the network structures and the loss function in detail. Further, we will introduce the space-heuristic path planning algorithm named CI-RRT* based on MPC. Finally, we will propose a novel utility function for goal selection which is suitable for efficient robot autonomous exploration.

3.1 Occupancy Map Prediction

To enable the robot to infer the occupancy of unknown environments, we design the map prediction GANs shown in Fig. 1 and Fig. 2. We design the generative network based on the U-Net architecture [26], and use a skip connection between the encoder and the decoder to break through the bottleneck. The generator first encode and downsample the training data, and then use several residual blocks to characterize features. Then, the generator decode the features through upsampling to generate a prediction map. When designing the discriminative network, to consider contextual information, we downsample the generated fake map to discriminate it on different scales. As shown in Fig. 2, for discriminators of different scales, we implement each network by stacking convolutional layers. To prepare the training data, we first crop maps with a size of $6\,\mathrm{m} \times 6\,\mathrm{m}$ to a size of $4\,\mathrm{m} \times 4\,\mathrm{m}$. Then we use the latter as the source map and the former as the target map to obtain a series of paired maps $\{(s_i, t_i)\}$. We train GANs so that

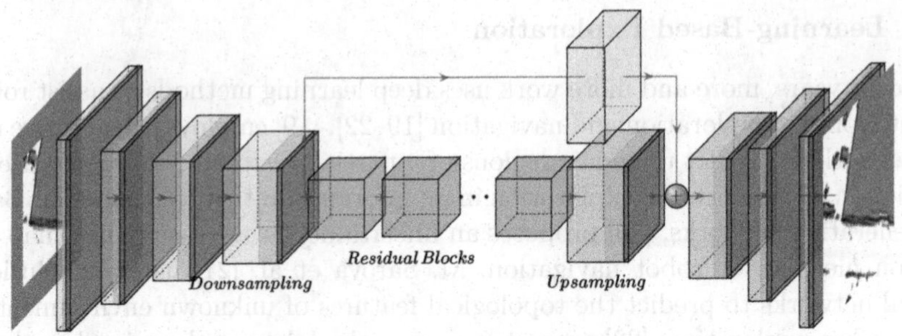

Fig. 1. The illustration of generative network architecture. The source map is downsampled first. Then a series of residual blocks are used to characterize the features. Finally, the prediction map is generated through upsampling. We use a skip connection between the encoder and the decoder to break through the bottleneck.

the map generated by the generator can pass the review of the discriminators. We hope that the generative network can complete the clipped part of the target map as realistically as possible. Furthermore, we expect it to be able to predict the occupancy of unknown areas outside the target map.

3.2 Adversarial Loss and Feature Extraction Loss

Different from the traditional GANs, our model consists of one generator G and multiple discriminators D_k. For our task, the objective of the generator G is to translate source maps s_i to more complete maps, while the discriminators D_k aims to distinguish target maps t_i from the translated ones. Based on the given training dataset $\{(s_i, t_i)\}$, our GANs aim to model the conditional distribution of target maps given the source maps via the following minimax game:

$$\min_{G} \max_{D_k} \sum_{k=1}^{n} \mathcal{L}_{GAN}(G, D_k), \tag{1}$$

where n represents the number of discriminators and the objective function $\mathcal{L}_{GAN}(G, D_k)$[1] is given by

$$\mathbb{E}_{(s,t)}[logD_k(s,t)] + \mathbb{E}_s[log(1 - D_k(s, G(s)))]. \tag{2}$$

Using discriminators of different scales to supervise the training process helps to improve the predictive performance of the generator. In addition, as shown in Fig. 2, to make the training process more stable, we extract features from multiple layers of each discriminator. Our GANs learn to match these intermediate representations from the target and the synthesized map. Specifically, we denote the feature extractor from the input to the top i layer in the k-th discriminator

[1] we denote $\mathbb{E}_s \triangleq \mathbb{E}_{s \sim p_{data}(s)}$ and $\mathbb{E}_{(s,t)} \triangleq \mathbb{E}_{(s,t) \sim p_{data}(s,t)}$.

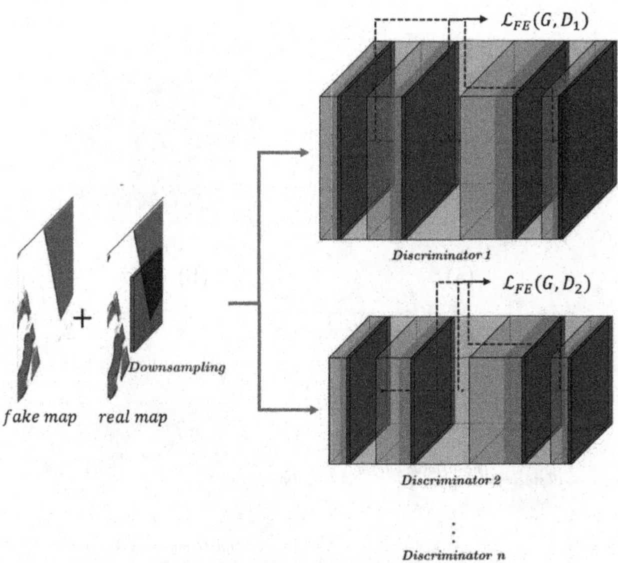

Fig. 2. The illustration of discriminators of different scales. We concatenate the fake map with the real map and double the number of channels. Then we downsample the concatenated map and the sampling factor is 0.5. Further, we input concatenated maps with different scales into discriminators of different scales. In addition, we calculate the feature extraction loss based on the features extracted from each layer of each discriminator.

as D_k^i. We use feature extractors D_k^i with different layers to calculate the loss, which makes the discriminator more robust. Therefore, the corresponding loss is defined as:

$$\mathcal{L}_{FE}(G, D_k) = \mathbb{E}_{(s,t)} \sum_{i=1}^{T} \frac{1}{N_i} [||D_k^i(s,t) - D_k^i(s, G(s))||_1], \tag{3}$$

where T is the number of layers and N_i represents the number of elements in each layer. Finally, the objective function is determined as:

$$\min_G((\max_{D_k} \sum_{k=1}^{n} \mathcal{L}_{GAN}(G, D_k)) + \lambda \sum_{k=1}^{n} \mathcal{L}_{FE}(G, D_k)), \tag{4}$$

where the loss related to feature extraction $\mathcal{L}_{FE}$ is minimized during the training process. λ is used to adjust the importance of the two losses.

3.3 CI-RRT* Algorithm

By restricting Informed-RRT* to the MPC, we propose CI-RRT* for autonomous exploration path planning. The process of constructing a MPC is shown in

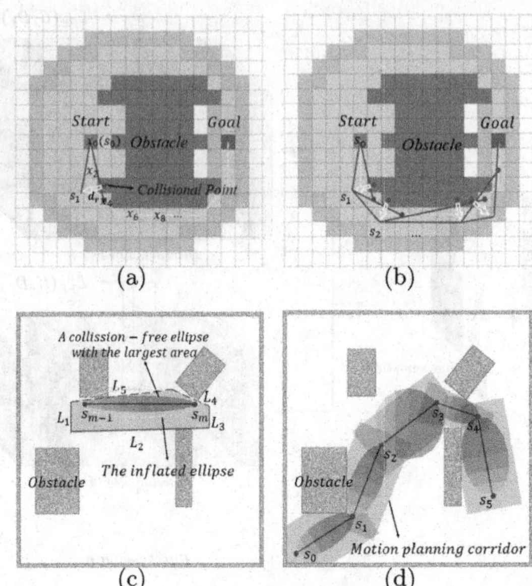

Fig. 3. The path searched by the A* algorithm is shown as the yellow path in (a) and (b). The red polyline path in (b) is the path obtained by pushing the A* path away from the obstacle. (c) illustrates the expansion process of the ellipse. L_4 and L_5 are the tangents between the expanding ellipse and the obstacle. L_1, L_2 and L_3 are the sides of the rectangle used to limit the MPC. A complete MPC is shown in (d). (Color figure online)

Algorithm 1. *Polygon_Sampling*(A, b)

```
CornerPoints ← ∅, Triangles ← ∅
P_I ← IntersectionPoint(A, b), N ← P_I.size()
for iter = 0, 1, ..., N − 1 do
    if A · P_I[iter] − b ≤ 0 then
        CornerPoints.push_back(P_I[iter])
    end if
end for
Triangles ← DividePolygon(CornerPoints)
M ← Triangles.size()
Polygonarea ← AreaSum(Triangles, M − 1)
randnum ← rand(0, Polygonarea)
index ← −1
for j = 0, 1, ..., M − 1 do
    if randnum ≤ Triangles[0].Area then
        index = 0
    else if AreaSum(Triangles, j) < random ≤ AreaSum(Triangles, j + 1) then
        index = j
    end if
end for
P_r ← TriangleSample(Triangles[index].V_1, Triangles[index].V_2)
return P_r
```

Algorithm 2. *MPC_Generation*(P_{A^*},d_r)

```
C_A ← ∅, C_b ← ∅
N ← P_A*.size(), S ← {P_A*.front()}
for iter = 0, 1, ..., N − 1 do
    l_d ← Line(S.back(), P_A*[iter])
    if IsCollision(l_d) then
        p_c ← CollisionPoint(l_d)
        V_u ← VerticalUnitVec(l_d, p_c)
        p_s ← PushAwayObs(p_c, V_u, d_r)
        S.push_back(p_s)
    end if
end for
S.push_back(P_A*.back())
for j = 0, 1, ..., S.size() − 2 do
    A_j, b_j ← GetPolyhedron(S[j], S[j + 1])
    C_A.push_back(A_j)
    C_b.push_back(b_j)
end for
return C_A, C_b
```

Algorithm 1 and Fig. 3. We first use the A* algorithm to search for a path $X = \{x_0, x_1, x_2, ..., x_{N-1}\}$ from the starting position to the goal in the grid map, as shown in the yellow route in Fig. 3(a). Then, we traverse the entire path from the starting position x_0, and stop when the line segment with x_0 and the traversed point as the endpoints collides with the obstacle. We calculate the unit vector perpendicular to the line segment and push the collision point p_c away in the direction of the unit vector. The waypoint pushed away is shown as s_1 in Fig. 3(a), and the distance pushed away is determined by d_r. Next, as shown in Fig. 3(b), we repeat the above operations with waypoint s_i as one of the endpoints of the line segment, until the traversal reaches the goal. Then, we obtain a safer polyline segment path $S = \{s_0, s_1, s_2, ..., s_{M-1}\}$.

As shown in Fig. 3(c), for each path segment, we use the line segment as the long axis of a ellipse and expand the ellipse. In the process of expansion, the tangents between the ellipse and the surrounding obstacles are stored one by one. Furthermore, we use fixed-size rectangles to limit the MPC to prevent the infinite growth of the area caused by the expansion of the ellipse. Finally, we obtain the polygonal area bounded by the tangents and the rectangle. As shown in Fig. 3(d), a series of polygons are overlapped and connected to form a MPC. Algorithm 1 returns the matrices and vectors corresponding to the sides of the polygons.

To ensure the probabilistic completeness of the RRT-based algorithm, it is necessary to achieve uniform sampling in the MPC. We describe the polygon uniform sampling algorithm in Algorithm 2. We decompose the sampling problem of the MPC into a polygon sampling problem, and further into a triangle sampling problem. We first calculate all the intersection points P_I of the straight lines corresponding to the edges, and then select the vertices *CornerPoints* of

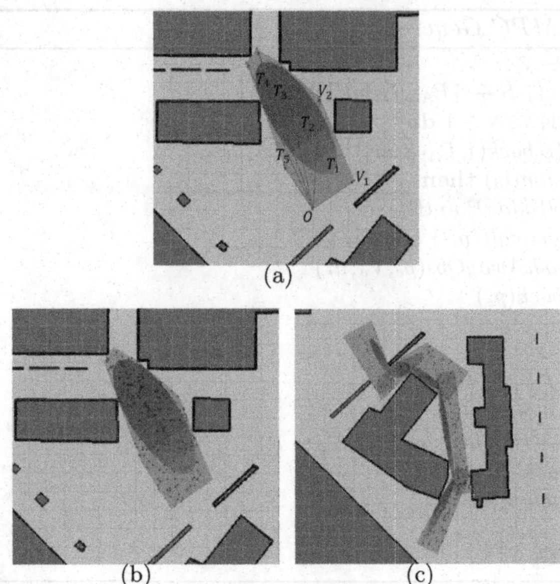

Fig. 4. (a) illustrates the vertices of a polygon and how to divide the polygon into an ordered triangle sequence T_i. V_1 and V_2 are vectors corresponding to two sides of the triangle T_1. (b) and (c) respectively show the sampling of 200 points in a single polygon and a MPC.

the polygon. Subsequently, we divide the polygon into triangles based on the vertices. Based on the area of the triangles, by judging which interval the random number *randnum* falls in, a triangle is randomly selected for further sampling. Finally, the function *TriangleSample*() is used to randomly sampling inside the triangle. For the case in Fig. 4(a), the sampling point P_r in T_1 is obtained by:

$$P_r = \begin{cases} O + \alpha V_1 + \beta V_2 & ,\alpha + \beta < 1 \\ O + (1-\alpha)V_1 + (1-\beta)V_2 & ,\alpha + \beta > 1. \end{cases} \tag{5}$$

where α and β are random numbers between 0 and 1.

As shown in Fig. 5(a) and (b), the Informed-RRT* algorithm gradually fine-tune the path by continuously narrowing a heuristic search domain. Our CI-RRT* algorithm restricts the heuristic sampling of Informed-RRT* to the MPC. CI-RRT* takes less time to sample the shortest path and reduces sampling operations. The preliminary sampled RRT path in the MPC is shown in Fig. 5(c). By fine-tuning in the common area of the heuristic search domain and the MPC, the final RRT path is shown in Fig. 5(d).

3.4 Utility Function for Goal Selection

We formulate the utility function for goal selection by considering the trade-off between path length and information gain. However, this goal selection strategy

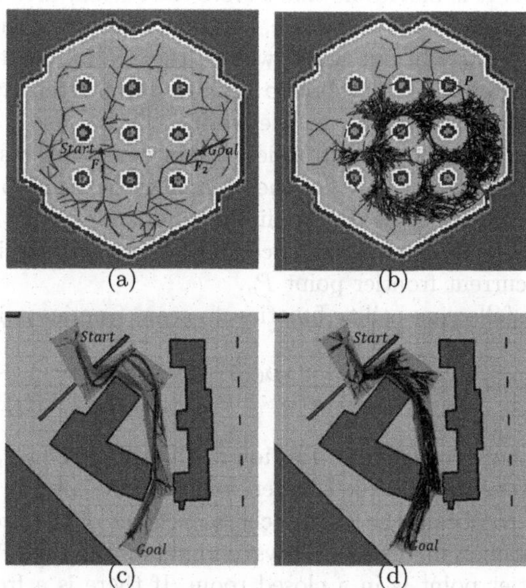

Fig. 5. (a) and (b) illustrate the preliminary path searched by the RRT* and the optimizing process of the Informed-RRT*, respectively. The red path in (c) is the preliminary path sought by CI-RRT*. (d) illustrates the sampling process of CI-RRT* and the optimized result based on the preliminary path. (Color figure online)

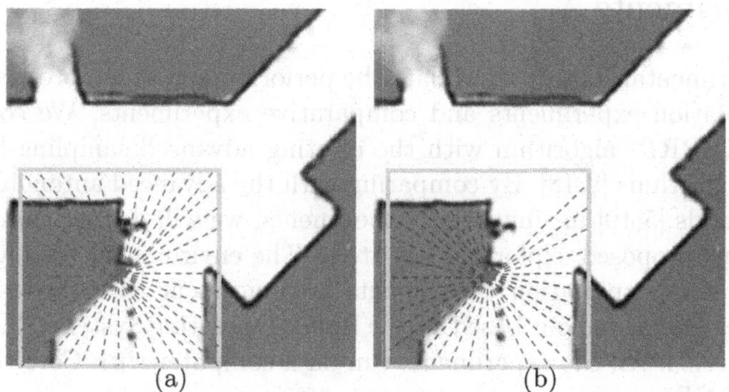

Fig. 6. The green points in (a) and (b) are current frontier point P_n. (a) illustrates the sparse raycasting operation. With the frontier point as the center, a 360° sparse raycasting is performed to count the number of unknown grids in the current submap. The number of unknown grids is used as an estimate of the information gain of the frontier point. (b) illustrates the use of sparse raycasting to obtain the spatial characteristics in the submap. The red rays indicate occluded rays and unknown rays. (Color figure online)

is greedy in nature. It is inevitable that before the current area is fully explored, the robot turns to other areas to explore, causing the robot to run back and forth and explore inefficiently. In our work, we use the spatial characteristics of the predicted submaps to guide the robot to explore step by step, avoiding repeated back and forth movements. Refer to the work in [5], we use a global RRT and a local RRT for global and local frontier points detection, respectively. Then, based on the current map, we predict the occupancy of the map in a square area centered on each frontier point. By predicting and completing the existing map, as shown in Fig. 6(a), we use sparse raycasting to estimate the information gain $Gain(P_n)$ of the current frontier point P_n.

We utilize the following utility function to score frontier points:

$$S(n) = Gain(P_n) - e^{\lambda D(x_{cur}, P_n)} + \frac{\delta N_r}{D(x_{cur}, P_n)}. \tag{6}$$

where x_{cur} represents the current position of the robot. $D(x_{cur}, P_n)$ represents the path length of the robot to the frontier point P_n. As shown in Fig. 6(b), we use sparse raycasting to obtain the spatial characteristics of the predicted submap. We count the number of occluded rays and unknown rays N_r to determine whether the frontier point is in a closed room. If there is a frontier point in a closed room in the current exploration area, we encourage the robot to further explore it completely. Subsequently, the robot will step by step to explore each frontier point under the drive of information gain. λ and δ in the utility function are used to weigh the importance of each item. The robot always explores the frontier point with the highest score first.

4 Experiments

In the experimental stage, we evaluate the performance of map prediction GANs through ablation experiments and comparative experiments. We compare the proposed CI-RRT* algorithm with the existing advanced sampling-based path planning algorithms [9,18]. By comparing with the advanced autonomous exploration methods [5,19] in simulation experiments, we will demonstrate the superiority of our proposed exploration strategy. The environment setting details of the autonomous exploration experiments are consistent with those in [5]. The experiments are carried out based on the Robot Operating System (ROS) Kinetic on Ubuntu16.04. All the experiments run on a computer with Core i7-9700K at 4.9 GHz, 16 GB memory.

4.1 Evaluation for Map Prediction

We prove the superiority of the proposed GANs for map prediction through comparative experiments and ablation experiments. We use the cartographer framework and the method described in Sect. 3.1 to obtain training data and test data. By cropping large-scale maps, we obtain 6000 pairs of training data and 2000 pairs of test data and there is no intersection between them. We use

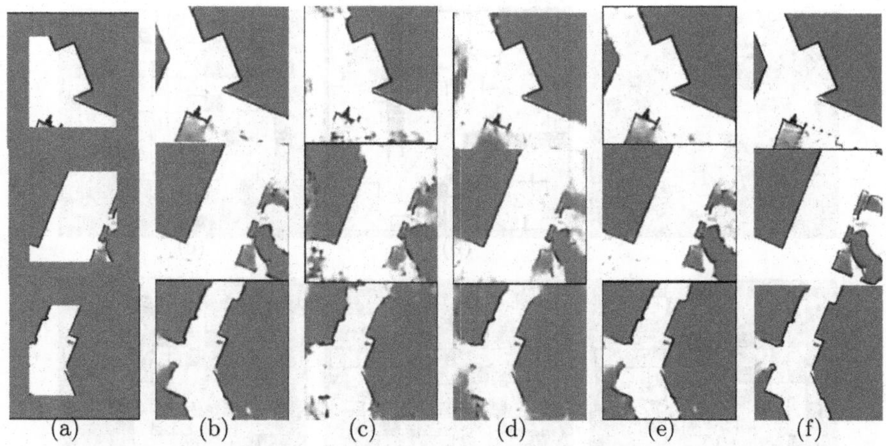

Fig. 7. (a) illustrates three input maps. (b) illustrates the prediction results using feature extraction loss and two discriminators. (c) illustrates the prediction results of the method in [20]. (d) illustrates the prediction results using feature extraction loss but without using feature extraction loss. (e) illustrates the prediction results using only one discriminators and using feature extraction loss. (f) illustrates the target maps.

a NVIDIA GeForce GTX 1080 Ti graphics card for networks training. With the same training data, we train each network for 200 epochs with a batch size of 8. We use the Adam optimizer with an initial learning rate of 0.005 and momentum parameters $\beta = 0.9$. As shown in Fig. 7, we compare with the map prediction method in [20] in three representative map scenarios. In contrast, our GANs (Fig. 7(b)) pay more attention to the details of the maps and can accurately predict the obstacles and corners in the maps. Although the method in [20] can predict the approximate occupancy of the environment, it ignores the details of obstacles (Fig. 7(c)). To highlight the advantages of discriminators of different scales, we compared the use of one discriminator (Fig. 7(d)) and two discriminators (Fig. 7(b)). To show the effect of feature extraction loss, we compared the case of using (Fig. 7(b)) and not using (Fig. 7(e)) feature extraction loss. Comparative experimental results show that discriminators of different scales and feature extraction loss can help improve the accuracy of map prediction.

4.2 Evaluation for CI-RRT*

As shown in Fig. 8(a), (b) and (c), we conduct evaluation experiments in three different scenarios. All the three maps have a resolution of 0.05 m. The dimensions of the three maps are respectively 1000×898, 775×746, and 769×729. (a), (b) and (c) in Fig. 8 respectively show the path planning performance of CI-RRT* in three maps. For qualitative analysis, we respectively show the path planning performance of Informed-RRT* [9], MOD-RRT* [18] and CI-RRT* in Scenario3. In contrast, Informed-RRT* and MOD-RRT* consume more computing resources for sampling, while CI-RRT* samples fewer nodes. For quantitative

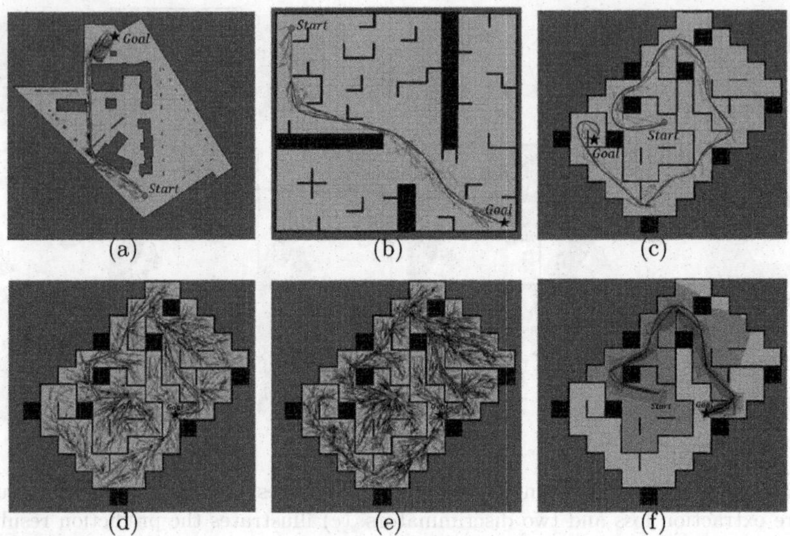

Fig. 8. (a), (b) and (c) respectively illustrate the path planning performance of the CI-RRT* algorithm in three representative maps. (d), (e) and (f) respectively illustrate the path planning comparison of Informed-RRT*, MOD-RRT* and CI-RRT* in Scenario 3. The red polyline paths are the RRT paths, and the blue curves are the results of further optimization. The pink area in (f) is the area traversed by A* algorithm. (Color figure online)

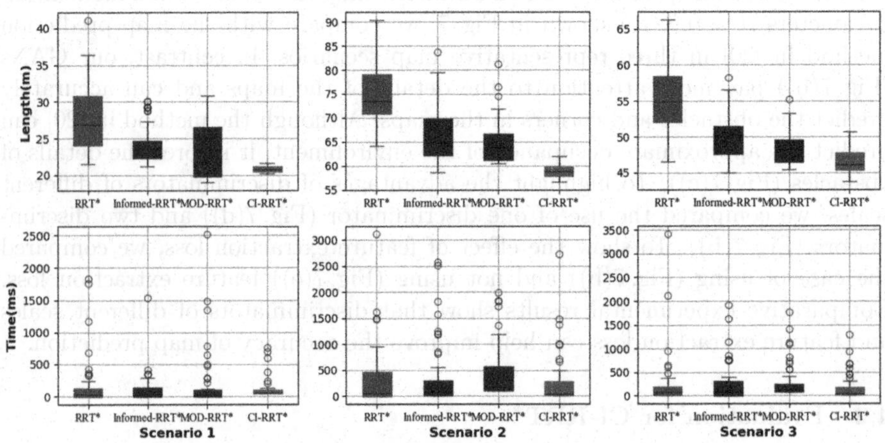

Fig. 9. The illustration of specific experimental data.

analysis, we choose path length, computation time, and success rate as indicators to evaluate the performance of the algorithms. Take the RRT* algorithm as a reference, we simulate each algorithm 50 times in each scenario. To achieve real-time path planning, we assume that the path planning task fails when the computation takes more than 4 s. The specific experimental data are shown in Table 1 and Fig. 9. It is obvious that the CI-RRT* algorithm consumes less average time to search for shorter paths. In addition, CI-RRT* can guarantee a 100% success rate.

Table 1. Performance comparison of different algorithms

Scenario	Method	Path length (m)			Time (ms)			S. rate (%)
		Max	Avg	Min	Max	Avg	Min	Avg
Scenario1	RRT*	41.04	27.56	21.55	1828.23	494.84	0.17	98
	Informed-RRT* [9]	30.08	24.11	20.67	1535.01	276.81	0.29	98
	MOD-RRT* [18]	30.77	23.03	**18.80**	2815.83	316.72	**0.11**	98
	CI-RRT*	**21.58**	**20.80**	20.23	**760.60**	**112.53**	44.96	**100**
Scenario2	RRT*	90.31	74.70	64.08	3113.59	484.12	1.24	98
	Informed-RRT* [9]	83.63	66.13	58.13	**2572.10**	682.11	1.55	96
	MOD-RRT* [18]	81.25	64.77	60.17	2724.72	496.63	**0.60**	**100**
	CI-RRT*	**61.94**	**58.73**	**56.56**	2643.32	**259.74**	98.52	**100**
Scenario3	RRT*	66.26	55.36	46.54	3423.04	245.88	1.52	98
	Informed-RRT* [9]	58.30	49.93	45.02	1919.78	245.03	3.28	96
	MOD-RRT* [18]	54.21	47.17	44.17	**1289.79**	320.25	**0.94**	98
	CI-RRT*	**50.63**	**46.62**	**43.72**	2042.44	**156.88**	67.45	**100**

Table 2. Exploration statistic in different scenes

Scene	Method	Exploration time (s)			Path length (m)		
		Max	Min	Avg	Max	Min	Avg
Scene1	MRRT [5]	94	68	78.4	68	57	64.4
	LMP [19]	89	53	67.5	62	51	53.7
	Ours	**72**	**46**	**57.1**	**58**	**43**	**48.8**
Scene2	MRRT [5]	91	59	65.4	64	51	57.6
	LMP [19]	82	48	60.4	56	45	47.1
	Ours	**54**	**36**	**38.1**	**37**	**29**	**31.9**

4.3 Comparative Simulation Experiments

Integrating map prediction and CI-RRT* algorithm into the autonomous exploration framework proposed in [5], we compare the new framework with the methods MRRT [5] and LMP [19]. In terms of frontier points detection, we set the stride length of global RRT and local RRT to 0.8 m and 0.3 m, respectively. We set the parameter d_r in Algorithm 2 to 1 m. We set the parameters λ and δ in the utility function to 0.5 and 25. The maximum speed of the robot is

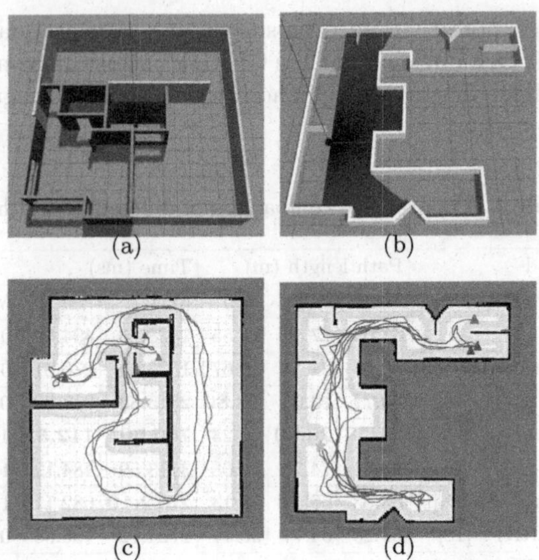

Fig. 10. (a) and (b) respectively show two representative simulation environments. (c) and (d) respectively show the trajectories of the three methods of autonomous exploration. The **star** represents the start position. The **triangles** represent the end positions. The pairing relationship between trajectories and strategies: **red-MRRT**, **green-LMP**, **blue-Ours**. (Color figure online)

0.25 m/s, and the field of view of the lidar is $[-90°, 90°]$. The angular interval of the sparse raycasting is set to 5°. We evaluate different methods with the length of the exploration path and the exploration time as evaluation indicators. We simulate each method 10 times in two representative scenes. The specific experimental data are shown in Table 2. Figure 10 shows the motion trajectories when different exploration strategies are adopted. As shown in Fig. 10(a), Scene1 is a circular indoor scene with a scale of 15 m × 15 m. According to the motion trajectories of the robot, the autonomous exploration method we proposed has explicit exploration goals. As shown in Table 2, our method has certain advantages in terms of exploration time and path length.

As shown in Fig. 10(b), Scene2 is a long and narrow corridor with a scale of 11.5 m × 11.5 m. In scenes similar to Scene2, the greedy exploration of MRRT and LMP caused the robot to move back and forth inefficiently. Even if LMP has the ability of map prediction, it can only improve the exploration time and path length slightly. Benefit from the utility function we proposed, we make full use of the spatial characteristics of the prediction map to ensure the robot explore efficiently and step by step.

5 Conclusions

To achieve efficient autonomous exploration of robots, we propose a method of using GANs to predict and complete the incomplete map. Based on the predicted submaps, we use sparse raycasting to calculate the information gain of the frontier points. To further utilize the structural characteristics of the predicted submaps, we propose a novel utility function to evaluate the exploration goals. Furthermore, we combined the motion planning corridor with the Informed-RRT* algorithm to propose a space-heuristic path planning algorithm named CI-RRT*. By integrating the above methods into the framework of robot autonomous exploration, we compare it with the existing advanced methods. The experimental results reflect the superiority of our methods.

Acknowledgement. This work is supported in part by the Fundamental Research Funds for the Central Universities of Central South University (China) under Grant $No.\,2021zzts0753$ and $No.\,2021zzts0735$.

References

1. Wang, T., Lu, Y., Wang, J., et al.: EIHDP: edge-intelligent hierarchical dynamic pricing based on cloud-edge-client collaboration for IoT systems. IEEE Trans. Comput. **14**, 1285–1298 (2021)
2. Wang, T., Luo, H., Zeng, X., et al.: Mobility based trust evaluation for heterogeneous electric vehicles network in smart cities. IEEE Trans. Intell. Transp. Syst. **22**(3), 1797–1806 (2020)
3. Klimentjew, D., Arli, M., Zhang, J.: 3D scene reconstruction based on a moving 2D laser range finder for service-robots. In: IEEE International Conference on Robotics and Biomimetics (ROBIO), pp. 1129–1134 (2009)
4. Neumann, T., Ferrein, A., Kallweit, S., Scholl, I.: Towards a mobile mapping robot for underground mines. In: Proceedings of the PRASA, RobMech and AfLaI Internatioinal Joint Symposium, Cape Town, South Africa (2014)
5. Umari, H., Mukhopadhyay, S.: Autonomous robotic exploration based on multiple rapidly-exploring randomized trees. In: 2017 IEEE/RSJ International Conference on Intelligent Robots and Systems (IROS). IEEE (2017)
6. Wang, C., Chi, W., Sun, Y., et al.: Autonomous robotic exploration by incremental road map construction. IEEE Trans. Autom. Sci. Eng. **16**(4), 1720–1731 (2019)
7. Goodfellow, I., Pouget-Abadie, J., Mirza, M., et al.: Generative adversarial nets. In: Advances in Neural Information Processing Systems 27 (2014)
8. Wang, T.C., Liu, M.Y., Zhu, J.Y., et al.: High-resolution image synthesis and semantic manipulation with conditional GANs. In: Proceedings of the IEEE Conference on Computer Vision and Pattern Recognition, pp. 8798–8807 (2018)
9. Gammell, J.D., Srinivasa, S.S., Barfoot, T.D.: Informed RRT*: optimal sampling-based path planning focused via direct sampling of an admissible ellipsoidal heuristic. In: 2014 IEEE/RSJ International Conference on Intelligent Robots and Systems, pp. 2997–3004. IEEE (2014)
10. Gao, F., et al.: Online safe trajectory generation for quadrotors using fast marching method and bernstein basis polynomial. In: 2018 IEEE International Conference on Robotics and Automation (ICRA). IEEE (2018)

11. Liu, J., Lv, Y., Yuan, Y., et al.: A prior information heuristic based robot exploration method in indoor environment. In: 2021 IEEE International Conference on Real-time Computing and Robotics (RCAR), pp. 129–134. IEEE (2021)
12. Cieslewski, T., Kaufmann, E., Scaramuzza, D.: Rapid exploration with multi-rotors: a frontier selection method for high speed flight. In: 2017 IEEE/RSJ International Conference on Intelligent Robots and Systems (IROS), pp. 2135–2142. IEEE (2017)
13. Zhou, B., et al.: FUEL: fast UAV exploration using incremental frontier structure and hierarchical planning. IEEE Robot. Autom. Lett. **6**(2), 779–786 (2021)
14. Bircher, A., Kamel, M., Alexis, K., Oleynikova, H., Siegwart, R.: Receding horizon "next-best-view" planner for 3D exploration. In: 2016 IEEE International Conference on Robotics and Automation (ICRA), pp. 1462–1468. IEEE (2016)
15. Schmid, L., Pantic, M., Khanna, R., Ott, L., Siegwart, R., Nieto, J.: An efficient sampling-based method for online informative path planning in unknown environments. IEEE Robot. Autom. Lett. **5**(2), 1500–1507 (2020)
16. Karaman, S., Frazzoli, E.: Sampling-based algorithms for optimal motion planning. Int. J. Robot. Res. **30**, 846–894 (2011)
17. Coelho, F.O., Carvalho, J.P., Pinto, M.F., Marcato, A.L.: Direct-DRRT*: a RRT improvement proposal. In: 2018 13th APCA International Conference on Automatic Control and Soft Computing (CONTROLO), pp. 154–158 (June 2018)
18. Qi, J., Yang, H., Sun, H.: MOD-RRT*: a sampling-based algorithm for robot path planning in dynamic environment. IEEE Trans. Ind. Electron. **68**, 7244–7251 (2021). https://doi.org/10.1109/TIE.2020.2998740
19. Shrestha, R., Tian, F.P., Feng, W., et al.: Learned map prediction for enhanced mobile robot exploration. In: 2019 International Conference on Robotics and Automation (ICRA), pp. 1197–1204. IEEE (2019)
20. Katyal, K., Popek, K., Paxton, C., et al.: Uncertainty-aware occupancy map prediction using generative networks for robot navigation. In: 2019 International Conference on Robotics and Automation (ICRA), pp. 5453–5459. IEEE (2019)
21. Saroya, M., Best, G., Hollinger, G.A.: Online exploration of tunnel networks leveraging topological CNN-based world predictions. In: 2020 IEEE/RSJ International Conference on Intelligent Robots and Systems (IROS), pp. 6038–6045. IEEE (2020)
22. Reinhart, R., Dang, T., Hand, E., et al.: Learning-based path planning for autonomous exploration of subterranean environments. In: 2020 IEEE International Conference on Robotics and Automation (ICRA), pp. 1215–1221. IEEE (2020)
23. Niroui, F., Zhang, K., Kashino, Z., et al.: Deep reinforcement learning robot for search and rescue applications: exploration in unknown cluttered environments. IEEE Robot. Autom. Lett. **4**(2), 610–617 (2019)
24. Zhu, D., Li, T., Ho, D., et al.: Deep reinforcement learning supervised autonomous exploration in office environments. In: 2018 IEEE International Conference on Robotics and Automation (ICRA), pp. 7548–7555. IEEE (2018)
25. Li, H., Zhang, Q., Zhao, D.: Deep reinforcement learning-based automatic exploration for navigation in unknown environment. IEEE Trans. Neural Netw. Learn. Syst. **31**(6), 2064–2076 (2019)
26. Ronneberger, O., Fischer, P., Brox, T.: U-Net: convolutional networks for biomedical image segmentation. In: Navab, N., Hornegger, J., Wells, W.M., Frangi, A.F. (eds.) MICCAI 2015. LNCS, vol. 9351, pp. 234–241. Springer, Cham (2015). https://doi.org/10.1007/978-3-319-24574-4_28

Service Dependability and Security Algorithms

Edge DDoS Attack Detection Method Based on Software Defined Networks

Gangsheng Ren, Yang Zhang, Shukui Zhang(✉), and Hao Long

School of Computer Science and Technology, Soochow University, Suzhou, China
20195227071@stu.suda.edu.cn, zhangsk@suda.edu.cn

Abstract. Edge computing extends the traditional cloud computing architecture by using the computing and storage resources on the edge of the network, making people's work and life more convenient. However, these devices at the edge of the network are widely distributed and the environment is relatively complex. Attackers use these vulnerable IoT devices to build botnets to initiate distributed denial of service attacks, posing a serious threat to the normal use of such networks. In response to this problem, we propose an anomaly detection framework based on software-defined networking (SDN). The edge controller in the SDN network is used to obtain the flow information and extract the features of the flow. The XGBoost algorithm optimized by genetic algorithm (GA-XGBoost) we proposed is used to classify and detect the flow. Experimental results show that compared with other algorithms of the same type, our proposed algorithm has a lower false alarm rate and higher accuracy.

Keywords: Edge computing · Software-defined network (SDN) · Distributed denial of service attack (DDoS) · Anomaly detection · Machine learning

1 Introduction

With the development of the Internet of Things technology, more and more smart devices are deployed in the network to provide services, and the era of the Internet of Everything has arrived. In this case, the cloud computing model has been unable to efficiently process the data generated by edge devices. The edge computing model is designed to solve this problem. The idea of edge computing is to run computing tasks on computing resources close to the data source, which can better support IoT applications, and edge computing has attracted widespread attention and research in the academic community.

This work was supported by the National Natural Science Foundation of China (61572340), Advance Research Fund (No. 61403120402), Priority Academic Program Development of Jiangsu Higher Education Institutions (PAPD); Collaborative Innovation Center of Novel Software Technology and Industrialization of Universities in Jiangsu Province.

Y. Lai et al. (Eds.): ICA3PP 2021, LNCS 13155, pp. 597–611, 2022.
https://doi.org/10.1007/978-3-030-95384-3_37

Distributed Denial of Service (DDoS) has always been one of the main security threats to the Internet [1–3]. Although edge computing solves many problems of traditional cloud computing models, it also faces many security threats [4–6]. DDoS is one of the most serious security threats to edge computing. DDoS aims to control a large number of botnets to exhaust the victims' resources. In traditional networks, hosts and laptops are the main targets for attackers to build botnets. However, with the rapid development of the Internet of Things, edge devices have also become the target of building a botnet. Software defined networking (SDN) is an effective network architecture to solve this problem. As a new type of network technology, SDN realizes network virtualization. It separates the data domain from the control domain of the network, simplifies the network architecture, and realizes a "software management network". Its programmability makes the network more flexible, and it has been widely studied and applied [7]. Although edge computing technology can reduce the traffic pressure of massive data and reduce network delay, it also makes the management of edge networks more difficult. A large number of edge devices are connected to the network, making DDoS attacks more severe. SDN realizes centralized management and control of the network. The management and scheduling capabilities of SDN can help manage edge network resources in edge computing, and further enhance the applicability of edge computing.

The OpenFlow protocol is a communication interface standard between the controller and the switch in the SDN network architecture. The OpenFlow switch matches and forwards data packets according to the flow table, which is published and updated by the controller. The flow table is composed of flow table entries, which include flow matching fields, flow statistics, and a set of rules. The flow table can be changed dynamically, making the network highly scalable and flexible. Based on these characteristics of the SDN network and the OpenFlow protocol, we can easily determine whether there is abnormal traffic in the network. SDN-based DDoS attack detection has been extensively studied [8–11].

The main contributions of this paper can be summarized as follows:

1) We propose a model to detect DDoS attacks on edge devices in SDN, and use edge controller to analyze traffic and make decisions to achieve real-time detection and response.
2) By using the OpenFlow protocol, we can extract the six-dimensional feature vector as the input of the model to detect and determine whether a DDoS attack occurs.
3) We propose an XGBoost detection algorithm based on information entropy features, and optimize it by genetic algorithm, the detection rate and accuracy of the model are improved, and the false alarm rate of the model is reduced.

The paper consists of five sections. Section 2 presents related works on SDN-based DDoS attacks. Section 3 introduces the method we proposed for DDoS attack detection. Section 4 performs simulation experiments on the algorithm we proposed and compares it with other algorithms. Section 5 concludes the entire paper.

2 Related Works

DDoS detection methods are mainly divided into the following three types: detection methods based on statistical information, strategy and machine learning [12]. Information entropy is an important concept of information theory, which can reflect the degree of randomness of network data. In the detection method based on statistical information, the network condition can be judged by analyzing the information entropy characteristics of the network traffic. Mousavi et al. [9] proposed an early detection algorithm for DDoS attacks. By comparing the entropy value obtained by calculating the destination address of network data packets with a threshold, the algorithm can determine whether a DDoS attack has occurred within 250 data packets. Kalkan et al. [13] proposed a method based on joint entropy to detect DDoS attacks. By extracting network packet characteristics to generate a feature matrix, and calculating the joint entropy between different feature combinations to determine whether a DDoS attack has occurred.

In the detection method based on machine learning, the detection framework usually consists of a traffic collection module, a feature extraction module and a classifier module. In [14], the authors captured the source IP address, source port, destination IP address, destination port, packet size, and timestamp of all IP data packets sent by smart home devices and performed feature extraction, using machine learning to classify normal traffic and abnormal traffic. Bhunia et al. [8] proposed a framework for early detection of traffic anomalies at the edge of the network. By finding anomalies at the network edge nodes and responding in time, it is beneficial to reduce the impact of attack traffic. Dong et al. [15] proposed an improved KNN algorithm to detect DDoS attacks by extracting 4-tuple characteristics of network traffic, including the number of data packets, stream duration, stream bytes, and data packet rate. Tan et al. [2] proposes a combined machine learning algorithm based on a combination of K-means and KNN, which detects DDoS attacks by extracting the average number of bytes of the flow, the average duration of the flow, the percentage of symmetric flow, the rate of change of single flow, and the percentage of low data packet flow. Sahoo et al. [16] proposed an improved support vector machine (SVM) model, which reduced the feature flow dimension through principal component analysis, and optimized SVM parameters by genetic algorithm. In order to verify the effectiveness of their model, they used it to test the NSLKDD dataset, and the results show that the detection rate of this method is higher than that of other SVM models. However, the detection effect of this method on actual SDN experimental environmental data is unknown. Zhou et al. [10] proposed a DDoS detection algorithm based on XGBoost to test the KDD99 data set. The results show that compared with other traditional machine learning algorithms, this method has a higher detection rate and a relatively faster detection rate. However, this method also faces the problem that the detection effect of other experimental environmental data is unknown. Yang et al. [11] compared the detection effects based on machine learning and neural network by collecting the traffic in the network as a data set, and the results showed that both have a higher detection rate, but the false alarm rate based on the neural network detection algorithm is higher.

Overall, although there are various detection algorithms for DDoS attacks, they are mainly based on information entropy and machine learning technology. Information entropy can well reflect the randomness of the tested sequence, and the method based on information entropy has a fast detection rate. However, the determination of the threshold often requires a lot of preparatory work, and the actual experience of the researcher is critical to success. Moreover, the pure use of detection methods based on information entropy has the characteristics of poor scalability and low detection rate compared with detection methods based on machine learning. Based on machine learning algorithms, XGBoost has been widely used in many competitions and project development and has achieved very good results. XGBoost can efficiently process large amounts of data in SDN. Therefore, we propose a DDoS detection method combining information entropy and machine learning technology.

3 Anomaly Detection Model

3.1 Detection Framework

We propose a detection framework whose components include IoT devices, OpenFlow switches, and SDN edge controllers, as shown in Fig. 1. IoT devices include smart phones, smart homes and so on. These devices can easily be used to build botnets. OpenFlow switchs are responsible for forwarding data packets. SDN-based IoT gateways (SDNIGs) are a subset of OpenFlow switches and they are regarded as IoT gateways. Edge controller can provide high reliability and services. It knows the complete topology of the current network, can publish forwarding rules to OpenFlow switches, and provide computing capabilities for the edge.

In the SDN environment, OpenFlow switches are responsible for collecting and forwarding data, and periodically send relevant information about the current flow to the controller according to the OpenFlow protocol. The central controller RYU maintains the topology information of the entire network and the flow table rules of the switches, and monitors and manages the entire SDN network by controlling the switches.

When we detect anomaly flows, in order to obtain flow information, the controller needs to continuously send requests to the switch. This task is completed by the flow collection module. After the feature extraction module obtains the collected flow, it extracts flow information features. Finally, the extracted flow characteristics are input to the anomaly detection module to complete the flow detection, and we use the GA-XGBoost algorithm as the anomaly detection module algorithm. Therefore, in order to detect anomaly flows, we propose a detection process based on the three modules of the SDN controller. The anomaly detection in the SDN environment is completed through the cooperation of the controller, the switch and the three modules.

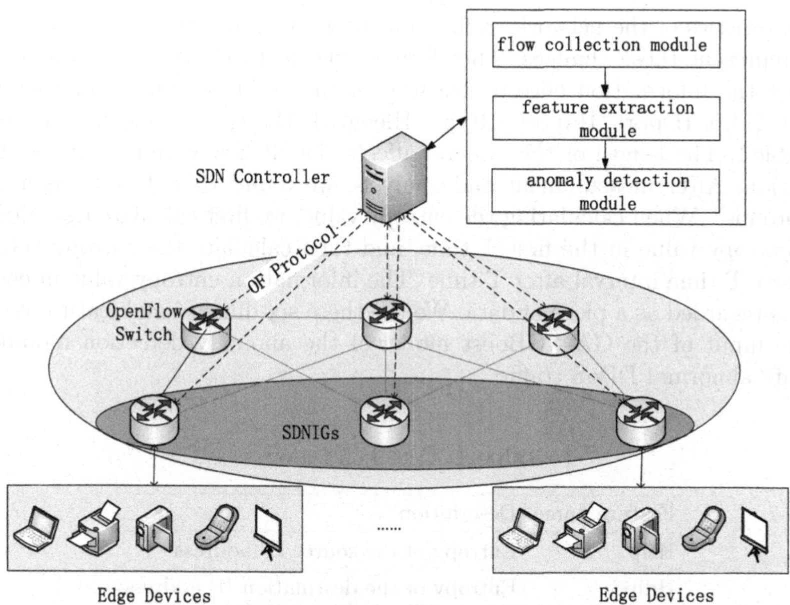

Fig. 1. Detection framework

3.2 Feature Extraction Based on Information Entropy

Information entropy can reflect the randomness of variable values. When the value of the variable is more random, the value of information entropy is larger. On the contrary, when the value of the variable tends to be single, the value of information entropy is smaller. The information entropy is expressed as follows:

$$H(x) = -\sum_{i=1}^{N} \left(\frac{n_i}{S}\right) lb \left(\frac{n_i}{S}\right) \tag{1}$$

In Eq. (1), the value set of sample x is $\{x_1, x_2, ..., x_n\}$, n_i is the number of occurrences of the ith sample in the sample data. S is the number of occurrences of all samples, that is, $\frac{n_i}{S}$ is the probability of occurrence of the ith sample in the sample data. Information entropy can reflect the stability of a system. The system is the most stable when $n_1 = n_2 = \cdots = n_N$, and the entropy value is 0. When the sample distribution is the most dispersed, that is, all samples have different values, and the entropy The maximum value is lbS. It can be seen that the value range of information entropy is $[0, lbS]$.

The main purpose of DDoS attacks is resource occupation and consumption. When an attacker attacks a network target through DDoS, the destination address, length, and protocol of the data packet vary in a single range, and the source address, source port, and destination port vary widely [17,18]. We can judge whether there is an attack in the network through the changes of the above characteristics of the network traffic in a short time, and the entropy value can well represent these changes. Therefore, in the feature extraction module, we extract the information entropy features of the traffic within a time interval: {Hsip, Hdip, Hsport, Hdport, Htype, Hlength}, the specific meaning is shown in Table 1. The length of the interval affects the efficiency and accuracy of the detection. After measurement and analysis, we found that T = 10 s is a good compromise. When calculating the entropy value, we first calculate the information entropy value in the first T time, and then calculate the entropy value in the next T time interval after T time. The information entropy value in each T time is regarded as a piece of data. We use these six-dimensional feature vectors as the input of the GA-XGBoost model of the anomaly detection module to identify abnormal DDoS traffic.

Table 1. Detection feature

Feature name	Description
Hsip	Entropy of the source IP address
Hdip	Entropy of the destination IP address
Hsport	Entropy of the source port
Hdport	Entropy of the destination port
Htype	Entropy of packet protocol
Hlength	Entropy of the packet length

3.3 Detection Method Based on GA and XGBoost(GA-XGBoost)

XGBoost Algorithm. The base classifier used by eXtreme Gradient Boosting (XGBoost) [19] is the Classification and Regression Trees (CART) model, so the XGBoost algorithm can solve regression and classification problems. Because the edge controller has limited processing capabilities, the detection model deployed on the edge controller should be relatively simple. XGBoost is an optimized distributed gradient boosting library, which can overcome the limited calculation speed and accuracy, and can quickly and accurately solve the prediction and classification problems. The XGBoost model has the characteristics of low computational complexity, fast running speed and high accuracy, which can effectively detect abnormal network behaviors while reducing the burden on the edge controller. Therefore, we deploy a XGBoost model in the edge controller to detect abnormal network traffic.

XGBoost continuously forms new decision trees to fit the residuals of previous predictions, so that the residuals between the predicted values and the true values are continuously reduced, thereby improving the prediction accuracy of the model. The algorithm performs a second-order Taylor expansion of the loss function on the basis of the Gradient Boosting Decision Tree (GBDT), and also adds a regular term to the loss function. XGBoost has made significant improvements to the traditional GBDT algorithm in terms of computing speed, generalization performance and scalability.

The objective function of the XGBoost algorithm is as follows:

$$F_{obj}(\theta) = L(\theta) + \Omega(\theta) \tag{2}$$

where,

$$L(\theta) = l\left(\hat{y}_i, y_i\right) \tag{3}$$

$$\Omega(\theta) = \gamma T + \frac{1}{2}\lambda\|\omega\|^2 \tag{4}$$

The objective function of the XGBoost algorithm is composed of two parts, $L(\theta)$ and $\Omega(\theta)$, and θ refers to various parameters in the formula. $L(\theta)$ refers to the differentiable convex loss function that distinguishes the difference between the predicted value $\hat{y}_i$ and the target value y_i. Its function is to make the predicted value of the model more closely fit the true value. $\Omega(\theta)$ represents the regular term in the objective function, where γ and λ represent the regular term coefficients to prevent the decision tree from being too complicated, T refers to the number of leaf nodes, and γT can prevent overfitting. Compared with the traditional algorithm GBDT, XGBoost adds $\frac{1}{2}\lambda\|\omega\|^2$ to the objective function, and ω represents the weight of the leaf node. Adding this item can strengthen the generalization ability of the model.

Each round of training of the XGBoost algorithm is generated iteratively on the basis of the previous round. The objective function of its tth iterative model is as follows:

$$F_{obj}^{(t)} = \sum_{i=1}^{n} l\left(y_i, \hat{y}_i^{(t-1)} + f_t\left(x_i\right)\right) + \Omega\left(f_t\right) \tag{5}$$

where $\hat{y}_i^{(t-1)}$ is the predicted value of the ith sample at the $t-1^{\text{th}}$ iteration, $f_t\left(x_i\right)$ is the predicted value at the tth iteration; and $\Omega\left(f_t\right)$ is the regular term of the objective function.

The second expansion of Taylor formula on Eq.(5) is as follows:

$$F_{obj}^{(t)} \cong \sum_{i=1}^{r}\left[\left(\sum_{i\in I_j} g_i\right)\omega_j + \frac{1}{2}\left(\sum_{i\in I_j} h_i + \lambda\right)\omega_j^2\right] + \gamma T \tag{6}$$

where, I_j is the sample set of the jth leaf node; g_i is the first derivative of the sample x_i; ω_j represents the output value of the jth leaf node; and h_i is the second derivative of the sample x_i.

It can be concluded from Eq. (6) that the objective function is a convex function, and the optimal ω_j^* can be obtained to make the objective function reach the optimal solution:

$$\omega_j^* = -\frac{\sum_{i\in I_j} g_i}{\sum_{i\in I_j} h_i + \lambda} \tag{7}$$

$$F_{obj}^{(t)} = -\frac{1}{2}\sum_{j=1}^{T}\frac{\left(\sum_{i\in I_j} g_i\right)^2}{\sum_{i\in I_j} h_i + \lambda} + \gamma T \tag{8}$$

Equation (8) can evaluate the quality of a tree model. The smaller the value, the better the tree model. The optimal tree structure is gradually generated by adding partitions to existing leaf nodes each time, and the gain of splitting is as follows:

$$\text{Gain} = \frac{1}{2}\left[\frac{G_L^2}{H_L+\lambda} + \frac{G_R^2}{H_R+\lambda} - \frac{(G_L+G_R)^2}{H_L+H_R+\lambda}\right] - \gamma \tag{9}$$

When the splitting gain is continuously smaller than the fixed value or the number of splits reaches the specified maximum depth, the splitting stops, and the final classification model is obtained.

GA-XGBoost Algorithm. Genetic Algorithm (GA) is designed and proposed according to the evolutionary laws of organisms in nature. It is a computational model that simulates Darwin's natural selection and Mendel's theory of group genetics [20]. It is a kind of random adaptive global search algorithm. Individuals in the population search for the optimal solution in the solution space through repeated selection, crossover and mutation operations, and continuous evolution, which can solve many practical problems. The algorithm has been successfully applied to optimization problems and machine learning.

Although the XGBoost algorithm has excellent performance in many aspects, there are still many problems, such as a large number of model parameters, slow convergence and easy to fall into a locally optimal solution. During the training of the XGBoost model, the larger the learning rate eta value, the more likely it will not converge, the greater the gamma of the minimum loss function required for node splitting, the more conservative the algorithm. At the same time, other parameters in the model, such as the maximum depth of the tree max_depth, the minimum sum of the weights min_child_weight, and the L2 regularization term lambda, will all have an impact on the learning and classification performance of the model. Different combinations of XGBoost hyperparameters will get different evaluation scores, and it will be more troublesome to optimize multiple hyperparameters at the same time using global searching. Therefore, we use GA to optimize the above five hyperparameters of the XGBoost model at the same time, and select the best parameter combinations of the XGBoost model to improve the performance of model classification. The optimization parameters are shown in Table 2.

Table 2. List of optimization parameters

Parameter name	Parameter meaning
eta	Step size shrinkage used in update to prevents overfitting
gamma	Minimum loss reduction required to make a further partition on a leaf node of the tree
max_depth	Maximum depth of a tree
Min_child_weight	Minimum sum of instance weight needed in a child
lambda	L2 regularization term on weights. Increasing this valuewill make model more conservative

We set the parameters of the initial population according to the characteristics of the genetic algorithm. The size of the initial population is set to 50. The crossover probability is set to 0.7, which not only facilitates the forward search of the population, but also protects the individuals with high fitness values in the population. The mutation probability is set to 0.05, which not only avoids falling into random search, but also ensures that the population can produce new individual genes. The number of iterations of the algorithm is set to 100. The flowchart of GA optimized XGBoost model is shown in Fig. 2.

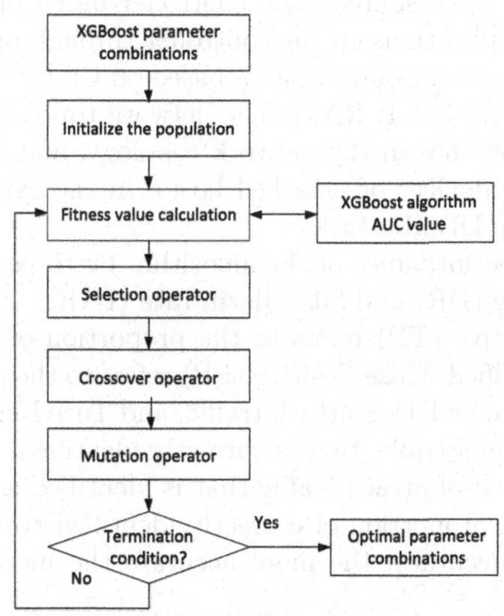

Fig. 2. GA-XGBoost algorithm flow chart

The steps are as follows:

Step1: Initialize the population. Each individual in the population represents a set of hyperparameters of the XGBoost model. During initialization, each individual generates five default hyperparameters of the XGBoost algorithm. The individual encoding method uses binary encoding, and the hyperparameters are shown in Table 2.

Step2: The score output by the XGBoost algorithm represents the fitness value of each individual, that is, the evaluation score of the XGBoost model of each group of hyperparameters is obtained, and the evaluation method uses the AUC value of the XGBoost model.

Step3: According to the selection operator, the hyperparameter combination represented by the individual is optimized and evolved, and the better individual in the model is selected.

Step4: The population performs crossover and mutation operations on individuals according to the crossover operator and mutation operator to obtain the next generation of population genes.

Step5: Determine whether the conditions for the termination of the loop are met. If not, go back to step 2. Otherwise, exit the loop, output the best individual, and the algorithm ends.

4 Performance Evaluation

4.1 Experimental Environment and Evaluation Index

We use the mininet platform to build an SDN network and select Ryu as the controller. Ryu is an open source SDN controller based on python. Users can program their own applications on the controller through python language. The operating system is an open source system based on Ubuntu 18.04, completed on a host with i5 CPU and 16 GB RAM. The network topology is shown in Fig. 3. There are 13 SDN switches in the network topology, and 10 edge switches are connected to 10 edge devices or attacked hosts. In the experiment, we use the scapy tool to simulate DDoS attacks.

We evaluate the performance of the algorithm by three indicators: accuracy (ACC), detection rate (DR) and false alarm rate (FAR).

Where, Ture Positive (TP) refers to the proportion of DDoS attack traffic that is correctly identified, False Positive (FP) refers to the proportion of normal traffic that is identified as DDoS attack traffic, and Ture Negative (TN) refers to the proportion of normal traffic that is correctly identified. False Negative (FN) refers to the proportion of attack traffic that is identified as normal traffic.

ACC refers to the proportion of correctly identified traffic in the total data set. The higher the accuracy, the more accurate the model's effect on traffic classification.

$$ACC = \frac{TP + TN}{TP + TN + FP + FN} \tag{10}$$

DR is the ratio of the number of correctly identified attack traffic to the number of all attack traffic in the sample. The higher the detection rate, the higher the accuracy of the model in identifying DDoS attacks.

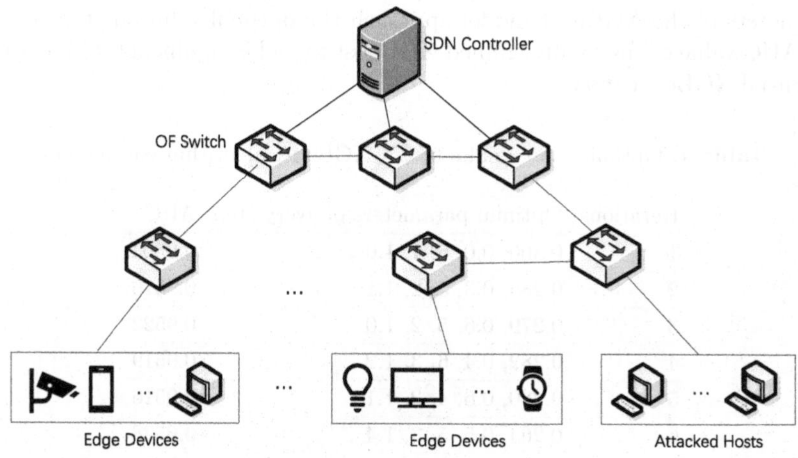

Fig. 3. The topology diagram

$$DR = \frac{TP}{TP + FN} \tag{11}$$

FAR is the ratio of normal traffic that is identified as attack traffic to all normal traffic in the sample. The lower the false alarm rate, the better the model classification effect.

$$FAR = \frac{FP}{FP + TN} \tag{12}$$

4.2 Detection Results and Analysis

The experimental data set is collected from the SDN network. We collected 20,000 traffic data, including 8,000 normal traffic and 12,000 DDoS attack traffic. Take 14,000 traffic as the training data set and 6000 traffic as the test set, as shown in Table 3.

Table 3. Data sets used in simulations

Category	Normal	DDOS
Training data set	5600	8400
Testing data set	2400	3600

In order to verify the optimization effect of GA on XGBoost, we use GA to optimize the five XGBoost parameters on the data set we collected. Through the iteration of the genetic algorithm, a set of optimal parameter combinations (0.221, 1.4, 6, 6, 2.2) are finally obtained. The parameter combination search process is shown in Table 4. It can be seen that after GA optimization, the parameters of the XGBoost model approach the optimal solution step by step. The AUC value of the finally trained XGBoost model is significantly higher than the initial XGBoost model.

Table 4. Optimization results of GA-XGBoost algorithm parameters

Iterations	Optimal parameters of every step	AUC
1	0.300, 0.0, 6, 1, 1.0	0.9504
2	0.286, 0.3, 5, 2, 1.1	0.9516
3	0.279, 0.6, 5, 2, 1.0	0.9522
4	0.282, 0.4, 6, 3, 1.2	0.9519
5	0.280, 0.6, 4, 2, 1.1	0.9516
6	0.264, 0.5, 5, 3, 1.4	0.9538
7	0.269, 0.7, 5, 2, 1.3	0.9561
8	0.261, 1.1, 4, 2, 1.4	0.9557
9	0.236, 1.3, 5, 5, 1.8	0.9612
10	0.223, 1.0, 4, 6, 1.3	0.9630
...	...	...
21	0.230, 1.3, 5, 7, 2.4	0.9657
22	0.213, 1.5, 5, 5, 2.1	0.9668
23	0.225, 1.4, 6, 7, 2.2	0.9671
...	...	...
Optimum	0.221, 1.4, 6, 6, 2.2	0.9673

Then, we compared the detection effect with other methods. Use the data set we collected to compare experiments with the XGBoost model in [10] and the SVM model in [16]. At the same time, the traditional machine learning algorithm KNN is also compared. The results are shown in Table 5. It can be seen that the GA-XGBoost detection model we proposed is better than the comparison algorithms in accuracy, detection rate and false alarm rate. The detection rate of the GA-XGBoost model can reach 95.73%, and the false alarm rate is significantly lower than other algorithms. The XGBoost algorithm uses the idea of boosting. Boosting is a method of combining multiple weak classifiers to form a strong classifier, so as to obtain a more ideal learning effect than a single learner. XGBoost uses a decision tree as its base learner. When training each weak classifier, it optimizes the objective function by continuously fitting and reducing the residual error of the previous model, so that each new classifier constructed focuses on making up for the error of the previous model and helping the final formation more accurate classification effect. Moreover, the combination of genetic algorithm and XGBoost can give full play to the greater advantages of XGBoost. Therefore, the model proposed in this paper can detect DDoS attacks more accurately than other comparative models.

Table 5. The comparison between algorithms %

Model	ACC	DR	FAR
GA-XGBoost	96.89	95.73	3.28
XGBoost	95.53	94.82	4.03
SVM	94.67	93.32	4.46
KNN	94.51	92.78	5.04

In order to further verify the performance of our proposed algorithm, the comparison accuracy rate is shown in Fig. 4(a) when the sample size of the data set is different. It can be seen that the accuracy of our proposed algorithm is higher than other algorithms in the case of different data sets. When the data set size is 20000, the accuracy rate of our proposed GA-XGBoost algorithm is about 96.89%, while the accuracy rates of XGBoost, SVM and KNN are 95.53%, 94.67% and 94.51% respectively, which are all lower than the GA-XGBoost algorithm.

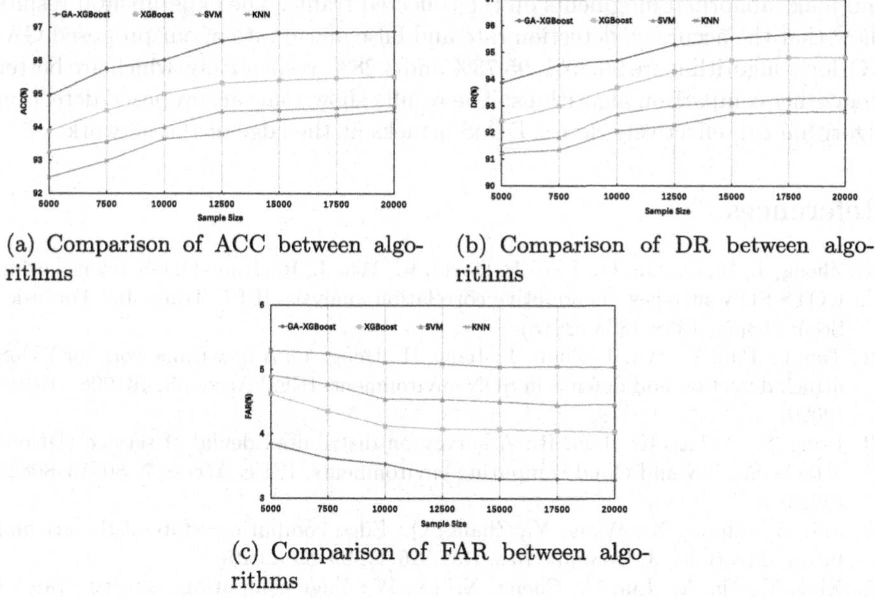

(a) Comparison of ACC between algorithms

(b) Comparison of DR between algorithms

(c) Comparison of FAR between algorithms

Fig. 4. Comparison between different algorithms

At the same time, we also compared the detection rates of these algorithms in different sizes of data sets, and the results are shown in Fig. 4(b). It can be seen that when the data set size is 20000, the detection rate of our proposed GA-XGBoost algorithm is about 95.73%, while the detection rates of XGBoost,

SVM and KNN are 94.82%, 93.32% and 92.78%, which are all lower than GA-XGBoost algorithm.

Finally, we compared the false alarm rates of these algorithms in different sizes of data sets, and the results are shown in Fig. 4(c). It can be seen that when the data set size is 20000, the false alarm rates of XGBoost, SVM and KNN are 4.03%, 4.46% and 5.04% respectively, which are all higher than the 3.28% of GA-XGBoost algorithm. And our proposed algorithm has the lowest false positive rate under different size data sets.

5 Conclusion

We propose a detection method for DDoS attacks on edge devices. This method uses a software-defined network to manage the Internet of Things devices at the edge of the network, and deploys the detection system in the controller. When a DDoS attack occurs on the network, the attacker will send a large number of data packets, and extract the flow characteristic information through the flow table of the switch. In response to these attacks, the GA-XGBoost algorithm is proposed. Collect the real traffic of the SDN network on the Mininet simulation platform, and make abnormal judgments on the collected traffic. The experimental results show that the accuracy, detection rate and false alarm rate of our proposed GA-XGBoost algorithm are 96.89%, 95.73% and 3.28%, respectively, which are better than other comparison algorithms. The results show that our proposed detection algorithm can effectively detect DDoS attacks at the edge of the network.

References

1. Zheng, J., Li, Q., Gu, G., Cao, J., David, K., Wu, J.: Realtime DDoS defense using COTS SDN switches via adaptive correlation analysis. IEEE Trans. Inf. Forensics Secur. **13**(7), 1838–1853 (2018)
2. Tan, L., Pan, Y., Wu, J., Zhou, J., Jiang, H., Deng, Y.: A new framework for DDoS attack detection and defense in SDN environment. IEEE Access **8**, 161908–161919 (2020)
3. Dong, S., Abbas, K., Jain, R.: A survey on distributed denial of service (DDoS) attacks in SDN and cloud computing environments. IEEE Access **7**, 80813–80828 (2019)
4. Shi, W., Zhang, X., Wang, Y., Zhang, Q.: Edge computing: state-of-the-art and future directions. J. Comput. Res. Dev. **56**(1), 69–89 (2019)
5. Xiao, Y., Jia, Y., Liu, C., Cheng, X., Lv, W.: Edge computing security: state of the art and challenges. Proc. IEEE **107**(8), 1608–1631 (2019)
6. Li, H., Wang, L.: Online orchestration of cooperative defense against DDoS attacks for 5G MEC. In: IEEE Wireless Communications and Networking Conference (WCNC), pp. 1–6 (2018)
7. Zhang, C., Cui, Y., Tang, H., Wu, J.: State-of-the-art survey on software-defined networking (SDN). J. Softw. **26**, 62–81 (2015)
8. Bhunia, S., Gurusamy, M.: Dynamic attack detection and mitigation in IoT using SDN. In: International Telecommunication Networks and Applications Conference (ITNAC), pp. 1–6 (2017)

9. Mousavi, S., St-Hilaire, M.: Early detection of DDoS attacks against SDN controllers. In: International Conference on Computing, Networking and Communications (ICNC), pp. 77–81 (2015)
10. Zhuo, C., Fu, J., Cheng, Y., Xin, G., Liu, W., Peng, J.: Xgboost classifier for DDoS attack detection and analysis in SDN-based cloud. In: IEEE International Conference on Big Data and Smart Computing (BigComp), pp. 251–256 (2018)
11. Yang, Y., Wang, J., Zhai, B., Liu, J.: IoT-based DDoS attack detection and mitigation using the edge of SDN. In: Vaidya, J., Zhang, X., Li, J. (eds.) CSS 2019. LNCS, vol. 11983, pp. 3–17. Springer, Cham (2019). https://doi.org/10.1007/978-3-030-37352-8_1
12. Dayal, N., Maity, P., Srivastava, S., Khondoker, R.: Research trends in security and DDoS in SDN. Secur. Commun. Netw. **9**, 6386–6411 (2016)
13. Kalkan, K., Altay, L., Gür, G., Alagöz, F.: Joint entropy-based DDoS defense scheme in SDN. IEEE J. Sel. Areas Commun. **36**(10), 2358–2372 (2018)
14. Doshi, R., Apthorpe, N., Feamster, N.: Machine learning DDoS detection for consumer internet of things devices. In: IEEE Security and Privacy Workshops (SPW), pp. 29–35 (2018)
15. Dong, S., Sarem, M.: DDoS attack detection method based on improved KNN with the degree of DDoS attack in software-defined networks. IEEE Access **8**, 5039–5048 (2020)
16. Sahoo, K.S., Tripathy, B.K.: An evolutionary SVM model for DDoS attack detection in software defined networks. IEEE Access **8**, 132502–132513 (2020)
17. Yao, L., Dong, P., Zhang, H.: Distributed denial of service attack detection based on object character in software defined network. J. Electron. Inf. Technol. **39**(2), 381–388 (2017)
18. Liu, X., Liu, P., Xu, H., Zhu, X.: Software defined internet of things based DDoS attack detection method. J. Comput. Appl. **40**(3), 753–759 (2020)
19. Chen, T., Guestrin, C.: Xgboost: a scalable tree boosting system. In: ACM SIGKDD International Conference on Knowledge Discovery and Data Mining, pp. 785–794 (2016)
20. Ma, Y., Wen, X.: Research progress of genetic algorithm. Appl. Res. Comput. **29**(4), 1201–1206 (2012)

GradMFL: Gradient Memory-Based Federated Learning for Hierarchical Knowledge Transferring Over Non-IID Data

Guanghui Tong, Gaolei Li(✉), Jun Wu(✉), and Jianhua Li

School of Electronic Information and Electrical Engineering, Shanghai Jiao Tong University, and Shanghai Key Laboratory of Integrated Administration Technologies for Information Security, Shanghai 200240, China
{ghtong,gaolei_li,junwuhn,lijh888}@sjtu.edu.cn

Abstract. The massive datasets are often collected under non-IID distribution scenarios, which enforces existing federated learning (FL) frameworks to be still struggling on the model accuracy and convergence. To achieve heterogeneity-aware collaborative training, the FL server aggregates gradients from different clients to ingest and transfer common knowledge behind non-IID data, while leading to information loss and bias due to statistical weighting. To address the above issues, we propose a Gradient Memory-based Federated Learning (GradMFL) framework, which enables Hierarchical Knowledge Transferring over Non-IID Data. In GradMFL, a data clustering method is proposed to categorize Non-IID data to IID data according to the similarity. And then, in order to enable beneficial knowledge transferring between hierarchical clusters, we also present a multi-stage model training mechanism using gradient memory, constraining the updating directions. Experiments on solving a set of classification tasks based on benchmark datasets have shown the strong performance of good accuracy and high efficiency.

Keywords: Federated learning · Gradient memory · Hierarchical clustering · Knowledge transferring · Non-IID data

1 Introduction

Due to the growing data volume, powerful computation capability, and development of the deep neural network, artificial intelligence (AI) has achieved great progress in recent years. It emerges as a promising technology to promote academic advance, industrial reformation and information security [1,2]. Most AI solutions are centralized where users transmit all of their collected data to a central data server or a cloud, which brings privacy concerns, latency, and bandwidth constraints. In contrast, the distributed architecture is a more privacy-preserving and efficient choice. Federated learning (FL) [3] is an emerging modality of distributed machine learning which allows for the collaboration of parties

Y. Lai et al. (Eds.): ICA3PP 2021, LNCS 13155, pp. 612–626, 2022.
https://doi.org/10.1007/978-3-030-95384-3_38

engaged in learning under the coordination of a parameter server. It provides considerable data privacy and low-cost machine learning models which is useful in many fields such as digital twin networks [4]. Under FL, local clients download a global model from the central parameter server, respectively perform local training with its own datasets, and then upload the updated model parameters instead of sensitive raw data to the server. The central server called aggregator combines these parameters using technologies such as Federated Averaging [5] to obtain a more effective global machine learning model, and then redistributes the aggregated parameters to local clients for the next global training iterations. The process is repeated until the communication round is over.

Clients generate and collect data in a non-IID manner, which violates the frequently-used IID assumption and rises challenges in terms of modeling heterogeneous data and convergence analysis of associated training procedures [6]. When naive weighted aggregation algorithms (such as Federated Averaging) are adopted on non-IID data, the global learning model converges to a stationary point of a mismatched objective function, which can be arbitrarily distinguished with the true objective, so-called objective inconsistency [7]. To handle these issues and improve convergence efficiency, a hierarchical clustering algorithm [8] is applied to transfer non-IID data into a set of IID data clusters according to similarity which is derived by the distance of clients. Those neighboring clients are merged as a cluster and assigned a specific task, each of which respectively represents a small-scale federated system. However, different clusters are separate from each other, and beneficial knowledge transferring is not allowed between them.

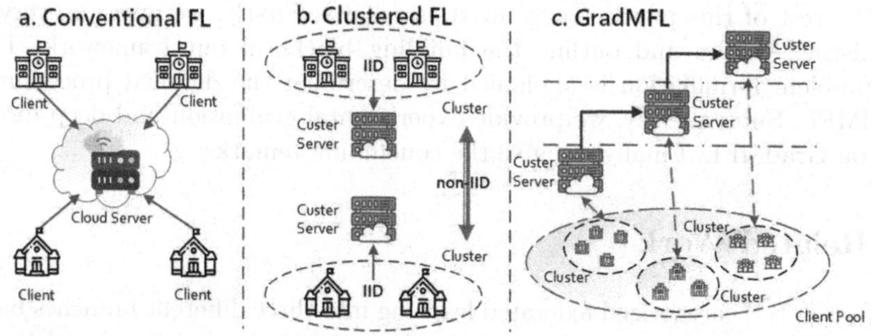

Fig. 1. Comparison of conventional FL, clustered FL, and proposed GradMFL.

In this case, a hierarchical architecture is constructed where the divided clusters of similar data distribution are organized into horizontal layers and different layers are permitted to exchange their knowledge. However, when a cluster is eager for acquiring knowledge from others and carries out a learning algorithm on a sequence of other clients' tasks, the problem of loss of knowledge concerning previous tasks called catastrophic forgetting happens because parameters and semantic representations learned from current tasks drift to the direction of past tasks [9]. Inspired by continual learning which has proposed various solutions to alleviate catastrophic forgetting, gradient memory [10] is used to store past samples and constrain the updating direction of training gradients to maintain good predictions on all tasks. Figure 1 depicts a comparison between conventional FL, clustered FL, and our GradMFL. In conventional FL, a single central server communicates with all clients accommodating heterogeneous data. To improve convergence efficiency and model performance, clustered FL divides different clients into groups holding IID data but these groups are isolated from each other. To solve these issues, our GradMFL partitions related participants into clusters in hierarchy and presents a multi-stage model training mechanism based on gradient memory to enable beneficial knowledge transferring between hierarchical clusters.

The main contributions of this paper are summarized as follows:

1) A Gradient Memory-based Federated Learning (GradMFL) framework is proposed, which improves the performance of heterogeneous federated learning by categorizing clients hierarchically according to similarity and allows beneficial knowledge transferring between clusters.
2) We present a collaborative training strategy for completing a series of tasks in hierarchical clusters, by introducing the use of gradient memory, which effectively alleviates catastrophic forgetting in hierarchical knowledge transferring.
3) Extensive numerical experiments conducted on Permuted MNIST and Fashion MNIST demonstrate our GradMFL scheme has a strong performance of good accuracy and high efficiency.

The rest of this paper is organized as follows. Firstly, we give an overview of existing studies and outline the building blocks in our framework. Then the problem formulation is explained by describing the detailed procedure of GradMFL. Subsequently, we provide experimental evaluation and deep discussion on GradMFL. Finally, we give the concluding remarks.

2 Related Work

Yang et al. [11] categorized federated learning into three different branches based on the distribution characteristics of the datasets: Horizontal Federated Learning, Vertical Federated Learning and Federated Transfer Learning. These standard federated learning architectures may severely degrade the performance of the model when training statistical heterogeneous data across user devices [12]. There exists plenty of literature in federated learning that attempts to promote the progress of the statistical heterogeneity problem. Smith et al. [13] introduced a multi-task learning framework and developed MOCHA meanwhile providing theoretical convergence guarantees on heterogeneous data. Sattler et al. [14] grouped the clients into multiple clusters with joint data distributions according to geometric properties of the FL loss. The most similar study is Briggs et al. [15], which separated the training clients based on the similarity of local updates. However, different from the existing studies, we use the data attributes to compute the similarity and provides knowledge transferring among clusters.

In order to capture knowledge from other clusters, continual learning provides a solution for learning a series of tasks with constrained resources such as limited memory and computation power [16]. It aims at transferring prior knowledge to current tasks to enforce the model to have a powerful prediction ability on all learned tasks. One of the serious challenges is catastrophic forgetting of the prior knowledge, which is mainly led by newly learned knowledge altering the previous parameters in the limited and fixed neural networks [17]. In the literature, there are several methods to mitigate catastrophic forgetting, such as parameter regularization [18], expansion of network [19] and adaptation with gradient memory [20]. Here, we are mostly interested in gradient memory which has a good performance based on limited resources.

Paz et al. [10] proposed gradient episodic memory (GEM) which requests that the direction of the gradients updating should be the alike as that of previous tasks. When the angle between the current gradient and previous gradients is more than 90°, GEM conducts a QP optimization on the local gradients. Because the computation of gradients and QP optimization process are needed for all previous tasks, it is time costly. Averaged GEM proposed by Chaudhry et al. [20] is on the basis of GEM and only requires no increase in losses for certain tasks, which has a better trade-off between accuracy and efficiency.

3 Preliminary

3.1 Federated Learning

A general FL system consists of one parameter server and N training clients. Denoting D_i as the local dataset at the training client C_i, where $i \in \{1, 2, \ldots, N\}$. An active training client, participating in the local training process, needs to update its parameters to minimize the loss function F [21]. At the parameter server, the goal is to learn a model over the data that reside at the N associated training clients. Formally, the parameter server aggregates model weights received from the N training clients as:

$$\mathbf{w_t} = \mathbf{w_{t-1}} - \gamma \sum_{i=1}^{N} p_i \mathbf{g}_i \tag{1}$$

where g_i is the gradient vector updated by the i-th training client, w is the weight vector after model aggregating at the parameter server, γ represents the learning rate, N is the number of training clients, $p_i = \frac{|\mathcal{D}_i|}{|\mathcal{D}|} \geq 0$ with $\sum_{i=1}^{N} p_i = 1$. Such a federated learning problem can be formulated as:

$$\mathbf{w}^* = \arg\min_{\mathbf{w}} \sum_{i=1}^{N} p_i F_i(\mathbf{w}, \mathcal{D}_i) \tag{2}$$

As a continuous iterative learning process, the FL keeps the global learning model updating across all training clients. Before the start of each training round, one participating user will download the global model parameters from the parameter server. And then, the participating user trains a local learning model and sends the updated model gradients into the parameter server. The central parameter server computes the latest global model parameter by aggregating the local updates from the selected users [22]. The training rounds will continue to loops unless the end condition is satisfied.

3.2 Gradient Memory

A multi-task machine learning model may suffer from the catastrophic forgetting problem, that means the learning model performs well on current tasks but not very well on previous tasks. As mentioned before, there are several approaches to alleviate the catastrophic forgetting problem. In this paper, we use the GEM [10], which mitigates the catastrophic forgetting problem by using an episodic memory $\mathcal{M}$ for storing the samples of previous tasks to guarantee the loss of previous tasks not to increase.

The GEM minimizes the loss on the current task, while avoiding the losses on the episodic memories of task $k < t$ increasing. The loss function on the memory is given by $\ell\left(f_\theta, \mathcal{M}_k\right) = \frac{1}{|\mathcal{M}_k|} \sum_{(x_i,k,y_i)\in\mathcal{M}_k} \ell\left(f_\theta\left(x_i, k\right), y_i\right)$. This allows GEM to maintain effective prediction on the past tasks. Formally, at task t, GEM solves for the following inequality constraints:

$$\begin{aligned} &\text{minimize}_\theta\ \ell\left(f_\theta(x,t), y\right) \\ &\text{subject to}\ \ell\left(f_\theta, \mathcal{M}_k\right) \leq \ell\left(f_\theta^{t-1}, \mathcal{M}_k\right)\ \text{ for all } k < t \end{aligned} \tag{3}$$

Where f_θ^{t-1} is the network trained till task t − 1. The angle between the loss gradient vector of previous tasks $G = -\left(g_1, \cdots, g_{t-1}\right)$ and the proposed update g_t is then computed. If the angle is greater than 90°, which refers that the loss in previous tasks increases, it projects the gradient to the closest gradient $\tilde{g}$ and keeps the angle less than 90°. Formally, the optimization problem is given by:

$$\begin{aligned} &\text{minimize}_{\tilde{g}}\ \frac{1}{2}\|g - \tilde{g}\|_2^2 \\ &\text{subject to}\ \langle\tilde{g}, g_k\rangle \geq 0 \text{ for all } k < t \end{aligned} \tag{4}$$

The projected gradient $\tilde{g}$ is finally given by $\tilde{g} = G^\top v^\star + g$, where $v^\star$ is the optimal value of the following QP problem:

$$\begin{aligned} &\text{minimize}_v\ \frac{1}{2}v^\top GG^\top v + g^\top G^\top v \\ &\text{subject to}\ v \geq 0 \end{aligned} \tag{5}$$

4 Problem Formulation

In this paper, we consider a common distributed machine learning scenario involving multiple parties. Each participant has their own dataset, predictable models and is willing to share data. The central server can coordinate these participants to make full use of their data. In the traditional FL, there is only one central parameter server communicating with N clients. We extend this definition to apply a hierarchical FL architecture and use the method of gradient memory to connect multiple layers, thus referring to it as Gradient Memory-based Federated Learning for Hierarchical Knowledge Transferring (GradMFL). In GradMFL, there are multiple servers, respectively controlling relevant participants, which are organized in a chain structure. A restricted version of GradMFL, considering only two layers, has been presented in Fig. 2. The use of multiple parameter servers is important because the tasks are diverse related to the different datasets and only one server cannot meet all of these distinct and scattered requirements. Our goal is to design an efficient federated learning mechanism, which can take better advantage of distributed data and avoid repetitive model training from scratch.

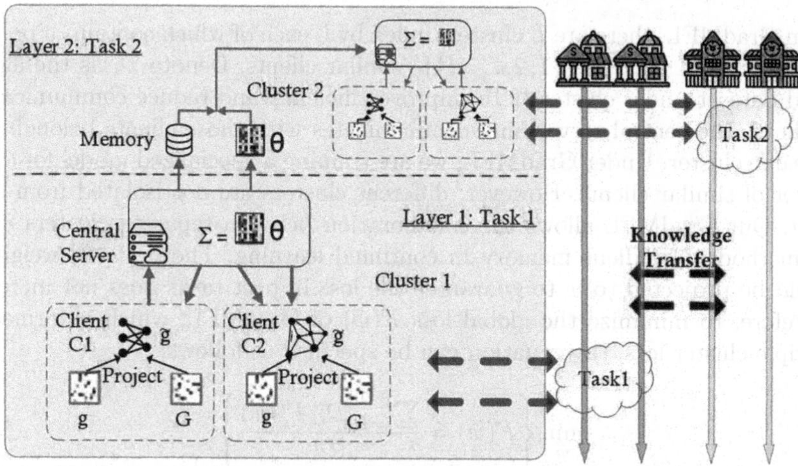

Fig. 2. Illustration of Proposed GradMFL Framework. The close clients are clustered into two layers, and then the lower system trains the specialized task based on gradient memory and sends model parameters and supplementary memory to the upper layer for its training.

The system model considered in this paper is introduced in detail in this subsection. There exist a set of n parties $\mathcal{P} = P_1, P_2, \ldots, P_n$. For any party $\mathcal{P}_i$, it holds a local dataset $\mathcal{D}_i$. Each of the n parties agrees on sharing its data to construct a better global model. Let $L = l_1, l_2, \ldots, l_m$ be the clusters merging the most similar clients, instead of assembling all the parties in a FL architecture, we divide the participated users into different groups according to their data. Denote $D_{i,j} = \{\boldsymbol{x}_k, y_k\}_{k=1}^{|D_{i,j}|}$ as a dataset held by the j-th participants belonging to the cluster i where x_k is the k-th input training sample, corresponding with the label y_k, $|D_{i,j}|$ is the total number of training samples. we define the loss function as $f_k(\boldsymbol{w}) = l(\boldsymbol{x}_k, y_k, \boldsymbol{w})$, which is the prediction error on the k-th data sample when the parameter of the model is w. The training process to minimize the total empirical loss $F_j^i(\boldsymbol{w})$ based on training dataset $D_{i,j}$ can be specified as follows:

$$F_j^i(\boldsymbol{w}) = \frac{\sum_{k=1}^{|D_{i,j}|} l(\boldsymbol{x}_k, y_k, \boldsymbol{w})}{|D_{i,j}|} = \frac{\sum_{k=1}^{|D_{i,j}|} f_k(\boldsymbol{w})}{|D_{i,j}|} \tag{6}$$

In a cluster l_i of our GradMFL, the datasets distributed on total N_i clients is denoted as $\{\mathcal{D}_{i,j}\}_{j=1}^{N_i}$, with $\cup_{j=1}^{N_i} \mathcal{D}_{i,j} = \mathcal{D}_i$ and the parameter sever cannot access these data directly. Thus the global loss is not computed by the central server, but derived in the form of a weighted average of distributed local loss functions $F_j^i(\boldsymbol{w})$ on local datasets $D_{i,j}$. Specifically, the global loss function $F_i(\boldsymbol{w})$ of a cluster l_i and local loss functions $F_j^i(\boldsymbol{w})$ are given by:

$$F^i(\boldsymbol{w}) = \frac{\sum_{j=1}^{N_i} |\mathcal{D}_{i,j}| F_j^i(\boldsymbol{w})}{|\mathcal{D}_i|}, \quad F_j^i(\boldsymbol{w}) = \frac{\sum_{k \in \mathcal{D}_{i,j}} f_k(\boldsymbol{w})}{|\mathcal{D}_{i,j}|} \tag{7}$$

In GradMFL, there are L clusters index by l, each of which contains a parameter server and $N_i, (i \in \{\{1, 2, \ldots, l\}\})$ similar clients. Denote $\mathcal{D}_i$ as the aggregated dataset under cluster l. To improve efficiency and reduce communication overhead, the central server only communicates with those clients belonging to the same cluster. Under GradMFL, we are training a specialized model for every cluster of similar clients. However, different clusters are not isolated from each other. Our GradMFL allows the collaboration between separate clusters using the method of gradient memory in continual learning. The updated weight w should be projected to $\tilde{w}$ to guarantee the loss in past tasks does not increase. Therefore, to minimize the global loss $F(\tilde{w})$ of GradMFL, which is formed of multiple cluster loss, the equation can be specified as follows:

$$\min_{\tilde{w}} \left\{ F(\tilde{w}) = \frac{\sum_{i=1}^{l} |\mathcal{D}_i| F^i(\tilde{w})}{|\mathcal{D}|} \right\} \tag{8}$$

5 Proposed GradMFL Framework

In large-scale networks, where many clients vary enormously depending on how their data are distributed and what categories data belong to, the communication efficiency and model convergence performance may be prohibitively poor. To this end, we propose our GradMFL framework, where data providers are clustered according to their data distribution and categories, and allow knowledge transferring between different cluster models with the help of gradient memory. The GradMFL method proposed in this paper is composed of three sections. In the beginning, we give a description of the clustering strategy determining how to divide clusters. Subsequently, a collaborative training strategy of optimizing the local cluster model meanwhile maintaining the past cluster model performance is examined. Finally, the detailed procedure of GradMFL in this research is described.

5.1 Hierarchical Clustering Under Non-IID

This section explains how to categorize participants under Non-IID into collaboration clusters. It can be accomplished by clustering the clients depending on the dataset similarities to train specialized models for clusters of related clients. Since we cannot determine how many clusters to divide beforehand, a clustering algorithm to automatically compute the number of clusters and their relevant parties is important [15]. We consider a hierarchical clustering algorithm applied for our GradMFL, which has the ability to generate specific clusters and allocate all the parties to the most suitable clusters under the premise of unknown cluster numbers and data distribution. Another advantage of hierarchical clustering is that the input order of samples is insensitive and it can construct a hierarchical structure of similarity between participants.

In our research, we choose a hierarchical clustering method that repeats the process of merging the separate clusters until the stopping condition is satisfied. For a more clear description, we provide an agglomerative hierarchical clustering procedure in Algorithm 1. At first, all the samples are regarded as their own initial single cluster. Subsequently, the distance between every pair of clusters is measured to estimate the similarity between them. The most similar couple clusters are merged to form a new cluster. The procedure of above calculating similarity and merging is repeated until the end. We introduced a hyperparameter of a distance threshold in advance to decide when to terminate the clustering process. By this means, the most similar clients of IID are gathered in the same clusters, and the different clusters keep a considerable distance.

Algorithm 1. Hierarchical Clustering of GradMFL

Input: Client set $D = \{\boldsymbol{D}_1, \boldsymbol{D}_2, \ldots, \boldsymbol{D}_m\}$; Cluster distance metric function Dis; Distance threshold t; the minimum cluster distance d;
Output: Divided clusters $\mathcal{C} = \{C_1, C_2, \ldots, C_k\}$

for $j = 1, 2, \ldots, m$ **do**
  $C_j = \{\boldsymbol{D}_j\}$
end for
while $d < t$ **do**
  d = $\min\{Dis\,(C_i, C_j)\}$, $i, j \in 1, 2, \ldots, k$
  Merge the closest cluster C_{i*} and C_{j*} : $C_{i^*} = C_{i^*} \bigcup C_{j^*}$
  k = k-1;
  Reassign sequence number $1, 2, \ldots, k$
end while

In the above procedure, we use client distance to measure the similarities among different clients. Hence, choosing an appropriate distance metric to compute the similarity between clients is of significance. Based on the Coupled

Object Similarity [23], we define our distance metric to measure the similarity of the data providers. Given two client samples d_1, d_2, the relevance distance between them is defined as

$$Dis(\mathbf{d_1}, \mathbf{d_2}) = \sum_{j=1}^{n} \delta_j^A (d_{1j}, d_{2j}) \tag{9}$$

where δ_j^A is the Coupled Attribute Value Similarity defined in [23] in accordance with coupled data similarities.

When a cluster contains multiple clients, it is important to determine choosing which client in the cluster to calculate the distance between clusters. According to different definitions of the similarity between clusters, there exist three techniques for measuring the distance between two united clusters. Single linkage and complete linkage only consider certain characteristic data but ignore the characteristics of the data within the clusters. Consequently, we prefer average linkage which takes the average of the distances between two clusters. It is more reasonable compared with the other two methods.

5.2 Local Training via Gradient Memory

In this section, the learning process in the hierarchical clusters is described. One advantage of our proposal is that it allows the beneficial transferring of knowledge between clusters. For a single cluster, which consists of a central parameter server and multiple similar clients, it can be considered as a small-scale federated learning system. The learning process repeats local training and parameter aggregation until model convergence is achieved. Each cluster is assigned a specific task model. However, when a cluster is eager to learn a new task of other

Algorithm 2. Federated Cluster Model Updating Strategy based on Gradient Memory

Input: The communication round r, the participants set $\mathcal{P}$ in the cluster t, the dataset $\mathcal{D}$, the training algorithm f_θ and loss function ℓ;
Output: The parameter global model θ;

for each round r = $1, 2, \ldots$ **do**
    for each $P_i \in \mathcal{P}$ **do**
        for (x,y) in $\mathcal{D}_i$ **do**
            $g_{t,i} \leftarrow \nabla_\theta \ell(f_\theta, (x, y)_i)$
            $g_k \leftarrow \nabla_\theta \ell(f_\theta, \mathcal{M}_k)$ for all $k < t$
            $\tilde{g}_{t,i} \leftarrow$ Project $(g_{t,i}, g_1, \ldots, g_{t-1})$
            $\theta_i \leftarrow \theta_i - \alpha \cdot \tilde{g}_{t,i}$
        end for
        P_i uploads θ_i
    end for
    $\mathcal{A}$ aggregates $\theta \leftarrow \sum_{i=1}^{|\mathcal{N}|} \theta_i$ and adds some samples to memory $\mathcal{M}$
end for

clusters, catastrophic forgetting occurs which hinders the knowledge transferring between clusters. As an alternative, the combination of all the distributed datasets and the redesign of a new joint training model may be a solution, but it results in low communication efficiency and low model accuracy. Accordingly, a better method of using gradient memory for clusters to transfer knowledge is taken.

We utilize GEM in our hierarchical federated learning, a continual learning method, for the purpose of optimizing model parameters in local clusters while maintaining correct predictions of previous tasks. An episode memory $\mathcal{M}_t$, which stores the samples from task t, is used to alleviate forgetting. The scalar product between the previous loss gradient vector and the local updated gradient is computed, which is equivalent to loss in past tasks. If the value is negative, the loss in some previous tasks increases after the parameters update. Hence, the proposed gradient should be projected to the closest gradient to make sure the value between the new gradient and the previous sample gradient is all positive. The project gradient is given by $\tilde{g} = G^\top v^\star + g$, where g is the gradient of local samples, and $G = -(g_1, \ldots, g_{t-1})$, is the gradients of samples of previous tasks $1, 2, \ldots, t-1$ and $v^\star$ is the optimal value of the following problem:

$$\begin{aligned} &\text{minimize}_v \ \tfrac{1}{2} v^\top G G^\top v + g^\top G^\top v \\ &\text{subject to} \ \ v \geq 0 \end{aligned} \tag{10}$$

The cluster model updating strategy is shown in Algorithm 2. Denote L as the request submitter cluster which learns a continuum task $T_1, T_2, \ldots, T_{k-1}$. The cluster L is assigned with a task t and the rest of $t-1$ tasks is in the charge of other $t-1$ clusters. At first, the server $\mathcal{A}$ in the cluster L distributes the learning model f_t and $\mathcal{M}_t$ to all the clients. Then, the clients compute the gradients $g_1, g_2, \ldots, g_{t-1}$ of previous samples in the $t-1$ tasks, its own gradient vector derived g_t by local dataset. Subsequently, the local clusters update with the projected gradient $\tilde{g}$ and send the updated parameters θ to the central server for aggregation. The process of local update with gradient memory and central aggregation and distribution repeats until the model is convergent.

5.3 Workflow of Proposed GradMFL

The detailed steps of our GradMFL scheme are as follows.

1) Initialization: As mentioned before, determining the partner clients for collaborative training is beneficial for better convergence in fewer training rounds and higher model accuracy. In this research, a hierarchical clustering based on similarity distance is executed to cluster all the participated parties under Non-IID into various groups of strong connections according to their distance towards each other. The more similar the two datasets are, the closer their distance is.

   Let P_i denote the i-th participant, C_i the i-th cluster, $Dis(C_i, C_j)$ the distance between i-th, j-th cluster. We also set the hyperparameter t as a limit to end the clustering. All the participants are partitioned into their own cluster in the beginning. For two clusters C_i, C_j, the distance between them $Dis(C_i, C_j)$ is derived by averaging the distances between the every two clients in C_i, C_j. Then, the minimum distance Dis_{min} between the clusters is selected to compare with the threshold t. If the distance of the closest clusters is less, these clusters are merged as a new cluster. The process of merging repeats until all the distances of clusters are greater than the threshold. It is a great selection of a hierarchical clustering algorithm under this circumstance, which can group the clients quickly and yield optimal results in the case of small computational complexity.

2) Launching Requests: Let $R = r_1, r_2, \ldots, r_t$ be the requests submitted by the clusters $C_1, C_2, \ldots, C_t$. We assume the request r_k is to learn a continuum of tasks $T_1, T_2, \ldots, T_{k-1}$ in the previous clusters in addition to its own task T_k. When a cluster launches a request, it searches the related clusters for collaborative learning.
3) Allocating Clusters In Hierarchy: A hierarchical cluster architecture is built by allocating the cluster C_i in the layer $\mathcal{L}_i$ according to the learning sequence. The hierarchy contains all the relevant clusters, each of which is assigned a specific model. The training process in the hierarchy starts from the bottom layer $\mathcal{L}_1$ to upper layers $\mathcal{L}_2, \ldots, \mathcal{L}_t$. Every two subsequent layers can have communication of exchanging their models to the upper layer, meanwhile pushing the last m samples to the memory $\mathcal{M}$.
4) Updating Cluster Models: The clusters at upper layers can capture knowledge from lower trained models. For a request r_t, we assume the previous clusters $C_1, C_2, \ldots, C_{t-1}$ at layers $\mathcal{L}_1, \mathcal{L}_2, \ldots, \mathcal{L}_{t-1}$ has finished their training and the memory $\mathcal{M}$ is representative of samples from the past clusters. In cluster C_t, the clients start to train the local model with its own dataset. The client $\mathcal{P}$ inputs the samples to derive the loss and gradients. Moreover, to make sure the loss of previous samples $(\mathcal{M}_1, \mathcal{M}_1, \ldots, \mathcal{M}_{t-1})$ not increase, the extra angle between the gradients of previous samples and local inputs are needed. The local gradients will be projected if the direction between them is different. After client local training, the clients send the updated projected gradients $\tilde{g}$ to the central server for aggregation.
   For the request r_{t+1}, it is also the same transmission process where the cluster C_t uploads its model parameters to C_{t+1}, adds m samples to the memory $\mathcal{M}$ and the cluster C_{t+1} repeats the cluster model updating process as described above. The training in a whole hierarchy is a continual learning process. It is noteworthy that the samples stored in the memory contain no sensitive information because the cluster has implemented the data masking.

6 Experiment

In this section, we present simulation results of GradMFL to verify the observations from the performance analysis and illustrate the advantages of the GradMFL framework.

6.1 Experimental Setup

We validate our GradMFL on the image classification problem of identifying objectives from pixel data. The publicly available MNIST dataset is chosen for this purpose in our experiment. This dataset includes 60,000 training samples and 10,000 testing instances of handwritten digits. Each example is a 28×28 grey-scale image of a digit between 0 and 9. The more complex Fashion-MNIST dataset is also considered which consists of 70,000 different Zalando's fashion gray-scale images of 10 categories [24]. Meanwhile, a variant of MNIST and Fashion-MNIST called Permuted MNIST and Permuted Fashion-MNIST is used to add more classes of heterogeneous data. The input pixels of training samples are permuted at a certain order in all images of this dataset. We use a fully connected network that is composed of three hidden layers, each including 100 ReLU units. In this setting, we consider 10 tasks, corresponding with 10 Permuted MNIST or Fashion MNIST, each of which is permuted randomly. For simplicity, the training data of each dataset is split into 20 disjoint sets. Each client holds a divided dataset, which contains 3,000 training samples and 500 testing instances. After clustering, the parties holding the homogeneous data are merged as a new cluster in hierarchy. For the local training on clients, we employ mini-batch Stochastic Gradient Descent (SGD) and set the batch size for training to $\beta = 64$, the learning rate to 0.1. The size of memory storing previous samples is set in different values to explore its impact.

For evaluation, we assume test sets in all previous tasks are accessible. Let acc_j be the accuracy evaluated on the test set of task T_j. After finishing learning the task T_k, the average performance on all k tasks is given as $ACC = \frac{1}{k}\sum_{j=1}^{k} acc_j$.

6.2 Result and Discussion

The first conclusion to verify is that more samples in the memory can improve the model performance in all clusters. In Fig. 3(a) and 3(b), we show how the average accuracy changes as the capacity of the memory vary. We see that the average accuracy with the existence of the memory is always high when the tasks continue in the hierarchical clusters, which demonstrates that knowledge is effectively transferred between clusters, and forgetting is alleviated in our framework. More importantly, we see that as the capacity of the memory expands, the average accuracy increases. This is due to the fact that the current task can learn more knowledge from the past samples stored in the memory.

Further, we examine how the performance of the past tasks changes during the number of clients in a cluster varies, to see how the federated setting improves the convergence and accuracy. Figure 3(c) and 3(d) show the performance of GradMFL based on 1, 2, 5, 10, 15 clients in a cluster when the memory size is stationary 3000 samples. If only one cluster exists, it is equivalent to conventional continual learning which we observe suffers from poor performance because of insufficient samples and the under-fitting of the model. In contrast, our framework shows higher average accuracy in all tasks. This is mainly because

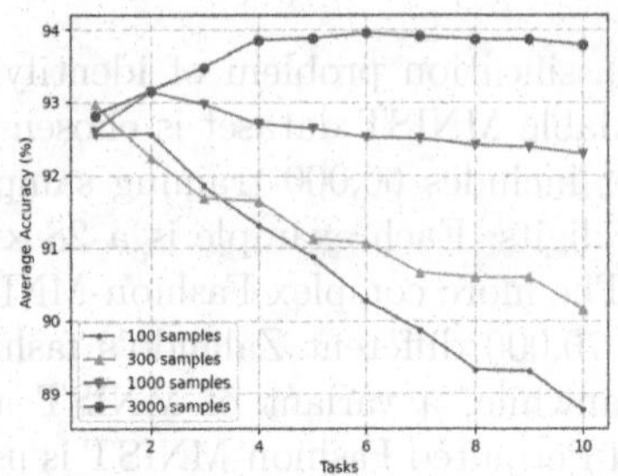

(a) Average accuracy changes as the memory size varies on Permuted MNIST

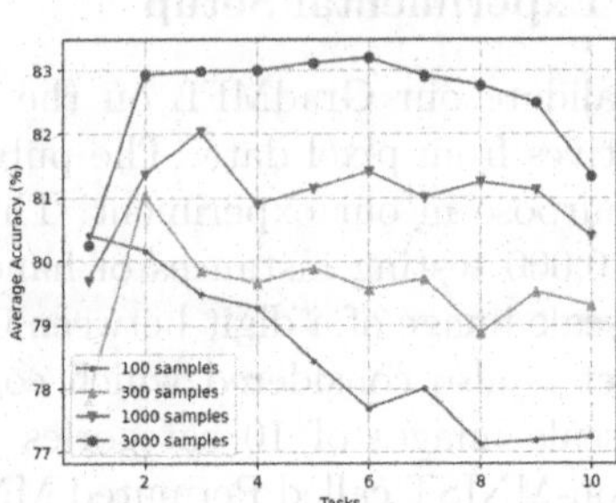

(b) Average accuracy changes as the memory size varies on Permuted Fashion-MNIST

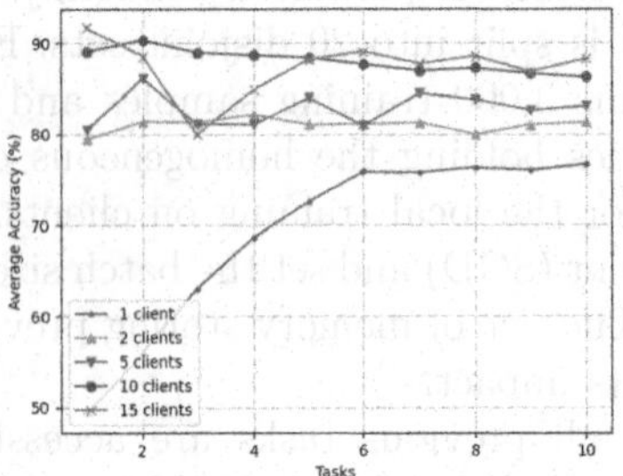

(c) Average accuracy changes as client number varies on Permuted MNIST

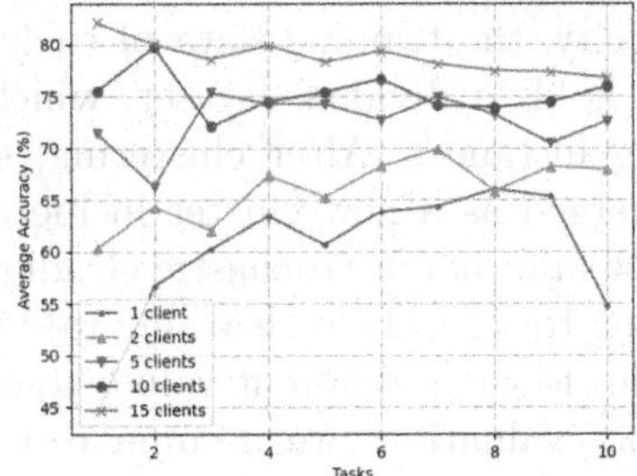

(d) Average accuracy changes as client number varies on Permuted Fashion-MNIST

Fig. 3. Average Accuracy changes as the memory size and client numbers vary.

each task is processed in a collaborative federated setting, where the sharing of distributed datasets provides enough training data. We also observe that the performance improves rapidly as the number of clients increases at first. But when the number of clients continues to increase, the performance improves at a lower step and then remains almost unchanged.

7 Conclusion

In this paper, we proposed the GradMFL framework that performs a continuum of learning tasks in the hierarchical federated learning setting, to enable strong convergence and energy-efficient. In this method, the hierarchical clustering method is first introduced to cluster the clients at different levels hierarchically according to the similarity of their datasets, without considering the optimal size and numbers of clusters in advance. The update of the inner cluster model is effective due to the homogeneous data distribution after clustering. Further, we present a collaborative learning algorithm for training a series of tasks in hierarchical clusters, by introducing the use of gradient memory, to allow beneficial knowledge transferring between these clusters. The experimental validation under various datasets shows that our GradMFL framework accomplishes good accuracy and high efficiency.

Acknowledgement. This work is supported by National Natural Science Foundation of China under Grant No. U20B2048 and 61972255, Shanghai Sailing Program under Grant No. 21YF1421700, Special Fund for Industrial Transformation and Upgrading Development of Shanghai Under Grant No. GYQJ-2018-3-03 and Shanghai Municipal Science and Technology Major Project under Grant 2021SHZDZX0102.

References

1. Yan, Z., Wu, J., Li, G., Li, S., Guizani, M.: Deep neural backdoor in semi-supervised learning: threats and countermeasures. IEEE Trans. Inf. Forensics Secur. **16**, 4827–4842 (2021). https://doi.org/10.1109/TIFS.2021.3116431
2. Huang, X., Leng, S., Maharjan, S., Zhang, Y.: Multi-agent deep reinforcement learning for computation offloading and interference coordination in small cell networks. IEEE Trans. Veh. Technol. **70**(9), 9282–9293 (2021). https://doi.org/10.1109/TVT.2021.3096928
3. Yang, Q., Liu, Y., Cheng, Y., Kang, Y., Chen, T., Yu, H.: Federated learning. Synth. Lect. Artif. Intell. Mach. Learn. **13**(3), 1–207 (2019)
4. Wu, Y., Zhang, K., Zhang, Y.: Digital twin networks: a survey. IEEE Internet Things J. **8**(18), 13789–13804 (2021). https://doi.org/10.1109/JIOT.2021.3079510
5. McMahan, H.B., Moore, E., Ramage, D., Arcas, B.A.Y.: Federated learning of deep networks using model averaging. CoRR abs/1602.05629 (2016)
6. Kairouz, P., et al.: Advances and open problems in federated learning. arXiv preprint arXiv:1912.04977 (2019)
7. Wang, J., Liu, Q., Liang, H., Joshi, G., Poor, H.V.: Tackling the objective inconsistency problem in heterogeneous federated optimization. arXiv preprint arXiv:2007.07481 (2020)
8. Nielsen, F.: Hierarchical clustering. In: Introduction to HPC with MPI for Data Science. UTCS, pp. 195–211. Springer, Cham (2016). https://doi.org/10.1007/978-3-319-21903-5_8
9. Pfülb, B., Gepperth, A., Abdullah, S., Kilian, A.: Catastrophic forgetting: still a problem for DNNs. In: Kůrková, V., Manolopoulos, Y., Hammer, B., Iliadis, L., Maglogiannis, I. (eds.) ICANN 2018. LNCS, vol. 11139, pp. 487–497. Springer, Cham (2018). https://doi.org/10.1007/978-3-030-01418-6_48
10. Lopez-Paz, D., Ranzato, M.: Gradient episodic memory for continual learning. In: Advances in Neural Information Processing Systems 30: Annual Conference on Neural Information Processing Systems 2017, Long Beach, CA, USA, 4–9 December 2017, pp. 6467–6476 (2017)
11. Yang, Q., Liu, Y., Chen, T., Tong, Y.: Federated machine learning: concept and applications. ACM Trans. Intell. Syst. Technol. **10**(2), 12:1-12:19 (2019). https://doi.org/10.1145/3298981
12. Li, T., Sahu, A.K., Zaheer, M., Sanjabi, M., Talwalkar, A., Smith, V.: Federated optimization in heterogeneous networks. arXiv preprint arXiv:1812.06127 (2018)
13. Smith, V., Chiang, C.K., Sanjabi, M., Talwalkar, A.: Federated multi-task learning. arXiv preprint arXiv:1705.10467 (2017)

14. Sattler, F., Müller, K.R., Samek, W.: Clustered federated learning: model-agnostic distributed multitask optimization under privacy constraints. IEEE Trans. Neural Netw. Learn. Syst. **32**(8), 3710–3722 (2020)
15. Briggs, C., Fan, Z., Andras, P.: Federated learning with hierarchical clustering of local updates to improve training on non-IID data. In: 2020 International Joint Conference on Neural Networks (IJCNN), pp. 1–9. IEEE (2020)
16. Delange, M., et al.: A continual learning survey: defying forgetting in classification tasks. IEEE Trans. Pattern Anal. Mach. Intell. (2021)
17. Kemker, R., McClure, M., Abitino, A., Hayes, T., Kanan, C.: Measuring catastrophic forgetting in neural networks. In: Proceedings of the AAAI Conference on Artificial Intelligence, vol. 32 (2018)
18. Kirkpatrick, J., et al.: Overcoming catastrophic forgetting in neural networks. CoRR abs/1612.00796 (2016)
19. Xiao, T., Zhang, J., Yang, K., Peng, Y., Zhang, Z.: Error-driven incremental learning in deep convolutional neural network for large-scale image classification. In: Proceedings of the 22nd ACM International Conference on Multimedia, pp. 177–186 (2014)
20. Chaudhry, A., Ranzato, M., Rohrbach, M., Elhoseiny, M.: Efficient lifelong learning with A-GEM. arXiv preprint arXiv:1812.00420 (2018)
21. Pan, Q., Wu, J., Bashir, A.K., Li, J., Yang, W., Al-Otaibi, Y.D.: Joint protection of energy security and information privacy for energy harvesting: An incentive federated learning approach. IEEE Trans. Ind. Inform. (2021). https://doi.org/10.1109/TII.2021.3105492
22. Zhang, C., Xie, Y., Bai, H., Yu, B., Li, W., Gao, Y.: A survey on federated learning. Knowl.-Based Syst. **216**, 106775 (2021)
23. Li, M., Li, J., Ou, Y., Zhang, Y., Luo, D., Bahtia, M., Cao, L.: Coupled K-nearest centroid classification for non-IID data. In: Nguyen, N.T., Kowalczyk, R., Corchado, J.M., Bajo, J. (eds.) Transactions on Computational Collective Intelligence XV. LNCS, vol. 8670, pp. 89–100. Springer, Heidelberg (2014). https://doi.org/10.1007/978-3-662-44750-5_5
24. Xiao, H., Rasul, K., Vollgraf, R.: Fashion-MNIST: a novel image dataset for benchmarking machine learning algorithms. arXiv preprint arXiv:1708.07747 (2017)

Linear Coded Federated Learning

Yingyao Yang[1], Jin Wang[1(✉)], Kejie Lu[2], Jianping Wang[3], and Zhaobo Lu[1]

[1] College of Computer Science and Technology, Soochow University, Suzhou 215006, China
{20195227058,20195227041}@stu.suda.edu.cn, wjin1985@suda.edu.cn

[2] Computer Science and Engineering, University of Puerto Rico at Mayagüez, Mayagüez, PR, USA
kejie.lu@upr.edu

[3] Computer Science, City University of Hong Kong, Hong Kong, China
jianwang@cityu.edu.hk

Abstract. In recent years, *federated learning* (FL) has attracted a lot of attention as a new edge computing paradigm for *artificial intelligence* (AI). FL facilitates multiple edge devices to collaboratively train a global model without leaking local data of any participant. In a typical edge computing scenario, the participants in FL are heterogeneous and can be composed of personal computers, smartphones, Internet of Things devices, network devices, *etc.*. In this heterogeneous setting, the slowest client in each training round becomes the bottleneck, which may limit the overall convergence speed and accuracy of the global model. To address this issue, one possible solution is to outsource the computing task in the slowest client to faster devices, which requires data transmission from the slowest client to other selected clients. Certainly, sending the original data set is not an option due to the privacy requirements. Therefore, in this paper, we propose an efficient *linear coded federated learning* (LCFL) framework to (1) speed up the convergence speed of heterogeneous FL and (2) protect the data privacy of the participants. Within the proposed framework, we design a *collaborative client selection* (CCS) algorithm that can select appropriate clients and assign the computation task of the slowest client to those selected devices. Finally, we build a practical experimental platform and conduct numerous experiments to evaluate the proposed LCFL framework from different aspects. The experimental results demonstrate that the proposed LCFL scheme can reduce the training time up to 93.73% when the participants have a large difference in terms of computing capability.

Keywords: Federated learning · Linear coding · Edge computing · Scheduling algorithm

1 Introduction

In the past few years, *federated learning* (FL) has attracted significant attention as a new edge computing paradigm for *artificial intelligence* (AI) [1,2]. FL usually consists of a server and multiple clients, each of which stores local data for

Y. Lai et al. (Eds.): ICA3PP 2021, LNCS 13155, pp. 627–644, 2022.
https://doi.org/10.1007/978-3-030-95384-3_39

training. FL can train the global model by scheduling clients to train the model locally without leaking local data of a participant to the server or other participants. To achieve the high accuracy of the global model, model training in FL needs multiple rounds.

Data privacy and the performance of training (convergence speed, accuracy, *etc.*) are major concerns in FL. In general, to ensure data privacy in FL, the local data of each client will not be transmitted to the server or other clients. Nevertheless, the participants of FL in a typical edge computing scenario are usually heterogeneous and can be composed of *personal computers* (PCs), smartphones, *Internet of Things* (IoT) devices, *etc.* In such a heterogeneous scenario, the relatively slow clients, namely stragglers, become the bottleneck, which limits the overall convergence speed and accuracy of the global model [3,4].

Since the stragglers must participate in the FL to increase the model accuracy, one remedy solution is to outsource the computing tasks to other clients. To this end, although a few existing *distributed machine learning* (DML) platforms can provide outsourcing support, they cannot protect data privacy [5]. Moreover, most existing privacy-preserved outsourcing schemes are based on homomorphic encryption or differential privacy, which may have performance issues [6–9]. Firstly, homomorphic encryption leads to high computation overheads, especially for a straggler with weak computation power, such as IoT devices [6,7]. Secondly, differential privacy-based schemes disturb the data by adding noise to protect the outsourced data. However, the disturbance may mask the true value [8].

To address the above issues, *coded distributed computing* (CDC) can be applied because it can not only mitigate the negative effects of stragglers, but also provide the privacy of the computing data in DML [10,11]. Recently, a linear coding approach has been introduced to train linear models in FL, in which the decoding process is not necessary. In [10], Dhakal *et al.* applied such a linear coding scheme to train a *linear regression* model. Their scheme can protect data privacy by linearly encoding the original data (both inputs and outputs) and does not require the decoding process with an assumption that the coding error is negligible. In [11], Prakash *et al.* extended the model in [10] to train a classification model, in which the data in all clients are first transformed by using a *random Fourier feature mapping* (RFFM) kernel [12], then the transformed data can be used to train a linear regression model with regularization. Although these two schemes are viable, the encoding process may introduce errors that affect model accuracy. Moreover, the approach cannot be applied to more general nonlinear models, such as *convolutional neural networks* (CNNs) in deep learning. In this paper, we aim to develop a coded computing scheme for FL that uses linear encoding to protect data privacy and obtain accurate training results. Therefore, our scheme can train both linear models and nonlinear models, which is more general.

In addition to the privacy of participants' data, the scheduling of clients in each round is another important issue in FL [13–16]. Although the server can schedule all clients to train the model in each round [13], such a simple scheduling

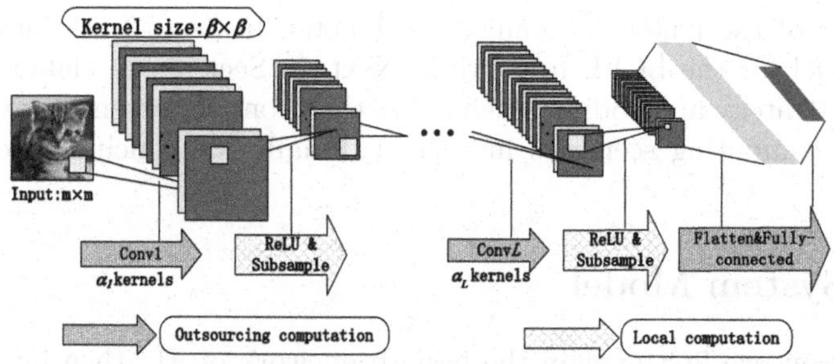

Fig. 1. CNN model structure

scheme may have a poor performance when participants are heterogeneous and there are many stragglers. For heterogeneous scenarios, the client selection ratio is considered for each training round as a hyper-parameter [14]. For example, in [15], Kairouz *et al.* considered a large-scale FL scenario with 10^6 to 10^{10} clients, in which they selected 50 to 5000 clients in each training round. Besides applying the client selection ratio, another idea is to monitor the running speed of each client and schedule slow clients less frequently [16].

In this paper, we propose an efficient *linear coded federated learning* (LCFL) framework for heterogeneous computing environments to (1) speed up the convergence speed of heterogeneous FL and (2) protect the data privacy of the participants. Within the LCFL framework, a training round is partitioned into a sequence of stages, each of which processes inputs and generates outputs. Clearly, such a sequential process is common to most learning frameworks, *e.g.*, the CNN model illustrated in Fig. 1. In each stage, the LCFL framework determines whether outsourcing is necessary. If outsourcing is not necessary, then the computation can be done locally, *e.g.*, the *rectified linear unit* (*ReLU*) function. However, when outsourcing is necessary, *e.g.*, for a convolution process in CNN, our work has the following contributions:

- We study and design a novel FL framework based on linear coding for a heterogeneous computing environment.
- Within the framework, we propose a *collaborative client selection* (CCS) algorithm, with which an appropriate number of the straggler's computing tasks can be outsourced to accelerate the training process.
- We design and implement a practical experimental platform that includes resource-limited clients, such as Raspberry Pi. Experimental results show that our system can significantly reduce the training time in practical scenarios with heterogeneous clients.

In this paper, we implement the proposed LCFL scheme in practical scenarios to complete the training of CNN models for image classification tasks using real a data set. To the best of the authors' knowledge, no previous work has been conducted to apply linear coding to nonlinear neural networks in FL.

The rest of the paper is organized as follows. In Sect. 2, we introduce the system model for the LCFL framework. Next, in Sect. 3, we elaborate on the scheduling strategy and coding scheme. We then conduct extensive experiments in realistic computing scenarios in Sect. 4. Finally, we conclude the paper in Sect. 5.

2 the System Model

In this section, we first explain the basic framework of FL, then introduce the coded distributed computing for CNN, and finally show the linear coded federated learning model (Table 1).

Table 1. Notations

Notations	Meaning
$\mathbf{A}_i$	The global model of i-th training round
$\mathbf{b}_i$	The biases of j-th convolutional layer in i-th training round
$\mathbf{C}$	The coding matrix
$\mathbf{E}_j^i$	The weights of j-th convolutional layer in i-th training round
g_l^i	The data transmission speed of l-th collaborative client in i-th training round
M	The number of blocks in blocking code
n	The total number of clients participating in FL
$\mathbf{Q}_k$	The matrix of data set owned by k-th client
$\mathbf{S}$	The set of all clients
$\mathbf{S}_i$	The set of clients scheduled in i-th training round
$\mathbf{S}_{c_i}$	The set of collaborative clients used in the i-th training round
u_k	The amount of data owned by the k-th client
$\mathbf{V}$	The set of all clients' speeds
v_l^i	The computational speed of l-th collaborative client in i-th training round
$\overline{\mathbf{W}}_i$	The set of the number of tasks to be undertaken by each client in a collaborative computing task
ξ	The number of collaborative clients

2.1 Federated Learning

In this sub-section, we use mathematical formulas to describe the process of CNN in heterogeneous FL as shown in Fig. 1. The CNN model consists of L convolutional layers, L subsampling layers, and one fully connected layer. FL completes the training of global model $\mathbf{A}$ by scheduling multiple rounds of clients. The system contains n heterogeneous clients $\mathbf{S} = \{s_1, \cdots, s_n\}$, and the detailed process of FL is as follows:

- **Model Target Confirmation**: Before scheduling clients to train, the server randomly generates or pre trains the global model $\mathbf{A}_0$ according to deep learning problems, *e.g.*, image classification [18].
- **Clients Selection**: At the beginning of i-th training round, the server sends the model $\mathbf{A}_{i-1}$ to the selected subset of all clients, denoted as $\mathbf{S}_i = \{s_i^1, \cdots, s_i^{h_i}\}$, in which h_i represents the number of selected clients in i-th training round.
- **Local Training**: In i-th training round, each selected client s_i^k trains model using local data. Let $\mathbf{E}_i^j$ and $\mathbf{b}_i^j$ be the weights and biases of the L-th convolutional layers and one fully connected layer, $j \in \{1, \cdots, L+1\}$. Set $\mathbf{X}_{j-1}$ as the input and $\mathbf{Y}_j$ as the output of j-th layer. Especially, $\mathbf{X}_0$ is the local data set. The core functions in forward propagation are expressed as follows:

$$Convolutional\ layers: \quad \mathbf{Y}_j = \mathbf{X}_{j-1} \odot \mathbf{E}_i^j + \mathbf{b}_i^j, j \in \{1, \cdots, L\} \quad (1)$$

$$Fully\ connected\ layer: \quad \mathbf{Y}_{L+1} = \mathbf{E}_i^{L+1}\mathbf{X}_L + \mathbf{b}_i^{L+1} \quad (2)$$

  $\odot$ in Eq. (1) represents the convolutional operation. The error Δ_i^k can be obtained by calculating the forward propagation calculation results and the real value of the label set using cross-entropy. After backward propagation, the gradient ∇_i^k of i-th training round can be obtained.
- **Model Aggregation**: At the end of the i-th training round, the server aggregates the gradients calculated by all selected clients $\mathbf{S}_i$ to obtain the global gradient $\nabla_i = \frac{\sum_{\rho=1}^{h_i} \nabla_i^{\rho}}{h_i}$. Since the gradient is consistent with the dimension of the model parameters, the model $\mathbf{A}_{i-1}$ can be updated using $\mathbf{A}_i = \mathbf{A}_{i-1} - r\nabla_i$, where r is the learning rate.
- **Model Verification**: The server tests the test set using the updated model $\mathbf{A}_i$. If the loss and accuracy meet expectations, the server sends the model $\mathbf{A}_i$ to all clients $\mathbf{S}$ and finishes the training. Otherwise, $(i+1)$-th training round will continue the training based on the model $\mathbf{A}_i$.

The above is the training process of traditional FL under heterogeneous background. Throughout the forward propagation, convolution operation and fully connected operation in Eq. (1) and Eq. (2) are the most computationally demanding functions. Similarly, these two functions also occupy huge computational power in backward propagation. Therefore, complex tasks need to be outsourced to speed up the training.

2.2 Coded Distributed Computing for CNN

In this sub-section, we mainly discuss how to apply linear coding to the training of CNN.

In the literature, linear coding has been applied to matrix multiplication in different scenarios. As mentioned in Sect. 2.1, convolution and fully connected operations in forward propagation and backward propagation need to be outsourced. The convolution operation can be converted to matrix multiplication without error [17]. The weights $\mathbf{E}_i^j$ and biases $\mathbf{b}_i^j$ $(j \in \{1, \cdots, L\})$ of the convolution layer in the i-th training round can be converted to $\widehat{\mathbf{E}}_i^j$ and $\widehat{\mathbf{b}}_i^j$ through matrix transformation. Equation (1) can be described in the form of matrix multiplication, which is given as $\widehat{\mathbf{Y}}_j = \widehat{\mathbf{X}}_{j-1}\widehat{\mathbf{E}}_i^j + \mathbf{b}_i^j, j \in \{1, \cdots, L\}$. Similarly, the calculation result $\widehat{\mathbf{Y}}_j$ can be transformed into $\mathbf{Y}_j$ via matrix transformation [17].

A large number of data matrices need to be divided into multiple blocks of the same size before being encoded. We define an operator for block matrix multiplication in the following Definition 1.

Definition 1 ***(Block matrix multiplication*** $\circledast$***).*** *Assume that* $\boldsymbol{C}$ *is an* $f \times f$ *dimensional matrix,* $\boldsymbol{Q}$ *is an* $h \times g$ *dimensional matrix,* h *is divisible by* f*,* $\boldsymbol{Q}$ *is divided into* f *blocks by rows* $\boldsymbol{Q} = [\boldsymbol{Q}_1^\top, \cdots, \boldsymbol{Q}_f^\top]^\top$*, and the dimension of each block is* $\frac{h}{f} \times g$*, then we can have the* $\frac{h}{f} \times g$ *dimensional matrix* $\boldsymbol{I} = \boldsymbol{C} \circledast \boldsymbol{Q} = [\boldsymbol{I}_1^\top, \cdots, \boldsymbol{I}_f^\top]^\top$*, in which each* $\frac{h}{f} \times g$ *dimensional sub-matrix:*

$$\boldsymbol{I}_i = \boldsymbol{C}_i \circledast \boldsymbol{Q} = \sum_{k=1}^{f} \boldsymbol{C}_{i,k}\boldsymbol{Q}_k, \forall i \in \{1, \cdots, f\}. \tag{3}$$

The coding matrix $\mathbf{C}$ must be full-ranked. Otherwise, the inverse matrix of coding matrix $\mathbf{C}$ will not exist and the raw result can not be obtained.

Next, we explain how the coded computing can be performed by a group of collaborative clients. Suppose there are ξ collaborative clients selected for helping the straggler in the i-th round. We divide data $\mathbf{X}$ into M $(M \geqslant \xi)$ blocks, i.e., $\mathbf{X} = \left[\mathbf{X}_1{}^\top, \cdots, \mathbf{X}_M{}^\top\right]^\top$. The number of blocks is defined as $\mathbf{W}_i = \left[w_1^i, \cdots, w_\xi^i\right]$, where w_l^i $(l \in \{1, \cdots, \xi\})$ represents the number of data blocks to be allocated to the scheduled l-th collaborative client and $\sum_{\rho=1}^{\xi} w_\rho^i = M$. Based on the speed of each collaborative $\mathbf{V} = \{v_1, \cdots, v_\xi\}$, we can get $w_l^i = \frac{v_l}{\sum_{\rho=1}^{\xi} v_\rho} M$. Regardless of redundancy, the specific coding computing process in the j-th $(j \in \{1, \cdots, L\})$ convolutional layer is as follows:

- **Data encoding**. In the j-th convolutional layer of the i-th training round, the data $\mathbf{X}$ and weights $\mathbf{E}_j^i$ are first converted to the $\widehat{\mathbf{X}}$ and $\widehat{\mathbf{E}}_j^i$ required for matrix multiplication [17]. We use the coding matrix $\mathbf{C}$ with dimension $M \times M$ to encode block data $\widehat{\mathbf{X}}$ to get $\widehat{\mathbf{X}}^* = \mathbf{C} \circledast \widehat{\mathbf{X}}$, where $\widehat{\mathbf{X}}^* = \left[\widehat{\mathbf{X}}_1^{*\top}, \cdots, \widehat{\mathbf{X}}_M^{*\top}\right]^\top$. Then, according to the number of data blocks allocated in $\mathbf{W}_i$, the M block coded data are divided into $\widehat{\mathbf{X}}_{\mathbf{c}}^* = [\widehat{\mathbf{X}}_{\mathbf{c}_1}^{*\top}, \cdots, \widehat{\mathbf{X}}_{\mathbf{c}_\xi}^{*\top}]^\top$. Finally, we send $\widehat{\mathbf{X}}_{\mathbf{c}}^*$ and $\widehat{\mathbf{E}}_j^i$ to ξ collaborative clients.
- **Local convolution calculation**. In the i-th training round, the l-th collaborative client uses the received $\widehat{\mathbf{E}}_j^i$ and data $\widehat{\mathbf{X}}_{\mathbf{c}_l}^*$ to complete the convolution operation in matrix multiplication form and obtains $\widehat{\mathbf{X}}_{\mathbf{c}_l}^*\widehat{\mathbf{E}}_j^i$.

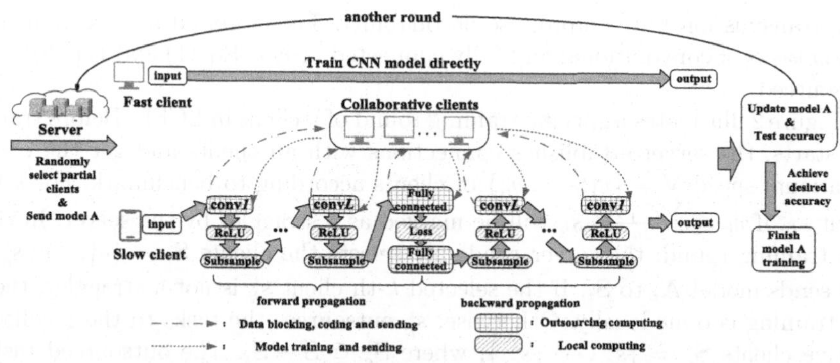

Fig. 2. System overall model structure

– **Result aggregation**. Aggregation of the computational results of all collaborative clients yields $\left[(\widehat{\mathbf{X}}^*_{\mathbf{c}_1}\widehat{\mathbf{E}}^i_j)^\top, \cdots, (\widehat{\mathbf{X}}^*_{\mathbf{c}_\xi}\widehat{\mathbf{E}}^i_j)^\top\right]^\top$. We get the correct result by decoding the aggregation result and completes the summation of the biases $\mathbf{b}^i_j$ and decoding results, which is given as

$$\begin{aligned}\widehat{\mathbf{Y}} &= \mathbf{C}^{-1}\left[(\widehat{\mathbf{X}}^*_{\mathbf{c}_1}\widehat{\mathbf{E}}^i_j)^\top, \cdots, (\widehat{\mathbf{X}}^*_{\mathbf{c}_\xi}\widehat{\mathbf{E}}^i_j)^\top\right]^\top + \mathbf{b}^i_j \\ &= \mathbf{C}^{-1}\left[(\widehat{\mathbf{X}}^*_1\widehat{\mathbf{E}}^i_j)^\top, \cdots, (\widehat{\mathbf{X}}^*_M\widehat{\mathbf{E}}^i_j)^\top\right]^\top + \mathbf{b}^i_j \\ &= \mathbf{C}^{-1}\mathbf{C} \circledast \left[(\widehat{\mathbf{X}}_1\widehat{\mathbf{E}}^i_j)^\top, \cdots, (\widehat{\mathbf{X}}_M\widehat{\mathbf{E}}^i_j)^\top\right]^\top + \mathbf{b}^i_j = \widehat{\mathbf{X}}\widehat{\mathbf{E}}^i_j + \mathbf{b}^i_j.\end{aligned}$$

Similarly, $\widehat{\mathbf{Y}}$ can be converted to $\mathbf{Y}$ [17].

The above is the outsourcing process of one convolution operation as shown in Eq. (1). Also, this process applies to linear operations such as fully connected operations.

2.3 Linear Coded Federated Learning System Model

In this sub-section, we explain the computing steps that can be accelerated by outsourcing with coded computing.

To enhance the expressive power of neural networks, almost all feature extraction layers need to use nonlinear functions in the activation process. However, the coded data cannot be decoded correctly after being processed by the nonlinear activation functions. Therefore, it is necessary to retrieve the calculation results of the previous layer before the model passes through the activation function. Similarly, the use of subsampling layers can significantly reduce the amount of data transmission but compromise decodability. Therefore, all matrix multiplication tasks of convolutional and fully connected layers (Eq. (1) and Eq. (2)) are outsourced.

Figure 2 illustrates a specific training round of a client in LCFL. Before training starts, the server establishes connections with all clients and get the computational speed $\mathbf{V} = \{v_1, \cdots, v_n\}$ of clients according to benchmark. For k-th client s_k, if $v_k \leqslant \frac{\sum_{h=1}^{n} v_h}{2n}$, s_k will be marked as a straggler by the server. In the i-th training round, the server randomly selects the clients $\mathbf{S}_i = \{s_1^i, \cdots, s_{d_i}^i\}$ and sends model $\mathbf{A}_i$ to $\mathbf{S}_i$. If the selected k-th client s_k^i is not a straggler, then the training is done locally. Otherwise, s_k^i outsources the tasks to the ξ collaborative clients $\mathbf{S}_\mathbf{c}^i = \{s_{c_1}^i, \cdots, s_{c_\xi}^i\}$, where $\mathbf{S}_\mathbf{c}^i \subseteq \mathbf{S} - \mathbf{S}_i$. The outsourced tasks include convolutional operation and fully connected operation. s_k^i encodes the data in forward propagation and then outsources it, while it will not encode the data in backward propagation.

In the i-th training round, the server sends the current global model $\mathbf{A}_{i-1}$ to all the clients in $\mathbf{S}_i$. Also, the server sends the relevant information of clients in $\mathbf{S}_\mathbf{c}^i$, *e.g.*, addresses and speeds, to s_k^i if s_k^i is a straggler. The whole process of one training round for a straggler is divided into *Training Fragments* (TFs) by the activation layers. The specific steps of a TF are as follows:

- **Client selection and task volume determination**. The straggler determines the selection of a set of collaborative clients and the number of tasks to be outsourced to each collaborative client based on the speeds of ξ free clients given by the server. We describe the CCS algorithm combined with the coding, including the number of collaborative clients and the amount of work each collaborative client undertaken in Sect. 3.
- **Data coding and outsourcing**. After straggler finishes blocking and encoding local data, it outsources the encoded data to the collaborative clients. The linear encoding scheme is proposed to encode the outsourced data which is processed simultaneously with the coded data transmission from the straggler to the collaborative clients.
- **Model computation and result aggregation**. After each collaborative client has completed the computation task, these results are returned to the straggler. The straggler integrates all the results received and decodes the coded results by linear decoding.

After completing all TFs in i-th round of FL, the global model $\mathbf{A}_{i-1}$ can be updated to $\mathbf{A}_i$.

Outsourcing computing can greatly improve the efficiency of computing, and the strategy of keeping part of the data by the outsourcer can further reduce the communication time. If the scheduled client $\mathbf{S}_i$ in i-th training round contains more than one straggler, the task of each straggler is outsourced for computation. Due to space limitations, the multi-stragglers collaborative outsourcing computation will be presented in the next paper.

3 Efficient and Secure Federated Learning

In this section, we present the CCS algorithm to solve the slowdown problem of stragglers for the LCFL. In the CCS algorithm, we determine the collaborative clients and the amount of computation undertaken by each client in the outsourcing computation of a certain training round.

Firstly, we choose ξ collaborative clients in the task outsourcing of a straggler. In the i-th training round, the server selects partial clients $\mathbf{S}_i$ from all clients $\mathbf{S}$. Suppose that s_k^i is a straggler and we need to select top ξ fastest clients from $\mathbf{S}-\mathbf{S}_i$ as collaborative clients to help compute. The speed is derived from a combination of computational speed and data transmission speed and is expressed as the time consumed by the client to compute and transmit fixed data, which is given as $\frac{1}{v_l^i}+\frac{1}{g_l^i}$, where v_l^i and g_l^i denote the computational speed of the l-th collaborative client and transmission speed from l-th collaborative client to the server in i-th training round respectively. v_l^i and g_l^i are initially given by the server based on each client's benchmark and are updated within the training based on the history.

Then we calculate the data volume of all TFs in one round according to the CNN model. The main basic parameters of the CNN model are as follows: (1) The size of one input image is $m\times m$. (2) The j-th ($j \in \{1,\cdots,L\}$) convolutional layer has α_j convolution kernels, where the dimension of each kernel is $\beta\times\beta$. (3) The dimension of the subsampling layer is $\gamma\times\gamma$. The use of each convolutional kernel in j-th convolutional layer yields a feature map with the same dimensionality as the input of j-th convolutional layer. For an $m\times m$ picture, the dimension of input of j-th convolutional layer is $\frac{m}{\gamma^{j-1}}\times\frac{m}{\gamma^{j-1}}$ according to Eq. (1), and the size of the output of j-th convolutional layer is $\alpha_j\frac{m}{\gamma^{j-1}}\frac{m}{\gamma^{j-1}}=\alpha_j\frac{m^2}{\gamma^{2(j-1)}}$. The features obtained from the convolution are outsourced after the subsampling layer. The size of the data outsourced after the j-th convolutional layer is $\alpha_j\frac{m}{\gamma^{j-1}}\frac{1}{\gamma}\frac{m}{\gamma^{j-1}}\frac{1}{\gamma}=\alpha_j\frac{m^2}{\gamma^{2j}}$. Therefore, the total amount of data to be sent in forward propagation is $(1+\sum_{\rho=1}^{L}\frac{\alpha_\rho}{\gamma^{2\rho}})m^2$. In backward propagation, the errors after upsampling and activation layers need to be outsourced, and the total amount of data to be sent in backward propagation is $\sum_{\rho=1}^{L}\frac{\alpha_\rho}{\gamma^{2(\rho-1)}}m^2$. For the k-th client s_k^i in i-th training round, the total amount of data transmitted in forward propagation D_f^i and backward propagation D_b^i is

$$\begin{aligned}D_t^i=D_f^i+D_b^i &=(1+\sum_{\rho=1}^{L}\frac{\alpha_\rho}{\gamma^{2\rho}})m^2+\sum_{\rho=1}^{L}\frac{\alpha_\rho}{\gamma^{2(\rho-1)}}m^2\\ &=(1+\sum_{\rho=1}^{L}\frac{(1+\gamma^2)\alpha_\rho}{\gamma^{2\rho}})m^2.\end{aligned} \tag{4}$$

The total data transmission time for s_k^i is $T_{d_k}=\frac{D_t^i}{g_k^i}$ and can not be ignored. Therefore, T_{d_k} can be used by s_k^i to calculate partial data. Let $\mathbf{W}_i=$

$\left[w_0^i, w_1^i, \cdots, w_\xi^i\right]$, where w_l^i ($l \in \{0, \cdots, \xi\}$) represents the number of sub tasks to be allocated to the scheduled l-th collaborative client $s_{c_l}^i$ and ξ represents the total number of collaborative clients scheduled in i-th training round. In particular, v_0^i, g_0^i and w_0^i are the computational speed, transmission speed and the amount of data calculated of the straggler s_k^i.

Based on the ratio between the speed of s_k^i and the total speeds of all collaborative clients $\mathbf{S}_c^i$, the respective amount of computation that the straggler s_k^i and each collaborative client $s_{c_l}^i$ will undertake is given as

$$w_l^i = \frac{v_l^i}{\sum_{\rho=0}^{\xi} v_\rho^i} u_k, l \in \{0, \cdots, \xi\}, \tag{5}$$

where u_k is the total amount of data owned by s_k^i. In a practical system, since data transmission and computation are asynchronous, s_k^i can use the data transmission time to complete partial data computation locally. The data transmission time is given as $\frac{\sum_{\rho=1}^{\xi} w_\rho^i D_t}{g_0^i}$ and the amount of this part of data is given as $\Psi = \frac{\sum_{\rho=1}^{\xi} w_\rho^i D_t^i}{g_0^i} v_0^i$. Since s_k^i calculates some additional data, the total amount of data used by the collaborative clients is reduced accordingly. The amount of data allocated to each collaborative client is updated to

$$w_l^i = \frac{v_l^i(u_k - \Psi)}{\sum_{\rho=0}^{\xi} v_\rho^i}, l \in \{1, \cdots, \xi\}. \tag{6}$$

Similarly, the amount of data calculated by s_k^i is updated as

$$w_0^i = \frac{v_0^i}{\sum_{\rho=0}^{\xi} v_\rho^i} u_k + \frac{\sum_{\rho=1}^{\xi} v_\rho^i}{\sum_{\rho=0}^{\xi} v_\rho^i} \Psi. \tag{7}$$

It can be compared that the additional amount of data computed by s_k^i after the update of w_0^i is less than the initially given Ψ value by $\frac{v_0^i}{\sum_{\rho=0}^{\xi} v_\rho^i} \Psi$. This part of the data volume is the quantity error caused by the dynamic update of $w_l^i (l \in \{1, \cdots, \xi\})$ and Ψ to each other, which has little impact on the total running time.

To ensure the privacy of data in the outsourcing process, data should be encoded before outsourcing. Because the block size of the encoding object must be the same, the original data in s_k^i should be divided into $M (M \in \{\xi, \cdots, u_k\})$ blocks of the same size, and then be combined after being encoded according to the size w_l^i required by l-th collaborative client $s_{c_l}^i$. Since w_l^i may be a decimal, which will cause the total number of blocks M to be very large to meet the exact number of blocks required by w_l^i, we need to update w_l^i according to M. In the selected ξ collaborative clients, the speed is from fast to slow, and the

Algorithm 1: Collaborative Client Selection (*CCS*)

Input: u_k, M, n, V, G, ξ
Output: $\overline{w}_l^i$

$l = 0$;
while $l \leqslant \xi$ **do**
  $w_l^i = \frac{v_l^i}{\sum_{\rho=0}^{\xi} v_\rho^i} u_k$;
  $l = l + 1$;
$\Psi = \frac{\sum_{\rho=1}^{\xi} w_\rho^i D_t^i}{g_0^i} v_0^i$;
$l = 1$;
while $l \leqslant \xi$ **do**
  $w_l^i = \frac{v_l^i (u_k - \Psi)}{\sum_{\rho=0}^{\xi} v_\rho^i}$;
  $l = l + 1$;
$w_0^i = \frac{v_0^i}{\sum_{\rho=0}^{\xi} v_\rho^i} u_k + \frac{\sum_{\rho=1}^{\xi} v_\rho^i}{\sum_{\rho=0}^{\xi} v_\rho^i} \Psi$;
for *l in range (2,ξ)* **do**
  $\overline{w}_l^i = \lfloor \frac{w_l^i}{u_k} M \rfloor$;
$\overline{w}_0^i = \lfloor w_0^i \rfloor$;
$\overline{w}_1^i = M - \overline{w}_0^i - \sum_{\rho=2}^{\xi} \overline{w}_\rho^i$;
return $\overline{w}_l^i$;

number of blocks w_l^i of $s_{c_l}^i$ is updated using *CCS* algorithm, which is given as $\overline{w}_l^i = \lfloor \frac{w_l^i}{u_k} M \rfloor, l \in \{2, \cdots, n\}$. Meanwhile, $\overline{w}_0^i = \lfloor w_0^i \rfloor$ and $\overline{w}_1^i = M - \overline{w}_0^i - \sum_{\rho=2}^{\xi} \overline{w}_\rho^i$. If the required number of blocks can not be obtained exactly for l-th ($l \in \{2, \cdots, n\}$) client, the fastest collaborative client $s_{c_1}^i$ will calculate the extra blocks. The update of w_0^i does not change with M, only rounding down to ensure the integer, the impact on the overall system time is almost zero. At this time, the total transmission time for s_k^i is

$$T_P = \frac{\sum_{\rho=1}^{\xi} \overline{w}_\rho^i D_t^i}{g_0^i}. \tag{8}$$

Next, we consider the size of the total block number M. Because there is no redundancy, the number of rows and columns of coding matrix $\mathbf{C}$ is the same as the total number of blocks. Suppose that the time for a linear calculation of m^2 data is T_c. According to the value of $\overline{\mathbf{W}}_i$, M_{max} is obtained to enable $s_{c_l}^i$ to get exactly $\overline{w}_l^i$ blocks. For M, $(0 \leqslant M \leqslant M_{max})$, the increased calculation time is the additional transmission time and local computation time required for the additional amount of data computed by $s_{c_1}^i$, which is given as

Scenario \ Number \ Type	PCs	Raspberry Pi
Scenario 1	8	2
Scenario 2	9	1

(a) Experimental scenarios

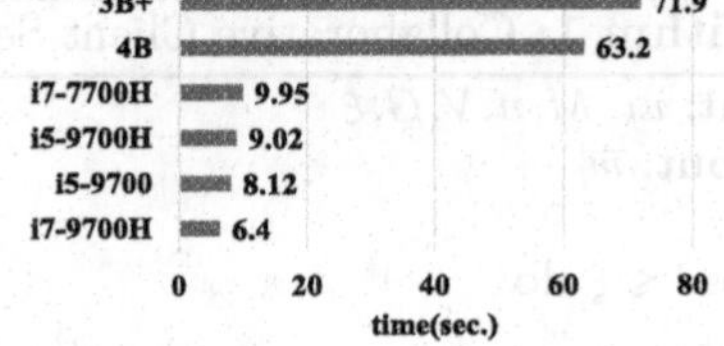

(b) Single round training time of various clients

Fig. 3. Experiment environment

$$
\begin{aligned}
T_E = & \frac{M - \overline{w}_0^i - \sum_{\rho=2}^{\xi} \lfloor \frac{w_\rho^i}{u_k} M \rfloor - (M_{max} - \overline{w}_0^i - \sum_{\rho=2}^{\xi} \lfloor \frac{w_\rho^i}{u_k} M_{max} \rfloor)}{g_0^i} \\
& + \frac{M - \overline{w}_0^i - \sum_{\rho=2}^{\xi} \lfloor \frac{w_\rho^i}{u_k} M \rfloor - (M_{max} - \overline{w}_0^i - \sum_{\rho=2}^{\xi} \lfloor \frac{w_\rho^i}{u_k} M_{max} \rfloor)}{v_1^i} \\
= & (M - M_{max} - \sum_{\rho=2}^{\xi} \lfloor \frac{w_\rho^i}{u_k} M \rfloor + \sum_{\rho=2}^{\xi} \lfloor \frac{w_\rho^i}{u_k} M_{max} \rfloor)(\frac{1}{g_0^i} + \frac{1}{v_1^i}). \qquad (9)
\end{aligned}
$$

Also, the reduction of the number of blocks will reduce the coding time, and this reduced time is $T_M = \frac{(D_f^i + D_b^i)(M_{max} - M)T_c}{m^2}$. Finally, M is obtained to minimize $T_E - T_M$.

4 Practical Experiments

In this section, we evaluate the CCS algorithm from real experiments. We mainly evaluate the performance of LCFL from the following criteria: 1) model accuracy and 2) training completion time.

4.1 Experiment Environment Configuration

To meet the universality, our neural network will use CPU for training and inferring, including three desktop computers with i5-9700 CPU, three laptops with i7-9700H CPU, two laptops with i5-9700H CPU, two laptops with i7-7700H CPU, one Raspberry Pi 4B and one Raspberry Pi 3B+. The experimental environment parameters and settings are shown in Fig. 3. The neural network model uses CNN without a deep learning framework, and the data set is MNIST handwritten numeral set. In each experiment, the server will schedule ten of the above devices for global model training, and each participating device will have one-tenth of the MNIST data set, that is, 6000 images. The model structure is shown in Fig. 1 and the server schedules 30% of the total number of clients each time, that is, 3 clients for training.

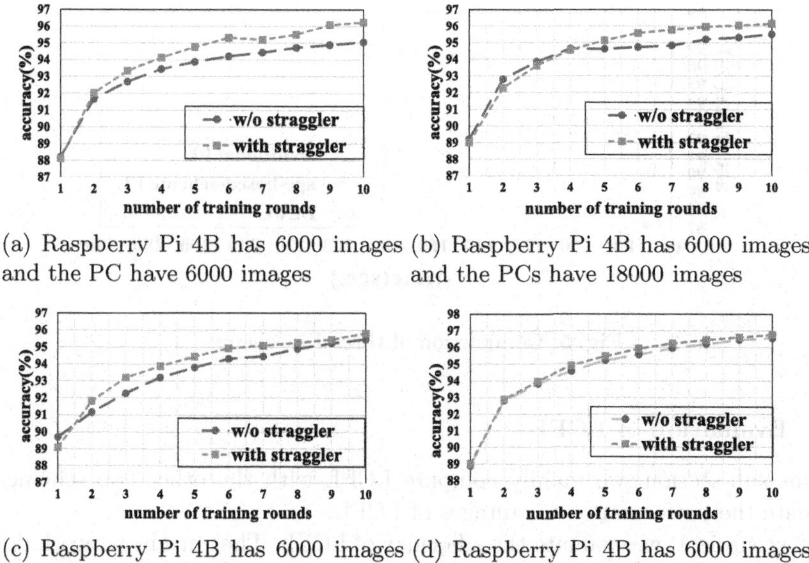

(a) Raspberry Pi 4B has 6000 images and the PC have 6000 images (b) Raspberry Pi 4B has 6000 images and the PCs have 18000 images

(c) Raspberry Pi 4B has 6000 images and the PCs have 24000 images (d) Raspberry Pi 4B has 6000 images and the PCs have 54000 images

Fig. 4. Influence of data quantity on accuracy with the same initial model

4.2 Accuracy of LCFL

In this sub-section, our experimental results are obtained by averaging the results from 10 experiments. First of all, we use all kinds of clients to train the neural network. Figure 3(b) shows the time required for each kind of client to complete the training of 6000 image data set. As an IoT device, Raspberry Pi takes six to seven times as long as PCs in one training round. There is no doubt that the Raspberry Pi is a straggler in the system.

Next, we test the influence of the amount of data on the accuracy of the random model. We vary the amount of data by controlling the number of PCs involved in the training. The same initial model is used to test the accuracy and ten rounds are included in one experiment. In each experiment, we set up two comparisons: (1) Scheduling all PCs and Raspberry Pi 4B to train. (2) Ignoring the Raspberry Pi 4B and scheduling all PCs to train. As shown in Fig. 4, ignoring the data in the straggler will inevitably lead to the decline of accuracy, especially when the proportion of data in the straggler is relatively large as shown in Fig. 4(a) and Fig. 4(b). When the proportion of data in straggler is small, whether to use these data for training, the difference is very small as shown in Fig. 4(d). The main reason for the small difference in accuracy is that the MNIST data set is shuffled of ten types of data, and each client has the data for all categories. If the data category of each client is incomplete, its impact on accuracy will be more obvious. However, the lack of straggler's data will inevitably reduce the accuracy of the model.

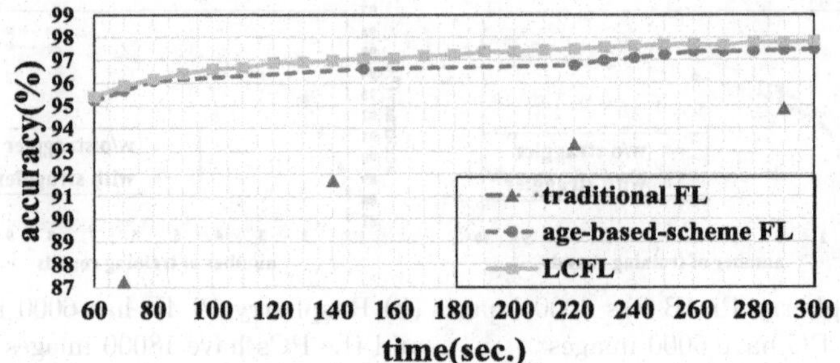

Fig. 5. Comparison of three FL schemes

4.3 Evaluation of LCFL

In this sub-section, we mainly compare LCFL with the other two schemes to evaluate the performance advantages of LCFL.

In scenario 1, we evaluate the efficiency of LCFL. The current network download speed is 7.8 Mb/s and the network upload speed is 6.9 Mb/s. In this experiment, ten devices in scenario 1 will use the image data set to train the model, and the accuracy will be given through testing the image test set.

Set the number of collaborative clients of one straggler to 2, we compare LCFL with traditional FL and age-based-scheme FL, and the experimental results are shown in Fig. 5. The traditional FL schedules all the clients to participate in a training round [13]. Age-based-scheme FL schedules partial clients in each round while ensuring that stragglers will be scheduled in a certain round [16]. It can be seen from Fig. 5 that LCFL is the most efficient scheme. When using the small image data set MNIST and the training time reaches 300 s, the accuracy of LCFL scheme is 97.854%. Under the same experimental conditions, the accuracy of age-based-scheme FL is 97.4519%, however LCFL can achieve 97.462% using only 220 s. In this experiment, LCFL is 26.7% faster than age-based-scheme FL. The main reason is that age-based-scheme FL is slowed down by straggler from 90 s to 210 s. The traditional FL is also dragged down by stragglers, which takes 70 to 80 s to complete a training round, resulting in the slowest convergence of the global model.

In scenario 1, we control the number of collaborative clients and network speed to compare age-based-scheme FL and LCFL. As shown in Fig. 6(a) and Fig. 6(b), when the network speed is 7.8 Mb/s, with the increase of the number of collaborative clients, LCFL can obtain higher accuracy at the same time. We use Tencent computer housekeeper[1] to limit the network speed to 3 Mb/s and 1 Mb/s. When the number of collaborative clients is 2, with the decrease of network speed, the training speed of LCFL also decreases as shown in Fig. 6(c) and Fig. 6(d) and the slope of the line between 100 s and 180 s in Fig. 6(d) is less than that in Fig. 6(c). However, the performance of LCFL is better than that of age-based-scheme FL under every setting. The main reason is that LCFL can

[1] https://guanjia.qq.com/.

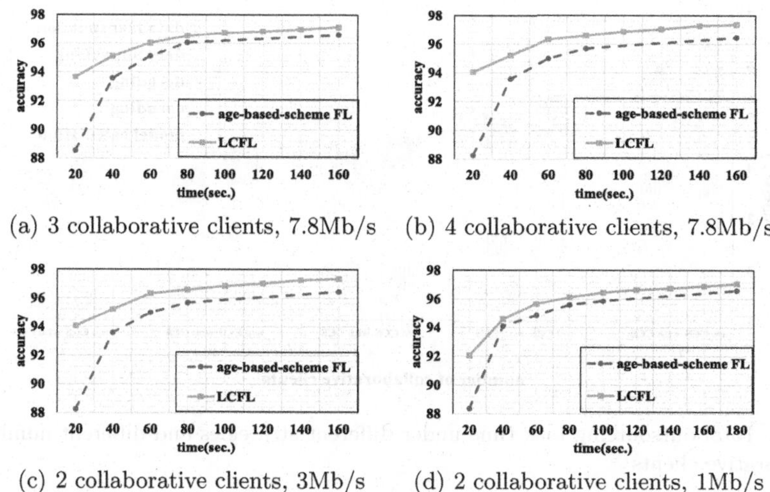

(a) 3 collaborative clients, 7.8Mb/s (b) 4 collaborative clients, 7.8Mb/s

(c) 2 collaborative clients, 3Mb/s (d) 2 collaborative clients, 1Mb/s

Fig. 6. Comparison between age-based-scheme FL and LCFL under different number of collaborative clients or network speed

greatly improve the speed of stragglers. In Fig. 6(d), LCFL's training speed is 22.2% faster than that of age-based-scheme FL. And LCFL's training speed is even 61.2% faster than that of age-based-scheme FL in Fig. 6(b).

4.4 Evaluation of CCS

In this sub-section, we evaluate how CCS algorithm can effectively reduce the training completion time.

We run the experiments in scenario 2 with the number of collaborative clients as a variable. The completion time is divided into five parts: data transmission, data encapsulation and unpacking, encoding, decoding and model calculation, where data encapsulation is the process of encapsulating the data matrix into a fixed format for transmission, and data unpacking is the process of converting the fixed format data into matrices. Figure 7 shows the completion time of outsourcing tasks under different strategies and different numbers of collaborative clients for Raspberry Pi 4B in scenario 2.

In the case of the same number of collaborative clients, the gradual addition of CCS algorithm can significantly reduce the training time. Compared with the scheme without CCS, the use of CCS algorithm can reduce 7.56% of the total time. The CCS algorithm can reduce the overall time by reducing the amount of outsourced data. When the number of collaborative clients is 2, the total time is reduced by 3.95% and when the number of collaborative clients is 6, the encoding and decoding time is even reduced by 8.746% after using the LPC scheme. With the increase of the number of collaborative clients, the time of model calculation is greatly reduced, and the efficiency improvement brought by CCS is more obvious. From the average total completion time in Fig. 7, LFCL

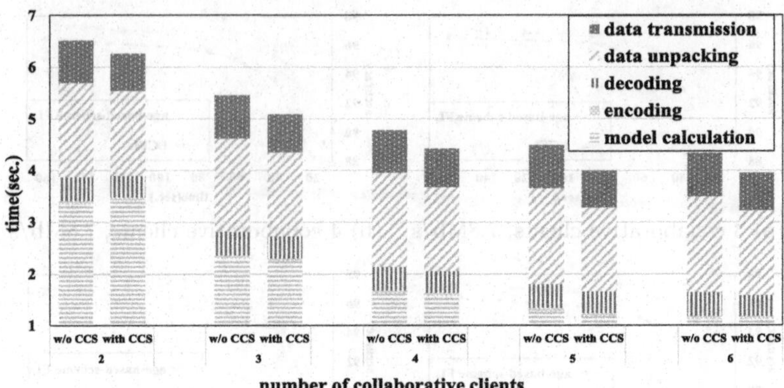

Fig. 7. Total outsourcing task time under different strategies and different number of collaborative clients

can reduce the training time of Raspberry Pi 4B by 93.73% compared with that Raspberry Pi 4B performs an independent calculation.

4.5 Simulation Conclusion

From the above simulations, we have the following conclusions.

- Traditional FL is often dragged by stragglers. In the proposed LCFL scheme, the task of stragglers is skillfully outsourced, and the drag effect no longer exists. Coding also provides weak security protection for data outsourcing.
- The proposed LCFL scheme can achieve good performance when the scale of training is below medium scale. For small data set MNIST, when the number of collaborative clients $\xi = 5$, the speed improvement and resource consumption achieve the best balance.
- The slower the speed of straggler in FL, the greater the speed increase ratio of LCFL. The PCs used in the experiments are 7 to 9 times faster than Raspberry Pi, and LCFL can reduce the training time of Raspberry Pi by 90%.

5 Conclusion

In this paper, we have proposed an efficient *linear coded federated learning* (LCFL) framework to (1) speed up the convergence speed of heterogeneous FL and (2) protect the data privacy of the participants. To design the proposed framework, we firstly design a *collaborative client selection* (CCS) algorithm that can select appropriate clients and assign the computation task of the slowest client to those selected devices. Finally, we build a practical experimental platform and conducted numerous experiments to evaluate the proposed LCFL framework from different aspects. The experimental results demonstrate that the proposed LCFL scheme can reduce the training time up to 93.73% when the participants have a large difference in terms of computing capability. In the future, our work will not be limited to focus on the impact of slow clients on the total training time, but will consider outsourcing the data of more clients to further accelerate the training of the federated learning.

Acknowledgment. This work is supported in part by National Natural Science Foundation of China (62072321, 61972272), Six Talent Peak Project of Jiangsu Province (XYDXX-084), China Postdoctoral Science Foundation (2020M671597), Jiangsu Postdoctoral Research Foundation (2020Z100), Suzhou Planning Project of Science and Technology (SNG2020073, SS202023, SYG202024), Tang Scholar of Soochow University, Collaborative Innovation Center of Novel Software Technology and Industrialization, and Soochow University Interdisciplinary Research Project for Young Scholars in the Humanities.

References

1. Li, T., Sahu, A., Talwalkar, A., Smith, V.: Federated learning: challenges, methods, and future directions. In Proceedings of the IEEE Signal Processing Magazine, vol. 37, no. 3, pp. 50–60 (2020)
2. Wang, T., et al.: Privacy-enhanced data collection based on deep learning for Internet of vehicles. IEEE Trans. Ind. Inf. **16**(10), 6663–6672 (2019)
3. Li, T., Sahu, A., Zaheer, M., Sanjabi, M., Talwalkar, A., Smith, V.: Federated optimization in heterogeneous networks. arXiv preprint arXiv:1812.06127 (2018)
4. Caldas, S., Konečny, J., McMahan, H., Talwalkar, A.: Expanding the reach of federated learning by reducing client resource requirements. arXiv preprint arXiv:1812.07210 (2018)
5. Verbraeken, J., Wolting, M., Katzy, J., Kloppenburg, J., Verbelen, T., Rellermeyer, J.: A survey on distributed machine learning. ACM Comput. Surv. (CSUR) **53**(2), 1–33 (2020)
6. Al-Rubaie, M., Chang, J.: Privacy-preserving machine learning: threats and solutions. IEEE Secur. Priv. **17**(2), 49–58 (2019)
7. Yang, H., Lee, J.: Secure distributed computing with straggling servers using polynomial codes. IEEE Trans. Inf. Forensics Secur. **14**(1), 141–150 (2018)
8. Geyer, R., Klein, T., Nabi, M.: Differentially private federated learning: a client level perspective. arXiv preprint arXiv:1712.07557 (2017)
9. Wu, Y., Huang, H., Wu, N., Wang, Y., Bhuiyan, M.Z.A., Wang, T.: An incentive-based protection and recovery strategy for secure big data in social networks. Inf. Sci. **508**, 79–91 (2020)
10. Dhakal, S., Prakash, S., Yona, Y., Talwar, S., Himayat, N.: Coded federated learning. In: Proceedings of the IEEE Globecom Workshops (GC Wkshps), pp. 1–6 (2019)
11. Prakash, S., Dhakal, S., Akdeniz, M., Avestimehr A., Himayat, N.: Coded computing for federated learning at the edge. arXiv preprint arXiv:2007.03273 (2020)
12. Rahimi, A., Recht, B.: Random features for large-scale kernel machines. NIPS **3**(4) (2007)
13. Zhao, Z., Feng, C., Yang, H., Luo, X.: Federated-Learning-enabled intelligent fog radio access networks: fundamental theory, key techniques, and future trends. IEEE Wirel. Commun. **27**(2), 22–28 (2020)

14. Nishio, T., Yonetani, R.: Client selection for federated learning with heterogeneous resources in mobile edge. In: Proceedings of the IEEE International Conference on Communications (ICC), Shanghai, China, pp. 1–7 (2019)
15. Kairouz, P., et al.: Advances and open problems in federated learning. arXiv preprint arXiv:1912.04977 (2019)
16. Yang, H., Arafa, A., Quek, T., Poor, H.: Age-based scheduling policy for federated learning in mobile edge networks. In: Proceedings of the IEEE International Conference on Acoustics, Speech and Signal Processing (ICASSP), pp. 8743–8747 (2020)
17. Chellapilla, K., Puri, S., Simard, P.: High performance convolutional neural networks for document processing. In: Proceedings of the Tenth International Workshop on Frontiers in Handwriting Recognition, Suvisoft (2006)
18. Muller, M.: A note on a method for generating points uniformly on N-dimensional spheres. Commun. ACM **2**(4), 19–20 (1959)

Verifiable Dynamic Searchable Symmetric Encryption with Forward Privacy in Cloud-Assisted E-Healthcare Systems

Haitang Lu[1], Jie Chen[2(✉)], and Kai Zhang[3]

[1] East China Normal University, Shanghai 200062, China
51194506024@stu.ecnu.edu.cn

[2] Shanghai Key Laboratory of Trustworthy Computing, East China Normal University, Shanghai 200062, China
s080001@e.ntu.edu.sg

[3] Shanghai University of Electric Power, Shanghai 201306, China
kzhang@shiep.edu.cn

Abstract. The integration of Internet of Things (IoT) and cloud computing is transforming traditional healthcare systems into cloud-assisted e-healthcare systems. In a cloud-assisted e-healthcare system, patients can upload their personal health information (PHI) files to the cloud, from where different healthcare service providers can obtain appropriate information to determine the patients' health status. However, this paradigm shift has raised many security and privacy concerns: data sharing, data tampering and information leakage. To address the above challenges, in this paper, we propose a verifiable dynamic searchable symmetric encryption (DSSE) scheme with forward privacy for e-healthcare systems, which enables different doctors to access and search PHI files in a secure and efficient manner. Forward privacy is achieved by maintaining state chains on the cloud server, while the verification of searched healthcare data comes from homomorphic MAC (HomMAC) technique. Detailed security analysis and simulations on real-world and simulated datasets demonstrate the practical efficiency of the proposed scheme in real-world e-healthcare applications.

Keywords: Searchable symmetric encryption · Multi-user · E-healthcare · Forward privacy · Verifiability

1 Introduction

With the fast development of cloud computing and Internet of Things (IoT), the traditional healthcare services industry has experienced a transformation, which has enabled flexible and efficient cloud-assisted e-healthcare systems [2,15,18]. In a typical cloud-assisted e-healthcare system, the wearable and implantable sensors (e.g., smart watches) regularly collect health data (e.g., heart rate, breathing rate, etc.) from patients at home. This information is aggregated into a personal health information (PHI) file at the IoT gateway, and then sent to the cloud

Y. Lai et al. (Eds.): ICA3PP 2021, LNCS 13155, pp. 645–659, 2022.
https://doi.org/10.1007/978-3-030-95384-3_40

server for storage. In this way, doctors can monitor patients' health status precisely in real time by submitting queries to the cloud server, so as to provide timely health advice and diagnosis plans. However, the security and privacy of these PHI files are major concerns in such systems due to malicious attacks, software vulnerabilities or accidental errors.

While simply encrypting PHI files before uploading to the cloud can ensure the privacy of e-healthcare systems, it makes searching for PHI files particularly challenging. Searchable symmetric encryption (SSE) [11], which allows the cloud server to search encrypted files using search tokens generated by the data users, provides a promising solution to the above problem. In the past few years, static SSE schemes [4,7] have been proposed, which are obviously not suitable for e-healthcare systems, as patients need to update their health data regularly. Therefore, dynamic searchable symmetric encryption (DSSE) schemes [6,13,21] have been later proposed. A DSSE scheme with forward privacy is highly desirable to prevent the cloud server from inferring sensitive information related to patients such as activity patterns, eating habits, etc. in e-healthcare systems. Forward privacy is a notion introduced in [12], which guarantees that the adversary cannot determine whether newly added data contains the keywords searched in the past. Moreover, forward privacy guarantees the privacy of query keywords against file injection attacks [19].

An essential function of e-healthcare systems is data sharing, which enables multiple doctors to perform searches on encrypted PHI files shared by patients [8,10,14,16,17]. However, existing schemes are either interactive (requiring the data owner to stay online to generate search tokens for the data users) or rely on bilinear pairing to achieve authorization. Therefore, how to design an effective multi-user scheme is still an open problem. Last but not least, the verifiability of e-healthcare systems is also critical, as any incorrect search results could negatively impact patients' health, such as incorrect diagnosis plans, thereby highlighting the need to deploy a verification process into the system [9,17,18].

Motivated by the above problems, we propose a verifiable DSSE scheme with forward privacy for e-healthcare systems. The main challenges are as follows. 1) How to design an efficient DSSE scheme in multi-user settings? 2) How to achieve forward privacy? 3) How to ensure that search results are verifiable rather than tampered or partial results? Our contributions can be summarized as follows.

- We propose a privacy-preserving DSSE scheme for e-healthcare systems, where PHI files are stored in encrypted form on the cloud server. Our scheme is able to achieve sub-linear search efficiency and forward privacy by maintaining the state chain for each keyword on the cloud server.
- The system aims at the application scenario where PHI files are generated from a single patient's health data and shared among multiple doctors, we utilize key distribution technique rather than key sharing to authorize doctors in multi-user settings. In addition, we design a verification process based on homomorphic MAC (HomMAC) technique that enables doctors to verify search results, which is suitable for doctors with resource-limited devices.

- We compare our DSSE scheme with some related schemes in terms of computation efficiency, and also implement keyword search based on real-world and simulated data. The results demonstrate that our scheme is not only feasible but also efficient. We also present security analysis to show forward privacy and the verifiability of our scheme.

2 Problem Formulation

2.1 System Model

Our system model consists of five entities as shown in Fig. 1: data owner (patient), an IoT gateway, a cloud server, data users (doctors) and a trusted authority (TA). The data owner is the patient whose health status is monitored by wearable devices and sent to the IoT gateway. The IoT gateway is the data aggregator that aggregates the collected data received from the patient periodically into PHI files, extracts keywords, and encrypts PHI files and keywords, then it uploads them to the cloud for storage. TA exchanges status information with the IoT gateway, distributes private keys for authorized data users, and also manages the search authorization of data users. When the data users (i.e., doctors) request the patient's health data, they first send queries to TA for checking the search authorization. If the check passes, TA sends search queries to the cloud server. The cloud server responds to search queries from TA, and sends the search results to the authorized data users. When receiving the search results from the cloud server, the data users who provide healthcare services for patients can verify the correctness of the search results.

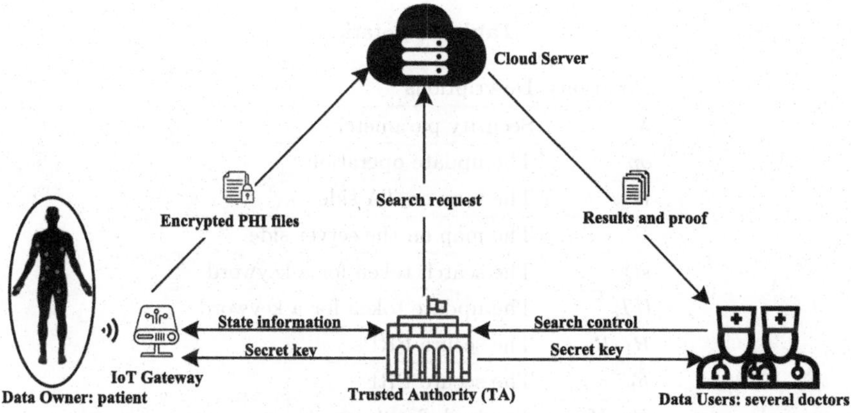

Fig. 1. The system model of our scheme

2.2 Threat Model

The cloud server in this model is considered as honest-but-curious (the cloud server faithfully executes the protocol but tries to infer information from the available data). However, the cloud server may be prone to external threats, and might return incorrect or incomplete search results to the data users (i.e., numerous doctors) for saving computational resources. Therefore, the data users should be able to verify the integrity of received PHI files. Note that we also assume that there is no collusion between the data users and the cloud server, or between the data users.

Similar to most keyword search schemes, TA and the IoT gateway in our model are fully trusted, which execute specified operations properly and do not leak any sensitive information to others. More specifically, TA generates and distributes keys to other entities, while the IoT gateway is responsible for generating encrypted PHI files. Moreover, the data users are also fully trusted as they are trusted healthcare practitioners to evaluate the health data.

3 Preliminaries

3.1 Notations

Let $x \rightarrow X$ denote uniformly sampling an element x from a set X and $|X|$ denote the number of elements in X; $\{0,1\}^n$ denotes the set of binary strings of length n while {0,1}* is the set of binary strings of arbitrary length; || denotes the concatenation of two strings; PPT is probabilistic polynomial time. Access is a list of authorized users, and the entries of the list are tuples (u, k_u), where u denotes the identifier of the authorized data user, and k_u denotes the private key of the authorized data user. Notations used in the proposed scheme are given in Table 1.

Table 1. Notations

Notations	Descriptions
λ	Security parameter
op	The update operation
W	The map on TA side
T	The map on the server side
st_c	The search token for a keyword
UT_c	The update token for a keyword
R_1, R_2	The secure PRPs
F_t	The secure PRF
H_1, H_2	The hash functions
R	The set of search results
$negl(\lambda)$	Negligible functions in λ
val	The value of verify process
Tag_w	The tag list of keyword w
c	The number of a keyword updates

3.2 Homomorphic MAC

The main idea of homomorphic MAC (HomMAC) is that the data user first uses the secrect key sk to authenticate a set of data $(m_1, m_2..., m_n)$ with corresponding labels $(\tau_1, \tau_2, ..., \tau_n)$, which can obtain a set of signatures $(\sigma_1, \sigma_2, ..., \sigma_n)$. Then, any entity can homomorphically execute a circuit f over $(m_1, m_2..., m_n)$ and $(\sigma_1, \sigma_2, ..., \sigma_n)$ to obtain a computation result χ and a new proof σ, respectively. Finally, the data user can use the proof σ to verify the correctness of χ. A HomMAC [5] includes four algorithms working as follows.

- $Setup(1^\lambda) \rightarrow (sk, ek)$: For a security parameter 1^λ, the algorithm outputs the secret key sk and evaluation key ek needed in the scheme.
- $Auth(sk, \tau, m) \rightarrow \sigma$: This algorithm inputs the secret key sk, a label τ, and a message $m \in \mathcal{M}$, where $\mathcal{M}$ is considered as n-dimensional vectors, and it outputs the corresponding tag σ.
- $Eval(ek, f, \sigma_1, ...\sigma_n) \rightarrow \sigma$: This algorithm inputs the evaluation key ek, a circuit $f : \mathcal{M}^n \rightarrow \mathcal{M}$ composed of addition and multiplication gates, and a vector of tags $(\sigma_1, ...\sigma_n)$, it outputs the proof σ used to verify the result χ of message set $(m_1, m_2..., m_n)$.
- $Ver(\chi, sk, \sigma, P) \rightarrow 0$ *or* 1: This algorithm inputs the secret key sk, a program $P = (f, \tau_1, \tau_2, ..., \tau_n)$, computation result χ of message set $(m_1, m_2..., m_n)$ and its proof σ, it outputs 0 (reject) or 1 (accept).

3.3 Security Definitions

Forward Privacy. We define the leakage functions that are used inside the definitions. For a list of search query Q, the entries in Q are $q_t = (t, w)$, the search pattern [3] is defined as $sp(w) = \{t : (t, w) \in Q\}$, and the access pattern is defined as $ap(q_t) = \{(1, DB_1(w_1)), ..., (t, DB_t(w_t))\}$. The integer t is a timestamp, initially set to 0, and which is incremented at each query. Moreover, the query pattern is defined as $qp(w) = \{t : (t, w) \in Q\}$ for all queries Q.

A DSSE scheme is forward private if for an update query $q_t = (t, op, ind)$, the leakage function can be written as: $\mathcal{L}_{Update}(t, op, ind) = \mathcal{L}'(t, op, ind, |\{w\}|)$, where $\mathcal{L}'$ is a stateless function, $|\{w\}|$ is the number of distinct keywords.

Reliability. Given a valid search result $R(w)$, the adversary $\mathcal{A}$ wins if she can forge invalid $R^*(w)$ that will pass *HomMAC.Ver* algorithm. We say that a verifiable DSSE scheme satisfies reliability if for any PPT adversary $\mathcal{A}$, $\mathbf{Adv}_{\mathcal{A}}(\lambda) = Pr(\mathcal{A}\ wins)$ is negligible for any search queries.

Security of DSSE Scheme. The definition is formulated using two games: *REAL* and *IDEAL*. The former is executed using our scheme, while the latter is simulated using the leakage as defined: $\mathcal{L} = (\mathcal{L}_{Setup}, \mathcal{L}_{Search}, \mathcal{L}_{Update})$ [7]. More precisely, $\mathcal{A}$ chooses a security parameter λ, then the game runs $Setup(\lambda$,U) or $\mathcal{S}(\mathcal{L}_{Setup}$(DB)), and returns initialized data structure to $\mathcal{A}$, $\mathcal{A}$ makes a polynomial number of adaptive queries q, and for each of them, the game then runs $Search(w, u, k_u, W, T)$ and $Update(sk, ind, op, W, T)$, or $\mathcal{S}(\mathcal{L}_{Search}(q))$ and $\mathcal{S}(\mathcal{L}_{Update}(q))$, then gives the generated transcript to $\mathcal{A}$. Eventually, $\mathcal{A}$ outputs a bit b $\in \{0,1\}$ as the output of the game. We say that our scheme is $\mathcal{L}$-adaptively-secure DSSE scheme if for any PPT adversary $\mathcal{A}$, there exists an efficient simulator $\mathcal{S}$ (with the input $\mathcal{L}$) such that:

$$|Pr[REAL^{\sum}_{\mathcal{A}}(\lambda) = 1] - Pr[IDEAL^{\sum}_{\mathcal{A},\mathcal{S}}(\lambda) = 1]| \leq \text{negl}(\lambda).$$

4 Our DSSE Construction

In this section, we first give the main idea of our DSSE scheme, and then demonstrate its concrete construction.

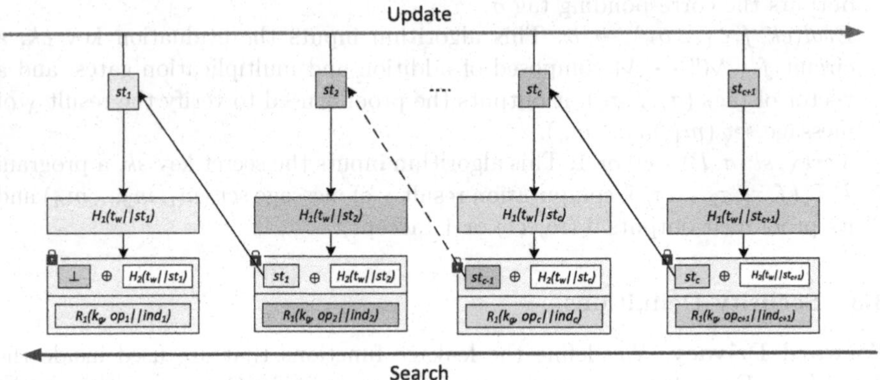

Fig. 2. The state chain of our scheme

4.1 Main Idea

In this paper, we construct the state chain (see Fig. 2) to achieve forward privacy. More specifically, when the IoT gateway wants to update keyword/file pair (w, ind), it firstly randomly generates the current search token st_{c+1}. Using st_{c+1}, the IoT gateway generates an update token $H_1(t_w||st_{c+1})$, the encrypted value $e_{c+1} = R_1(k_g, op||ind)$ and the mask token $H_2(t_w||st_{c+1}) \oplus st_c$. Moreover, the Iot gateway also generate t_w using pseudorandom function F_t. Finally, the IoT gataway sends them to the cloud server. However, due to the randomness of search tokens and the usage of pseudorandom function, the server cannot access the next search token st_{c+1} and learn any information about the keyword. In this way, we can achieve forward privacy.

To solve key sharing and key exposure problems in multi-user settings, TA distributes a randomly generated private key k_u to each authorized data user u, and stores (u, k_u) in list Access. Specifically, when a data user u wants to perform searching, TA firstly checks whether u is an authorized user by looking up list Access, and if the check passes, sends the search query to the cloud server. Note that we require that illegal and revoked users cannot search encrypted PHI files. Therefore, TA is responsible for generating private keys for authorized users, key distribution, adding and revoking data users, which can reduce the interaction between the IoT gateway and the data users, and avoids key exposure.

To achieve efficient verification, the IoT gateway computes a list of tags Tag_w for each keyword during update process. When the operation *op* is add, the IoT gateway uses *HomMAC.Auth* algorithm to generate a tag σ corresponding to (w, op, ind) pair, and adds it into Tag_w; when the operation *op* is delete, the IoT gateway also generates a tag σ corresponding to (w, op, ind) pair, and deletes it from the tag list Tag_w that is identical to it. In this way, Tag_w stores the tags that have not been deleted. In addition, the cloud server uses *HomMAC.Eval* algorithm to compute auxiliary information (σ', χ) after searching, where σ' is a new tag proof and χ is the computation result based on R, and sends them to the data user for verifying along with the search results R. Finally, the data user uses *HomMAC.Ver* algorithm to verify auxiliary information (σ', χ).

4.2 The Concrete Construction

Now, we are ready to present our proposed DSSE construction $\sum$ = (*Setup, Adduser, Search, Update, Revokeuser, Verify*). Suppose that H_1: $\{0,1\}^* \rightarrow \{0,1\}^n$ and H_2: $\{0,1\}^* \rightarrow \{0,1\}^\lambda$ are two keyed hash functions, F_t is PRF and R_1, R_2 are PRPs.

Scheme Initialization. *Setup*(λ, U) $\rightarrow$ (sk, pp, W, T, Access): TA takes the security parameter λ as input, and randomly generates keys $k_t, k_g \in \{0,1\}^\lambda$; (sk, ek) for *HomMAC.Setup* algorithm. For each user u in user list U, TA distributes private keys $k_u \in \{0,1\}^\lambda$, and stores (u, k_u) in list Access. Moreover, it also initializes an empty state information table W as TA storage and an empty table T as the server storage, respectively. Finally, it outputs parameters $pp = (F_t, R_1, R_2, H_1, H_2, ek, k_t, k_g)$.

Adding Authorized Data User. *Adduser*$(u, \lambda) \rightarrow$ Access: TA computes the private key $k_u \in \{0,1\}^\lambda$ using the data user identifier u (i.e., the doctor), then sends k_u to the data user u, and (u, k_u) for updating Access = Access $\cup$ (u, k_u).

File Encrypting at IoT Gateway. *Update*$(sk, ind, op, W, T) \rightarrow (W, T)$: When updating a PHI file with identifier *ind* into scheme, for each keyword in *ind*, it proceeds the following operations (see Algorithm 1).

1. The wearable devices collect health data from the data owner (i.e., the patient), and forward it to the IoT gateway.
2. The IoT gateway aggregates the health data into a PHI file *ind*, and extracts keywords from *ind* to generate a keyword set w_{ind}, where $w_{ind} = (w_1, w_2, ..., w_m)$ is sent to TA to get the state information $(st_c, c, Tag_w) \leftarrow W[w]$. TA then sends $\{sk, (st_{c_1}, c_1, Tag_{w_1}), (st_{c_2}, c_2, Tag_{w_2}), ...\}$ to the IoT gateway for updating.

3. The IoT gateway randomly generates a new search token $st_{c+1} \leftarrow \{0,1\}^{\lambda}$, and then gets the update token $UT_{c+1} \leftarrow H_1(t_w||st_{c+1})$ where $t_w \leftarrow F_t(k_t, w)$. Moreover, the IoT gateway encrypts (op, ind) to get the encrypted value $e_{c+1} \leftarrow R_1(k_g, op||ind)$, and generates the mask token $C_{st_c} \leftarrow H_2(t_w||st_{c+1}) \oplus st_c$.
4. Next, the IoT gateway generates a tag $\sigma_{c+1} \leftarrow HomMAC.Auth(sk, \tau, ind)$ based on (w, ind, op), and then updates the tag list Tag_w based on operations.
5. Finally, the IoT gateway sends $(UT_{c+1}, (e_{c+1}, C_{st_c}))$ to the cloud server to update $T[UT_{c+1}] \leftarrow (e_{c+1}, C_{st_c})$, and $(w, (st_{c+1}, c{+}1, Tag_w))$ to TA to update $W[w] \leftarrow (st_{c+1}, c+1, Tag_w)$.

Algorithm 1. File Encrypting

Update(ind, op, W, T, sk)

TA:

for each $w \in$ ind **do**
  $(st_c, c, Tag_w) \leftarrow W[w]$
end for
Send $\{sk, (st_{c_1}, c_1, Tag_{w_1}), ...\}$ to the IoT gateway

IoT gateway:

for each $w \in$ ind **do**
  $t_w \leftarrow F_t(k_t, w)$
  if$(st_c, c) = \perp$ **then**
    $c \leftarrow 0, st_c = \perp, \sigma_w \leftarrow \emptyset$
  end if
  $st_{c+1} \leftarrow \{0,1\}^{\lambda}$
  $\sigma_{c+1} \leftarrow HomMAC.Auth(sk, \tau, ind)$
  if $op = add$ **then**
    $Tag_w = Tag_w \cup \{\sigma_{c+1}\}$
  else
    $Tag_w = Tag_w \backslash \{\sigma_{c+1}\}$
  $UT_{c+1} \leftarrow H_1(t_w||st_{c+1})$
  $e_{c+1} \leftarrow R_1(k_g, op||ind)$
  $C_{st_c} \leftarrow H_2(t_w||st_{c+1}) \oplus st_c$
  Send $(UT_{c+1}, e_{c+1}, C_{st_c})$ to the server
  Send $(st_{c+1}, c+1, Tag_w)$ to TA
  Server:
  $T[UT_{c+1}] \leftarrow (e_{c+1}, C_{st_c})$
  TA:
  $W[w] \leftarrow (st_{c+1}, c+1, Tag_w)$
end for

Verifying the Authorization of Data User. To search for the keyword w (e.g., *blood pressure: 100*) in the patient PHI files, the data user u (i.e., the doctor) generates a query token $QT \leftarrow R_2(k_u, w)$ using PRP R_2, and sends it to TA. TA firstly uses list Access to check whether the data user u has authorization to search the encrypted PHI files, and if the check passes, TA then decrypts the query token to obtain the keyword $w \leftarrow R_2^{-1}(k_u, QT)$ and $t_w \leftarrow F_t(k_t, w)$, and then sends (t_w, st_c, c, Tag_w) to the server.

Keyword Searching at Server. *Search*$(w, u, k_u, W, T) \rightarrow (\chi, \sigma, R)$: The keyword search process is performed as follows (see Algorithm 2).

1. The server firstly computes update tokens $UT_i \leftarrow H_1(t_w||st_i)$ using st_c to retrieve all previous e_i and search tokens $st_{i-1} \leftarrow H_2(t_w, st_i) \oplus C_{st_{i-1}}$. Note that, in order to save the server storage, for every search, the server can remove all entries corresponding to w, and store the final result for next searching $E \leftarrow E \cup \{e_i\}$, $T[UT_c] \leftarrow (E, \perp)$.

2. Moreover, the server decrypts all e_i to compute $(op||ind) \leftarrow R_1^{-1}(k_g, e_i)$, and deletes file identifiers where the operation is del, which can obtain the search results R.
3. The server generates auxiliary information (σ', χ) by using *HomMAC.Eval* where $\sigma \leftarrow HMAC.Eval(ek, f, Tag_w)$, $\chi \leftarrow f(R[1], R[2], ...R[size])$, and returns (χ, σ, R) to the data user (i.e., the doctor).

Algorithm 2. Keyword Searching

Search(w, u, k_u, W, T)
TA:
$(st_c, c, Tag_w) \leftarrow W[w]$
$t_w \leftarrow F_t(k_t, w)$
if$(st_c, c) = \perp$ **then**
  return $\emptyset$
end if
Send (t_w, st_c, c, Tag_w) to the server
Server:
$R, R_d, E \leftarrow \emptyset$
for $i = c$ *to* 0 **do**
  $UT_i \leftarrow H_1(t_w||st_i)$
  $(e_i, C_{st_{i-1}}) \leftarrow T[UT_i]$
  $E \leftarrow E \cup \{e_i\}$
  $(op||ind) \leftarrow R_1^{-1}(k_g, e_i)$
  remove $T[UT_i]$
  if $op = del$ **then**
    $R_d \leftarrow R_d \cup \{ind\}$
  else if $op = ind$ **then**
    if $ind \in R_d$ **then**
      $R_d \leftarrow R_d \backslash \{ind\}$
    else
      $R \leftarrow R \cup \{ind\}$
  end if
  if $C_{st_{i-1}} = \perp$ **then**
    break
  end if
  $st_{i-1} \leftarrow H_2(t_w, st_i) \oplus C_{st_{i-1}}$
end for
$T[UT_c] \leftarrow (E, \perp)$
$\sigma \leftarrow HMAC.Eval(ek, f, Tag_w)$
$\chi \leftarrow f(R[1], R[2], ...R[size])$
Send (R, σ, χ) to the data user

Revoking Authorized Data User. *Revokeuser*$(u) \rightarrow$ Access: If a data user u wants to revoke his or her search authorization (e.g., the doctor leaving), TA will be responsible for this revocation. Given a data user identifier u, TA updates authorized user list Access = Access$\backslash\{(u, k_u)\}$.

Verifying Searched Results at Data User. *Verify*$(w, \chi, \sigma, sk) \rightarrow val$: When the results returned by the server are obtained, the data user (i.e., the doctor) should verify them to prevent incorrect diagnosis. TA firstly sends *sk* to the data user for verification. The data user uses *HomMAC.Ver* algorithm $val \leftarrow HomMAC.Ver(sk, \chi, \mathcal{P}, \sigma)$ to verify the correctness of the returned results. If the *val* is 0, the results is rejected, otherwise, the results is accepted.

5 Security Analysis

We firstly give the proofs for our construction to satisfy forward privacy. In addition, we prove that our construction satisfies reliability.

Theorem 1. *(Adaptive security): Let F_t be secure PRF, R_1, R_2 be secure PRPs, H_1, H_2 be hash functions modeled as random oracles. Define leakage as:*

$$\mathcal{L}_{DSSE} = \begin{cases} \mathcal{L}_{Setup} = \perp \\ \mathcal{L}_{Search}(w) = (ap(w), qp(w)) \\ \mathcal{L}_{Update}(t, op, ind) = |\{w\}| \end{cases}$$

then DSSE is $\mathcal{L}$-adaptively secure DSSE with forward privacy.

Proof. We construct a series of indistinguishable games, and we show that they cannot be distinguished from the previous one. The first game G_0 is identical to the real-world game while the last game is identical to the idea-world game. Finally, we can obtain the fact that the real-world game and the ideal-world are indistinguishable.

Game G_0: G_0 is identical to the real-world game $REAL_{\mathcal{A}}^{\sum}(\lambda)$:

$$Pr[REAL_{\mathcal{A}}^{\sum}(\lambda) = 1] = Pr[G_0 = 1].$$

Game G_1: In G_1, it replaces every call to F_t by using table Key_t. If there exists an adversary $\mathcal{A}$ that can distinguish between G_0 and G_1, we can build an adversary $\mathcal{B}_1$ that can distinguish F_t from a truly random function. Thus, we can say that:

$$|Pr[G_0 = 1] - Pr[G_1 = 1]| \leq \mathbf{Adv}_{F_t, \mathcal{B}_1}^{PRF}(\lambda).$$

Game G_2: In G_2, it replaces every call to R_2 by using the table Key_u. If there exists an adversary $\mathcal{A}$ that can distinguish between G_1 and G_2, we can build an adversary $\mathcal{B}_2$ that can distinguish R_1 from a truly random permutation. Thus, we can say that:

$$|Pr[G_1 = 1] - Pr[G_2 = 1]| \leq \mathbf{Adv}_{R_2, \mathcal{B}_2}^{PRP}(\lambda).$$

Game G_3: In G_3, it replaces every call to R_1 by using the table Key_g. The probability that two e are equal is at most $\frac{q^2}{2^{2\lambda}}$, where the number of queries to R_1 is q, which is a polynomial in the security parameter. If there exists an adversary $\mathcal{A}$ that can distinguish between G_2 and G_3, we can build an adversary $\mathcal{B}_3$ that can distinguish R_1 from a truly random permutation. Thus, we can say that:

$$|Pr[G_2 = 1] - Pr[G_3 = 1]| \leq \mathbf{Adv}_{R_1, \mathcal{B}_3}^{PRP}(\lambda) + \frac{q^2}{2^{2\lambda}}.$$

Game G_4: In G_4, it randomly picks a string from $\{0,1\}^{\lambda}$ rather than calling hash function H_1 to generate update token UT, and stores string in the table $\text{UT}[H_1(t_w||st_{c+1})]$ in the *Update*. The entry in $\text{UT}[H_1(t_w||st_{c+1})]$ will be returned to the adversary if he or she queries H_1 with $H_1(t_w||st_{c+1})$, the random oracle is programmed in the *Search*, and we use table H_1 to keep track of the transcript. Instead of storing the randomly picked string, it first checks whether there was a query to H_1 with input $H_1(t_w||st_{c+1})$ has happened. If the check is true, then $H_1(t_w||st_{c+1})$ is stored in $\mathrm{UT}[H_1(t_w||st_{c+1})]$ else the random picked string is stored in $\mathrm{UT}[H_1(t_w||st_{c+1})]$. The probability that an adversary guesses the right st_{c+1} is $\frac{p}{2^\lambda}$, where the number of queries is p. Thus, we can say that:

$$|Pr[G_3 = 1] - Pr[G_4 = 1]| \leq \frac{p}{2^\lambda}.$$

Game G_5: In G_5, we model H_2 as a random oracle which is similar to H_1 in G_5. Thus, we can say that:

$$|Pr[G_4 = 1] - Pr[G_5 = 1]| \leq \frac{p}{2^\lambda}.$$

Game G_6: In G_6, we initialize a global counter t in the *Setup* and it will increase in the *Update* to record the number of update operations. Now we can show that G_6 cannot distinguish from G_5.

Simulator. The simulator $\mathcal{S}$ is quite similar to G_6 except two differences. Firstly, the simulator $\mathcal{S}$ inputs the update history $ap(w)$ of keyword w. Secondly, the simulator $\mathcal{S}$ can map the keyword w to $\bar{w} \leftarrow min\ sp(w)$ for the searched keyword query w. Finally, we have

$$Pr[REAL_{\mathcal{A}}^{\Sigma}(\lambda) = 1] - Pr[IDEAL_{\mathcal{A},\mathcal{S}}^{\Sigma}(\lambda) = 1]$$
$$\leq \mathbf{Adv}_{F_t,\mathcal{B}_1}^{PRF}(\lambda) + \mathbf{Adv}_{R_2,\mathcal{B}_2}^{PRP}(\lambda) + \mathbf{Adv}_{R_1,\mathcal{B}_3}^{PRP}(\lambda) + \frac{q^2}{2^{2\lambda}} + \frac{2p}{2^\lambda}.$$

Theorem 2. *(Reliability): Let* Π = *(HomMAC.Setup, HomMAC.Auth, HomMAC.Eval, HomMAC.Ver) be a secure homomorphic MAC scheme. Then, our construction satisfies reliability.*

Proof. Our construction employs the homomorphic MAC techhnique from HomMAC in [5], thus it has similar properties: authentication correctness, evaluation correctness, succinctness and authenticator security. Suppose that there exists an adversary $\mathcal{A}$ who breaks the reliability for some search queries $w_1, ..., w_q$.

From our assumption, $\mathcal{A}$ returns $R^*(w)$ such that $R^*(w)$ is invalid and *Verify*$(w, \chi, \sigma, W, sk) \rightarrow 1$ with no-negligible probability. However, the generated tag σ_w in update process using *HomMAC.Auth* algorithm has authentication correctness, then the computational result χ and a new proof σ are generated using *HomMAC.Eval* algorithm, which has evalutation correctness. Therefore, if *HomMAC.Ver* algorithm returns accept, then we can say that the returned result satisfies reliability.

6 Experiment Evaluation

The scheme is implemented in JAVA, using the JPBC libraries for cryptographic operations: HMAC-SHA-1 for PRF F_t and PRPs R_1, R_2; SHA-256 for hash functions H_1 and H_2. We leverage the machine with four core Intel Core i7-10510U CPU 2.30 GHz processor, running Ubuntu 14.04 LTS, with 16 GB RAM to simulate algorithms. We use the Indian Liver Patient Dataset dataset as real-world dataset [1]. The dataset has 11 attribute (i.e., 11 keywords) and 583 instances, we treat each instance as a PHI file. We also use a simulated dataset to evaluate our scheme, where each PHI file consists of 18 pairs of attributes in the format of attribute:value, which each pair is considered as a single keyword (e.g., $w =$ *blood pressure*: 100)[1]. The number of PHI files in the simulated dataset ranges from 1,000 to 11,000, and the number of keyword/file pairs ranges from 10 to 10^5. The comparison with existing DSSE schemes is given in Table 2.

Table 2. Comparison with existing DSSE schemes

Scheme	Forward privacy	Multi-user	Verifiable	Search time	Update time
Sophos [3]	✓	×	×	$O(a_w)$	O(1)
FDSSE [21]	✓	✓	×	$O(a_w)$	O(1)
MFS [16]	✓	✓	×	$O(n_w)$	O(1)
VFDSSE [20]	✓	×	✓	$O(a_w)$	O(1)
Ours	✓	✓	✓	$O(a_w)$	O(1)

6.1 Setup

Figure 3(a) shows the time cost of initialization phase, which is linear to the number of data users, since Setup process consists of generating scheme parameters, HomMAC keys and multi-user keys. The time cost of Setup process is less than 1 s when the number of data users is less than 100.

6.2 Update

The experiments in Fig. 3(b) shows the update time of our scheme compared with MFS [16] and FDSSE [21] under the simulated dataset. We can observe that the update overhead increases linearly with the number of keyword/file pairs for all three, and the major factor of the update time is the computation time of the new update token. MFS has a heavier overhead on the update protocol due to bilinear pairing computation, while our scheme and FDSSE just rely on hash functions to generate update tokens. Updates in our scheme only needs a constant amount of computation at the client and the server for each update operation, which is close to the existing efficient forward private DSSE schemes for single keyword search [6,21].

[1] Available: https://www.clouddx.com/downloads/Heart-Friendly-Report-2015-12-24-092313.pdf.

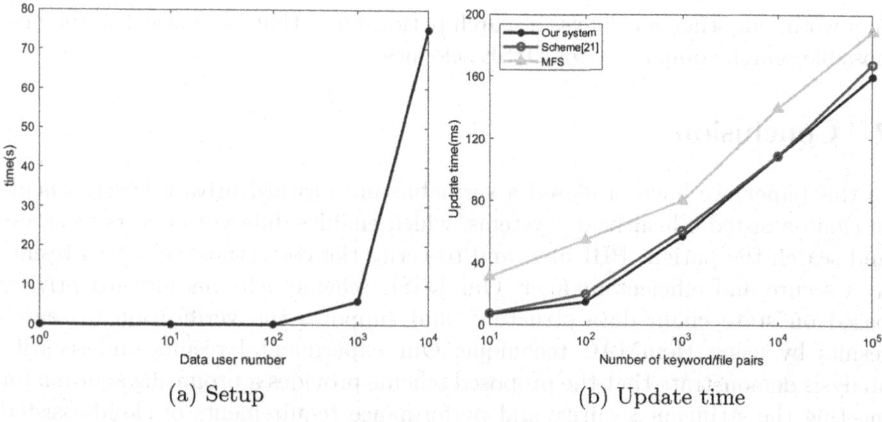

(a) Setup (b) Update time

Fig. 3. Setup and update phase

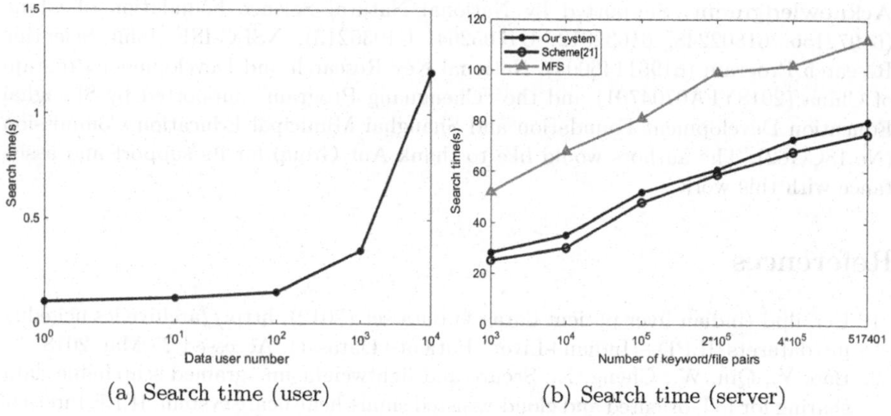

(a) Search time (user) (b) Search time (server)

Fig. 4. Search phase

6.3 Search

We evaluate two parts of *Search* process: the user-side search and the server-side search. From Fig. 4(a), we can see that the search time growth on the user side is linear when the number of data users grows. When there are fewer than 100 users, it takes less than 0.13 s. Figure 4(b) shows the time cost of *Search* process under the real-world dataset. We can say that search time for all three increases with the number of keyword/file pairs. Specifically, MFS generates the largest search overhead among the three schemes because of bilinear pairing computation (i.e. $O(n_w)$), where n_w is the size of the result set, while FDSSE and our scheme only use hash operations. The search complexity of our scheme is linear with the number of update operations involving the search keyword w (i.e. $O(a_w)$). Note that the best possible search complexity for DSSE schemes [6] is linear with the number of matching files with search keyword (i.e. $O(a_w)$).

In a word, our scheme achieves a search performance that is identical to the best possible search complexity for DSSE schemes.

7 Conclusion

In this paper, we have proposed a verifiable and forward private DSSE scheme in cloud-assisted e-healthcare systems, which enables different doctors to access and search the patient PHI files, and to verify the correctness of search results in a secure and efficient manner. Our DSSE scheme achieves forward privacy based on state chain data structure, and supports the verification of search results by using HomMAC technique. Our experimental results and security analysis demonstrate that the proposed scheme provides a promising solution for meeting the stringent security and performance requirements of cloud-assisted e-healthcare systems.

Acknowledgment. Supported by National Natural Science Foundation of China (61972156, 61802248, 61632012, U1705264, U1936213), NSFC-ISF Joint Scientific Research Program (61961146004), National Key Research and Development Program of China (2018YFA0704701) and the "Chenguang Program" supported by Shanghai Education Development Foundation and Shanghai Municipal Education Commission (No.18CG62). The authors would like to thank Ant Group for its support and assistance with this work.

References

1. Uci. ilpd (indian liver patient dataset) data set (2012). http://archive.ics.uci.edu/ml/datasets/ILPD+(Indian+Liver+Patient+Dataset), Accessed 25 May 2018
2. Bao, Y., Qiu, W., Cheng, X.: Secure and lightweight fine-grained searchable data sharing for IoT-oriented and cloud-assisted smart healthcare system. IEEE Internet Things J. **9**(4), 2513–2526 (2021)
3. Bost, R.: $\sum o\varphi o\varsigma$: forward secure searchable encryption. In: Proceedings of the 2016 ACM SIGSAC Conference on Computer and Communications Security, CCS 2016, pp. 1143–1154 (2016)
4. Cash, D., Jarecki, S., Jutla, C., Krawczyk, H., Roşu, M.-C., Steiner, M.: Highly-scalable searchable symmetric encryption with support for Boolean queries. In: Canetti, R., Garay, J.A. (eds.) CRYPTO 2013. LNCS, vol. 8042, pp. 353–373. Springer, Heidelberg (2013). https://doi.org/10.1007/978-3-642-40041-4_20
5. Catalano, D., Fiore, D.: Practical homomorphic MACs for arithmetic circuits. In: Johansson, T., Nguyen, P.Q. (eds.) EUROCRYPT 2013. LNCS, vol. 7881, pp. 336–352. Springer, Heidelberg (2013). https://doi.org/10.1007/978-3-642-38348-9_21
6. Chamani, J.G., Papadopoulos, D., Papamanthou, C., Jalili, R.: New constructions for forward and backward private symmetric searchable encryption. In: Proceedings of the 2018 ACM SIGSAC Conference on Computer and Communications Security, CCS 2018, pp. 1038–1055 (2018)
7. Curtmola, R., Garay, J.A., Kamara, S., Ostrovsky, R.: Searchable symmetric encryption: improved definitions and efficient constructions. In: Proceedings of the 13th ACM Conference on Computer and Communications Security, CCS 2006, pp. 79–88 (2006)

8. Du, L., Li, K., Liu, Q., Wu, Z., Zhang, S.: Dynamic multi-client searchable symmetric encryption with support for Boolean queries. Inf. Sci. **506**, 234–257 (2020)
9. Ge, X., Yu, J., Chen, F., Kong, F., Wang, H.: Towards verifiable phrase search over encrypted cloud-based IoT data. IEEE Internet Things J. **8**, 12902–12918 (2021)
10. Kiayias, A., Oksuz, O., Russell, A., Tang, Q., Wang, B.: Efficient encrypted keyword search for multi-user data sharing. In: Askoxylakis, I., Ioannidis, S., Katsikas, S., Meadows, C. (eds.) ESORICS 2016. LNCS, vol. 9878, pp. 173–195. Springer, Cham (2016). https://doi.org/10.1007/978-3-319-45744-4_9
11. Song, D.X., Wagner, D.A., Perrig, A.: Practical techniques for searches on encrypted data. In: Proceedings of 2000 IEEE Symposium on Security and Privacy, SP 2000, pp. 44–55. IEEE (2000)
12. Stefanov, E., Papamanthou, C., Shi, E.: Practical dynamic searchable encryption with small leakage. In: Proceedings of 21st Annual Network and Distributed System Security Symposium, NDSS (2014)
13. Sun, S., et al.: Practical backward-secure searchable encryption from symmetric puncturable encryption. In: Proceedings of the 2018 ACM SIGSAC Conference on Computer and Communications Security, CCS 2018, pp. 763–780 (2018)
14. Tong, Y., Sun, J., Chow, S.S., Li, P.: Cloud-assisted mobile-access of health data with privacy and auditability. IEEE J. Biomed. Health Inf. **18**(2), 419–429 (2013)
15. Wang, K., Chen, C.M., Tie, Z., Shojafar, M., Kumar, S., Kumari, S.: Forward privacy preservation in IoT enabled healthcare systems. IEEE Trans. Ind. Inf. **18**, 1991–1999 (2021)
16. Wang, Q., Guo, Yu., Huang, H., Jia, X.: Multi-user forward secure dynamic searchable symmetric encryption. In: Au, M.H., Yiu, S.M., Li, J., Luo, X., Wang, C., Castiglione, A., Kluczniak, K. (eds.) NSS 2018. LNCS, vol. 11058, pp. 125–140. Springer, Cham (2018). https://doi.org/10.1007/978-3-030-02744-5_9
17. Xu, C., Wang, N., Zhu, L., Sharif, K., Zhang, C.: Achieving searchable and privacy-preserving data sharing for cloud-assisted e-healthcare system. IEEE Internet Things J. **6**(5), 8345–8356 (2019)
18. Yang, L., Zheng, Q., Fan, X.: RSPP: a reliable, searchable and privacy-preserving e-healthcare system for cloud-assisted body area networks. In: IEEE INFOCOM 2017-IEEE Conference on Computer Communications, pp. 1–9. IEEE (2017)
19. Zhang, Y., Katz, J., Papamanthou, C.: All your queries are belong to us: the power of file-injection attacks on searchable encryption. In: 25th USENIX Security Symposium, USENIX Security 2016, pp. 707–720. USENIX Association (2016)
20. Zhang, Z., Wang, J., Wang, Y., Su, Y., Chen, X.: Towards efficient verifiable forward secure searchable symmetric encryption. In: Sako, K., Schneider, S., Ryan, P.Y.A. (eds.) ESORICS 2019. LNCS, vol. 11736, pp. 304–321. Springer, Cham (2019). https://doi.org/10.1007/978-3-030-29962-0_15
21. Zuo, C., Sun, S.-F., Liu, J.K., Shao, J., Pieprzyk, J.: Dynamic searchable symmetric encryption with forward and stronger backward privacy. In: Sako, K., Schneider, S., Ryan, P.Y.A. (eds.) ESORICS 2019. LNCS, vol. 11736, pp. 283–303. Springer, Cham (2019). https://doi.org/10.1007/978-3-030-29962-0_14

Security Analysis of Poisoning Attacks Against Multi-agent Reinforcement Learning

Zhiqiang Xie, Yingxiao Xiang, Yike Li, Shuang Zhao, Endong Tong(✉), Wenjia Niu(✉), Jiqiang Liu, and Jian Wang

Beijing Key Laboratory of Security and Privacy in Intelligent Transportation, Beijing Jiaotong University, Beijing 100044, China
{edtong,niuwj}@bjtu.edu.cn

Abstract. As the closest machine learning method to general artificial intelligence, multi-agent reinforcement learning (MARL) has shown great potential. However, there are few security studies on MARL, and related security problems also appear, especially the serious misleading caused by the poisoning attack on the model. The current research on poisoning attacks for reinforcement learning mainly focuses on single-agent setting, while there are few such studies for multi-agent RL. Hence, we propose an analysis framework for the poisoning attack in the MARL system, taking the multi-agent soft actor-critic algorithm, which has the best performance at present, as the target of the poisoning attack. In the framework, we conduct extensive poisoning attacks on the agent's state signal and reward signal from three different aspects: the modes of poisoning attacks, the impact of the timing of poisoning, and the mitigation ability of the MARL system. Experiment results in our framework indicate that 1) compared to the baseline, the random poisoning against state signal reduces the average reward by as high as -65.73%; 2) the timing of poisoning has completely opposite effects on reward-based and state-based attacks; and 3) the agent can completely alleviate the toxicity when the attack interval is 10000 episodes.

Keywords: Reinforcement learning · Multi-agent system · Soft actor-critic · Poisoning attack · Security analysis

1 Introduction

As a branch of artificial intelligence, reinforcement learning based on reward maximization is the most popular method to achieve general artificial intelligence. With the development of computing power, we have witnessed many great successes in reinforcement learning in recent years, such as AlphaGo Zero [1], autonomous vehicles [2], and unmanned aerial vehicles [3]. However, the problems are complex and changeable in reality. The intelligent decision-making of a single agent can no longer meet the needs, and multi-agent intelligent decision-making has gradually become the mainstream. In the multi-agent setting, agents need to compete or cooperate in the dynamically changing environment. In order to learn more effectively, the agent needs to observe the state

Y. Lai et al. (Eds.): ICA3PP 2021, LNCS 13155, pp. 660–675, 2022.
https://doi.org/10.1007/978-3-030-95384-3_41

of the environment and pay attention to the state of other agents. So, in this case, the easiest way is to take the state of all other agents and the state of the environment as a joint state space, and use this as the agent's observation. However, due to the constant changes of other agents' policies, the overall environment is neither stable nor Markovian. In the current mainstream works, a more advanced architecture is the centralized training and distributed execution (CTDE) [4]. In the CTDE architecture, agents' critics are located in the central nodes and the agent executes the action through a distributed actor. This architecture can solve partial observability problems and meanwhile avoid huge input and output space dimensions caused by centralized execution. Hence, the CTDE architecture is used in this paper as the setting of the poisoned MARL algorithm.

With the advancement of the MARL field, hidden security issues have gradually emerged. Like federated learning, multi-agent systems may also suffer malicious damage such as poisoning attacks and privacy inference attacks. Poisoning attacks are usually easier to implement than privacy inference attacks, and the contamination of data by poisoning attacks is more likely to cause irreversible consequences. For example, in an unmanned driving system, if the observation value of the agent is contaminated, the consequences will be disastrous. In the MARL framework, agents receive partial state signals and reward signals from the environment through sensors, and agents also need to communicate with each other to obtain information about other agents. In the process of data transmission, the adversary can manipulate the signal through data injection attacks or prevent the agent from receiving the correct signal by adding malicious interference to the channel. For example, because the sensor has the limited computing resource and storage space, the adversary is likely to attack the sensor and cause its normal data to be contaminated. In our previous introduction, we have mentioned that agents need to perceive the states of other agents to make decisions; this will lead to another attack method. The malicious agent may deliberately send false state information to mislead the normal agent. Because the feedback value from the environment is maliciously manipulated, the MARL algorithm is either unable to learn the policy or is misled to learn an incorrect policy. Under malicious attacks, the failure of the MARL algorithm will bring unpredictable consequences. For example, in autonomous driving, wrong observation data will cause serious car accidents. Therefore, it is very necessary to research poisoning attacks in the MARL system.

The defense mechanism is necessary for MARL, and understanding poisoning attacks is essential to develop effective defense mechanisms and is an important step toward trustworthy and safe MARL. In our work, we propose a comprehensive analysis framework, which conducts security analysis on three different aspects of poisoning attacks for state and reward signals. First, we explore the effects of different poisoning attack modes. Second, we study the impact of attack timing on poisoning by changing the time period of poisoning. Finally, we analyze the mitigation ability of the model by changing the time interval of attacks. We have a more comprehensive understanding of them through multi-dimensional analysis of poisoning attacks, so we can adopt appropriate defense methods in the future to decrease the security risk.

The rest of the paper is structured as follows: In Sect. 2, we discuss related work. The threat model is proposed in Sect. 3. Section 4 conduct the security analysis mechanism and reports our findings. Finally, Sect. 5 concludes the work of this paper.

2 Related Work

MARL is a long-term development field. In recent years, deep learning methods have become more and more popular, breakthroughs have been made in this field. In [5], AlphaStar developed by Vinyals et al. showed its strength beyond the top human players in StarCraft 2. And OpenAI Five [6] also proved that MARL has unlimited potential by defeating the Dota 2 world champion team OG.

Nevertheless, there are still many challenges in the MARL field. Many MARL methods have non-stationarity problems and scalability problems, and at the same time are easily troubled by partial observability. Although these problems have not been completely resolved, there are already some solutions with good performance. In terms of scalability, the earliest work is [7], which supports two independent DQN agents. Wai et al. [8] proposed to use the gradient and reward information of neighbors to accelerate the convergence of the local model. Abouheaf and Gueaieb proposed an online adaptive reinforcement learning method to solve large-scale optimization problems [9]. Regarding the problem of non-stationarity, the authors in [10] solved the non-stationarity caused by experience replay, and the work [11] reduced the non-stationarity significantly through centralized critics. The multi-Agent deep deterministic policy gradient method proposed by the authors in [12] solves the non-stationarity and partial observation of the system. In a recent research work [13], a soft actor-critic under the multi-agent setting is proposed. It solves non-stationarity and partial observation through the CTDE architecture and introduces soft value to motivate the agent to explore the unknown space. And in the comparative experiment of this paper, the performance of its agent exceeds other algorithms. Therefore, the multi-agent soft actor-critic algorithm is selected as the target of the poisoning attack in our analysis framework.

At present, the research work in the MARL field generally focuses on the performance of the model, and there are few studies on its security. Performance improvements are important for reinforcement learning, but all these efforts may be in vain if the adversary successfully attacks the agent. At present, security research in the field of reinforcement learning mainly focuses on a single agent. The research work in [14–17] revolves around white-box attacks under a single agent setting, but this often requires the attacker to understand reinforcement learning algorithms. And the attack cost is relatively high. Compared with the above attack methods, poisoning attacks are usually easier to achieve. In [18], the authors proposed a policy poisoning framework for batch reinforcement learning, and experiments show that the attack can change the learner's policy with a small cost. In [19], Zhang et al. showed that effective attacks can be found empirically using reinforcement learning techniques. In [20], Rakhsha et al. studied the security threats of forcing the agent to execute the wrong strategy by poisoning the environment in reinforcement learning. With the evolution of technology, multi-agent reinforcement learning is the general trend in the future. However, the related security works are still relatively small. And research on poisoning attacks in the MARL system is even rarer, so its security analysis is imminent. Therefore, we propose a multi-dimensional analysis framework for poisoning attacks in the MARL system in this paper. We carried out extensive experiments from three aspects: the mode, timing, and frequency of poisoning. The study of malicious attacks in the MARL system is a crucial way to further develop defense mechanism.

3 Threat Model

3.1 Preliminaries

Markov Games. In this work, we consider a Markov decision process (MDPs) [21] under a multi-agent setting, which is called the partially observable Markov game. A Markov game of N agents is defined by a set of states S, a set of actions $A = \{A_1, \ldots, A_N\}$ and a set of observations $O = \{O_1, \ldots, O_N\}$. Each agent will continue to learn the local policy function, $\pi_i : O_i \rightarrow P(A_i)$ which is actually a mapping between the agent's observations and the actions performed. After the action is executed, the next state will be generated according to the state transition function $T : S \times A_1 \times \cdots \times A_N \rightarrow P(S)$ and the agent will receive a corresponding partial observationO_i. The cumulative reward that the ith agent gets at time T is $R_i = \sum_{t=0}^{T} \gamma^t r_{it}$, where γ represents discount factor that determines how much the agent prefers long-term gain or short-term gain. Each agent must learn as much as possible a better strategy to maximize the expected discounted return $J_i(\pi_i) = \mathbb{E}_{a_1 \sim \pi_1, \ldots, a_N \sim \pi_N, s \sim T}\left[\sum_{t=0}^{\infty} \gamma^t r_{it}(s_t, a_{1t}, \ldots, a_{Nt})\right]$.

Policy Gradient Algorithms. In policy-based reinforcement learning tasks, we use deep neural networks $\pi_\theta(a|s)$ to approximate the policy function where θ is the neural network parameter. The main idea of the Policy Gradient method [Policy gradient methods for reinforcement learning with function approximation] is to continuously adjust θ to maximize the objective function $J(\pi_\theta)$. The gradient of the policy network can be written as the following form:

$$\nabla_\theta J(\pi_\theta) = \mathbb{E}_{a_{t'} \sim \pi_\theta, s_{t'} \sim T}\left[\nabla_\theta \log \pi_\theta(a_t|s_t) \sum_{t'=t}^{\infty} \gamma^{t'-t} r_{t'}(s_{t'}, a_{t'})\right] \tag{1}$$

Actor-Critic and Soft Actor-Critic. The main idea of the Actor-Critic algorithm [22] is to combine a value-based approach and a policy-based approach. This method can avoid the situation where the variance is too large when only using the policy gradient algorithm, so it holds the promise of delivering faster convergence. One specific instance of actor-critic methods learns a function to estimate expected discounted returns, given a state and action, $Q_\psi(s_t, a_t) = \mathbb{E}\left[\sum_{t'=t}^{\infty} \gamma^{t'-t} r_{t'}(s_{t'}, a_{t'})\right]$, learned through off-policy temporal-difference learning by minimizing the regression loss:

$$\mathcal{L}_Q(\psi) = \mathbb{E}_{(s,a,r,s') \sim D}\left[\left(Q_\psi(s, a) - y\right)^2\right] \tag{2}$$

where

where $y = r(s, a) + \gamma \mathbb{E}_{a' \sim \pi(s')}\left[Q_{\overline{\psi}}\left(s', a'\right)\right]$, and $Q_{\overline{\psi}}$ is the target Q-value function, which is simply an exponential moving average of the past Q-functions and D is a replay buffer that stores past experiences.

Soft Actor-Critic (SAC) [23] is constructed under the framework of maximum entropy reinforcement learning, with the purpose of randomizing the policy. The maximum entropy of the policy also means that the exploration of the policy space and trajectory space is more sufficient than the deterministic algorithm. For states where there is more than one optimal action, the algorithm can output the probability distribution of one action instead of a certain action. The SAC learns a soft value function by modifying the policy gradient to incorporate an entropy term:

$$\nabla_\theta J(\pi_\theta) = \mathbb{E}_{s\sim D, a\sim\pi}\left[\nabla_\theta \log(\pi_\theta(a|s))\left(-\alpha\log(\pi_\theta(a|s)) + Q_\psi(s,a) - b(s)\right)\right] \quad (3)$$

where $b(s)$ is a state-dependent baseline (for the Q-value function) and α is the temperature parameter determining the balance between maximizing entropy and rewards. The loss function for temporal-difference learning of the value function is also revised accordingly with a new target:

$$y = r(s,a) + \gamma \mathbb{E}_{a'\sim\pi_{\overline{\theta}}(s')}\left[Q_{\overline{\psi}}\left(s',a'\right) - \alpha\log\left(\pi_{\overline{\theta}}\left(a'|s'\right)\right)\right] \quad (4)$$

3.2 Poisoning Attack

Under malicious attacks on reward signals or state signals, the agent will fail to perform reasonable actions while the remaining aspects of the MDP framework stay the same. Our proposed threat model is as follows:

Knowledge of the Attacker. The attacker knows the multi-agent SAC algorithm. In the state-based poisoning attack, the attacker has access to the environmental state sensor of the victim agents. In the state-based poisoning attack, the attacker has access to the critic of the victim agents, which is located at the central node.

Available Actions of the Attacker. Due to the limitation of computing power and storage space of sensors, they are usually more vulnerable to attacks. Therefore, the attacker is allowed to arbitrarily modify the state signal o_i sent by the sensor to $\overline{\mathrm{o_i}}$. Although the central node has more capabilities than sensors, it may also compromise under DOS attacks or illegal operations by insiders. Therefore, in the reward-based poisoning attack, the attacker can send a false reward signal $\overline{\mathrm{r_j}}$ to the critic of agent i.

Attacker's Goals. The attacker's task is to design forged signals based on his information structure and his available actions to achieve certain goals. In a multi-agent game, the two parties are in a competitive relationship. One party needs to get as many rewards as possible and reduce the rewards of the other party. The purpose of the attacker is to interfere with the opponent's decision-making by poisoning to reduce the opponent's reward and win the game.

As shown in Fig. 1, the overall architecture of the multi-agent SAC algorithm is set to CTDE. The central node will receive the signals uploaded by all agents and train critics, while the agents at the edge use local actors to perform actions. The critic of each agent needs to consider the action-value function of all agents simultaneously under the multi-agent setting. Although this can improve the performance of the agent, it also gives the attacker an opportunity. We propose two types of poisoning attacks on the system: reward-based, and state-based. Next, we will introduce the details of these two types of attacks.

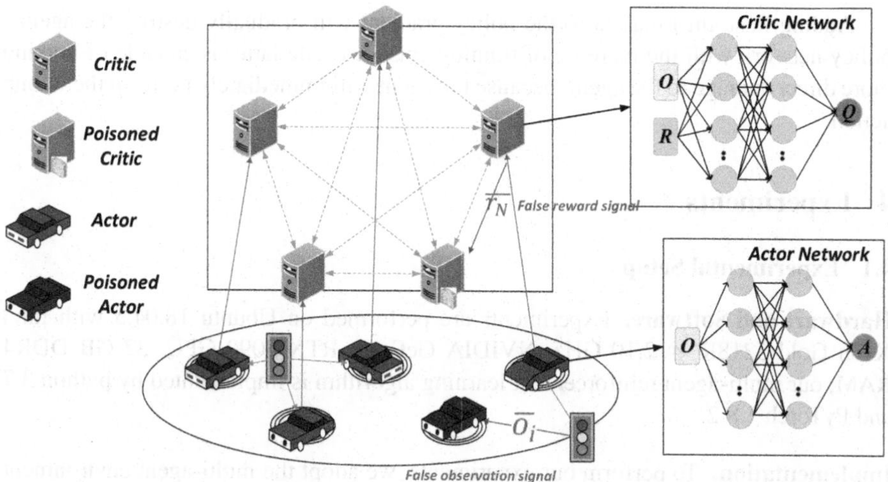

Fig. 1. The CTDE architecture in the multi-agent SAC algorithm.

(1) *Reward-based poisoning attack*. Since the reward function is fixed, malicious agents generally cannot tamper with the reward values of other agents. But malicious agents can send false signals when broadcasting their rewards. As the gradient descent in each round, the false signal will increase the error between the Q function and the actual value. For example, the malicious agent may deny that it has received a large reward, so the normal agent's decision will go astray. Under this attack, the loss function of the victim agent takes the following form:

$$\mathcal{L}_Q(\psi) = \sum_{i=1}^{N-1} \mathbb{E}_{\left(o_i, a_i, r_i, o_i'\right)\sim D}\left[\left(Q_{i\psi}(o_i, a_i) - y_i\right)^2\right] + \mathbb{E}_{\left(o_i, a_i, r_i, o_i'\right)\sim D}\left[\left(Q_{i\psi}(o_i, a_i) - \overline{\mathrm{y_N}}\right)^2\right] \tag{5}$$

where $\overline{\mathrm{y_N}} = \overline{r_N} + \gamma \mathbb{E}_{a_i' \sim \pi_{\overline{\theta}}\left(o_i'\right)}\left[Q_{\overline{\psi}}\left(o_i', a_i'\right) - \alpha \log\left(\pi_{\overline{\theta}}\left(a_i' | o_i'\right)\right)\right]$.

(2) *State-based poisoning attack*. Malicious agents can send false self-states or tamper with state signals sent to normal agents (such as changing the location information), resulting in the normal agent being unable to obtain the true environmental observation and therefore taking wrong actions. In a state-based poisoning attack, both the agent's policy gradient and the actions it performs will be affected:

$$\nabla_\theta J(\pi_\theta) = \mathbb{E}_{a_{it'} \sim \pi_\theta, o_{it'} \sim T}\left[\nabla_\theta \log \pi_\theta(a_{it} | \overline{\mathrm{o_{it}}}) \sum_{t'=t}^{\infty} \gamma^{t'-t} r_{t'}\left(o_{it'}, a_{it'}\right)\right] \tag{6}$$

$$P(a_i) = \pi_{\overline{\theta}}(a_i | \overline{\mathrm{o_i}}) \tag{7}$$

Adding false information to the policy gradient will gradually destroy the agent's policy network with the increase of training episodes. The latter is capable of making more direct damage to the agent, because the agent will immediately perform the wrong action.

4 Experiments

4.1 Experimental Setup

Hardware and Software. Experiments are performed on Ubuntu 18.04.5 with Intel Xeon Gold 5218R @ 2.10 GHz, NVIDIA GeForce RTX 3090 GPU, 32 GB DDR4 RAM, our multi-agent reinforcement learning algorithm is implemented by python 3.7 and PyTorch 1.8.2.

Implementation. To perform our experiments, we adopt the multi-agent environment proposed in GitHub project [24], which is a simple multi-agent particle world with a continuous observation and discrete action space, along with some basic simulated physics. Our experiment is conducted in a game in this environment. Below we will provide detailed information for this game.

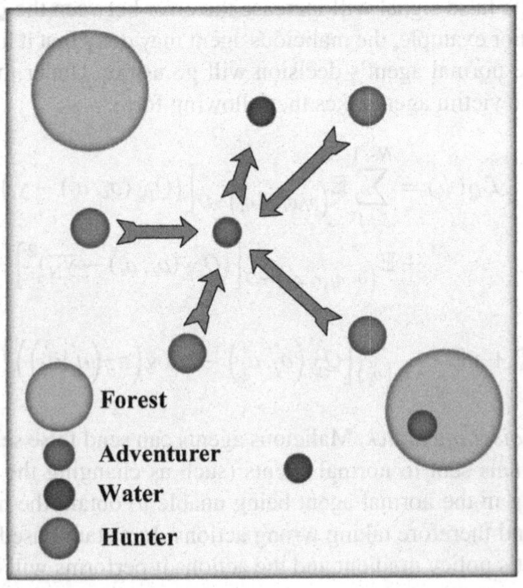

Fig. 2. An illustration of the *Looking for water*.

Looking for Water. As shown in Fig. 2, there are 6 agents in this game. Two of the green balls are faster. They are rewarded by approaching the blue ball, which is the water source. And the other four red balls are slower. They need to chase the green ball to get a reward. The two big green balls represent the forests. When the green balls enter the forests, the red balls cannot observe them. In this environment, only the agent (the green balls and the red balls) can move. After each episode, the environment will change randomly. In this game, both the green ball and the red ball need to learn to cooperate with each other to maximize the overall reward, while the green ball and the red ball are in a competitive relationship.

4.2 Experimental Results

In our analysis mechanism, we launch different poisoning attacks from three angles to explore the poisoning attacks in the MARL system more comprehensively. (1) We adopt three different modes of poisoning, namely random poisoning, reverse poisoning, and halved poisoning. The poisoning modes' details are described as Table 1. (2) We carry out poisoning attacks at different stages of the training process and observe its effects. (3) We perform poisoning attacks with different frequencies.

Table 1. Details of different modes of poisoning

Mode of poisoning	Detailed description
Random poisoning	Replace normal values with random numbers
Reverse poisoning	Replace normal values with opposite numbers
Halved poisoning	Cut the normal value to half

Attack Effects of Different Poisoning Modes

In this section, we use different poisoning modes to contaminate the received state signals and reward signals of the green ball. One of the four red balls is malicious, which will send false signals to the green ball.

As Fig. 3 shows, we can see the clear trend of the average rewards per episode under state-based attacks and reward-based attacks, respectively. The gray curve represents a normal agent that receives the correct state signals and reward signals, while the other three colors represent the reward trend under three poisoning modes, respectively.

In order to quantitatively evaluate the harmful effects of poisoning attacks on normal agents, we subtracted the average reward of the last 5000 episodes under the poisoning attack from normal one. The results are shown in Table 2.

In Fig. 3(a), we can see that random poisoning works best under state-based attacks, while halved poisoning has negligible impact on the agent. In addition, reversed poisoning can also have a greater negative impact, which is between the other two poisoning modes. So, it can be seen that if the state value is multiplied by a constant coefficient between 0–1, the agent can still learn a good policy. This is probably because the model has learned to observe the environment in a new mode. When the state signal is multiplied by 0.5, it can be regarded as a part of the environment compressed according to a

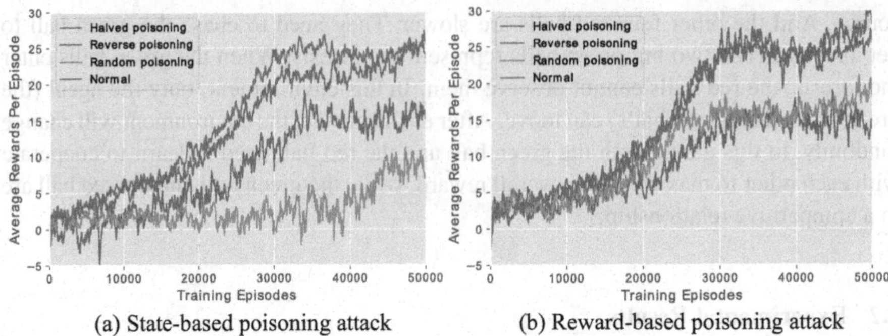

(a) State-based poisoning attack (b) Reward-based poisoning attack

Fig. 3. Use different methods of poisoning to attack the agent

Table 2. Decrease in average reward per episode caused by different poisoning modes

Mode of poisoning	State-based poisoning attack	Reward-based poisoning attack
Random poisoning	**−65.73%**	−28.30%
Reverse poisoning	−37.99%	**−41.83%**
Halved poisoning	−4.04%	+5.92%

fixed ratio. This part of the environment state that has been tampered with can still have a one-to-one correspondence with the real environment state. For example, when playing table tennis, even if the size of the ball is deliberately changed, it does not have a great impact on the players. When the state received by the agent is reversed, the reward of the agent can eventually converge to an acceptable interval. When performing reverse poisoning, the coordinates of some agents become opposite, but the agent can still learn to map the opposite state to the real state one by one. However, since the normal state mapping and the reverse mapping exist simultaneously, the agent needs to learn a policy model that can take into account both types of mapping at the same time. This situation causes a lot of difficulties for the agent's decision-making, so the reward value has a certain degree of decline. In random poisoning, since the state of each transmission is randomly generated, the agent cannot construct a one-to-one mapping between the false state and the real state. Therefore, the agent can only rely on the part of the real environmental observations to make decisions, so the final reward is significantly reduced compared to the normal situation.

Figure 3(b) shows the trend of the average reward per episode when malicious agents use reward-based attacks. In this case, the malicious agent publishes a false reward value, while the normal agent broadcasts the real reward value. We can see that under halved poisoning, the reward given to the agent is almost the same as the normal situation. This result is the same as in the state-based attack. Therefore, we can make a reasonable conjecture that the agent has learned to map the reward after halving to the real environment state in this case. In other words, the agent learned to make decisions in a new mode. The destructive effects caused by reverse and random poisoning are very obvious, and the attack effect of the former is slightly better. In reverse poisoning, when the malicious red ball hunts down the green ball, it will not inform others that it has obtained the capture reward but broadcasts an opposite number as the reward signal. In this way, another green ball will make an optimistic but wrong judgment about the current situation and make a wrong decision. Random poisoning is slightly worse than reverse poisoning. The reason may be that some of the values in the generated random numbers are not much different from the true values, so the misleading effect on normal agents is not as good as reverse poisoning.

The Impact of the Timing of Poisoning

In this section, we explore the attack effect of poisoning at different stages of training. The methods of poisoning used in this experiment are all reverse poisoning attacks.

The training of the model is conducted for a total of 50,000 episodes, and we divide the total number of episodes into five equal stages. Since the model has not learned any knowledge in the 0–10000 episodes, there is no need to carry out the poisoning experiment at this stage. We conduct poisoning attacks on episodes 10000–20000, episodes 20000–30000, episodes 30000–40000, and episodes 40000–50000. The variation of the average rewards per episode when using state-based poisoning attack and reward-based poisoning attack is shown in Fig. 4 and Fig. 5. The four subplots represent episodes of 10,000 to 20,000, 20,000 to 30,000, 30,000 to 40,000, and 40,000 to 50,000, respectively. Furthermore, by subtracting the average value of the reward before poisoning with the value after poisoning, we can more clearly see the different effects in different stages under poisoning attacks. Table 3 shows the percentage of average reward decrease after poisoning attacks.

Table 3. Decrease in average reward caused by poisoning in different time periods

Time of poisoning	State-based poisoning attack	Reward-based poisoning attack
10000–20000	−1.08%	−20.99%
20000–30000	−20.29%	**−35.38%**
30000–40000	−43.28%	−26.14%
40000–50000	**−53.46%**	−14.39%

In Fig. 4(a), we can see that the reward curve oscillates constantly, but the reward value has not dropped significantly during the poisoning stages. In Fig. 4(b), the curve begins to have an obvious decline accompanied by stronger oscillations than before. And in the later part of training, 30,000–40000 episodes and 40,000–50,000 episodes, the trend of the curve began to change drastically. In Fig. 4(c) and Fig. 4(d), we can see that the reward curve of an agent falls off a cliff immediately after being poisoned, and then goes into an oscillation. When we launched a state-based poisoning attack in the stage of 40,000–50,000 episodes, the average reward per episode of the agent dropped by −53.46% compared to normal conditions as shown in Table 3, which is more than 50 times better than the attack effect that launched the attack at the beginning.

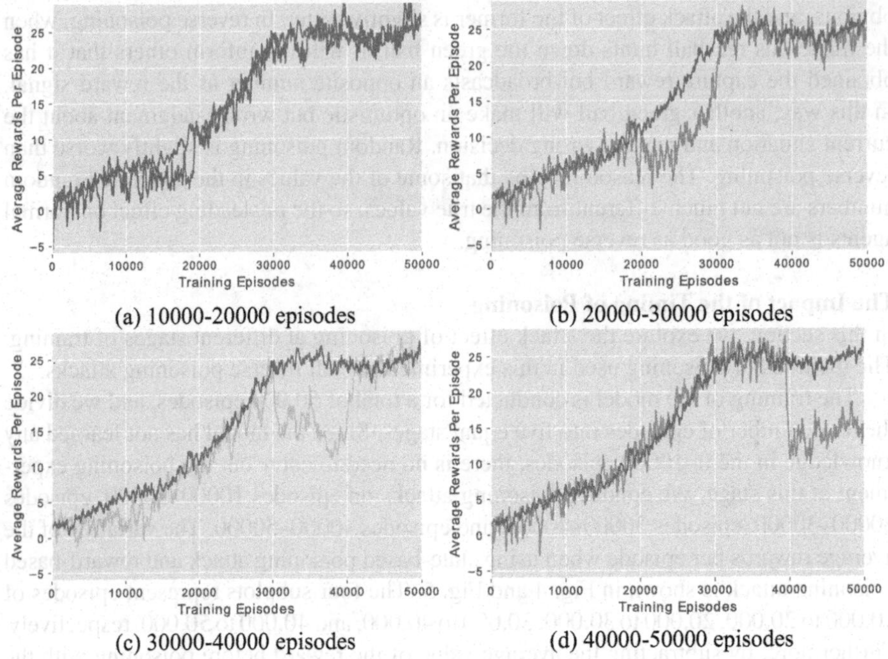

(a) 10000-20000 episodes

(b) 20000-30000 episodes

(c) 30000-40000 episodes

(d) 40000-50000 episodes

Fig. 4. State-based poisoning attacks at different stages of training

In summary, we can find that as the attack time period goes backward, the damage effect of poisoning on the model gradually increases. This is largely due to the different training levels of agents in different time periods. In the earlier stage of training, the agent is still in the stage of trial and its local strategy is very simple. Even if the state information given to the agent is true, it may not get great rewards. Therefore, when faced with maliciously tampered state information, the agent's reward drops slightly. In the later stage of training, the agent's strategy is already very mature, so it can respond reasonably to the given state. Since the agent has fully adapted to the normal state observation, the flip of the state signal will completely subvert the action that the agent should do. When the agent faces the false state signal, its actions will be completely opposite to the actions it should do. Therefore, when the agent suffers from an attack in the later stage, the average reward drops significantly.

Focusing on the reward-based poisoning attack, in Fig. 5 we have the following intuitive observations. In Fig. 5(a), the reward growth trend of the victim agent after 30,000 episodes has slowed down significantly, and at the end of the training, it has dropped by −20.99% compared to the normal situation. In Fig. 5(b), the reward gap between the victim agent and the normal agent reaches the maximum of −35.38%, and the reward of the agent even declines to some extent after 30,000 episodes. And when we launched a poisoning attack after 30,000 episodes, as shown in Fig. 5(c) and Fig. 5(d), we can see that as the timing of the attack is postponed, the impact on the agent is getting less and less. When we poisoned during 40,000 to 50,000 episodes, the agent's reward decreased by only −14.39%.

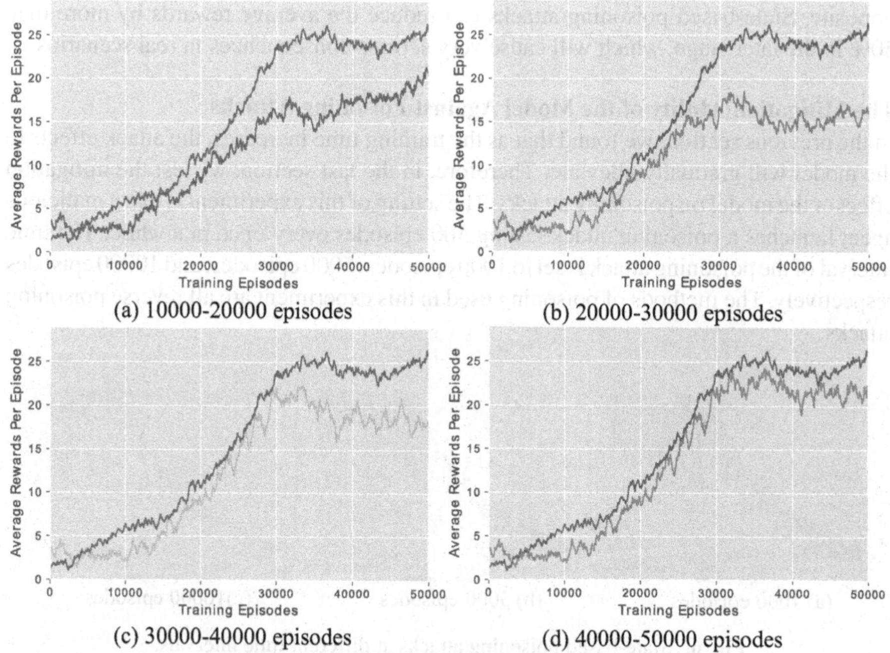

(a) 10000-20000 episodes (b) 20000-30000 episodes

(c) 30000-40000 episodes (d) 40000-50000 episodes

Fig. 5. Reward-based poisoning attacks at different stages of training

Through the above observations, we can find that the timing of the attack has completely different effects on the two types of poisoning modes. In a state-based poisoning attack, the agent is more severely affected in the later stage. The situation is just the opposite while in a reward-based poisoning attack. The agent's reward will gradually converge after 30,000 episodes under normal circumstances, so we can judge that the agent's policy has been almost perfect at the 30,000 episodes. However, because the victim agent received the wrong reward signal before 30,000 episodes, the rate of increase of its reward curve is much slower than that of the normal agent. Since the poisoning attack in Fig. 5(a) was initiated earlier than Fig. 5(b), the toxicity suffered by the agent in the former has been relieved to a certain extent with training. Therefore, when the attack is launched between 20,000 and 30,000 episodes, the agent suffers a greater impact. In the later stage of training, the model of the agent is already relatively mature and possesses a certain degree of robustness. The wrong reward signal can only gradually deteriorate the model through many rounds of gradient descent, so the average reward per episode of the agent will not drop suddenly.

Through the experimental results in this section, we can intuitively see the obvious difference between the two types of poisoning attacks. Reward-based poisoning attacks are more destructive in the earlier stage, while state-based poisoning attacks are just the opposite. State-based poisoning attacks can reduce the average rewards by more than 50% in the later stage, which will cause very serious consequences in real scenarios.

The Mitigation Ability of the Model Against Poisoning Attacks

In the previous section, we found that as the training time increases, the attack effects to the model will gradually alleviate. Therefore, in the last section, we test the mitigation effect of the model on poisoning attacks. The setting of this experiment is that a malicious agent launches a poisoning attack lasting 500 episodes every once in a while. The time interval of the poisoning attack is set to 1000 episodes, 5000 episodes, and 10000 episodes respectively. The methods of poisoning used in this experiment are all reverse poisoning attacks.

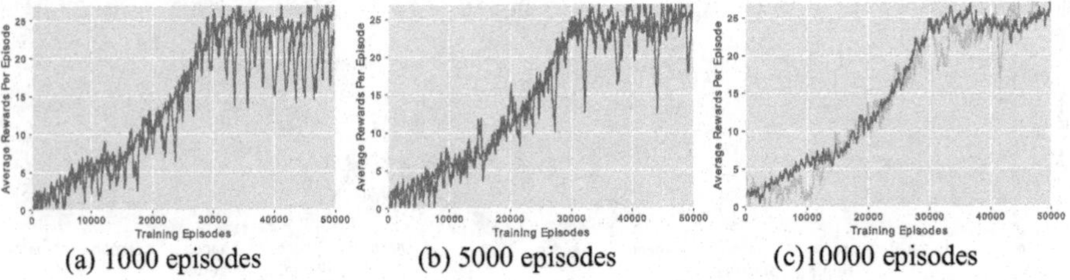

(a) 1000 episodes (b) 5000 episodes (c)10000 episodes

Fig. 6. State-based poisoning attacks at different time intervals.

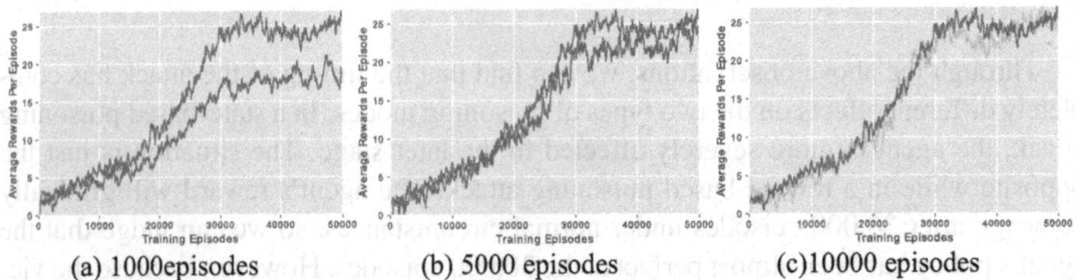

(a) 1000episodes (b) 5000 episodes (c)10000 episodes

Fig. 7. Reward-based poisoning attacks at different time intervals.

Figure 6 shows the average reward per episode of the agent under the state-based poisoning attack at different time intervals. We can see that when the attack interval is 1000 episodes, the agent's reward fluctuates very sharply. When the attack interval is 5000 episodes, the fluctuation of the curve is alleviated to a certain extent. And when the attack interval is 10000 episodes, the curve of the victim agent is almost the same as that of the normal agent. This result conveys an obvious message to us: when the frequencies of attacks is not high, the local model of the agent can completely eliminate the impact of poisoning attacks.

The training results of the agent, when it is subjected to different frequencies of reward-based poisoning attacks, are shown in Fig. 7. The experimental results obtained in this part are consistent with those in Fig. 6: as the attack frequency decreases, the impact on the agent is also decreasing. When the attack interval is 10,000 episodes, the victim agent has been able to completely eliminate the damage caused by the attack.

An excessively high attack frequency means an increase in cost and exposure risk for the adversary, but an excessively low attack frequency cannot achieve the desired destructive effect. Therefore, the adversary's attack frequency is a trade-off between the attack cost and the attack effect. In addition, for an experienced adversary, a reward-based poisoning attack may be a better choice. We will find that sending false state signals frequently will cause a large vibration of the reward as shown in Fig. 6 and Fig. 7. The large vibration will cause malicious behavior that can be easily detected. However, reward-based poisoning attacks do not need to worry about this situation, because the trend of the reward curve is very stable. In this way, the adversary can evade the malicious behavior detection even if the attack frequency is high. Therefore, reward-based poisoning attacks can reach a trade-off between the excellent attack effects with low exposure risk.

5 Conclusion

In the current MARL field, there are few studies on poisoning attacks. Hence, we propose an analysis framework for the poisoning attacks in the MARL system. In the framework, we carry out multi-dimensional experiments on poisoning attacks against the multi-agent soft actor-critic algorithm. 1) We conduct a thorough analysis of the three modes of poisoning attacks (halved poisoning, reverse poisoning, and random poisoning) for state and reward signal. 2) We study the impact of the timing of poisoning on the agent by trying poisoning in different stages for the state- and reward-based poisoning attacks. 3) We analyze the mitigation ability of the MARL system for poisoning by changing the frequency of poisoning. Through the systematical analysis of poisoning attacks in this analysis framework, we can obtain the following conclusions. Reverse poisoning has the best poisoning effect under a reward-based poisoning attack, while random poisoning is even better under a state-based poisoning attack. Under the state-based poisoning attack, the damage to the agent increases as the attack timing moves back, while the reward-based poisoning attack results in the opposite. As the attack time interval increases, the model can gradually alleviate the toxicity. When the attack interval is 10,000 episodes, the agent can completely alleviate the toxicity. Based on the above conclusions, we can make a more reasonable choice of the mode, timing, and frequency of poisoning, to achieve a better attack effect. This work is the primary research to develop an effective defense mechanism against poisoning attacks in the MARL system.

Acknowledgment. The work was supported by the National Natural Science Foundation of China under Grant Nos. 61972025, 61802389, 61672092, U1811264, and 61966009, the National Key R&D Program of China under Grant Nos. 2020YFB1005604 and 2020YFB2103802.

References

1. Silver, D., et al.: Mastering the game of go without human knowledge. Nature **550**(7676), 354–359 (2017)
2. Xu, Z., Chen, J., Tomizuka, M.: Guided policy search model-based reinforcement learning for urban autonomous driving. arXiv preprint arXiv:2005.03076 (2020)

3. Yang, B., Liu, M.: Keeping in touch with collaborative UAVs: a deep reinforcement learning approach. In: IJCAI, pp. 562–568 (2018)
4. Chen, G.: A new framework for multi-agent reinforcement learning–centralized training and exploration with decentralized execution via policy distillation. In: Proceedings of the 19th International Conference on Autonomous Agents and MultiAgent Systems, pp. 1801–1803 (2020)
5. Vinyals, O., et al.: Grandmaster level in starcraft ii using multi-agent reinforcement learning. Nature **575**(7782), 350–354 (2019)
6. Berner, C., et al.: Dota 2 with large scale deep reinforcement learning. arXiv preprint arXiv: 1912.06680 (2019)
7. Tampuu, A., et al.: Multiagent cooperation and competition with deep reinforcement learning. PloS one **12**(4), e0172395 (2017)
8. Wai, H.T., Yang, Z., Wang, Z., Hong, M.: Multi-agent reinforcement learning via double averaging primal-dual optimization. In: NeurIPS (2018)
9. Abouheaf, M., Gueaieb, W.: Multi-agent reinforcement learning approach based on reduced value function approximations. In: 2017 IEEE International Symposium on Robotics and Intelligent Sensors (IRIS), pp. 111–116. IEEE (2017)
10. Foerster, J., et al.: Stabilising experience replay for deep multi-agent reinforcement learning. In: International Conference on Machine Learning, pp. 1146–1155. PMLR (2017)
11. Foerster, J., Farquhar, G., Afouras, T., Nardelli, N., Whiteson, S.: Counterfactual multi-agent policy gradients. In: Proceedings of the AAAI Conference on Artificial Intelligence, vol. 32 (2018)
12. Lowe, R., Wu, Y., Tamar, A., Harb, J., Abbeel, P., Mordatch, I.: Multi-agent actor-critic for mixed cooperative-competitive environments. In: Proceedings of the 31st International Conference on Neural Information Processing Systems, pp. 6382–6393 (2017)
13. Iqbal, S., Sha, F.: Actor-attention-critic for multi-agent reinforcement learning. In: International Conference on Machine Learning, pp. 2961–2970. PMLR (2019)
14. Huang, S., Papernot, N., Goodfellow, I., Duan, Y., Abbeel, P.: Adversarial attacks on neural network policies. arXiv preprint arXiv:1702.02284 (2017)
15. Xiang, Y., Niu, W., Liu, J., Chen, T., Han, Z.: A pca-based model to predict adversarial examples on q-learning of path finding. In: 2018 IEEE Third International Conference on Data Science in Cyberspace (DSC), pp. 773–780. IEEE (2018)
16. Lin, Y.C., Hong, Z.W., Liao, Y.H., Shih, M.L., Liu, M.Y., Sun, M.: Tactics of adversarial attack on deep reinforcement learning agents. arXiv preprint arXiv:1703.06748 (2017)
17. Pan, X., Wang, W., Zhang, X., Li, B., Yi, J., Song, D.: How you act tells a lot: Privacy-leaking attack on deep reinforcement learning. In: Proceedings of the 18th International Conference on Autonomous Agents and MultiAgent Systems, pp. 368–376 (2019)
18. Ma, Y., Zhang, X., Sun, W., Zhu, X.: Policy poisoning in batch reinforcement learning and control. arXiv preprint arXiv:1910.05821 (2019)
19. Zhang, X., Ma, Y., Singla, A., Zhu, X.: Adaptive reward-poisoning attacks against reinforcement learning. In: International Conference on Machine Learning, pp. 11225–11234. PMLR (2020)
20. Rakhsha, A., Radanovic, G., Devidze, R., Zhu, X., Singla, A.: Policy teaching in reinforcement learning via environment poisoning attacks. arXiv preprint arXiv:2011.10824 (2020)
21. Littman, M.L.: Markov games as a framework for multi-agent reinforcement learning. In: Machine Learning Proceedings 1994, pp. 157–163. Elsevier (1994)
22. Konda, V.R., Tsitsiklis, J.N.: Actor-critic algorithms. In: Advances in Neural Information Processing Systems, pp. 1008–1014 (2000)

23. Haarnoja, T., Zhou, A., Abbeel, P., Levine, S.: Soft actor-critic: off-policy maximum entropy deep reinforcement learning with a stochastic actor. In: International Conference on Machine Learning, pp. 1861–1870. PMLR (2018)
24. Multi-Agent Particle Environment. https://github.com/openai/multiagent-particle-envs, Accessed 20 Juy 2021

A Blockchain-Based Proxy Oriented Cloud Storage Public Audit Scheme for Low-Performance Terminal Devices

Mande Xie[1], Qiting Zhao[2], and Haibo Hong[2](✉)

[1] School of Information and Electric Engineering, Zhejiang Gongshang University, Hangzhou, China

[2] School of Computer and Information Engineering, Zhejiang Gongshang University, Hangzhou, China

Abstract. Due to the advantages of cloud computing, more and more people tend to store their data to cloud server (CS). Cloud storage audit scheme can help data owner (DO) to confirm the integrity of their stored data, but the existing schemes still have some limitations. Therefore, in this paper, we propose a blockchain-based proxy oriented cloud storage public audit scheme (BBPO-PA) for low performance terminal devices. Firstly, we employ a proxy that is authorized by DO. The proxy can process and upload ciphertext files for DO. Secondly, we introduce blockchain into our scheme and utilize smart contracts instead of centralized trusted third party audit (TPA) to improve the reliability and stability of audit results. Subsequently, the security analysis shows that our scheme has achieved the designed security goals. Finally, the performance evaluation demonstrates that our scheme has significant advantages over other schemes, in particular significantly reducing the resource overheads of DO.

Keywords: Public auditing · Blockchain · Smart contract · Proxy · Low performance terminal device

1 Introduction

With the rapid development of cloud computing, more and more individuals/companies tend to outsource their data to CS for reducing their local resource overheads. Although cloud computing has brought great convenience, it indeed causes many security issues like privacy leakage of DO. Therefore, many researchers have made great efforts to deal with these tricky problems [1–4].

To verify the integrity of outsourced data, researchers have proposed a number of cloud storage audit schemes. The existing schemes are mainly divided into two categories: privacy audit scheme and public audit scheme. In the case of privacy audit scheme, it is primarily the interaction between DO and CS to perform validation operations. While in a public audit scheme, a TPA interacts with the CS on behalf of DO, it performs audit tasks and returns audit results

Y. Lai et al. (Eds.): ICA3PP 2021, LNCS 13155, pp. 676–692, 2022.
https://doi.org/10.1007/978-3-030-95384-3_42

to DO. But at the same time, TPA has also brought several troubles. In existing schemes, TPA is generally considered to be fully trusted, but in real application scenarios, TPA is always malicious.

1.1 Contributions

In this paper, we propose a novel blockchain-based proxy oriented cloud storage public audit scheme for low-performance terminal devices. Specifically, our contribution mainly includes the following three aspects:

1. Firstly, our scheme introduces a trusted proxy authorized by DO. The proxy generates the tag for DO's file and finally uploads the file and tag to CS. This approach reduces the considerable computational overhead of DO during the tag generation phase.
2. Secondly, we design a blockchain-based smart contract platform for our scheme. During the audit phase, DO does not need to interact with CS, which periodically sends proof information to the smart contract for auditing. The smart contract performs audits and generates audit results instead of centralized TPA. This approach reduces a lot of communication and computing overhead for DO during the audit phase.
3. Finally, we carry out a security analysis on our scheme, and the security analysis shows that our scheme can achieve the expected security goals. In addition, we evaluate the performance of our scheme and compare it with other classic schemes. The comparison results show that our scheme has great advantages in efficiency and functionality.

1.2 Related Work

Currently, how to ensure the integrity of data stored in untrusted CS is an important security issue. Many researchers have put forward several efficient schemes to tackle this core security problem. In 2007, Ateniese et al. [5] proposed a provable data possession (PDP) model. Afterwards, some PDP schemes with data dynamics are proposed in [6,7]. Also in 2007, Juels et al. [8] raised the proofs of retrievability (POR) model. Subsequently, Shacham et al. [9] improved the POR scheme to support public audit.

After the work of Shacham et al., many experts and scholars have done a lot of research in this field and devised many public audit schemes [10,11], which are usually based on public key infrastructure (PKI). Afterwards, many researchers brought forward a series of identity-based public audit schemes [12,13], which greatly reduced the communication and computing burdens of all entities in the system.

In these public audit schemes, a trusted TPA usually took the place of DO to verify the integrity of the data, thereby reducing the computational and communication overhead of DO. However, in a real application scenario, the auditor is untrusted. Due to the inherent characteristics of blockchain, many experts and scholars have proposed public audit schemes based on blockchain to eliminate the negative impacts brought by TPA [14,15].

1.3 Organization

The rest of this paper is organized as follows. In Sect. 2, we introduce the background and preliminary related to our scheme. In Sect. 3, we give the formal definition, threat model and design goals of our scheme. In Sect. 4, we describe the specific structure of our scheme in detail. Then, we present the security analysis and experimental performance of our scheme in Sect. 5 and Sect. 6, respectively. Finally, we come to the conclusion in Sect. 7.

2 Preliminaries

2.1 Bilinear Pairings [5]

Suppose that G and G_T are both multiplicative cyclic groups with prime order q and g is the generator of group G, then $e : G \times G \rightarrow G_T$ is a bilinear map if it satisfies the following three properties:

1. Bilinearity: $\forall g_1, g_2 \in G$ and $a, b \in \mathcal{Z}_q^*$, $e(g_1^a, g_2^b) = e(g_1, g_2)^{ab}$.
2. Computability: $\forall g_3, g_4 \in G$, there exits an efficiently computable algorithm for computing map $e(g_3, g_4)$.
3. Non-degenerate: $\forall g_5, g_6 \in G$, $e(g_5, g_6) \neq 1$.

2.2 Intractable Assumption

Definition 1. *(Computational Diffie-Hellman (CDH) problem) Suppose G is a multiplicative cyclic group of prime order q, let g be the generator of G. Then, computational Diffie-Hellman (CDH) problem is defined as follows: for a given triple (g, g^a, g^b), to compute g^{ab}, where $a, b \in \mathcal{Z}_q^*$. F*

For any PPT adversary A, the probability of computing g^{ab} is negligible. In sequel, we design the new scheme based on the intractability of CDH problem.

2.3 Blockchain and Smart Contract

In 2008, Satoshi Nakamoto [16] firstly came up with the concept of bitcoin, which can conduct peer-to-peer cash transactions without relying on trusted central authorities. And blockchain is the underlying technology of Bitcoin. In blockchain, the hash value of the current block participates in the generation of next block, thus forming a logical connection between blocks. Blockchain is stored and managed by multiple nodes, and blocks can be added only after consensus is reached between nodes through a consensus algorithm.

Subsequently, researchers introduced smart contracts into blockchain, further improving the functionality of blockchain. In short, a smart contract is the executable code stored in a blockchain. Ethereum [17] is a blockchain platform that uses smart contracts. The Ethereum blockchain is shown in Fig. 1. In Ethereum, each smart contract has a unique contract account, and external accounts call programs by interacting with the contract account.

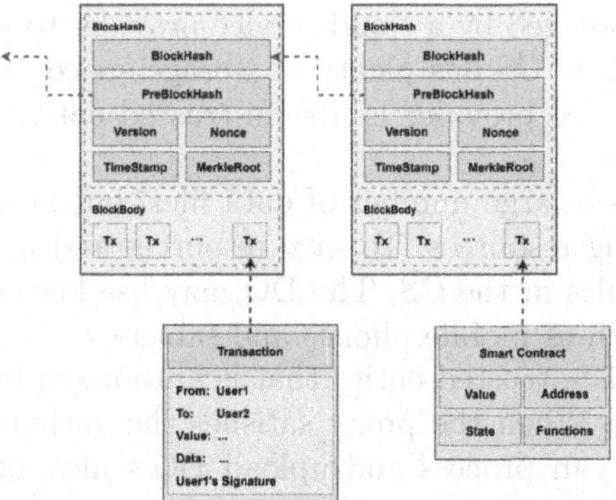

Fig. 1. Ethereum blockchain.

3 System Model, Threat Model and Design Goals

3.1 System Model

The system model of our scheme is shown in Fig. 2. Our scheme consists of four entities: **Key Generation Center (KGC)**, **Data Owner (DO)**, **Proxy**, and **Cloud Server (CS)**.

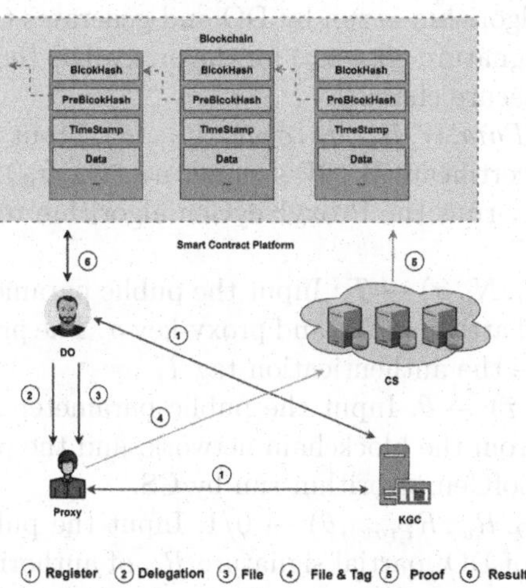

Fig. 2. System model of our scheme.

- **KGC:** The KGC is managed by a fully trusted authority. The KGC will generate system public parameters and corresponding keys according to the identity information of the entity.

- **CS:** The CS is managed by a cloud service provider to provide cloud storage services for DO. The CS has plenty of storage space and computing power, but the cloud service provider is a semi-trusted entity that can cause data corruption or loss.
- **DO:** The DO has a large number of data files, but he/she has limited storage and computing resources. To save resources and improve efficiency, DO chooses to store files in the CS. The DO may use low computing power terminal devices, such as mobile phones and tablets.
- **Proxy:** A proxy is a trusted entity that is authorized by DO to process and upload DO's files. When the proxy satisfies the authorization certificate W issued by DO, it can process and upload DO's files; otherwise, it can't do anything.

Ours scheme consists of the following seven algorithms: **Setup**, **KeyGen**, **Auth**, **ProxyKeyGen**, **TagGen**, **ProofGen** and **Audit**.

Setup$(k) \to (Para, s)$: Input the security parameter k, and the Setup algorithm outputs the system public parameter $Para$, and the master private key s of the KGC.

KeyGen$(Para, s, ID) \to (sk_{ID})$: Input the public parameter $Para$, the master private key s, and the identity ID of the entity U, the KeyGen algorithm generates the secret key $sk_{ID} = (R_{ID}, \sigma_{ID})$ for U. Then, the KGC sends the sk_{ID} to U via secure channel.

Auth$(Para, ID_{DO}, \sigma_{DO}, ID_{Proxy}) \to (W, R_w, \sigma_w)$: Input the public parameter $Para$, DO's identity ID_{DO}, DO's private key σ_{DO}, and the proxy's identity ID_{Proxy}. The Auth algorithm is run by DO and generates the authorization certificate W and W's signature (R_w, σ_w) for the proxy. DO then sends (W, R_w, σ_W) to the proxy over a secure channel.

ProxyKeyGen$(Para, W, R_w, \sigma_w, \sigma_{Proxy}) \to (\sigma)$: Input the public parameter $Para$, authorization certificate W, W's signature (R_w, σ_w), and proxy's private key σ_{Proxy}. The proxy runs the ProxyKeyGen algorithm to generate the proxy-key σ for the proxy.

TagGen$(Para, B_i, N_i, \sigma) \to T_i$: Input the public parameter $Para$, data block B_i, data block related attribute N_i, and proxy-key σ. The proxy runs the TagGen algorithm to generate the authentication tag T_i

ProofGen$(Para, \tau) \to \theta$: Input the public parameter $Para$ and the public state information τ from the blockchain network, and the proof information θ is generated by the ProofGen algorithm run by CS.

Audit$(Para, R_{DO}, R_w, R_{Proxy}, \theta) \to 0/1$: Input the public parameter $Para$, the public key R_{DO} of DO, partial signature R_w of authorization certificate W, the public key R_{Proxy} of the proxy, and the proof information θ. The Audit result $res \in \{0, 1\}$ is output by the Audit algorithm run by the smart contract SC_0, where 0 indicates error and 1 indicates correct.

3.2 Threat Model and Design Goals

In the threat model, we consider CS to be the primary adversary, which may have the following malicious behaviors. Firstly, CS may deliberately delete outsourced data that is rarely accessed by DO to save its storage space. Secondly, there may be a sudden external hacker attack/internal software or hardware failure, and the data stored by CS may be tampered with, lost or deleted. However, in order to protect its reputation, CS may deliberately hide these unexpected situations from DO. In order to achieve a secure and efficient cloud storage audit scheme, our scheme should achieve the following design goals.

Functionality:

- Proxy data processing and uploading: The proxy can process DO's files to generate corresponding authentication tags and upload the files and authentication tags to CS.
- Public auditability: Nodes in the blockchain network can always recalculate audit results based on audit information stored in the blockchain.
- Decentralization: Use a decentralized smart contract instead of a centralized TPA to verify the validity of proof information from CS for DO.
- Non-interactive: There is no interaction between DO and CS during the entire audit phase.
- Automatic arbitration: Smart contracts are automatically arbitrated based on the validity of the proof information from CS.
- Traceability: Historical audit information is stored in blockchain, including relevant public information, proof information and audit results.

Security:

- Correctness: When KGC, DO, proxy, and CS are all honest and execute the process correctly, the proof information generated by CS can always pass the audit of the smart contract.
- Storage correctness: CS will only pass the audit of smart contract if it correctly stores DO's data and authentication tags.

Efficiency:

- The terminal device used by DO may be low computing power. Our scheme can be applied to all kinds of resource-constrained terminal devices, and the low performance terminal devices will not have any negative impact on execution efficiency of our scheme.

4 Proposed Scheme

In this section, we firstly design a fair payment smart contract platform based on blockchain, and finally describe the concrete structure of our scheme in detail.

4.1 A Fair Payment Smart Contract Platform Based on Blockchain

We implement this new cloud storage fair payment model by using blockchain and smart contracts. When DO sends the file to the proxy, it submits a smart contract SC_{DO} to the smart contract platform, see Fig. 3. SC_{DO} guarantees that a service fee will be sent from DO's account to CS's account if the proof information submitted by the CS passes the smart contract SC_0. When CS receives the file and authentication tag from the proxy, it submits a smart contract SC_{CS} to the smart contract platform, see Fig. 4. SC_{CS} guarantees that if the proof information submitted by the CS fails to pass the smart contract SC_0, a compensation payment will be sent from CS's account to the account of DO. CS periodically invokes the smart contract SC_0 for service fees, see Fig. 5. SC_0 will verify proof information θ and obtain the audit result res with the parameters of SC_{DO}, SC_{CS} and θ. If $res = 1$ and SC_0 activates SC_{DO}, DO pays the service fee to the CS; If $res = 0$, SC_0 activates SC_{CS}, and the CS pays compensation to DO. Finally, SC_0 generates a transaction TX that includes the public information, θ, and res related to this audit.

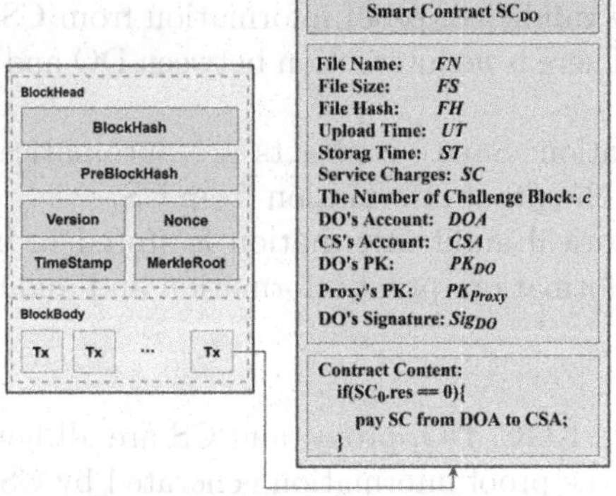

Fig. 3. Smart contract SC_{DO}.

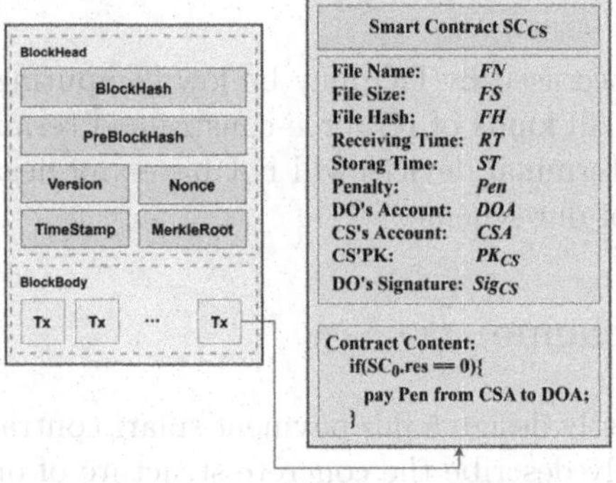

Fig. 4. Smart contract SC_{CS}.

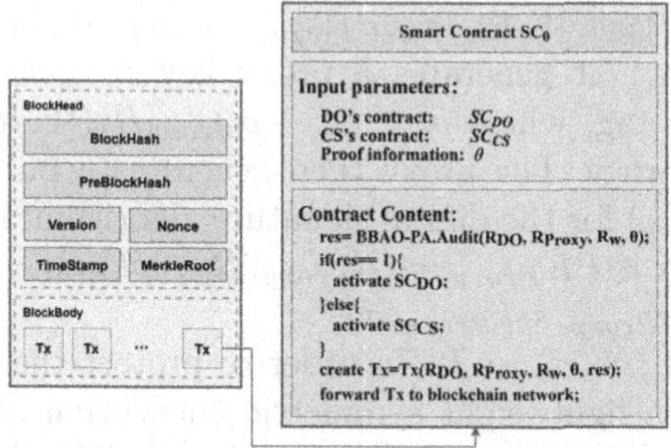

Fig. 5. Smart contract SC_0.

4.2 Construction of Our Scheme

Our BBPO-AO scheme is mainly on the basis of scheme [18], but there are several significant differences as follows. Firstly, there is no interaction between DO and CS during the audit phase of our scheme. Secondly, we utilize smart contracts to help DO validate proof information from CS, enabling public audits.

- **Setup**$(k) \rightarrow (Para, s)$: According to the security parameter k, the KGC selects two multiplicative cyclic group G and G_T of the same prime order q and a bilinear mapping $e: G \times G \rightarrow G_T$. Let g be the generator of group G. KGC selects two cryptographic hash functions $H_1: \{0,1\}^* \rightarrow \mathcal{Z}_q^*$ and $H_2: \{0,1\}^* \rightarrow G$. KGC selects a pseudo-random function $f(\cdot): \{0,1\}^* \rightarrow \mathcal{Z}_q^*$ and a pseudo-random permutation $\pi(\cdot): \{0,1\}^* \rightarrow \{1, 2, \ldots, n\}$. The KGC selects a random value $s \in \mathcal{Z}_q^*$ as its master private key, and then computes $P = g^s$. Hence, the system public parameter $Para = \{G, G_T, q, g, H_1, H_2, f(\cdot), \pi(\cdot), P\}$. The master private key s is kept by KGC.
- **KeyGen**$(Para, s, ID) \rightarrow (sk_{ID})$: Input the identity ID of entity U, and KGC selects a random value $r_{ID} \in \mathcal{Z}_q^*$ and calculates U's secret key $sk_{ID} = (R_{ID}, \sigma_{ID})$, where $R_{ID} = g^{r_{ID}}$, $\sigma_{ID} = r_{ID} + sH_1(ID, R_{ID})$. The KGC sends $sk_{ID} = (R_{ID}, \sigma_{ID})$ to U over a secure channel as U's secret key. If U is DO, he/she selects a random value $u \in G$ and generates a public/private key pair (spk_{DO}, ssk_{DO}) for digital signature algorithm. Finally, the full public key of DO is $PK_{DO} = (R_{DO}, spk_{DO}, u)$, and the full private key is $SK_{DO} = (\sigma_U, ssk_{DO})$.
- **Auth**$(Para, ID_{DO}, \sigma_{DO}, ID_{Proxy}) \rightarrow (W, R_w, \sigma_w)$: DO generates a certificate of authorization W for the proxy. W contains some necessary information, such as DO's identity ID_{DO}, the proxy's identity ID_{Proxy}, W's validity period $ProxyTime$ and the name of the file $FileName$. thus, $W = ID_{DO}||ID_{Proxy}||ProxyTime||FileName$. The proxy cannot process and upload DO's files unless the proxy satisfies W. DO chooses a random value $r_w \in \mathcal{Z}_q^*$, and computes W's signature (R_w, σ_w), where $R_w = g^{r_w}$, $\sigma_w = r_w + \sigma_{DO} H_1(W, R_w)$. Finally, DO sends W and W's signature (R_w, σ_w) to the proxy and CS via a secure communication channel.

- **ProxyKeyGen**$(Para, W, R_w, \sigma_w, \sigma_{Proxy}) \rightarrow (\sigma)$: When the proxy received a triple (W, r_w, σ_w), it generates a proxy key σ that is used to generate an authentication tag, where $\sigma = \sigma_w + \sigma_{Proxy} H_1(W, R_w)$. σ is held confidentially by the proxy. The proxy then generates a public/private key pair $(spk_{Proxy}, ssk_{Proxy})$ for the digital signature algorithm. Finally, the full public key of the proxy is $PK_{Proxy} = (R_{Proxy}, R_w, spk_{Proxy})$, and the full private key is $SK_{DO} = (\sigma_{Proxy}, ssk_{Proxy}, \sigma)$.
- **TagGen**$(Para, B_i, N_i, \sigma) \rightarrow T_i$: In order to protect the privacy of plaintext files $\hat{F}$, DO uses a lightweight symmetric encryption algorithm to encrypt plaintext files $\hat{F}$ and obtains ciphertext files F. DO divides F into n data blocks according to the size of F, that is, $F = (B_1, B_2, \ldots, B_n)$, where B_i represents the i-th data block of F, and $B_i \in \mathcal{Z}_q^*$. Then, DO sends $\{(B_i, N_i) \mid 1 \leq i \leq n\}$ to the proxy in a secure communication channel, where $N_i = FileName \| i$. Finally, DO generates a smart contract SC_{DO}, which contains information about contracts between DO and CS, and uploads SC_{DO} to the smart contract platform.
  When the proxy meets the conditions in the authorization certificate W, it can process and upload DO's files. When the proxy receives $\{(B_i, N_i) \mid 1 \leq i \leq n\}$ from DO, it computes the authentication tag T_i for each data block B_i, where $T_i = (H_2(N_i) u^{B_i})^{\sigma}$. The proxy then uploads all the data blocks and authentication tags $\{(B_i, T_i), 1 \leq i \leq n\}$ to CS over a secure communication channel.
  When CS receives the data blocks and certification tags $\{(B_i, T_i) \mid 1 \leq i \leq n\}$, it firstly checks whether the proxy meets the W. If yes, CS goes to the next step. Otherwise, CS refuses to accept these files. If all the verification passes, CS generates the smart contract SC_{CS} and sends it to the smart contract platform. SC_{CS} contain information about the contract between the DO and the CS.
- **ProofGen**$(Para, \tau) \rightarrow \theta$: CS periodically interacts with the smart contract SC_0 to get the service fee. Firstly, CS obtains the current time t and the public state information τ in blockchain network. τ contains the latest block hash at time t and other public information, so τ is not controlled by CS. Note: τ is determinate when t is determinate.
  CS gets the number of blocks to be challenged c from a smart contract SC_{DO} issued by DO. For $i \in [1, c]$, CS computes $v_i = \pi(i \| \tau)$, where $v_i \epsilon [1, n]$, v_i is the index of challenged data block. And CS calculates the $a_i = f(i \| \tau)$, where $a_i \in \mathcal{Z}_q^*$, a_i as the coefficient required for generating an authentication tag T_i. CS then calculates T and $\bar{F}$, Where $T = \prod_{1 \leq i \leq c} T_{v_i}^{a_i}$ and $\bar{F} = \sum_{1 \leq i \leq c} a_i B_{v_i}$ Finally, CS sends the proof information $\theta = (T, \bar{F}, t, \tau)$, the contract address of SC_{DO} and SC_{CS} as parameters to the smart contract SC_0.
- **Audit**$(Para, R_{DO}, R_w, R_{Proxy}, \theta) \rightarrow 0/1$: When the smart contract SC_0 receives the proof information $\theta = (T, \bar{F}, t, \tau)$, the contract address of SC_{DO} and SC_{CS}, it automatically executes Audit algorithm to obtain the audit result *res*.
  SC_0 firstly verifies the authenticity of the public state information τ through

the timestamp t. If τ is invalid, it directly returns the audit result $res = 0$. If τ is valid, for $1 \leq i \leq c$, SC_0 begins to compute:

$$v_i = \pi(i\|\tau), a_i = f(i\|\tau)$$

$$h_{v_i} = H_2(N_{v_i}), N_{v_i} = FileName\|v_i$$

Where c and *FileName* are obtained from SC_{DO}. And then SC_0 computes

$$\bar{R} = R_w(R_{DO}R_{Proxy}P^{H_1(ID_{DO},R_{DO})+H_1(ID_{Proxy},R_{Proxy})})^{H_1(W,R_w)}$$

Finally, SC_0 verifies whether the following equation holds:

$$e(T,g) \stackrel{?}{=} e(\sum_{i=1}^{c} h_{v_i}^{a_i} u^{\bar{F}}, \bar{R}) \tag{1}$$

If the Eq. (1) is true, it means that the data of DO is intact and returns $res = 1$; otherwise, this means that DO's data is corrupted and returns the audit result $res = 0$.

- **Modification**: If DO wants to change block B_i to B_i', he/she does as follows:
  1. DO sends a request $\Gamma = (\Omega = (M, FileName, i, B', t), Sig_{ssk_{DO}}(\Omega))$ to the proxy, where M represents the modification operation and the t represents a timestamp.
  2. When the proxy receives the request Γ, it verifies whether $Sig_{ssk_{DO}}(\Omega)$ is a valid signature on Ω. If the signature is valid, the proxy generates the tag T_i' in the same way as in *TagGen* and sends $(\Omega, Sig_{ssk_{DO}}(\Omega), T_i', Sig_{ssk_{Proxy}}(T_i'))$ to CS.
  3. When CS receives $(\Omega, Sig_{ssk_{DO}}(\Omega), T_i', Sig_{ssk_{Proxy}}(T_i'))$, it verifies the validity of signatures $Sig_{ssk_{DO}}(\Omega)$ and $Sig_{ssk_{Proxy}}(T_i')$. If all authentications pass, CS recovers (B_i, T_i) from the storage space and changes (B_i, T_i) to (B_i', T_i').

5 Security Analysis

5.1 Correctness

Theorem 1. *If the KGC, CS, DO, and the proxy are honest and follow the prescribed process, then, the proof information θ of CS can pass the audit of smart contract SC_0.*

Proof. The correctness of BBAO-PO scheme comes from the followings:

$$\begin{aligned}
e(T,g) &= e(\prod_{i=1}^{c} T_{v_i}^{a_i}, g) \\
&= e(\prod_{i=1}^{c}((H_2(N_{v_i})u^{B_{v_i}})^{\sigma})^{a_i}, g) \\
&= e(\prod_{i=1}^{c} H_2(N_{v_i})^{a_i} u^{B_{v_i} a_i}, g^{\sigma}) \\
&= e(\prod_{i=1}^{c} H_2(N_{v_i})^{a_i} u^{\Sigma_{i=1}^{c} B_{v_i} a_i}, g^{\sigma}) \\
&= e(\prod_{i=1}^{c} H_2(N_{v_i})^{a_i} u^{\bar{F}}, g^{\sigma}) \\
&= e(\prod_{i=1}^{c} h_{v_i}^{a_i} u^{\bar{F}}, g^{\sigma})
\end{aligned}$$

On the other hand, we have:

$$\begin{aligned}
g^{\sigma} &= g^{\sigma_w + \sigma_{Proxy} H_1(W,R_w)} \\
&= g^{\sigma_w} g^{\sigma_{Proxy} H_1(W,R_w)} \\
&= R_w (R_{DO} P^{H_1(ID_{DO},R_{DO})})^{H_1(W,R_W)} (R_{Proxy} P^{H_1(ID_{Proxy},R_{Proxy})})^{H_1(W,R_w)} \\
&= R_w (R_{Proxy} R_{DO} P^{H_1(ID_{DO},R_{DO}) + H_1(ID_{Proxy},R_{Proxy})})^{H_1(W,R_w)} \\
&= \bar{R}
\end{aligned}$$

As a result, $e(T,g) = e(\prod_{i=1}^{c} h_{v_i}^{a_i} u^{\bar{F}}, \bar{R})$. Therefore, the correctness of our scheme is proved. □

5.2 Unforgeability

Theorem 2. *If the CDH problem holds in group G, then our scheme has unforgerability in the random oracle model.*

Proof. The *TagGen* stage in our scheme is the same as that in scheme [18], so the proof process is the same as that in [18]. Hence, we only present the part with differences. In scheme [18], u is calculated by hashing function h. In our scheme, u is chosen randomly from the group G. In the random oracle, there is no difference between the output value of the hash function and the randomly selected value of the group G. When CS is malicious, the scheme [18] achieves unforgeability in the random oracle, so our proposal also achieves unforgeability in the random oracle. Specifically, if the challenged data blocks of the outsourced files stored in CS are corrupted, CS cannot generate any valid proof information to pass data integrity verification. □

5.3 Proxy-Protection

Theorem 3. *If the CDH problem holds in group G, then our scheme has the characteristics of proxy-protection in the random oracle model.*

Proof. Since the *TagGen* stage in our scheme is the same as that scheme [18], the proof process is the same as that in scheme [18]. According to the security analysis in scheme [18], we can know that scheme [18] is existentially proxy-protection in the random oracle model, so our scheme is also existentially proxy-protection in the random oracle model. That means, if the DO is malicious, it cannot impersonate the proxy's identity to generate authentication tags.

5.4 Non-Interactive

Theorem 4. *When the input values of the pseudo-random function $f(\cdot)$ and the pseudo-random permutation $\pi(\cdot)$ are pseudo-random, our scheme is non-interactive.*

Proof. The algorithm *ProofGen* in our scheme is almost the same as that in [14], because scheme [14] is non-interactive, so our scheme is also non-interactive. The specific proof process is as follows.

Our scheme differs from the interactive public audit model of scheme [14] in that there is no need for any interaction between CS and DO during the audit phase. To achieve this, we use a pseudo-random permutation of $\pi(\cdot)$ to generate an index set for the challenged data blocks, and a pseudo-random function $f(\cdot)$ to generate a coefficient set for generating proof information. Because the input values of $\pi(\cdot)$ and $f(\cdot)$ contain the public state information τ at the current time t, τ contains the latest block hash at time t and other public information, so it is not controlled by CS. Due to the randomness of τ, the index and coefficient sets generated in this way are no different from those randomly generated by the validator, so our scheme is non-interactive. □

6 Performance Evaluation

6.1 Functional Comparison

In order to display the superiority of our scheme, we compare our scheme with the other two schemes in functionality. As describes in Table 1, scheme [18] only supports public auditing, proxy data processing and uploading, and scheme [12] only implement public auditing. While our scheme has achieved all functionality including many inherent properties of blockchain such as decentralization, traceability, non-interaction and automatic arbitration. Moreover, since DO does not need to participate in tag generation and cost any computing and communication overhead during the audit phase, our scheme is very friendly to low-performance terminal devices.

Table 1. Functionality comparison

Scheme	[12]	[18]	Ours
Public auditability	√	√	√
Proxy data processing and uploading	×	√	√
Decentration	×	×	√
Non-interactive	×	×	√
Automatic arbitration	×	×	√
Traceability	×	×	√
Low performance end device friendliness	×	×	√

6.2 Theoretical Analysis

In this part, we mainly reveal the computational cost of our scheme. In bilinear pairing-based scheme, the computational overhead mainly comes from pairing operation, multiplication operation and exponential operation on group G. The computational overhead of our scheme is shown in Table 2, where n represents the number of data blocks of the file, c represents the number of challenged data blocks, M and E represent the multiplication operation and exponential operation on group G, respectively, and BP represents the bilinear pairing operation.

Table 2. Analysis of computational overhead

Stage	Computational cost
Setup	E
KeyExtra	$E + M$
Auth	$E + M$
ProxyKeyGen	M
TagGen	$2nE + nM$
ProofGen	$cE + (c - 1)M$
Audit	$(c + 3)E + (c + 3)M + 2BP$

It can be clearly seen from Table 3 that the computational cost of our scheme mainly comes from *TagGen*, *ProofGen* and *Audit*. In the *TagGen* phase, the proxy helps DO generate the authentication tag, which performs 2n exponentials and n multiplications on group G. In the *ProofGen* phase, CS performs c exponentials and (c-1) multiplications on group G. In the *Audit* phase, smart contract SC_0 performs (c+3) multiplications and exponentials on group G, and 2 pairing operations. As is depicted in Table 3, we compare our scheme with scheme [18] in terms of entity's computational overhead. DO only needs to perform 1 exponential and 1 multiplication on group G to generate the authorization certificate for the proxy in the *Auth* phase. Therefore, our scheme does not require much computational overhead from DO.

6.3 Experimental Analysis

In this section, we simulate all stages of our scheme, and conduct a deeper efficiency analysis of our scheme. Specifically, we carry out this experiment in a laptop with a 2.1 GHz AMD Rayon 5 4600U CPU, 16 GB of RAM and Windows 10 operating system based on Java Pairing Based Cryptography (JPBC) library version 2.0.0 [19] and select the Type-A pairing parameters of order 160 bit. In order to make the performance evaluation more convincing, we choose the average of 20 experiments as the final presentation results.

Table 3. Comparison of computational overhead

Scheme	[18]	Ours
DO's computational cost	$(c+4)E+(c+4)M+2BP$	$E+M$
Proxy's computational cost	$2nE+(n+1)M$	$2nE+(n+1)M$
CS's computational cost	$cE+(c-1)M$	$cE+(c-1)M$
Smart contract's computational cost	×	$(c+3)E+(c+3)M+2BP$

In the tag generation phase, the proxy generates the corresponding authentication tag for DO's data block. In Fig. 6, we describe the computation time required for the proxy to process different numbers of data blocks in the tag generation phase, with the number of data blocks represented on the X-axis and the computation time required on the Y-axis. We can see that the computation time increases linearly with the data block. The results show that when DO has a large amount of data, the time required to generate the corresponding authentication tag is also large.

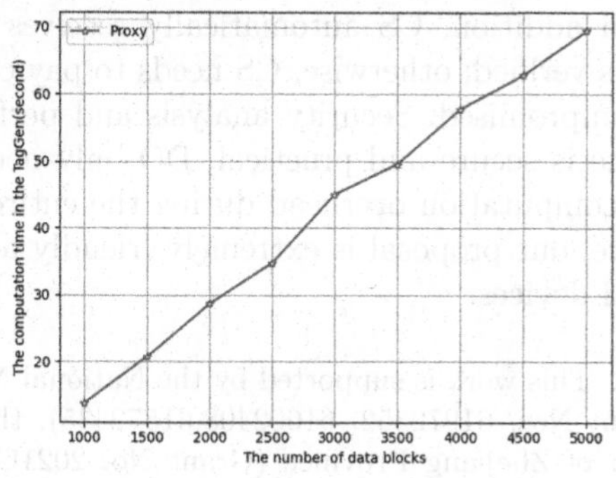

Fig. 6. The computation time in the tag generation phase

In the audit phase, CS generates the proof information and the smart contract checks the validity of the proof information. Figure 7 demonstrates the computation time required for CS and smart contracts to process different amounts of challenged data blocks. The experimental results show that although the tag generation stage and audit stage have high computational overhead, they do not require the participation of DO, so our scheme is very friendly to DO with low performance terminal devices.

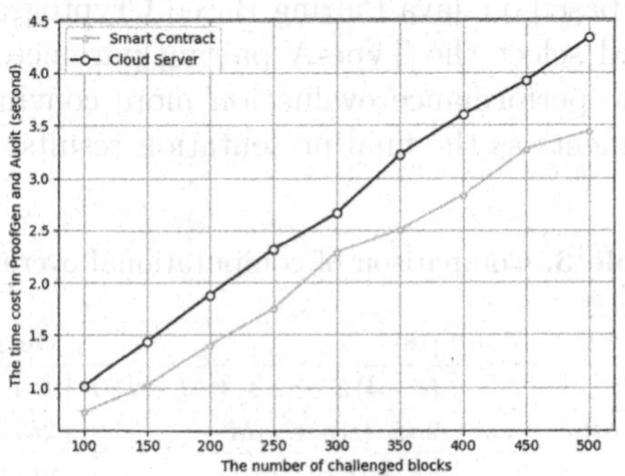

Fig. 7. The computation time in the proof information generation and audit phase

7 Conclusions

In this paper, we propose a blockchain-based proxy oriented cloud storage public audit scheme for low-performance terminal devices. In the tag generation phase, DO only sends the documents to the proxy, which generates authentication tag and uploads encrypted documents to CS. In the audit stage, with the help of blockchain and smart contract, DO does not need to interact with CS. CS will send proof information to the smart contract, and the smart contract will verify the validity of the proof information. DO only needs to check the audit results in the blockchain. In addition, CS automatically receives a service fee when DO's data integrity is verified; otherwise, CS needs to pay compensation to DO whose integrity is compromised. Security analysis and performance evaluation show that our scheme is secure and practical. DO only needs to cost minimal communication and computation overhead during the entire implementation of the scheme. Therefore, our proposal is extremely friendly and practical for low performance terminal devices.

Acknowledgements. This work is supported by the National Natural Science Foundation of China (Grant Nos. 61972352, 61602408,61572435), the Key Research and Development Program of Zhejiang Province (Grant No. 2021C03150) and Zhejiang Provincial Natural Science Foundation of China under Grant (Nos. LY19F020005, LY18F020009).

References

1. Wang, T., et al.: Privacy-enhanced data collection based on deep learning for internet of vehicles. IEEE Trans. Ind. Inf. **16**(10), 6663–6672 (2019)
2. Wang, T., Jia, W., Xing, G., Li, M.: Exploiting statistical mobility models for efficient wi-fi deployment. IEEE Trans. Veh. Technol. **62**(1), 360–373 (2012)
3. Chen, M., Wang, T., Ota, K., Dong, M., Zhao, M., Liu, A.: Intelligent resource allocation management for vehicles network: an a3c learning approach. Comput. Commun. **151**, 485–494 (2020)
4. Xie, M., Ruan, Y., Hong, H., Shao, J.: A cp-abe scheme based on multi-authority in hybrid clouds for mobile devices. Futur. Gener. Comput. Syst. **121**, 114–122 (2021)
5. Ateniese, G., et al.: Provable data possession at untrusted stores. In: Proceedings of the 14th ACM Conference on Computer and Communications Security, pp. 598–609 (2007)
6. Erway, C.C., Küpçü, A., Papamanthou, C., Tamassia, R.: Dynamic provable data possession. ACM Trans. Inf. Syst. Secur. (TISSEC) **17**(4), 1–29 (2015)
7. Esiner, E., Küpçü, A., Özkasap, Ö.: Analysis and optimization on FlexDPDP: a practical solution for dynamic provable data possession. In: Al-Saidi, A., Fleischer, R., Maamar, Z., Rana, O.F. (eds.) Analysis and optimization on flexdpdp: A practical solution for dynamic provable data possession. LNCS, vol. 8993, pp. 65–83. Springer, Cham (2015). https://doi.org/10.1007/978-3-319-19848-4_5
8. Juels, A., Kaliski Jr, B.S.: Pors: proofs of retrievability for large files. In: Proceedings of the 14th ACM Conference on Computer and Communications Security, pp. 584–597 (2007)
9. Shacham, H., Waters, B.: Compact proofs of retrievability. In: Pieprzyk, J. (ed.) ASIACRYPT 2008. LNCS, vol. 5350, pp. 90–107. Springer, Heidelberg (2008). https://doi.org/10.1007/978-3-540-89255-7_7
10. Cash, D., Küpçü, A., Wichs, D.: Dynamic proofs of retrievability via oblivious ram. J. Cryptol. **30**(1), 22–57 (2017)
11. Zhang, J., Tang, W., Mao, J.: Efficient public verification proof of retrievability scheme in cloud. Clust. Comput. **17**(4), 1401–1411 (2014). https://doi.org/10.1007/s10586-014-0394-8
12. Wang, H.: Identity-based distributed provable data possession in multicloud storage. IEEE Trans. Serv. Comput. **8**(2), 328–340 (2014)
13. Zhang, J., Dong, Q.: Efficient id-based public auditing for the outsourced data in cloud storage. Inf. Sci. **343**, 1–14 (2016)
14. Wang, H., Qin, H., Zhao, M., Wei, X., Shen, H., Susilo, W.: Blockchain-based fair payment smart contract for public cloud storage auditing. Inf. Sci. **519**, 348–362 (2020)
15. Yuan, H., Chen, X., Wang, J., Yuan, J., Yan, H., Susilo, W.: Blockchain-based public auditing and secure deduplication with fair arbitration. Inf. Sci. **541**, 409–425 (2020)
16. Nakamoto, S.: Bitcoin: a peer-to-peer electronic cash system. Decentralized Business Review, p. 21260 (2008)
17. Buterin, V., et al.: Ethereum white paper. GitHub Repository **1**, 22–23 (2013)

18. Wang, H., He, D., Tang, S.: Identity-based proxy-oriented data uploading and remote data integrity checking in public cloud. IEEE Trans. Inf. Forensics Secur. **11**(6), 1165–1176 (2016)
19. De Caro, A., Iovino, V.: jpbc: Java pairing based cryptography. In: Proceedings of the 16th IEEE Symposium on Computers and Communications, ISCC 2011, pp. 850–855. IEEE, Kerkyra, Corfu, Greece, June 28–1 July (2011)

Sunspot: A Decentralized Framework Enabling Privacy for Authorizable Data Sharing on Transparent Public Blockchains

Yepeng Ding(✉) and Hiroyuki Sato

The University of Tokyo, Tokyo, Japan
{youhoutei,schuko}@satolab.itc.u-tokyo.ac.jp

Abstract. Blockchain technologies have been boosting the development of data-driven decentralized services in a wide range of fields. However, with the spirit of full transparency, many public blockchains expose all types of data to the public such as Ethereum. Besides, the on-chain persistence of large data is significantly expensive technically and economically. These issues lead to the difficulty of sharing fairly large private data while preserving attractive properties of public blockchains. Although direct encryption for on-chain data persistence can introduce confidentiality, new challenges such as key sharing, access control, and legal rights proving are still open. Meanwhile, cross-chain collaboration still requires secure and effective protocols, though decentralized storage systems such as IPFS bring the possibility for fairly large data persistence. In this paper, we propose Sunspot, a decentralized framework for privacy-preserving data sharing with access control on transparent public blockchains, to solve these issues. We also show the practicality and applicability of Sunspot by MyPub, a decentralized privacy-preserving publishing platform based on Sunspot. Furthermore, we evaluate the security and privacy of Sunspot through theoretical analysis and experiments.

Keywords: Data sharing · Privacy preservation · Access control · Blockchain · Decentralized storage · Digital rights management

1 Introduction

Worldwide decentralization of data persistence and sharing is advancing with the evolution of public blockchain technologies, as can be seen by the boom of decentralized applications (DApp), especially decentralized finance (DeFi) [39] in recent years. Different from decentralized payment platforms such as

This research was partially supported by KAKENHI (Grant-in-Aid for JSPS Fellows) 21J21087 and KAKENHI (Grants-in-Aid for Scientific Research) (C) 19K11958.

Y. Lai et al. (Eds.): ICA3PP 2021, LNCS 13155, pp. 693–709, 2022.
https://doi.org/10.1007/978-3-030-95384-3_43

Bitcoin [24][1], DeFi has wider applicability that provides various financial products and services such as lending [3], insurance [34], trading, and investment [5]. Although the energy consumption of DeFi compared to traditional finance is still under debate, DeFi has presented a set of attractive properties including accessibility, automation, transparency, interoperability, finality, borderlessness, and innovativeness. These properties are generally achieved through on-chain data persistence and sharing supported by transparent public blockchains such as Ethereum [36] that are permissionless and fully disclosing all types of data. But it would still be limited if data are finance-oriented only.

With the maturity of the ERC-721 [10], the application based on non-fungible token (NFT) techniques has enlarged applicability to all types of digital assets in a wide range of fields including art, law, finance, entertainment, as well as personal data. An NFT is generally used to represent legal rights such as ownership of a digital asset, though is possible for a traditional asset. The proofs of these legal rights possess beneficial properties of public blockchains such as tamperproofing, transparency, traceability, and high availability due to its nature of solidifying proof data on public blockchains. However, proof of legal rights does not imply prevention of infringement of these rights because off-chain assets may still face security and privacy issues.

Although on-chain storage of digital assets is theoretically feasible, it is too expensive to store large data directly on public blockchains due to the consensus difficulty. Consequently, current public blockchains have strict data size limitations. While widespread centralized solutions such as self-hosted storage and cloud storage [37] have lower costs, some properties like integrity and availability are hard to implement. As a compromise solution, public blockchains that leverage peer-to-peer file sharing protocols have been experimented in industry to store large data without centralized mechanisms and authorities such as InterPlanetary File System (IPFS) [4]. With elaborated incentive mechanisms and cryptographic support, some blockchains have made significant progress to attract participants such as the Filecoin network that is built on top of IPFS. These systems enhance the security of data storage including integrity, durability, and availability but still do not solve privacy issues because all data are stored publicly without any confidentiality protection at default. In other words, anyone with a public link can fully acquire that data. Even if data are encrypted by data providers before storage, it is challenging to share decryption keys and authorize data consumers without central entities.

From the background above, we summarize and formalize challenges of decentralized data sharing as follows, where $\mathbf{D}$ is the shared data set, $\mathbf{R}$ the legal right set, $\mathbf{U}$ the user set.

Challenge 1 [Proof of legal rights]. $\mathcal{P}_r(d, r) = u$ if and only if u has the legal right r, where $\mathcal{P}_r : \mathbf{D} \times \mathbf{R} \to \mathbf{U}, d \in \mathbf{D}, r \in \mathbf{R}$.

---

[1] For brevity, when mentioning the name of a cryptocurrency in this paper, we refer to its underlying blockchain. For instance, "Bitcoin" refers to the blockchain built on the Bitcoin Backbone Protocol [11].

Challenge 2 [Fairly large data storage]. $\exists d \in \mathbf{D}, \mathit{Size}(d) > 1\text{MB}$.

Challenge 3 [Interoperability]. The system is compatible with various blockchains and allows multiple heterogeneous blockchains to implement functionalities with minimum compatibility issues.

Challenge 4 [Privacy preservation]. $\forall d \in \mathbf{D}$, d is encrypted.

Challenge 5 [Access control]. $\mathcal{P}_a(u, d) = \mathit{Dec}(d)$ if and only if u has the access privilege of d, where $\mathcal{P}_a : \mathbf{U} \times \mathbf{D} \rightarrow \{\mathit{Dec}(d) : d \in \mathbf{D}\}$.

Motivated by the above issues and challenges, we propose a framework named Sunspot to enable privacy and provide access control for flexible data sharing on transparent public blockchains. Sunspot has a decentralized architecture and supports multiple access control mechanisms based on customizable identity management models. We generalize its protocols to make Sunspot compatible with various public blockchains and decentralized storage systems. Besides, Sunspot has a cross-chain collaboration mechanism that bridges blockchains and decentralized storage systems, as well as a multi-chain mechanism that allows heterogeneous blockchains to present unified characteristics.

We summarize our main contributions of this paper and Sunspot as follows.

1. We present Sunspot, a decentralized privacy-preserving data sharing framework that solves the challenges above. To the best of our knowledge, Sunspot is the first framework that focuses on preserving privacy for data sharing on transparent public blockchains.
2. We generalize protocols and provide support for multiple access control mechanisms to enable flexible data sharing in various scenarios. Particularly, we show an identity management model and a management-free model for identification based on two different access control mechanisms.
3. We present MyPub, a decentralized privacy-preserving publishing platform based on Sunspot, to show the practicality and applicability of Sunspot.
4. We prove the security and privacy properties of Sunspot and demonstrate them with experiments from the perspective of attackers.

2 Related Work

The privacy issues coming with full transparency are rising concerns in society and endangering legacy of some decentralized services due to tightening of regulations such as the General Data Protection Regulation (GDPR) [27]. In such a context, many enterprise products turn into solutions based on private blockchains and consortium blockchains such as HyperLedger Fabric [2] and XRP Ledger [29]. In some scenarios, redeveloping an open-source public blockchain as a privacy-preserving blockchain also becomes a solution.

For existing transparent public blockchains, solutions based on zero-knowledge proof (ZKP) [12] are attracting attention. As an early exploration, Zerocoin [22] provides a feasible solution to Bitcoin's pseudonymity by zero-knowledge proofs for set membership. Security such as prevention of double spending is assured without revealing the transaction. But the original implementation of Zerocoin significantly increased computation cost and slowed down the network efficiency. Later, Zcash, the first widespread blockchain based on the Zero-Knowledge Succinct Non-Interactive Argument of Knowledge (zk-SNARKs) [14], was launched on top of the Zerocoin protocol. Zcash enhances privacy by making the sender, recipient and amount completely private via encryption as well as improves network performance. However, it is merely a privacy protocol in the Bitcoin network that can neither support smart contracts nor store large data.

Besides ZKP-based approaches, Monero [31], which is based on the CryptoNote [32], provides another solution based on Ring Confidential Transactions (RingCT) [25]. It uses a variant of linkable ring signature [20] to allow a member of a group to stay in anonymity while signing messages on behalf of the group. However, the same with ZKP-based approaches, the smart contract is not supported. These privacy-preserving blockchains are generally used for decentralized payment only and not for extended functionalities such as generalized data sharing.

As for blockchain-based data sharing, many solutions have been proposed in different scenarios. For AI-powered network operations, Zhang et al. proposed a mutual trust data sharing framework [40] to break data barriers between different operators. This framework contains three layers: system management layer, storage layer, and user layer. Only the system management layer adopts a blockchain for control, while large data are stored in the cloud. Consequently, it is still vulnerable to threats to cloud storage and threatened by the trustiness of cloud service providers. In work [38], a framework named BBDS is proposed for data sharing of electronic medical records (EMR). It adopts a decentralized access control mechanism to secure medical records. However, BBDS is based on a permissioned blockchain and also relies on cloud storage. And privacy issues are not addressed including data encryption and key sharing. Recently, Derepo [8], a distributed data repository, is proposed to preserve the privacy of medical data with decentralized access control and homomorphic encryption (HE). Although the HE scheme proves to be effective for solving privacy issues, it leads to a significant performance issue.

To the best of our knowledge, there is no existing work that tackles the privacy issues for decentralized and general-purpose data sharing on transparent public blockchains.

3 Sunspot

Sunspots are dark areas on the surface of the Sun. Analogically, our framework enables private data sharing "in the dark", i.e., without disclosing any meaningful information from the shared data to the unauthorized public, on fully transparent public blockchains. Meanwhile, it is built on top of public blockchains to preserve their security properties.

Note, in the rest of the paper, we simply use the word *blockchain* when referring to the phrase *transparent public blockchain*. In Sunspot, we introduce new terminologies to clarify non-standardized concepts. We categorize blockchains into two types: control chain and storage chain. A blockchain can be a control chain if and only if it supports the smart contract, event mechanism, and NFT protocol. A storage chain is a blockchain that supports decentralized large data storage, which usually adopts a hybrid architecture. A blockchain can be both a control chain and a storage chain with the only conceptual difference. Notably, with the feature of the multi-chain compatibility of Sunspot, a control chain or a storage chain can be composed of a set of heterogeneous blockchains.

3.1 Architecture

Sunspot involves two main types of roles: organizer and participant. Organizers are responsible for initialization, deployment, and optionally for identification that is a continuous work to endorse participants' identities. In this paper, we will show an identity management model based on a fine-grained access control mechanism in this section, a management-free model that depends on a payment-based access control mechanism in Sect. 4, and a self-sovereign identity model in Sect. 6.

Participants in Sunspot can be data providers, data consumers, or both. We call participants who are willing to share data as data providers while those who are willing to acquire data as data consumers. There is no direct interaction between a provider and a consumer. Sunspot encapsulates all operations and exposes interfaces to participants in a decentralized manner, which means it has a high probability that the interaction target is not the intended target but a set of unknown nodes in the blockchain network that are even not aware of the functionality.

As shown in Fig. 1, Sunspot contains five components: Distributor, ID Manager, Cipher Suite, Control Linker, and Storage Linker. From the view of deployment, except Distributor and ID Manager that are deployed on the control chain, other components are initialized on the storage chain.

Distributor. Distributor is a smart contract on the control chain that implements the NFT protocol. It provides core functionalities to solidify the legal rights of data by minting a unique NFT as its meta-information. A data owner uses Distributor to share data while a data consumer to query and get the meta-information of the shared data.

ID Manager. ID Manager is used for participant registration and identity query, which is a smart contract on the control chain. Participant registration integrates identification information such as wallet addresses of different blockchains and authorization information. For the identity management model in this section, Ciphertext-Policy Attribute-Based Encryption (CP-ABE) [35] is used for fine-grained access control. Hence, authorization information is a private key that describes the attributes of a participant. And identity query is a read-only operation for verification, which is implemented in a query engine.

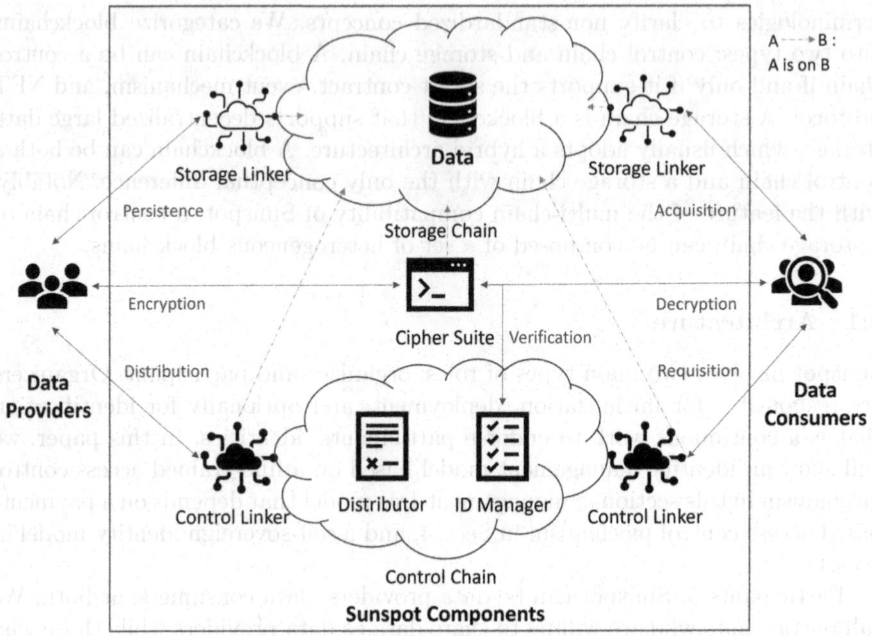

Fig. 1. The architecture of Sunspot.

Cipher Suite. Cipher Suite plays a pivotal role in enabling privacy and securing access control. Its main functionalities are encrypting data and generating decryptors. It is designed to be secure and implemented with a set of security measures such as obfuscation [13] to ensure control-flow integrity (CFI) [1], type safety, and memory safety.

Control Linker. Control Linker is the gateway to the control chain that encapsulates interactions such as signing transactions and making the remote procedure call (RPC). A common interface is implemented to facilitate the redevelopment and multi-chain support.

Storage Linker. Similar to the Control Linker, Storage Linker is the gateway to the storage chain. Particularly, it encapsulates interactions with the storage chain such as data persistence and data retrieval.

3.2 Protocol

Environment Assumption. We assume there is an organizer set $\mathbf{O}$ and a participant set $\mathbf{P}$. In the worst case, $g \leq |\mathbf{O}|$ of organizers are available, and at least one organizer of the available ones is honest and never colludes with the others.

The control chain $\mathfrak{C}$ and the storage chain $\mathfrak{S}$ already exist. For clarity, we assume $\mathfrak{C}$ is merely a blockchain. And so is $\mathfrak{S}$. We will illustrate the multi-chain case in Sect. 3.3.

Blockchains selected for $\mathfrak{C}$ and $\mathfrak{S}$ enable Transport Layer Security (TLS) for the JSON-RPC request/response communication channel.

Two master key pairs used to initialize Sunspot, $(\mathrm{MPK}^c, \mathrm{MSK}^c, \mathrm{MWA}^c)$ for interacting with $\mathfrak{C}$, $(\mathrm{MPK}^s, \mathrm{MSK}^s, \mathrm{MWA}^s)$ for interacting with $\mathfrak{S}$, are created, where MPK is a master public key, MSK a master secret key, and WMA a master wallet address. Besides, MSK^c and MSK^s are encoded as integers in $\mathbb{Z}/m\mathbb{Z}$ and divided into $|\mathbf{O}|$ shares respectively based on the Shamir's Secret Sharing (SSS) [30] with $(g, |\mathbf{O}|)$-threshold. And each share for each of them is distributed to a unique organizer.

All participants already have key pairs for interacting with $\mathfrak{C}$ and $\mathfrak{S}$. $\forall P_i \in \mathbf{P}$, P_i has a pair $(\mathrm{PK}_i^c, \mathrm{SK}_i^c, \mathrm{W}_i^c)$ for $\mathfrak{C}$ and $(\mathrm{PK}_i^s, \mathrm{SK}_i^s, \mathrm{WA}_i^s)$ for $\mathfrak{S}$, where PK, SK, WA, denote public key, secret key, and wallet address respectively. For Ethereum, SK is generated as random 256 bits usually by SHA-256 [6]. PK is derived from SK by the Elliptic Curve Digital Signature Algorithm (ECDSA) [16] for the elliptic curve secp256k1. And WA is created by applying the Keccak-256 [9] to PK.

Initialization. This protocol is used for initializing Sunspot.

1. Deploy Distributor and ID Manager contracts to $\mathfrak{C}$ with MSK^c and get addresses C_{dis} and C_{id} as their identifiers.
2. Run *Setup*(λ, Γ) algorithm of CP-ABE [35] to produce public parameter $\hat{\mathrm{PK}}$ and a master key $\hat{\mathrm{MK}}$, where λ is the security parameter and Γ is the attribute space;
3. Store $\hat{\mathrm{PK}}$ and $\hat{\mathrm{MK}}$ to state variables of C_{dis} by invoking *SetABE*$(\hat{\mathrm{PK}}, \hat{\mathrm{MK}})$;
4. Configure and compile Cipher Suite, Control Linker, and Storage Linker with parameters C_{dis} and C_{id}. Store them to $\mathfrak{S}$ with MSK^s and get addresses S_{ciph}, S_{cl}, S_{sl} as their identifiers;
5. Emit a *Storage Registered Event*$(\mathrm{MWA}^s, S_{ciph}, S_{cl}, S_{sl})$ on $\mathfrak{C}$ with MSK^c;
6. Make $\hat{\mathrm{PK}}$, $\hat{\mathrm{MK}}$, S_{ciph}, S_{cl}, S_{sl} public to participants.

Registration. Users need to execute this protocol to get qualified as participants or rejected. Suppose an arbitrary user $u \in \mathbf{U}$ requests to get registered as a participant.

1. u submits a structured record Rec_u with proofs to organizers who are responsible for identity management. Rec_u consists of WA_u^c, WA_u^s and a set of attributes $\mathbf{A}$;
2. Run *KeyGeneration*$(\hat{\mathrm{MK}}, \mathbf{A})$ algorithm of the CP-ABE to produce a private key χ_u;
3. A *Registered Event*$(\mathrm{WA}_u^c, \mathrm{WA}_u^s, \chi_u)$ is emitted on $\mathfrak{C}$ by organizers via C_{id} to indicate $u \in \mathbf{P}$ if Rec_u is approved.

Solidification. Data providers use this protocol to solidify their data while preserving privacy through three sub-protocols: Encryption, Persistence, and Distribution.

Suppose a data provider $P_d \in \mathbf{P}$ wants to share data D in private.

Encryption. After obtaining Cipher Suite from S_{ciph}, execute this protocol to produce encrypted data *Enc*(D) and corresponding decryptor *Dec*.

1. Generate a secret key $\kappa = Bcrypt(\mathrm{SK}_d^c \oplus \epsilon)$ where *Bcrypt* is a password scheme [26], ϵ is a nonce that is generated from a cryptographically-secure pseudo-random number generator [33], and $\oplus$ denotes an operation to obfuscate SK_d^c and ϵ such as concatenation and bitwise exclusive or;
2. Get *Enc*(D) by encrypting D with κ based on the AES-GCM-SIV scheme (AES) [15];
3. Run $Encrypt(\hat{\mathrm{PK}}, M, \gamma)$ algorithm of CP-ABE to produce an access control policy ρ, where M is a random challenge message and γ is an access structure over Γ;
4. Generate decryptor *Dec* with κ, ρ and M.

Persistence. This sub-protocol is executed via S_{sl} to store *Enc*(D) and *Dec* on $\mathfrak{S}$ and produces metadata address S_μ where storing the meta-information of D.

1. Get persistent addresses S_{data} and S_{dec} after storing *Enc*(D) and *Dec* to $\mathfrak{S}$ with SK_d^s signing transactions;
2. Generate a structured metadata μ that at least contains S_{data}, $\mathcal{H}(Enc(D))$, S_{dec}, and $\mathcal{H}(Dec)$ where $\mathcal{H}: \{0,1\}^* \to \{0,1\}^l$ is a cyrptographic hash function with a fixed size l.
3. Get persistent address S_μ after storing μ to $\mathfrak{S}$ with SK_d^s.

Distribution. This sub-protocol solidifies S_μ as an NFT on $\mathfrak{C}$ through S_{cl}. After this protocol, P_d can make S_μ public to share D in a decentralized and private way.

1. Generate a unique token id ι;
2. Invoke $Mint(\mathrm{WA}_d^c, \iota)$, a method in the NFT protocol, of C_{dis} with SK_d^c;
3. Update ID mapping $\mathcal{M}_{id}$ with $S_\mu \mapsto \iota$;
4. Emit a $Distributed\ Event(\mathrm{WA}_d^c, \iota)$ on C_{dis}.

Authorization. Data consumers use this protocol to get authorized and acquire decrypted data through four sub-protocols: Requisition, Verification, Acquisition, and Decryption.

Suppose a data consumer $P_c \in \mathbf{P}$ knows S_μ and wants to acquire decrypted data D.

Requisition. In this protocol, P_c explicitly makes a request via S_{cl}. After this protocol, P_c knows ι.

1. Validate identity existence by ensuring a *Registered Event*($\text{WA}_c^c, \text{WA}_c^s, *$) exists, where WA_c^c is derived from SK_c^c, WA_c^s from SK_c^s. If the validation result is $\perp$, terminate the protocol with *Non-existence* exception. Otherwise, continue the protocol;
2. Get ι by $\mathcal{M}_{id}(S_\mu)$;
3. Emit a *Requested Event*(WA_c^c, ι) with SK_c^c.

Acquisition. By this sub-protocol, P_c gets decryptor *Dec* via S_{sl}.

1. Get μ from S_μ;
2. Get *Dec* from $\mu[dec] \implies S_{dec}$ and verify its integrity by $\mathcal{H}(Dec)$.

Verification. This sub-protocol is enforced by *Dec* in memory to authorize a request and produce verification result $\eta \in \{\top, \perp\}$.

1. Validate identity existence by ensuring a *Registered Event*($\text{WA}_c^c, \text{WA}_c^s, *$) exists, where WA_c^c is derived from SK_c^c, WA_c^s from SK_c^s. If the validation result is $\perp$, terminate the protocol with *Non-existence* exception. Otherwise, continue the protocol;
2. Validate ownership by $\text{WA}_c^c = \textit{OwnerOf}(\iota)$, where *OwnerOf* is a method in the NFT protocol, to produce the validation result η_0;
3. Validate request existence to produce η_1 by ensuring a *Requested Event* (WA_c^c, ι) exists;
4. Match the challenge message M with the result of *Decrypt*($\hat{\text{PK}}, \rho, \chi_c$) algorithm of the CP-ABE to produce η_2;
5. Emit a *Verified Event*($\text{WA}_c^c, \iota, \eta$) with SK_c^c, where $\eta = \eta_0 \vee \eta_1 \wedge \eta_2$.

Decryption. This sub-protocol follows the *Verification* protocol to decrypt *Enc*(*data*), which is also enforced by *Dec* in memory.

1. Validate the verification result by ensuring a *Verified Event*($\text{WA}_c^c, \iota, \top$) exists, where WA_c^c is derived from SK_c^c;
2. Get μ from *TokenURI*(ι), a method in the NFT protocol;
3. Get *Enc(D)* from $\mu[data] \implies S_{data}$ and verify its integrity by $\mathcal{H}(Enc(D))$;
4. Restore *Enc*(D) to D with κ.

3.3 Multi-Chain Support

Sunspot allows a heterogeneous structure for a control chain and a storage chain. In a heterogeneous control chain, there is a master blockchain and a set of peer blockchains. And a storage chain can consist of a set of heterogeneous peer blockchains. In Sunspot, we formulate a Chain Manager and additional protocols to enable multi-chain compatibility on top of fundamental protocols illustrated in Sect. 3.2.

Chain Manager. Chain Manager is a DApp for registering information about blockchain clusters used to constitute the control chain or the storage chain. Its main functionality is to permanently store addresses of smart contracts deployed on all control-oriented blockchains and addresses of other components deployed on storage-oriented blockchains.

Environment Assumption. We assume the same environment illustrated in Sect. 3.2 with additional assumptions as follows.

- The control chain $\mathfrak{C}$ consists of a master blockchain $\mathfrak{M}$ and a finite set of peer blockchains $\mathbf{N}_c = \{\mathfrak{N}_0^c, \mathfrak{N}_1^c, \dots\}$.
- Chain Manager C_{cm} is deployed on $\mathfrak{M}$ with a key pair ($\text{MPK}^m, \text{MSK}^m$, MWA^m).
- The storage chain $\mathfrak{S}$ consists of a finite set of peer blockchains $\mathbf{N}_s = \{\mathfrak{N}_0^s, \mathfrak{N}_1^s, \dots\}$.

Chain Initialization. This protocol is executed on $\mathfrak{M}$ via C_{cm} to register $\mathbf{N}_c \cup \mathbf{N}_s$. It contains two sub-protocols: *Control Chain Registration* and *Storage Chain Registration.*

Control Chain Registration. This sub-protocol registers all blockchains in $\mathbf{N}_c$ to produce an address set $\mathbf{C}$ of deployed contracts.

We present the main process for registering $\mathfrak{N}_i^c \in \mathbf{N}_c$. This process is repeated for all elements in $\mathbf{N}_c$.

1. Get addresses C_{dis}^i and C_{id}^i by deploying Distributor and ID Manager optimized for $\mathfrak{N}_i$ to $\mathfrak{N}_i$ with MSK_i^c;
2. Emit a *Control Registered Event*$(i, \text{MWA}_i^c, C_{dis}^i, C_{id}^i)$ on $\mathfrak{M}$ via C_{cm} with MSK^m;
3. Append (C_{dis}^i, C_{id}^i) to $\mathbf{C}$.

Storage Chain Registration. This sub-protocol registers all blockchains in $\mathbf{N}_s$.

We present the main process for registering $\mathfrak{N}_j^s \in \mathbf{N}_s$. This process is repeated for all elements in $\mathbf{N}_s$.

1. Configure and compile Cipher Suite, Control Linker, and Storage Linker with $\mathbf{C}$. Store them to $\mathfrak{N}_j^s$ with MSK_j^s and get addresses S_{ciph}^j, S_{cl}^j, S_{sl}^j as their identifiers;
2. Emit a *Storage Registered Event*$(j, \text{MWA}_j^s, S_{ciph}^j, S_{cl}^j, S_{sl}^j)$ through C_{cm} with MSK^m.

Chain Interaction. For any protocol that has interactions with the control chain, it interacts with the target underlying blockchain through the Chain Manager C_{cm}.

Suppose a protocol calls a function of C_{dis}^i. After constructing $\mathbf{C}$ by querying *Control Registered Events* of $\mathfrak{M}$, the protocol can identify the blockchain deploying C_{dis}^i and invoke that function.

For interacting with the storage chain, it is the same with the single-chain case with the only difference of Application Programming Interfaces (APIs).

4 MyPub

MyPub[2] is a decentralized privacy-preserving publishing platform based on Sunspot. Creators can publish their work such as digital paintings, writings, and music, without disclosing all contents on MyPub and share the metadata freely. To acquire the full contents of a work, customers can choose to buy the ownership or pay for the right to use. All operations of MyPub are enforced on blockchains via variants of protocols of Sunspot without central authorities. Based on Sunspot, MyPub can protect legal rights including the copyright, ownership, and right to use based on properties such as confidentiality, integrity, availability, and transparency.

MyPub adopts Ethereum as its control chain and Filecoin as its storage chain. To adapt to its business logic, MyPub simplifies but concretizes Sunspot protocols and adopts a management-free identity model that depends on a payment-based access control mechanism. This model does not require organizers to endorse participants' identities. Instead, it is based on the intrinsic transaction mechanism of the blockchain. For instance, to obtain the right to use of a work, the customer executes the *Requisition* protocol with a specific amount of cryptocurrency made by the owner. During the execution of the *Verification* protocol, the decryptor validates payment proofs, instead of authenticating the customer according to the registered attributes, to determine whether to decrypt the data. Payment proofs are forgery resistant because they have been already accepted in a blockchain network. Consequently, the payment-based access control mechanism is fully decentralized and does not require authorities for endorsements.

5 Security and Privacy Analysis

5.1 Assumption

We make reasonable assumptions below.

Blockchain Safety. Given a blockchain in Sunspot, most of the nodes, the number of which is greater than the threshold to make the consensus mechanism function well, are trustworthy. By assuming blockchain safety, any blockchain used in a system based on Sunspot preserves properties of a transparent public blockchain including on-chain data integrity and availability.

RPC Safety. Given a blockchain in Sunspot, its API at least enables TLS for RPCs. By assuming RPC safety, any blockchain used in a system based on Sunspot preserves connection integrity and RPC availability.

[2] https://github.com/yepengding/MyPub.

Control-Flow Integrity. Cipher Suite preserves control-flow integrity [1]. Cipher Suite in Sunspot is implemented by the Rust [21], a programming language that provides guarantees for memory safety [17] through an ownership-based resource management model, without using *unsafe* and *external* blocks. Besides, the implementation follows the security through obscurity to protect sensitive function flows and constants.

Organizer Trust. By adopting $(g, |\mathbf{O}|)$-threshold SSS, at least g organizers out of $\mathbf{O}$ are always available and constitute an available set $\mathbf{O}_g \subseteq \mathbf{O}$. $\forall o \in \mathbf{O}_g$, o can collude with at most g' organizers where $g' < g-1$. This assumption ensures there does not exist an organizer can dominate the identity management of a system based on Sunspot.

5.2 Evaluation Model

We use the same notation system illustrated in the sections above. To facilitate reading, we repeat some notations as follows.

For sets, we denote the participant set as $\mathbf{P}$, the shared data set as $\mathbf{D}$. Two types of blockchains in Sunspot are the control chain $\mathfrak{C}$ and the storage chain $\mathfrak{S}$.

Property 1 (Data Privacy). $\forall d \in \mathbf{D}$, $p \in \mathbf{P}$ can decrypt d if and only if $\eta^p = \top$.

Proof. In Sunspot, data privacy is ensured by confidentiality. Data are encrypted through the *Encryption* protocol before being stored on $\mathfrak{S}$, which introduces confidentiality for the stored data.
$\Rightarrow$. In the *Encryption* protocol, d is encrypted via the AES with secret key κ obfuscated and hard-coded in its decryptor. By the *Control-Flow Integrity* assumption, κ is not disclosed to any $p \in \mathbf{P}$, of which the safety is guaranteed. Hence, executing the *Authorization* protocol is the only way for p to decrypt d. In the sub-protocol *Verification*, η^p is the only condition to determine the enforcement of the *Decryption* protocol. Besides, η^p becomes tamper-evident and immutable after corresponding *Verified Event* is emitted according to the *Blockchain Safety.* If p can decrypt d, η^p must be satisfied.

$\Leftarrow$. In the *Verification* protocol, $\eta^p = \top$ holds in two cases: $\eta_0^p = \top$ or $\bigwedge_{i=1}^{2} \eta_i^p = \top$. The first case implies p is the owner of d, while the second case implies the attributes of p satisfies the access control policy ρ_d, which is ensured by the CP-ABE. Based on the assumptions above, p can decrypt d in either case when $\eta^p = \top$.

This completes the proof.

Property 2 (Identity Privacy). $\forall p \in \mathbf{P}$ with a set of attributes $\mathbf{A}_p$, $\mathbf{A}_p$ cannot be disclosed.

Proof. Sunspot protects the privacy of identities in all types of identity management models. Since identity management depends on its underlying access control mechanisms, two cases in Sunspot are proved below.

Case 1 (Fine-Grained Mechanism). In the *Registration* protocol, the private key χ_p that describes $\mathbf{A}_p$ is generated based on the CP-ABE and persist on $\mathfrak{C}$. χ_p is the only public information associated with $\mathbf{A}_p$. Since $\mathbf{A}_p$ cannot be disclosed with χ_p according to [35], the identity privacy is protected.

Case 2 (Payment-Based Mechanism). This mechanism does not require identity information other than the wallet address embedded in a payment transaction. Since the wallet address does not directly link to the real identity, the identity information cannot be disclosed.

This completes the proof.

Property 3 (Data Availability). $\forall d \in \mathbf{D}$, if the token id ι of d is known, d is always available.

Proof. The metadata μ of d is always available via calling *TokenURI*(ι) of the Distributor C_{dis} according to the *Blockchain Safety* and *RPC Safety* assumptions. On-chain data integrity implies μ and S_{data} implied from $\mu[data]$ are tamper-proof. Based on the *Blockchain Safety*, d is always available via its address S_{data}.

5.3 Experiments

To demonstrate the properties are effective to protect Sunspot from threats, we conduct two representative experiments from the perspective of attackers in a simulation environment where we set up an Ethereum and a Filecoin network with TLS enabled, as well as 10 account pairs (one for each network) for organizers, 10 for data providers, and 10 for data consumers. The *Initialization* and *Registration* protocols are executed after the environment setup.

Man-in-the-Middle Attack. We simulate a middleman who aims to deceive the verification process of the decryptor by hijacking the connections between the decryptor and the blockchain network. After penetrating our deliberately developed vulnerable network, all packets are successfully intercepted by the middle man. However, without the private key, the middle man can neither know the transmitting data nor forge a connection to compromise the *Verification* protocol.

We also formulate an advanced attack by making the middleman pretend to be a legal node in the network with a self-signed certificate. However, the decryptor throws an *Unauthenticated Error* immediately after detecting an unknown certificate authority in the first process of the *Verification* protocol.

Return-Oriented Programming. To test the control-flow integrity, we use return-oriented programming (ROP) [28] to analyze the decryptor and try to find gadgets to bypass the *Verification* protocol by modifying the control flow. We decode the decryptor into a file in the LLVM assembly language format [19].

However, the file is over 2 million lines with unreadable flows and values due to the obfuscation. And our experiments fail to exploit it with static and runtime analyzers such as [18].

6 Discussion

We claim that Sunspot provides a feasible solution to the issues and challenges described in Sect. 1.

Challenge 1 [Proof of legal rights]. Metadata are stored on the control chain as NFTs. The proofs of legal rights derived from on-chain data are immutably verifiable due to the public blockchain properties. For instance, in Mypub, a creator of an NFT holds copyright and initial ownership while a customer can hold the right to use of some data after the payment, i.e., getting authorized. These rights are permanently traceable and verifiable.

Challenge 2 [Fairly large data storage]. In Sunspot, the control chain collaborates with the storage chain to enable fairly large data storage. The storage chain is implemented by decentralized storage systems to store large data. Besides, a cross-chain collaboration mechanism is implemented based on protocols in Sect. 3.2 to bridge the control chain and the storage chain.

Challenge 3 [Interoperability]. First, Sunspot protocols and access control mechanisms are generalized and not blockchain-specific. Therefore, Sunspot can support all types of blockchains that satisfy the conditions defined in Sect. 3. Furthermore, Sunspot enables multi-chain compatibility and allows heterogeneous structures by the Chain Manager and additional protocols illustrated in Sect. 3.3. In this manner, a chain component can consist of multiple heterogeneous blockchains to be scalable and dependable.

Challenge 4 [Privacy preservation]. In Sunspot, identity privacy and data privacy are preserved. We provide proofs in Sect. 5.2. Participants' attributes are stored on chain but described by a private key generated in the *Registration* protocol. And the shared data are fully encrypted via the *Encryption* protocol before being stored on the storage chain.

Challenge 5 [Access control]. Currently, two access control mechanisms are mechanized in Sunspot including a fine-grained mechanism and a payment-based mechanism. Based on these two mechanisms, Sunspot supports an identity management model based on the combination of the CP-ABE and blockchains that is described in Sect. 3 and a management-free model purely enforced on the blockchain that is described in Sect. 4.

However, the identity management model based on the fine-grained access control mechanism highly depends on the *Control-Flow Integrity* assumption. Although the management-free model is enforced on the blockchain, it lacks support for complex access control policies. Therefore, the implementation of the self-sovereign identity (SSI) [23] model is on our schedule. With the support of the SSI, participants can manage their own identities based on blockchains and present on-demand verifiable presentations for authentication and authorization.

Besides, Sunspot-based systems are also threatened by the security issues of blockchains such as smart contract vulnerabilities [7] and the Sybil attack [41].

7 Conclusion

In this paper, we have presented Sunspot, a decentralized privacy-preserving framework with multiple access control mechanisms, to solve the key challenges of data sharing on transparent public blockchains including proof of legal rights, fairly large data storage, interoperability, privacy preservation, and access control. To enlarge the applicability of Sunspot, we have generalized its protocols and made it support multi-chain and heterogeneous structures. Besides, we have presented MyPub, a decentralized privacy-preserving publishing platform based on Sunspot to demonstrate Sunspot in practice. We have also proved the security and privacy properties of Sunspot including data privacy, identity privacy, and data availability. For security in practice, we have conducted experiments from the perspective of attackers to launch representative attacks to demonstrate the effectiveness of Sunspot. Furthermore, we have discussed the methods mechanized in Sunspot of solving the key challenges of data sharing on transparent public blockchains, as well as the limitations and further improvements.

References

1. Abadi, M., Budiu, M., Erlingsson, U., Ligatti, J.: Control-flow integrity principles, implementations, and applications. ACM Trans. Inf. Syst. Secur. (TISSEC) **13**(1), 1–40 (2009). iSBN: 1094-9224 Publisher: ACM New York, NY, USA
2. Androulaki, E., et al.: Hyperledger fabric: a distributed operating system for permissioned blockchains. In: Proceedings of the Thirteenth EuroSys Conference, pp. 1–15 (2018)
3. Bartoletti, M., Chiang, J.H., Lafuente, A.L.: SoK: lending pools in decentralized finance. In: Bernhard, M., Bracciali, A., Gudgeon, L., Haines, T., Klages-Mundt, A., Matsuo, S., Perez, D., Sala, M., Werner, S. (eds.) FC 2021. LNCS, vol. 12676, pp. 553–578. Springer, Heidelberg (2021). https://doi.org/10.1007/978-3-662-63958-0_40
4. Benet, J.: Ipfs-content addressed, versioned, p2p file system. arXiv preprint arXiv:1407.3561 (2014)
5. Chen, Y., Bellavitis, C.: Blockchain disruption and decentralized finance: the rise of decentralized business models. J. Bus. Venturing Insights **13**, e00151 (2020), iSBN: 2352-6734
6. Dang, Q.H.: Secure hash standard (2015)
7. Destefanis, G., Marchesi, M., Ortu, M., Tonelli, R., Bracciali, A., Hierons, R.: Smart contracts vulnerabilities: a call for blockchain software engineering? In: 2018 International Workshop on Blockchain Oriented Software Engineering (IWBOSE), pp. 19–25. IEEE (2018)

8. Ding, Y., Sato, H.: Derepo: a distributed privacy-preserving data repository with decentralized access control for smart health. In: 2020 7th IEEE International Conference on Cyber Security and Cloud Computing (CSCloud)/2020 6th IEEE International Conference on Edge Computing and Scalable Cloud (EdgeCom), pp. 29–35. IEEE (2020)
9. Dworkin, M.J.: SHA-3 standard: Permutation-based hash and extendable-output functions (2015)
10. Entriken, W., Shirley, D., Evans, J., Sachs, N.: Erc-721 non-fungible token standard. Ethereum Foundation (2018)
11. Garay, J., Kiayias, A., Leonardos, N.: The bitcoin backbone protocol: analysis and applications. In: Oswald, E., Fischlin, M. (eds.) EUROCRYPT 2015. LNCS, vol. 9057, pp. 281–310. Springer, Heidelberg (2015). https://doi.org/10.1007/978-3-662-46803-6_10
12. Goldreich, O., Oren, Y.: Definitions and properties of zero-knowledge proof systems. J. Cryptol. **7**(1), 1–32 (1994). iSBN: 1432-1378
13. Goldwasser, S., Rothblum, G.N.: On best-possible obfuscation. J. Cryptology **27**(3), 480–505 (2013). https://doi.org/10.1007/s00145-013-9151-z
14. Groth, J., Kohlweiss, M., Maller, M., Meiklejohn, S., Miers, I.: Updatable and universal common reference strings with applications to zk-SNARKs. In: Shacham, H., Boldyreva, A. (eds.) CRYPTO 2018. LNCS, vol. 10993, pp. 698–728. Springer, Cham (2018). https://doi.org/10.1007/978-3-319-96878-0_24
15. Gueron, S., Langley, A., Lindell, Y.: AES-GCM-SIV: Specification and Analysis. IACR Cryptol. ePrint Arch. **2017**, 168 (2017)
16. Johnson, D., Menezes, A., Vanstone, S.: The elliptic curve digital signature algorithm (ECDSA). Int.J Inf. Secur. **1**(1), 36–63 (2001)
17. Jung, R., Jourdan, J.H., Krebbers, R., Dreyer, D.: RustBelt: securing the foundations of the Rust programming language. In: Proceedings of the ACM on Programming Languages 2(POPL), pp. 1–34. ACM, New York (2017), iSBN: 2475-1421
18. Křoustek, J., Matula, P., Zemek, P.: Retdec: an open-source machine-code decompiler, December 2017
19. Lattner, C., Adve, V.: LLVM: a compilation framework for lifelong program analysis & transformation. In: International Symposium on Code Generation and Optimization, CGO 2004, pp. 75–86. IEEE (2004)
20. Liu, J.K., Wei, V.K., Wong, D.S.: Linkable spontaneous anonymous group signature for ad hoc groups. In: Wang, H., Pieprzyk, J., Varadharajan, V. (eds.) ACISP 2004. LNCS, vol. 3108, pp. 325–335. Springer, Heidelberg (2004). https://doi.org/10.1007/978-3-540-27800-9_28
21. Matsakis, N.D., Klock, F.S.: The rust language. ACM SIGAda Ada Lett. **34**(3), 103–104 (2014). iSBN: 1094-3641
22. Miers, I., Garman, C., Green, M., Rubin, A.D.: Zerocoin: anonymous distributed e-cash from bitcoin. In: 2013 IEEE Symposium on Security and Privacy, pp. 397–411. IEEE (2013)
23. Mühle, A., Grüner, A., Gayvoronskaya, T., Meinel, C.: A survey on essential components of a self-sovereign identity. Comput. Sci. Rev. **30**, 80–86 (2018). iSBN: 1574-0137
24. Nakamoto, S.: Bitcoin: A peer-to-peer electronic cash system. Technical report, Manubot (2019)
25. Noether, S.: Ring signature confidential transactions for monero. IACR Cryptol. ePrint Arch. **2015**, 1098 (2015)
26. Provos, N., Mazieres, D.: Bcrypt algorithm. In: USENIX (1999)

27. Regulation, G.D.P.: Regulation EU 2016/679 of the European Parliament and of the Council of 27 April 2016. Official Journal of the European Union (2016)
28. Roemer, R., Buchanan, E., Shacham, H., Savage, S.: Return-oriented programming: systems, languages, and applications. ACM Trans. Inf. Syst. Secur. (TISSEC) **15**(1), 1–34 (2012). iSBN: 1094-9224
29. Schwartz, D., Youngs, N., Britto, A.: The ripple protocol consensus algorithm. Ripple Labs Inc White Paper **5**(8), 151 (2014)
30. Shamir, A.: How to share a secret. Commun. ACM **22**(11), 612–613 (1979). iSBN: 0001-0782
31. Sun, S.-F., Au, M.H., Liu, J.K., Yuen, T.H.: RingCT 2.0: a compact accumulator-based (linkable ring signature) protocol for blockchain cryptocurrency Monero. In: Foley, S.N., Gollmann, D., Snekkenes, E. (eds.) ESORICS 2017. LNCS, vol. 10493, pp. 456–474. Springer, Cham (2017). https://doi.org/10.1007/978-3-319-66399-9_25
32. Van Saberhagen, N.: CryptoNote v 2.0 (2013)
33. Vazirani, U.V., Vazirani, V.V.: Efficient and secure pseudo-random number generation (extended abstract). In: Blakley, G.R., Chaum, D. (eds.) CRYPTO 1984. LNCS, vol. 196, pp. 193–202. Springer, Heidelberg (1985). https://doi.org/10.1007/3-540-39568-7_17
34. Wan, Z., Guan, Z., Cheng, X.: Pride: a private and decentralized usage-based insurance using blockchain. In: 2018 IEEE International Conference on Internet of Things (iThings) and IEEE Green Computing and Communications (GreenCom) and IEEE Cyber, Physical and Social Computing (CPSCom) and IEEE Smart Data (SmartData), pp. 1349–1354. IEEE (2018)
35. Waters, B.: Ciphertext-policy attribute-based encryption: an expressive, efficient, and provably secure realization. In: Catalano, D., Fazio, N., Gennaro, R., Nicolosi, A. (eds.) PKC 2011. LNCS, vol. 6571, pp. 53–70. Springer, Heidelberg (2011). https://doi.org/10.1007/978-3-642-19379-8_4
36. Wood, G.: Ethereum: a secure decentralised generalised transaction ledger. Ethereum Project Yellow Paper **151**(2014), 1–32 (2014)
37. Wu, J., Ping, L., Ge, X., Wang, Y., Fu, J.: Cloud storage as the infrastructure of cloud computing. In: 2010 International Conference on Intelligent Computing and Cognitive Informatics, pp. 380–383. IEEE (2010)
38. Xia, Q., Sifah, E.B., Smahi, A., Amofa, S., Zhang, X.: BBDS: blockchain-based data sharing for electronic medical records in cloud environments. Information **8**(2), 44 (2017)
39. Zetzsche, D.A., Arner, D.W., Buckley, R.P.: Decentralized finance. J. Financ. Regul. **6**(2), 172–203 (2020). iSBN: 2053-4841
40. Zhang, G., Li, T., Li, Y., Hui, P., Jin, D.: Blockchain-based data sharing system for ai-powered network operations. J. Commun. Inf. Networks **3**(3), 1–8 (2018). iSBN: 2509-3312
41. Zhang, S., Lee, J.H.: Double-spending with a sybil attack in the bitcoin decentralized network. IEEE Trans. Ind. Inf. **15**(10), 5715–5722 (2019). iSBN: 1551-3203

A Novel Protection Method of Continuous Location Sharing Based on Local Differential Privacy and Conditional Random Field

Linghe Zhu[1], Haibo Hong[1(✉)], and Mande Xie[2]

[1] School of Computer and Information Engineering, Zhejiang Gongshang University, Hangzhou, China

[2] School of Information and Electric Engineering, Zhejiang Gongshang University, Hangzhou, China

Abstract. At present, the protection of user's location privacy (especially mobile user's location privacy) is widely concerned by academia and industry. Many experts and scholars have proposed several secure and efficient solutions to this problem. However, in this context, if a mobile user wants to obtain the services she/he wants, she/he needs to share her/his location information continuously with an untrusted third-party in user's locations. This will cause privacy and security issue. To tackle this problem, in this paper, we apply local differential privacy (LDP) in supporting continuous location sharing among mobile users. Firstly, we put forward a new idea of using conditional random field (CRF) in model user's mobility. Then, we combine δ-location set and ε-LDP and advance a mechanism to support continuous location sharing. Finally, we conduct experiments on real data set to evaluate our proposed mechanism. The results show that our mechanism is more effective than planar isotropic mechanism (PIM).

Keywords: Privacy protection · Location privacy · Continuous location sharing · Local differential privacy · Conditional random field

1 Introduction

Nowadays, people are more and more dependent on smart devices, smart wearable devices that record the location of households can provide accurate services to households according to the location information of households. However, the location information of households is completely mastered by service providers, and users are gradually beginning to worry about whether their information will be maliciously used or sold by service providers. In the past decade, many location protection algorithms have been proposed, but these algorithms either have too high computational complexity or cannot provide sufficient privacy protection with strict mathematical proof. The main idea of these algorithms is to map the

Y. Lai et al. (Eds.): ICA3PP 2021, LNCS 13155, pp. 710–725, 2022.
https://doi.org/10.1007/978-3-030-95384-3_44

real location of user to a dummy location for achieving privacy protection. Specifically, most of these algorithms only consider the static scene, that is, assuming that the user is static in a location or only consider a timestamp, and then execute the protection algorithm to generate a dummy location. However, users are mostly movable in most instances and need to share their locations with the server in real time. Therefore, the adversaries can infer the real location without considering the time relationship generated by multiple timestamp locations.

To capture the time correlation in multiple locations of the user, previous algorithms mainly adopt Markov model. However, the prediction accuracy of the low-order Markov model is not high [1,2], also the generation models such as the hidden Markov model cannot fully utilize the context information of the current location. In order to solve the above problems, in this paper, we take advantage of conditional random field model to predict the location trajectory of moving objects. In this way, we can achieve higher prediction accuracy.

Meanwhile, emergence of differential privacy [3] provides strict mathematical proof for privacy protection, and it has gradually attracted wide attention. Differential privacy was originally used to protect the personal information of the database for data release, and was later applied to the field of location protection.

Contributions. In this paper, in order to make users be assured to disturb their data, we employ local differential privacy instead of centralized differential privacy. We have investigated the issue of continuous location uploading by using LDP. In Fig. 1, a user disturbs each timestamp location of the real location trajectory by LDP mechanism, and then sends the disturbed location to the service provider. Our main contributions are as follows:

1. Firstly, we emply a conditional random field to model the user's mobility, and then implement a protection mechanism that satisfies LDP on the δ-location set. The problem of ignoring the location context in the Markov model is avoided, the prior distribution probability is more accurate, and the corresponding δ-location set is also more accurate, so the utility is improved.
2. Secondly, we combine generalized randomization response (GRR) with randomized aggregatable privacy preserving ordinal responses (RAPPOR) to build a set of protection mechanisms that support continuous location sharing for movable users based on the perturbation algorithm. The user dynamically selects the corresponding algorithm according to their privacy needs, which improves the effectiveness of the mechanism.
3. Thirdly, we conduct experimental evaluation on real trajectory datasets, and compare our mechanism with PIM mechanism. The results show that our protection mechanism can provide better utility while ensuring privacy.

2 Related Works

2.1 Location Privacy

In the field of location privacy protection, a large number of scholars have given relevant research and protection methods, a series of work including privacy stan-

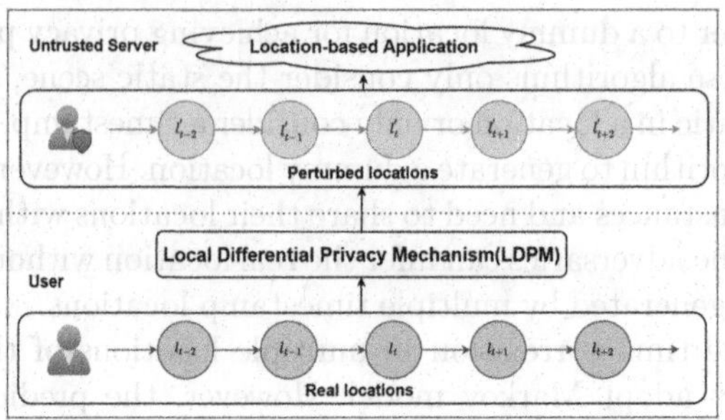

Fig. 1. Continuous location sharing based on LDP.

dards, location privacy protection methods have emerged [4–6]. Generally speaking, location privacy protection methods include mix zone [7], k-anonymity [8], virtual location [9], and differential privacy [10]. The main protection method of location privacy protection mechanism is the disturbance method, such as spatial cloaking, which maps the real location of the user to one or more dummy locations and achieves the purpose that the adversaries cannot distinguish the real location. In 2013, Andrés et al. [10] proposed the concept of geo-indistinguishability. However, the geo-indistinguishability mechanism does not take into account the time correlation between user locations, and geo-indistinguishability mechanism is more about static scenes, but in reality users tend to be moving, either walking or travelling by means of transport [11–14], which leads to the failure to effectively protect location privacy in the scenario of continuous location sharing.

Before this, many works took Markov model with the mobility of users and predict the location or trajectory of users. In the continuous location privacy protection scenario, Ardagna et al. [15] put forward a novel method to protect the location. They firstly used Markov chain to model the user's mobile preferences, and then devised a location perturbation algorithm. Similarly, PIM algorithm in [16] utilized Markov model to model user's mobility and captured the temporal correlation between user locations. However, the Markov model can not effectively use the context information of the location when predicting the location or trajectory, and the prediction performance is insufficient. Hence, we take advantage of the CRF prediction model to model the user's mobility in this paper.

2.2 Local Differential Privacy

In order to meet the needs of users to disturb their privacy information locally, local differential privacy (LDP) was proposed in [17]. The main protection method of LDP is the randomized response mechanism (RR) [18]. However, RR is limited by binary variables, so Kariouz et al. [19] raised a K-randomized response (KRR) mechanism that can be used in multivariate scenarios. Subsequently, Wang et al. [20] advanced the generalized random response (GRR) as the generalized version of RR and KRR. Also, in [21], they carried out a series of experimental analysis for the frequency estimation problem in LDP. The results prove that if the value range $d < 3e^{\varepsilon} + 2$, then generalized random response has better performance than other random response mechanisms, which also provides the basis for the mechanism in our scheme and the parameter of the corresponding LDP. In addition, we also provide the RAPPOR disturbance mechanism, and users can choose independently according to the current privacy preference. In the field of location privacy, there have been some work to design protection mechanisms based on local differential privacy. For the first time, LDP was applied to spatial location data aggregation in [22]. Sei and Ohsuga [23]proposed a Bayesian-based multiple dummies mechanism for combating untrusted servers. This method can effectively provide location privacy for a single user under the premise of satisfying LDP, but this method can only protect the static location of users. We use LDP to protect moviable users under the condition of continuous location sharing, preventing adversaries from inferring the real location of users through the correlation between user locations.

3 Preliminaries

Some important symbol tags are summarized in Table 1.

Table 1. Denotation.

s_i	A grid after the target area is divided, $i = 1, 2, ..., n$
s, l	The location represented by state coordinate system and map coordinate system
s^*, l^*	Real location of users in state and map coordinates
z	The perturbed location published by users
p_t^-, p_t^+	Prior and posterior probabilities
$\Delta X, \lvert\Delta X\rvert$	δ-location set and its size
$P^u_{i,j}$	User preference,the probability of user u_i accessing location l_j
$P^c_{i,j}(c_j)$	Transition probability of user u_i moving from one cluster c_m to another cluster c_j

3.1 Two Coordinate Systems

In this paper,we represent a location by emply different coordinate systems, namely the state coordinate system and the map coordinate system. The former is to use the RAPPOR method to generate the perturbed location in the mechanism proposed later, and the latter is used in the map model. Let $S = \{s_1, \ldots, s_n\}$ represent the target area, meaning that the target area is divided into some grids according to the finest grain, and each grid is represented by s_i. s_i can be regarded as m-dimensional unit vector whose element at the ith position is 1, the other element is 0. Each grid represents the location of a user. At the same time, the user's location can also be represented by geometric coordinates, that is, the map coordinates are represented by longitude and latitude. That means, the user's location l is a two-dimensional vector, the first element is $l[1]$ and another is $l[2]$, as described as Fig. 2. One user is now at location s_{10}, the user's detailed coordinates can be expressed as follows:

$$\boldsymbol{s} = \boldsymbol{s_{10}} = [0\,0\,0\,0\,0\,0\,0\,0\,0\,1\,0\,\ldots\,0] \quad \boldsymbol{l} = [3,2] \; \textit{with} \; \boldsymbol{l}[1] = 3 \; \textit{and} \; \boldsymbol{l}[2] = 2$$

As the user moves, the user's moving location trajectory can be represented by the map coordinate system $l_1, \ldots, l_t$ or the state coordinate system $s_1, \ldots, s_t$.

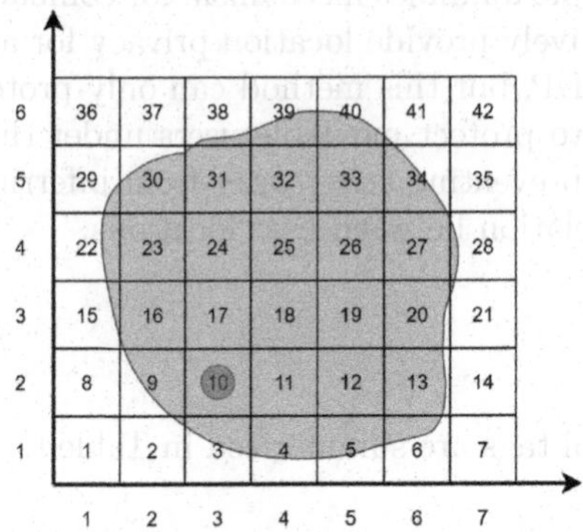

Fig. 2. Coordinate systems.

3.2 Mobility and Inference Model

We make use of the conditional random field [24] to model the event correlation and semantic context between user locations. The user's movable mode and the surrounding road network can be captured by the model. Before this, some scholars took Markov model as a prediction model, but Markov model has some limitations in its predictability [25]. In this paper, we combine context-based location clustering [26], user preferences with posterior probability to synthesize the feature function of conditional random field to predict the future location.

User Preference: By detecting the user's historical movement correlation, we can capture the user's location preference. $P^u_{i,j}$ denotes the probability of user u_i accessing location l_j. Suppose that $v_{k,i}(j)$ denotes a visit of user u_k and u_i to location l_j, if user u_i and user u_k have been to location l_j, then $v_{k,i}(j)$ equals 1, otherwise 0, then the user preference probability $P^u_{i,j}$ can be given by the following equation

$$P^u_{i,j} = \frac{\sum_{u_k} \rho_{k,i} \cdot v_{k,i}(j)}{\sum_{u_k} \rho_{k,i}}, \tag{1}$$

where $p_{k,i}$ represents the access correlation between user u_k and u_i. We randomly select the locations visited by n users u_k and u_i, and then calculate the Pearson correlation

$$\rho_{k,i} = \frac{\sum ki - \frac{(\sum k)(\sum i)}{n}}{\sqrt{\left(\sum k^2 - \frac{(\sum k)^2}{n}\right)\left(\sum i^2 - \frac{(\sum i)^2}{n}\right)}} \tag{2}$$

Prior and Posterior Probabilities: Since the conditional random field model is assumed to be public, adversaries can reason the user's current location based on the locations previously published. At time point t, we use p_t^- and p_t^+ to represent the prior probability and the posterior probability of the user before and after publishing the perturbed location, respectively. The prior probability of location at timestamp t is as follows

$$\boldsymbol{p}_t^-[i] = \Pr\left(\mathbf{s}_t^* = \mathbf{s}_i \mid \boldsymbol{z}_{t-1}, \cdots, \boldsymbol{z}_1\right) \tag{3}$$

After the user publishes the perturbed location z_t, the adversary obtains the perturbed location z_t by observation, and then uses Bayesian inference to calculate the posterior probability of the user location. For each grid s_i,

$$\boldsymbol{p}_t^+[i] = \Pr\left(\boldsymbol{s}_t^* = \boldsymbol{s}_i \mid \boldsymbol{z}_t\right) = \frac{\Pr\left(\boldsymbol{z}_t \mid \boldsymbol{s}_t^* = \boldsymbol{s}_i\right) \boldsymbol{p}_t^-[i]}{\sum_j \Pr\left(\boldsymbol{z}_t \mid \mathbf{s}_t^* = \boldsymbol{s}_j\right) \boldsymbol{p}_t^-[j]} \tag{4}$$

Historical Context-based Cluster Transitions: Assam et al. [26] proposed cluster geographical location according to the semantic context of location, and this attribute can capture context-based conversion frequency and behavior according to the context information of movable users. We use the cluster name of the user's current location instead of the user's location to generate a user's moving sequence. Let $T = \{c_1, c_2, \cdots, c_n\}$ represent a context-based ordered sequence of location clustering, where c_i represents the corresponding clustering name. Suppose that $P_{i,j}^c(c_j)$ represents the transition probability based on location context, that is the probability user u_i moves from one cluster c_m to another cluster c_j, where location $l_m \in c_m$, $l_j \in c_j$. The formula is as follows :

$$P_{i,j}^c(c_j) = \frac{T(c_m, c_j)}{T(C^i)} \tag{5}$$

where $T(c_m, c_j)$ represents the number of times that user u_i moves from cluster c_m to cluster c_j, and $T(C^i)$ represents the total number of times that user u_i moves between all clusters.

δ-Location Set: Xiao et al. [16] proposed the δ-location set, which help us determine the user's all potential location set at any time point, while hiding the user's real location in this location set. We use Eq. 3 to express the prior probability of the location at the time point t. To get the user's possible location set, To get all the potential locations for the user, we first select the appropriate value for parameter δ. Formal definition is that the location in δ-location set should reach the minimum number under the following conditions, that is, the sum of prior probabilities of these locations is greater than or equal to $1 - \delta$

$$\Delta\mathbf{X}_t = \min\left\{\boldsymbol{s}_i \mid \sum_{\boldsymbol{s}_i} \boldsymbol{p}_t^-[i] \geqslant 1 - \delta\right\} \tag{6}$$

Suppose a target area is composed of $[s_1, s_2, s_3, s_4, s_5, s_6]$ and their prior probability is $p_t^- = [0.1, 0.3, 0.25, 0.1, 0.05, 0.2]$, then there are

$$\Delta\mathbf{X} = \begin{cases} [s_1, s_2, s_3, s_6] & \delta = 0.2 \\ [s_2, s_3, s_6] & \delta = 0.3 \end{cases}$$

If $\delta = 0$, then $\Delta\mathbf{X}$ will contain all possible locations. When generating δ-location set, if the inference model is Markov model, it is possible that the user's real location is not in the δ-location set due to its limited prediction accuracy. This phenomenon is called "drift" phenomenon. However, the reasoning model we use is the conditional random field model, and its prediction performance is greatly improved compared with the Markov model. In spite of this, if there is a "drift" phenomenon in the experiment, we directly add the real location of the user to the δ-location set. Therefore, comparing with [16], our method can obtain better utility while ensuring privacy.

In fact, our intention is not to obtain the prediction results, but to obtain the prior probability of the user location of the next timestamp. Of course, the location with the maximum prior probability will be considered as the prediction location of the next timestamp. Here, we use CRF to model and calculate the prior probability $\boldsymbol{p}_t^-$ of the user location at time t as follows:

$$P_t^- = \Lambda_u P_{i,j}^u + \Lambda_f P_{t-1}^+ + \Lambda_c P_{i,j}^c (c_j) \tag{7}$$

Among them, $P_{i,j}^u, P_{t-1}^+, P_{i,j}^c (c_j)$ are user preferences, posterior probability of user location and user mobility attributes based on context clustering, respectively. $\Lambda_u, \Lambda_f, \Lambda_c$ represent the weight of user preference probability, posterior probability and context-based access probability, respectively. CRF is responsible for determining these weights based on the user's historical mobile data.

3.3 Local Differential Privacy

Here, we employ local differential privacy [17]. Its formal definition is as follows:

Definition 1 *[17]. Given a randomized algorithm M, Range(A) represents all possible results of algorithm M, and then $\varepsilon \geq 0$, for any two locations $\boldsymbol{l}, \boldsymbol{l}'$ in the region $\mathcal{L}$ and all $\boldsymbol{O} \subseteq$ Range(A), if algorithm M satisfies the following inequality, then algorithm M satisfies ε-Local differential privacy.*

$$\Pr(M(\boldsymbol{l}) \in \boldsymbol{O}) \leqslant e^{\varepsilon} \cdot \Pr(M(\boldsymbol{l}') \in \boldsymbol{O})$$

3.4 Privacy Definition

Similar to [22], in order to avoid the adversary finding the user's real location, we hide the user's real location in a security area, that is, the δ-location set mentioned in the previous section. In other words, we combine LDP and δ-location set to define locally differential privacy, therefore, it is difficult for the adversary to find the real location of the user within the location set. Formally definition is as follows:

Definition 2. *Exists a privacy protection algorithm M, for any two locations l_t, l_t' in the δ-location set, if for arbitrary output $z_t \in \text{Range}(M)$, the following always established*

$$\Pr\left(M\left(l_t\right)=z_t\right) \leqslant e^{\varepsilon_t} \cdot \Pr\left(M\left(l_t^{'}\right)=z_t\right),$$

then the algorithm M satisfies ε-LDP.

In this way, the user's real location is protected in δ-location set, and the perturbed location published in each timestamp satisfies LDP, and supports privacy location sharing.

4 Location Protection Algorithm

4.1 Framework

We firstly show the overall framework of the location protection algorithm, whose pseudocode is shown in Algorithm 1.

Algorithm 1: Framework

Data: $\varepsilon_t, \delta_t, CRF, \mathbf{p}_{t-1}^{+}, l_t^*, s_t^*$
Result: Algorithm $1(\varepsilon_{t+1}, \delta, CRF, \mathbf{p}_t^{+}, l_{t+1}^*, s_{t+1}^*)$

$P_t^- \leftarrow \Lambda_u P_{i,j}^u + \Lambda_f P_{t-1}^{+} + \Lambda_c P_{i,j}^c\left(c_j\right)$;
if *the location information requires to be uploaded* **then**
  Calculate out $\Delta\mathbf{X}_t$ based on P_t^-;
  if $l_t^* \notin \Delta\mathbf{X}_t$ **then**
    put l_t^* in $\Delta\mathbf{X}_t$;
  end
  $z_t \leftarrow$Algorithm $2(\varepsilon_t, \Delta\mathbf{X}_t, l_t^*/s_t^*)$;
  Derive postererior probability P_t^+ by Equation 4 ;
end
return *Algorithm* $1(\varepsilon_{t+1}, \delta, CRF, \mathbf{p}_t^{+}, l_{t+1}^*, s_{t+1}^*)$;
go to next timestamp ;

We explain the overall framework in detail: At the beginning of the algorithm, we firstly calculate the prior probability vector P_t^- of each timestamp t. If the user's current location needs to be shared to the third party server, we construct the corresponding δ-location set according to the calculated P_t^-. There is a possibility of "drift" at this time, adding the user's real location directly to the δ-location set. Then, Algorithm 2 satisfying LDP is applied to generate the perturbed location z_t of the current timestamp, and Algorithm 2 will be shown in the next section. At the same time, the perturbed location z_t is used to update the posterior probability P_t^+, so that in the next time $t+1$, prior probability P_{t+1}^- is updated based on the value of the posterior probability P_t^+, and then the above process is repeated. Note that in each timestamp, user can define ε_t, δ_t according to the current privacy requirements. For convenience, we assume that each timestamp parameter ε_t, δ_t is the same.

4.2 Location Release Algorithm

There are two basic methods to implement LDP, GRR and RAPPOR. In terms of privacy budget, there are some differences in performance between the two perturbation methods [27]: Given privacy budget $\varepsilon = \ln k$, Algorithm 2 uses RAPPOR when privacy budget is low, and Algorithm 2 uses GRR when privacy budget is high. In this paper, we both utilize GRR and RAPPOR. The experimental results show that the dynamic selection of perturbation algorithm according to the size of privacy budget will obtain better utility. Therefore, in this paper, we let $k = |\Delta\mathbf{X}_t|$, if $\varepsilon < \ln k$, we use RAPPOR; if privacy budget $\varepsilon > \ln k$, we use GRR. We provide users with two different location perturbation algorithms, so that users can choose the appropriate location perturbation method according to their current location privacy requirements. In the following section, we firstly describe these two algorithms, and then in the end of this section, we give the pseudocode of Algorithm 2.

Generalized Randomized Response. In the scene of location privacy, the number of locations is more than two, so $d = |D| > 2$, then the location perturbation method is defined as

$$\Pr\left[M_{\mathrm{GRR}}(l^*) = z\right] = \begin{cases} p = \frac{e^{\varepsilon}}{e^{\varepsilon}+d-1}, & \text{if } z = l^* \\ q = \frac{1-p}{d-1} = \frac{1}{e^{\varepsilon}+d-1}, & \text{if } z \neq l^* \end{cases} \tag{8}$$

That is to say, GRR will use probability $\mathrm{e}^{\varepsilon_t} / (\mathrm{e}^{\varepsilon_t} + |\Delta\mathbf{X_t}| - 1)$ to return the user's real location as the generated perturbed location; otherwise, it will use probability $1/(\mathrm{e}^{\varepsilon_t} + |\Delta\mathbf{X_t}| - 1)$ to randomly select a location in $\{\Delta\mathbf{X}_t/l^*\}$ as the generated perturbed location.

RAPPOR. We use state coordinates to represent the user's location, $S = \{s_1, \ldots, s_n\}$ to represent the target area, n to represent the total number of grids in the target area, s_i to represent the grid where the user is currently located, and then use L to represent the user's location, L can be regarded as n-dimensional unit vector whose element at the ith position is 1, the other element is 0. The formal definition is expressed as follows:

$$L_j = \begin{cases} 1, & j = i \\ 0, & \text{otherwise} \end{cases} \tag{9}$$

Here, L_j represents the jth bit of L.

With the user's location coordinates, we can disturb the user's real location. Under the RAPPOR, L becomes the following form by randomized response perturbation:

$$P\left(U_k = x\right) = \begin{cases} 0.5f, & x = 0 \\ 0.5f, & x = 1 \\ 1 - f, & x = L_k \end{cases} \tag{10}$$

The above process is called permanent randomized response. The f on the right side of the equation is a system parameter that can control the degree of privacy, and its value ranges from 0 to 1. It's showed that the closer its value is to 1, the higher the degree of privacy protection of the protection mechanism is, the lower the corresponding data utility is.

After generating a permanent randomized response U, it is the instantaneous randomized response process. RAPPOR adds additional noise to each bit of U, so that it has more randomness. The formal expression is as follows:

$$P(S_k = 1) = \begin{cases} q, & \text{if } U_k = 1 \\ p, & \text{if } U_k = 0 \end{cases} \tag{11}$$

The above equation means that for each bit of the new disturbance information S, if the value of the corresponding bit in the corresponding permanent random response U is 1, then the probability of value of the corresponding bit in S is q. If the value of the corresponding bit in the corresponding permanent random response U is 0, the probability of the value of the corresponding bit in S being 1 is p, where q and p are the system parameters in RAPPOR, and the value range from 0 to 1. Finally, we take the grid location closest to the user's real location in the perturbation location information as the perturbed location. Find the nearest one to the real location in all bits with a value of 1 in the perturbation information S, and then use the value of this bit to 1 and the other bits to zero to generate the perturbed location $\boldsymbol{z}$. Figure 3 shows the system structure of RAPPOR. Also have the following equation [28]:

$$\begin{gathered} \varepsilon = h \log \left(\frac{q^*(1-p^*)}{p^*(1-q^*)} \right) \\ q^* = P(S_i = 1 \mid L_i = 1) = \tfrac{1}{2} f(p+q) + (1-f)q \\ p^* = P(S_i = 1 \mid L_i = 0) = \tfrac{1}{2} f(p+q) + (1-f)p \end{gathered}$$

then the previous random perturbation process satisfies ε-differential privacy.

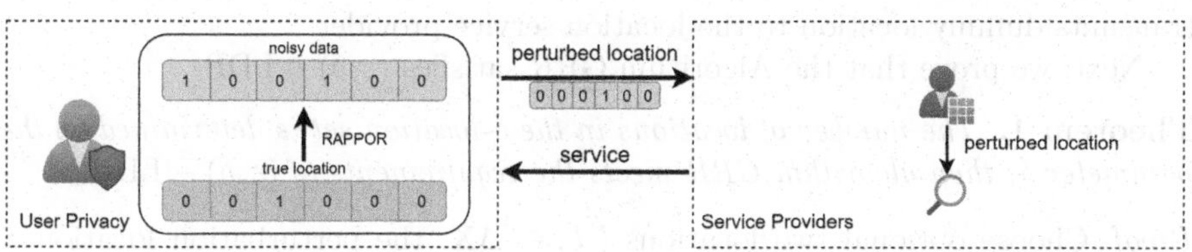

Fig. 3. Architecture of RAPPOR.

The pseudocode of Algorithm 2 is as follows:

The input of Algorithm 2 is the privacy budget ε_t of the current timestamp, δ-location set $\Delta\mathbf{X}_t$ and the real location l_t^*/s_t^* of the user. The output result is the user location $\boldsymbol{z}_t$ released after disturbance.

Line 1 determines the user's privacy requirements for the current location based on the size of the δ-location set generated in Algorithm 1, that is, $k =$

Algorithm 2: Location Release Algorithm

Data: $\varepsilon_t, s_t^*, l_t^*, \Delta\mathbf{X}_t$
Result: z_t
1 $k \leftarrow |\Delta\mathbf{X}_t|$;
2 **if** $\varepsilon_t > \ln k$ **then**
3 $\quad b \leftarrow \text{Bern}\left(e^{\varepsilon_t}/(e^{\varepsilon_t} + |\Delta\mathbf{X}_t| - 1)\right)$;
4 $\quad$ **if** $b = 1$ **then**
5 $\quad\quad z_t \leftarrow l^*$;
6 $\quad$ **else**
7 $\quad\quad z_t \leftarrow$ Uniform $(\{\Delta\mathbf{X}_t/l^*\})$;
8 $\quad$ **end**
9 **else**
10 $\quad$ generate unit vector L by Equation 9;
11 $\quad$ generate permanent random response U by Equation 10;
12 $\quad$ generate instantaneous random response S by Equation 11;
13 $\quad$ select the closest location from s^* in S as z_t ;
14 **end**
15 **return** z_t ;

$|\Delta\mathbf{X}_t|$, then select the appropriate location perturbation algorithm. Lines 2–8, if $\varepsilon > \ln k$, the real location of the user is perturbed by the GRR method, the GRR takes the probability $e^{\varepsilon_t}/(e^{\varepsilon_t} + |\Delta\mathbf{X}_t| - 1)$ to return the real location of the user as the generated perturbed location, and the generated perturbed location is in δ-location set. Otherwise, GRR will randomly select a location in $\{\Delta\mathbf{X}_t/l^*\}$ as the generated perturbed location. In lines 10–13, if $\varepsilon < \ln k$, the RAPPOR method is selected as the location disturbance algorithm. Firstly, the unit vector L is generated according to the real location of the user, and then the permanent random response U is generated. Finally, the instantaneous random response S is generated, and the grid location closest to the real location of the user in the disturbance location information S is taken as the perturbed location. Line 15 transmits dummy location to the location service provider.

Next, we prove that the Algorithm GRR satisfies $(\varepsilon, \delta) - \text{LDP}$.

Theorem 1. *The number of locations in the δ-location set is determined by the parameter δ, then algorithm GRR meets the requirements of* $(\varepsilon, \delta) - \text{LDP}$.

Proof. Choose optional two locations $l, l^{'} \in \Delta\mathbf{X}$, the perturbation location z published by GRR, we can obtain the following inequality

$$\frac{\Pr\left[M_{\text{GRR}}(\boldsymbol{l}) = \boldsymbol{z}\right]}{\Pr\left[M_{\text{GRR}}(\boldsymbol{l'}) = \boldsymbol{z}\right]} \leqslant \frac{p}{q} = \frac{\frac{e^{\varepsilon}}{(e^{\varepsilon}+|\Delta\mathbf{X}|-1)}}{\frac{1}{(e^{\varepsilon}+|\Delta\mathbf{X}|-1)}} = e^{\varepsilon}$$

Therefore, GRR satisfies $(\varepsilon, \delta) - \text{LDP}$. □

Finally, because algorithm RAPPOR satisfies $\varepsilon - \text{LDP}$, so Algorithm 2 satisfies $\varepsilon - \text{LDP}$, so continuous location sharing Algorithm 1 meets the requirement of $(\varepsilon, \delta) - \text{LDP}$.

5 Experimental Evaluation

We will conduct an experimental evaluation of our proposed location privacy protection mechanism in this section. We implement the experiments on a desktop with Windows 10 operating system, 2.90 GHz AMD Ryzen 7 4800H CPU and 16-GB RAM. All of algorithms are implemented by Python. In the experiment, we take PIM mechanism [16] as a benchmark to compare its performance with our proposed mechanism.

5.1 Experimental Setup

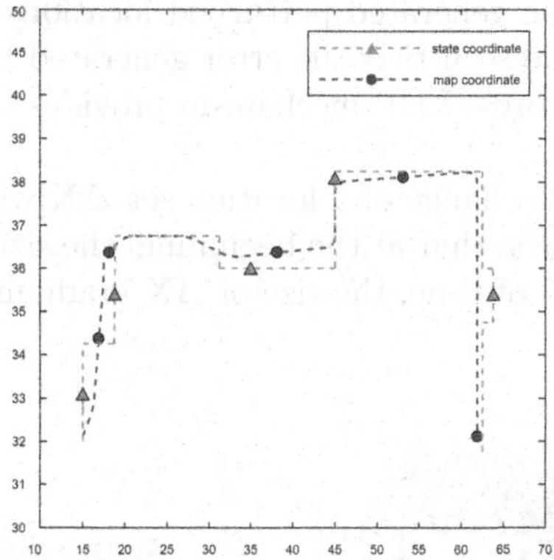

Fig. 4. Real trace.

Dataset. We use Geolife [29] as the experimental data set, which is a GPS trajectory data set of 182 users collected by Microsoft Asia Research Institute within five years. The user trajectory in this dataset is a sequence of points based on timestamp. Each point in the sequence includes latitude, longitude and altitude of the user's current location. The location points in each sequence are divided by time or distance, such as every 1–5 s or 5–10 m record a location point. In this dataset, we select all user trajectories near the Beijing Third Ring to generate CRF model, and then divide target region into grids of size $0.34 \times 0.34\ \text{km}^2$.

Metrics. We mainly use two related indicators to evaluate the performance of GRR, RAPPOR and PIM. First of all, the privacy budget ε and δ-location set used in each timestamp of the user trajectory are closely related to our privacy definition. So we evaluate the size of δ-location set $\Delta\mathbf{X}$ to observe how $\Delta\mathbf{X}$ changes over time. Then, We regard the distance between real and dummy locations as utility factor, reflecting the error of the mechanism.

5.2 Performance over Time

Concretely, we randomly select a user trajectory from the Geolife data set, which is composed of 500 timestamps. We assume $\varepsilon = 1$ and $\delta = 0.01$ on each timestamp, and then compare PIM with our mechanism. The state and map coordinates of the real location we selected are shown in Fig. 4, map coordinates are represented by latitude and longitude, and state coordinates are represented by grids divided by the target area, so there are some differences. We run two mechanisms on this trajectory 10 times, and then take their average. (a) and (b) in Fig. 5 show the location points of each timestamp after disturbed through PIM and LDP (for convenience, the mechanism in this paper is referred to as LDP). It is obvious that the location points disturbed by LDP mechanism is closer to its real location points. This is more obvious in (d), where denotes the distance between the generated perturbed locations and its real locations. By comparison, it can be seen that the error generated by LDP is smaller than that of PIM. In other words, LDP mechanism provides better utility than PIM mechanism.

Figure 5(c) shows the change of δ-location set $\Delta\mathbf{X}$ with timestamp for both mechanisms. It can be seen that at the beginning, the size of $\Delta\mathbf{X}$ varies dramatically, but after a period of time, the size of $\Delta\mathbf{X}$ gradually becomes stable. The

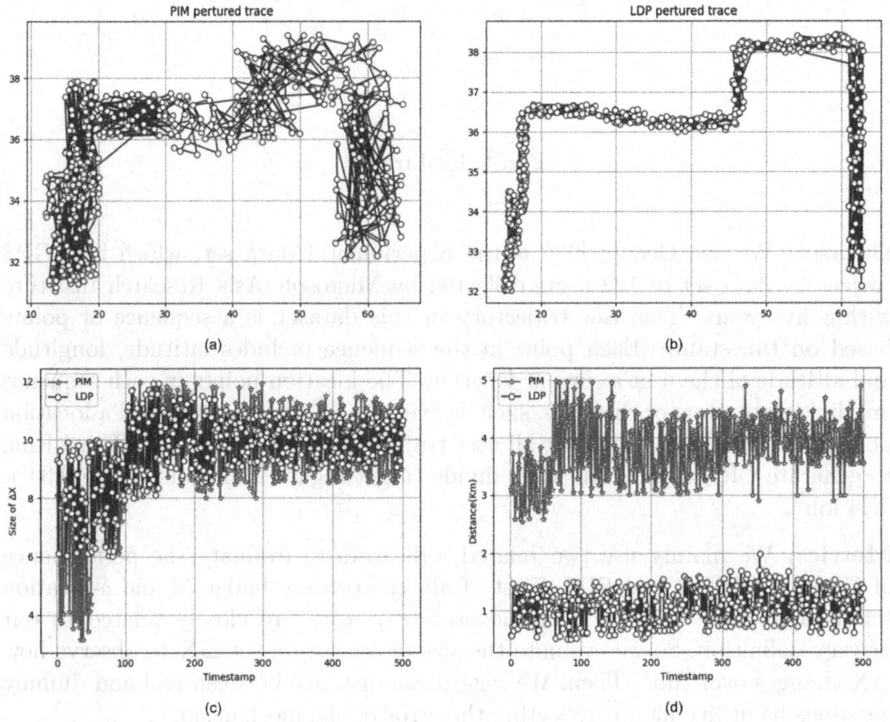

Fig. 5. Performance: (a) (b) Released traces, (c) Size of $\Delta\mathbf{X}$, (d) Distance over time.

reasoning mechanism determines the prior probability distribution at the current time point, the selection of δ-location set is related to the reasoning mechanism. Each choice of δ-location set will increase the probability of locations in $\Delta\mathbf{X}$, and also reduces the probability of other locations. Over time, $\Delta\mathbf{X}$ will become more stable. Finally, we can find that the size of $\Delta\mathbf{X}$ generated by two mechanisms is very close, indicating that the effect of parameter δ on the two different mechanisms is almost the same.

6 Conclusion

In this paper, we study the continuous location sharing mechanism for movable users. Firstly, we use the conditional random field to model the time correlation among multiple locations. Then, this paper propose a new location privacy protection mechanism by combining δ-location set with local differential privacy. This mechanism combines GRR and RAPPOR, allowing users to dynamically select required mechanisms according to their current location privacy requirements. Finally, we design a continuous location sharing framework based on the new mechanism and demonstrate the performance of new framework on real data set. The results show that our mechanism is more effective than PIM.

Acknowledgement. This work is supported by the National Natural Science Foundation of China (Grant Nos. 61602408, 61972352, 61572435), the Key Research and Development Program of Zhejiang Province (Grant No. 2021C03150) and Zhejiang Provincial Natural Science Foundation of China under Grant (Nos. LY19F020005, LY18F020009).

References

1. Krumm, J.: A markov model for driver turn prediction. In: Society of Automotive Engineers (SAE) 2008 World Congress, April 2008. SAE 2008 World Congress, April 2008. https://www.microsoft.com/en-us/research/publication/markov-model-driver-turn-prediction/, lloyd L. Withrow Distinguished Speaker Award
2. Götz, M., Nath, S., Gehrke, J.: Maskit: privately releasing user context streams for personalized mobile applications. In: Proceedings of the 2012 ACM SIGMOD International Conference on Management of Data, pp. 289–300 (2012)
3. Alvim, M.S., Chatzikokolakis, K., McIver, A., Morgan, C., Palamidessi, C., Smith, G.: Differential privacy. Presented at the (2020). https://doi.org/10.1007/978-3-319-96131-6_23
4. Sun, G., Song, L., Liao, D., Yu, H., Chang, V.: Towards privacy preservation for "check-in" services in location-based social networks. Inf. Sci. **481**, 616–634 (2019)
5. Peng, T., Liu, Q., Meng, D., Wang, G.: Collaborative trajectory privacy preserving scheme in location-based services. Inf. Sci. **387**, 165–179 (2017)
6. Location Privacy in Mobile Applications. SCSSN. Springer, Singapore (2018). https://doi.org/10.1007/978-981-13-1705-7_6

7. Freudiger, J., Shokri, R., Hubaux, J.-P.: On the optimal placement of mix zones. In: Goldberg, I., Atallah, M.J. (eds.) PETS 2009. LNCS, vol. 5672, pp. 216–234. Springer, Heidelberg (2009). https://doi.org/10.1007/978-3-642-03168-7_13
8. Gedik, B., Liu, L.: Protecting location privacy with personalized k-anonymity: architecture and algorithms. IEEE Trans. Mob. Comput. **7**(1), 1–18 (2007)
9. Wu, X., Li, S., Yang, J., Dou, W.: A cost sharing mechanism for location privacy preservation in big trajectory data. In: 2017 IEEE International Conference on Communications (ICC), pp. 1–6. IEEE (2017)
10. Andrés, M.E., Bordenabe, N.E., Chatzikokolakis, K., Palamidessi, C.: Geo-indistinguishability: differential privacy for location-based systems. In: Proceedings of the 2013 ACM SIGSAC Conference on Computer & Communications security, pp. 901–914 (2013)
11. Wang, T., Cao, Z., Wang, S., Wang, J., Qi, L., Liu, A., Xie, M., Li, X.: Privacy-enhanced data collection based on deep learning for internet of vehicles. IEEE Trans. Industr. Inf. **16**(10), 6663–6672 (2019)
12. Wang, T., Jia, W., Xing, G., Li, M.: Exploiting statistical mobility models for efficient wi-fi deployment. IEEE Trans. Veh. Technol. **62**(1), 360–373 (2012)
13. Chen, M., Wang, T., Ota, K., Dong, M., Zhao, M., Liu, A.: Intelligent resource allocation management for vehicles network: an a3c learning approach. Comput. Commun. **151**, 485–494 (2020)
14. Xie, M., Ruan, Y., Hong, H., Shao, J.: A cp-abe scheme based on multi-authority in hybrid clouds for mobile devices. Futur. Gener. Comput. Syst. **121**, 114–122 (2021)
15. Ardagna, C.A., Livraga, G., Samarati, P.: Protecting privacy of user information in continuous location-based services. In: 2012 IEEE 15th International Conference on Computational Science and Engineering, pp. 162–169. IEEE (2012)
16. Xiao, Y., Xiong, L.: Protecting locations with differential privacy under temporal correlations. In: Proceedings of the 22nd ACM SIGSAC Conference on Computer and Communications Security, pp. 1298–1309 (2015)
17. Duchi, J.C., Jordan, M.I., Wainwright, M.J.: Local privacy and statistical minimax rates. In: 2013 IEEE 54th Annual Symposium on Foundations of Computer Science, pp. 429–438. IEEE (2013)
18. Warner, S.L.: Randomized response: a survey technique for eliminating evasive answer bias. J. Am. Stat. Assoc. **60**(309), 63–69 (1965)
19. Kairouz, P., Oh, S., Viswanath, P.: Extremal mechanisms for local differential privacy. Adv. Neural. Inf. Process. Syst. **27**, 2879–2887 (2014)
20. Wang, T., Blocki, J., Li, N., Jha, S.: Locally differentially private protocols for frequency estimation. In: 26th {USENIX} Security Symposium ({USENIX} Security 17), pp. 729–745 (2017)
21. Wang, T., Li, N., Jha, S.: Locally differentially private frequent itemset mining. In: 2018 IEEE Symposium on Security and Privacy (SP), pp. 127–143. IEEE (2018)
22. Chen, R., Li, H., Qin, A.K., Kasiviswanathan, S.P., Jin, H.: Private spatial data aggregation in the local setting. In: 2016 IEEE 32nd International Conference on Data Engineering (ICDE), pp. 289–300. IEEE (2016)
23. Sei, Y., Ohsuga, A.: Differential private data collection and analysis based on randomized multiple dummies for untrusted mobile crowdsensing. IEEE Trans. Inf. Forensics Secur. **12**(4), 926–939 (2016)
24. McCallum, A.: Efficiently inducing features of conditional random fields. arXiv preprint arXiv:1212.2504 (2012)
25. Song, C., Qu, Z., Blumm, N., Barabási, A.L.: Limits of predictability in human mobility. Science **327**(5968), 1018–1021 (2010)

26. Assam, R., Seidl, T.: Context-based location clustering and prediction using conditional random fields. In: Proceedings of the 13th International Conference on Mobile and Ubiquitous Multimedia, pp. 1–10 (2014)
27. Kairouz, P., Bonawitz, K., Ramage, D.: Discrete distribution estimation under local privacy. In: International Conference on Machine Learning, pp. 2436–2444. PMLR (2016)
28. Erlingsson, Ú., Pihur, V., Korolova, A.: Rappor: randomized aggregatable privacy-preserving ordinal response. In: Proceedings of the 2014 ACM SIGSAC Conference on Computer and Communications Security, pp. 1054–1067 (2014)
29. Zheng, Y., Xie, X., Ma, W.Y., et al.: Geolife: A collaborative social networking service among user, location and trajectory. IEEE Data Eng. Bull. **33**(2), 32–39 (2010)

An Intelligent Allocation Mechanism Based on Ethereum Blockchain in Microgrid

Yingming Zeng[1,2], Liyu Deng[1], and Haibin Zhang[1](✉)

[1] School of Cyber Engineering, Xidian University, Xi'an 710071, Shaanxi, China
hbzhang@mail.xidian.edu.cn

[2] Beijing Institute of Computer Technology and Applications, Beijing 100854, China

Abstract. The data security and the efficiency of energy scheduling are two main challenges for the practical application of microgrid. The existing researches have disadvantages of a single system component, unreasonable scheduling, and the lack of analysis on the main grid. In light of this, we formalize an efficient microgrid system based on Ethereum blockchain. We first use the blockchain network to upload the IoT data and predict the total load in some region and the energy generation of renewable energy, and then complete the deployment of energy based on the predicted results under the restrictions of relevant resources. In particular, we introduce a credit bidding mechanism that is based on Etheruem smart contracts to optimize energy allocation. It maximizes the proportional-fairness participation of all parties and avoids the waste of energy. In addition, we adopt group signature to preserve the privacy of user data. Simulation results show that the proposed scheme can significantly reduce users' cost, increase the profit rate, enforce proportional fairness, and improve the operation stability of the main grid.

Keywords: Microgrid · Blockchain · Intelligent scheduling · Credit bidding · Group signature

1 Introduction

In recent years, the global energy demand has been growing exponentially, which also exacerbates environmental pollution. Thus, it is of great necessity to solve the problem of how to optimize the use of energy, minimize production costs [1]. Smart energy grid is produced to make the energy system more efficient by reducing the peakto-average load, minimizing the production cost and incorporating Renewable Energy Sources (RES). Specifically, microgrid is a small self-reliant electricity grid that produces and distributes energy in a limited area, such as

Supported by the National Natural Science Foundation of China under Grant 61771373.

Y. Lai et al. (Eds.): ICA3PP 2021, LNCS 13155, pp. 726–740, 2022.
https://doi.org/10.1007/978-3-030-95384-3_45

a village or industrial area. Besides, microgrid can also transfer electric power with a larger traditional grid without outside electrical connection [2].

Intelligent allocation of microgrid is to make reasonable control strategy for smart equipment of microgrid based on predicted results, which includes peak clipping, valley filling, load shifting, strategic load growth, strategic conservation, and flexible load shape [3]. Tushar [4] proposed a real-time decentralized demand-side management (RDCDSM) system, which adjusted real-time housing loads according to the predicted total load to track the future planned energy generation. Chaouachi [5] used fuzzy logic expert system for battery scheduling, and AI method to reduce pollutant emissions while increasing economic costs. Hittinger [2] proposed a high-resolution Energy System Model (ESM), which adds several important aspects in battery modeling including temperature effects, rate-based variable efficiency, and operational modeling of capacity fade to optimize the system design and reduce the electricity cost.

However, the existing intelligent scheduling systems are lack of considering multiple RES and batteries schedule, and don't analyze the impact that microgrid users connect to the main grid, and it is unacceptable to reduce the cost by controlling users' electricity habit, which will affect users' electricity experience. Tushar [6] applied a mixed integer linear programming method (MILP) to solve the electricity consumption scheduling problem. However, to make MILP works efficiently, the batteries and RES constraints must be formulated as linear formulation, which is far away from reality facts. A. Rabiee [7] proposed a two-stage optimization model to optimize scheduling of microgrid, which considering of electrical vehicles and responsive loads of wind and PV units uncertainties. Its limitation is that the charging/discharging schedule of the vehicles is determined based on the mean load. L. Ju [8] introduced multi-agent system (MAS) for intelligent scheduling model which considering different operation strategies of energy storage device and traditional RES, but lacked of experiments to show how to achieve control coordination between microgrids and the main power grid.Besides, it is also wise to analyze the impact that microgrid users connect to the main grid, because the main grid supplier disagree to things that are only beneficial to the users.

In addition, the large amount of energy data are the basis for the efficient operation of the microgrid system, so it is of great importance to ensure the data security, especially the sensors frequently lose either user's privacy or data integrity [9]. The traditional centralized energy management system relies on the trusted third party to ensure system security, which can cause additional costs and new problems. In terms of data security, the traditional centralized energy management system relies on the trusted third party to ensure system security, which can cause additional costs and new problems [10].

To solve these problems, we propose an intelligent allocation mechanism based on blockchain, which allows programs to execute in a distributed management manner. Blockchain is a distributed database that can receive any information, including records, events, transactions, etc. It updates the data information by specific rules, uses the cryptographic hash function to calculate the block header, and derives a string of numbers uniquely identified by each block. At the same time, each block contains the hash value of previous block at its head, and its hash value is also affected by the feature of the parent block's hash value.

In addition, encryption technique is widely used in blockchain network, which provides authority for all interactions in the network. The smart contract [6] is stored in the blockchain digitally, which means that the storage, reading and execution processes are transparent, traceable and tamper-proof by using the blockchain. All of these advantages make the blockchain an ideal functional module in the IoT layout. Based on these characteristics of blockchain technology, there are many studies attempt to integrate blockchain technology into microgrid [11]. However, this powerful technology still isn't adopted in distributed allocation platform to realize the peaking and filling of the power curve of the regional main grid and balanced configuration of various power sources. Thus, in this paper, the proposed system is built on the blockchain technique and the corresponding cryptography security mechanism. The main contributions of this paper are summarized as follows

- We formalize a microgrid model based on the blockchain, which breaks the trust barrier among different enterprise users, improves the operation efficiency and stability of the whole system. Except for the environmental benefit, the system significantly reduces the cost of each participant. All agents of blockchain store dataset copies of microgrid and communicate with each other by smart contract, whose updates including real-time power generation, historical change records of agents and power demand prediction for the next day.
- Introducing a credit exchange principle for impelling Distributed Energy Resources (DERs) to fairly participate in voltage regulation. In the scheduling phase, we divide the DERs into two subsets A and B with A participating in the scheduling phase and B not participating. When joining B, DERs can ask for a credit, which will be paid by the DERs in A operating at their full capacity. In turn, the credit decrease of the DERs that have not participated in B ultimately forces them to join B in future. The credit statuses of all DERs are tracked by using a blockchain protocol.
- Temperature, wind intensity and light intensity are token into account in the power prediction. Besides, we present a heuristic algorithm for intelligent allocation to reduces users' cost compared to traditional power system, and make the main grid achieve higher profit rate and more stable.
- Using the group signature algorithm to ensures the privacy of users' data and preserves the tracking ability (abnormal detection) of the system.

Problem statement is presented in Sect. 2. Section 3 shows our system model. Problem formulation and implementation are shown in Sect. 4. Section 5 discusses the experimental results. And Sect. 6 concludes the paper and suggests the direction for future work.

2 Problem Statement

The operation process of the basic model is to obtain the users' electricity data, predict the electricity demand in the coming days with certain prediction algorithm, and then allocate the energy resources (including RES and traditional energy). However, the existing microgrid researches rarely analyze the energy storage equipment allocation, RES management and the impact on the main grid (including power load and economic benefits), which makes it impossible to optimize regional power resources to the utmost. Besides, due to the lack of an efficient and credible platform, the microgrid cannot really smooth the power load of the main grid (achieve by "shaving the peaks and filling the valley").

To these existing problems of the traditional microgrid, we formalize a smart microgrid system based on the blockchain platform in this paper.

First, we build a *management hub* node on the traditional IoT, which is responsible for format conversion of sensor data in a certain region (transforming data of different communication protocols into JSONRPC messages that can be identified by blockchain nodes). Then, we predict users' demand and the energy generation capacity of the enterprises for electricity in the next 24 h by using recorded historical data, which is called the prediction phase. Besides, meteorological factors are taken into account in this phase, and present a heuristic algorithm for intelligent allocation. Finally, based on the predicted results and the characteristics of resources, we distribute the energy generation and energy storage equipment in different time periods, which is called the scheduling phase. Specifically, we introduce a principle based on credit exchange to inspire the Distributed Energy Resources (DERs) for fairly participating in voltage regulation. In the proposed framework, the DERs installed on the same distribution feeder stipulate a smart contract, determine which units will act as voltage regulator over the control periods based on their available credit statuses and the adopted economic strategy.

3 System Model

Figure 1 shows the formalized system which composed of the hardware such as RES, i.e., wind turbines and photovoltaics (PV) generators, energy storage equipments, and the main grid.

3.1 Mathematical Model of System Components

Renewable Energy. In the proposed model, the system is allowed to start and stop wind turbine and PV. Equation (1) describes the power generation model

$$\begin{cases} p_{wd}^t = X_{wd}^t \cdot P_{wd_{on}}^t \\ p_{pv}^t = X_{pv}^t \cdot P_{pv_{on}}^t \\ X_{wd}^t, X_{pv}^t \in \{0,1\} \end{cases} \tag{1}$$

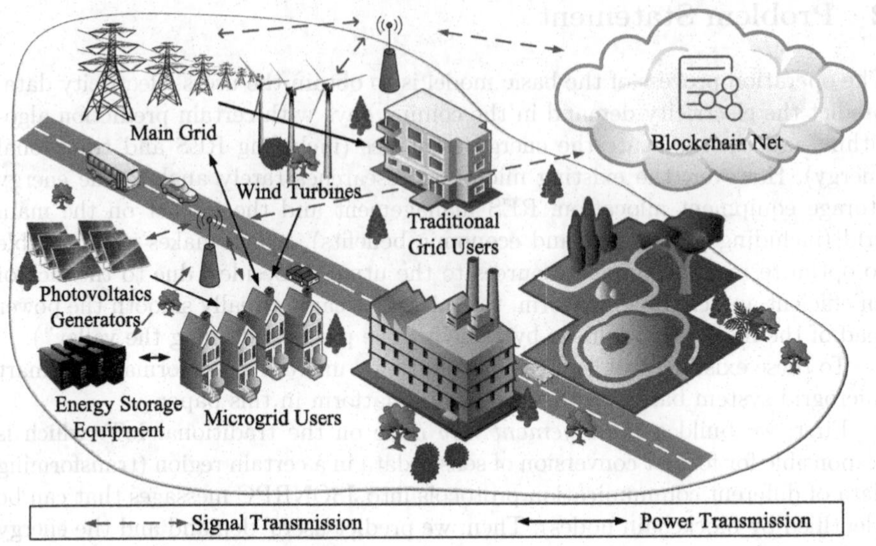

Fig. 1. Application scenario of smart microgrid.

where $P^t_{wd_{on}}$, $P^t_{pv_{on}}$ denote the on-state wind energy, the on-state PV energy. $X^t_{wd}$and X^t_{pv} represent the wind start-stop factor and the PV start-stop factor in which 1 means on and 0 means off.

Energy Storage Equipments. We assume the charging and discharging power of the battery are constant. The state of charge (SOC) $S^{t+1} = S^t + P^t_{bt} \cdot \Delta t/E_b$ denotes the ratio of the remaining battery energy to the battery capacity during the period $\Delta t + 1$. We consider that the life of the energy storage equipment is related to the charge and discharge power and the SOC. Moreover, the energy state of the battery should be equal at the beginning and end of the scheduling. So the storage equipment needs to meet the constraints described by Eq. (2)–(4).

$$S_{min} \leqslant S^t \leqslant S_{max} \tag{2}$$

$$0 \leqslant |P^t_{bt}| \leqslant 0.2E_b \tag{3}$$

$$S^0 = S^N \tag{4}$$

where S_{min}, S_{max}, E_b, P^t_{bt} denote the minimum SOC value, the maximum SOC value, the battery capacity, the charging or discharging power of the battery at time t. Equation (3) limits the maximum charge or discharge power of the equipment to 20% of its capacity per time interval.

Power Exchange with Main Grid and Power Conservation. Equation (5) describes the constraint of power conservation in the microgrid system.

$$P^t_{load} + P^t_{bt} = P^t_{wd} + P^t_{pv} + P^t_{gd} \tag{5}$$

where P^t_{load} ,P^t_{wd}, P^t_{pv} denotes the load power of the user at time t, the power generated by the wind turbines at time t, the PV power at time t, P^t_{bt} denotes the charge or discharge power of the battery at time t with the positive indicating that battery is charging and negative indicating discharging. $P^t_{gd} \in [P_{gmin}, P_{gmax}]$ denotes the exchange power between the microgrid and the main grid at time t, in which the positive value means the microgrid is purchasing the electricity from the main grid and the negative means selling.

3.2 The Protocol of Blockchain Model

We formalize an efficient distributed management model using smart contracts (a smart contract is a computer program that can be executed by a blockchain agent) and the group signature algorithm [12].

In the blockchain, all agents store identical copies of the database in a list which contained blocks. Each block stores a set of smart contract updates including real-time power generation, credit score and historical change records of the nodes, as well as the power demand data for the next cycle in the prediction phase. In addition, to ensure the energy data is reliable, each agent can carry out the process of outlier correction before data transmission [13]. The new state is communicated through a smart contract update. To avoid the proliferation of different contract states, the blockchain in parallel executes the update verification process via block generation. Besides, it ensures that only verified updates are included in the blockchain. The detailed process is described as follows:

1. Every agent is provided with a local copy of the blockchain and a buffer named 'mempool'. When smart contract updates are produced, they will be sent to the neighbors.
2. In parallel, agents also generate blocks, i.e., working to solve the computational puzzle by the blockchain protocol and obtain the Proof of Work (POW). When an agent obtains POW and generates a new block, it is filled with the smart contract updates present in its mempool and transmits the new block to other agents.
3. Upon the reception of the block, a neighbouring agent verifies that it contains a valid POW and updates by checking the provided solution of the puzzle. If the verification succeeds, the agent adds the block to its blockchain, and propagates it further to its neighbors. If fails, the validation results continue to be propagated to other agents.
4. When all agents have received and verified the block, they have the same updated version of the blockchain. The agents also remove from their mempools the smart contract updates included in the received and verified blocks.

Figure 2 gives the architecture of the proposed system.

Wireless Sensor Network. Wireless sensor network (WSN) mainly contains smart meters, energy storage devices (equipped with communication modules), communicable energy monitoring sensors(fixed in wind or PV energy generation

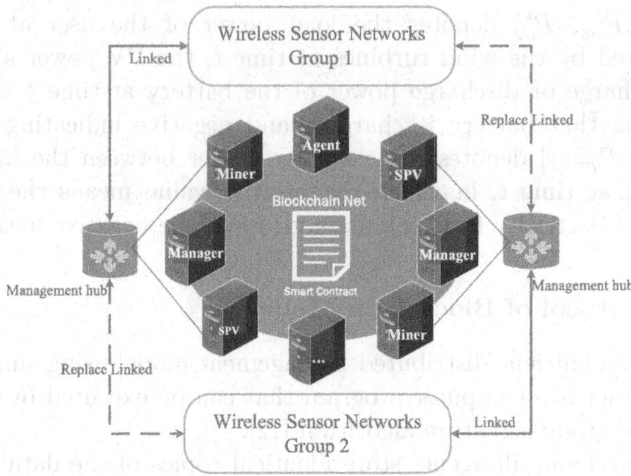

Fig. 2. The architecture of the proposed blockchain plat operation.

equipments) and other IoT terminals (to monitor temperature or wind speed). According to previous work on WSN's vulnerability [14], once the worm attack spreads in WSN, the whole network possibly collapses. Thus we should take the security of WSN into consideration. In some way, put the sensor data into blockchain is a good idea. Most block information in blockchain will increase as time elapsing. Thus, the IoT terminals and the *management hub* are not nodes in the blockchain network because of their memory and computing power limit. In addition, IoT terminals and the *management hub* use CoAP [15] as communication protocol, which means every terminal can communicate with multiple *management hubs* to implement view replacement in group signature.

Management Hub. The duty of a management hub node is to format conversion of sensor data (such as smart meters) in a certain region. Firstly, it links the terminal and the blockchain node (such as *miner*). Besides, it transforms the data of IoT terminals from CoAP coding to JSONRPC messages that can be recognized by blockchain nodes. Secondly, it serves as a group management node to achieve the privacy protection, anomaly detection and processes the data of users and enterprises within the microgrid. Finally, the *Management hub* is responsible for verifying the relevant permissions of the IoT terminals including uploading and querying data.

Manager. A *manager* is a node which can interact with smart contract to implement the registration of terminal and set the relevant access control permission in the blockchain network. In details, the *manager* informs several *management hubs* of the permission and identity plate of a new added IoT terminal, it will also inform the IoT terminal of *management hubs*' location.

Traditional Blockchain Node. Miner and Simplified Payment Verification (SPV) are traditional nodes in the blockchain network. The miner packs the relevant data (e.g. the credit score recorded in the smart contract, the corresponding historical records and power-related data)within a period into a block. The SPV node can provide agents with the query of the data stored in the blockchain. For example, the SPV node can find the transaction records when the agent uses the others energy storage equipment to complete a full load charge and discharge once a day.

Differenced with the traditional blockchain application, we have made different innovations in model structure, consensus and encryption technique. Firstly, *management hub* node can avoid a large number of IoT terminals connecting directly to the blockchain network, which can reduce the pressure of terminal components upgrade. On the other hand, sub-networks based on different protocols are allowed to access the blockchain network, which greatly improves the scalability and compatibility of the model. Secondly, to balance the transparent transmission requirements of blockchain network and the privacy of user data, we use the group signature mechanism in the *management hub* node to confuse the correspondence between data and users. In addition, the *management hub* node keeps key pairs of group members. In this way, the system can trace the source of the problem terminal when a terminal fails.

4 Problem Formulation and Implementation

This section mainly introduces the process of intelligent scheduling, i.e., power prediction and intelligent allocation, and the core content of the blockchain framework in the form of smart contracts, i.e., the credit mechanism. Credit bidding is proposed to meet the demand for electricity and maximizing the interests of enterprises by better realizing the fair-proportionate participation of various power enterprises.

4.1 Power Prediciton

Considering the influence of meteorological factors, we can predict power data for the upcoming day using the model expressed as

$$\hat{y}^t = f_\theta(y^{t-1}, y^{t-2}..., y^{t-N}; M^{t-1}, M^{t-2}..., M^{t-L}) \tag{6}$$

where y^{t-i} represents the real value of power at time $t-i$, $\hat{y}^t$ is the predicted value at time t, M^{t-i} denotes the real meteorological data at time $t-i$, f_θ represents the prediction function with parameter θ which can be estimated by historical load power, wind turbine power, and PV power data.

This paper uses an improved radial basis function neural network-based model with an Error feedback scheme (IRBFNN-EF, which is poposed in [16]) for forecasting short-term power.

4.2 Intelligent Allocation

Base on prediction results, we can obtain a set of operation decision sequence $D = \{P^t_{wd}, P^t_{pv}, P^t_{bt}\}$ of relevant energy equipments to enable users to minimize the cost of electricity in the next day.

The total cost of microgrid can be calculated by $C = C_{reb} + C_{stor} + C_{gd}$. In addition, the RES cost can be obtained by

$$C_{reb} = \sum_{t=1}^{N} [U_{wd} \cdot P^t_{wd} + U_{pv} \cdot P^t_{pv}] \cdot \Delta t \tag{7}$$

where U_{wd} and U_{pv} denote the unit prices of wind and PV energy generation. Energy storage cost can be calculated by

$$C_{stor} = U_{bt} \sum_{t=1}^{N} |P^t_{bt}| \cdot \Delta t \tag{8}$$

where U_{bt} denotes the unit price of energy storage cost.

The energy exchange cost between the microgrid and the main grid can be obtained by

$$C_{gd} = \sum_{t=1}^{N} C^t_{gd} = \begin{cases} \sum_{t=1}^{N} U^t_{pgd} \cdot |P^t_{bt}| \cdot \Delta t, P^t_{bt} \geqslant 0 \\ \sum_{t=1}^{N} -U^t_{sgd} \cdot |P^t_{bt}| \cdot \Delta t, P^t_{bt} < 0 \end{cases} \tag{9}$$

where C^t_{gd} represents the electricity exchange cost between the microgrid and the main grid at period t, U^t_{pgd} (U^t_{sgd}) is purchasing (selling) price from the main grid at period t.

The microgrid economic scheduling problem can be expressed as the multi-stage decision optimization problem (MSDOP) shown by Eq. (15)

$$\min_{D} C' = C + \beta \cdot |S^N - S^0| \tag{10}$$

which aims at minimizing the cost C' of electricity for users in the upcoming day, where $\beta \cdot |S^N - S^0|$ represents the punishment for the difference of the initial and final state for the battery SOC. We use Simulated Annealing (SA) to deal with this NP-Hard Problem.

4.3 Credit Bidding Strategy

Growing complexity of systems and heterogeneous networking will enlarge the destructive effect of compromised or malicious sensor nodes [17]. In this paper, we introduce credit bidding mechanism to balance the participants' economy benefits during scheduling. All nodes participate in intelligent scheduling according to the bidding results. The detailed process of bidding results generation is describes as follows:

- First, in each round of power allocation, each DER node will be distributed the same credit score at the beginning of a day cycle for election participation.
- Then, the campaign score is encrypted and transmitted to other nodes. When all nodes receive this information, the key is sent to other nodes to ensure that all nodes have the same bidding table. All nodes select the node with the highest payment credit score according to the bidding, and record the credit value changes of relevant nodes (e.g. payment nodes and other nodes not selected due to low bidding score) in the local mempool buffer.
- Ultimately, all nodes have the same updated credit table. In addition, the node that calculates the new block first gets a certain credit score reward, and the score of the nodes which are verified to be invalid will be deducted.

A feasible bidding rule is given below:

There are a total of M companies participating in the auction within a bidding period T. At the beginning of the distribution, each service provider is assigned an initial credit score of $Cred_i(0)$, which indicates the reputation of the i-th company at time 0). In the t period, based on the results of the power forecast, they will calculate the proportion of the profit that can be obtained in the $t+1$ period to the total profit in all future periods, which will also be the proportion of the credit points they have to pay when bidding in the t period. Expressed as

$$Pay_i(t) = Cred_i(t)\frac{U(t+1)\tilde{P}(t+1)}{\sum_{j=t+1}^{T} U(j)\tilde{P}(j)} \tag{11}$$

where $U(t)$ represents the unit price at time t, and $\tilde{P}$ represents the power prediction value, which may be wind power or photoelectric power. 1 Then, the computer program will randomly bid according to the reputation scored by each company. The selected probability is proportional to the reputation they voted, which is expressed as

$$Prob_i(t) = \frac{Pay_i(t)}{\sum_{i=1}^{M} Pay_i(t)} \tag{12}$$

where $Prob_i(t)$ represents the probability that company i is selected at time t. The credit points paid of the successfully bidder for this round will be deducted while the credit points paid by other unsuccessfully bidder will be returned. By deducting more credits during the period of more profit, this rule avoids the phenomenon that individual companies obtain higher profit.

5 Experimental Results

We conduct experiments on real data from the relevant historical data of a certain region of China. Results are obtained by a computer with Intel Core i7 and an ubuntu 16.04 version. The blockchain software installed in the DERs controller supports a private Ethereum blockchain [18], which provides the possibility of writing complex smart contracts, and it is simulated using EthereumJS testrpc.

Table 1. Electricity purchase prices with different times

Time	0:00∼8:00	8:00∼12:00	12:00∼14:00	14:00∼18:00	18:00∼24:00
Selling Price($/kWh)	0.032	0.095	0.062	0.095	0.062
Purchase Price($/kWh)	0.037	0.120	0.078	0.120	0.078

Table 2. Parameters of the intelligent scheduling system

Main grid	RES		Battery			
Power limit	WP price	PV price	Storage price	Capacity	SOC range	Powerlimit
[−180 kW,180 kW]	0.077 $/kWh	0.110 $/kWh	0.015 $/kWh	600 kWh	[0.3,0.95]	[−120 kW,120 kW]

5.1 Performance Evaluation of Intelligent Allocation

We carry out the intelligent allocation of microgrid based on the predicted results by regulating the time-phase pricing scheme of the main grid and various parameters of the system which are shown in Table 1 and Table 2. The experimental results are shown in Fig. 3.

Figure 3(a) shows the power load curves of the main grid using respectively the traditional way and proposed model. The 'Intelligent microgrid (Predicted)' label represents the power load curve of the main grid derived from the call scheduling algorithm based on the prediction results. The 'Intelligent microgrid (Actual)' label represents the actual load curve of the main grid for the second day. We can clearly see the "peak load shifting" effect of the scheduling system on the traditional load curve.

Figure 3(b) and Fig. 3(c) show the decision results of the scheduling. We can see that RES generation equipment is turned on when the cost of the grid is higher than the energy cost of RES, otherwise it will be shut down. Moreover, we can see that the battery is charged during the period when the electricity price is low, and it is discharged when electricity price is high. After further observation, we can find that only a small amount of electric energy is released during the peak period from 8:00–12:00, while a large amount of electric energy is released during the period from 14:00–18:00. The reason is that discharging takes place when RES is scarce due to the limited battery capacity. And we can see that at the beginning and end of the scheduling period, the SOC values of the batteries are equal (0.4).

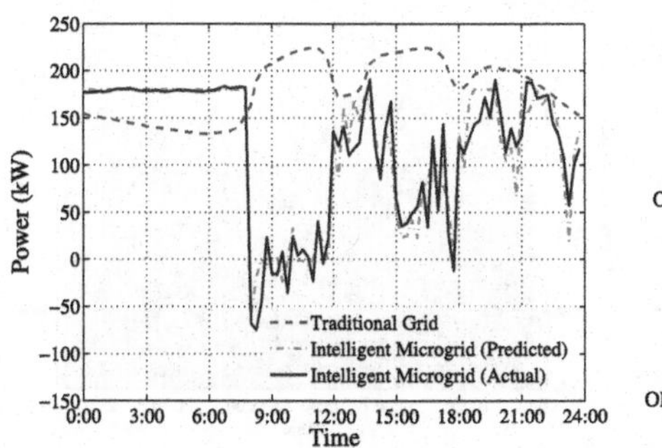

(a) Comparison of main grid load curves between traditional power grid and the intelligent microgrid.

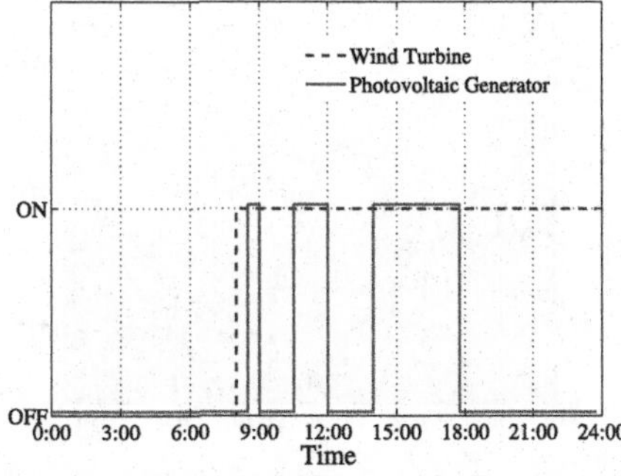

(b) RES intelligent scheduling strategy.

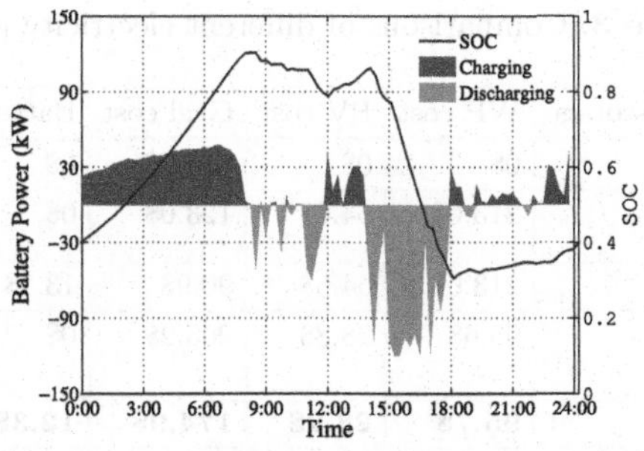

(c) Battery intelligent scheduling strategy.

Fig. 3. Performances of the intelligent scheduling.

5.2 Bidding Strategy Based on Intelligent Contract

To evaluate the performance of our proposed bidding scheme, we compare it with the way of randomly selecting a company to give services. The comparison results are shown in Fig. 4, we can see that the profit proportion of Company2 is far less than that of Company1 in the ordinary bidding strategy because Company2 have bad luck in the high profit period of 8:00–13:00. In contrast, Company3 don't gain an advantage in the period of higher profits for our bidding strategy, but it has a greater advantage in the future period because it spend less credit points, and those three companies obtain more average profits finally.

5.3 Cost Comparison of Different Energy Plans

To further illustrate the effectiveness of the proposed model, we select five different energy supply options for comparison, the results of which are shown in Table 3. We can see that the cost of electricity will increase, which will not be accepted by users if RES is used without reasonable scheduling. After introducing the energy storage equipment, the cost of electricity will not drop significantly

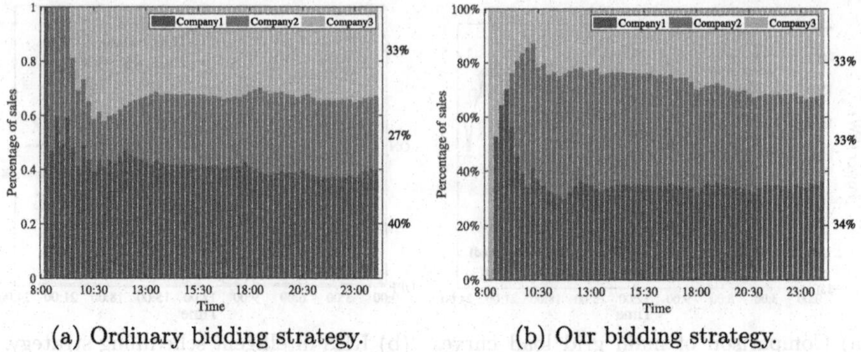

(a) Ordinary bidding strategy. (b) Our bidding strategy.

Fig. 4. The comparison of bidding strategy

Table 3. Comparisons of different electricity plans

RES	Energy storage	WP cost	PV cost	Grid cost	Battery cost	Average cost
No	No	0$	0$	364.7$	0$	0.084$/kWh
Fully accepted	No	213.0$	64.6$	123.0$	0$	0.093$/kWh
	Yes	213.0$	64.6$	90.9$	13.1$	0.087$/kWh
Selective acceptance	No	95.6$	28.8$	203.2$	0$	0.075$/kWh
	Yes	**95.7$**	**28.2$**	**174.9$**	**12.3$**	**0.071$/kWh**

due to its limited capacity and its own cost. From the last line of Table 3, we can see that the proposed model has the minimum electricity cost, which is 0.013 $/kWh lower than the average cost of the traditional energy supply mode.

5.4 Main Grid Side Economic Analysis

We evaluate the impact of microgrid users connecting to the main grid (we assume that the total number of users is 1000), the results of which is shown in Fig. 5.

It can be found that when 40% of the microgrid users are connected, the peak voltage of the main grid load decreases by 20%. Considering the system stability, the operating cost of the main grid can be reduced by about 25%. From Table 3, we can see that the switching cost of electric energy between the user and the main grid can be reduced by about 52% after the adoption of smart microgrid system, that is, the energy supply income of the main grid operator can be reduced by about 21% (since $52\% * 40\% = 21\%$). So the profit rate of the main grid operators have been improved. However, we find that the value of $|Peak\text{-}Valley|$ will increase when the proportions of microgrid users is too large. We suspect that the fluctuations in the load curve are due to a large number of RES connections, which will be discussed in our future works.

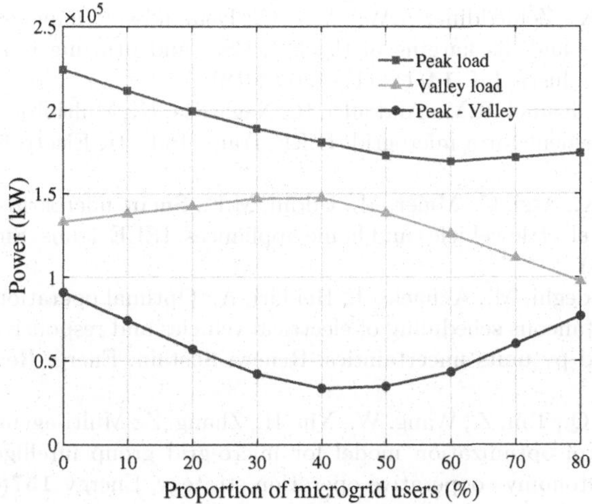

Fig. 5. Impact of microgrid users connecting to the main grid.

6 Conclusions

In this paper, we formalized a distributed energy management system based on blockchain. The system first collects wireless sensor networks data, stores the data on the blockchain network by group signature mechanism and consensus algorithm, and then adjusts energy resources in some region to realize the optimized energy allocation and efficient regional energy consumption based on prediction results and pricing rules for the main gird. Also, it can reduce the relevant economic costs of main grid and residents. Moreover, our proposed bidding strategy based on credit score makes each DER supplier obtain more fair competition. It also realized high security, robustness (the implementation of decentralization, and greatly promotion of system's ability to resist malicious attacks). In addition, the flexibility of the system architecture design (including the establishment of the management hub node, the registration of devices and group signature) make the system be able to detect and process anomalies. However, the anomaly detection is reserved which will be one of our future works.

References

1. Payne, J.E.: A survey of the electricity consumption-growth literature. Appl. Energy **87**(3), 723–731 (2010)
2. Hittinger, E., Wiley, T., Kluza, J., Whitacre, J.: Evaluating the value of batteries in microgrid electricity systems using an improved energy systems model. Energy Convers. Manage. **89**, 458–472 (2015)
3. Khan, A.R., Mahmood, A., Safdar, A., Khan, Z.A., Khan, N.A.: Load forecasting, dynamic pricing and DSM in smart grid: a review. Renew. Sustain. Energy Rev. **54**(C), 1311–1322 (2016)

4. Tushar, M.H.K., Zeineddine, A.W., Assi, C.: Demand-side management by regulating charging and discharging of the EV, ESS, and utilizing renewable energy. IEEE Trans. Industr. Inf. **14**(1), 117–126 (2018)
5. Chaouachi, A., Kamel, R.M., Andoulsi, R., Nagasaka, K.: Multiobjective intelligent energy management for a microgrid. IEEE Trans. Industr. Electron. **60**(4), 1688–1699 (2013)
6. Tushar, M.H.K., Assi, C., Maier, M., Uddin, M.F.: Smart microgrids: optimal joint scheduling for electric vehicles and home appliances. IEEE Trans. Smart Grid **5**(1), 239–250 (2014)
7. Rabiee, A., Sadeghi, M., Aghaeic, J., Heidari, A.: Optimal operation of microgrids through simultaneous scheduling of electrical vehicles and responsive loads considering wind and pv units uncertainties. Renew. Sustain. Energ. Rev. **57**, 721–739 (2016)
8. Ju, L., Zhang, Q., Tan, Z., Wang, W., Xin, H., Zhang, Z.: Multi-agent-system-based coupling control optimization model for micro-grid group intelligent scheduling considering autonomy-cooperative operation strategy. Energy **157**(C), 1035–1052 (2018)
9. Tian Wang, Md., Bhuiyan, Z.A., Wang, G., Qi, L., Jie, W., Hayajneh, T.: Preserving balance between privacy and data integrity in edge-assisted internet of things. IEEE Internet Things J. **7**(4), 2679–2689 (2020)
10. Du, X., Xiao, Y., Ci, S., Guizani, M., Chen, H.-H.: A routing-driven key management scheme for heterogeneous sensor networks. In: 2007 IEEE International Conference on Communications, pp. 3407–3412 (2007)
11. Goranović, A., Meisel, M., Fotiadis, L., Wilker, S., Treytl, A., Sauter, T.: Blockchain applications in microgrids an overview of current projects and concepts. In: IECON 2017–43rd Annual Conference of the IEEE Industrial Electronics Society, pp. 6153–6158 (2017)
12. Chaum, D., van Heyst, E.: Group signatures. In: Davies, D.W. (ed.) EUROCRYPT 1991. LNCS, vol. 547, pp. 257–265. Springer, Heidelberg (1991). https://doi.org/10.1007/3-540-46416-6_22
13. Sun, L., Zhou, K., Zhang, X., Yang, S.: Outlier data treatment methods toward smart grid applications. IEEE Access **6**, 39849–39859 (2018)
14. Tian, W., et al.: Propagation modeling and defending of a mobile sensor worm in wireless sensor and actuator networks. Sensors **17**(1), 139 (2017)
15. Shelby, Z.: The constrained application protocol (coap). Draft-ietfcore-coap-08, **47**(5), 9–11 (2010)
16. Chang, G.W., Lu, H.J., Chang, Y.R., Lee, Y.D.: An improved neural network-based approach for short-term wind speed and power forecast. Renewable Energy **105**(MAY), 301–311 (2017)
17. Wang, T., Luo, H., Zeng, X., Yu, Z., Liu, A., Sangaiah, A.K.: Mobility based trust evaluation for heterogeneous electric vehicles network in smart cities. IEEE Trans. Intell. Transp. Syst. **22**(3), 1797–1806 (2020)
18. Wood, D.D.: Ethereum: a secure decentralised generalised transaction ledger (2014)

Data Science

Multi-layer Adaptive Sampling for Per-Flow Spread Measurement

Boyu Zhang[1], Yang Du[1](✉), He Huang[1], Yu-E Sun[2], Guoju Gao[1], Xiaoyu Wang[1], and Shiping Chen[3](✉)

[1] School of Computer Science and Technology, Soochow University, Suzhou, Jiangsu, China
{duyang,huangh,gjgao,xywang21}@suda.edu.cn

[2] School of Rail Transportation, Soochow University, Suzhou, Jiangsu, China
sunye12@suda.edu.cn

[3] School of Optical-Electrical and Computer Engineering, University of Shanghai for Science and Technology, Shanghai, China
chensp@usst.edu.cn

Abstract. Per-flow spread measurement in high-speed networks, which aims to estimate the number of distinct elements of each flow, plays an important role in many practical applications. Most existing solutions adopt compact data structures (i.e., sketches) to share memory units among flows so that they can fit in limited on-chip memory, resulting in low estimation accuracy for small flows. Unlike sketch-based solutions, non-duplicate sampling measures per-flow spreads by sampling each distinct element with the same sampling probability. However, it ignores that, compared to small flows, large flows only need lower sampling probabilities to achieve the same relative estimation error, wasting significant on-chip memory for large flows. This paper presents multi-layer adaptive sampling to complement the prior work by assigning lower probabilities to larger flows. The proposed framework employs a multi-layer model to sample distinct elements, ensuring that most small flows will stay in lower layers and large flows will get to higher layers. Besides, higher layers are designed with smaller overall probabilities to ensure that larger flows have lower sampling probabilities. Experimental results based on real Internet traces show that, compared to the state-of-the-art method, our solution can reduce up to 86% average relative errors for per-flow spread estimation and reduce the FPRs and FNRs of flow misclassification by around one to two magnitudes.

Keywords: Traffic measurement · Spread estimation · Sampling · Adaptive sampling · High-speed networks

1 Introduction

Per-flow traffic measurement, which aims to collect traffic statistics of high-speed network flows, can provide indispensable information to many practical applications, e.g., anomaly detection, network billing, traffic engineering, and content

Y. Lai et al. (Eds.): ICA3PP 2021, LNCS 13155, pp. 743–758, 2022.
https://doi.org/10.1007/978-3-030-95384-3_46

access profiling [2,4,6,9–11,13–26]. Here, a flow is a set of packets sharing the same flow label, where flow labels are pre-defined based on application interests. This paper focuses on the per-flow spread measurement problem [6,9–11,14–16,21], a particularly challenging traffic measurement task which estimates the number of distinct elements (*i.e.*, spread) in each flow. Here, the elements can also be flexibly defined according to application interests. For example, when detecting scan attackers, we can define the flow labels as the source addresses and let the destination addresses be the elements. Then, we can report the sources with abnormally large spreads (*i.e.*, connected to an abnormally large number of destinations) as scan attackers [9,11,15,16]. Notice that an element may appear multiple times in the packet stream. Thus, per-flow spread measurement needs to filter duplicate elements from packet stream and hence faces more challenges than traditional tasks like per-flow size measurement (only needs to count the number of packets) [2,17,22,24–26].

One challenge for per-flow spread measurement is the extremely high line speed of high-speed networks, which can reach up to 400 Gbps or higher. Another challenge is that precise duplicate filtering requires significant memory space to record the element information and identify duplicates. In other words, implementing a precise spread estimator requires large memory bandwidth and large memory capacity to meet the challenges raised by line speed and duplicate filtering. Unfortunately, existing memory modules cannot provide large bandwidth and large capacity at the same time. Specifically, on-chip memory (SRAM) is fast but only has a limited size (usually less than 8.25 MB) [20]; off-chip memory (DRAM) is large enough to hold the entire data stream, but its update speed is much slower than the line rate.

To address the above challenges, various compact data structures, *i.e.*, sketches, have been proposed to estimate per-flow spreads [13,15,19,21–23]. However, these sketch-based methods mainly focus on reducing their memory requirements to fit in the limited on-chip memory, which need to mix up different flows' traffic information and hence can hardly achieve high accuracy for small flows. Moreover, they are only suitable for offline queries since they need to scan hundreds or thousands of bits/registers to answer a query for a flow's spread. A different strategy to implement the spread estimator is non-duplicate sampling [18], which adopts an on-chip/off-chip model to address the limitations of sketch-based solutions. Specifically, it uses on-chip memory to filter duplicate elements and then samples and sends a portion of distinct elements to off-chip memory for recording, which solves the mismatch between the off-chip memory speed and line rate in high-speed networks. Moreover, like the sampling-based methods for size measurement [5,7,8,12], non-duplicate sampling maintains an individual off-chip counter for each flow too, avoiding the space sharing noise. However, non-duplicate sampling has the problem that it uses the same sampling probability for all flows. As pointed out in [3,12], compared to small flows, larger flows only need lower sampling probabilities to guarantee the same relative estimation error. Consequently, the estimation accuracy of large flows is much higher than the application requirements, which indicates a waste of on-chip memory resources.

We aim to complement the prior work by assigning larger flows with lower sampling probabilities so that we can satisfy the estimation accuracy requirements using less on-chip memory. Also, we want to improve the estimation accuracy of small flows, which is usually ignored in existing spread estimators. We want to stress that an accurate estimation for small flows is indispensable to certain applications, such as detecting stealthy denial-of-quality attacks or stealthy scanners [4,15]. However, the assignment of adaptive sampling probabilities is not a trivial task since packet streams arrive in a one-pass manner: we cannot identify if an incoming flow has a large or small spread.

This paper presents multi-layer adaptive sampling, a novel framework that uses a multi-layer model to sample larger flows with lower probabilities. In the beginning, the elements of all flows are mapped to the bit arrays in the first layer. Then, the bit arrays containing large flows will get full, and the following elements that have been sampled by these bit arrays need to get to the next layer for sampling again. Hence, larger flows need to pass more layers' sampling and will have lower overall sampling probabilities. Besides, this framework adopts an on-chip/off-chip model. It places an multi-layer sampling module on the network processor chip to catch up with the line rate in high-speed networks and an off-chip recording module to store traffic information. Thanks to the large space of off-chip memory, the off-chip recording module can store each flow's data separately, preventing the noise introduced by space sharing. The experimental results based on real Internet traffic traces show that our design can have much lower estimation errors for small and medium flows than the prior art.

2 Problem Statement

In high-speed networks, a flow is a set of packets sharing the same flow label, where flow labels are pre-defined subsets of the header fields based on application interests. The task of per-flow spread measurement is to estimate the spread (*i.e.*, number of distinct elements) of each flow. The definitions of elements can also be flexibly defined to meet different applications' requirements. For example, to detect the victims of DDoS attacks, we can define the destination addresses as flow labels and define the source addresses as elements.

Notice that an element may be carried by multiple packets that arrive at different times, making the per-flow spread measurement problem more challenging than the traditional per-flow size measurement task. Consider a flow with 1000 packets carrying the same element. The size and spread of this flow are 1000 and 1 separately. Per-flow size measurement only needs to count the number of packets to get the flow's size. However, for per-flow spread measurement, it has to "remember" the arrived elements and filter the duplicates to get the flow's spread. Notice that, due to the limited on-chip memory, it is impractical for per-flow traffic measurement to precisely measure each flow's size/spread in high-speed networks. Considering the packet stream as a set of flows $\mathcal{F} = \{f_1, f_2, f_3, ...\}$ where the actual size/spread of flow f_i is n_{f_i}, the outputs of per-flow traffic measurement are the estimations $\{\widehat{n_{f_1}}, \widehat{n_{f_2}}, \widehat{n_{f_3}}, ...\}$.

3 Related Work

Most solutions for per-flow spread measurement are sketch-based [13,15,19,21–23], which employ highly compact data structures (*i.e.*, sketches) to store traffic information. These designs focus on reducing their memory requirements so that they can fit in the limited on-chip memory. However, to do so, they always have to adopt an aggressive space sharing schema where different flows' information are mixed together, which causes two main limitations. First, they can hardly achieve high accuracy for small flows since small flows' recording information can easily be overwritten by large flows during the space sharing process. Second, their estimation formulas are expensive to compute since they have to scan hundreds or thousands of bits/registers to answer a query for a flow's spread [13,15,19, 21,23]. Hence, they have to periodically send their sketch data structures to an offline analysis server to make calculations and answer queries, which is not suitable for online tasks.

To address the above limitations, recent work proposes a different strategy called non-duplicate sampling [18]. Non-duplicate sampling is a sampling-based design that employs an on-chip/off-chip model. Specifically, it places a sampling module on the network processor chip to filter the duplicate elements and sample each distinct element with a pre-defined sampling probability. And, the sampled elements will be recorded in the recording module placed in off-chip memory. This design solves the mismatch between the off-chip memory speed and line rate in high-speed networks. Besides, thanks to the large space of off-chip memory, each flow can have an individual counter to store its traffic information, which prevents the noise introduced by space sharing. This design can also support online queries since off-chip memory speed is fast enough for most online query tasks.

However, non-duplicate sampling uses the same sampling probability for all flows, regardless of flow spreads. As pointed out in [3,12], due to the characteristic of sampling, the relative estimation errors of large flows are much smaller than that of small flows when using the same sampling probability. As a result, large flows will have a much higher estimation accuracy than the application requirements, which is unnecessary and indicates a waste of on-chip memory resources. This motivates us to explore an adaptive sampling method, which can sample larger flows with lower sampling probabilities.

4 Multi-layer Adaptive Sampling

4.1 Overview

As shown in Fig. 1, multi-layer adaptive sampling can be divided into two modules: sampling module and recording module. In order to catch up with the line rate, the sampling module is deployed on the network processor chip of a router (or gateway) and employs high-speed on-chip memory such as SRAM to process the arrival packets in real time. In the sampling module, we use several contiguous physical memory blocks as layers. Each physical memory block is implemented as a number of bit arrays, and each bit array has a counter to record the number of '1's in it. Time is divided into epochs. All the bit arrays and counters are initialized to zeros at the beginning of each measurement epoch. In the recording module, which is implemented in off-chip memory such as DRAM, each flow has an individual counter array to record the number of elements that are sampled and send recording information from each layer.

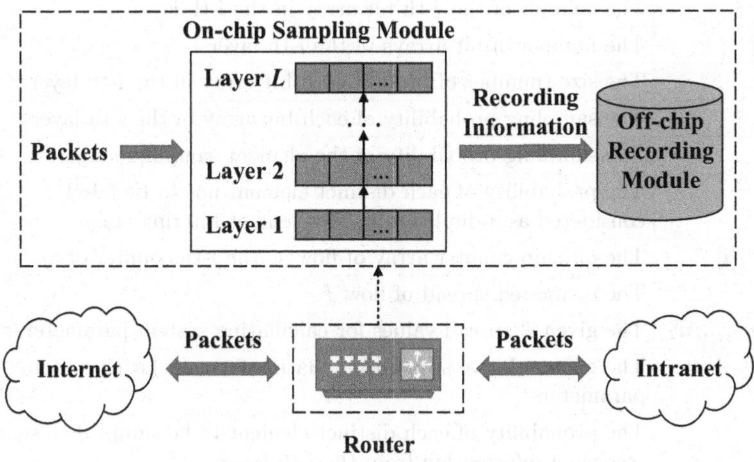

Fig. 1. System model.

The objective of multi-layer adaptive sampling is to let larger flows have lower sampling probabilities so that the on-chip memory occupied by large flows can be saved. To do so, we let elements be sampled layer by layer. In the beginning, all the elements are mapped to the bit arrays in the first layer. Soon, large flows' bit arrays will get full. Then, the following elements that have passed the sampling of these bit arrays need to get to the next layer for sampling again. In other words, small flows will stay in the lower layers, and larger flows will get to higher layers. Hence, larger flows need to pass more layers' sampling and will have lower overall sampling probabilities.

Next, we give the details of multi-layer adaptive sampling. Also, we list the variables and parameters used in this paper in Table 1.

4.2 Sampling

The sampling module employs an architecture of multiple layers. Each layer is implemented on a contiguous physical memory block and includes a number of bit arrays with the same size (*i.e.*, number of bits). Notice that each bit array has a counter to record the number of '1's in it. For simplicity, we use B and C to separately represent all the bit arrays and counters in the sampling module. And, $B[i][j]$ and $C[i][j]$ are denoted as the j-th bit array and its counter in

Table 1. Notations

Symbols	Meaning
L	The number of layers
B	All the bit arrays in the on-chip sampling module
C	All the counters in the on-chip sampling module
$B[i][j]$	The j-th bit array in the i-th layer
$C[i][j]$	The counter of the j-th bit array in the i-th layer
$N[i]$	The number of bit arrays in the i-th layer
$S[i]$	The size (number of bits) of each bit array in the i-th layer
$P[i]$	The sampling probability of each bit array in the i-th layer
p'	The sampling probability of the element sampling stage
p''	The probability of each distinct element not to be falsely considered as a duplicate in the element filtering stage
$c_f, c_f[i]$	The off-chip counter array of flow f, the i-th counter of c_f
$\widehat{n_f}$	The estimated spread of flow f
$n_1, n_2, ..., n_L$	The given L spread values for calculating system parameters
σ_r	The given relative standard deviation for calculating system parameters
p_i	The probability of each distinct element to be sampled to send recording information from the i-th layer
S_{ti}	The total size (number of bits) of all bit arrays in the i-th layer
C_{ti}	The sum of all the counters in the i-th layer
N_i	The number of distinct elements processed by the i-th layer

the i-th layer separately. Before the measurement, we have to set the following system parameters: the number of layers L, the number of bit arrays in each layer, the size of bit arrays in each layer, and the sampling probability in each layer. We use $N[i]$, $S[i]$, and $P[i]$ to denote the number of bit arrays, the size of each bit array, and the sampling probability in the i-th layer separately. Next, we give the details of the sampling module.

The algorithm of the sampling module is shown in Algorithm 1. For each arrival packet, we extract the flow label f and element e, and initialize the current layer number i to 1 in the beginning. Then, we process this packet layer by layer. When the packet gets to the i-th layer, we first perform a hash $j = H(f \oplus R[i]) \bmod N[i]$ to find the bit array $B[i][j]$ corresponding to flow f in the i-th layer, where H is a hash function, R is an array containing pre-defined random numbers, and $\oplus$ is the XOR operator. Notice that we will terminate the estimation epoch if $i > L$. Next, we employ a sampling method that is modified based on non-duplicate sampling [18] to sample elements using the bit array $B[i][j]$. Like non-duplicate sampling, our sampling method can be divided into two stages: element sampling and element filtering.

Element Sampling Stage. As shown in Line 6 to 8 in Algorithm 1, in this stage, we need to get the counter value $c = C[i][j]$ and sample each distinct element with a probability of

$$p' = \frac{S[i]}{S[i] - c} P[i]. \tag{1}$$

The value of p' is actually determined by the second stage, and we will explain it when presenting the second stage. To implement this sampling, we perform a hash $r = H'(f \oplus e \oplus R[i]) \bmod X$, where X is a large enough constant number within the range of the hash function H'. If $\frac{r}{X} < p'$, the element e is selected. Notice that, the packet processing is ended if the first stage does not select the element, *i.e.*, only the selected elements will be sent to the second stage for duplicate filtering.

Element Filtering Stage. This stage is to filter the duplicates. As shown in Line 9 to 18 in Algorithm 1, for each element e sent from the first stage, we first perform a hash $h = H(f \oplus e \oplus R[i]) \bmod S[i]$ to map it to the h-th bit $B[i][j][h]$ in the bit array $B[i][j]$. There are two cases to consider: (1) If $B[i][j][h] = 0$, we consider the element e has never been recorded by this stage before. Next, we need to check if the bitmap $B[i][j]$ has been full. If $\frac{S[i]-c}{S[i]} > P[i]$, we say this bitmap is not full. In this case, we set $B[i][j][h]$ to 1 and increase $C[i][j]$ by 1. Then, we send recording information $<f, i>$ to the recording module in off-chip memory. Otherwise, we cannot directly set the bit $B[i][j][h]$ to 1. Instead, we increase i by one and send this element to the next layer for sampling (Line 5 to 18 in Algorithm 1). (2) If $B[i][j][h] = 1$, we consider the element e has been recorded by this stage before. In other words, we consider the current element e as a duplicate. Hence, we will not send any recoding information to the recording module. However, the bit $B[i][j][h]$ may be set by another element, which means we will make false positives in such cases. Obviously, the false positive probability is $\frac{c}{S[i]}$. Therefore, the probability of each distinct element not to be considered as a duplicate is

$$p'' = 1 - \frac{c}{S[i]} = \frac{S[i] - c}{S[i]}. \tag{2}$$

We can easily find that the probability p' and p'' have the relationship of $p'p'' = P[i]$, which is the reason why we set p' to $\frac{S[i]}{S[i]-c}P[i]$. Since $p' \leq 1$ and p' is increasing as c increases, $\frac{S[i]-c}{S[i]}$ cannot be less than $P[i]$ according to (1). Therefore, we say a bitmap in the i-th layer has been full when $\frac{S[i]-c}{S[i]} \leq P[i]$.

Discussion on Our Sampling Strategy. In our strategy, the elements that are not selected by the element sampling stage will not be sent to the element filtering stage. In other words, we will not set any bit in the bit arrays to 1 for the elements discarded by the first stage, which can save a large amount of on-chip memory. Hence, our sampling method can have a much higher sampling probability than non-duplicate sampling using the same memory, especially when the occupied on-chip memory is very limited. Though this strategy makes some discarded elements still have a chance to be selected by the first stage, which makes the estimation results have a tiny upper deviation. However, since the upper deviation is too tiny, when using the same on-chip memory, our sampling method can still have a much higher estimation accuracy than non-duplicate sampling due to its higher sampling probability.

Algorithm 1. Algorithm of the sampling module

Input: number of layers L, bitmaps B, counters C, numbers of bitmaps in the layers N, sizes of bitmaps in the layers S, sampling probabilities in the layers P, constant number X, random number array R

Initialize B and C to 0;
for each packet in packet stream **do**
  Extract flow label f and element e;
  for $i = 1, 2, ..., L$ **do**
    Set $j = H(f \oplus R[i]) \bmod N[i]$;
    Set $c = C[i][j]$;
    Set $r = H'(f \oplus e \oplus R[i]) \bmod X$;
    if $\frac{r}{X} \leq \frac{S[i]}{S[i]-c} P[i]$ **then**
      Set $h = H(f \oplus e \oplus R[i]) \bmod S[i]$;
      if $B[i][j][h] = 1$ **then**
        break;
      else
        if $\frac{S[i]-c}{S[i]} \leq P[i]$ **then**
          continue;
        else
          Set $B[i][j][h] = 1$;
          Send $<f, i>$ to the off-chip recording module;
          break;
        end if
      end if
    end if
  end for
end for

4.3 Recording

In the off-chip recording module, each flow has an individual counter array with L counters. The i-th counter is to record the number of elements sending recording information from the i-th layer. Specifically, each time a piece of recording information $<f, i>$ arrives, it means a distinct element of flow f is sampled at the i-th layer. Therefore, we find the counter array c_f of flow f and increase the i-th counter $c_f[i]$ by 1.

4.4 Online Query Process

Due to the architecture of multi-layer adaptive sampling, the estimation process is quite straightforward. Hence, multi-layer adaptive sampling can answer online queries in real time. Recall that each flow has an individual counter array in the off-chip recording module. Thus we only need to read the counter array c_f to get the traffic information of the queried flow f. Each time $c_f[i]$ is increased by 1, there must be a piece of recording information $<f, i>$ sent from the i-th layer. To send recording information from the i-th layer, each distinct element has to be sampled from the first layer to the i-th layer, and the probability is approximately $\prod_{l=1}^{i} P[l]$. Therefore, we estimate the spread $\widehat{n_f}$ of flow f as the following equation:

$$\widehat{n_f} = \sum_{i=1}^{L} \frac{c_f[i]}{\prod_{l=1}^{i} P[l]}. \quad (3)$$

5 System Parameters

In this section, we show how to configure the system parameters of multi-layer adaptive sampling. Since the employed sampling method may sample elements using a probability higher than the defined sampling probability, and the multi-layer model is complex, we cannot give a precise optimal parameter setting schema. However, we can give a suggestive parameter setting schema, which shows a good performance in the experiments. We use the relative standard deviation as an indication of the relative error of estimated spreads. Given a relative standard deviation σ_r and L spread values $n_1, n_2, ..., n_L$ where we make flows have the same relative standard deviation, we will divide L layers, and give the number of bit arrays $N[i]$, the size of bit arrays $S[i]$, and the sampling probability $P[i]$ in the i-th layer, where $n_1 < n_2 < ... < n_L$ and $1 \leq i \leq L$.

Obviously, the number of sampled elements obeys the binomial distribution. Thus, the variance of the number of sampled elements is $D(X) = np(1-p)$, and the mathematical expectation is $E(X) = np$, where X is a random variable representing the number of sampled elements, n is the flow spread, and p is the sampling probability. Let Y represent the estimated spread, and we have $Y = \frac{X}{p}$. Therefore, the variance and mathematical expectation of the estimated spread are $D(Y) = D(\frac{X}{p}) = \frac{n(1-p)}{p}$ and $E(Y) = E(\frac{X}{p}) = n$. Then, we can find that the relative standard deviation of the estimated spread is

$$\sigma_r = \frac{\sqrt{D(Y)}}{E(Y)} = \sqrt{\frac{1-p}{np}}. \quad (4)$$

According to Eq. (4), we can find that, given the same sampling probability, the relative standard deviation is getting smaller as the flow spread increases. Thus, the estimation accuracy of large flows is always much higher than the application requirements if we use the same sampling probability for both small and large flows. We want to decrease large flows' sampling probabilities and increase small flows' sampling probabilities. To do so, we let flows with spreads

$n_1, n_2, ..., n_L$ have the same relative standard deviation σ_r. Then, as the following equation, we can calculate the sampling probability p_i that a distinct element, which belongs to a flow with spread n_i, sends recording information to off-chip memory.

$$p_i = \frac{1}{n_i \sigma_r^2 + 1}, 1 \leq i \leq L. \tag{5}$$

Obviously, $p_1 > p_2 > ... > p_L$ since $n_1 < n_2 < ... < n_L$. Next, we divide L layers, and the sampling probability of the bit arrays in the i-th layer is

$$P[i] = \begin{cases} \frac{p_i}{p_{i-1}} & i \geq 2 \\ p_i & i = 1 \end{cases}. \tag{6}$$

Since our multi-layer model uses the overflow of bit arrays to detect flow spreads, small flows will be maintained in lower layers, and larger flows will get to higher layers. We suppose that flows less than n_2 will be maintained in the first layer, flows not less than n_L will be maintained in the last layer, and flows within $[n_i, n_{i+1})$ will be maintained in the i-th layer, where $2 \leq i \leq L-1$. Notice that this supposition is not precise. We just use this supposition to calculate the suggestive parameters.

Next, we calculate the total size of all bit arrays in each layer. Consider all the bit arrays in the i-th layer as a whole bit array, and let S_{ti}, C_{ti}, $P[i]$ be the size, number of '1's, and sampling probability of the whole bit array separately. As mentioned in Sect. 4, we consider the whole bit array is not full when

$$\frac{S_{ti} - C_{ti}}{S_{ti}} > P[i]. \tag{7}$$

We suppose that each layer only processes the elements of the flows maintained in that layer. Let N_1 be the total spread of flows less than n_2, N_L be the total spread of flows not less than n_L, and let N_i be the total spread of flows within $[n_i, n_{i+1})$, where $2 \leq i \leq L-1$. These values can be estimated using historical statistics. Thus, the i-th layer will process N_i distinct elements. If each distinct element has a probability of p_i to be sampled to sending recording information from the i-th layer, there are expected to be $N_i p_i$ elements to be sampled, which means C_{ti} is expected to be $N_i p_i$. Combine Eq. (7), we have

$$S_{ti} > \frac{N_i p_i}{1 - P[i]}. \tag{8}$$

Notice that, in multi-layer sampling, the actual sampling probability of each bit array in the i-th layer may be a little larger than $P[i]$. Also, the probability of a distinct element to be sampled and send recording information from the i-th layer may be larger than p_i. Besides, the elements processed by a layer are not equally assigned to the bit arrays in the layer, which means some bit arrays will process fewer elements while others will process more elements and overflow, which will waste a portion of memory. Thus, $\lceil \frac{N_i p_i}{1-P[i]} \rceil$ bits are not enough for the i-th layer to process N_i distinct elements. According to our experience, we suggest giving the last layer more space to prevent overflow but not giving other layers more memory. Using five minutes of per-destination traffic data downloaded from CAIDA [1], we find that about twice of $\lceil \frac{N_i p_i}{1-P[i]} \rceil$ bits are appropriate for the total bit array size of the last layer. Thus, we can set the total bit array size of each layer as follows:

$$\begin{cases} S_{ti} = \lceil \frac{N_i p_i}{1-P[i]} \rceil & i < L \\ S_{ti} \approx 2\lceil \frac{N_i p_i}{1-P[i]} \rceil & i = L \end{cases}. \tag{9}$$

Next, we need to determine the single bit array size in each layer. According to our experience, the single bit array sizes have little effect on the performance if we do not set them to extreme values, and we suggest setting the bit array size in the following manner. Obviously, the last layer only needs one bit array. Thus, the bit array size $S[L]$ is equal to S_{tL}. For other layers, we suppose the largest flow that can be maintained in the i-th layer has a spread of $n_{i+1} - 1$. Therefore, the minimum size of the bit array in the i-th layer should be $\frac{(n_{i+1}-1)P[i]}{1-P[i]}$. Then, we suggest setting the size of the bit array in the i-th layer as follows:

$$S[i] = \begin{cases} \lceil \frac{(n_{i+1}-1)P[i]}{1-P[i]} \rceil & i < L \\ S_{ti} & i = L \end{cases}. \tag{10}$$

Finally, we can get the number of bit arrays in each layer as follows:

$$N[i] = \lceil \frac{S_{ti}}{S[i]} \rceil \tag{11}$$

Given L spread values $n_1, n_2, ..., n_L$ and a relative standard deviation σ_r, we can calculate the system parameters $N[i]$, $S[i]$, and $P[i]$ according to Eq. (5), (6), and (9)–(11). Using these system parameters, we can let larger flows be sampled with lower sampling probabilities. Though these suggestive system parameters are not optimal, they show a good performance in the experiments.

6 Experimental Evaluation

6.1 Experiment Setup

In this section, we evaluate our estimator's performance using real Internet traffic traces. We use five minutes of data downloaded from CAIDA [1] as our dataset. It contains 513889 per-destination flows and 152163629 packets. In Table 2, we present the distribution of flow spreads. We can find that large flows take up a large portion of distinct elements though they have tiny numbers.

Table 2. The distribution of per-destination flows in different spread intervals.

Spread interval	[1, 200)	[200, 500)	[500, 1000)	[1000, +∞)	All
Number of flows	511610	1391	596	292	513889
Proportion of flows	0.9956	0.0027	0.0012	0.0006	1.0000
Number of distinct elements	1561482	426674	400675	761909	3150740
Proportion of distinct elements	0.4956	0.1354	0.1272	0.2418	1.0000

Table 3. The system parameters of MLAS.

σ_r	0.2			0.15		
Parameters	$i = 1$	$i = 2$	$i = 3$	$i = 1$	$i = 2$	$i = 3$
$S[i]$	70	90	75000	120	150	140000
$N[i]$	3572	445	1	3750	400	1
$P[i]$	0.111	0.432	0.501	0.182	0.451	0.524

In the following, we compare multi-layer adaptive sampling (denoted as MLAS for short) and non-duplicate sampling [18] (denoted as NDS) in terms of per-flow spread estimation accuracy first. Then, we compare MLAS and NDS on flow misclassification in terms of false-positive rate (FPR) and false-negative rate (FNR). To calculate the system parameters of MLAS, we set the same relative standard deviation for flows of spreads 200, 500, and 1000. The values of σ_r are set to 0.2 and 0.15, and the on-chip memory (including the on-chip counters, denoted as m) occupied by MLAS are 0.047 MB and 0.081 MB separately. The specific system parameters of MLAS are shown in Table 3. To make a fair comparison, the on-chip memory occupied by NDS is the same as that of MLAS, and the parameters of NDS are set to the optimal values according to [18].

6.2 Comparison in Terms of Estimation Accuracy

We first compare MLAS and NDS in terms of per-flow spread estimation accuracy. The estimation results of MLAS and NDS are shown in Fig. 2, Fig. 3 and Table 4. Figure 2 and Fig. 3 are scatter diagrams in log scale, where each point represents a flow with its x coordinate being the actual spread and its y coordinate being the estimated spread. The closer a point gets to the equality line $y = x$, the more accurate the estimation is. Table 4 shows the average relative errors (ARE) in different spread intervals.

Obviously, the estimation results of MLAS are much more accurate than that of NDS when using the same on-chip memory, especially when the on-chip memory is very limited. For example, NDS must use a very small sampling probability of 0.00034 to fit in the 0.047 MB on-chip memory. In this case, a flow's estimated spread will be 2941.18 even if only one element of the flow is sampled. Therefore, we can find a blank area where estimated spreads are less than 2941.18 in Fig. 3(a). However, due to the multi-layer model and our sampling method,

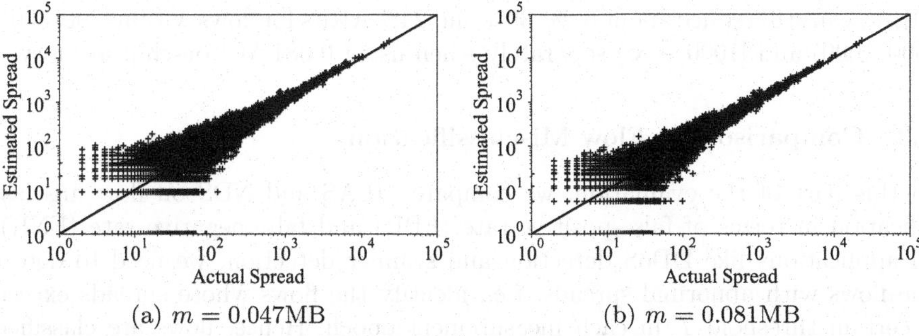

(a) $m = 0.047\text{MB}$ (b) $m = 0.081\text{MB}$

Fig. 2. Spread estimation accuracy of MLAS when the relative standard deviation is 0.2 ($m = 0.047$ MB) and 0.15 ($m = 0.081$ MB).

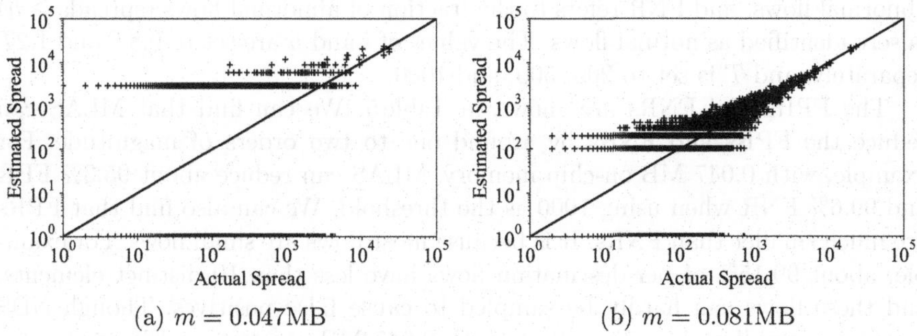

(a) $m = 0.047\text{MB}$ (b) $m = 0.081\text{MB}$

Fig. 3. Spread estimation accuracy of NDS when the occupied memory is 0.047 MB ($p = 0.00034$) and 0.081 MB ($p = 0.00969$).

MLAS can have much larger sampling probabilities for small and medium flows than NDS, which leads to a higher estimation accuracy. Though our sampling method can cause an upper deviation, the upper deviation is so tiny that it only has little effect on the estimation results. As shown in Table 4, the AREs of MLAS can reduce 83%, 86% and 84% of that of NDS for flows within $[200, 500)$, $[500, 1000)$ and $[1000, +\infty)$ separately. The difference between MLAS and NDS gets smaller when their occupied memory increases. However, compared to NDS,

Table 4. The AREs of NDS and MLAS.

Memory	Algorithm	$[1, 200)$	$[200, 500)$	$[500, 1000)$	$[1000, +\infty)$
0.047 MB	NDS	1.053	1.753	1.583	1.053
	MLAS	0.993	0.299	0.220	0.172
0.081 MB	NDS	1.084	0.472	0.303	0.200
	MLAS	0.911	0.225	0.186	0.152

MLAS can still reduce about 52%, 39% and 24% AREs for flows within [200, 500), [500, 1000) and [1000, +∞) separately when using 0.081 MB on-chip memory.

6.3 Comparison on Flow Misclassification

In this part of the evaluation, we compare MLAS and NDS on flow misclassification in terms of false-positive rate (FPR) and false-negative rate (FNR). In applications like DDoS detection and scanner detection, we need to detect the flows with abnormal spreads, *i.e.*, identify the flows whose spreads exceed a certain threshold T in each measurement epoch. Hence, flows are classified into normal flows and abnormal flows according to their estimated spreads. Due to the estimation errors, we may make false positives and false negatives. FPR refers to the fraction of normal flows (spreads $\leq l$) that are falsely classified into abnormal flows, and FNR refers to the fraction of abnormal flows (spreads $\geq h$) falsely identified as normal flows. The values of l and h are set to $0.8T$ and $1.2T$ separately, and T is set to 200, 500, and 1000.

The FPRs and FNRs are shown in Table 5. We can find that MLAS can reduce the FPRs and FNRs by around one to two orders of magnitude. For example, with 0.047 MB on-chip memory, MLAS can reduce about 95.3% FPR and 99.6% FNR when using 1000 as the threshold. We can also find that FPRs are much smaller than FNRs. It is because most flows are small flows. For example, about 99.15% of per-destination flows have less than 10 distinct elements, and these flows can hardly be sampled to cause false positives. Though NDS loses most small and medium flows with 0.047 MB memory and hence has an FPR of 0.00099 when the threshold is 200, MLAS can still reduce 19.2% FPR.

Table 5. The FPRs and FNRs of flow misclassification.

T		200		500		1000	
Memory	Algorithm	FPR	FNR	FPR	FNR	FPR	FNR
0.047 MB	NDS	0.00099	0.80318	0.00124	0.66665	0.00148	0.44535
	MLAS	0.00080	0.01640	0.00020	0.00938	0.00007	0.00197
0.081 MB	NDS	0.00304	0.09652	0.00039	0.10218	0.00010	0.04648
	MLAS	0.00047	0.00758	0.00010	0.00338	0.00003	0.00047

7 Conclusion

This paper proposes an efficient adaptive sampling framework called multi-layer adaptive sampling for per-flow spread measurement. Based on the multi-layer model and our sampling method, multi-layer adaptive sampling can sample larger flows with lower sampling probabilities, and small flows can be sampled with a higher probability than non-duplicate sampling. The experimental results based on real Internet traffic traces demonstrate that multi-layer adaptive sampling can achieve a much higher estimation accuracy and lower flow misclassification rates than non-duplicate sampling with the same on-chip memory. Our future work is to extend traffic measurement research on other application environments, such as internet of things and software defined network.

Acknowledgments. The corresponding authors of this paper are Yang Du and Shiping Chen. The research of authors was supported by National Natural Science Foundation of China under Grant No. 62072322, No. 61873177, and No. U20A20182, Natural Science Foundation of Jiangsu Province under Grant No. BK20210706, and Jiangsu Planned Projects for Postdoctoral Research Funds under Grant No. 2021K165B. The research of Guoju Gao was partially supported by the National Natural Science Foundation of China (NSFC) under Grant No. 62102275, the NSF of Jiangsu in China under Grant No. BK20210704, and the NSF of the Jiangsu Higher Education Institutions of China under Grant No. 21KJB520025.

References

1. CAIDA: The CAIDA UCSD Anonymized Internet Traces 2016. http://www.caida.org/data/passive/passive_2016_dataset.xml. Accessed 28 Jul 2019
2. Chen, M., Chen, S., Cai, Z.: Counter tree: a scalable counter architecture for per-flow traffic measurement. IEEE/ACM Trans. Networking **25**(2), 1249–1262 (2017). https://doi.org/10.1109/TNET.2016.2621159
3. Choi, B.Y., Park, J., Zhang, Z.L.: Adaptive random sampling for traffic load measurement. In: IEEE International Conference on Communications, 2003, ICC 2003, vol. 3, pp. 1552–1556. IEEE (2003)
4. Dai, H., Shahzad, M., Liu, A.X., Li, M., Zhong, Y., Chen, G.: Identifying and estimating persistent items in data streams. IEEE/ACM Trans. Networking **26**(6), 2429–2442 (2018)
5. Dimitropoulos, X., Hurley, P., Kind, A.: Probabilistic lossy counting: an efficient algorithm for finding heavy hitters. ACM SIGCOMM Comput. Commun. Rev. **38**(1), 7–16 (2008)
6. Du, Y., Huang, H., Sun, Y.E., Chen, S., Gao, G.: Self-adaptive sampling for network traffic measurement. In: IEEE INFOCOM 2021-IEEE Conference on Computer Communications, pp. 1–10. IEEE (2021)
7. Duffield, N., Lund, C., Thorup, M.: Learn more, sample less: control of volume and variance in network measurement. IEEE Trans. Inf. Theory **51**(5), 1756–1775 (2005)
8. Duffield, N., Lund, C., Thorup, M., Thorup, M.: Flow sampling under hard resource constraints. In: ACM SIGMETRICS Performance Evaluation Review, vol. 32, pp. 85–96 (2004)
9. Estan, C., Varghese, G.: New directions in traffic measurement and accounting: focusing on the elephants, ignoring the mice. ACM Trans. Comput. Syst. (TOCS) **21**(3), 270–313 (2003)
10. Hao, F., Kodialam, M., Lakshman, T.: ACCEL-RATE: a faster mechanism for memory efficient per-flow traffic estimation. In: ACM SIGMETRICS Performance Evaluation Review, vol. 32, pp. 155–166 (2004)

11. Heule, S., Nunkesser, M., Hall, A.: HyperLogLog in practice: algorithmic engineering of a state of the art cardinality estimation algorithm. In: Proceedings of of the 16th International Conference on Extending Database Technology (EDBT 2013), pp. 683–692 (2013)
12. Hu, C., Wang, S., Tian, J., Liu, B., Cheng, Y., Chen, Y.: Accurate and efficient traffic monitoring using adaptive non-linear sampling method. In: IEEE INFOCOM 2008-The 27th Conference on Computer Communications, pp. 26–30. IEEE (2008)
13. Huang, H., et al.: You can drop but you can't hide: *k*-persistent spread estimation in high-speed networks. In: Proceedings of the IEEE Conference on Computer Communications (INFOCOM 2018). pp. 1889–1897 (2018)
14. Kumar, A., Xu, J., Wang, J.: Space-code bloom filter for efficient per-flow traffic measurement. IEEE J. Sel. Areas Commun. **24**(12), 2327–2339 (2006)
15. Li, T., Chen, S., Luo, W., Zhang, M.: Scan detection in high-speed networks based on optimal dynamic bit sharing. In: Proceedings of the IEEE Conference on Computer Communications (INFOCOM 2011), pp. 3200–3208 (2011)
16. Lieven, P., Scheuermann, B.: High-speed per-flow traffic measurement with probabilistic multiplicity counting. In: Proceedings of the IEEE Conference on Computer Communications (INFOCOM 2010), pp. 1–9 (2010)
17. Lu, Y., Montanari, A., Prabhakar, B., Dharmapurikar, S., Kabbani, A.: Counter braids: a novel counter architecture for per-flow measurement. ACM SIGMETRICS Perform. Eval. Rev. **36**(1), 121–132 (2008)
18. Sun, Y.E., Huang, H., Ma, C., Chen, S., Du, Y., Xiao, Q.: Online spread estimation with non-duplicate sampling. In: Proceedings of IEEE INFOCOM 2020, pp. 2440–2448 (2020)
19. Xiao, Q., Qiao, Y., Zhen, M., Chen, S.: Estimating the persistent spreads in high-speed networks. In: Proceedings of the IEEE 22nd International Conference on Network Protocols (ICNP 2014), pp. 131–142 (2014)
20. Yang, T., Xu, J., Liu, X., Liu, P., Wang, L., Bi, J., Li, X.: A generic technique for sketches to adapt to different counting ranges. In: Proceedings of the IEEE Conference on Computer Communications (INFOCOM 2019), pp. 2017–2025 (2019)
21. Yoon, M., Li, T., Chen, S., Peir, J.K.: Fit a compact spread estimator in small high-speed memory. IEEE/ACM Trans. Networking (TON) **19**(5), 1253–1264 (2011)
22. Zhou, Y., Zhang, Y., Ma, C., Chen, S., Odegbile, O.O.: Generalized sketch families for network traffic measurement. In: Proceedings of the ACM on Measurement and Analysis of Computing Systems (POMACS), vol. 3, no. 3, p. 51 (2019)
23. Zhou, Y., Zhou, Y., Chen, M., Chen, S.: Persistent spread measurement for big network data based on register intersection. In: Proceedings of the ACM on Measurement and Analysis of Computing Systems, vol. 1, p. 15 (2017)
24. Zhou, Y., Zhou, Y., Chen, M., Xiao, Q., Chen, S.: Highly compact virtual counters for per-flow traffic measurement through register sharing. In: Proceedings of the IEEE GLOBECOM 2016, pp. 1–6 (2016)
25. Zhou, Y., Zhou, Y., Chen, S., Zhang, Y.: Per-flow counting for big network data stream over sliding windows. In: Proceedings of the IEEE/ACM IWQoS 2017, pp. 1–10 (2017)
26. Zhou, Y., Zhou, Y., Chen, S., Zhang, Y.: Highly compact virtual active counters for per-flow traffic measurement. In: Proceedings of the IEEE Conference on Computer Communications (INFOCOM 2018) (2018)

Transformer-Based Rating-Aware Sequential Recommendation

Yang Li[1], Qianmu Li[1(✉)], Shunmei Meng[1], and Jun Hou[2,3]

[1] Nanjing University of Science and Technology, Nanjing, China
{liyang1998,qianmu,mengshunmei}@njust.edu.cn
[2] School of Social Science,
Nanjing Vocational University of Industry Technology, Nanjing, China
[3] Intelligent Manufacturing Department, Wuyi University, Jiangmen, China

Abstract. As a significant application of Big Data, recommender system can effectively solve information overload. The user's behavior sequence forms massive data and has excellent mining value. Sequential recommendation is to extract user's features in massive sequential data and predict the next interaction based on the user's recent temporal behavior. Currently, recurrent neural networks (RNN) and Graph Neural Networks (GNN) take on the role of item embedding in sequential recommendation and have shown adequate performance. However, such RNN based model and GNN based model cannot deeply mine the complex behavior sequence and neglect user preference like rating information. Inspired by the popular Transformer, we adopt the Transformer encoder layer to process sequence and represent item embedding by multi-head attention. Meanwhile, rating information is integrated into weight calculation when we represent the user preference with self-attention. Weight with rating not only retains the structural information of sequence but also combines the user's preferences. What's more, we consider global and local preferences to formulate hybrid performance and make recommendations in Top-N. For persuasiveness, we conduct experiments on large real-world datasets, and our model performs better in most cases on two datasets compared to state-of-the-art methods.

Keywords: Recommender system · Sequential recommendation · Transformer · Attention mechanism · Big data application · Data analytics

1 Introduction

Recommender system is a typical application of Big Data, which can efficiently perform data analysis. Service providers make personalized recommendations for users based on a large number of historical behavior records. As a kind of serialized information in Big Data, behavior record contains the user's preferences to some extends [1]. Accordingly, sequential recommendation aims to extract users' features from massive sequential data and make predictions in the next stage [2].

Considering the timing and consistency of the sequence, the research of sequential recommendations has become a hot issue. Basically, such as Markov-chain-based methods [3] and Socially-Aware Personalized Markov Chains [4] are classic Markov models,

Y. Lai et al. (Eds.): ICA3PP 2021, LNCS 13155, pp. 759–774, 2022.
https://doi.org/10.1007/978-3-030-95384-3_47

which predict users' subsequent behavior according to previous behavior with a basic probability model. Due to the development of deep learning, many scholars have recently combined neural networks with sequential recommendation. In recent years, models based on recurrent neural networks (RNNs) have been widely used. Practically, the recurrent neural network was proposed to apply in the session-based sequential recommendation [5]. Moreover, Neural Attentive Session-based Recommendation (NARM) [6] integrates user's global preferences and local preferences by constructing two RNNs layers.

However, limitations are visible when using RNNs merely to handle complicated relationships between sequences. The representation of the relationship between items is in a predicament by the model based on RNNs, considering the complex and uncertain transitions of items. Hence, attention mechanisms and Graph Neural Networks (GNN) take on the role of RNNs to some extends. The most famous model is Session-Based Recommendation with Graph Neural Networks (SRGNN) [7], which employs GNN to capture embedding of items in session graph and apply attention layers to calculate the weight between adjacent nodes [8]. Target Attentive Graph Neural Networks (TAGNN) [9] and Star Graph Neural Networks with Highway Network (SRNN- HN) [10] make resemble simulate and have their different innovation.

Despite the satisfactory results achieved by the methods above, existing work still has limitations and deficiencies in some ways. Firstly, previous work like GRU4Rec [5], SRGNN [7] and TAGNN [9] both employ Gate Recurrent Unit (GRU) to process the transitions in the session. It is worth noting that user interaction used in the above three models is concise, and it is hard to mine the transfer of user preferences [11]. In other words, the performance of these models in dealing with complex sequences is questionable. It was clearly pointed out that the performance of complex sequences was not satisfactory in the ablation experiment of SRGNN [7]. Secondly, the focus of sequential recommendation is to capture the transition of consecutive items in the sequence. However, it does not mean we can ignore original user preferences information such as ratings. Few works consider both sequence information and rating information in sequential recommendation simultaneously. Some methods, such as TAGNN [9] and SRNN-HN [10], lack rating preference because of the dataset without rating. The other methods, like MFGAN [12] and SSE-PT [13], neglect rating information when modeling users' preferences. So we argue that those methods are still in their infancy and it remains to be improved.

In the light of this challenge, we propose a Transformer-based Rating-Aware Sequential Recommendation (TRA-SR) to tackle the related issues and overcome the limitations mentioned above. In this work, we use transformer [14] encoder layer instead of GRU to model the complex transition pattern in the session, capturing more delicate item representation than the GRU-based model. Additionally, to better use the rating information, we combine rating information with a soft-attention mechanism as the weight for the item in the session. In this way, we can learn the adaptive weights with user preference [15]. Finally, we represent user hybrid preference by item embedding and mixed weight for the recommendation.

To summarize, this work makes the following contributions:

- We utilize the Transformer encoder layer to capture the transition in the user interaction session and achieve item embedding. We improve the efficiency of embedding through the multi-head attention mechanism.
- To the best of our knowledge, we are the few to apply rating information in sequential recommendations. As to details, we propose a self-attention mechanism to learn the weight between items in session and apply the user's ratings of items through a linear layer as another weight. By fusing these two weights, user preferences can be better expressed.
- For persuasiveness, we conduct experiments on two real-world datasets, MovieLens and Last-FM, empirically. We compare the performance against the state-of-the-art baselines by evaluating in metrics HR@20, MRR@20 and NDCG@20.

2 System Overview and Preliminary

2.1 Problem Statement

Sequential recommendation aims to predict the user's next click according to the previous, ongoing interactions. Let $V = \{v_1, v_2, \dots v_{|V|}\}$ denotes all items that the user has interacted with in the sequence where $|V|$ is the length of the whole sequence. We focus on the sub-sequence $V_{sub} = \{v_1, v_2, \dots, v_t, \dots, v_k\}$, which consists of k interactions, and v_t is the t-th item in the subsequence. We aim to analyze the first k interactive records in the sequence and predict the item v_{k+1} that the user will click. In detail, we calculate the probability of all the items in sequence $y = \{y_1, y_2, \dots y_{|V|}\}$ where y_i represent the likelihood of the candidate item v_i. Finally, items with Top-K likelihood will be recommended for the target user.

2.2 System Overview

The workflow of Transformer-based Rating-Aware Sequential Recommendation (TRA-SR) is plotted in Fig. 1. TRA-SR model consists of item embedding, preference representation and prediction layer. First, items v_i in sequence are represented in d-dimension embedding $\mathbf{h}_i \in \mathbb{R}^d$ through the Transformer Encoder Layer. Then we calculate the weight between items in the sequence and combine it with the normalized rating information to achieve comprehensive weight. Next, we get hybrid preference from global and local preferences generated from the item embedding and mixed weight. Finally, sessions have been represented in user preference, and we calculate the score of candidate items through the prediction layer and make Top-N recommendations according to probability ranking.

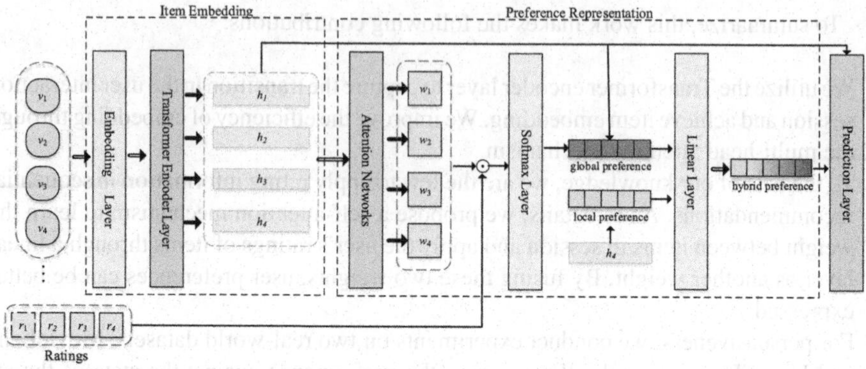

Fig. 1. The architecture of TRA-SR

2.3 Multi-head Attention

Transformer we adopt is widely applied in sequences, and it consists of an encoder layer and decoder layer. For item embedding, we take the encoder layer in our model. Moreover, the kernel of the encoder layer is multi-head attention. It performs three separate linear layers on $\mathbf{X}_{\text{embedding}}$ and assigns three weights in terms of Q, K, V.

$$Q, K, V = Linear\left(\mathbf{X}_{\text{embedding}}\right) = \mathbf{X}_{\text{embedding}}\mathbf{W}_{Q,K,V} \tag{1}$$

where $\mathbf{W}_Q, \mathbf{W}_K, \mathbf{W}_V \in \mathbb{R}^{d \times d}$ and d is the embedding dimension. Q, K, V are three matrix which has the same dimension as $\mathbf{X}_{\text{embedding}}$ and respectively represent query, key and value.

Self-attention is a particular case of multi-head attention mechanism. The attention mechanism can be understood as calculating the correlation. According to the more prominent the dot product of the two vectors, the more similar, they are. So, we calculate QK^T to find the attention matrix, then the V is weighted according to the attention matrix:

$$\mathrm{f}_{\text{Attention}}(Q, K, V) = \text{softmax}\left(\frac{QK^T}{\sqrt{d_k}}\right)V \tag{2}$$

where $\sqrt{d_k}$ is the scaling factor and d_k is the same as dimension d, transforming the attention matrix into normal distribution and Attention$(Q, K, V) \in \mathbb{R}^{n \times d}$. Besides, the softmax function here is used to weight normalization.

Furthermore, multi-head attention divides attention matrix into multiple parts from the embedding dimension, and one part corresponds to one head. So the embedding dimension d must be able to divide the number of heads h. Each head searches independently, and they are contacted by linear layer. And The formula is as follows:

$$\begin{aligned} \mathrm{f}_{\text{MultiHead}}(Q, K, V) &= \text{Concat}\,(head_1, \ldots, head_h)\mathbf{W}^O \\ head_{\text{i}} &= \mathrm{f}_{\text{Attention}}\left(Q_i\mathbf{W}_i^Q, K_i\mathbf{W}_i^K, V_i\mathbf{W}_i^V\right) \end{aligned} \tag{3}$$

where $\mathbf{W}_i^Q, \mathbf{W}_i^K, \mathbf{W}_i^V \in^{d/h \times d/h}$ are the parameter matrices and $Q_i, K_i, V_i \in \mathbb{R}^{n \times h \times (d/h)}$ because they're the division of dimension.

3 Methodology

As indicated in Fig. 1, TRA-SR consists of three parts: Item embedding, preference representation and prediction. Furthermore, we will elaborate on the model from these three aspects in this section.

3.1 Item Embedding

For the session $S = \{v_1, v_2, \ldots, v_k\}$ we get, we need to get the item embedding for subsequent use. In our work, we employ Transformer one encoder layer to get latent embedding of items $\mathbf{H} = \{\mathbf{h}_1, \mathbf{h}_2, \ldots, \mathbf{h}_k\}$. This step mainly includes positional encoding, multi-head self-attention, residual connection and feed-forward network.

Initial Embedding
We first put each item in the session through the embedding layer and transform it into a dense initialization vector. In other words, each item in the session is transformed into a vector in the d dimension as a look-up dictionary.

$$\mathbf{X}_{\text{embedding}} = \mathit{EmbeddingLookup}(\mathbf{X}) \tag{4}$$

where $\mathbf{X} \in \mathbb{R}^n$ and $\mathbf{X}_{\text{embedding}} \in \mathbb{R}^{n \times d}$, n is the items number.

Positional Encoding
The position information is crucial in the sequence. Since the recursion and convolution of loop structure are abandoned in the calculation of transformer, so it needs to be added artificially and provide position information to identify the sequence relationship. We achieve this module as following [14]:

$$\begin{aligned} PE_{\sin}(index, 2i) &= \sin\left(index/10000^{2i/d}\right) \\ PE_{\cos}(index, 2i+1) &= \cos\left(index/10000^{2i/d}\right) \\ \mathbf{X}_{\text{PE}}(1) &= [PE_{\sin}(0), PE_{\cos}(0), \ldots, PE_{\sin}(d/2), PE_{\cos}(d/2)] \end{aligned} \tag{5}$$

where $index$ is the position and i is the dimension of positional encoding. We calculate the positional encoding of each item, and we can get $\mathbf{X}_{\text{PE}} \in \mathbb{R}^{n \times d}$. Due to the same dimension, if we set i as item embedding d, we add them together to get the new embedding $\mathbf{X}_{\text{PE}}$, which contains the position information.

Residual Connection
We put the output of the previous step through multi-head attention and get $\mathbf{X}_{\text{MHA}}$ which has the exact dimension of $\mathbf{X}_{\text{embedding}}$. Then add them up to make the residual connection.

To detail, the input of the previous layer **X** is added to the output of the previous layer *SubLayer*(**X**), which is **X** + *SubLayer*(**X**). At last, we perform Layer-Normalization on the output. It will normalize the hidden layer in the form of standard normal distribution and accelerate the convergence:

$$\begin{aligned} \mathbf{X}_{\text{attention}} &= \mathbf{X}_{\text{embedding}} + \mathbf{X}_{\text{PE}} + \mathbf{X}_{\text{MHA}} \\ \mathbf{X}_{\text{attention}} &= LayerNorm(\mathbf{X}_{\text{attention}}) \end{aligned} \tag{6}$$

where $\mathbf{X}_{\text{embedding}}$ is item embedding, and $\mathbf{X}_{\text{PE}}$ is positional encoding and $\mathbf{X}_{\text{MHA}}$ is the output of multi-head attention.*LayerNorm* function is defined as follow:

$$\begin{aligned} \sigma_j^2 &= \frac{1}{m}\sum_{i=1}^{m}\left(x_{ij} - \frac{1}{m}\sum_{i=1}^{m} x_{ij}\right)^2 \\ LayerNorm(x) &= a \odot \frac{x_{ij} - \mu_i}{\sqrt{\sigma_i^2 + \epsilon}} + \beta \end{aligned} \tag{7}$$

where μ_i is the mean value of the row of the matrix and σ_j^2 is the variance of the row of the matrix, it subtracts the mean value of this row from each element of each row, and divide by the standard deviation of this row to get the normalized value.

Feed-Forward Network (FNN)

The feed-forward network is a two-layer linear mapping and then passes through the activation function:

$$\begin{aligned} \mathbf{X}_{\text{hidden}} &= \text{ReLU}(Linear(Linear(\mathbf{X}_{\text{attention}})) \\ &= \max(0, \mathbf{X}_{\text{attention}}\mathbf{W}_1 + \mathbf{b}_1)\mathbf{W}_2 + \mathbf{b}_2 \end{aligned} \tag{8}$$

where $\mathbf{W}_1, \mathbf{W}_2 \in \mathbb{R}^{d \times d}$ are weight parameter of the linear layer and $\mathbf{b}_1, \mathbf{b}_2$ are bias parameter. ReLu is activation function. $\mathbf{X}_{\text{attention}}$ is the output of the last step.

3.2 Preference Representation

In this component, we concentrate on exploring the correlation between items in the session. Similar to previous work, we try to use an attention mechanism to measure an item's importance in the session and express user preference directly by nodes involved in the session. User's rating information is integrated into the attention mechanism, making weight more reasonable and explainable at the feature level. The preference we consider can be classified by global preference and local preference [11]. In detail, for a session $S = \{v_1, v_2, \ldots, v_k\}$, we simply take $\mathbf{h}_k$, the embedding of the last item v_k, as local preference. Meanwhile, the global preference can be represented by items in the session by comprehensive weight sum. The weight we need to use combines two types of information: the correlation learned from the sequence itself, and the other is users' rating information. Most of the previous work only considered the former and ignored the scoring information. Due to the sequence complexity, the correlation between items in session can be realized by soft-attention mechanism. For the session $S = \{v_1, v_2, \ldots, v_k\}$, the latent embedding can be expressed in $\mathbf{X}_{\text{hidden}} = \{\mathbf{h}_1, \mathbf{h}_2, \ldots, \mathbf{h}_k\}$. Linear layers are adapted to learn the similarity between v_i and v_k:

$$\gamma_i = \mathbf{W}_0^T \sigma(\mathbf{W}_1 \mathbf{h}_i + \mathbf{W}_2 \mathbf{h}_k + \mathbf{b}) \tag{9}$$

where $\mathbf{W}_0 \in \mathbb{R}^d, \mathbf{W}_1, \mathbf{W}_2 \in \mathbb{R}^{d \times d}$ are weight parameters, and γ_i represents the correlation between v_i and v_k. The correspond weight can be defined as $\boldsymbol{\gamma} = \{\gamma_1, \gamma_2, \ldots \gamma_k\}$.

As for the rating information $\mathbf{r} = \{r_1, r_2, \ldots r_k\}$, we first perform the normalization operation to prevent data overflow caused by excessive ratings. Then we multiply it with the weight obtained in the previous step as the input of softmax function:

$$\boldsymbol{\beta} = \text{softmax}(\mathbf{r} \odot \boldsymbol{\gamma}) = \{\beta_i, \beta_2, \ldots \beta_k\} \tag{10}$$

where β_i is the comprehensive weight of v_i in the session, thus, we can calculate global preference $\mathbf{p}_g$ in the form of weighted sums:

$$\mathbf{p}_g = \sum_{i=1}^{k} \beta_i \mathbf{h}_i = \sum_{i=1}^{k} \frac{\exp(r_i \gamma_i)}{\sum_j \exp(r_j \gamma_j)} \mathbf{h}_i \tag{11}$$

Till now, we can compute the hybrid preference $\mathbf{p}_h$ by the global preference $\mathbf{p}_g$ and local preference $\mathbf{h}_k$ with the help of linear layer:

$$\mathbf{p}_h = \mathbf{W}_3 \text{Concat}(\mathbf{p}_g, \mathbf{h}_k) + \mathbf{b}_1 \tag{12}$$

where $\mathbf{W}_3 \in \mathbb{R}^{d \times 2d}$ is the weight parameter which can transform the $2d$ dimension into d dimension and $\mathbf{b}_1 \in \mathbb{R}^{d \times 1}$ is the bias parameter.

3.3 Prediction

After acquiring item embedding and the user's hybrid preference, we can make recommendations by computing the probabilities on all candidate items $v_i \in V$. It should be noted that the candidate items here refer to all items in V rather than part of the items in session S. We calculate the scores $\tilde{\mathbf{y}}$ of item in V by multiplying the hybrid preference with item embedding $\mathbf{h}_i$ as follows:

$$\tilde{\mathbf{y}} = \mathbf{p}_h^{\mathrm{T}} \mathbf{h}_i \tag{13}$$

where $\mathbf{p}_h$ is the hybrid preference and $\mathbf{h}_i$ is i-th candidate item embedding.

Although the score $\tilde{\mathbf{y}}$ can be treated as the basis for Top-N ranking already, considering convergence and distinction, softmax function is adopted here to normalize the score for the probabilities $\hat{\mathbf{y}}$ of items:

$$\hat{\mathbf{y}} = \frac{\exp(\tilde{\mathbf{y}}_i)}{\sum_i \exp(\tilde{\mathbf{y}}_i)} \tag{14}$$

This kind of predictive recommendation problem is essentially a two-category problem. In the case of dichotomy, there are only two cases where the model needs to predict the result at the end: presence and absence. Therefore, the cross-entropy function should be employed in the loss function L:

$$L = \frac{1}{|V|} \sum_i -\left[\mathbf{y}_i \cdot \log(\hat{\mathbf{y}}_i) + (1 - \mathbf{y}_i) \cdot \log(1 - \hat{\mathbf{y}}_i)\right] \tag{15}$$

where $\hat{\mathbf{y}}_i$ is probability of item v_i being selected and $\mathbf{y}_i$ means the label that whether item v_i in the ground-truth, so $\mathbf{y}_i = 1$ if the item is the target item, otherwise $\mathbf{y}_i = 0$.

Furthermore, we use the learning rate decay mechanism when the networks are in Back Propagation training. It will cause oscillations in the search process, causing the parameters to linger near the extreme optimal value if the learning rate is large.

4 Experiment

This section explains experiment settings and then reports the detailed comparison with other state-of-the-art algorithms. Our experiments are designed empirically to answer the following research questions:

RQ1: What's the performance of our model compared to state-of-the-art algorithms?
RQ2: Whether scoring information can improve the accuracy of the model?
RQ3: Whether the model can ensure stable performance under different max lengths?

4.1 Datasets

We evaluate the performance of TRA-SR and other algorithms on two real-world datasets, **Last-FM** and **MovieLens**.

- **Last-FM:** Last-FM HetRec 2K dataset contains a set of 2000 user social networks, tags, listening times and music artists' listening information[16].
- **MovieLens:** A benchmark dataset is widely used in recommender systems. It consists of a large number of detailed ratings and we select ML-1M in this experiment.

For Last-FM and MovieLens, we filter out the items which appear less than five times and users' sequences whose length is smaller than 2, following [7, 17]. We determine the sequence order of actions by timestamp in datasets. Significantly, Last-FM does not have the same obvious rating information as MovieLens. However, the number of times users listen to songs is in the raw dataset. In our experiment, we regard the number of listening as the user's rating, considering it is explicit feedback. Moreover, we implement the data augment because every user has only one sequence in datasets, and it's definitely insufficient for train data. Similar to [9, 10], we split the sequence into several subsets and corresponding labels. For sequence $S = \{v_1, v_2, \ldots v_k\}$,we split it into $\{([v_1], v_2), ([v_1, v_2], v_3), \ldots, ([v_1, \ldots v_{k-1}], v_k)\}$ where $[v_1, \ldots v_{k-1}]$ is training session and v_k is the target label. Thus, a sequence of k length can be conducted into $k - 1$ sub-sequence. Due to the augment, the total number of the session exponentially increases. Significantly, the average sequence length in MovieLens is longer than 100, so we use the latest 50 actions for all datasets and explore the influence of different lengths on the model later. The statistics of datasets after preprocessing are presented in Table 1.

We randomly split the first 70% interactions in sequence as the train set for each user and the recent 30% as the test set for each dataset. From the train set, we take the last 10% as the validation set to tune parameters. Both train set and test set undergoing data augment to achieve session and corresponding label.

Table 1. Datasets statistics

Statistics	Last-FM	ML-1M
#users	1892	6040
#items	4031	3290
#avg actions/user	100.79	165.59
#avg actions/session	14.19	22.47
#sessions in train set	30262	183248
# sessions in test set	10497	76121

4.2 Baselines

To show the performance of our method, we compared the proposed TRA-SR with several representative state-of-the-art methods: **S-POP** is non-personalized method that recommends Top-N popular ranking item, and the popularity is defined as the frequency. **Item-KNN** [18] is a method that recommends items based on the similarity of the last item user interacts in. **GRU4REC** [5] employs RNNs network to model item information and adopts a session-parallel mini-batch training process. **NARM** [6] employs GRU to model the user sequential behavior and utilizes the attention mechanism in the local encoder. **SRGNN** [7] adopts gated graph neural networks to capture the transition in the session and represent users' preference in terms of item latent vector. **TAGNN** [9] improves SRGNN by learning interest representation which varies with different target items and increases expression of the model.

4.3 Evaluation Metrics

To distinguish the performance of the proposed model and baseline method, we adopt three regular Top-K metrics and set K = 20 in our experiment. We evaluate the performance of models under HR@K, MRR@K and NDCG@K respectively.

4.4 Implementation Detail

For our proposed TRA-SR, we use one Transformer encoder layer in item embedding. And the main parts of the model are implemented in PyTorch, and the optimizer is Adam. We set the initial learning rate at 0.001 and decay by 0.1 every three steps. The $\ell 2$ normalization penalty is set to 0.00001. The embedding dimension is fixed on 100, and the batch size is 100 for all models in comparison. The number of heads in multi-heads is 5. And the other hyper-parameters are tuned on the validation set. Here is a statement that the implements of baselines refer to the open-source framework in the recommendation system RecBole [19].

Table 2. Overall Performance with all methods on two datasets The best performance in each column is boldfaced, and the second-best is in underline. Improvements over the second-best method are presented in the last row.

Models	ML-1M			Last-FM		
	HR@20	MRR@20	NDCG@20	HR@2	MRR@20	NDCG@20
S-POP	5.89	1.15	2.15	7.74	2.17	3.38
Item-KNN	16.12	4.38	6.93	35.70	19.47	23.24
GRU4REC	19.54	5.60	8.61	42.31	29.36	32.29
NARM	20.07	5.89	8.93	38.70	25.35	28.33
SRGNN	23.40	7.19	10.71	54.50	41.10	44.12
TAGNN	<u>25.76</u>	<u>7.77</u>	<u>11.68</u>	<u>58.03</u>	<u>47.68</u>	<u>50.02</u>
TRA-SR	**27.91**	**7.89**	**12.23**	**61.37**	**52.55**	**54.56**
Improv	8.34%	1.54%	4.49%	5.75%	10.21%	9.07%

4.5 Performance Comparison

The performances of all methods on two datasets are plotted in Table 2. It can easily distinguish from Table 2 that the proposed TRA-SA has the best performance over baselines, especially on Last-FM.

Observing the second-best method in baselines across two datasets, classic methods without neural networks are unsatisfactory such as S-POP and Item-KNN. In line with expectations, traditional models cannot process sequence and poorly mine the information in the sequence. Due to the similarity of model construction, the performance of

GRU4REC and NARM are at the same level, especially on ML-1M dataset. As a further expansion, SRGNN added the degree of the graph in GRU training, and the improvement was achieved. TAGNN, the most robust baseline, improves SRGNN by adding target item embedding in session representation and obtains the narrow advancement on ML-1M dataset and more steps on Last-FM. Unlike the above methods, our model employs the Transformer encoder layer rather than GRU in item embedding and receives a certain improvement. The improvement from TRA-SR against baselines roughly comes from two components. Transformer encoder layer can better conduct item embedding, and rating information can improve weights as auxiliary data. We will elaborate on them in a later ablation study.

It is a remarkable fact that all methods perform better on Last-FM than ML-1M, and it's consistent with the paper [20], considering that we have a similar way of processing data. Compared to ML-1M, our method improves more against TAGNN on Last-FM. Through data analysis, we find the repeatability makes a tremendous difference. In MovieLens, items in the sequence are unique from each other. That means no duplicates in user interactions. As to Last-FM, the user can listen to one song many times at different timestamps so that the item may occur one more time in sequence. This difference will affect the model in two ways. First, items with high occurrences will have a higher weight through attention mechanism calculation. Second, the Transformer may have greater adaptability on sequence with repeated nodes than GRU due to the positional encoding.

4.6 Ablation Study

Utility of Rating Information

To answer **RQ2**, we experiment on the model without rating information and the model without attention-based weight. For persuasiveness, we also added the two best baseline methods TAGNN and SRGNN, in this comparison. The results in HR@20, MRR@20, NDCG@20 are plotted in Fig. 2. TRA-NR means the proposed model without rating information, and TRA-NA means the proposed model without attention-based weight.

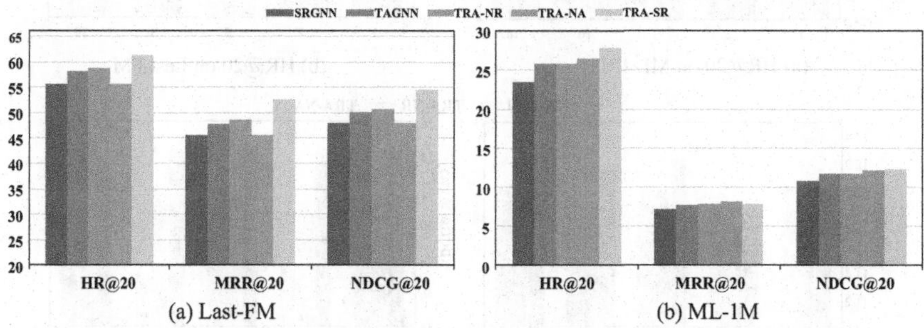

(a) Last-FM (b) ML-1M

Fig. 2. Performance of different components.

As shown in Fig. 2, the rating information can improve the accuracy on two datasets, especially on Last-FM. It's consistent with our original intention that rating information is significant for weight calculation in session representation. The weight learned by the attention mechanism can only reflect the internal relationship of the sequence itself. While rating information can naturally bring user's preference in the sequence in the weight learning. Both components are essential, and combining these two types of information can make the presentation better. From Fig. 2, we can observe that the performance of the proposed model with only one type of weight is not satisfactory on the two datasets under two metrics. And the performance of TRA-NR is not good as TRA-SR, which indicates the rating information plays a significant role in the components of our model. As to Transformer, both SRGNN and TAGNN are implemented by GRU to capture item embedding, and their performance is not good as TRA-NR and TRA-SA. The model without rating information as an auxiliary is better than the strongest baseline TAGNN on Last-FM, and it's at the same level as TAGNN on ML-1M. It confirms that the Transformer encoder layer makes an outstanding contribution to the proposed model, and it's excellent at processing sequences with duplicate items.

Influence of Multi-head Number

As indicated in the methodology, we adopt multi-head attention in the item embedding layer. Here we explore the influence of different head numbers on the results. In Transformer, multi-head attention essentially cuts dimension into several parts. So the embedding dimension must be divided by the head numbers. We empirically observe the multi-head effect within {1, 2, 5, 10, 20}. And the results are plotted in Fig. 3. For credible results and the proposed model TRA-SR, we also carry out experiments on TRA-NA and TRA-NR, which keeps the Transformer component as well.

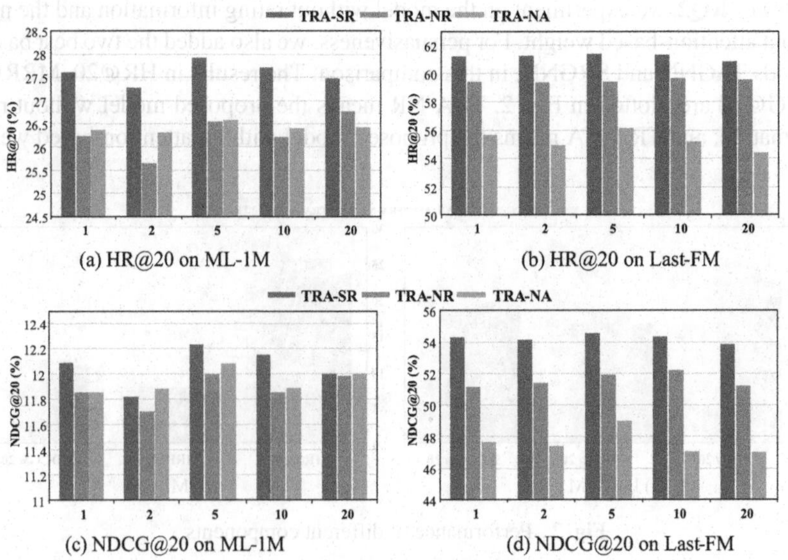

(a) HR@20 on ML-1M (b) HR@20 on Last-FM

(c) NDCG@20 on ML-1M (d) NDCG@20 on Last-FM

Fig. 3. Performance under different head numbers

In Fig. 3, we could find head numbers slightly impact the overall performance of all of the models. In general, the proposed TRA-SR shows better adaptability on different numbers of heads compared to TRA-NA and TRA-NR. From the perspective of HR@20 and NDCG@20, TRA-SR reaches the peak with five heads on two datasets. Meanwhile, it gets the worst performance with two heads on ML-1M and 20 on Last-FM, respectively. And the performance of the model with one head is not as good as the model with five heads on two datasets. It confirms that multi-head can indeed improve the expression of items and help the network capture more affluent features.

Influence of Sequence Length

To answer **RQ3**, we experimented with different max lengths of sequence on two datasets and investigated the sequence length's influence. We compare the proposed model with the best two baselines TAGNN and SRGNN, and the comparison under different max lengths is presented in Fig. 4. In terms of trends, the effect of our model gradually deteriorates with the increase of sequence length on two datasets.

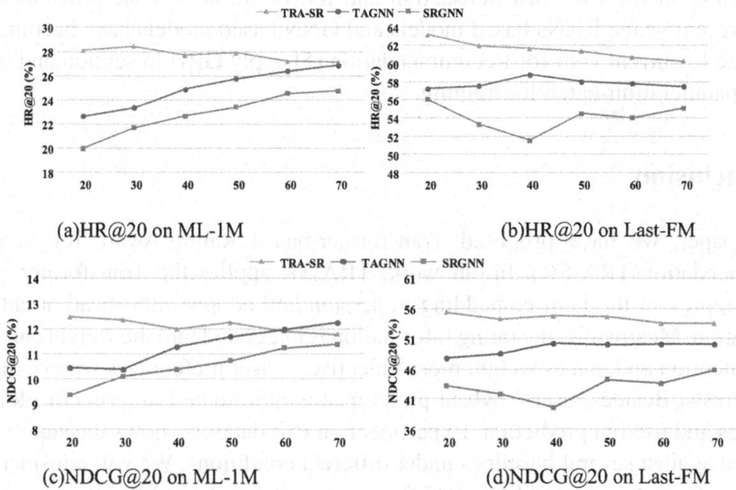

Fig. 4. Performance under different max lengths

On ML-1M, TAGNN and SRGNN, which adopt GRU demonstrate better processing capabilities for longer sequences, but it hasn't shown this trend on Last-FM yet. Our proposed model has an obvious advantage over TAGNN and SRGNN within all the sequence lengths on Last-FM. On the other hand, TRA-SR shows greater effectiveness than TAGNN on ML-1M from length 20 to 60 in terms of HR@20 despite slight backwardness at the length of 70. Overall, our model performs better in most cases on two datasets compared to the strongest baseline.

5 Related Work

5.1 General Recommendation

Collaborative Filtering is the most successful algorithms applied in the recommender system. Classic models such as Matrix Factorization [21, 22] methods seek to represents users' preference and items' properties by matrix. Recently, Neural Collaborative Filtering [23] employs multi-layer perceptron for matrix decomposition. Although CF based on hybrid technology has been applied in various scenarios [24], it's not suitable for capturing sequential patterns due to the lack of modeling sequential interaction.

5.2 Sequential Recommendation

Sequential recommendation seeks to capture sequential patterns among successive items. The fundamental method is Markov Chain based model, such as Factorizing Personalized Markov Chains [3]. Hierarchical Representation Model (HRM) analyses sequential information from the last transaction and moves further on the previous method [25]. In recent years, RNNs based models and GNN based models have become dominant. Gate Recurrent Unit for Recommendation [5] apply GRU in session and utilize a session-parallel mini-batch for training.

6 Conclusion

In this paper, we have proposed Transformer-based Rating-Aware for Sequential Recommendation(TRA-SR). In our work, TRA-SR applies the Transformer encoder layer to represent the item embedding in session and adopts multi-head attention for optimization. Meanwhile, the rating information is integrated into the weight calculation of self-attention and makes weight more reflective of user preferences when expressing user interests. Besides, users' hybrid preference is represented in terms of global and local ones and used in prediction. Experiment on two datasets shows the superiority of the model against several baselines under different conditions. We will consider fusing semantic information as auxiliary item feature into the model in future work.

Acknowledgements. This work was supported in part by the National Key R&D Program (2020YFB1804604), the 2020 Industrial Internet Innovation and Development Project from Ministry of Industry and Information Technology ,Jiangsu Province Modern Education Technology Research Project (84365); Scientific research project of Nanjing Vocational University of Industry Technology(2020SKYJ03).

References

1. Wu, L., Yu, H.-F., Rao, N., Sharpnack, J., Hsieh, C.: Graph DNA: Deep neighborhood aware graph encoding for collaborative filtering. In: The 23rd International Conference on Artificial Intelligence and Statistics, 2020, Online, pp. 776–787. PMLR (2019)

2. Wang, B., Cai, W.: Knowledge-enhanced graph neural networks for sequential recommendation. Inf. **11** (2020)
3. Rendle, S., Freudenthaler, C., Schmidt-Thieme, L.: Factorizing personalized Markov chains for next-basket recommendation. In: Proceedings of the 19th International Conference on World Wide Web, WWW 2010, pp. 811–820. ACM (2010)
4. Cai, C., He, R., McAuley, J.: SPMC: Socially-aware personalized Markov chains for sparse sequential recommendation. In: IJCAI International Joint Conference on Artificial Intelligence, pp. 1476–1482. ijcai.org (2017)
5. Hidasi, B., Karatzoglou, A., Baltrunas, L., Tikk, D.: Session-based recommendations with recurrent neural networks. In: 4th International Conference on Learning Representations, ICLR 2016 - Conference Track Proceedings (2016)
6. Li, J., Ren, P., Chen, Z., Ren, Z., Lian, T., Ma, J.: Neural attentive session-based recommendation. In: International Conference on Information and Knowledge Management, Proceedings, pp. 1419–1428. ACM (2017)
7. Wu, S., Tang, Y., Zhu, Y., Wang, L., Xie, X., Tan, T.: Session-based recommendation with graph neural networks. In: 33rd AAAI Conference on Artificial Intelligence, AAAI 2019, pp. 346–353. Press (2019)
8. Kang, W.C., McAuley, J.: Self-attentive sequential recommendation. In: Proceedings - IEEE International Conference on Data Mining, ICDM, pp. 197–206. IEEE (2018)
9. Yu, F., Zhu, Y., Liu, Q., Wu, S., Wang, L., Tan, T.: TAGNN: target attentive graph neural networks for session-based recommendation. In: SIGIR 2020 - Proceedings of the 43rd International ACM SIGIR Conference on Research and Development in Information Retrieval, pp. 1921–1924 (2020)
10. Pan, Z., Cai, F., Chen, W., Chen, H., De Rijke, M.: star graph neural networks for session-based recommendation. In: International Conference on Information and Knowledge Management, Proceedings, pp. 1195–1204 (2020)
11. Liu, Q., Mokhosi, R., Zeng, Y., Zhang, H.: STAMP: short-term attention/memory priority model for session-based recommendation. In: Proceedings of the ACM SIGKDD International Conference on Knowledge Discovery and Data Mining, pp. 1831–1839. ACM (2018)
12. Ren, R., Liu, Z., Li, Y., Zhao, W.X., Wang, H., Ding, B., Wen, J.R.: Sequential Recommendation with Self-Attentive Multi-Adversarial Network. In: SIGIR 2020 - Proceedings of the 43rd International ACM SIGIR Conference on Research and Development in Information Retrieval, pp. 89–98. ACM (2020)
13. Wu, L., Li, S., Hsieh, C.J., Sharpnack, J.: SSE-PT: sequential recommendation via personalized transformer. In: RecSys 2020 - 14th ACM Conference on Recommender Systems, pp. 328–337 (2020)
14. Vaswani, A., et al.: Attention is all you need. In: Advances in Neural Information Processing Systems, pp. 5999–6009 (2017)
15. Oliveira, J., Nogueira, M., Ramos, C., Renna, F., Ferreira, C., Coimbra, M.: Using soft attention mechanisms to classify heart sounds. In: Proceedings of the Annual International Conference of the IEEE Engineering in Medicine and Biology Society, EMBS, pp. 6669–6672. IEEE (2019)
16. Da Conceição Moreira, P.S., Tsunoda, D.F.: LAST.FM songs database: a database for musical genre classification. In: IC3K 2018 - Proceedings of the 10th International Joint Conference on Knowledge Discovery, Knowledge Engineering and Knowledge Management, pp. 253–260. SciTePress (2018)
17. Xu, C., Feng, J., Zhao, P., Zhuang, F., Wang, D., Liu, Y.: Long- and short-term self-attention network for sequential recommendation. Neurocomputing **423**, 580–589 (2021)
18. Sarwar, B., Karypis, G., Konstan, J., Riedl, J.: Item-based collaborative filtering recommendation algorithms. In: Proceedings of the 10th International Conference on World Wide Web, WWW 2001, pp. 285–295. ACM (2001)

19. Zhao, W.X., Mu, S., Hou, Y.: RecBole: towards a unified, comprehensive and efficient framework for recommendation algorithms. In: The 30th International Conference on Information and Knowledge Management, Virtual Event, 2021. pp. 4653–4664. ACM (2021)
20. Lonjarret, C., Auburtin, R., Robardet, C., Plantevit, M.: Sequential recommendation with metric models based on frequent sequences. Data Min. Knowl. Disc. **35**(3), 1087–1133 (2021). https://doi.org/10.1007/s10618-021-00744-w
21. Siy, P.W., et al.: Matrix factorization techniques for analysis of imaging mass spectrometry data. In: 8th IEEE International Conference on BioInformatics and BioEngineering, BIBE 2008, pp. 1–6. IEEE (2008)
22. Zhang, X.Q., Liu, X.X., Guo, J., Liu, B.Y., Gan, D.G.: A matrix factorization based recommendation algorithm for science and technology resource exploitation. In: Proceedings of IEEE/ACS International Conference on Computer Systems and Applications, AICCSA, pp. 1–6. IEEE (2020)
23. He, X., Liao, L., Zhang, H., Nie, L., Hu, X., Chua, T.S.: Neural collaborative filtering. In: 26th International World Wide Web Conference, WWW 2017, pp. 173–182. ACM (2017)
24. Meng, S., Gao, Z., Li, Q., Wang, H., Dai, H.N., Qi, L.: Security-Driven hybrid collaborative recommendation method for cloud-based iot services. Comput. Secur. **97**, 101950 (2020)
25. Wang, P., Guo, J., Lan, Y.: Learning hierarchical representation model for next basket recommendation. In: SIGIR 2015 - Proceedings of the 38th International ACM SIGIR Conference on Research and Development in Information Retrieval, pp. 403–412. ACM (2015)

An Effective Single-Pass Approach for Estimating the Φ-quantile in Data Streams

Zhengyuan Xue(✉)

Henan University of Technology, Zhengzhou 450001, China
xuezy@haut.edu.cn

Abstract. Random sampling is a common method to deal with large-scale data sets and in particular to deal with data streams. However, the accuracy of this method decreases greatly with the reduction of sampling size when dealing with quantile-related problems. It is known that $\Omega(\varepsilon^{-1} \log \log \delta^{-1})$ space are required, to estimate the rank of any query item up to additive error εn with probability of at least $1 - \delta$, where n is the total number of items in the data stream. In many cases, however, we cannot predict the scale of a data stream in advance. This paper proposes a novel approach for estimating the Φ-quantile item in a data stream. The Φ-quantile result of the approach has high accuracy, and the required extra space is very little. Specifically, the general idea is as follows: firstly, we properly divide the items in the data stream into several groups, then the extreme values in each group are properly selected, and the estimated quantiles are finally obtained based on summarizing each extreme value. In the above process, we carefully calculate the proper number of groups, the extreme value of each group, and design the summarizing of the extreme values, to estimate the Φ-quantile in the data stream with a single-pass. At the same time, ensuring that the expected rank of the result item is exactly equal to the rank of the desired item. A preliminary analysis of the cost of the proposed approach is also given. Moreover, an extensive empirical study using synthetic data sets is reported to verify the effectiveness of the methods.

Keywords: Quantile estimation · Φ-quantile · Data stream · Single pass

1 Introduction

In the rapid development of big data era, the explosive growth in data size brings us a great challenge to deal with these data. Rigorous accuracy is less important than getting a quick grasp of their broad outlines or progress over time [1]. Given a large amount of data, a first and foundational problem is to describe the data distribution.

Quantiles are the most commonly used non-parametric representation for data distributions. Imagine you are given a really large stream of data items, e.g., queries on Google searches last month, sales of Walmart during the Christmas season, or online network flow monitoring data of a university. Note that you probably do not know the

This work is supported by the High level talent foundation of Henan University of Technology under Grant 2018BS057.

Y. Lai et al. (Eds.): ICA3PP 2021, LNCS 13155, pp. 775–789, 2022.
https://doi.org/10.1007/978-3-030-95384-3_48

size of the data stream in advance. To quickly grasp the outline of the large data, now your goal is to efficiently find out or to estimate the "entry score" of the top 1% items, i.e., the *1%-quantile* item.

Quantile computation is arguably one of the most fundamental problems in data analysis. Computing the quantiles has significant practical importance. In the Sawzall language that is the basis for all of Google's log data analysis, quantile computation is one of the seven basic operators defined (the others including sum, max, top-k, and count-distinct, etc.) [2]. The quantiles also play an important role in network health monitoring [3, 4] and data collection in wireless sensor networks [5, 6], etc.

Given a data set, it seems not difficult to find out the Φ-quantile. This is true when the data scale is not large enough. However, in the case of a data stream with a large data scale, only a single pass over the data is possible, and the quantile is computed using in-memory structures that are not able to store the entire data seen so far. Hence, quantile computation in a data stream is generally approximate, with a provable guarantee on the quality of approximation.

Although people no longer seek an exact result, but only an approximate result (e.g., ε-approximate quantile), it is still difficult to make a great breakthrough in storage space. The state-of-the-art work in [7] proved that, even though we only desire an estimated result of the ε-approximate Φ-quantile, at least $\boldsymbol{\Omega}(\varepsilon^{-1} \log \log \delta^{-1})$ space is required, to make sure the error not bigger than εn with probability at least $1-\delta$. However, this result is still not satisfactory, as ε and δ are usually very small in practical applications. To get more practical and feasible methods in terms of space requirements, this paper proposes a novel approximate Φ-quantile algorithm. The algorithm needs little space, while the accuracy of the results is guaranteed.

The problem can be defined as follows. Given a large scale data stream D with unknown data size of items, and a fixed ratio Φ ($0 < \Phi \leq 1/2$), we need to estimate the Φ-quantile item in D with a single-pass, such that the expected rank of the result item is exactly equal to the rank of the desired item, and the sample size is as small as possible. The rank of item x is the number of stream items x_i such that $x_i \geq x$.

To the best of our knowledge, there is no other work gives an available and effective solution to the proposed problem. In this paper, we propose a simple strategy that requires significantly less memory than the existed methods [7–9] for solving the Φ-quantile problem.

The main challenges mainly include three aspects:

- The first challenge lies in that the data size n of the data stream is too large (n is not known in advance). Thus we cannot store all items in the data stream for further processing.
- The second challenge is, the desired quantile is w.r.t. Φ, a fixed ratio, rather than a fixed quantity of items. With an unknown data size, the existing selection algorithms (e.g., Quickselect [10], Priority queues [11], and DC-Top-k [12], etc.) no longer works.
- The last major challenge is, whether there exists a method that requires minimal space to fit the bill? Moreover, how can we get it and how to balance the required space and the accuracy of the algorithm?

Of course, the first challenge can be alleviated by referring to the *Reservoir sampling* method [13] to get an approximate solution. But then a serious question arises due to the second challenge. Folklore analysis shows that if the sample has size $\mathbf{O}(\varepsilon^{-2}\log\delta^{-1})$, the Φ-quantile of the sample is an ε-approximate quantile of the input data set with probability of at least $1-\delta$ [9, 14]. Note that this value $\mathbf{O}(\varepsilon^{-2}\log\delta^{-1})$ seems larger than the state-of-the-art lower bound $\mathbf{O}(\varepsilon^{-1}\log\log\delta^{-1})$. Is there an estimator whose expected rank is Φn, and can be reliably computed using very little or optimal space? How many items should we sample, and what should we further do after the sampling? Note that computing the maximum (or minimum) in a data stream requires only $\mathbf{O}(1)$ space. Thus, if we can get a sample in which the expected rank of the maximum is exactly Φn, we may obtain the desired result.

In summary, the contributions in this paper are as follows:

- Firstly, we give the minimal sample size when using the sampling method to find the Φ-quantile item in a data stream with an unknown data scale.
- Secondly, to achieve a more accurate estimation result, we propose a novel approach for estimating the Φ-quantile item. As a result, the estimation is more accurate, and the required extra space is very little.
- Thirdly, we test the proposed algorithm with large amount of synthetic data by extensive experiments to verify the effectiveness of the proposed algorithm.

The rest of the paper is organized as follows. Section 2 introduces the related work. And then, to answer the above questions, we first propose an important conclusion about estimating the k-th largest in n items and give the proof in Sect. 3. We also give the method and related conclusions for finding the minimal sample size in Sect. 3. On this basis, Sect. 4 further give the overall solution to the proposed problem. Section 5 conducts some experiments, and Sect. 6 concludes the paper.

2 Related Work

Given a stream of n items, the Φ-quantile of the stream is the item whose rank in this stream equals Φn. Due to the particularity of the data stream, it is often allowed to traverse the data only once, i.e., single-pass. Moreover, the data size is particularly large, or the data size is unknown, and the data cannot be fully loaded into memory. Researchers usually can only get an approximate result instead of an accurate result. Specifically, the ε-approximate Φ-quantile problem asks for some single-pass algorithm such that after reading a data stream of n items, we can find, for a fixed Φ $(0 < \Phi \leq 1/2)$, an item x whose rank is in $[(\Phi - \varepsilon)n, (\Phi + \varepsilon)n]$. We call x an ε-approximate Φ-quantile of the data stream. Note that we do not know the data size n in advance. The main challenge is to use as little memory as possible.

To solve the ε-approximate quantiles problem, existed work can mainly categorizes into two classes [14–16], i.e., the deterministic algorithm [17, 18], where an algorithm returns a quantile between $\Phi - \varepsilon$ and $\Phi + \varepsilon$, and the randomized algorithm [5, 7, 8, 19, 20], where an algorithm may return an incorrect quantile (i.e., exceeding the stated ε error) with a small probability.

Hung *et al.* [17] investigate the lower bound of the deterministic algorithm. As a result, they proved that any comparison-based algorithm needs $\mathbf{\Omega}(\varepsilon^{-1}\log\varepsilon^{-1})$ space for finding ε-approximate quantiles. The authors of [18] propose a deterministic algorithm that requires $\mathbf{O}(\varepsilon^{-1}\log(n\varepsilon))$ space. This is the best known deterministic algorithm for the ε-approximate Φ-quantile problem [7].

Some researchers focus on the randomized algorithm in randomly ordered streams [8, 19]. They show that any algorithm computing the median of a stream presented in random order, using $\log(n)$ space, requires optimal $\mathbf{\Omega}(\log\log n)$ passes. Guha *et al.* [8] also give some other results in the random-order model. Felber *et al.* [20] propose a randomized algorithm that using $\mathbf{O}(\varepsilon^{-1}\log\varepsilon^{-1})$ space to solve the ε-approximate Φ-quantile problem with success probability $1-\mathrm{e}^{-\mathrm{poly}}(\varepsilon^{-1})$. Huang *et al.* [5] focus on the communication-efficiently computing approximate quantiles in large-scale sensor networks, and present a sampling-based quantile computation algorithm with less communication. Recently, Karnin *et al.* [7] obtained an $\mathbf{O}(\varepsilon^{-1}\log\log\delta^{-1})$ space algorithm and proved that this is the exact lower bound of the randomized algorithm, which is considered as the state-of-the-art work.

There are some other works related to our quantiles topic. [21] estimates quantiles from the union of historical and streaming data. [22] applies the existed deterministic stream quantile algorithm to Monte Carlo simulation in order to estimate quantile with limited memory. Based on the existed quantile algorithms, [23] proposes space-efficient algorithms for performing the Kolmogorov-Smirnov test on streaming data. Moreover, [24] studies the moment estimators for quantile approximation from an empirical perspective, which is somewhat weak in theory. [25] proposes a technique for dynamic tracking of quantiles of data streams, and [26] proposes a technique for data-center telemetry monitoring. They both focus on the dynamics of data.

Readers can see [15, 16] for surveys of existed quantile algorithms in data streams. The surveys provide a theoretical comparisons and extensive experimental analysis. On the whole, there is no other work that gives an available and effective solution to the proposed problem in *Introduction* Section. In this paper, we propose a simple strategy that requires significantly less memory than the existed methods for solving the Φ-quantile problem. On this basis, we further propose an effective single-pass approach for estimating the Φ-quantile in data streams.

3 Finding the Minimal Sample Size

3.1 Estimating the *k*-th Largest Item

To get the minimal sample size when using the sampling method to find the Φ-quantile item in a data stream with an unknown data scale, we first propose a theorem and prove the result. On this basis, further results are given.

Theorem 1. *Suppose the input* $\mathbf{R}_n$ *is consists of n numbers. If we randomly sample* $x = (n-k+1)/k$ *items (denoted as* $\mathbf{R}_x$*) from* $\mathbf{R}_n$*, then the expected rank-k largest will be* d_{max}*, the maximum in* $\mathbf{R}_x$*.*

Proof. Denote by $\boldsymbol{R}_x$ the set of x items randomly sampled from $\boldsymbol{R}_n$ ($\boldsymbol{R}_x \subset \boldsymbol{R}_n$, $1 <= x < n$), and $d_{\max}$ as the maximum in $\boldsymbol{R}_x$. For other items in the set $\boldsymbol{R}_n - \boldsymbol{R}_x$, we need to find out the number of items which are larger than $d_{\max}$.

Suppose $\boldsymbol{R}_n' = \{e_1, e_2, \ldots, e_n\}$ is a rearrangement of $\boldsymbol{R}_n$, and with the items ordered randomly. This assumption will not affect the result even if the items in $\boldsymbol{R}_n$ are not ordered randomly, since we only need to know the quantity of the items larger than $d_{\max}$ in $\boldsymbol{R}_n - \boldsymbol{R}_x$, and $\boldsymbol{R}_x$ is also sampled randomly. Thus, in the rest of the proof, we consider $\boldsymbol{R}_n$ as $\boldsymbol{R}_n'$ with the items ordered randomly.

Denote by m_j the j-th largest item in $\boldsymbol{R}_n'$. Then the probability of the i-th item e_i is exactly m_j

$$P(e_i = m_j) = (n-1)!/n! = 1/n (1 <= i <= n, <= j <= n) \quad (1)$$

For any item $e_i \in \boldsymbol{R}_n - \boldsymbol{R}_x$, i.e., $e_i \in \boldsymbol{R}_n' - \boldsymbol{R}_x$, we get the probability of $e_i > d_{\max}$

$$\begin{aligned} P(e_i > d_{max}) &= \sum_{j=1}^{n-x} P(e_i = m_j) P(e_i > d_{max} | e_i = m_j) \\ &= (1/n) * \sum_{j=1}^{n-x} A_{n-j}^{x} A_{n-1-x}^{n-1-x} / (n-1)! \\ &= (n-1-x)!x!/n! \sum_{j=1}^{n-x} C_{n-j}^{x} \\ &= (n-1-x)!x!/n! \, C_n^{x+1} \\ &= 1/(x+1) \end{aligned} \quad (2)$$

where $\boldsymbol{A}$ denotes the number of permutations, e.g.,

$$A_n^m = n!/(n-m)!$$

and $\boldsymbol{C}$ denotes the number of combinations, e.g.,

$$C_n^m = A_n^m / m!$$

According to (1), for any item $e_i \in \boldsymbol{R}_n - \boldsymbol{R}_x$, e_i is larger than $d_{\max}$ with the probability of $p = 1/(x+1)$. Denote by Y the number of items larger than $d_{\max}$ in $\boldsymbol{R}_n - \boldsymbol{R}_x$. Clearly, we know that the random variable Y is binomial distributed, i.e., $Y \sim B(n-x, p)$. Denote by $E(Y)$ the Expected value of Y, then $E(Y) = (n-x)p$. Let $E(Y) = k-1$ to make $d_{\max}$ the k-th largest on average, we get $x = (n-k+1)/k$.

In conclusion, we need to randomly sample $x = (n-k+1)/k$ items from n items and then find out the maximum $d_{\max}$ in the sample. From the perspective of mathematical expectation, $d_{\max}$ is the k-th largest in the n items (there are $k-1$ items larger than $d_{\max}$ on average). ■

3.2 Estimating the Φ-quantile by Minimal Sample Size

According to Theorem 1, to estimate the k-th largest from n items, we need to randomly sample $(n - k + 1)/k$ items. Thus, to estimate the "entry score" of the top 1% (Φ = 1%), i.e., the $(n/100)$-th largest item, we need to randomly sample about $(1/\Phi - 1) = 99$ items.

Above all, the minimal sampling method to estimate the Φ-quantile in a data stream using single-pass is as follows.

(1) Randomly sample the items from the data stream with a sample size of $1/\Phi$. Specifically, we should store the first $1/\Phi$ items into a "reservoir"; then we select the i-th item with a probability of $p_i = 1/(\Phi i)$ $(i > 1/\Phi)$, and it should randomly replace one of the $1/\Phi$ items in the "reservoir", provided that the i-th item is indeed selected, or else we continue to deal with the next item.
(2) When the sampling is completed (the data stream is over), we find out the maximum of the $1/\Phi$ items in the "reservoir", $d_{\max}$. Thus, $d_{\max}$ is considered as an estimate of the Φ-quantile item in the data stream.

Cost Analysis: Note that step (1) is a sampling procedure, and it is almost real-time w.r.t the data stream since we only need to execute a random function at most two times when a data comes [13]. Thus we only focus on the cost in step 2. Clearly, in step 2 we only need to compare the items for $(1/\Phi - 2)$ times to find the maximum as the final result. So the cost of step 2 in the method is only $(1/\Phi - 2)$ comparison operations between items.

In contrast, the state-of-the-art work [7] proved that, to get an estimated result of the ε-approximate Φ-quantile, at least $\Omega(\varepsilon^{-1} \log\log \delta^{-1})$ space is required, to make sure the error not bigger than εn with probability at least $1 - \delta$.

It is not difficult to find that the above sampling method realizes the minimal sampling using single-pass to solve the quantile problem in the data stream. The extra space used by this method is very small, but unfortunately, the accuracy of the quantile result obtained by this sampling method is not high (this is also confirmed in the following experiments). Therefore, although this result has theoretical significance, it has little practical application value.

To better estimate the quantile in the large-scale data stream with an unknown scale that using single-pass, we further explore and propose a practical and feasible approach considering both the space used by the algorithm and the accuracy of the result. This is partially based on the conclusions in Theorem 1. Most importantly, we abandon the sampling strategy and propose a completely new approach.

4 Proposed Approach for Φ-quantile

In order to introduce the approach more clearly, first, we give a new theorem, i.e., Theorem 2.

Theorem 2. *Suppose the input $\boldsymbol{R}_n$ is consists of n numbers. If we randomly divide the data into k groups on average and set the maximums in the k groups respectively as $max_1, max_2, max_3, \ldots, max_k$. Denote by threshold the minimum of the k maximums, i.e., threshold $= min(max_1, max_2, max_3, \ldots, max_k)$. Denote by Q the number of items in the rest (n-k) items greater than threshold. Then the Expected value of Q (denoted by E(Q)) satisfies $E(Q) \approx kH_k$. H_k is the k-th harmonic number.*

Proof. Suppose that $\boldsymbol{R}_n = \{e_1, e_2, \ldots, e_n\}$ are ordered randomly. Denote by m_j the j-th largest item in $\boldsymbol{R}_n$. According to (1) in Theorem 1, the probability of the i-th item e_i is exactly m_j

$$P(e_i = m_j) = 1/n(1 <= i <= n, 1 <= j <= n) \tag{3}$$

Denote $\boldsymbol{R}_s$ as a set of s items, $\boldsymbol{R}_s \subset \boldsymbol{R}_n$, and rm_s as the maximum in $\boldsymbol{R}_s$ ($1 <= s < n$). According to (2) in Theorem 1, For any item $e_i \notin \boldsymbol{R}_s$ (i.e., $e_i \in \boldsymbol{R}_n - \boldsymbol{R}_s$, totally n-s), we get the probability of $e_i > rm_s$

$$P(e_i > rm_s) = 1/(s+1) \tag{4}$$

Let $N = n/k$ stands for the number of items in each group. Denote $\boldsymbol{G}_t$ as a set of t groups totally tN items, $\boldsymbol{G}_t \subset \boldsymbol{R}_n$, and gm_t as the maximum in $\boldsymbol{G}_t$ ($1 <= t < k$). According to (4), for any item $e_i \notin \boldsymbol{G}_t$ (i.e., $e_i \in \boldsymbol{R}_n - \boldsymbol{G}_t$, totally n-tN), we can get the probability of $e_i > gm_t$

$$P(e_i > gm_t) = 1/(tN+1) \tag{5}$$

Denote by Q_t the number of all the items which are bigger than gm_t, we know that the random variable Q_t is binomial distributed, i.e., $Q_t \sim B((k-t)N, 1/(tN+1))$, according to the above analysis. Thus,

$$E(Q_t) = (k-t)N/(tN+1) \approx k/t - 1 \tag{6}$$

Denote by Q' the number of all the items that are bigger than *threshold*, we know that $Q' = k+Q$. According to the *Inclusion-exclusion Principle*, we get

$$\begin{aligned} E(Q') &= C_k^1 E(Q_1) - C_k^2 E(Q_2) + C_k^3 E(Q_3) - \ldots + (-1)^k C_k^{k-1} E(Q_{k-1}) \\ &= \sum_{t=1}^{k-1} (-1)^{t+1} C_k^t E(Q_t) \approx \sum_{t=1}^{k-1} (-1)^{t+1} C_k^t (k/t - 1) \\ &= k \sum_{t=1}^{k-1} (-1)^{t+1} C_k^t /t - \sum_{t=1}^{k-1} (-1)^{t+1} C_k^t \\ &= k(H_{k-1} + 1) - 1 - (-1)^k \end{aligned} \tag{7}$$

According to $Q' = k+Q$, we get the Expected value of Q

$$E(Q) = E(Q') - k = k\mathrm{H}_{k-1} - 1 - (-1)^k \approx kH_k \tag{8}$$

where H_k denotes the k-th harmonic number

$$H_k = \sum_{i=1}^{k} \frac{1}{i} = \ln_k + \gamma + o(1)$$

and γ is the Euler constant. ■

Given a data stream of n items, then the Φ-quantile is the item whose rank in this stream equals Φn. Suppose we evenly divide all the items in the data stream into k' groups, then find out the maximum in each group, and denote by *threshold*' the minimum of the k' maximums. Based on the conclusion of Theorem 2, we know that the mathematical expectation of the rank of *threshold*' is $E(r_{threshold'}) = k'H_{k'} + k'$. Let $E(r_{threshold'}) = k'H_{k'} + k' = \Phi n$, and figure out the appropriate number of groups k' $= f(\Phi n)$, the data size of each group *num*' $= n/k' = (H_{k'} + 1)/\Phi$. We can see that the number of groups k' and the data size of each group *num*' are determined by n and Φ. When the items in a data stream are evenly divided into $k' = f(\Phi n)$ groups, then the corresponding *threshold*' can be an estimation of the Φ-quantile in the data stream.

However, since it is generally difficult to predict the size of the data stream in advance, the exact value of n can not be obtained until the end of the data stream, which results in the k' value (i.e., the proper number of groups to be divided). The *num*' value (i.e., the proper data size of each group) cannot be obtained in advance.

Next, we further design a series of strategies to solve this problem. The adjusted approach does not need to depend on the total size (i.e., n) of the data stream in advance.

Obviously, in the above method, the appropriate number of groups $k' < \Phi n$, and the data size of each group *num*' $= n/k' > 1/\Phi$. Therefore, we use $num_0 = 1/\Phi$ items as the basic unit. Thus, there are $num'/num_0 = (n/k')/(1/\Phi) = (\Phi n)/k' = (k'H_{k'} + k')/k' = H_{k'} + 1$ basic units in a group of items.

Specific solutions are given as follows:

(1) Start with receiving the items in the data stream in turn, save a maximum in each basic unit (i.e., $num_0 = 1/\Phi$ items) until the data stream is complete. Denote by n the total size of the data stream, then the total number of maximums is Φn;
(2) Let $k'H_{k'} + k' = \Phi n$ and figure out the value of k'; then find out the maximum in each group, that is, merge $H_{k'} + 1$ maximums of $H_{k'} + 1$ basic units into one maximum in turn;
(3) There are now k' maximums, one in each group. Find out *threshold*', the minimum among the k' maximums, which is approximately the Φ-quantile required in the data stream.

See Algorithm 1 for the details of the proposed approach.

Algorithm 1 : Algorithm for estimating the *Φ*-quantile in a data stream

```
Input: r[0..n-1]: the n source items in the data stream, one after another
       Φ: the quantile in the data stream we need
  Output: Φ_quantile: The item of Φ quantile in the data stream
  1 num0 ← 1/Φ;
  2 i ← 0;
  3 for uld ← 0 to i/num0 do
  4   max[uld] ← r[uld * num0];
  5   for i ← uld*num0 to uld*num0+num0-1 do
  6     if r[i] == END then
  7       break;
  8     elseif r[i] > max[uld] then
  9       max[uld] ← r[i];
  10      i++;
  11    else
  12      i++;
  13    endif
  14  endfor
  15 endfor
  16 n ← i;
  17 k'← Root (k'H_k'+k'= n*Φ);
  18 num1 ← H_k'+1;
  19 for gld ← 0 to k'-1 do
  20   Max[gld] ← max[gld * num1];
  21   for i ← gld*num1 to gld*num1+num1-1 do
  22     if max[i] > Max[gld] then
  23       Max[gld] ← max[i];
  24     endif
  25   endfor
  26 endfor
  27 threshold' ← Max[0];
  28 for gld ← 1 to k'-1 do
  29   if Max[gld] < threshold' then
  30     threshold' ← Max[gld];
  31   endif
  32 endfor
  33 Φ_quantile ← threshold';
  34 return Φ_quantile;
```

In Algorithm 1, lines 1–15 implement Step(1) to find each maximum *max*[] within each basic unit. num_0 denotes the number of items in each basic unit, variable *i* counts the amount of items received from the data stream, *uld* denotes the number of a basic unit, and END denotes the end flag of the data stream. Lines 16–26 implement Step(2) to find the maximum in each group, that is, to find the maximum in those basic units contained in the group. *n* is the total size of the data stream, *k'* is the number of groups to be divided, **Root** is the root of the equation, num_1 is the number of basic units contained in each group, and *gld* is the number of a group. By summarizing the num_1 basic units in each group, the maximum *Max*[] in each group is deduced. Lastly, lines 27–34 implement Step(3) by finding the minimum in *Max*[] as *threshold'*, which is the proper estimation of *Φ_quantile*.

The Number of Comparisons in Algorithm 1: After an in-depth analysis of Algorithm 1, it can be found that the purpose of comparison operations in the algorithm is to get the maximal or the minimal item. Thus, each comparison operation discards one item (the comparison operations in Step(1)(2) discard the smaller item each time, and the comparison operations in Step(3) discard the larger item each time). Therefore, the total number of comparison operations in the algorithm is $n - 1$.

Extra Space Required by Algorithm 1: In Step(1) of the algorithm, the maximums of each basic unit are successively retained, totaling Φn. In Step(2), the quantity of the reserved items is further reduced based on the items retained in Step(1), and in Step(3), the quantity of the reserved items is further reduced based on the items retained in Step(2). Therefore, the extra space required by the algorithm is Φn.

5 Experimental Analysis and Discussions

In this section, we test the performance of the minimal random sampling method and the proposed Algorithm 1. We use synthetic data sets for experiments. All the items in the data stream with double precision are generated by the computer randomly, with a range of 1 to 10^8. The data size of the simulated data stream is $n = 10^7$ (n is unknown in advance). Experiments were made with $\Phi = 10^{-4}, 10^{-3}, 10^{-2}, 10^{-1}$, respectively. That is, four upper quantiles were tested. The results were 500 runs of the algorithm in each experiment.

Table 1 lists the mean value and standard deviation of the experimental results of 500 runs by the two algorithms, i.e., the minimal random sampling method and the Algorithm 1 proposed in this paper.

Table 1. Mean value and standard deviation

Φ	Minimal sampling		Algorithm 1	
	Mean value	Std. deviation	Mean value	Std. deviation
10^{-4}	1043	1050.60	930	214.81
10^{-3}	9067	9052.64	9431	1619.13
10^{-2}	92462	92537.32	95685	12798.20
10^{-1}	857809	788732.49	929814	85445.41

The results of each 500 runs are shown in Fig. 1, Fig. 2, Fig. 3, Fig. 4, Fig. 5, Fig. 6, Fig. 7 and Fig. 8. Among them, Fig. 1, Fig. 2, Fig. 3 and Fig. 4 estimates the Φ-quantile using the minimal random sampling method when $\Phi = 10^{-4}$, $10^{-3}, 10^{-2}, 10^{-1}$, respectively. Figure 5, Fig. 6, Fig. 7 and Fig. 8 estimates the Φ-quantile using Algorithm 1 when $\Phi = 10^{-4}$, $10^{-3}, 10^{-2}, 10^{-1}$, respectively. In all of the experiments, the data size of the data stream is unknown in advance.

From the experimental results in Table 1 and Fig. 1, Fig. 2, Fig. 3 and Fig. 4, it can be seen that although the result of the minimal random sampling method is equal to the expected value theoretically, and the same is generally true of the results in the experiment, the deviation of each result value is too large to be practical. The accuracy of the results is far from practical application. Thus, from the results of 500 runs, this random sampling method has little use value except for theoretical significance.

From the experimental results in Table 1 and Fig. 5, Fig. 6, Fig. 7 and Fig. 8, it can be seen that the proposed Algorithm 1 achieves excellent performance in the experiments, and the 500 result data in the experiment are significantly concentrated near the real quantile, which is consistent with the theoretical analysis.

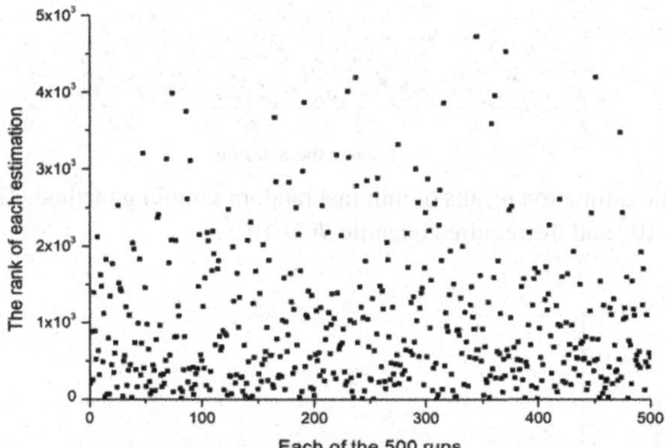

Fig. 1. The quantile estimation results of minimal random sampling method. The data size of the data stream is $n = 10^7$ and the required quantile $\Phi = 10^{-4}$.

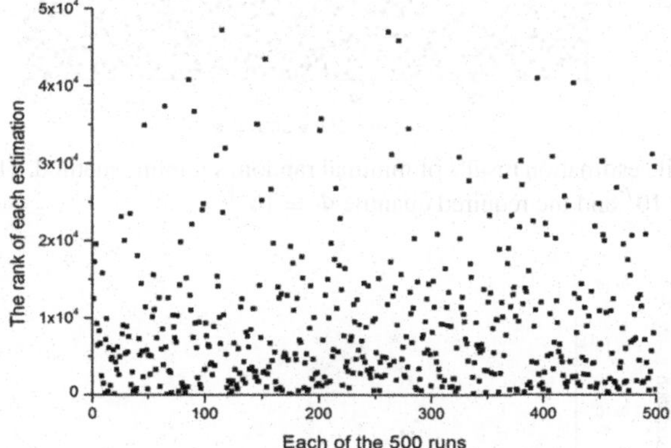

Fig. 2. The quantile estimation results of minimal random sampling method. The data size of the data stream is $n = 10^7$ and the required quantile $\Phi = 10^{-3}$.

In addition, we observed the following phenomena: (1) With a fixed data size and the quantile $\Phi = 10^{-4}$, 10^{-3}, 10^{-2}, 10^{-1}, respectively in turn, as the Φ value increases, the algorithm performance seems to be getting better, and the estimated data result will be more and more concentrated. The minimal sampling method does not seem to perform as well. (2) We noticed that over half of the 500 experimental results are higher than the expected quantile rank, that is, most estimates are slightly greater than the real quantile

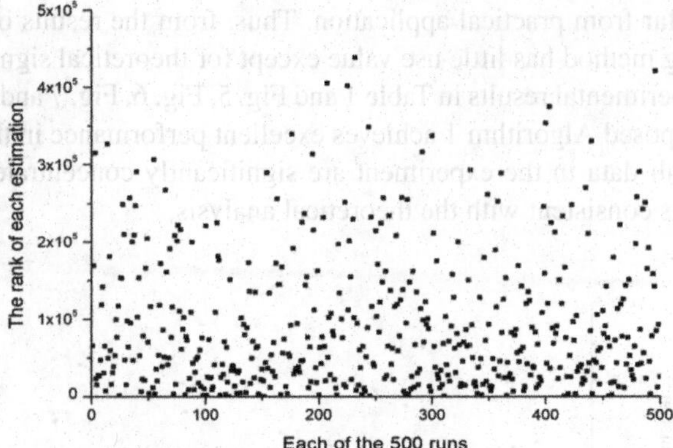

Fig. 3. The quantile estimation results of minimal random sampling method. The data size of the data stream is $n = 10^7$ and the required quantile $\Phi = 10^{-2}$.

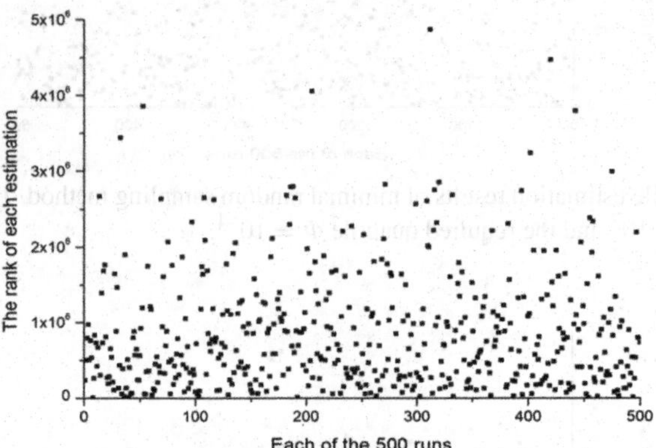

Fig. 4. The quantile estimation results of minimal random sampling method. The data size of the data stream is $n = 10^7$ and the required quantile $\Phi = 10^{-1}$.

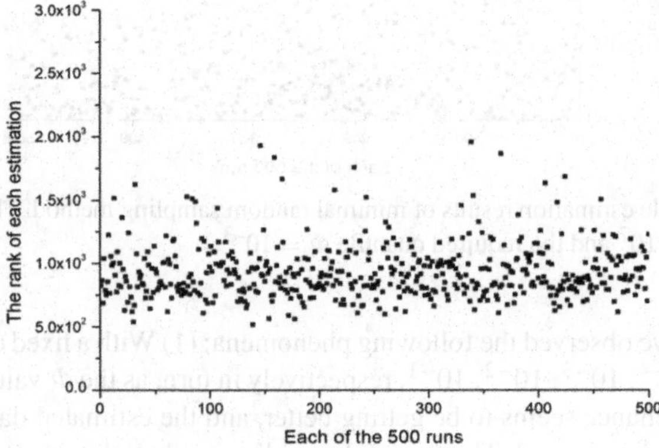

Fig. 5. The quantile estimation results of Algorithm 1. The data size of the data stream is $n = 10^7$ and the required quantile $\Phi = 10^{-4}$.

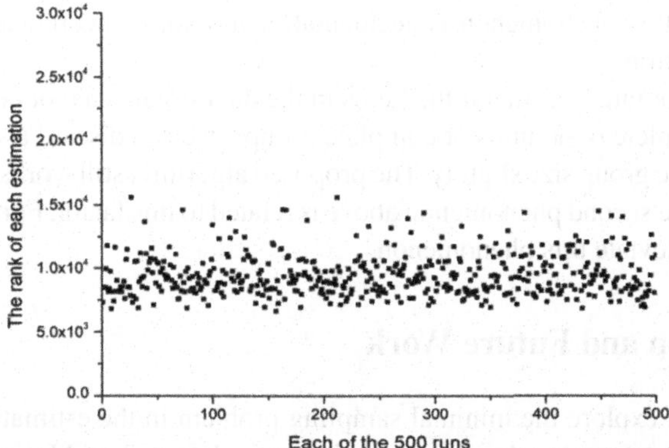

Fig. 6. The quantile estimation results of Algorithm 1. The data size of the data stream is $n = 10^7$ and the required quantile $\Phi = 10^{-3}$.

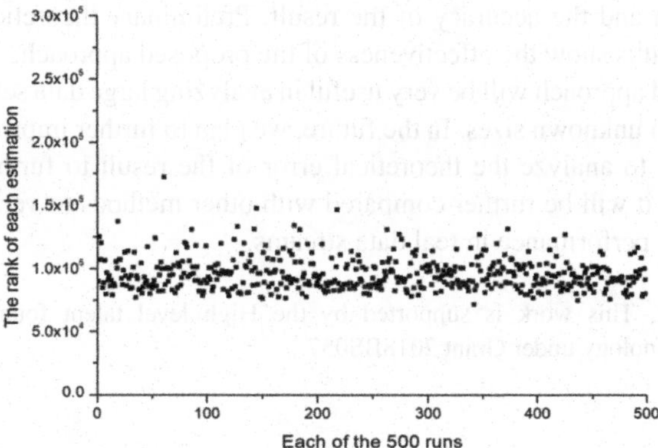

Fig. 7. The quantile estimation results of Algorithm 1. The data size of the data stream is $n = 10^7$ and the required quantile $\Phi = 10^{-2}$.

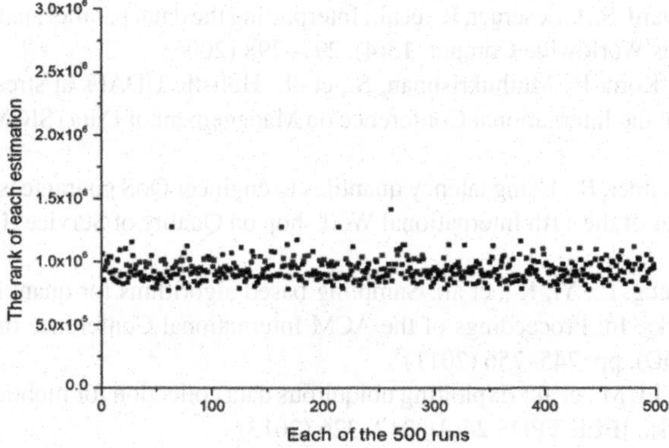

Fig. 8. The quantile estimation results of Algorithm 1. The data size of the data stream is $n = 10^7$ and the required quantile $\Phi = 10^{-1}$.

items. Whether this phenomenon is accidental or has some theoretical support needs further investigation.

It should be pointed out that if the items in the data stream cannot be evenly divided into several complete basic units or complete groups, we can discard a small number of items or adjust the group size slightly. The proposed algorithm still works. It is tentatively suggested that the second phenomenon above is related to this factor. Further exploration is expected to alleviate this phenomenon.

6 Conclusion and Future Work

In this paper, we explore the minimal sampling problem in the estimation approach to solve the problem of getting the Φ-quantile item in a data stream. Moreover, to estimate the Φ-quantile in the data stream with a single-pass, while ensuring that the expected rank of the result item is exactly equal to the rank of the desired item, we further explore and propose a practical and feasible approach considering both the space used by the algorithm and the accuracy of the result. Preliminary theoretical analysis and experimental results show the effectiveness of the proposed approach.

The proposed approach will be very useful in analyzing large data sets, especially for data streams with unknown sizes. In the future, we plan to further improve the proposed methods and try to analyze the theoretical error of the result to further optimize the algorithm. Also, it will be further compared with other methods to verify and improve the experimental performance in real data streams.

Acknowledgment. This work is supported by the High level talent foundation of Henan University of Technology under Grant 2018BS057.

References

1. Mayer-Schönberger, V., Cukier, K.: Big Data: a revolution that will transform how we live, work, and think, Houghton Mifflin Harcourt, p. 31 (2013). ISBN 9780544002692
2. Pike, R., Dorward, S., Griesemer, R., et al.: Interpreting the data: parallel analysis with sawzall. Dynamic Grids Worldwide Comput. **13**(4), 277–298 (2005)
3. Cormode, G., Korn, F., Muthukrishnan, S., et al.: Holistic UDAFs at streaming speeds. In: Proceedings of the International Conference on Management of Data (SIGMOD), pp. 35–46. ACM (2004)
4. Fiedler, U., Plattner, B.: Using latency quantiles to engineer QoS guarantees for web services. In: Proceedings of the 11th International Workshop on Quality of Service (IWQoS), pp. 345–362 (2003)
5. Huang, Z., Wang, L., Yi, K.,, et al.: Sampling based algorithms for quantile computation in sensor networks. In: Proceedings of the ACM International Conference on Management of Data (SIGMOD), pp. 745–756 (2011)
6. Li, Z., Liu, Y., Li, M., et al.: Exploiting ubiquitous data collection for mobile users in wireless sensor networks. IEEE TPDS **24**(2), 312–326 (2013)
7. Karnin, Z., Lang, K., Liberty, E.: Optimal quantile approximation in streams. In: Proceedings of the IEEE 57th Annual Symposium on Foundations of Computer Science (FOCS), pp. 71–78. IEEE (2016)

8. Guha, S., Mcgregor, A.: Stream order and order statistics: quantile estimation in random-order streams. SIAM J. Comput. **38**(5), 2044–2059 (2008)
9. Manku, G.S., Rajagopalan, S., Lindsay, B.G.: Random sampling techniques for space efficient online computation of order statistics of large datasets. ACM SIGMOD Rec. **28**(2), 251–262 (1999)
10. Hoare, C.A.R.: Algorithm 65: find. Commun. ACM **4**(7), 321–322 (1961)
11. Sedgewick, R., Wayne, K.: Algorithms, Fourth Edition. Addison-Wesley. Section 2.4: Priority Queues, pp. 308–335 (2011). ISBN 978-0-321-57351-3
12. Xue, Z., Li, R., Zhang, H., et al.: DC-Top-k: a novel Top-k selecting algorithm and its parallelization. In: Proceedings of the 45th International Conference on Parallel Processing (ICPP), pp. 370–379. IEEE (2016)
13. Vitter, J.S.: Random sampling with a reservoir. ACM Trans. Math. Softw. **11**(1), 37–57 (1985)
14. Buragohain, C., Suri, S.: Quantiles on streams. In: Encyclopedia of Database Systems, pp. 2235–2240 (2009)
15. Wang, L., Luo, G., Yi, K., et al.: Quantiles over data streams:an experimental study. In: Proceedings of the International Conference on Management of Data (SIGMOD), pp. 737–748. ACM (2013)
16. Luo, G., Wang, L., Yi, K., Cormode, G.: Quantiles over data streams: experimental comparisons, new analyses, and further improvements. VLDB J. **25**(4), 449–472 (2016). https://doi.org/10.1007/s00778-016-0424-7
17. Hung, R.Y.S., Ting, H.F.: An $\Omega(1/\varepsilon \log 1/\varepsilon)$ space lower bound for finding ε-approximate quantiles in a data stream. In: Proceedings of the International Conference on Frontiers in Algorithmics, pp. 89–100. Springer (2010)
18. Greenwald, M., Khanna, S.: Efficient online computation of quantile summaries. In: Proceedings of the ACM International Conference on Management of Data (SIGMOD), pp. 58–66. ACM (2001)
19. Chakrabarti, A., Jayram, T.S., Cu, M.: Tight lower bounds for selection in randomly ordered streams. In: Proceedings of the 19th ACM-Siam Symposium on Discrete Algorithms (SODA), pp. 720–729 (2008)
20. Felber, D., Ostrovsky, R.: A randomized online quantile summary in $O(1/\varepsilon \log 1/\varepsilon)$ words. In: Proceedings of the 18th International Workshop on Approximation Algorithms for Combinatorial Optimization Problems (APPROX 2015), pp. 775–785 (2015)
21. Singh, S.A., Srivastava, D., Tirthapura, S.: Estimating quantiles from the union of historical and streaming data. Proc. VLDB Endow. **10**(4), 433–444 (2016)
22. Wang, W., Ching, W., Wang, S., et al.: Quantiles on stream: an application to Monte Carlo simulation. J. Syst. Sci. Inf. **4**(4), 334–342 (2016)
23. Lall, A.: Data streaming algorithms for the Kolmogorov-Smirnov test. In: Proceedings of the IEEE International Conference on Big Data (ICBD), pp. 95–104. IEEE (2015)
24. Mitchell, R., Frank, E., Holmes, G.: An empirical study of moment estimators for quantile approximation. ACM Trans. Database Syst. **46**(1), 1–21 (2021)
25. Tiwari, N., Pandey, P.C.: A technique with low memory and computational requirements for dynamic tracking of quantiles. J. Signal Process. Syst. **91**(5), 411–422 (2018). https://doi.org/10.1007/s11265-017-1327-6
26. Lim, G., Hassan, M.S., Jin, Z., et al.: Approximate quantiles for datacenter telemetry monitoring. In: Proceedings of the 36th IEEE International Conference on Data Engineering (ICDE), pp. 1914–1917. IEEE (2020)

Fed-Tra: Improving Accuracy of Deep Learning Model on Non-iid in Federated Learning

Wenjie Xiao[1], Xuehai Tang[1,2(✉)], Biyu Zhou[1], Wang Wang[1,2], Yangchen Dong[1], Liangjun Zang[1], Jizhong Han[1], and Songlin Hu[1]

[1] Institute of Information Engineering, Chinese Academy of Sciences, Beijing, China
{xiaowenjie,tangxuehai,zhoubiyu,wangwang,dongyangchen,zangliangjun, hanjizhong,husonglin}@iie.ac.cn

[2] University of Chinese Academy of Sciences, Beijing, China

Abstract. Federated Learning (FL) has received more and more attention from researches and industries in that it can break the data island while protecting the data privacy. However, the original federated average algorithm proposed in FL ignores the difference in distribution of data from multiple participants (widespread in reality), which seriously undermine the performance of deep learning models. It is more serious when the gap of data-volume is large. In this paper, we propose a novel and universal federated learning method, named Fed-Tra, to effectively weaken biases in the model training to build high-precision models. Fed-Tra does this with dynamically adjusting the weight of local training samples for each round in all participants. Moreover, our evaluation on real-world datasets shows that the Fed-Tra achieves nearly +28% improvement of F1-Score compared with the original federated average algorithm.

Keywords: Federated learning · Distributed machine learning · Weight adjustment · Data privacy

1 Introduction

With the success of AlphaGo, deep learning has shown its powerful capabilities in almost every industry and different walks of life, including electric business, finance, etc. Deep learning relies on large quantities of highly-qualified annotated data to achieve precise prediction. For example, ResNet-152 [1] only have about 4% error in Image-net, which is an image database containing over 1,000,000 images in 1000 object class.

However, it is difficult or even impossible to acquire large scale annotated data to support the training of deep neural networks. To obtain large amounts of labeled data, domain experts have to spend much time to annotate the raw data, which costs too much. At the meantime, the labeled data in different enterprises cannot to be shared to protect user privacy due to the regulations of

Y. Lai et al. (Eds.): ICA3PP 2021, LNCS 13155, pp. 790–803, 2022.
https://doi.org/10.1007/978-3-030-95384-3_49

General Data Protection Regulation (GDPR) [2]. And because of this, the data in different enterprises become the data islands, which aggravates this problem.

Recent years, the Federated-Learning (FL) [3,4,26] has been proposed to make use of data from different participants (enterprises or organizations) to train a high precision model without violating privacy of user. While the original federated average algorithm (core of FL) used to merge gradients from multiple participants in FL, it assumes that the data in different participants is independent identically distributed (i.i.d) [8], which is not reasonable. In fact, the data of multiple participants are usually non-i.i.d [6,24]. The gradient aggregation method based on the naive assumption leads to the suboptimal performance of final prediction model due to the divergence of distribution of data from multiple participants. What's worse, the large difference of samples-volume will introduce more biases in the federated-model training between enterprises.

Recent research on federated learning has focused on reducing communication overhead [10] and protecting privacy [12–14], but a few existing works to tackle the problem of non-i.i.d, especially under the situation where there is a large difference in data-volume. Krishna Pillutla proposed Robust Aggregation algorithms [16] to address the differences of data distribution relying on a robust secure aggregation oracle based on the geometric median. Astraea [18], a self-balancing federated learning framework, was proposed by Moming Duan in 2019 to improve classification accuracy of mobile deep learning applications by global data distribution based data augmentation. However, they perform poorly in this scene where the ratio of data is large between multiple enterprises. Yang [19] also proposed Federated transfer learning (FTL), which firstly applied feature-representation transfer learning technology to address this problem across silos. However, this framework is needed to make the pre-processing of samples aligned for structured data, not applicable for unstructured data.

In view of the above challenges, we propose a new and general federated learning method for silos, which assigns effective weights to samples to alleviate the difference of data distribution among multiple enterprises. It is still meaning for the large gap of data-volume. Specifically, local model is updated with leveraging Stochastic Gradient Descent (SGD) algorithm on the local weighted sample, which is assigned sample weights based on classification error rate of local model. In this way, the distribution divergence of data across various enterprises is narrowed. After then, all local models are sent to third-party trusted server and aggregated into the global model which will be returned to locality to train iteratively until a excellent precision is reached. We evaluate Fed-Tra on real-world datasets from different enterprises, including news text-sets, image-sets and structural data. The results confirm that Fed-Tra outperforms 3 state-of-the-art federated learning methods. And Fed-Tra can effectively improve model precision by 26.59% (mean value on text and image) and F1-Score by 19.79% (mean value on text and image).

The contributions of this paper are as follows:

- We systematically study how the large difference of daga-volume intensified affects the performance of deep learning model on Non-iid in FL.

- We propose a universal FL method facing silos which can overcome the impact of distribution divergence of data from multiple participants and achieve precise prediction even when the ratio of data from other participants to the local is very large.
- We have performed experiments on 3 types of real-world datasets, including news text-sets, image-sets and structural data. And the results confirm the effectiveness Fed-Tra in improving the performance of federated model for multiple types of datasets.

This paper is organized as follows: we first elaborate the background and motivation of our work in Sect. 2. In Sect. 3, we introduce the related work. Section 4 describes the details of our proposed method Fed-Tra. We discuss reports the experiments and corresponding results in Sect. 5. Lastly, Sect. 6 concludes this paper.

2 Background and Motivation

2.1 Background

When it comes to applications of AI in real-world scenarios, it is often the case that corporations only possess low-quality and insufficient labeled data. This case motivates enterprises possessing less labeled data to train a neural network model towards high accuracy with employing federated learning. Traditional federated learning enables participating enterprises train their data locally and exchange the model parameters rather than local data to establish a common model. But it is clearly known that local models are also different due to the different distribution and quantity of local data. Federated Averaging is a widely used model aggregation algorithm in FL. This learning algorithm applies for the following target function (1).

$$f(w) = \sum_{k=1}^{K} \frac{n_k}{n} F_k(w) \tag{1}$$

where f(w) is the global loss function and $F_k(w)$ is the local loss. n_k denotes the set of indexes of data points on client k.

This algorithm enables all clients calculate and update their weights by the local update $w^k \leftarrow w^k - \eta \bigtriangledown F_k(w^k)$ in parallel, then a trusted server collects the updates of clients and aggregates them into average of the results with applying the update $w_{t+1} \leftarrow w_t - \eta \sum_{k=1}^{K} \frac{n_k}{n} w^k$.

2.2 Motivation

However, federated average algorithm relies on stochastic gradient descent (SGD), which should ensure that the local gradient is an unbiased estimate of the full gradient in the scenario where the data is evenly distributed among enterprises. Unlike the common training dataset, the data distribution of different enterprises is always imbalanced in practice. Recently Li [5] further examined the influence of non-IID data on model performance and show empirically and theoretically that heterogeneous data will slow down convergence of the mode. In addition, Yue Zhao [29] indicated that when the distribution of data is imbalanced, for each participant, due to the distance between the data distribution, the divergence between participants becomes much larger and accumulates very fast, which makes the divergence between participants much larger. In addition, as is seen by function (1), what the contribution of local gradient in the aggregated global model is related to the number of local samples. If the data amount of one party is too little, the model parameters of this party will be seriously disturbed by the model parameters of other participants with different data distribution, which will introduce biases in the model training and sharply reduces the accuracy of federated learning model.

We proved the existence of this problem experimentally. MNIST and USPS datasets (collected by different organizations) are used to simulate this scenario where the volume gap is increased. The proportion of annotation sample is 1:1, 1:10, 1:30, 1:100, 1:300 (two participants). Furthermore, three sets of news texts from different companies are simulated with quantitative proportion of 1:1:1, 1:8:10, 1:20:50, 1:40:100, 1:60:1000 (three participants).

Table 1. Federated model accuracy on imbalanced annotation-data with two parties.

Sample proportion	Series accuracy
1:1	0.9642
1:10	0.7375
1:30	0.6837
1:100	0.5421
1:300	0.4691

The test top-1 accuracy on five sample proportion is shown in Table 1 and Table 2. Experimental results show that as the difference in the quantity ratio increases among enterprises with different data distribution, the accuracy of the model is greatly reduced. Qualitatively, it is clear that the accuracy on sample proportion 1:1 and 1:300 is 96.42% and 46.91% in Table 1. In addition, experimental results with three parties in Table 2 shows that when sample proportion from 1:1:1 to 1:60:1000, 54% reduction in accuracy is observed.

3 Related Work

In 2017 McMahan [26] proposed the term federated learning, including the model aggregation algorithm FedAvg, which is used to update the model locally by the end-user of Android mobile phone with satisfying data privacy. However, federated learning still has potential challenges. In terms of privacy, Wang [28]

Table 2. Federated model accuracy on imbalanced annotation-data with three parties.

Sample proportion	Accuracy
1:1:1	0.9841
1:8:10	0.8261
1:20:50	0.7538
1:40:100	0.6447
1:60:1000	0.3472

attempted to explore the user-level privacy leakage faced with a malicious server in federated learning. In terms of communication efficiency, Mehryar Mohri proposed a new framework of agnostic federated learning [30] and Mikhail Yurochkin [31] developed a federated learning method based on Bayesian non-parametric. To address data heterogeneity, Liping Li et al. proposed Byzantine-Robust [17] Stochastic Aggregation methods. In application, Hyesung Kim [21] also combined on-device federated learning mode with blockchain and Qian, Yongfeng [22] adopted federated learning to address service placement with users' privacy and limited resources for mobile edge computing.

With the increasing awareness of large companies compromising on data security and user privacy, the emphasis on data privacy and security has become a worldwide major issue [23]. Recently, Yang Qiang [3] proposed the novel concept of federated learning facing the scene of cross-silo, which can be divided into three categories based on the distribution characteristics of the data:

- *Horizontal Federated Learning*: Horizontal federated learning [3], or called with sample-based federated learning, is introduced in the scenarios where two companies locate in different regions which result in having very different user groups, but their business is very similar. A typical architecture for a horizontal federated learning system is likely to McMahan proposed the term federated learning for Android phone model updates. In this system, participants collaboratively learn a model with the help of a true parameter or cloud server in which it allows models built at client devices to security aggregate at the server site to build a global federated model. The process of model building ensures that there is no data leakage.
- *Vertical Federated Learning*: Vertical federated learning [3], or called with feature-based federated learning, is applicable to the cases in which two different companies have the same Users ID come from the same city, but they are different kinds of companies. In this system, it aggregates these different features and computing the training loss and gradients in a privacy-preserving manner to build a model with data from both parties collaboratively.
- *Federated Transfer Learning*: Federated Transfer Learning [3,20] applies to the scenarios that the locations and types of business differ between enterprises, which causes that two datasets are different not only in samples but also in feature space. In this case, transfer learning techniques can be applied

to provide solutions for the entire sample and feature space in FL. For example, Yang Liu [19] proposed Federated Transfer Learning (FTL) framwork.

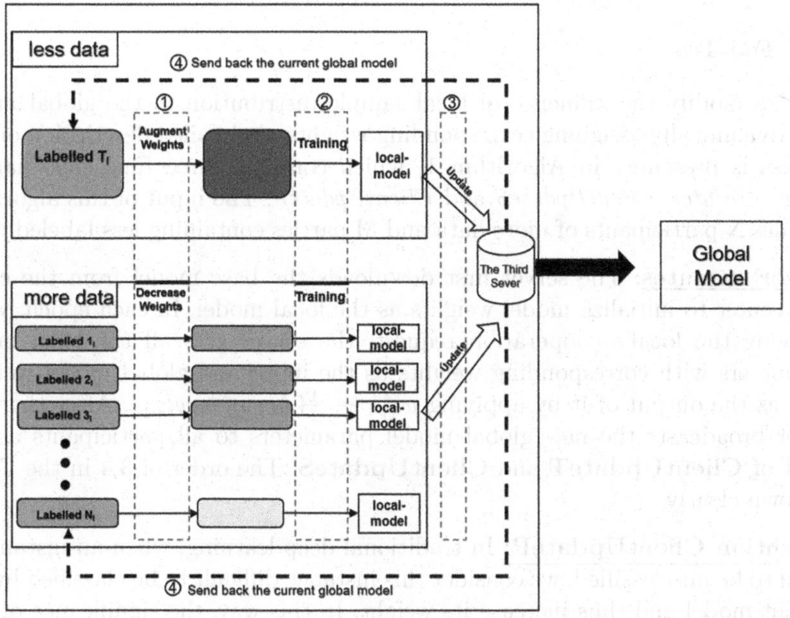

Fig. 1. The overall workflow of Fed-Tra

4 Method

To solve the precision reduction of federated learning on No-IID data, we present the Fed-Tra, a novel federated learning method based on dynamic sample weight. So that sample distribution in the federated network shifts towards more uniformity. Fed-Tra walks the reader through the order of training sample processed in this section.

The workflow of Fed-Tra is shown in Fig. 1. Fed-Tra contains three parts, including the third-party server, one party of less data and one party of more data. The local firstly dynamically assigns corresponding weights to every sample using the global model from the previous round. The performance of the federated model depends on all local models, each of which is effected by sample distribution and data-volume in the local. To narrow the gap of the distribution of accuracy in the federated network from the sample level, the sample weights of parties with poor labeled-data should be augmented to increase the influence of these samples for the global model. Further, the sample weights of organizations with more data may be decreased. After then, we train the local model on weighted samples and upload local model parameters to the third-party server.

It securely averages local model parameters into the current global model. Then this model will be sent back to the local and start a new iteration. Through this way, after many rounds of training, this method can train a model with strong generalization ability and the precision is improved.

4.1 Fed-Tra

Fed-Tra modify the influence of local sample distribution to the global model with dynamically assigning corresponding weights. A detailed description of this process is presented in Algorithm 1, which contains three functions, namely *Server executes*, *ClientUpdateS* and *ClientUpdateP*. The input of this algorithm includes N participants of more data and M parties containing less labeled data.

Server executes: The server first downloads the base model from the cloud data center to initialize model weights as the local model. In each epoch, when receiving the local's cooperation request, the server gets all the local model parameters with corresponding weights as the input and global model weights w_{t+1} as the output of it by applying $w_{t+1} \leftarrow \frac{1}{2}(Array + w_{t+1}^T)$. After then the server broadcasts the new global model parameters to all participants as the input of **ClientUpdateT** and **ClientUpdateS**. The order of 3,4 in the Fig. 1 is shown clearly.

Funchtion ClientUpdateP: In traditional deep learning, when an instance is found to be misclassified, we consider this instance difficult to be classified by the current model and thus increase its weight. In this way, the significance of this instance will become greater in the next iteration. To begin with, client T receives base model from the server as initialize local model and then give instances a initialized sample weights distribution to performs the iteration process. In each cycle, we dynamically set the new weights based on the previous iteration results and calculate the error rate of current local model on labeled-instances. To bridge the gap, if client T instances are misclassified, we increase the weights of these instances to pay more attention in the next iteration through multiplying these instances by $a_i^{t+1} = a_i^t \varphi_t^{|y(x_i^T) - Q_t^T(x_i^T)|}$. Noted $y(x_i^T)$ is the true label of each instance x_i^T in client T. After setting weights, clients are equivalent to get the new sample sets used to update local model weights.

Funchtion ClientUpdateS: Each participants firstly receives base model and initialize sample weights distribution used to train local model. In each iterations, we will decrease the weight of that instance in the next iteration to reduce its influences on the global model. If instances of party of more data are misclassified, they are considered to be different from the distribution of less data. Due to this party with more training samples, it has a greater impact on global model parameters than parties of less training samples. As is seen in Figure 1, it is clear that we decrease the weights of these instances to reduce their influences on the global model in the next iteration. Specifically, we multiply the instances by $a_i^{t+1} = a_i^t \varphi^{|y(x_i^s) - Q_t^s(x_i^s)|}$.

Algorithm 1. Fed-boost Algorithm

Input: More-data parties S={1_l,2_l,...,N_l};less labeled-data parties P={1_l,2_l,...,M_l}; The maximum number of iterations,K.
Output: Final-classifier

initialize model-weights w_0
function SERVER EXECUTES($\{w_1, w_2, ..., w_N\}; w_T$)
  k = 1
  for $t = 1, ..., K$ **do**
    for $s = 1, 2, ..., N$ **do**
      $w_{t+1}^s \leftarrow ClientUpdateS(w_t^s)$
      $Array[k++] \leftarrow w_{t+1}^s$
    end for
    for $p = 1, 2, ..., M$ **do**
      $w_{t+1}^p \leftarrow ClientUpdateT(w_t^p)$
      $Array[a++] \leftarrow w_{t+1}^p$
      $w_{t+1} \leftarrow \frac{1}{2}(Array[k++] + Array[a++])$
    end for
    return w_{t+1} ▷ Global model weights

    function $ClientUpdateP(w_t^p)$
      $Q_t^p \leftarrow w_t^p$ ▷ Update local model using w_t^p
      $\epsilon_t = \sum_{i=1}^{m} \frac{|y(x_i^p) - Q_t^M(x_i^p)| a_i^t}{\sum_{i=1}^{m} a_i^t}$ ▷ x_i^p is sample of client T
      $\varphi_t = \epsilon_t/(1 - \epsilon_t)$ ▷ Required to be less than 1/2
      $a_i^{t+1} = a_i^t \varphi_t^{|y(x_i^p) - Q_t^p(x_i^p)|}$ ▷ Update the sample weight distributions
      $T_i \leftarrow a_i^{t+1}$ ▷ i is the sample amount of client T
      $Q_p^{t+1} \leftarrow Q_p^t(p_i)$ ▷ Train local model
      $w_{t+1}^p \leftarrow Q_p^{t+1}$
      return w_{t+1}^p ▷ T client sent local model to the server
    end function

    function $ClientUpdateS(w_t^s)$
      $Q_t^s \leftarrow w_t^s$ ▷ Update local model using global model weights
      $\varphi = \frac{1}{\sqrt{2 \ln 2/K + 1}}$
      $a_i^{t+1} = a_i^t \varphi^{|y(x_i^s) - Q_t^s(x_i^s)|}$ ▷ x_i^s is each sample of client, s={1,2...,N}
      $s_i \leftarrow a_i^{t+1}$ ▷ i is the sample amount of client s
      $Q_{t+1}^s \leftarrow Q_t^s(s_i)$ ▷ Train local model
      $w_{t+1}^s \leftarrow Q_{t+1}^s$
      return w_{t+1}^s ▷ S={1_l,2_l,...,N_l} all sent local model to the server
    end function

After several iterations, we can get a shared model with strong generalization ability with employing this method.

5 Experiments

In this section, extensive experiments are conducted to evaluate the improved precision of Fed-Tra on multiple types of datasets. We first describe the methodology, followed by experimental results of comparing Fed-Tra against 3 state-of-the-art federated learning methods with respect to the evaluation metrics.

5.1 Methodology

We first describe the characteristics of the real-world datasets and related base-model. Then we introduce our competitive methods of model training. Finally, the experiment setup will be portrayed.

Datasets and Models. To evaluate the effectiveness and universality of Fed-Tra on various tasks, we will use different types of datasets and the corresponding model in our experiments. We adopt three widely used real-world datasets, including: 1) THUCNews, TOUTIAO and SouGou (news text-sets); 2) MNIST and USPS (image-sets); 3) Default-of-Credit-Card-Clients (structured datasets for banks).

THUCNews, TOUTIAO and SouGou which are both Chinese news text datasets from different organizations are utilized for text classification. And Text-CNN model is selected as a base-classifier in our federated learning experiments. THUCNews contains 740,000 news documents generated by filtering the historical data of Sina News and are composed of 14 classification categories. TOUTIAO dataset contains 382,688 entries, distributed in 15 classifications. Besides, SouGou provides nearly 100,000 SoHu news texts containing 11 categories.

MNIST and USPS are both ten categories of handwritten digital picture datasets which are generated from different organizations. They both contain 60000 training sample and 10000 testing sample. For image classification, we choose the CNN model as its classifier.

The "Default-of-Credit-Card-Clients" dataset consists of credit card records with 30,000 samples with users' default payments as labels [10], including user demographics, history of payments, and bill statements, etc.

Comparative Generative Models. To verify the effectiveness of our proposed Fed-Tra, we compare our approach against three state-of-the-art federated learning methods.

Table 3. Setting of imbalanced dataset.

3*DataType	*SouGou → TOUTIAO&THUCNews*			*MNIST → USPS*		Default-of-Credit-Card-Clients	
	SouGou	TOUTIAO	THUCNews	MNIST	USPS	Target	Source
Train	10000	135000	270000	2000	60000	600	18000
Test	93000	None	None	10000	None	10000	None
2*DataType	*SouGou → TOUTIAO&THUCNews*			*MNIST → USPS*			
Train	2000	135000	270000	800	60000		
Test	93000	None	None	10000	None		

- **FedAvg**. This algorithm is the first proposed in federated learning, which has been applied to Google keyboard for improving query suggestions. It runs E steps of SGD in parallel on a small sampled subset of devices and then averages the resulting model updates via a central server once in a while. In comparison with SGD and its variants, FedAvg performs more local computation and less communication.
- **Robust Aggregation**. The Robust Aggregation is relies on a robust secure aggregation oracle based on FedAvg algorithm, which returns a robust aggregate using a constant number of calls to a regular non-robust secure average oracle. This way aggregates model parameters in a robust manner to narrow the difference of local samples and then address problems of imbalanced datasets.
- **Astrea**. The Astrea, a self-balancing federated learning framework, reduce the imbalances by global data augmentation and mediator rescheduling which aims to improve classification accuracy of mobile deep learning model. Astrea achieves this by reducing global imbalance with data augmentation. And then for averaging the local imbalance, it creates the mediator to reschedule the training of clients based on KullbackLeibler divergence (KLD) of their data distribution.

Experiment Setup. We simulate multiple parties in federated learning by splitting dataset in all parties. Specifically, one of three parties contains a small amount of labeled samples, others own more non-iid data. We reconstructed new subset with four categories from three news text-sets. The size of new subsets are 1) THUCNews:270000; 2) TOUTIAO:170000; 3) SouGou:100000. Note that the categories are the same between these datasets. In terms of image datasets, we randomly select a small number of labeled samples from training set of MNIST as one party with less labeled-datsets. Default-of-Credit-Card-Clients is split directly into two parties. The detail settings are shown in Table 3. The differences of sample size are more larger in the bottom half of Table 3 than the top half. Note that The right side of the arrow represents enterprises with more annotated data (introduce samples) and the left side is enterprises with less annotated data (target samples). Besides, local batch size and the maximum number of training epochs are set to 128 and 2000, respectively.

5.2 Evaluation Comparison

We evaluate the performance of our method with the metrics including precision, recall and F1-score in our experiments. Table 4 and Table 5 provides experimental results with differences in sample size.

Table 4. Comparing metric of different data type (Sample size is the top half of Table 3).

2*DataType	$SouGou \rightarrow TOUTIAO\&THUCNews$)			$MNIST \rightarrow USPS$)		
	Precision	Recall	F1-Score	Precision	Recall	F1-Score
FedAvg	0.6254	0.5877	0.6407	0.6018	0.7028	0.6447
Robust Aggregation	0.8861	0.5877	0.7431	0.8619	0.6901	0.7055
Astrea	0.8172	0.6221	0.8016	0.8831	0.6951	0.7779
Fed-Tra	**0.9109**	**0.8142**	**0.8493**	**0.8486**	**0.7625**	**0.8219**

Table 5. Comparing metric of different data type (Sample size is the bottom half of Table 3).

2*DataType	$SouGou \rightarrow TOUTIAO\&THUCNews$			$MNIST \rightarrow USPS$		
	Precision	Recall	F1-Score	Precision	Recall	F1-Score
FedAvg	0.4672	0.3087	0.3766	0.2782	0.6018	0.3805
Robust Aggregation	0.7326	0.4654	0.5608	0.7075	0.6318	0.5617
Astrea	0.6999	0.5036	0.6099	0.7888	0.4568	0.5785
Fed-Tra	**0.8786**	**0.6951**	**0.7955**	**0.7961**	**0.7028**	**0.7483**

The Effectiveness of Fed-Tra. As is shown, all these federated learning methods can improve the precision and F1-Score compared with the original FedAvg. Among them, Fed-Tra has the best performance and it is effective for all types of datasets, no matter the text-sets or image-sets. Fed-Tra performs better than other methods for a key reason that the weight-value method fine-grained narrows the difference of sample distribution between all participants by assigning effective sample weight to samples of relevant participants. As seen in Table 4 comparing with FedAvg, Fed-TrA improves respectively the precision and F1-Score by about 28.55% and 20.86% on text-sets, 24.63% and 18.72% on image-sets. In addition, Fed-Tra also has the higher F1-Score than Robust Aggregation and Astrea by up about 10% and 4%.

As is shown in Tabel 5, as the increase of the ratio of tsample-volume among enterprises, Fed-Tra still performs better than other methods. The F1-Score of Robust Aggregation and Astrea are reduced by about 18.67% compared with Table 4. The reason for the sharp drop in accuracy is that samples distribution in the part with more labeled-data more seriously affect enterprises with less labeled-data. However, Fed-Tra reduced the rate by only 8.59% in F1-Score. So as the experiments with image datasets. So, Fed-Tra is more robust than others when the ratio of data among enterprises is very large.

Table 6. Comparing accuracy of same datasets (FATE's FTL vs Fed-TrA).

1*Dataset	Default-of-credit-card-clients
	Accuracy
FATE's FTL	0.9124
Fed-Tra	**0.8968**

Otherwise, comparing with FATE's FTL, Fed-Tra is less than FATE's FTL by about 2% on the same structured dataset in Table 6. The reason about the loss of accuracy is that FATE's FTL employs preprocessing of sample alignment and based-feature transfer learning with direct exchange of intermediate results to update model weights. However, as is shown in Table 4 and Table 5, Fed-Tra can fit with different types of data.

6 Conclusion

Federated learning is a promising distributed machine learning framework with the advantage of privacy-preserving. However, FL does not handle Non-iid well. It is more serious when the gap of data-volume is large. To resolve this problem, we present Fed-Tra, a universal federated learning method, which aims to effectively narrow biases of sample distribution and improve the accuracy of federated model. We use horizontal federation learning to train model with protecting the security and privacy of cross-silo raw data. During the training process, we are to narrow the distances of sample distribution by assigning relevant weights to each labeled sample between multiple participants. Through experiments on real-world dataset, we evaluate our approach to demonstrate Fed-ATL outperforms 3 state-of-the-art federated learning methods on text-sets and image-sets and structural data.

Acknowledgments. We thank the anonymous reviewers for their help in improving our paper. This work was supported by Grant 2020YFB 1005402 from the National Key R&D Program of China.

References

1. He, K., Zhang, X., Ren, S., Sun, J.: Deep residual learning for image recognition. In: Proceedings of the IEEE Conference on Computer Vision and Pattern Recognition, pp. 770–778 (2016)
2. Voigt, P., von dem Bussche, A.: The EU General Data Protection Regulation (GDPR). Springer, Cham (2017). https://doi.org/10.1007/978-3-319-57959-7
3. Yang, Q., Liu, Y., Chen, T., Tong, Y.: Federated machine learning: concept and applications. ACM Trans. Intell. Syst. Technol. **10**(2), 1–19 (2019)
4. Aledhari, M., Razzak, R., Parizi, R.M., Saeed, F.: Federated learning: a survey on enabling technologies, protocols, and applications. IEEE Access **8**, 140699–140725 (2020)

5. Li, X., Huang, K., Yang, W., Wang, S., Zhang, Z.: On the convergence of fedavg on non-iid data. arXiv preprint arXiv:1907.02189 (2019)
6. Tian, L., Anit Kumar, S., Manzil, Z., Maziar, S., Ameet, T., Virginia, S.: Federated optimization in heterogeneous networks. arXiv preprint arXiv:1812.06127 (2018)
7. Liu, F., Wu, X., Ge, S., Fan, W., Zou, Y. : Federated learning for vision-and-language grounding problems. In: Proceedings of the AAAI Conference on Artificial Intelligence (2018), pp. 11572–11579 (2020)
8. Hsieh, K., Phanishayee, A., Mutlu, O., Gibbons, P.: The non-iid data quagmire of decentralized machine learning. In: International Conference on Machine Learning, pp. 4387–4398 (2020)
9. Christopher, B., Zhong, F., Andras, P.: Federated learning with hierarchical clustering of local updates to improve training on non-IID data. In: 2020 International Joint Conference on Neural Networks (IJCNN), pp. 1–9 (2020)
10. Luping, W., Wei, W., Bo, L.I.: CMFL: mitigating communication overhead for federated learning. In: 2019 IEEE 39th International Conference on Distributed Computing Systems (ICDCS), pp. 954–964 (2029)
11. Konečnỳ, J.: Federated learning: strategies for improving communication efficiency. arXiv preprint arXiv:1610.05492 (2016)
12. Stacey, T., et al.: A hybrid approach to privacy-preserving federated learning. In: Proceedings of the 12th ACM Workshop on Artificial Intelligence and Security, pp. 1–11(2019)
13. Liu, X., Li, H., Xu, G., Lu, R., He, M.: Adaptive privacy-preserving federated learning. Peer-to-peer Netw. Appl. **13**(6), 2356–2366 (2020). https://doi.org/10.1007/s12083-019-00869-2
14. Lu, Y., Huang, X., Dai, Y., Maharjan, S., Zhang, Y.: Blockchain and federated learning for privacy-preserved data sharing in industrial IoT. IEEE Trans. Ind. Inf. **16**(6), 4177–4186 (2019)
15. Chuan, M.: On safeguarding privacy and security in the framework of federated learning. IEEE Network **34**(4), 242–248 (2020)
16. Pillutla, K., Kakade, S.M., Harchaoui, Z.: Robust aggregation for federated learning. arXiv preprint arXiv:1912.13445 (2019)
17. Li, L., Xu, W., Chen, T., Giannakis, G.B., Ling, Q.: RSA: byzantine-robust stochastic aggregation methods for distributed learning from heterogeneous datasets. In: Proceedings of the AAAI Conference on Artificial Intelligence, pp. 1544–1551 (2019)
18. Chuan, M., et al.: Astraea: self-balancing federated learning for improving classification accuracy of mobile deep learning applications. In: 2019 IEEE 37th International Conference on Computer Design (ICCD), pp. 246–254 (2019)
19. Liu, Y., Kang, Y., Xing, C., Chen, T., Yang, Q.: A secure federated transfer learning framework. IEEE Intell. Syst. **35**(4), 70–82 (2019)
20. Chen, Y., Qin, X., Wang, J., Yu, C., Gao, W.: Fedhealth: a federated transfer learning framework for wearable healthcare. IEEE Intell. Syst. **35**(4), 83–93 (2020)
21. Kim, H., Park, J., Bennis, M., Kim, S.-L.: Blockchained on-device federated learning. IEEE Commun. Lett. **24**(6), 1279–1283 (2019)
22. Qian, Y., Hu, L., Chen, J., Guan, X., Hassan, M.M., Alelaiwi, A.: Privacy-aware service placement for mobile edge computing via federated learning. Inf. Sci. **505**, 562–570 (2019)
23. Li, T., Sahu, A.K., Talwalkar, A., Smith, V.: Federated learning: challenges, methods, and future directions. IEEE Sig. Process. Mag. **37**(3), 50–60 (2020)

24. Sattler, F., Wiedemann, S., Müller, K.R., Samek, W.: Robust and communication-efficient federated learning from non-iid data. IEEE Trans. Neural Netw. Learn. Syst. **31**(9), 3400–3413 (2019)
25. Tran, N.H., Bao, W., Zomaya, A., Nguyen, M.N., Hong, C.S.: Federated learning over wireless networks: optimization model design and analysis. In: IEEE INFOCOM 2019-IEEE Conference on Computer Communications, pp. 1387–1395 (2020)
26. McMahan, B., Moore, E., Ramage, D., Hampson, S., y Arcas, B.A.: Communication-efficient learning of deep networks from decentralized data. In: Artificial Intelligence and Statistics, pp. 1273–1282 (2020)
27. Lyu, L., Yu, H., Yang, Q.: Threats to federated learning: a survey. arXiv preprint arXiv:2003.02133 (2020)
28. Wang, Z., Song, M., Zhang, Z., Song, Y., Wang, Q., Qi, H.: Beyond inferring class representatives: User-level privacy leakage from federated learning. In: IEEE INFOCOM 2019-IEEE Conference on Computer Communications [U+FF0C, pp. 2512–2520 (2019)
29. Zhao, Y., Li, M., Lai, L., Suda, N., Civin, D., Chandra, V.: Federated learning with non-iid data. arXiv preprint arXiv:1806.00582 (2018)
30. Mohri, M., Sivek, G., Suresh, A.T.: Agnostic federated learning. In: International Conference on Machine Learning, pp. 4615–4625 (2019)
31. Yurochkin, M., Agarwal, M., Ghosh, S., Greenewald, K., Hoang, N., Khazaeni, Y.: Bayesian nonparametric federated learning of neural networks. In: International Conference on Machine Learning, pp. 7252–7261 (2019)

Author Index